Linear Model Theory

Dale L. Zimmerman

Linear Model Theory

With Examples and Exercises

Springer

Dale L. Zimmerman
Department of Statistics and Actuarial
Science
University of Iowa
Iowa City, IA, USA

ISBN 978-3-030-52065-6 ISBN 978-3-030-52063-2 (eBook)
https://doi.org/10.1007/978-3-030-52063-2

Mathematics Subject Classification: 62J05, 62J10, 62F03, 62F10, 62F25

This Springer imprint is published by the registered company Springer Nature Switzerland AG.
The registered company address is: Gewerbestrasse 11, 6330 Cham, Switzerland

To my wife, Bridget,
and our children, Nathan, Joshua, Bethany,
Anna, and Abby

Preface

Several excellent books exist on the theory of linear models (I won't list them here, for fear of omitting someone's favorite!). So why did I write another one? Primarily for two reasons. First, while the existing books do a fine job (imho) of presenting a broad range of appropriate content, they tend to be too light on examples and exercises for many students to become proficient at obtaining concrete results for specific linear models. For example, a student who has taken a linear models course using an existing book undoubtedly can tell you that the best linear unbiased estimator (BLUE) of an estimable function $\mathbf{c}^T\boldsymbol{\beta}$, under a linear model $\mathbf{y} = \mathbf{X}\boldsymbol{\beta} + \mathbf{e}$ with Gauss–Markov assumptions on the errors, is $\mathbf{c}^T\hat{\boldsymbol{\beta}}$, where $\hat{\boldsymbol{\beta}}$ is any solution to the equations $\mathbf{X}^T\mathbf{X}\boldsymbol{\beta} = \mathbf{X}^T\mathbf{y}$; but in my experience, that student will often struggle to use that knowledge to obtain a simplified expression for, say, the BLUE of a factor-level difference in a two-way main effects model. Second, it has long been my view [see, for example, Zimmerman (2014)] that existing linear models texts give relatively too much attention to estimation and hypothesis testing of parameters, and not enough attention to predictive inference for random variables. This book, in contrast, features many more examples and exercises illustrating concrete results for specific linear models than other texts (hence, the subtitle "With Examples and Exercises"), and it puts nearly as much emphasis on prediction as on estimation and hypothesis testing.

There are 113 examples, to be exact, scattered throughout the book, which constitute more than 25% of the content. To be clear, the examples are not analyses of data but are either toy examples illustrating a theoretical concept or specializations of theorems or general expressions to specific cases of linear models. In that respect, they are similar to examples that one might find in a typical introductory mathematical statistics textbook. A separate list of the examples is included after the table of contents, for ease of reference. Also interspersed throughout the book are 12 "Methodological Interludes." These are descriptions of specific methodologies that the reader is likely to have encountered previously in an applied regression course, and they are placed immediately after the requisite theory is presented. The descriptions are relatively brief, focusing on theoretical justification rather than implementation. Such topics as variance inflation factors, added variable plots, and influence diagnostics are among those featured in these interludes. A separate listing of these is also provided.

The book contains 296 numbered exercises, located at the end of every chapter save the first. Be forewarned: some of the exercises have multiple parts and are quite lengthy. Some are proofs of theorems not supplied in the text, but most are similar to the book's examples in the sense that they are specializations of general results to specific linear models. A handful require the use of a computer, but none involves the analysis of actual data. Solutions to the exercises are available in a companion volume (Zimmerman 2020), which students and instructors (even those using a different text) may find useful. I am not aware of any other graduate-level text on the theory of linear models that has an accompanying published solutions volume.

The book focuses primarily on what I consider to be the core material of linear model theory: best linear unbiased point estimation and prediction of (linearly) estimable/predictable functions of parameters and random variables, and related inferences (interval estimation and prediction, and hypothesis testing) under an assumption of multivariate normality. Some topics that are included in other books on the theory of linear models, but not central to these core themes, are not found in this book. Examples of such topics are generalized linear models, nonlinear least squares, robust regression, nonparametric regression, principal components regression, ridge regression, Bayesian methods, and inference based on elliptical distributions. Various regression diagnostics, which also receive considerable attention in other books, are described only briefly here via the aforementioned Methodological Interludes. Nevertheless, there are topics that receive rather more attention in this book than in most others. These include estimability, reparameterization, partitioned analyses of variance, constrained least squares estimation, prediction (as noted previously), inference for models with a singular variance–covariance matrix, effects of model misspecification, likelihood-based methods of inference for variance components, and empirical best linear unbiased estimation and prediction.

A unique feature of the order of presentation of topics in the book is that distribution theory (pertaining specifically to the multivariate normal and related distributions) is not introduced until the 14th chapter (out of 17). Until then, none of the results presented require that the distribution of the model errors (and other random effects, if any) have any particular form.

The book is intended for use as a textbook for a one-semester course taken by second-year graduate students in statistics, biostatistics, and related fields. By their second year, most such students will have completed a yearlong course in probability and mathematical statistics using a textbook such as *Statistical Inference* by George Casella and Roger Berger. Thus, I assume that the reader is familiar with univariate and multivariate probability theory, including such topics as cumulative distribution functions, probability density functions, moments, moment generating functions, independence of random variables and random vectors, bias, mean squared error, maximum likelihood estimation, confidence intervals, and hypothesis testing. Because phrases like "with probability one" and "a set of probability zero" are also used in *Statistical Inference*, I presume that my (infrequent) usage of these phrases herein will not be a stumbling block. It is not assumed that the reader has taken a course in measure-theoretic probability.

Two additional prerequisites of the book are an upper-undergraduate or first-year graduate course in applied regression analysis and an undergraduate course in linear algebra. Chapter 2 of the book briefly reviews selected topics in linear and matrix algebra but does so very quickly and with little context provided, which is why some prior exposure to linear algebra is essential.

I have used this book, in the form of a continually evolving set of "lecture notes," for a one-semester, four-credit course at the University of Iowa that I have taught every other year, on average, for more than 30 years. Usually I have just enough time to cover all the topics. Instructors who use the book for a one-semester three-credit course may find that challenging. If so, I would suggest omitting those portions that present estimation and/or prediction for constrained models (Chap. 10) or models with singular variance–covariance matrices (Sects. 11.2 and 13.3). Additional candidates for omission are Sect. 11.3 on the equivalence of ordinary least squares estimation and best linear unbiased estimation, Chap. 12 on the effects of model misspecification, and Chap. 17 on empirical best linear unbiased estimation and prediction.

I owe debts of gratitude to many individuals. First and foremost are two giant figures in the field of linear model theory: Oscar Kempthorne and David Harville. I was privileged to take courses from both and to have David Harville as my Ph.D. advisor while I was a graduate student at the Iowa State University in the 1980s. Their influence on me pervades many aspects of this book, from the overall approach taken, to the choice of topics, and undoubtedly even to the presentation style. I wish to thank all the students who have taken my linear models course over the years. Each successive generation provided feedback, which improved the quality of the book over time and led to my decision to incorporate so many examples. I particularly wish to thank two recent students, Jun Tang and Rui Huang, for their careful proofreading of portions of the manuscript and for checking the solutions to the exercises, saving me from the embarrassment of numerous mistakes and typos (I'm sure some still exist, however, for which I am solely responsible!). Thanks also to two anonymous reviewers for their comments and suggestions and to Veronika Rosteck at Springer for her guidance through the publication process.

Iowa City, IA, USA Dale L. Zimmerman

Reference

Zimmerman, D. L. (2014). Comment on "The need for more emphasis on prediction: A non-denominational model-based approach." *The American Statistician, 68*, 85–86.

Contents

List of Examples

List of Methodological Interludes

A Brief Introduction

1

This book deals with a class of models that pertain to data consisting of the values of $p + 1$ variables $y, x_1, x_2, \ldots, x_p$ taken on each of n "observational units." We refer to y as the *response variable* (regardless of whether it actually makes sense to think of y as a response to anything) and to $x_1, \ldots, x_p$ as the *explanatory variables* (regardless of whether they actually help to explain anything). Let $y_i, x_{i1}, x_{i2}, \ldots, x_{ip}$ represent the observed values of these variables for the ith observational unit. Let us assemble these into an n-dimensional column vector $\mathbf{y}$ and n p-dimensional row vectors $\mathbf{x}_1^T, \ldots, \mathbf{x}_n^T$ as follows:

$$\mathbf{y} = \begin{pmatrix} y_1 \\ \vdots \\ y_n \end{pmatrix}, \quad \mathbf{x}_1^T = (x_{11}, \ldots, x_{1p}), \quad \ldots, \quad \mathbf{x}_n^T = (x_{n1}, \ldots, x_{np}).$$

Let us further assemble the row vectors $\mathbf{x}_1^T, \ldots, \mathbf{x}_n^T$ into a matrix $\mathbf{X}$:

$$\mathbf{X} = \begin{pmatrix} \mathbf{x}_1^T \\ \vdots \\ \mathbf{x}_n^T \end{pmatrix} = \begin{pmatrix} x_{11} & x_{12} & \ldots & x_{1p} \\ \vdots & \vdots & \vdots & \vdots \\ x_{n1} & x_{n2} & \ldots & x_{np} \end{pmatrix}.$$

It is often of interest to study whether, and how, y may be related to the explanatory variables $x_1, \ldots, x_p$. One approach, among many, to studying this relationship is based on modeling it by a linear model. We are adopting a *linear model* when we act as though the observed values of y are related to the observed values of $x_1, \ldots, x_p$ via the equations

$$y_i = \mathbf{x}_i^T \boldsymbol{\beta} + e_i \quad (i = 1, \ldots, n)$$

D. L. Zimmerman, *Linear Model Theory*,
https://doi.org/10.1007/978-3-030-52063-2_1

or equivalently

$$\mathbf{y} = \mathbf{X}\boldsymbol{\beta} + \mathbf{e}, \tag{1.1}$$

where $\boldsymbol{\beta} = (\beta_j)$ is a p-vector of unknown parameters that is contained in some parameter space and $\mathbf{e} = (e_i)$ is a n-vector of unobservable "errors" or "residuals." We are adopting a *statistical linear model* when, in addition, we regard the errors as realizations of n random variables. Thus, under a statistical linear model the elements of both $\mathbf{e}$ and $\mathbf{y}$ are realizations of random variables, but only $\mathbf{y}$ is observed. The elements of $\mathbf{X}$ are observed but, in contrast to the elements of $\mathbf{y}$, are not regarded as realizations of random variables. The elements of $\boldsymbol{\beta}$ are not observed, and may be regarded as either random (corresponding to a Bayesian approach) or nonrandom (corresponding to a frequentist approach). Throughout this book, we adopt the frequentist perspective.

As just noted, it is assumed that the explanatory variables are nonrandom. Arguably, in some settings it may be just as reasonable to regard one or more of the explanatory variables as random as it is to regard the response variable as random; as an example consider a situation in which y is the weight, and a single explanatory variable, x, is the height, of a person. In such a situation the model will be interpreted as a model for the conditional distribution of $\mathbf{y}$ given the observed value of $\mathbf{X}$, and all inferences made on the basis of this model are conditional (on this $\mathbf{X}$) rather than unconditional.

We give the following names to $\mathbf{y}$, $\mathbf{X}$, $\boldsymbol{\beta}$, and $\mathbf{e}$: $\mathbf{y}$ is the *response vector*, $\mathbf{X}$ is the *model matrix*, $\boldsymbol{\beta}$ is the *parameter vector*, and $\mathbf{e}$ is the *error vector* or *residual vector*; the last two terms are used interchangeably.

Throughout, we consider only those statistical linear models for which the expectations of the errors exist and are equal to zero. Thus, for example, we do not consider models in which the errors have Cauchy distributions. Henceforth, we use the term *linear model* without modifiers to refer to any statistical linear model in which $\boldsymbol{\beta}$ is nonrandom and the errors all have mean zero. Because the errors of a linear model so defined have mean zero, $\mathrm{E}(y_i) = \mathbf{x}_i^T\boldsymbol{\beta}$ for all i, or equivalently, $\mathrm{E}(\mathbf{y}) = \mathbf{X}\boldsymbol{\beta}$.

The only parameters that appear explicitly in the linear model equation (1.1) are the elements of $\boldsymbol{\beta}$. However, the joint distribution of $\mathbf{e}$, and hence the model, may have additional parameters. For example, if it is assumed that all of the e_i's have a common unknown variance σ^2, then the model also includes σ^2 as a parameter. In fact, the e_i's for the simplest linear models we consider have such a common variance, i.e., they are *homoscedastic*; furthermore, they are uncorrelated. More general models may allow the e_i's to have different variances (i.e., to be *heteroscedastic*), or to be correlated, and may thus have additional variance–covariance parameters. In some of these models, correlation among the e_i's may arise as a result of them being linear combinations of other mean-zero, uncorrelated random variables with finite variances.

Often, but not always, the parameter space for $\boldsymbol{\beta}$ is taken to be $\mathbb{R}^p$. When this is so, we say that the linear model is *unconstrained* (in $\boldsymbol{\beta}$). Restricting $\boldsymbol{\beta}$ to belong to some subset of $\mathbb{R}^p$ results in a *constrained* linear model. We restrict attention herein to linear equality constraints on $\boldsymbol{\beta}$. There may also be constraints on the parameters describing the joint distribution of the errors. For example, if it is assumed that all of the e_i's have a common unknown variance σ^2, then obviously σ^2 must be nonnegative.

In what sense is a linear model linear? The model is said to be linear because $\mathrm{E}(y_i)$ is a linear combination of the unknown parameters. That is, for all i,

$$\mathrm{E}(y_i) = \mathbf{x}_i^T \boldsymbol{\beta} = \sum_{j=1}^{p} \beta_j x_{ij}.$$

The elements of $\mathbf{X}$ need not be linear functions of the explanatory variables actually observed. Some examples of $\mathrm{E}(y_i)$ in linear models are as follows:

- $\mathrm{E}(y_i) = \beta_1 + \beta_2 x_i + \beta_3 x_i^2$
- $\mathrm{E}(y_i) = \beta_1 + \beta_2 \exp(-x_{i1}) + \beta_3 \sin(x_{i2})$
- $\mathrm{E}(y_i) = \beta_1 x_{i1} + \beta_2 x_{i2} + \beta_3 x_{i1} x_{i2}$

If $\mathrm{E}(y_i)$ is a nonlinear, rather than linear, function of a p-vector of unknown parameters, the model $y_i = \mathrm{E}(y_i) + e_i$ is called a *nonlinear* model. There are some nonlinear models, known as *transformably linear models*, that can be made linear by a suitable transformation of $\mathrm{E}(y_i)$ (and possibly by a transformation of one or more of the explanatory variables as well). Two examples are the following:

- $\mathrm{E}(y_i) = \exp(\beta_1 + \beta_2 x_i)$
- $\mathrm{E}(y_i) = \frac{\beta_1 x_i}{\beta_2 + x_i}$

The transformations required to "linearize" these models are, respectively, as follows:

- $\log[\mathrm{E}(y_i)] = \beta_1 + \beta_2 x_i$
- $\frac{1}{\mathrm{E}(y_i)} = \beta_3 \left(\frac{1}{x_i}\right) + \beta_4$, where $\beta_3 = \frac{\beta_2}{\beta_1}$ and $\beta_4 = \frac{1}{\beta_1}$

Some statistical methods that have been developed for linear models may be adapted for use with transformably linear models.

Some nonlinear models, however, are not transformably linear. The following are two examples:

- $\mathrm{E}(y_i) = \beta_1 + \beta_2 \beta_3^{x_i}$
- $\mathrm{E}(y_i) = \beta_1 + \beta_2 x_i^{\beta_3}$

No transformation of $\mathrm{E}(y_i)$ can "linearize" these models. However, if β_3 were known, then both would be linear models. Such models are known as *conditionally linear models* because, conditioned on knowing one or more of the parameters, they are linear in the remaining parameters. Some methods that have been developed for linear models may be adapted for use with conditionally linear models also.

This book presents theory needed to solve three main inference problems for linear models. The first is that of making inferences about $\boldsymbol{\beta}$ or, more generally, linear functions of $\boldsymbol{\beta}$. The inferences to be considered include point estimation, interval (and regional) estimation, and hypothesis testing. The second problem is that of making inferences about unobserved random variables that either appear in the model (as components of $\mathbf{e}$) or are related to those that appear in the model. This problem is called *prediction* and involves both point and interval/regional prediction. The final problem is that of making inferences about parameters that characterize the variance–covariance structure of $\mathbf{e}$. Again, these inferences include point and interval estimation and hypothesis testing.

The order of presentation of linear model theory in the book is roughly aligned with the order in which the inference problems were just listed. However, there is a considerable amount of background material on matrix algebra, generalized inverses, solutions to linear systems of equations, and moments of random vectors and certain functions thereof that comes first. These are the topics of Chaps. 2, 3, and 4. Chapter 5 presents additional background on widely used mean structures and error structures of linear models. It is not until Chap. 6 that we begin to take up point estimation of linear functions of $\boldsymbol{\beta}$. That chapter considers which linear functions of $\boldsymbol{\beta}$ are estimable, i.e., able to be estimated unbiasedly by a linear function of the observations. Chapter 7 then presents the method of least squares estimation of estimable functions, and derives some of its properties under a model with homoscedastic and uncorrelated errors. Chapter 8 describes a geometric interpretation of least squares estimation and introduces the overall analysis of variance, a method for partitioning the variation among the observations to that which can be explained by the model and that which cannot. Chapter 9 takes the analysis of variance a step further by showing how the variation may be partitioned among the different explanatory variables in the model.

Chapters 10 and 11 extend least squares estimation of estimable functions and ANOVA to two more general types of linear models: constrained linear models and models for which the elements of $\mathbf{e}$ are heteroscedastic and/or correlated. The effects of model misspecification on least squares estimators are considered in Chap. 12. Chapter 13 presents point prediction; in particular, the best linear unbiased predictor (BLUP) of a so-called predictable linear function of $\boldsymbol{\beta}$ and unobserved random variables is derived under models such as those considered in Chap. 11.

Up to and including Chap. 13, no particular form of the distribution of $\mathbf{e}$ is assumed. In Chap. 14, the multivariate normal distribution and several distributions derived from it are introduced. An assumption that the distribution of $\mathbf{e}$ (or, in more complicated models, the joint distribution of all the random variables whose linear combination constitutes $\mathbf{e}$) is multivariate normal makes it possible to perform inferences beyond mere point estimation. Chapter 15 takes up interval estimation and hypothesis testing for estimable/predictable functions of $\boldsymbol{\beta}$ and unobservable random variables when all variance–covariance parameters except an overall scale parameter are known, while Chap. 16 presents the same topics for the variance–covariance parameters and functions thereof. Chapter 17 briefly summarizes what is known about inference for estimable/predictable functions of $\boldsymbol{\beta}$ and unobserved random variables when the variance–covariance parameters are unknown.

Selected Matrix Algebra Topics and Results 2

Before we can begin to tackle the inference problems described near the end of the previous chapter, we must first develop an adequate working knowledge of matrix algebra useful for linear models. That is the objective of this chapter. Admittedly, the topics and results selected for inclusion here are severely abridged, being limited almost exclusively to what will actually be needed in later chapters. Furthermore, for some of the results (particularly those that are used only once or twice in the sequel), little context is provided. For much more thorough treatments of matrix algebra useful for linear models and other areas of statistics, we refer the reader to the books by Harville (1997) and Schott (2016). In fact, for proofs not given in this chapter, we provide a reference to a proof given in one or both of those books.

2.1 Basic Notation and Terminology

A *matrix* $\mathbf{A}_{n\times m} = (a_{ij})$ is a rectangular $(n \times m)$ array of numbers. The numbers in the matrix are called *elements*. Throughout this book, we restrict attention exclusively to *real* matrices, i.e., matrices whose elements are real numbers. The *dimensions* of $\mathbf{A}$ are n (number of rows) by m (number of columns). An *n-vector* $\mathbf{a} = (a_i)$ is an $n \times 1$ matrix. (By convention, we take a vector to be a *column vector*; a matrix consisting of one row is called a *row vector*.) The *dimension* of a vector is the number of its elements. A 1×1 matrix is also called a *scalar*.

A *submatrix* of a matrix $\mathbf{A}$ is a matrix obtained by deleting entire rows and/or columns of $\mathbf{A}$. A *subvector* is a submatrix of a column or row vector. A matrix separated into subsets of contiguous columns and/or subsets of contiguous rows is called a *partitioned matrix*. The matrices $\mathbf{B}$, $\mathbf{C}$, $\mathbf{D}$, $\mathbf{E}$ comprising the partitioned matrices $\mathbf{A} = (\mathbf{B} \;\; \mathbf{C})$, $\mathbf{A} = \begin{pmatrix} \mathbf{B} \\ \mathbf{C} \end{pmatrix}$, and $\mathbf{A} = \begin{pmatrix} \mathbf{B} & \mathbf{C} \\ \mathbf{D} & \mathbf{E} \end{pmatrix}$ are called *blocks*.

If the dimensions n and m of an $n \times m$ matrix $\mathbf{A}$ are equal, we say that $\mathbf{A}$ is *square*. If $\mathbf{A} = (a_{ij})$ is square with n rows (and n columns), its elements $a_{11}, a_{22}, \ldots, a_{nn}$

D. L. Zimmerman, *Linear Model Theory*,
https://doi.org/10.1007/978-3-030-52063-2_2

are called *main diagonal* elements, and its remaining elements are referred to as *off-diagonal elements*. If all of the off-diagonal elements of an $n \times n$ matrix $\mathbf{A}$ are equal to zero, $\mathbf{A}$ is said to be a *diagonal* matrix, which may be written as $\mathbf{A} = \text{diag}(a_{11}, \ldots, a_{nn})$ or more simply as $\mathbf{A} = \text{diag}(a_1, \ldots, a_n)$ where $a_i \equiv a_{ii}$. If a square matrix $\mathbf{A}$ may be partitioned as

$$\mathbf{A} = \begin{pmatrix} \mathbf{A}_{11} & \mathbf{A}_{12} & \cdots & \mathbf{A}_{1t} \\ \mathbf{A}_{21} & \mathbf{A}_{22} & \cdots & \mathbf{A}_{2t} \\ \vdots & \vdots & \ddots & \vdots \\ \mathbf{A}_{t1} & \mathbf{A}_{t2} & \cdots & \mathbf{A}_{tt} \end{pmatrix}$$

where $\mathbf{A}_{ii}$ is a square matrix for each i and all the elements of $\mathbf{A}_{ij}$ are 0 for every $i \neq j$, then $\mathbf{A}$ is called a *block diagonal* matrix.

If $\mathbf{A} = (a_{ij})$ is an $n \times m$ matrix, the *transpose* of $\mathbf{A}$, written as $\mathbf{A}^T$, is the $m \times n$ matrix (a_{ji}). Thus if $\mathbf{A} = \begin{pmatrix} 1 & 2 & 3 \\ 4 & 5 & 6 \end{pmatrix}$, then

$$\mathbf{A}^T = \begin{pmatrix} 1 & 4 \\ 2 & 5 \\ 3 & 6 \end{pmatrix}.$$

Some additional frequently used notation and terminology are as follows:

- 0—scalar zero
- $\mathbf{0}_n$—zero (or *null*) n-vector
- $\mathbf{0}_{n \times m}$—$n \times m$ zero (or *null*) matrix
- $\mathbf{1}_n$—n-vector whose elements are all equal to one
- $\mathbf{J}_{n \times m} = \mathbf{1}_n \mathbf{1}_m^T$—$n \times m$ matrix whose elements are all equal to one
- $\mathbf{J}_n = \mathbf{J}_{n \times n} = \mathbf{1}_n \mathbf{1}_n^T$
- $\mathbf{I}_n$—$n \times n$ *identity matrix*, i.e., an $n \times n$ diagonal matrix whose main diagonal elements are all equal to one
- $\mathbf{u}_i^{(n)}$—ith *unit n-vector*, i.e., the ith column of $\mathbf{I}_n$

Usually, the context of a given situation makes clear the dimensions of these matrices, so we often omit the subscripts (or the superscript, in the case of $\mathbf{u}_i^{(n)}$).

It is assumed in this book that the reader knows the basic definitions and rules governing the addition of two matrices, the scalar multiplication of a matrix, and the multiplication of two matrices; if not, see Section 1.2 of Harville (1997) or Section 1.3 of Schott (2016). In regard to matrix multiplication, it is important to be aware of the following three facts. First, the matrix product $\mathbf{AB}$ is well defined only when $\mathbf{B}$ has the same number of rows as $\mathbf{A}$ has columns, in which case we say that $\mathbf{A}$ is conformal for post-multiplication by $\mathbf{B}$ (or equivalently, that $\mathbf{B}$ is conformal for pre-multiplication by $\mathbf{A}$). Second, matrix multiplication is not commutative, i.e., $\mathbf{AB}$

(when well defined) does not necessarily equal **BA**; in fact, **BA** need not be well defined, even when **AB** is. Third, post-multiplication of an $n \times m$ matrix **A** by an m-vector **b** is equivalent to forming a linear combination of the columns of **A** in which the elements of **b** are scalar coefficients, i.e.,

$$\mathbf{A}\mathbf{b} = (\mathbf{a}_1, \mathbf{a}_2, \ldots, \mathbf{a}_m) \begin{pmatrix} b_1 \\ b_2 \\ \vdots \\ b_m \end{pmatrix} = \sum_{j=1}^{m} b_j \mathbf{a}_j.$$

Similarly, pre-multiplication of an $n \times m$ matrix **A** by an n-dimensional row vector $\mathbf{c}^T$ is equivalent to forming a linear combination of the rows of **A** in which the elements of $\mathbf{c}^T$ are scalar coefficients, i.e.,

$$\mathbf{c}^T \mathbf{A} = (c_1, c_2, \ldots, c_n) \begin{pmatrix} \boldsymbol{\alpha}_1^T \\ \boldsymbol{\alpha}_2^T \\ \vdots \\ \boldsymbol{\alpha}_n^T \end{pmatrix} = \sum_{i=1}^{n} c_i \boldsymbol{\alpha}_i^T.$$

The result for **Ab** may be extended to the following result for the matrix product **AB**, where **A** is $n \times m$ and **B** is $m \times q$:

$$\mathbf{A}\mathbf{B} = (\mathbf{a}_1, \mathbf{a}_2, \ldots, \mathbf{a}_m) \begin{pmatrix} \mathbf{b}_1^T \\ \mathbf{b}_2^T \\ \vdots \\ \mathbf{b}_m^T \end{pmatrix} = \sum_{j=1}^{m} \mathbf{a}_j \mathbf{b}_j^T.$$

Theorem 2.1.1 *Let* **A** *and* **B** *represent* $n \times m$ *matrices. If* $\mathbf{Ax} = \mathbf{Bx}$ *for all* m*-vectors* **x**, *then* $\mathbf{A} = \mathbf{B}$.

Proof Write **A** and **B** in terms of their columns as $\mathbf{A} = (\mathbf{a}_1, \mathbf{a}_2, \ldots, \mathbf{a}_m)$ and $\mathbf{B} = (\mathbf{b}_1, \mathbf{b}_2, \ldots, \mathbf{b}_m)$. By hypothesis, $\mathbf{A}\mathbf{u}_i = \mathbf{B}\mathbf{u}_i$ for each unit m-vector $\mathbf{u}_i$ $(i = 1, \ldots, n)$, i.e., $\mathbf{a}_i = \mathbf{b}_i$. Thus, column by column, **A** and **B** are identical, implying that $\mathbf{A} = \mathbf{B}$. □

For any n-vector **a**, the *norm* or *length* of **a** is defined as $\|\mathbf{a}\| = (\mathbf{a}^T\mathbf{a})^{\frac{1}{2}}$. One measure of distance between two n-vectors **a** and **b** is $\|\mathbf{a} - \mathbf{b}\|$. In fact, this measure satisfies the definition of a distance metric on $\mathbb{R}^n$: it is nonnegative, it equals 0 if and only if $\mathbf{a} = \mathbf{b}$, it is symmetric in **a** and **b**, and it satisfies the triangle inequality. Note that $\|(1/n)^{\frac{1}{2}}\mathbf{1}_n\| = 1$ and $\|\mathbf{u}_i^{(n)}\| = 1$ for all $i = 1, \ldots, n$.

2.2 Linearly Dependent and Linearly Independent Vectors

Definition 2.2.1 A finite nonempty set of vectors $\{\mathbf{v}_1, \ldots, \mathbf{v}_m\}$ of the same dimension is said to be *linearly dependent* if real numbers $c_1, \ldots, c_m$, at least one of which is nonzero, exist such that $\sum_{i=1}^{m} c_i \mathbf{v}_i = \mathbf{0}$. If no such real numbers exist, the set of vectors is said to be *linearly independent.*

Example 2.2-1. Some Linearly Independent Sets and Linearly Dependent Sets of Vectors

It is clear from Definition 2.2.1 that a set consisting of a single arbitrary nonnull n-vector is linearly independent, and that a set consisting of any two n-vectors is linearly independent unless one of the vectors is null or each vector is a nonzero scalar multiple of the other. Thus, for example, $\{\mathbf{1}_n, \mathbf{u}_1^{(n)}\}$ is a linearly independent set (for $n > 1$), but $\{\mathbf{1}_n, -6\mathbf{1}_n\}$ is linearly dependent. It is also clear that for any n, $\{\mathbf{u}_1^{(n)}, \ldots, \mathbf{u}_n^{(n)}\}$ is a linearly independent set. However, $\{\mathbf{u}_1^{(n)}, \mathbf{u}_2^{(n)}, \ldots, \mathbf{u}_n^{(n)}, \mathbf{1}_n\}$ is linearly dependent because $\sum_{i=1}^{n} \mathbf{u}_i^{(n)} - \mathbf{1}_n = \mathbf{0}_n$. Similarly,

$$\left\{ \begin{pmatrix} 1 \\ 1 \\ 0 \\ 0 \end{pmatrix}, \begin{pmatrix} 0 \\ 0 \\ 1 \\ 1 \end{pmatrix}, \begin{pmatrix} 1 \\ 0 \\ 1 \\ 0 \end{pmatrix}, \begin{pmatrix} 0 \\ 1 \\ 0 \\ 1 \end{pmatrix} \right\}$$

is linearly dependent because, for example, the fourth vector is equal to the first plus the second, minus the third. ■

The following three theorems provide useful ways to determine if a set of n-vectors is linearly independent.

Theorem 2.2.1 *If $\mathbf{0}_n$ is an element of a finite set of n-vectors, then the set is linearly dependent.*

Proof Without loss of generality, assume that there are $m \geq 1$ vectors in the set, and that $\mathbf{0}_n$ is the first such vector. Then, for $c_1 \neq 0$ and $c_2 = c_3 = \cdots = c_m = 0$, we have $\sum_{i=1}^{m} c_i \mathbf{v}_i = \mathbf{0}$. □

Theorem 2.2.2 *A finite set of two or more n-vectors is linearly dependent if and only if at least one vector in the set may be expressed as a linear combination of the others.*

Proof Harville (1997, Lemma 3.2.1) or Schott (2016, Theorem 2.4). □

Theorem 2.2.3 *Let* $\{\mathbf{v}_1, \ldots, \mathbf{v}_m\}$ *be a set of n-vectors. If* $m > n$*, then the set is linearly dependent.*

Proof Harville (1997, Corollary 4.3.3) or Schott (2016, Theorem 2.6). □

2.3 Orthogonal Vectors

Definition 2.3.1 A finite set of vectors $\{\mathbf{v}_1, \ldots, \mathbf{v}_m\}$ (all of the same dimension) is said to be *orthogonal* if $\mathbf{v}_i^T\mathbf{v}_j = 0$ for all $i \neq j$. If, in addition, $\mathbf{v}_i^T\mathbf{v}_i = 1$ for all $i = 1, \ldots, m$, then the set is said to be *orthonormal*.

Observe that if $\{\mathbf{v}_1, \ldots, \mathbf{v}_m\}$ is an orthogonal set of nonnull vectors, then $\{(\mathbf{v}_1^T\mathbf{v}_1)^{-\frac{1}{2}}\mathbf{v}_1, \ldots, (\mathbf{v}_m^T\mathbf{v}_m)^{-\frac{1}{2}}\mathbf{v}_m\}$ is an orthonormal set.

Example 2.3-1. Some Sets of Orthogonal Vectors

The set of unit n-vectors $\{\mathbf{u}_1^{(n)}, \ldots, \mathbf{u}_n^{(n)}\}$ is orthogonal; in fact, it is orthonormal. As a second example, consider the set of 3-vectors $\{(1, 1, 0)^T, (1, -1, 0)^T, (0, 0, 1)^T\}$. It is easily verified that this set is orthogonal, but not orthonormal. The corresponding "normalized" set is $\{(\sqrt{2}/2, \sqrt{2}/2, 0)^T, (\sqrt{2}/2, -\sqrt{2}/2, 0)^T, (0, 0, 1)^T\}$.

Another set of orthogonal vectors is $\{\mathbf{1}_n, \mathbf{x}-\bar{x}\mathbf{1}_n\}$, where $\mathbf{x} = (x_i)$ is an arbitrary n-vector and $\bar{x} = (1/n)\sum_{i=1}^n x_i$; observe that $\mathbf{1}_n^T(\mathbf{x} - \bar{x}\mathbf{1}_n) = \mathbf{1}_n^T\mathbf{x} - \bar{x}\mathbf{1}_n^T\mathbf{1}_n = \sum_{i=1}^n x_i - n\bar{x} = 0$. ■

The following theorem describes a relationship between orthogonality and linear independence for a set of (nonnull) n-vectors.

Theorem 2.3.1 *Any orthogonal set of nonnull vectors is linearly independent.*

Proof Harville (1997, Lemma 6.2.1). □

The converse of Theorem 2.3.1 is false; a counterexample is the set $\{\mathbf{1}_n, \mathbf{u}_1^{(n)}\}$ (for $n > 1$).

2.4 Vector Spaces and Subspaces

Definition 2.4.1 Suppose that V_n is a nonempty set of n-vectors satisfying the following two properties:

1. If $\mathbf{v}_1 \in V_n$ and $\mathbf{v}_2 \in V_n$, then $\mathbf{v}_1 + \mathbf{v}_2 \in V_n$;
2. If $\mathbf{v}_1 \in V_n$ and $c \in \mathbb{R}$, then $c\mathbf{v}_1 \in V_n$.

Such a set V_n is said to be a *vector space*. A nonempty set of n-vectors S_n is a *subspace* of a vector space V_n if S_n is a vector space and S_n is a subset of V_n.

It follows easily from the second part of Definition 2.4.1 that $\mathbf{0}_n$ is an element of every n-dimensional vector space.

Example 2.4-1. Some Vector Spaces and Subspaces

Some easy-to-verify examples of vector spaces include $\{\mathbf{0}_n\}$, $\{c\mathbf{a} : c \in \mathbb{R}\}$ where $\mathbf{a}$ is any n-vector, $\{a\mathbf{v}_1 + b\mathbf{v}_2 : a \in \mathbb{R}, b \in \mathbb{R}\}$ where $\mathbf{v}_1$ and $\mathbf{v}_2$ are arbitrary n-vectors, and $\mathbb{R}^n$. Another example is $\{\mathbf{v} = (v_i) \in \mathbb{R}^2 : v_1 = -v_2\}$. Furthermore, $\{\mathbf{0}_n\}$ is a subspace of $\{c\mathbf{a} : c \in \mathbb{R}\}$, which in turn is a subspace of $\mathbb{R}^n$. In fact, $\{\mathbf{0}_n\}$ is a subspace of every n-dimensional vector space, and every n-dimensional vector space is a subspace of $\mathbb{R}^n$. ■

In some settings two n-dimensional vector spaces are of interest, but neither is a subspace of the other. Such vector spaces may have other kinds of relationships to each other, two of which are defined as follows.

Definition 2.4.2 Two n-dimensional vector spaces U_n and V_n are said to be *essentially disjoint* if they have only the null vector in common, i.e., if $U_n \cap V_n = \{\mathbf{0}_n\}$.

Definition 2.4.3 Two n-dimensional vector spaces U_n and V_n are said to be *orthogonal* to each other if every vector in U_n is orthogonal to every vector in V_n. An n-vector $\mathbf{a}$ is said to be *orthogonal* to a vector space V_n if the vector spaces $U_n \equiv \{c\mathbf{a} : c \in \mathbb{R}\}$ and V_n are orthogonal to each other.

Example 2.4-2. Some Essentially Disjoint and/or Orthogonal Vector Spaces

Define

$$U_2 = \left\{c\begin{pmatrix}1\\1\end{pmatrix} : c \in \mathbb{R}\right\}, \quad V_2 = \left\{c\begin{pmatrix}1\\2\end{pmatrix} : c \in \mathbb{R}\right\}, \quad W_2 = \left\{c\begin{pmatrix}1\\-1\end{pmatrix} : c \in \mathbb{R}\right\}.$$

Then U_2, V_2, and W_2 are all subspaces of $\mathbb{R}^2$; in fact, each corresponds to a line in $\mathbb{R}^2$ passing through the origin. In terms of Cartesian coordinates (x, y), U_2 is the line $y = x$, V_2 is the line $y = 2x$, and W_2 is the line $y = -x$. Each pair of lines intersect only at the origin, so the subspaces are essentially disjoint. Furthermore, U_2 and W_2 are orthogonal to each other, but the other pairs of vector spaces are not.

As another example, consider the vector spaces

$$U_4 = \left\{ \mathbf{v} : \mathbf{v} = c_1 \begin{pmatrix} 1 \\ 1 \\ 0 \\ 0 \end{pmatrix} + c_2 \begin{pmatrix} 0 \\ 0 \\ 1 \\ 1 \end{pmatrix}, \ c_1 \in \mathbb{R},\ c_2 \in \mathbb{R} \right\}$$

and

$$V_4 = \left\{ \mathbf{v} : \mathbf{v} = c_1 \begin{pmatrix} 1 \\ 0 \\ 1 \\ 0 \end{pmatrix} + c_2 \begin{pmatrix} 0 \\ 1 \\ 0 \\ 1 \end{pmatrix}, \ c_1 \in \mathbb{R},\ c_2 \in \mathbb{R} \right\}.$$

Although neither $(1, 1, 0, 0)^T$ nor $(0, 0, 1, 1)^T$ is a linear combination of $(1, 0, 1, 0)^T$ and $(0, 1, 0, 1)^T$, and neither $(1, 0, 1, 0)^T$ nor $(0, 1, 0, 1)^T$ is a linear combination of $(1, 1, 0, 0)^T$ and $(0, 0, 1, 1)^T$, U_4 and V_4 are not essentially disjoint, for they share the vector $(1, 1, 1, 1)$ (and all scalar multiples thereof) in common. U_4 and V_4 are not orthogonal to each other either, for

$$(1\ 1\ 0\ 0) \begin{pmatrix} 1 \\ 0 \\ 1 \\ 0 \end{pmatrix} \neq 0.$$

■

The following theorem describes a relationship between orthogonal vector spaces and essentially disjoint vector spaces.

Theorem 2.4.1 *If U_n and V_n are orthogonal vector spaces, then they are essentially disjoint.*

Proof Let $\mathbf{a}$ represent an n-vector belonging to both vector spaces, i.e., $\mathbf{a} \in U_n$ and $\mathbf{a} \in V_n$. Then, because U_n and V_n are orthogonal to each other, $\mathbf{a}^T\mathbf{a} = 0$, implying further (because $\mathbf{a}^T\mathbf{a}$ is the sum of squares of elements of $\mathbf{a}$) that $\mathbf{a} = \mathbf{0}_n$. □

The converse of Theorem 2.4.1 is false; a counterexample is provided by U_2 and V_2 in Example 2.4-2.

Definition 2.4.4 The set of vectors in a vector space V_n that are orthogonal to a subspace S_n of V_n is called the *orthogonal complement of* S_n (relative to V_n).

It may be shown (see the exercises) that the orthogonal complement of any subspace is itself a subspace. Thus, by Theorem 2.4.1, a vector space and its orthogonal complement are essentially disjoint. Observe that in Example 2.4-2, W_2 is the orthogonal complement of U_2 (relative to $\mathbb{R}^2$).

2.5 Spans, Spanning Sets, Bases, and Dimension

Definition 2.5.1 Let $\{\mathbf{v}_1, \ldots, \mathbf{v}_m\}$ be a finite nonempty set of vectors in a vector space V_n. If each vector in V_n can be expressed as a linear combination of the vectors $\mathbf{v}_1, \ldots, \mathbf{v}_m$, then the set $\{\mathbf{v}_1, \ldots, \mathbf{v}_m\}$ is said to *span*, and be a *spanning set* of, V_n, and we write $\text{span}\{\mathbf{v}_1, \ldots, \mathbf{v}_m\} = V_n$.

Definition 2.5.2 . A *basis* for a vector space V_n is a finite set of linearly independent vectors that span V_n.

A basis exists for every vector space except $\{\mathbf{0}_n\}$ (for every n). In general, a basis for any other vector space is not unique, but the number of vectors in a basis, called the *dimension* of the vector space, *is* unique; for a proof of this, see Harville (1997, Theorem 4.3.6). We write $\dim(V_n)$ for the dimension of a vector space V_n; by convention $\dim(\{\mathbf{0}_n\}) = 0$.

Example 2.5-1. Spans, Spanning Sets, Bases, and Dimension of Some Vector Spaces

For the vector space $\{c\mathbf{a}_n : c \in \mathbb{R}\}$, where $\mathbf{a}_n$ is any nonnull n-vector, $\{\mathbf{a}_n\}$ is a basis, and another is $\{3\mathbf{a}_n\}$; any basis has dimension 1. One spanning set of the same vector space is $\{\mathbf{a}_n, 2\mathbf{a}_n, -3\mathbf{a}_n\}$. Next consider

$$\mathcal{V}_4 = \text{span}\left\{\begin{pmatrix}1\\1\\0\\0\end{pmatrix}, \begin{pmatrix}0\\0\\1\\1\end{pmatrix}, \begin{pmatrix}1\\0\\1\\0\end{pmatrix}, \begin{pmatrix}0\\1\\0\\1\end{pmatrix}\right\}.$$

Any subset of three vectors in the given spanning set of $\mathcal{V}_4$ is a basis for $\mathcal{V}_4$ because such a subset is linearly independent and the excluded vector is a linear combination of the three vectors in the subset. Thus $\dim(\mathcal{V}_4) = 3$. Finally consider the vector space $\mathbb{R}^n$. A natural basis for $\mathbb{R}^n$ is the set of unit vectors $\{\mathbf{u}_1^{(n)}, \ldots, \mathbf{u}_n^{(n)}\}$ but there are many others, such as $\{\mathbf{u}_1^{(n)}, \mathbf{u}_1^{(n)}+\mathbf{u}_2^{(n)}, \ldots, \sum_{i=1}^n \mathbf{u}_i^{(n)}\}$. All bases of $\mathbb{R}^n$ comprise n vectors, so $\dim(\mathbb{R}^n) = n$. ■

2.6 Column, Row, and Null Spaces

Definition 2.6.1 Let $\mathbf{A} = (\mathbf{a}_1, \ldots, \mathbf{a}_m)$, where each $\mathbf{a}_j$ is an n-vector, and let $S = \{\mathbf{v} : \mathbf{v} = \sum_{j=1}^{m} c_j \mathbf{a}_j, c_j \in \mathbb{R}\}$, which can be written alternatively as $\{\mathbf{v} : \mathbf{v} = \mathbf{A}\mathbf{c}, \mathbf{c} \in \mathbb{R}^m\}$. S is a special vector space which is called the *column space of* $\mathbf{A}$, denoted by $\mathcal{C}(\mathbf{A})$. Similarly, if

$$\mathbf{A} = \begin{pmatrix} \boldsymbol{\alpha}_1^T \\ \vdots \\ \boldsymbol{\alpha}_n^T \end{pmatrix},$$

where each $\boldsymbol{\alpha}_i^T$ is an m-dimensional row vector, then the set of vectors $\{\mathbf{v}^T : \mathbf{v}^T = \sum_{i=1}^{n} d_i \boldsymbol{\alpha}_i^T, d_i \in \mathbb{R}\}$, which can alternatively be written as $\{\mathbf{v}^T : \mathbf{v}^T = \mathbf{d}^T\mathbf{A}, \mathbf{d} \in \mathbb{R}^n\}$, is called the *row space* of $\mathbf{A}$, denoted by $\mathcal{R}(\mathbf{A})$.

Observe that $\mathcal{C}(\mathbf{A})$ is the set of all n-vectors spanned by the columns of $\mathbf{A}$ and is a subspace of $\mathbb{R}^n$. Similarly, $\mathcal{R}(\mathbf{A})$ is the set of all m-dimensional row vectors spanned by the rows of $\mathbf{A}$ and is a subspace of $\mathbb{R}^m$. Observe also that $\mathbf{v} \in \mathcal{C}(\mathbf{A})$ if and only if $\mathbf{v} \in \mathcal{R}(\mathbf{A}^T)$, and $\mathbf{v} \in \mathcal{R}(\mathbf{A})$ if and only if $\mathbf{v} \in \mathcal{C}(\mathbf{A}^T)$.

Example 2.6-1. Column Spaces and Row Spaces of Selected Matrices

Observe that $\mathcal{C}(\mathbf{I}_n) = \mathcal{R}(\mathbf{I}_n) = \mathbb{R}^n$ and $\mathcal{C}(\mathbf{J}_n) = \mathcal{C}(\mathbf{1}_n) = \{c\mathbf{1}_n : c \in \mathbb{R}\}$. Also, if

$$\mathbf{A} = \begin{pmatrix} 1 & 1 & 0 \\ 1 & 1 & 0 \\ 1 & 0 & 1 \end{pmatrix},$$

then $\mathcal{C}(\mathbf{A}) = \{(a, b, c)^T : a = b\}$ and $\mathcal{R}(\mathbf{A}) = \{(a, b, c) : a = b + c\}$. ■

Theorem 2.6.1 *$\mathcal{C}(\mathbf{B})$ is a subspace of $\mathcal{C}(\mathbf{A})$ if and only if $\mathbf{AC} = \mathbf{B}$ for some matrix $\mathbf{C}$; likewise $\mathcal{R}(\mathbf{B})$ is a subspace of $\mathcal{R}(\mathbf{A})$ if and only if $\mathbf{DA} = \mathbf{B}$ for some matrix $\mathbf{D}$.*

Proof Harville (1997, Lemma 4.2.2). □

Definition 2.6.2 The *null space* of an $n \times m$ matrix $\mathbf{A}$, denoted by $\mathcal{N}(\mathbf{A})$, is the set of m-vectors $\mathcal{N}(\mathbf{A}) = \{\mathbf{v} : \mathbf{Av} = \mathbf{0}\}$.

It can be shown (see the exercises) that $\mathcal{N}(\mathbf{A})$ is a subspace of $\mathbb{R}^m$. Furthermore, there is a relationship between the null space and row space of a matrix, as described by the following theorem.

Theorem 2.6.2 *For any matrix* $\mathbf{A}$, $\mathcal{C}(\mathbf{A})$ *and* $\mathcal{N}(\mathbf{A}^T)$ *are orthogonal to each other. In fact,* $\mathcal{N}(\mathbf{A}^T)$ *is the orthogonal complement of* $\mathcal{C}(\mathbf{A})$. *Similarly,* $\mathcal{N}(\mathbf{A})$ *is the orthogonal complement of* $\mathcal{R}(\mathbf{A})$.

Proof Harville (1997, Lemma 12.5.2). □

2.7 Matrix Transpose and Symmetric Matrices

In Sect. 2.1 we defined the *transpose* of a matrix $\mathbf{A}$ and wrote it as $\mathbf{A}^T$. The following theorem gives some basic properties of the transpose.

Theorem 2.7.1 *If* $\mathbf{A}$ *is any matrix, then:*

(a) $(\mathbf{A}^T)^T = \mathbf{A}$;
(b) if $\mathbf{B}$ *has the same dimensions as* $\mathbf{A}$, *and* c *and* d *are arbitrary scalars, then* $(c\mathbf{A} + d\mathbf{B})^T = c\mathbf{A}^T + d\mathbf{B}^T$; *and*
(c) if $\mathbf{A}$ *is conformal for post-multiplication by* $\mathbf{B}$, *then* $(\mathbf{AB})^T = \mathbf{B}^T\mathbf{A}^T$.

Proof Proofs of parts (a) and (b) are trivial. The proof of part (c) is not much more difficult; one may be found in Harville (1997, Section 1.2d) or Schott (2016, Section 1.4). □

Definition 2.7.1 A (square) matrix $\mathbf{A}$ is said to be *symmetric* if $\mathbf{A}^T = \mathbf{A}$.

Example 2.7-1. Some Symmetric Matrices

By inspection, it can be seen that the following matrices are symmetric: any scalar, $\mathbf{0}_{n\times n}$, $\mathbf{I}_n$, $\mathbf{J}_n$, $a\mathbf{I}_n + b\mathbf{J}_n$ for any scalars a and b, and any diagonal matrix. Also, $\mathbf{aa}^T$ is symmetric for any n-vector $\mathbf{a} = (a_1, \ldots, a_n)^T$ because $\mathbf{aa}^T = (a_i a_j)$ and $a_i a_j = a_j a_i$ for all i and j. ■

If a matrix is square and symmetric, then each off-diagonal element below the main diagonal coincides with a particular off-diagonal element above the main diagonal. Thus, when displaying a square symmetric matrix in terms of its scalar elements, only the main diagonal elements and the off-diagonal elements to the right and above the main diagonal need to be included; the remainder of the display may be left blank. For example, a 3×3 symmetric matrix $\mathbf{A} = (a_{ij})$ may be completely specified by the representation

$$\mathbf{A} = \begin{pmatrix} a_{11} & a_{12} & a_{13} \\ & a_{22} & a_{23} \\ & & a_{33} \end{pmatrix}.$$

Theorem 2.7.2 *If* $\mathbf{A}$ *is any matrix, then* $\mathbf{A}^T\mathbf{A}$ *and* $\mathbf{A}\mathbf{A}^T$ *are symmetric matrices.*

Proof Using parts (c) and (a) of Theorem 2.7.1, we obtain $(\mathbf{A}^T\mathbf{A})^T = \mathbf{A}^T(\mathbf{A}^T)^T = \mathbf{A}^T\mathbf{A}$, showing that $\mathbf{A}^T\mathbf{A}$ is symmetric. A similar approach shows that $\mathbf{A}\mathbf{A}^T$ is symmetric also. □

2.8 The Rank of a Matrix

Definition 2.8.1 The *row rank* of a matrix $\mathbf{A}$ is defined as the dimension of $\mathcal{R}(\mathbf{A})$, and the *column rank* of $\mathbf{A}$ is defined as the dimension of $\mathcal{C}(\mathbf{A})$. The row rank of any matrix $\mathbf{A}$ is equal to the column rank of $\mathbf{A}$ [see Harville (1997, Section 4a) or Schott (2016, Section 2.5)], and their common value is simply called the *rank* of $\mathbf{A}$ and written as rank($\mathbf{A}$).

Example 2.8-1. Ranks of Selected Matrices

Using Definition 2.8.1, we see immediately that $\text{rank}(\mathbf{0}_{n\times m}) = 0$, $\text{rank}(\mathbf{I}_n) = n$, and $\text{rank}(\mathbf{J}_{n\times m}) = 1$. The rank of $\mathbf{I}_n + \mathbf{J}_n$ is not as obvious from the definition, so it will be considered later. Furthermore, if $\mathbf{a} = (a_i)$ is a nonnull n-vector, then $\text{rank}(\mathbf{a}\mathbf{a}^T) = 1$ because $\mathbf{a}\mathbf{a}^T = (a_1\mathbf{a}, a_2\mathbf{a}, \ldots, a_n\mathbf{a})$ and thus each column of $\mathbf{a}\mathbf{a}^T$ is an element of $\mathcal{C}(\mathbf{a})$, which has dimension 1. Finally, if

$$\mathbf{A} = \begin{pmatrix} 1 & 0 & 1 & 0 \\ 1 & 0 & 0 & 1 \\ 0 & 1 & 1 & 0 \\ 0 & 1 & 0 & 1 \end{pmatrix},$$

then $\text{rank}(\mathbf{A}) = 3$ because $\mathcal{C}(\mathbf{A})$ coincides with $\mathcal{V}_4$ in Example 2.5-1, where it was shown that $\dim(\mathcal{V}_4) = 3$. ■

If $\mathbf{A}$ is an $n\times m$ matrix, it is clear from Definition 2.8.1 that $\text{rank}(\mathbf{A}) \leq \min(n, m)$. Furthermore, because the dimension of $\mathcal{R}(\mathbf{A})$ is equal to the number of vectors in a basis for this vector space [with a similar result for the dimension of $\mathcal{C}(\mathbf{A})$], and such vectors are linearly independent, we have the following result.

Theorem 2.8.1 *Let* $\mathbf{A}$ *represent an* $n \times m$ *matrix. Then rank*($\mathbf{A}$) $= n$ *if and only if the rows of* $\mathbf{A}$ *are linearly independent, and rank*($\mathbf{A}$) $= m$ *if and only if the columns of* $\mathbf{A}$ *are linearly independent.*

The following theorems give several other important results pertaining to the ranks of various matrices.

Theorem 2.8.2 *For any matrix* $\mathbf{A}$, *rank*$(\mathbf{A}^T) = $ *rank*$(\mathbf{A})$.

Proof Left as an exercise. □

Theorem 2.8.3 *For any matrix* $\mathbf{A}$ *and any nonzero scalar c, rank*$(c\mathbf{A}) = $ *rank*$(\mathbf{A})$.

Proof Left as an exercise. □

Theorem 2.8.4 *For any matrices* $\mathbf{A}$ *and* $\mathbf{B}$ *such that* $\mathbf{AB}$ *is well defined,*

$$rank(\mathbf{AB}) \leq \min[rank(\mathbf{A}), rank(\mathbf{B})].$$

Proof Harville (1997, Corollary 4.4.5) or Schott (2016, Theorem 2.8a). □

Theorem 2.8.5 *Let* $\mathbf{A}$ *and* $\mathbf{B}$ *represent matrices having the same number of rows. If* $\mathcal{C}(\mathbf{B}) \subseteq \mathcal{C}(\mathbf{A})$ *and rank*$(\mathbf{B}) = $ *rank*$(\mathbf{A})$, *then* $\mathcal{C}(\mathbf{B}) = \mathcal{C}(\mathbf{A})$. *Similarly, let* $\mathbf{A}$ *and* $\mathbf{C}$ *represent matrices having the same number of columns. If* $\mathcal{R}(\mathbf{C}) \subseteq \mathcal{R}(\mathbf{A})$ *and rank*$(\mathbf{C}) = $ *rank*$(\mathbf{A})$, *then* $\mathcal{R}(\mathbf{C}) = \mathcal{R}(\mathbf{A})$.

Proof Harville (1997, Theorem 4.4.6). □

Theorem 2.8.6 *For any matrices* $\mathbf{A}$ *and* $\mathbf{B}$ *having the same number of rows, rank*$(\mathbf{A}, \mathbf{B}) \leq $ *rank*$(\mathbf{A}) + $ *rank*$(\mathbf{B})$, *with equality if and only if* $\mathcal{C}(\mathbf{A}) \cap \mathcal{C}(\mathbf{B}) = \{\mathbf{0}\}$. *Similarly, for any matrices* $\mathbf{A}$ *and* $\mathbf{B}$ *having the same number of columns,* $rank\begin{pmatrix}\mathbf{A}\\ \mathbf{B}\end{pmatrix} \leq rank(\mathbf{A}) + rank(\mathbf{B})$, *with equality if and only if* $\mathcal{R}(\mathbf{A}) \cap \mathcal{R}(\mathbf{B}) = \{\mathbf{0}\}$.

Proof Harville (1997, Theorem 17.2.4). □

Theorem 2.8.7 *For any matrices* $\mathbf{A}$ *and* $\mathbf{B}$ *of the same dimensions, rank*$(\mathbf{A} + \mathbf{B}) \leq $ *rank*$(\mathbf{A}) + $ *rank*$(\mathbf{B})$.

Proof Harville (1997, Corollary 4.5.9) or Schott (2016, Theorem 2.8b). □

Theorem 2.8.8 *For any matrix* $\mathbf{A}$, *rank*$(\mathbf{A}^T\mathbf{A}) = $ *rank*$(\mathbf{A})$.

Proof Harville (1997, Corollary 7.4.5) or Schott (2016, Theorem 2.8c). □

Definition 2.8.2 If $\mathbf{A}$ is an $n \times m$ matrix, and rank$(\mathbf{A}) = n$, then $\mathbf{A}$ is said to have *full row rank*; if rank$(\mathbf{A}) = m$, then $\mathbf{A}$ is said to have *full column rank*. A matrix is said to be *nonsingular* if it has both full row rank and full column rank.

Clearly, if $\mathbf{A}$ is nonsingular it must be square. Thus, by Definition 2.8.2 an $n \times n$ matrix $\mathbf{A}$ is nonsingular if and only if rank($\mathbf{A}$) $= n$. A square matrix that is not nonsingular is said to be *singular*.

Example 2.8-2. Some Nonsingular and Singular Matrices

Clearly $\mathbf{I}_n$ is nonsingular, as is any diagonal matrix whose main diagonal elements are all nonzero. Another example is

$$\begin{pmatrix} \sqrt{2}/2 & \sqrt{2}/2 & 0 \\ \sqrt{2}/2 & -\sqrt{2}/2 & 0 \\ 0 & 0 & 1 \end{pmatrix};$$

its columns are orthogonal (refer back to Example 2.3-1), so by Theorems 2.3.1 and 2.8.1 it is nonsingular. But $\mathbf{0}_{n\times n}$ and $\mathbf{J}_n$ have ranks 0 and 1, respectively, so $\mathbf{0}_{n\times n}$ is singular, and so is $\mathbf{J}_n$ unless $n = 1$. ■

Two important theorems in which nonsingular matrices and the ranks of other matrices are involved are as follows.

Theorem 2.8.9 *Let* $\mathbf{A}$ *represent any matrix. If* $\mathbf{B}$ *and* $\mathbf{C}$ *are nonsingular matrices conformal for post- and pre-multiplication, respectively, by* $\mathbf{A}$*, then rank*($\mathbf{A}$) $=$ *rank*($\mathbf{BA}$) $=$ *rank*($\mathbf{AC}$) $=$ *rank*($\mathbf{BAC}$).

Proof Harville (1997, Corollary 8.3.3). □

Theorem 2.8.10 *Let* $\mathbf{A}$ *represent an* $n \times m$ *nonnull matrix of rank* r*. Then nonsingular matrices* $\mathbf{P}$ *and* $\mathbf{Q}$ *of dimensions* $n \times n$ *and* $m \times m$*, respectively, exist such that*

$$\mathbf{PAQ} = \begin{pmatrix} \mathbf{I}_r & \mathbf{0} \\ \mathbf{0} & \mathbf{0} \end{pmatrix}.$$

Proof Harville (1997, Theorem 4.4.9). □

2.9 The Inverse of a Nonsingular Matrix

Definition 2.9.1 For any $n \times n$ nonsingular matrix $\mathbf{A}$, a unique $n \times n$ nonsingular matrix $\mathbf{C}$ exists such that $\mathbf{AC} = \mathbf{I}_n$ and $\mathbf{CA} = \mathbf{I}_n$. The matrix $\mathbf{C}$ is called the *inverse* of $\mathbf{A}$, and is denoted by $\mathbf{A}^{-1}$. If $\mathbf{A}$ is singular or nonsquare, it has no inverse.

Inverses play an important role in solving systems of linear equations. If a solution to the system of equations $\mathbf{Ax} = \mathbf{b}$ is desired (where $\mathbf{A}$ and $\mathbf{b}$ are a specified matrix and vector, respectively, and $\mathbf{x}$ is a vector of unknowns), then, provided that $\mathbf{A}$ is nonsingular, $\mathbf{A}^{-1}\mathbf{b}$ is a solution for $\mathbf{x}$.

We will not be concerned with numerical computation of the inverse of a nonsingular matrix, except for the 2×2 case: the inverse of a nonsingular 2×2 matrix $\begin{pmatrix} a & b \\ c & d \end{pmatrix}$ is $[1/(ad - bc)]\begin{pmatrix} d & -b \\ -c & a \end{pmatrix}$, for it can be easily verified that pre-multiplying either matrix by the other yields $\mathbf{I}_2$.

It follows immediately from the definition of the inverse that if $\mathbf{A} = (\mathbf{a}_1, \ldots, \mathbf{a}_n)$ is nonsingular, then $(\mathbf{A}^{-1})^{-1} = \mathbf{A}$ and $\mathbf{A}^{-1}\mathbf{a}_i = \mathbf{u}_i^{(n)}$ $(i = 1, \ldots, n)$. The following theorems document some additional important properties of the inverse or specify expressions for inverses of nonsingular matrices having certain structures.

Theorem 2.9.1 *For any nonsingular matrix* $\mathbf{A}$ *and any nonzero scalar* c, $(c\mathbf{A})^{-1} = (1/c)\mathbf{A}^{-1}$.

Proof $(c\mathbf{A})(1/c)\mathbf{A}^{-1} = c(1/c)\mathbf{AA}^{-1} = \mathbf{I}$. □

Theorem 2.9.2 *For any nonsingular matrix* $\mathbf{A}$, $(\mathbf{A}^T)^{-1} = (\mathbf{A}^{-1})^T$.

Proof By Theorem 2.8.2, $\mathbf{A}^T$ is nonsingular. By the symmetry of $\mathbf{I}$ and by Theorem 2.7.1c,

$$\mathbf{I} = \mathbf{A}^T(\mathbf{A}^T)^{-1} = [\mathbf{A}^T(\mathbf{A}^T)^{-1}]^T = [(\mathbf{A}^T)^{-1}]^T\mathbf{A},$$

implying (by post-multiplication of both sides by $\mathbf{A}^{-1}$) that $[(\mathbf{A}^T)^{-1}]^T = \mathbf{A}^{-1}$. The result follows by transposing both sides and using Theorem 2.7.1a. □

Corollary 2.9.2.1 *If* $\mathbf{A}$ *is nonsingular and symmetric, then* $\mathbf{A}^{-1}$ *is symmetric.*

Theorem 2.9.3 *Let* $\mathbf{A}$ *represent an* $n \times n$ *diagonal matrix, i.e.,* $\mathbf{A} = diag(a_1, \ldots, a_n)$. *Then,* $\mathbf{A}$ *is nonsingular if and only if* $a_i \neq 0$ *for all* $i = 1, \ldots, n$, *in which case* $\mathbf{A}^{-1} = diag(1/a_1, \ldots, 1/a_n)$.

Proof If $a_i \neq 0$ for all $i = 1, \ldots, n$, then $\mathbf{AA}^{-1} = \text{diag}[a_1(1/a_1), a_2(1/a_2), \ldots, a_n(1/a_n)] = \mathbf{I}$. If $a_i = 0$ for some i, then the ith row (and column) of $\mathbf{A}$ is a null vector, in which case $\mathbf{A}$ is not nonsingular. □

Theorem 2.9.4 *For any nonsingular matrices* $\mathbf{A}$ *and* $\mathbf{B}$ *of the same dimensions,* $(\mathbf{AB})^{-1} = \mathbf{B}^{-1}\mathbf{A}^{-1}$.

Proof $(\mathbf{AB})(\mathbf{B}^{-1}\mathbf{A}^{-1}) = \mathbf{A}(\mathbf{BB}^{-1})\mathbf{A}^{-1} = \mathbf{AA}^{-1} = \mathbf{I}$. □

Theorem 2.9.5 *Let* $\begin{pmatrix} \mathbf{A} & \mathbf{B} \\ \mathbf{C} & \mathbf{D} \end{pmatrix}$ *represent a partitioned matrix.*

(a) If $\mathbf{A}$ *is nonsingular and so is* $\mathbf{Q} = \mathbf{D} - \mathbf{CA}^{-1}\mathbf{B}$*, then*

$$\begin{pmatrix} \mathbf{A} & \mathbf{B} \\ \mathbf{C} & \mathbf{D} \end{pmatrix}^{-1} = \begin{pmatrix} \mathbf{A}^{-1} + \mathbf{A}^{-1}\mathbf{BQ}^{-1}\mathbf{CA}^{-1} & -\mathbf{A}^{-1}\mathbf{BQ}^{-1} \\ -\mathbf{Q}^{-1}\mathbf{CA}^{-1} & \mathbf{Q}^{-1} \end{pmatrix}.$$

(b) If $\mathbf{D}$ *is nonsingular and so is* $\mathbf{P} = \mathbf{A} - \mathbf{BD}^{-1}\mathbf{C}$*, then*

$$\begin{pmatrix} \mathbf{A} & \mathbf{B} \\ \mathbf{C} & \mathbf{D} \end{pmatrix}^{-1} = \begin{pmatrix} \mathbf{P}^{-1} & -\mathbf{P}^{-1}\mathbf{BD}^{-1} \\ -\mathbf{D}^{-1}\mathbf{CP}^{-1} & \mathbf{D}^{-1} + \mathbf{D}^{-1}\mathbf{CP}^{-1}\mathbf{BD}^{-1} \end{pmatrix}.$$

Proof Harville (1997, Theorem 8.5.11) or Schott (2016, Theorem 7.1). □

Definition 2.9.2 The matrix $\mathbf{Q} = \mathbf{D} - \mathbf{CA}^{-1}\mathbf{B}$ in part (a) of Theorem 2.9.5 is called the *Schur complement of* $\mathbf{A}$. Likewise, the matrix $\mathbf{P} = \mathbf{A} - \mathbf{BD}^{-1}\mathbf{C}$ in part (b) of Theorem 2.9.5 is called the *Schur complement of* $\mathbf{D}$.

Corollary 2.9.5.1 *If* $\begin{pmatrix} \mathbf{A} & \mathbf{0} \\ \mathbf{0} & \mathbf{B} \end{pmatrix}$ *is a block diagonal matrix, and* $\mathbf{A}$ *and* $\mathbf{B}$ *are both nonsingular, then*

$$\begin{pmatrix} \mathbf{A} & \mathbf{0} \\ \mathbf{0} & \mathbf{B} \end{pmatrix}^{-1} = \begin{pmatrix} \mathbf{A}^{-1} & \mathbf{0} \\ \mathbf{0} & \mathbf{B}^{-1} \end{pmatrix}.$$

Proof $\begin{pmatrix} \mathbf{A} & \mathbf{0} \\ \mathbf{0} & \mathbf{B} \end{pmatrix} \begin{pmatrix} \mathbf{A}^{-1} & \mathbf{0} \\ \mathbf{0} & \mathbf{B}^{-1} \end{pmatrix} = \begin{pmatrix} \mathbf{I} & \mathbf{0} \\ \mathbf{0} & \mathbf{I} \end{pmatrix} = \mathbf{I}.$ □

Theorem 2.9.6 *Let* $\mathbf{M} = \begin{pmatrix} \mathbf{A} & \mathbf{B} \\ \mathbf{C} & \mathbf{D} \end{pmatrix}$ *represent a partitioned matrix. If* $\mathbf{A}$ *is nonsingular, then rank*$(\mathbf{M})$ = *rank*$(\mathbf{A})$ + *rank*$(\mathbf{D} - \mathbf{CA}^{-1}\mathbf{B})$.

Proof Harville (1997, Theorem 8.5.10). □

Theorem 2.9.7 *Let* $\mathbf{A}$, $\mathbf{B}$, $\mathbf{C}$, *and* $\mathbf{D}$ *represent* $n \times n$, $n \times m$, $m \times m$, *and* $m \times n$ *matrices, respectively. Suppose that* $\mathbf{A}$ *and* $\mathbf{C}$ *are nonsingular. Then* $\mathbf{A} + \mathbf{BCD}$ *is nonsingular if and only if* $\mathbf{C}^{-1} + \mathbf{DA}^{-1}\mathbf{B}$ *is nonsingular, in which case*

$$\begin{aligned}(\mathbf{A} + \mathbf{BCD})^{-1} &= \mathbf{A}^{-1} - \mathbf{A}^{-1}\mathbf{B}(\mathbf{C}^{-1} + \mathbf{DA}^{-1}\mathbf{B})^{-1}\mathbf{DA}^{-1} \\ &= \mathbf{A}^{-1} - \mathbf{A}^{-1}\mathbf{BC}(\mathbf{C} + \mathbf{CDA}^{-1}\mathbf{BC})^{-1}\mathbf{CDA}^{-1}.\end{aligned}$$

Proof Harville (1997, Theorem 18.2.8) or Schott (2016, Theorem 1.9). □

The following important corollary to Theorem 2.9.7 is widely known as the Sherman–Morrison–Woodbury formula.

Corollary 2.9.7.1 *Let* $\mathbf{A}$ *represent an* $n \times n$ *nonsingular matrix, and let* $\mathbf{b}$ *and* $\mathbf{d}$ *represent n-vectors. If* $\mathbf{d}^T\mathbf{A}^{-1}\mathbf{b} \neq -1$*, then* $\mathbf{A} + \mathbf{bd}^T$ *is nonsingular and its inverse is given by*

$$(\mathbf{A} + \mathbf{bd}^T)^{-1} = \mathbf{A}^{-1} - (1 + \mathbf{d}^T\mathbf{A}^{-1}\mathbf{b})^{-1}\mathbf{A}^{-1}\mathbf{bd}^T\mathbf{A}^{-1}.$$

Proof Set $\mathbf{B} = \mathbf{b},\ \mathbf{C} = 1$, and $\mathbf{D} = \mathbf{d}^T$ in Theorem 2.9.7. □

Example 2.9-1. The Inverse of a Nonsingular Matrix of the Form $a\mathbf{I} + b\mathbf{J}$

Consider an $n \times n$ matrix of the form $a\mathbf{I}_n + b\mathbf{J}_n$. Such a matrix occurs frequently in the context of linear model theory. If the matrix is nonsingular, its inverse, interestingly, is of similar form. Specifically,

$$(a\mathbf{I}_n + b\mathbf{J}_n)^{-1} = (1/a)\mathbf{I}_n + \{-b/[a(a + nb)]\}\mathbf{J}_n, \tag{2.1}$$

as can be verified either directly or as a special case of Corollary 2.9.7.1. Observe that $\mathbf{J}_n$ (which corresponds to the case $a = 0,\ b = 1$) does not have an inverse, nor does $a\mathbf{I}_n - (a/n)\mathbf{J}_n = a[\mathbf{I}_n - (1/n)\mathbf{J}_n]$ (which corresponds to the case $b = -a/n$, where a is arbitrary). Thus, $a\mathbf{I}_n + b\mathbf{J}_n$ is nonsingular if and only if $a \neq 0$ and $b \neq -a/n$. ■

2.10 The Trace of a Square Matrix

Definition 2.10.1 The *trace* of a square matrix is the sum of the elements on its main diagonal. That is, if $\mathbf{A} = (a_{ij})$ is an $n \times n$ matrix, then its trace, written as tr($\mathbf{A}$), is

$$\text{tr}(\mathbf{A}) = \sum_{i=1}^{n} a_{ii}.$$

The following two theorems give some basic properties of the trace.

Theorem 2.10.1 *Let* $\mathbf{A}$ *represent any square matrix. Then:*

(a) $tr(\mathbf{A}^T) = tr(\mathbf{A})$;

(b) If **B** *is a (square) matrix of the same dimensions as* **A**, *and c and d are arbitrary scalars, then* $tr(c\mathbf{A} + d\mathbf{B}) = ctr(\mathbf{A}) + dtr(\mathbf{B})$.

Proof Proofs of both parts are trivial. □

Theorem 2.10.2 *If* **A** *is* $m \times n$ *and* **B** *is* $n \times m$, *then* $tr(\mathbf{AB}) = tr(\mathbf{BA})$.

Proof Harville (1997, Lemma 5.2.1) or Schott (2016, Theorem 1.3d). □

Consider how Theorem 2.10.2 might be extended to a result on the trace of the product of three matrices. That is, suppose that **A** is $n \times m$, **B** is $m \times q$, and **C** is $q \times n$, and consider tr(**ABC**). We may guess that tr(**ABC**) would equal tr(**CBA**); however, this cannot be right because the product **CBA** might not be well defined. Even in the special case $n = m = q$ for which **CBA** is defined, it is not generally true that tr(**ABC**) = tr(**CBA**). But by Theorem 2.10.2,

$$\text{tr}(\mathbf{ABC}) = \text{tr}[(\mathbf{AB})\mathbf{C}] = \text{tr}[\mathbf{C}(\mathbf{AB})] = \text{tr}(\mathbf{CAB}) = \text{tr}[(\mathbf{CA})\mathbf{B}] = \text{tr}[\mathbf{B}(\mathbf{CA})] = \text{tr}(\mathbf{BCA})$$

for arbitrary n, m, and q. Hence we have the following theorem.

Theorem 2.10.3 *If* **A** *is* $n \times m$, **B** *is* $m \times q$, *and* **C** *is* $q \times n$, *then* $tr(\mathbf{ABC}) = tr(\mathbf{CAB}) = tr(\mathbf{BCA})$.

The property described by Theorem 2.10.3 is often called the *cyclic property* of the trace because equality is preserved by "cycling" the last term in the product to the first, any number of times.

Theorem 2.10.4 *For any* $n \times m$ *matrix* **A**, $tr(\mathbf{A}^T\mathbf{A}) \geq 0$, *with equality if and only if* $\mathbf{A} = \mathbf{0}$.

Proof For $\mathbf{A} = (a_{ij})$, $\text{tr}(\mathbf{A}^T\mathbf{A}) = \sum_{j=1}^m \sum_{i=1}^n a_{ij}^2 \geq 0$. Furthermore, $\text{tr}(\mathbf{A}^T\mathbf{A}) = 0$ if and only if $a_{ij} = 0$ for all i and j; that is, if and only if $\mathbf{A} = \mathbf{0}$. □

Corollary 2.10.4.1 $\mathbf{A}^T\mathbf{A} = \mathbf{0}$ *if and only if* $\mathbf{A} = \mathbf{0}$.

2.11 The Determinant of a Square Matrix

The *determinant* of an $n \times n$ matrix **A**, written as $|\mathbf{A}|$, is a function that maps $n \times n$ matrices into $\mathbb{R}$. The determinant of a 1×1 matrix (i.e., a scalar) is merely that scalar, the determinant of a 2×2 matrix $\begin{pmatrix} a & b \\ c & d \end{pmatrix}$ is $ad - bc$, and the determinant of a 3×3 matrix $\begin{pmatrix} a & b & c \\ d & e & f \\ g & h & i \end{pmatrix}$ is $a(ei - fh) - b(di - fg) + c(dh - eg)$. The definition of this function for an arbitrary $n \times n$ matrix may be found in Harville (1997, Section 13.1). For our purposes, various properties of the determinant are much more important than the general definition. The theorems in this section give some of these properties.

Theorem 2.11.1 *The determinant of the diagonal matrix* $\mathbf{A} = diag(a_1, \ldots, a_n)$ *is* $\prod_{i=1}^{n} a_i$.

Proof Harville (1997, Corollary 13.1.2). □

Theorem 2.11.2 $|\mathbf{A}| = 0$ *if and only if* $\mathbf{A}$ *is singular.*

Proof Harville (1997, Theorem 13.3.7). □

Theorem 2.11.3 *For any square matrix* $\mathbf{A}$, $|\mathbf{A}^T| = |\mathbf{A}|$.

Proof Harville (1997, Lemma 13.2.1). □

Theorem 2.11.4 *For any* $n \times n$ *matrix* $\mathbf{A}$ *and any scalar* c, $|c\mathbf{A}| = c^n|\mathbf{A}|$.

Proof Harville (1997, Corollary 13.2.4). □

Theorem 2.11.5 *For any square matrices* $\mathbf{A}$ *and* $\mathbf{B}$ *of the same dimensions,* $|\mathbf{AB}| = |\mathbf{A}||\mathbf{B}|$.

Proof Harville (1997, Theorem 13.3.4) or Schott (2016, Theorem 1.7). □

Theorem 2.11.6 *For any nonsingular matrix* $\mathbf{A}$, $|\mathbf{A}^{-1}| = 1/|\mathbf{A}|$.

Proof By Theorems 2.11.5 and 2.11.1, $|\mathbf{A}^{-1}||\mathbf{A}| = |\mathbf{A}^{-1}\mathbf{A}| = |\mathbf{I}| = 1$, implying (with the aid of Theorem 2.11.2) that $|\mathbf{A}^{-1}| = 1/|\mathbf{A}|$. □

Theorem 2.11.7 *For any square partitioned matrices* $\begin{pmatrix} \mathbf{A} & \mathbf{B} \\ \mathbf{C} & \mathbf{D} \end{pmatrix}$ *and* $\begin{pmatrix} \mathbf{D} & \mathbf{C} \\ \mathbf{B} & \mathbf{A} \end{pmatrix}$ *for which* $\mathbf{A}$ *is nonsingular,*

$$\left|\begin{pmatrix} \mathbf{A} & \mathbf{B} \\ \mathbf{C} & \mathbf{D} \end{pmatrix}\right| = \left|\begin{pmatrix} \mathbf{D} & \mathbf{C} \\ \mathbf{B} & \mathbf{A} \end{pmatrix}\right| = |\mathbf{A}||\mathbf{D} - \mathbf{C}\mathbf{A}^{-1}\mathbf{B}|.$$

Proof Harville (1997, Theorem 13.3.8) or Schott (2016, Theorem 7.4). □

Theorem 2.11.8 *Let* $\mathbf{A}$, $\mathbf{B}$, $\mathbf{D}$, *and* $\mathbf{C}$ *represent* $n \times n$, $n \times m$, $m \times m$, *and* $m \times n$ *matrices, respectively. If* $\mathbf{A}$ *and* $\mathbf{D}$ *are nonsingular, then* $|\mathbf{A}+\mathbf{BDC}| = |\mathbf{A}||\mathbf{D}||\mathbf{D}^{-1}+\mathbf{CA}^{-1}\mathbf{B}|$.

Proof Harville (1997, Theorem 18.1.1). □

2.12 Idempotent Matrices

Definition 2.12.1 A square matrix $\mathbf{A}$ is said to be *idempotent* if $\mathbf{AA} = \mathbf{A}$.

Example 2.12-1. Some Idempotent Matrices

Observe that $\mathbf{0}_{n\times n}\mathbf{0}_{n\times n} = \mathbf{0}_{n\times n}$, so $\mathbf{0}_{n\times n}$ is idempotent. Similarly, $\mathbf{I}_n\mathbf{I}_n = \mathbf{I}_n$,

$$(1/n)\mathbf{J}_n(1/n)\mathbf{J}_n = (1/n^2)\mathbf{1}_n\mathbf{1}_n^T\mathbf{1}_n\mathbf{1}_n^T = (1/n^2)\mathbf{1}_n(n)\mathbf{1}_n^T = (1/n)\mathbf{J}_n,$$

and

$$\begin{aligned}[\mathbf{I}_n - (1/n)\mathbf{J}_n][\mathbf{I}_n - (1/n)\mathbf{J}_n] &= \mathbf{I}_n\mathbf{I}_n - \mathbf{I}_n(1/n)\mathbf{J}_n - (1/n)\mathbf{J}_n\mathbf{I}_n + (1/n)\mathbf{J}_n(1/n)\mathbf{J}_n \\ &= \mathbf{I}_n - (1/n)\mathbf{J}_n - (1/n)\mathbf{J}_n + (1/n)\mathbf{J}_n \\ &= \mathbf{I}_n - (1/n)\mathbf{J}_n,\end{aligned}$$

so $\mathbf{I}_n$, $(1/n)\mathbf{J}_n$ and $\mathbf{I}_n - (1/n)\mathbf{J}_n$ are idempotent as well. ■

Some consequences of idempotency are given by the theorems in this section.

Theorem 2.12.1 *If* $\mathbf{A}$ *is idempotent, then so are* $\mathbf{A}^T$ *and* $\mathbf{I} - \mathbf{A}$.

Proof By Theorem 2.7.1c, $\mathbf{A}^T\mathbf{A}^T = (\mathbf{AA})^T = \mathbf{A}^T$; furthermore, $(\mathbf{I}-\mathbf{A})(\mathbf{I}-\mathbf{A}) = \mathbf{I}-\mathbf{A}-\mathbf{A}+\mathbf{AA} = \mathbf{I}-\mathbf{A}$. □

Theorem 2.12.2 *If* $\mathbf{A}$ *is idempotent, then rank(*$\mathbf{A}$*) = tr(*$\mathbf{A}$*).*

Proof Harville (1997, Corollary 10.2.2) or Schott (2016, Theorem 11.1d). □

Theorem 2.12.3 *Let* $\mathbf{A}_1, \ldots, \mathbf{A}_k$ *represent* $n \times n$ *matrices, and define* $\mathbf{A} = \sum_{i=1}^k \mathbf{A}_i$. *If* $\mathbf{A}$ *is idempotent and rank(*$\mathbf{A}$*)* $= \sum_{i=1}^k$ *rank(*$\mathbf{A}_i$*), then* $\mathbf{A}_i$ *is idempotent for* $i = 1, \ldots, k$ *and* $\mathbf{A}_i\mathbf{A}_j = \mathbf{0}$ *for all* $j \neq i = 1, \ldots, k$.

Proof Harville (1997, Theorem 18.4.1) or Schott (2016, Theorem 11.8). □

Example 2.12-2. Illustrations of Theorems 2.12.2 and 2.12.3

We use the idempotent matrices in Example 2.12-1 to illustrate Theorems 2.12.2 and 2.12.3. First it can be verified easily that $\text{rank}(\mathbf{0}_{n\times n}) = \text{tr}(\mathbf{0}_{n\times n}) = 0$, $\text{rank}(\mathbf{I}_n) = \text{tr}(\mathbf{I}_n) = n$, $\text{rank}(\frac{1}{n}\mathbf{J}_n) = \text{tr}(\frac{1}{n}\mathbf{J}_n) = 1$, and $\text{rank}(\mathbf{I} - \frac{1}{n}\mathbf{J}_n) = \text{tr}(\mathbf{I} - \frac{1}{n}\mathbf{J}_n) = n - (1/n)n = n - 1$.

Next, consider $\mathbf{A}_1 = \frac{1}{n}\mathbf{J}_n$ and $\mathbf{A}_2 = \mathbf{I} - \frac{1}{n}\mathbf{J}_n$, whose sum is $\mathbf{A} \equiv \mathbf{A}_1 + \mathbf{A}_2 = \mathbf{I}_n$. Because $\mathbf{I}_n$ is idempotent and its rank (which is n) is equal to the sum of the ranks of $\frac{1}{n}\mathbf{J}_n$ and $\mathbf{I} - \frac{1}{n}\mathbf{J}_n$ (which are 1 and $n - 1$, respectively), the theorem informs us that $\frac{1}{n}\mathbf{J}_n$ and $\mathbf{I} - \frac{1}{n}\mathbf{J}_n$ are idempotent (which we already knew from Example 2.12-1) and that $\frac{1}{n}\mathbf{J}_n(\mathbf{I} - \frac{1}{n}\mathbf{J}_n) = \mathbf{0}$ (which may also be verified by direct calculation). ■

2.13 Orthogonal Matrices

Definition 2.13.1 A square matrix $\mathbf{P}$ is said to be *orthogonal* if $\mathbf{P}^T\mathbf{P} = \mathbf{I}$.

Observe that if $\mathbf{P}$ is orthogonal, then $\mathbf{P}^T = \mathbf{P}^{-1}$, implying further that $\mathbf{P}$ is nonsingular and that $\mathbf{P}\mathbf{P}^T = \mathbf{I}$. The column vectors of an orthogonal matrix are orthonormal, as are its row vectors.

The requirement that $\mathbf{P}$ be square in order that it be orthogonal is important. Nonsquare matrices $\mathbf{A}$ exist for which $\mathbf{A}^T\mathbf{A} = \mathbf{I}$, but they are not orthogonal matrices and it does not follow that $\mathbf{A}\mathbf{A}^T = \mathbf{I}$ for them.

Theorem 2.13.1 *If* $\mathbf{P}$ *is orthogonal, then either* $|\mathbf{P}| = 1$ *or* $|\mathbf{P}| = -1$.

Proof By Theorems 2.11.3 and 2.11.5,

$$|\mathbf{P}||\mathbf{P}| = |\mathbf{P}^T||\mathbf{P}| = |\mathbf{P}^T\mathbf{P}| = |\mathbf{I}| = 1,$$

implying that $|\mathbf{P}| = \pm 1$. □

2.14 Quadratic Forms

Definition 2.14.1 Let $\mathbf{x} = (x_1, \ldots, x_n)^T$ represent an n-vector. A function $f(\mathbf{x})$ of the form $f(\mathbf{x}) = \sum_{i=1}^n \sum_{j=1}^n a_{ij}x_ix_j$, or in matrix notation, $f(\mathbf{x}) = \mathbf{x}^T\mathbf{A}\mathbf{x}$, is called a *quadratic form* (in $\mathbf{x}$). The matrix $\mathbf{A}$ is called the matrix of the quadratic form.

Example 2.14-1. Some Quadratic Forms

Some examples of quadratic forms are

$$x_1^2 - x_2^2 = (x_1\ x_2)\begin{pmatrix} 1 & 0 \\ 0 & -1 \end{pmatrix}\begin{pmatrix} x_1 \\ x_2 \end{pmatrix}$$

and

$$2x_1^2 + 6x_1x_2 + 6x_2^2 = (x_1\ x_2)\begin{pmatrix} 2 & 3 \\ 3 & 6 \end{pmatrix}\begin{pmatrix} x_1 \\ x_2 \end{pmatrix}.$$

In each of these examples, the matrix of the quadratic form is not unique, for we may re-express the same quadratic forms as

$$x_1^2 - x_2^2 = (x_1\ x_2)\begin{pmatrix} 1 & -2 \\ 2 & -1 \end{pmatrix}\begin{pmatrix} x_1 \\ x_2 \end{pmatrix}$$

and

$$2x_1^2 + 6x_1x_2 + 6x_2^2 = (x_1\ x_2)\begin{pmatrix} 2 & 1 \\ 5 & 6 \end{pmatrix}\begin{pmatrix} x_1 \\ x_2 \end{pmatrix}.$$

Observe that the matrix of the quadratic form is symmetric in the first, but not the second, representation of each quadratic form. ■

Theorem 2.14.1 *If* $\mathbf{x}^T\mathbf{A}\mathbf{x}$ *is any quadratic form, then a unique symmetric matrix* $\mathbf{B}$ *exists such that* $\mathbf{x}^T\mathbf{A}\mathbf{x} = \mathbf{x}^T\mathbf{B}\mathbf{x}$ *for all* $\mathbf{x}$.

Proof $\mathbf{x}^T\mathbf{A}\mathbf{x} = (\mathbf{x}^T\mathbf{A}\mathbf{x})^T$ because a quadratic form is a scalar, and $(\mathbf{x}^T\mathbf{A}\mathbf{x})^T = \mathbf{x}^T\mathbf{A}^T\mathbf{x}$ by Theorem 2.7.1c. Thus for all $\mathbf{x}$,

$$\mathbf{x}^T\mathbf{A}\mathbf{x} = \frac{1}{2}\mathbf{x}^T\mathbf{A}\mathbf{x} + \frac{1}{2}\mathbf{x}^T\mathbf{A}^T\mathbf{x} = \mathbf{x}^T\left(\frac{1}{2}\mathbf{A} + \frac{1}{2}\mathbf{A}^T\right)\mathbf{x} = \mathbf{x}^T\mathbf{B}\mathbf{x},$$

say, where $\mathbf{B}$ is symmetric. To establish the uniqueness of $\mathbf{B}$, let $\mathbf{C}$ represent an arbitrary $n \times n$ symmetric matrix such that $\mathbf{x}^T\mathbf{B}\mathbf{x} = \mathbf{x}^T\mathbf{C}\mathbf{x}$ for all $\mathbf{x}$. Then for all $i = 1, \ldots, n$, $\mathbf{u}_i^T\mathbf{B}\mathbf{u}_i = \mathbf{u}_i^T\mathbf{C}\mathbf{u}_i$, i.e., $b_{ii} = c_{ii}$. Furthermore, for all $i > j = 1, \ldots, n$, $(\mathbf{u}_i - \mathbf{u}_j)^T\mathbf{B}(\mathbf{u}_i - \mathbf{u}_j) = (\mathbf{u}_i - \mathbf{u}_j)^T\mathbf{C}(\mathbf{u}_i - \mathbf{u}_j)$, i.e., $b_{ii} + b_{jj} - b_{ij} - b_{ji} = c_{ii} + c_{jj} - c_{ij} - c_{ji}$. By the equality of b_{ii} and c_{ii} for all i and the symmetry of $\mathbf{B}$ and $\mathbf{C}$, we obtain $-2b_{ij} = -2c_{ij}$, i.e., $b_{ij} = c_{ij}$. □

The upshot of Theorem 2.14.1 is that when representing a quadratic form we can always, without loss of generality, take the matrix of the quadratic form to be symmetric. We shall do this throughout, usually without explicitly saying so.

Definition 2.14.2 The quadratic form $\mathbf{x}^T\mathbf{A}\mathbf{x}$ is said to be *nonnegative definite* if $\mathbf{x}^T\mathbf{A}\mathbf{x} \geq 0$ for all $\mathbf{x} \in \mathbb{R}^n$. If a quadratic form $\mathbf{x}^T\mathbf{A}\mathbf{x}$ is nonnegative definite and $\mathbf{0}$ is the only value of $\mathbf{x}$ for which $\mathbf{x}^T\mathbf{A}\mathbf{x}$ equals 0, $\mathbf{x}^T\mathbf{A}\mathbf{x}$ is said to be *positive definite*; otherwise it is said to be *positive semidefinite*. Similarly, a quadratic form $\mathbf{x}^T\mathbf{A}\mathbf{x}$ is said to be *nonpositive definite* if $\mathbf{x}^T\mathbf{A}\mathbf{x} \leq 0$ for all $\mathbf{x} \in \mathbb{R}^n$. If a quadratic form $\mathbf{x}^T\mathbf{A}\mathbf{x}$ is nonpositive definite and $\mathbf{0}$ is the only value of $\mathbf{x}$ for which $\mathbf{x}^T\mathbf{A}\mathbf{x}$ equals 0, $\mathbf{x}^T\mathbf{A}\mathbf{x}$ is said to be *negative definite*; otherwise it is said to be *negative semidefinite*. If a quadratic form is neither nonnegative definite nor nonpositive definite, it is said to be *indefinite*.

In the theory of linear models, nonnegative definite and positive definite quadratic forms play a much larger role than other types of quadratic forms.

Example 2.14-2. Some Nonnegative Definite and Positive Definite Quadratic Forms

Consider the quadratic forms $\mathbf{x}^T\mathbf{I}_n\mathbf{x}$, $\mathbf{x}^T\mathbf{J}_n\mathbf{x}$, and $\mathbf{x}^T(\mathbf{I}_n - \frac{1}{n}\mathbf{J}_n)\mathbf{x}$, where $n \geq 2$. Observe that

$$\mathbf{x}^T\mathbf{I}_n\mathbf{x} = \mathbf{x}^T\mathbf{x} = x_1^2 + \cdots + x_n^2,$$

which is positive unless $\mathbf{x}$ is null. Thus $\mathbf{x}^T\mathbf{I}_n\mathbf{x}$ is positive definite. Next,

$$\mathbf{x}^T\mathbf{J}_n\mathbf{x} = \mathbf{x}^T\mathbf{1}\mathbf{1}^T\mathbf{x} = (x_1 + \cdots + x_n)^2,$$

which is nonnegative but can equal 0 for some nonnull $\mathbf{x}$ [e.g., $\mathbf{x} = (a, -a, \mathbf{0}_{n-2}^T)^T$ where $a \neq 0$]. Thus $\mathbf{x}^T\mathbf{J}_n\mathbf{x}$ is positive semidefinite. Last,

$$\mathbf{x}^T(\mathbf{I}_n - \frac{1}{n}\mathbf{J}_n)\mathbf{x} = \mathbf{x}^T\mathbf{x} - (1/n)\mathbf{x}^T\mathbf{1}\mathbf{1}^T\mathbf{x} = \sum_{i=1}^n x_i^2 - \left(\sum_{i=1}^n x_i\right)^2 \Big/ n = \sum_{i=1}^n (x_i - \bar{x})^2,$$

which is nonnegative but can equal 0 for some nonnull $\mathbf{x}$ (e.g., $\mathbf{x} = \mathbf{1}_n$). So $\mathbf{x}^T[\mathbf{I}_n - (1/n)\mathbf{J}_n]\mathbf{x}$ is positive semidefinite. ■

2.15 Nonnegative Definite, Positive Definite, and Positive Semidefinite Matrices

Definition 2.15.1 A symmetric matrix **A** is said to be *nonnegative definite, positive definite*, or *positive semidefinite* if the quadratic form $\mathbf{x}^T\mathbf{A}\mathbf{x}$ is nonnegative definite, positive definite, or positive semidefinite, respectively.

Example 2.15-1. Nonnegative Definiteness and Positive Definiteness of Matrices of the Form $a\mathbf{I} + b\mathbf{J}$

Consider $n \times n$ matrices of the form $a\mathbf{I}_n + b\mathbf{J}_n$, which are clearly symmetric. For what values of a and b is $a\mathbf{I}_n + b\mathbf{J}_n$ nonnegative definite, and for what values is it positive definite? To answer this question, consider the quadratic form $\mathbf{u}^T(a\mathbf{I}_n + b\mathbf{J}_n)\mathbf{u} = a\sum_{i=1}^n u_i^2 + b(\sum_{i=1}^n u_i)^2$, where $\mathbf{u} = (u_i)$ is an arbitrary nonnull n-vector. Let us partition the set of nonnull n-vectors into three subsets: $\mathcal{V}_1 = \{\mathbf{u} \neq \mathbf{0} : \mathbf{u}^T\mathbf{1} = 0\}$, $\mathcal{V}_2 = \{\mathbf{u} : \mathbf{u} = x\mathbf{1}$ for some scalar $x \neq 0\}$, and $\mathcal{V}_3 = \mathbb{R}^n \backslash (\mathbf{0} \cup \mathcal{V}_1 \cup \mathcal{V}_2)$. For $\mathbf{u} \in \mathcal{V}_1$, $\mathbf{u}^T(a\mathbf{I} + b\mathbf{J})\mathbf{u} = a\sum_{i=1}^n u_i^2$, which is greater than or equal to 0 if and only if $a \geq 0$ (and strictly greater than 0 if and only if $a > 0$). For $\mathbf{u} \in \mathcal{V}_2$, $\mathbf{u}^T(a\mathbf{I} + b\mathbf{J})\mathbf{u} = x^2 n(a + nb)$, which is greater than or equal to 0 if and only if $b \geq -a/n$ (and strictly greater than 0 if and only if $b > -a/n$). For $\mathbf{u} \in \mathcal{V}_3$, $\mathbf{u}^T(a\mathbf{I} + b\mathbf{J})\mathbf{u} = a\sum_{i=1}^n u_i^2 + b(\sum_{i=1}^n u_i)^2 > (a + nb)(\sum_{i=1}^n u_i)^2/n$ by the Cauchy–Schwartz inequality [i.e., $\sum_{i=1}^n u_i^2 > (\sum_{i=1}^n u_i)^2/n$ within this subset]. Thus, for $\mathbf{u} \in \mathcal{V}_3$, the quadratic form is strictly positive if and only if $b > -a/n$. Putting all cases together, we see that $a\mathbf{I}_n + b\mathbf{J}_n$ is nonnegative definite if and only if $a \geq 0$ and $b \geq -a/n$, and that $a\mathbf{I}_n + b\mathbf{J}_n$ is positive definite if and only if $a > 0$ and $b > -a/n$. ■

Many important results about nonnegative definite, positive definite, and positive semidefinite matrices are described by the theorems in this section.

Theorem 2.15.1 *If* **A** *is a nonnegative definite matrix, then all of its diagonal elements are nonnegative; if* **A** *is positive definite, then all of its diagonal elements are positive.*

Proof Left as an exercise. □

Theorem 2.15.2 *Let c represent a positive scalar. If* **A** *is a positive definite matrix, then so is c***A***; if* **A** *is positive semidefinite, then so is c***A***.*

Proof Left as an exercise. □

Theorem 2.15.3 *Let* $\mathbf{A}$ *and* $\mathbf{B}$ *represent* $n \times n$ *matrices. If* $\mathbf{A}$ *and* $\mathbf{B}$ *are both nonnegative definite, then so is* $\mathbf{A} + \mathbf{B}$*; if either one is positive definite and the other is nonnegative definite, then* $\mathbf{A} + \mathbf{B}$ *is positive definite.*

Proof Left as an exercise. □

Theorem 2.15.4 *The block diagonal matrix* $\begin{pmatrix} \mathbf{A} & \mathbf{0} \\ \mathbf{0} & \mathbf{B} \end{pmatrix}$ *is nonnegative definite if and only if both* $\mathbf{A}$ *and* $\mathbf{B}$ *are nonnegative definite, and is positive definite if and only if both* $\mathbf{A}$ *and* $\mathbf{B}$ *are positive definite.*

Proof Left as an exercise. □

Theorem 2.15.5 *If* $\mathbf{A}$ *is a positive definite matrix, then it is nonsingular; if* $\mathbf{A}$ *is positive semidefinite, then it is singular.*

Proof Harville (1997, Corollary 14.3.12). □

Theorem 2.15.6 *If* $\mathbf{A}$ *is a positive definite matrix, then so is* $\mathbf{A}^{-1}$.

Proof Harville (1997, Corollary 14.2.11). □

Theorem 2.15.7 *Let* $\mathbf{M} = \begin{pmatrix} \mathbf{A} & \mathbf{B} \\ \mathbf{B}^T & \mathbf{C} \end{pmatrix}$ *represent a symmetric partitioned matrix.*

(a) If $\mathbf{M}$ *is positive definite, then so are* $\mathbf{A}$, $\mathbf{C}$, $\mathbf{A} - \mathbf{B}\mathbf{C}^{-1}\mathbf{B}^T$, *and* $\mathbf{C} - \mathbf{B}^T\mathbf{A}^{-1}\mathbf{B}$.
(b) If $\mathbf{M}$ *is nonnegative definite and* $\mathbf{A}$ *is positive definite, then* $\mathbf{C}$ *and* $\mathbf{C} - \mathbf{B}^T\mathbf{A}^{-1}\mathbf{B}$ *are nonnegative definite.*

Proof A little manipulation of results proven by Harville (1997, Corollary 14.2.12 and Theorem 14.8.4) yields the result. □

Theorem 2.15.8 *If* $\mathbf{A}$ *is nonnegative definite, then* $|\mathbf{A}| \geq 0$*; if* $\mathbf{A}$ *is positive definite, then* $|\mathbf{A}| > 0$.

Proof Harville (1997, Theorem 14.9.1). □

Importantly, the converses of the assertions in Theorem 2.15.8 are false. Thus, the sign of the determinant does not definitively determine whether a square matrix is nonnegative definite (or positive definite).

Theorem 2.15.9 *Let* $\mathbf{A}$ *represent an* $n \times m$ *matrix of rank* r*. Then* $\mathbf{A}^T\mathbf{A}$ *is an* $m \times m$ *nonnegative definite matrix of rank* r*. Furthermore,* $\mathbf{A}^T\mathbf{A}$ *is positive definite if* $r = m$*, or is positive semidefinite if* $r < m$.

Proof For any m-vector $\mathbf{x}$, $\mathbf{x}^T(\mathbf{A}^T\mathbf{A})\mathbf{x} = (\mathbf{Ax})^T(\mathbf{Ax}) \geq 0$. This establishes that $\mathbf{A}^T\mathbf{A}$ is nonnegative definite, and by Theorem 2.8.8 rank$(\mathbf{A}^T\mathbf{A}) = r$. Furthermore, it follows from Corollary 2.10.4.1 that $\mathbf{x}^T(\mathbf{A}^T\mathbf{A})\mathbf{x}$ is equal to 0 if and only if $\mathbf{Ax} = \mathbf{0}$. But this is so for nonnull $\mathbf{x}$ if and only if the columns of $\mathbf{A}$ are linearly dependent, or equivalently (by Theorem 2.8.1) if and only if $r < m$. □

Corollary 2.15.9.1 *If* $\mathbf{A}$ *is an* $n \times n$ *symmetric idempotent matrix, then* $\mathbf{A}$ *is nonnegative definite.*

Throughout this book, there are numerous occasions where we want to determine whether a square matrix is positive definite or not. Some of the theorems already presented in this section can be helpful in this regard. For example, by Theorem 2.15.8, if the determinant of the matrix is negative, then the matrix is not positive definite. Or, by Theorem 2.15.9, if the matrix can be written in the form $\mathbf{A}^T\mathbf{A}$ for some $n \times m$ matrix $\mathbf{A}$ of rank m, then it is positive definite. The next two theorems can also help to determine whether a square matrix is positive definite.

Theorem 2.15.10 *Let* $\mathbf{A} = (a_{ij})$ *represent an* $n \times n$ *symmetric matrix with positive diagonal elements. If* $\mathbf{A}$ *is* diagonally dominant, *i.e., if* $|a_{ii}| > \sum_{j \neq i} |a_{ij}|$ *for* $i = 1, \ldots, n$, *then* $\mathbf{A}$ *is positive definite.*

Proof Harville (2001, Solution to Exercise 14.36). □

Theorem 2.15.11 *A symmetric matrix* $\mathbf{A} = (a_{ij})$ *is nonnegative definite if and only if*

$$a_{11} \geq 0, \qquad \begin{vmatrix} a_{11} & a_{12} \\ a_{21} & a_{22} \end{vmatrix} \geq 0, \qquad \ldots, \qquad |\mathbf{A}| \geq 0.$$

Furthermore, $\mathbf{A}$ *is positive definite if and only if strict versions of the same inequalities hold.*

Proof Harville (1997, Theorems 14.9.5 and 14.9.11). □

Recall that a real number a is nonnegative if and only if a unique nonnegative real number b exists such that $a = b^2$, and that such a b is called the nonnegative square root of a. Furthermore, b is positive if and only if a is positive. The next theorem is a generalization of these results to nonnegative matrices.

Theorem 2.15.12 $\mathbf{A}$ *is a nonnegative definite matrix of rank* r *if and only if a unique nonnegative definite (hence symmetric) matrix* $\mathbf{B}$ *of rank* r *exists such that* $\mathbf{A} = \mathbf{BB}$. *Furthermore,* $\mathbf{B}$ *is positive definite if and only if* $\mathbf{A}$ *is positive definite.*

Proof Harville (1997, Theorem 21.9.1 and surrounding discussion). □

Definition 2.15.2 The unique matrix $\mathbf{B}$ mentioned in Theorem 2.15.12 is called the *nonnegative definite square root* of $\mathbf{A}$ and is written as $\mathbf{A}^{\frac{1}{2}}$.

Next we give two useful corollaries to Theorem 2.15.12. The proof of the second corollary is trivial, hence not provided.

Corollary 2.15.12.1 *If* $\mathbf{A}$ *is an* $n \times n$ *nonnull nonnegative definite matrix and* $\mathbf{C}$ *is any matrix having n columns, then* $\mathbf{CAC}^T$ *is nonnegative definite.*

Proof Left as an exercise. □

Corollary 2.15.12.2 *The nonnegative definite square root of a symmetric and idempotent matrix* $\mathbf{A}$ *is* $\mathbf{A}$.

We end this section with four more theorems pertaining to nonnegative definite or positive definite matrices.

Theorem 2.15.13 *Let* $\mathbf{A}$ *and* $\mathbf{B}$ *represent nonnegative definite matrices. Then* $tr(\mathbf{AB}) \geq 0$*, with equality if and only if* $\mathbf{AB} = \mathbf{0}$.

Proof Left as an exercise. □

Theorem 2.15.14 *If* $\mathbf{A}$ *is an* $n \times n$ *nonnull nonnegative definite matrix of rank r, then an* $n \times r$ *matrix* $\mathbf{B}$ *of rank r exists such that* $\mathbf{A} = \mathbf{BB}^T$.

Proof Harville (1997, Theorem 14.3.7). □

Theorem 2.15.15 *If* $\mathbf{A}$ *is an* $n \times n$ *nonnull symmetric idempotent matrix of rank r, then an* $n \times r$ *matrix* $\mathbf{B}$ *of rank r exists such that* $\mathbf{A} = \mathbf{BB}^T$ *and* $\mathbf{B}^T\mathbf{B} = \mathbf{I}_r$.

Proof A proof can be constructed by combining the results of Theorems 21.5.7 and 21.8.3 of Harville (1997). □

Theorem 2.15.16 *Let* $\mathbf{A}$ *represent any* $n \times n$ *positive definite matrix, and let* $\mathbf{B}$ *represent any matrix with n columns. Then* $\mathbf{AB}^T(\mathbf{I} + \mathbf{BAB}^T)^{-1} = (\mathbf{A}^{-1} + \mathbf{B}^T\mathbf{B})^{-1}\mathbf{B}^T$.

Proof Clearly $(\mathbf{A}^{-1} + \mathbf{B}^T\mathbf{B})\mathbf{AB}^T = \mathbf{B}^T(\mathbf{I} + \mathbf{BAB}^T)$. Now because $\mathbf{A}$ is positive definite and $\mathbf{BAB}^T$ is nonnegative definite, by Theorems 2.15.3 and 2.15.6 both of $\mathbf{A}^{-1}+\mathbf{B}^T\mathbf{B}$ and $\mathbf{I}+\mathbf{BAB}^T$ are positive definite (hence nonsingular). Pre-multiplying both sides of the equality by $(\mathbf{A}^{-1} + \mathbf{B}^T\mathbf{B})^{-1}$ and post-multiplying both sides by $(\mathbf{I} + \mathbf{BAB}^T)^{-1}$ yield the result. □

2.16 The Generalized Cauchy–Schwartz Inequality

The following theorem gives a generalization of the well-known Cauchy–Schwartz inequality.

Theorem 2.16.1 (Generalized Cauchy–Schwartz Inequality) *For any n-vectors* $\mathbf{a}$ *and* $\mathbf{b}$ *and any positive definite matrix* $\mathbf{A}$,

$$(\mathbf{a}^T\mathbf{b})^2 \le (\mathbf{a}^T\mathbf{A}\mathbf{a})(\mathbf{b}^T\mathbf{A}^{-1}\mathbf{b}),$$

with equality holding if and only if $\mathbf{a}$ *is proportional to* $\mathbf{A}^{-1}\mathbf{b}$.

Proof Because $\mathbf{A}$ is positive definite, its inverse exists and is positive definite as well (Theorem 2.15.6). Thus, if $\mathbf{a}^T\mathbf{A}\mathbf{a} = 0$ then $\mathbf{a} = \mathbf{0}$, and similarly if $\mathbf{b}^T\mathbf{A}^{-1}\mathbf{b} = 0$ then $\mathbf{b} = \mathbf{0}$. Thus, the inequality is satisfied if either $\mathbf{a}^T\mathbf{A}\mathbf{a} = 0$ or $\mathbf{b}^T\mathbf{A}^{-1}\mathbf{b} = 0$. Alternatively, if $\mathbf{a}^T\mathbf{A}\mathbf{a} > 0$ and $\mathbf{b}^T\mathbf{A}^{-1}\mathbf{b} > 0$, then for any scalars c and d,

$$0 \le (c\mathbf{a} - d\mathbf{A}^{-1}\mathbf{b})^T\mathbf{A}(c\mathbf{a} - d\mathbf{A}^{-1}\mathbf{b}) = c^2\mathbf{a}^T\mathbf{A}\mathbf{a} + d^2\mathbf{b}^T\mathbf{A}^{-1}\mathbf{b} - 2cd\mathbf{a}^T\mathbf{b},$$

implying that

$$2cd\mathbf{a}^T\mathbf{b} \le c^2\mathbf{a}^T\mathbf{A}\mathbf{a} + d^2\mathbf{b}^T\mathbf{A}^{-1}\mathbf{b}.$$

Now take $c = (\mathbf{b}^T\mathbf{A}^{-1}\mathbf{b})^{\frac{1}{2}}$ and $d = -(\mathbf{a}^T\mathbf{A}\mathbf{a})^{\frac{1}{2}}$, yielding

$$-2(\mathbf{b}^T\mathbf{A}^{-1}\mathbf{b})^{\frac{1}{2}}(\mathbf{a}^T\mathbf{A}\mathbf{a})^{\frac{1}{2}}(\mathbf{a}^T\mathbf{b}) \le 2(\mathbf{a}^T\mathbf{A}\mathbf{a})(\mathbf{b}^T\mathbf{A}^{-1}\mathbf{b}),$$

which in turn implies that

$$\mathbf{a}^T\mathbf{b} \ge -(\mathbf{a}^T\mathbf{A}\mathbf{a})^{\frac{1}{2}}(\mathbf{b}^T\mathbf{A}^{-1}\mathbf{b})^{\frac{1}{2}}.$$

Similarly, taking $c = (\mathbf{b}^T\mathbf{A}^{-1}\mathbf{b})^{\frac{1}{2}}$ and $d = (\mathbf{a}^T\mathbf{A}\mathbf{a})^{\frac{1}{2}}$, we find that

$$\mathbf{a}^T\mathbf{b} \le (\mathbf{a}^T\mathbf{A}\mathbf{a})^{\frac{1}{2}}(\mathbf{b}^T\mathbf{A}^{-1}\mathbf{b})^{\frac{1}{2}}.$$

Squaring both sides of these last two inequalities establishes that lower bound on $(\mathbf{a}^T\mathbf{b})^2$ given in the theorem. To establish the equality condition, note that $(c\mathbf{a} - d\mathbf{A}^{-1}\mathbf{b})^T\mathbf{A}(c\mathbf{a} - d\mathbf{A}^{-1}\mathbf{b}) = 0$ if and only if $c\mathbf{a} - d\mathbf{A}^{-1}\mathbf{b} = \mathbf{0}$, or equivalently (because $c > 0$ and $d > 0$) if and only if $\mathbf{a}$ is proportional to $\mathbf{A}^{-1}\mathbf{b}$. □

Corollary 2.16.1.1 *For any n-vector* $\mathbf{b}$ *and any* $n \times n$ *positive definite matrix* $\mathbf{A}$,

$$\max_{\mathbf{a} \neq \mathbf{0}} \left(\frac{(\mathbf{a}^T \mathbf{b})^2}{\mathbf{a}^T \mathbf{A} \mathbf{a}} \right) = \mathbf{b}^T \mathbf{A}^{-1} \mathbf{b},$$

and the maximum is attained if and only if $\mathbf{a}$ *is proportional to* $\mathbf{A}^{-1}\mathbf{b}$.

2.17 Kronecker Products

Some model matrices and other matrices that arise in the theory of linear models have a structure given by the following definition.

Definition 2.17.1 The *Kronecker product* of an $n \times m$ matrix $\mathbf{A} = (a_{ij})$ and a $p \times q$ matrix $\mathbf{B} = (b_{ij})$, written as $\mathbf{A} \otimes \mathbf{B}$, is defined as the $np \times mq$ matrix

$$\mathbf{A} \otimes \mathbf{B} = \begin{pmatrix} a_{11}\mathbf{B} & a_{12}\mathbf{B} & \cdots & a_{1m}\mathbf{B} \\ a_{21}\mathbf{B} & a_{22}\mathbf{B} & \cdots & a_{2m}\mathbf{B} \\ \vdots & \vdots & & \vdots \\ a_{n1}\mathbf{B} & a_{n2}\mathbf{B} & \cdots & a_{nm}\mathbf{B} \end{pmatrix}.$$

The following theorems describe a number of useful properties of the Kronecker product. Proofs of the first four theorems are trivial, hence not included.

Theorem 2.17.1 *For any matrices* $\mathbf{A}$ *and* $\mathbf{B}$ *and any scalar* c, $(c\mathbf{A}) \otimes \mathbf{B} = \mathbf{A} \otimes (c\mathbf{B}) = c(\mathbf{A} \otimes \mathbf{B})$.

Theorem 2.17.2 *For any* $n \times m$ *matrices* $\mathbf{A}$ *and* $\mathbf{B}$ *and* $p \times q$ *matrices* $\mathbf{C}$ *and* $\mathbf{D}$, $(\mathbf{A}+\mathbf{B}) \otimes (\mathbf{C}+\mathbf{D}) = (\mathbf{A} \otimes \mathbf{C}) + (\mathbf{A} \otimes \mathbf{D}) + (\mathbf{B} \otimes \mathbf{C}) + (\mathbf{B} \otimes \mathbf{D})$.

Theorem 2.17.3 *For any matrices* $\mathbf{A}$, $\mathbf{B}$, *and* $\mathbf{C}$, $(\mathbf{A} \otimes \mathbf{B}) \otimes \mathbf{C} = \mathbf{A} \otimes (\mathbf{B} \otimes \mathbf{C})$.

Theorem 2.17.4 *For any matrices* $\mathbf{A}$ *and* $\mathbf{B}$, $(\mathbf{A} \otimes \mathbf{B})^T = \mathbf{A}^T \otimes \mathbf{B}^T$.

Theorem 2.17.5 *For any* $n \times m$ *matrix* $\mathbf{A}$, $p \times q$ *matrix* $\mathbf{B}$, $m \times u$ *matrix* $\mathbf{C}$, *and* $q \times v$ *matrix* $\mathbf{D}$, $(\mathbf{A} \otimes \mathbf{B})(\mathbf{C} \otimes \mathbf{D}) = (\mathbf{AC}) \otimes (\mathbf{BD})$.

Proof Harville (1997, Lemma 16.1.2) or Schott (2016, Theorem 8.2). □

Theorem 2.17.6 *For any square matrices* $\mathbf{A}$ *and* $\mathbf{B}$, $tr(\mathbf{A} \otimes \mathbf{B}) = tr(\mathbf{A}) \cdot tr(\mathbf{B})$.

Proof Harville (1997, Section 16.1) or Schott (2016, Theorem 8.3). □

Theorem 2.17.7 *For any* $n \times n$ *matrix* $\mathbf{A}$ *and* $p \times p$ *matrix* $\mathbf{B}$, $|\mathbf{A} \otimes \mathbf{B}| = |\mathbf{A}|^p |\mathbf{B}|^n$.

Proof Harville (1997, Section 16.3e) or Schott (2016, Theorem 8.6). □

Theorem 2.17.8 *For any matrices* $\mathbf{A}$ *and* $\mathbf{B}$, *rank*$(\mathbf{A} \otimes \mathbf{B}) =$ *rank*$(\mathbf{A}) \cdot$ *rank*$(\mathbf{B})$.

Proof Harville (1997, Section 16.1) or Schott (2016, Theorem 8.7). □

Theorem 2.17.9 *If* $\mathbf{A}$ *and* $\mathbf{B}$ *are nonsingular, then so is* $\mathbf{A} \otimes \mathbf{B}$, *and its inverse is* $(\mathbf{A} \otimes \mathbf{B})^{-1} = \mathbf{A}^{-1} \otimes \mathbf{B}^{-1}$.

Proof Using Theorem 2.17.5,

$$(\mathbf{A} \otimes \mathbf{B})(\mathbf{A}^{-1} \otimes \mathbf{B}^{-1}) = (\mathbf{A}\mathbf{A}^{-1}) \otimes (\mathbf{B}\mathbf{B}^{-1}) = \mathbf{I} \otimes \mathbf{I} = \mathbf{I}.$$

□

2.18 Vecs and Direct Sums

Two operators on matrices useful for their notational efficiency are the vec and the direct sum.

Definition 2.18.1 The *vec* of an $n \times m$ matrix $\mathbf{A} = (\mathbf{a}_1, \mathbf{a}_2, \ldots, \mathbf{a}_m)$ is the (nm)-vector obtained by stacking the columns of $\mathbf{A}$ on top of one another, i.e.,

$$\text{vec}(\mathbf{A}) = \begin{pmatrix} \mathbf{a}_1 \\ \mathbf{a}_2 \\ \vdots \\ \mathbf{a}_m \end{pmatrix}.$$

Obviously, if $\mathbf{A}$ is a symmetric matrix, then $\text{vec}(\mathbf{A}^T) = \text{vec}(\mathbf{A})$. It is just as obvious that $\text{vec}(\cdot)$ is a linear operator, i.e., for any matrices $\mathbf{A}$ and $\mathbf{B}$ of the same dimensions and any scalars c and d, $\text{vec}(c\mathbf{A} + d\mathbf{B}) = c\text{vec}(\mathbf{A}) + d\text{vec}(\mathbf{B})$.

Definition 2.18.2 The *direct sum* of arbitrary matrices $\mathbf{A}_1, \mathbf{A}_2, \ldots, \mathbf{A}_k$, written as $\mathbf{A}_1 \oplus \mathbf{A}_2 \oplus \cdots \oplus \mathbf{A}_k$ or as $\oplus_{i=1}^k \mathbf{A}_i$, is defined as the matrix

$$\begin{pmatrix} \mathbf{A}_1 & \mathbf{0} & \cdots & \mathbf{0} \\ \mathbf{0} & \mathbf{A}_2 & \cdots & \mathbf{0} \\ \vdots & \vdots & \ddots & \vdots \\ \mathbf{0} & \mathbf{0} & \cdots & \mathbf{A}_n \end{pmatrix}.$$

The matrices $\mathbf{A}_1, \ldots, \mathbf{A}_k$ need not be square for their direct sum to be well defined, nor must their dimensions be identical. If $\mathbf{A}_i$ is square for each i, then $\oplus_{i=1}^k \mathbf{A}_i$ is block diagonal. If $\mathbf{A}_1 = \mathbf{A}_2 = \cdots = \mathbf{A}_k = \mathbf{A}$, say, then $\oplus_{i=1}^k \mathbf{A}_i = \mathbf{I} \otimes \mathbf{A}$.

Theorem 2.18.1 *If $\mathbf{A}_1, \mathbf{A}_2, \ldots, \mathbf{A}_k$ are nonsingular, then so is $\oplus_{i=1}^k \mathbf{A}_i$, and its inverse is $\oplus_{i=1}^k \mathbf{A}_i^{-1}$.*

Proof Left as an exercise. □

2.19 Unconstrained Optimization of a Function of a Vector

The last three sections of this chapter describe some results that lie at the intersection of matrix algebra and multivariable calculus. In this section, we briefly discuss the problem of optimizing (minimizing or maximizing) a function of a vector. We begin with a definition of a certain vector that is often useful for solving such a problem.

Definition 2.19.1 Let $f(\mathbf{x})$ represent a real-valued differentiable function of an n-vector $\mathbf{x} = (x_1, \ldots, x_n)^T$. The *gradient* of f, denoted ∇f, is the n-vector of partial derivatives of $f(\mathbf{x})$ with respect to $\mathbf{x}$, i.e.,

$$\nabla f = \frac{\delta f(\mathbf{x})}{\delta \mathbf{x}} = \begin{pmatrix} \frac{\partial f(\mathbf{x})}{\partial x_1} \\ \frac{\partial f(\mathbf{x})}{\partial x_2} \\ \vdots \\ \frac{\partial f(\mathbf{x})}{\partial x_n} \end{pmatrix}.$$

For the functions that we wish to minimize in the course of developing linear model theory (the theory of least squares estimation in particular), the gradients of two particular types of functions are of considerable importance: a linear function $f(\mathbf{x}) = \mathbf{a}^T\mathbf{x}$, and a quadratic function (quadratic form) $f(\mathbf{x}) = \mathbf{x}^T\mathbf{A}\mathbf{x}$. (Here, $\mathbf{a}$ and $\mathbf{A}$ are a column vector and symmetric matrix, respectively, of fixed constants.) The following theorem is easily proven by writing the linear function and quadratic form as sums of products and then using partial differentiation results from calculus.

Theorem 2.19.1 *Let $\mathbf{x}$ represent an n-vector of variables.*

(a) If $f(\mathbf{x}) = \mathbf{a}^T\mathbf{x}$ where $\mathbf{a}$ is an n-vector of constants, then $\nabla f = \frac{\partial(\mathbf{a}^T\mathbf{x})}{\partial \mathbf{x}} = \mathbf{a}$.

(b) If $f(\mathbf{x}) = \mathbf{x}^T\mathbf{A}\mathbf{x}$ where $\mathbf{A}$ is an $n \times n$ symmetric matrix of constants, then $\nabla f = \frac{\partial(\mathbf{x}^T\mathbf{A}\mathbf{x})}{\partial \mathbf{x}} = 2\mathbf{A}\mathbf{x}$.

In standard multivariable calculus courses, it is shown that if $f(\mathbf{x})$ is defined on an open set, is differentiable at an interior point $\mathbf{x}_0$ of that set, and has a local extremum (minimum or maximum) at $\mathbf{x}_0$, then $\nabla f(\mathbf{x}_0) = 0$. This fact is the basis of a method for obtaining local and/or global extrema of the function. Specifically, the method is to determine an expression for $\nabla f(\mathbf{x})$, set it equal to 0, and solve the resulting equations for $\mathbf{x}$. However, while the condition that $\nabla f(\mathbf{x}_0) = 0$ is necessary for $f(\mathbf{x}_0)$ to be an extremum, it is not sufficient: solutions to the equations (known as stationary points) can be either a (local) minimizer, a (local) maximizer, or neither (a saddle point). Additional analysis is required to classify a stationary point to one of these categories and, if it is determined to be a local optimizer, to further determine whether it is a global optimizer. One method for classifying a stationary point that the reader may have encountered is based on the positive (or negative) definiteness of the matrix of second-order partial derivatives (assuming that they exist) at the point; this is an extension of the second derivative test used in conjunction with the optimization of a function of a single variable. However, for the applications of gradient-based optimization to linear models considered in this book, we use alternative methods for determining the nature of stationary points that do not require the calculation of second-order partial derivatives.

Example 2.19-1. Minimizing $f(\mathbf{x}) = \mathbf{x}^T\mathbf{x} - 2\mathbf{a}^T\mathbf{x}$

To illustrate the optimization method just described, suppose that we wish to minimize $f(\mathbf{x}) = \mathbf{x}^T\mathbf{x} - 2\mathbf{a}^T\mathbf{x}$ over $\mathbf{x} \in \mathbb{R}^n$. We begin by taking partial derivatives. Using Theorem 2.19.1 we obtain

$$\nabla f = 2\mathbf{x} - 2\mathbf{a}.$$

Setting these partial derivatives equal to 0 and solving yields a single stationary point $\mathbf{x} = \mathbf{a}$. Evaluating f at this point, we obtain

$$f(\mathbf{a}) = \mathbf{a}^T\mathbf{a} - 2\mathbf{a}^T\mathbf{a} = -\mathbf{a}^T\mathbf{a}.$$

Now let $\boldsymbol{\delta}$ represent any n-vector. Then

$$\begin{aligned} f(\mathbf{a} + \boldsymbol{\delta}) &= (\mathbf{a} + \boldsymbol{\delta})^T(\mathbf{a} + \boldsymbol{\delta}) - 2\mathbf{a}^T(\mathbf{a} + \boldsymbol{\delta}) \\ &= \mathbf{a}^T\mathbf{a} + \boldsymbol{\delta}^T\mathbf{a} + \mathbf{a}^T\boldsymbol{\delta} + \boldsymbol{\delta}^T\boldsymbol{\delta} - 2\mathbf{a}^T\mathbf{a} - 2\mathbf{a}^T\boldsymbol{\delta} \\ &= -\mathbf{a}^T\mathbf{a} + \boldsymbol{\delta}^T\boldsymbol{\delta}. \end{aligned}$$

Because $\boldsymbol{\delta}^T\boldsymbol{\delta} \geq 0$ for all $\boldsymbol{\delta}$, $\mathbf{a}$ is a global minimizer of $f(\mathbf{x})$. ■

2.20 Constrained Optimization of a Function of a Vector

Suppose that $f(\mathbf{x})$ and $g(\mathbf{x})$ are continuously differentiable real-valued functions of an n-vector $\mathbf{x}$ defined on an open set, and suppose further that $\nabla g(\mathbf{x}) \neq 0$ for all $\mathbf{x}$ in that set. In many multivariable calculus courses, it is noted that if an interior point $\mathbf{x}_0$ is a local or global minimizer (or maximizer) of $f(\mathbf{x})$ subject to constraints $g(\mathbf{x}) = k$ (where k is a specified constant), then a value λ_0 exists such that $(\mathbf{x}_0, \lambda_0)$ is a stationary point of the function

$$L(\mathbf{x}, \lambda) \equiv f(\mathbf{x}) - \lambda[g(\mathbf{x}) - k].$$

The function $L(\mathbf{x}, \lambda)$ is called the *Lagrangian function* associated with this constrained minimization (maximization) problem, and λ is called the *Lagrange multiplier*. This fact is the basis of a method, called the method of Lagrange multipliers, for obtaining local and/or global extrema of $f(\mathbf{x})$ subject to the constraints. Specifically, the method is to determine an expression for $\nabla L(\mathbf{x}, \lambda)$, set it equal to 0, and solve the resulting equations for $\mathbf{x}$ and λ. As with unconstrained optimization, however, additional analysis is required to determine whether a stationary point of $L(\mathbf{x}, \lambda)$ determined in this way is a local and/or global optimizer or a saddle point of $f(\mathbf{x})$ subject to the constraints. For applications to linear model theory considered in this book, we make this determination using methods that do not require the calculation of second-order partial derivatives.

Example 2.20-1. Minimizing $f(\mathbf{x}) = \mathbf{x}^T\mathbf{x}$ Subject to the Constraint $\mathbf{1}^T\mathbf{x} = 1$

To illustrate the method of Lagrange multipliers, suppose that we wish to minimize $f(\mathbf{x}) = \mathbf{x}^T\mathbf{x} = \sum_{i=1}^n x_i{}^2$ subject to the constraint $\mathbf{1}^T\mathbf{x} = \sum_{i=1}^n x_i = 1$. Accordingly, we form the Lagrangian function

$$L(\mathbf{x}, \lambda) = \sum_{i=1}^n x_i{}^2 - \lambda\left(\sum_{i=1}^n x_i - 1\right)$$

and take partial derivatives:

$$\frac{\partial L}{\partial x_i} = 2x_i - \lambda, \quad i = 1, \ldots, n,$$

$$\frac{\partial L}{\partial \lambda} = \sum_{i=1}^n x_i - 1.$$

Equating these partial derivatives to 0 yields the equations

$$x_i = \lambda/2, \quad i = 1, \ldots, n,$$
$$\sum_{i=1}^{n} x_i = 1.$$

Substituting the solution for x_i (in terms of λ) into the last equation yields $\sum_{i=1}^{n}(\lambda/2) = 1$, i.e., $n\lambda/2 = 1$, i.e., $\lambda = 2/n$. Substituting this solution for λ back into each of the first n equations yields $x_i = (2/n)/2$, i.e., $x_i = 1/n$. Thus $L(\mathbf{x}, \lambda)$ has a stationary point at $x_1 = x_2 = \cdots = x_n = 1/n$, $\lambda = 2/n$. Is $\mathbf{x}_0 \equiv (1/n, 1/n, \ldots, 1/n)^T$ a global minimizer of $f(\mathbf{x})$ subject to the constraints? To answer this, let $\boldsymbol{\delta} = (\delta_1, \ldots, \delta_n)^T$ represent an n-vector such that $\sum_{i=1}^{n} \delta_i = 0$, so that $\mathbf{x} + \boldsymbol{\delta}$ satisfies the constraint if and only if $\mathbf{x}$ does. Then

$$\sum_{i=1}^{n}\left(\frac{1}{n} + \delta_i\right)^2 = \sum_{i=1}^{n}\left(\frac{1}{n^2} + \frac{2\delta_i}{n} + \delta_i^2\right) = \sum_{i=1}^{n}\frac{1}{n^2} + \sum_{i=1}^{n}\delta_i^2 \geq \sum_{i=1}^{n}\frac{1}{n^2},$$

with equality holding if and only if $\delta_i = 0$ for all $i = 1, \ldots, n$. Thus $\mathbf{x}_0$ is indeed a global minimizer subject to the constraints. ■

If a constrained minimization (or maximization) problem has more than one constraint, then an additional Lagrange multiplier is introduced for each constraint. That is, if $\mathbf{g}(\mathbf{x}) = \mathbf{k}$ is a vector of constraints, then we minimize (or maximize) the Lagrangian function

$$L(\mathbf{x}, \boldsymbol{\lambda}) = f(\mathbf{x}) - \boldsymbol{\lambda}^T[\mathbf{g}(\mathbf{x}) - \mathbf{k}].$$

In this case, $\boldsymbol{\lambda}$ is called the *vector of Lagrange multipliers.*

2.21 Derivatives of Selected Functions of Matrices with Respect to a Scalar

Differentiation of certain functions of matrices is of importance in some aspects of linear model theory. We have already introduced (in Sect. 2.19) the derivatives of linear and quadratic forms of a vector with respect to that vector. Now we define the derivative of an arbitrary matrix with respect to a scalar and present a theorem giving expressions for such a derivative for special types of matrices or functions thereof.

Definition 2.21.1 Let $\mathbf{A} = (a_{ij})$ represent an $n \times m$ matrix whose elements are differentiable functions of a scalar variable t. The *derivative of* $\mathbf{A}$ *with respect to* t, written as $\frac{d\mathbf{A}}{dt}$, is the $n \times m$ matrix whose ijth element is $\frac{da_{ij}}{dt}$.

Theorem 2.21.1 *Let* **A** *and* **B** *represent two matrices whose elements are differentiable functions of a scalar variable t.*

(a) If **A** *and* **B** *have the same dimensions, then*

$$\frac{d}{dt}(\mathbf{A}+\mathbf{B}) = \frac{d\mathbf{A}}{dt} + \frac{d\mathbf{B}}{dt}.$$

(b) If **A** *has the same number of columns as* **B** *has rows, then*

$$\frac{d}{dt}(\mathbf{AB}) = \frac{d\mathbf{A}}{dt}\mathbf{B} + \mathbf{A}\frac{d\mathbf{B}}{dt}.$$

(c) If **A** *is nonsingular, then*

$$\frac{d}{dt}(\mathbf{A}^{-1}) = -\mathbf{A}^{-1}\frac{d\mathbf{A}}{dt}\mathbf{A}^{-1}.$$

(d) If **A** *is nonsingular, then*

$$\frac{d}{dt}(\log|\mathbf{A}|) = tr\left(\mathbf{A}^{-1}\frac{d\mathbf{A}}{dt}\right).$$

Proof Harville (1997, Lemmas 15.4.2 and Lemma 15.4.3, and Section 15.8). □

2.22 Exercises

1. Let V_n represent a vector space and let S_n represent a subspace of V_n. Prove that the orthogonal complement of S_n (relative to V_n) is a subspace of V_n.
2. Let **A** represent an $n \times m$ matrix. Prove that $\mathcal{N}(\mathbf{A})$ is a subspace of $\mathbb{R}^m$.
3. Prove Theorem 2.8.2.
4. Prove Theorem 2.8.3.
5. Prove Theorem 2.15.1.
6. Prove Theorem 2.15.2.
7. Prove Theorem 2.15.3.
8. Prove Theorem 2.15.4.
9. Prove Corollary 2.15.12.1.
10. Prove Theorem 2.15.13.
11. Prove Theorem 2.18.1.
12. Prove Theorem 2.19.1.

References

Harville, D. A. (1997). *Matrix algebra from a statistician's perspective*. New York, NY: Springer.
Harville, D. A. (2001). *Matrix algebra: Exercises and solutions*. New York, NY: Springer.
Schott, J. R. (2016). *Matrix analysis for statistics* (3rd ed.). Hoboken, NJ: Wiley.

Generalized Inverses and Solutions to Systems of Linear Equations 3

Throughout this book we will discover that in order to obtain "good" estimators of the elements of $\boldsymbol{\beta}$ (or functions thereof) under a linear model, and for other purposes as well, we need to solve various systems of linear equations of the general form

$$\mathbf{Ax} = \mathbf{b}.$$

Here $\mathbf{A}$ is a specified $n \times m$ matrix called the *coefficient matrix*, $\mathbf{b}$ is a specified n-vector called the *right-hand side vector*, and $\mathbf{x}$ is an m-vector of unknowns. Any vector of unknowns that satisfies the system is called a *solution*. A solution may or may not exist.

If $n = m$ and $\mathbf{A}$ is nonsingular, solving the system of equations $\mathbf{Ax} = \mathbf{b}$ is simple: $\mathbf{A}^{-1}\mathbf{b}$ is the one and only solution, as is easily verified. But if $\mathbf{A}$ is not square, or if $\mathbf{A}$ is square but singular, how to find a solution, if indeed one (or more) exists, is not as obvious. In this chapter, we describe a general method for solving such systems. This method is based on the notion of a generalized inverse. The first section of the chapter gives the definition of a generalized inverse, and the second describes how a generalized inverse may be used to solve a system of linear equations. The third section presents a number of results that will be useful for developing the theory of linear models in subsequent chapters, and the last defines and gives some properties of a special generalized inverse.

D. L. Zimmerman, *Linear Model Theory*,
https://doi.org/10.1007/978-3-030-52063-2_3

3.1 Generalized Inverses

Definition 3.1.1 A *generalized inverse* of an arbitrary matrix $\mathbf{A}$ is any matrix $\mathbf{G}$ for which $\mathbf{AGA} = \mathbf{A}$.

Example 3.1-1. Generalized Inverses of Some Special Matrices

Consider the following matrices: $\mathbf{I}_n$, $\mathbf{0}_{n\times m}$, $\mathbf{u}_i^{(n)}$, $\mathbf{1}_n$, and $\mathbf{aa}^T$ where $\mathbf{a}$ is any nonnull vector. Observe that

$$\begin{aligned}
\mathbf{I}_n\mathbf{I}_n\mathbf{I}_n &= \mathbf{I}_n, \\
\mathbf{0}_{n\times m}\mathbf{0}_{m\times n}\mathbf{0}_{n\times m} &= \mathbf{0}_{n\times m}, \\
\mathbf{u}_i^{(n)}\mathbf{u}_i^{(n)T}\mathbf{u}_i^{(n)} &= \mathbf{u}_i^{(n)} \cdot 1 = \mathbf{u}_i^{(n)}, \\
\mathbf{1}_n\left(\frac{1}{n}\mathbf{1}_n^T\right)\mathbf{1}_n &= \frac{1}{n}\mathbf{1}_n(\mathbf{1}_n^T\mathbf{1}_n) = \mathbf{1}_n, \\
\mathbf{aa}^T\left(\frac{1}{\|\mathbf{a}\|^4}\mathbf{aa}^T\right)\mathbf{aa}^T &= \frac{1}{\|\mathbf{a}\|^4}\mathbf{a}(\mathbf{a}^T\mathbf{a})(\mathbf{a}^T\mathbf{a})\mathbf{a}^T = \mathbf{aa}^T,
\end{aligned}$$

implying that $\mathbf{I}_n$, $\mathbf{0}_{m\times n}$, $\mathbf{u}_i^{(n)T}$, $\frac{1}{n}\mathbf{1}_n^T$, and $\frac{1}{\|\mathbf{a}\|^4}\mathbf{aa}^T$, respectively, are generalized inverses of these matrices. Also observe that

$$\begin{pmatrix} 1 & 1 \\ 1 & 1 \end{pmatrix}\begin{pmatrix} 1 & 0 \\ 0 & 0 \end{pmatrix}\begin{pmatrix} 1 & 1 \\ 1 & 1 \end{pmatrix} = \begin{pmatrix} 1 & 1 \\ 1 & 1 \end{pmatrix},$$

so $\begin{pmatrix} 1 & 0 \\ 0 & 0 \end{pmatrix}$ is a generalized inverse of $\mathbf{J}_2 = \begin{pmatrix} 1 & 1 \\ 1 & 1 \end{pmatrix}$. ■

Does a generalized inverse exist for every matrix $\mathbf{A}$? If $\mathbf{A}$ is nonsingular, then $\mathbf{AA}^{-1}\mathbf{A} = \mathbf{A}$, so $\mathbf{A}^{-1}$ is a generalized inverse of $\mathbf{A}$ in this case. In fact, if $\mathbf{A}$ is nonsingular, then $\mathbf{A}^{-1}$ is its only generalized inverse, for if $\mathbf{G}_1$ and $\mathbf{G}_2$ are two generalized inverses of such an $\mathbf{A}$, then $\mathbf{AG}_1\mathbf{A} = \mathbf{AG}_2\mathbf{A}$, implying (upon pre- and post-multiplying both sides by $\mathbf{A}^{-1}$) that $\mathbf{G}_1 = \mathbf{G}_2$.

If $\mathbf{A}$ is singular, however, then it can have more than one generalized inverse. For example, consider the matrix $\mathbf{J}_2$. It can be verified that the following matrices are all generalized inverses of $\mathbf{J}_2$:

$$\begin{pmatrix} 1 & 0 \\ 0 & 0 \end{pmatrix}, \quad \begin{pmatrix} 0 & 1 \\ 0 & 0 \end{pmatrix}, \quad \begin{pmatrix} 0.5 & 0 \\ 0.5 & 0 \end{pmatrix}, \quad 0.5\mathbf{I}_2, \quad 0.25\mathbf{J}_2. \tag{3.1}$$

(That the first matrix in this list is a generalized inverse of $\mathbf{J}_2$ was shown in Example 3.1-1.) Similarly, if $\mathbf{A}$ is nonsquare, it can have more than one generalized inverse. For example, it is easily verified that the following are all generalized inverses of the row vector $(1, 2)$:

$$\begin{pmatrix} 1 \\ 0 \end{pmatrix}, \quad \begin{pmatrix} 0 \\ 0.5 \end{pmatrix}, \quad \begin{pmatrix} -1 \\ 1 \end{pmatrix}.$$

Does a singular or nonsquare matrix always have a generalized inverse? If so, how many generalized inverses can it have? And is there a relatively simple method for computing some of these generalized inverses? The following theorem will provide answers to these questions.

Theorem 3.1.1 *Let* $\mathbf{A}$ *represent an arbitrary matrix of rank* r*, and let* $\mathbf{P}$ *and* $\mathbf{Q}$ *represent nonsingular matrices such that*

$$\mathbf{PAQ} = \begin{pmatrix} \mathbf{I}_r & \mathbf{0} \\ \mathbf{0} & \mathbf{0} \end{pmatrix}.$$

(Such matrices $\mathbf{P}$ *and* $\mathbf{Q}$ *exist by Theorem 2.8.10.) Then the matrix*

$$\mathbf{G} = \mathbf{Q} \begin{pmatrix} \mathbf{I}_r & \mathbf{F} \\ \mathbf{H} & \mathbf{B} \end{pmatrix} \mathbf{P},$$

where $\mathbf{F}$, $\mathbf{H}$, *and* $\mathbf{B}$ *are arbitrary matrices of appropriate dimensions, is a generalized inverse of* $\mathbf{A}$.

Proof Left as an exercise. □

Theorem 3.1.1 establishes that every matrix, regardless of its dimensions and/or rank, has at least one generalized inverse. It also reveals that a matrix has an infinite number of generalized inverses unless the matrix is nonsingular. (If a matrix is nonsingular, its ordinary inverse is its unique generalized inverse, as shown previously). Finally, Theorem 3.1.1 suggests a way to compute a generalized inverse of any matrix; we do not, however, describe an algorithm for doing so because we shall not be concerned with computing generalized inverses of arbitrary matrices.

Although a matrix $\mathbf{A}$ can have infinitely many generalized inverses, it is possible to express all of them in terms of just one, as indicated by the following theorem.

Theorem 3.1.2 *Let* $\mathbf{A}$ *represent an arbitrary* $n \times m$ *matrix, and let* $\mathbf{G}$ *represent a generalized inverse of* $\mathbf{A}$. *Then* $\mathbf{B}$ *is a generalized inverse of* $\mathbf{A}$ *if and only if*

$$\mathbf{B} = \mathbf{G} + \mathbf{U} - \mathbf{GAUAG}$$

for some $m \times n$ *matrix* $\mathbf{U}$.

Proof If $\mathbf{B} = \mathbf{G} + \mathbf{U} - \mathbf{GAUAG}$ for some $m \times n$ matrix $\mathbf{U}$, then

$$\mathbf{ABA} = \mathbf{AGA} + \mathbf{AUA} - \mathbf{AGAUAGA} = \mathbf{A} + \mathbf{AUA} - \mathbf{AUA} = \mathbf{A},$$

which demonstrates that $\mathbf{B}$ is a generalized inverse of $\mathbf{A}$. Conversely, suppose $\mathbf{B}$ is a generalized inverse of an $n \times m$ matrix $\mathbf{A}$. Then $\mathbf{ABA} = \mathbf{AGA}$, hence we have the identity

$$\mathbf{B} = \mathbf{G} + (\mathbf{B} - \mathbf{G}) - \mathbf{GA}(\mathbf{B} - \mathbf{G})\mathbf{AG}.$$

This expression for $\mathbf{B}$ has the form specified by the theorem, with $\mathbf{U} = \mathbf{B} - \mathbf{G}$. □

Subsequently, for any matrix $\mathbf{A}$ let $\mathbf{A}^-$ denote an arbitrary generalized inverse of $\mathbf{A}$.

3.2 Solving a System of Linear Equations via a Generalized Inverse

Before seeing how generalized inverses are involved in solving a system of linear equations, we must first realize that not every system of equations has a solution. For example, the system

$$\begin{pmatrix} 1 & 1 \\ 1 & 1 \end{pmatrix} \begin{pmatrix} x_1 \\ x_2 \end{pmatrix} = \begin{pmatrix} 0 \\ 1 \end{pmatrix}$$

has no solution, as it is not possible for $x_1 + x_2$ to be equal to 0 and 1 simultaneously.

Definition 3.2.1 The system of equations $\mathbf{Ax} = \mathbf{b}$ is said to be *consistent* if it has one or more solutions, i.e., if one or more values of $\mathbf{x}$ exist for which $\mathbf{Ax} = \mathbf{b}$.

Given a system of linear equations, how do we determine if it is consistent? One way to demonstrate consistency is, of course, to produce (somehow) a solution. However, finding a solution, even if one exists, may prove difficult. The notion of compatible equations, defined below, provides an alternative way to establish consistency.

Definition 3.2.2 The system of equations $\mathbf{Ax} = \mathbf{b}$ is said to be *compatible* if $\mathbf{v}^T\mathbf{b} = 0$ for every vector $\mathbf{v}$ for which $\mathbf{v}^T\mathbf{A} = \mathbf{0}^T$.

Theorem 3.2.1 *The system of equations* $\mathbf{Ax} = \mathbf{b}$ *is consistent if and only if it is compatible.*

Proof Suppose that the system of equations $\mathbf{Ax} = \mathbf{b}$ is consistent. Then a vector $\mathbf{x}^*$ exists such that $\mathbf{b} = \mathbf{Ax}^*$. Then, letting $\mathbf{v}$ represent any vector such that $\mathbf{v}^T\mathbf{A} = \mathbf{0}^T$, we have

$$\mathbf{v}^T\mathbf{b} = \mathbf{v}^T\mathbf{Ax}^* = (\mathbf{v}^T\mathbf{A})\mathbf{x}^* = \mathbf{0}^T\mathbf{x}^* = 0.$$

This establishes that consistency implies compatibility. For a proof of the converse, see Harville (1997, pp. 73–74). □

The upshot of Theorem 3.2.1 is that one can determine whether a system of equations is consistent merely by determining whether the system is compatible; one does not have to actually find a solution. For some systems, demonstrating compatibility is much easier than finding a solution.

We are now ready to establish the link between generalized inverses and solutions to a system of linear equations.

Theorem 3.2.2 *Let* $\mathbf{A}$ *represent any matrix. Then* $\mathbf{G}$ *is a generalized inverse of* $\mathbf{A}$ *if and only if* $\mathbf{Gb}$ *is a solution to the system of equations* $\mathbf{Ax} = \mathbf{b}$ *for every vector* $\mathbf{b}$ *for which the system is consistent.*

Proof Let $\mathbf{a}_i$ represent the ith column of $\mathbf{A}$. The system of equations $\mathbf{Ax} = \mathbf{a}_i$ clearly is consistent; hence, if $\mathbf{Gb}$ is a solution to the system of equations $\mathbf{Ax} = \mathbf{b}$ for every vector $\mathbf{b}$ for which the system is consistent, then $\mathbf{Ga}_i$ is a solution to the system $\mathbf{Ax} = \mathbf{a}_i$. That is, $\mathbf{AGa}_i = \mathbf{a}_i$ for all i, or equivalently $\mathbf{AGA} = \mathbf{A}$, establishing that $\mathbf{G}$ is a generalized inverse of $\mathbf{A}$. Conversely, suppose that $\mathbf{G}$ is any generalized inverse of $\mathbf{A}$ and that $\mathbf{b}$ is any vector for which the system of equations $\mathbf{Ax} = \mathbf{b}$ is consistent. Let $\mathbf{x}^*$ represent any solution to the system $\mathbf{Ax} = \mathbf{b}$. Then,

$$\mathbf{b} = \mathbf{Ax}^* = \mathbf{AGAx}^* = \mathbf{AGb} = \mathbf{A}(\mathbf{Gb}),$$

demonstrating that $\mathbf{Gb}$ is also a solution to this system. □

Theorem 3.2.3 *For any given generalized inverse* $\mathbf{A}^-$ *of* $\mathbf{A}$*, all solutions to the consistent system of equations* $\mathbf{Ax} = \mathbf{b}$ *are obtained by putting*

$$\mathbf{x} = \mathbf{A}^-\mathbf{b} + (\mathbf{I} - \mathbf{A}^-\mathbf{A})\mathbf{z}$$

and letting $\mathbf{z}$ *range over the space of all vectors (of appropriate dimension).*

Proof By Theorem 3.2.2, we have, for any $\mathbf{z}$ of appropriate dimension,

$$\mathbf{A}[\mathbf{A}^-\mathbf{b} + (\mathbf{I} - \mathbf{A}^-\mathbf{A})\mathbf{z}] = \mathbf{AA}^-\mathbf{b} = \mathbf{b},$$

showing that $\mathbf{A}^{-}\mathbf{b} + (\mathbf{I} - \mathbf{A}^{-}\mathbf{A})\mathbf{z}$ is a solution to the system. Conversely, for any solution $\mathbf{x}^*$ to the system of equations $\mathbf{Ax} = \mathbf{b}$,

$$\mathbf{x}^* = \mathbf{A}^{-}\mathbf{b} + \mathbf{x}^* - \mathbf{A}^{-}\mathbf{b} = \mathbf{A}^{-}\mathbf{b} + (\mathbf{I} - \mathbf{A}^{-}\mathbf{A})\mathbf{x}^*,$$

which has the specified form (with $\mathbf{z} \equiv \mathbf{x}^*$). □

Example 3.2-1. Solutions to the System of Equations $\mathbf{aa}^T\mathbf{x} = \mathbf{a}$

Let $\mathbf{a}$ represent a nonnull n-vector, and consider the system of equations $\mathbf{aa}^T\mathbf{x} = \mathbf{a}$. This system is consistent because $\frac{1}{\|\mathbf{a}\|^2}\mathbf{a}$ is a solution, as is easily verified. Thus, by Theorem 3.2.3, and using the generalized inverse of $\mathbf{aa}^T$ given in Example 3.1-1, all solutions to the system are of the form

$$\frac{1}{\|\mathbf{a}\|^4}\mathbf{aa}^T\mathbf{a} + (\mathbf{I} - \frac{1}{\|\mathbf{a}\|^4}\mathbf{aa}^T\mathbf{aa}^T)\mathbf{z}, \quad \mathbf{z} \in \mathbb{R}^n$$

or equivalently,

$$\frac{1}{\|\mathbf{a}\|^2}\mathbf{a} + (\mathbf{I} - \frac{1}{\|\mathbf{a}\|^2}\mathbf{aa}^T)\mathbf{z}, \quad \mathbf{z} \in \mathbb{R}^n.$$ ■

3.3 Some Additional Results on Generalized Inverses

This section gives some additional results on generalized inverses that will be used heavily throughout this book. These include results on the symmetry (or lack thereof) of a generalized inverse, some properties of generalized inverses of a matrix of the form $\mathbf{A}^T\mathbf{A}$, and results for generalized inverses of partitioned matrices and sums of matrices. For still more results on generalized inverses, the interested reader may consult Rao and Mitra (1971) and Harville (1997).

Theorem 3.3.1 *$\mathbf{G}$ is a generalized inverse of $\mathbf{A}$ if and only if $\mathbf{G}^T$ is a generalized inverse of $\mathbf{A}^T$.*

Proof By Theorem 2.7.1c, transposing both sides of $\mathbf{AGA} = \mathbf{A}$ yields $\mathbf{A}^T\mathbf{G}^T\mathbf{A}^T = \mathbf{A}^T$ (and vice versa). The result follows immediately by Definition 3.1.1. □

Corollary 3.3.1.1 *If $\mathbf{A}$ is symmetric, then $(\mathbf{A}^{-})^T$ is a generalized inverse of $\mathbf{A}$.*

It is important to recognize that Corollary 3.3.1.1 does not say or imply that a generalized inverse of a symmetric matrix is symmetric. Consider, for example, the symmetric matrix $\mathbf{J}_2$, several generalized inverses of which were listed in (3.1). Of the listed generalized inverses, some plainly are symmetric but some are not. On the other hand, if $\mathbf{A}$ is symmetric it has at least one symmetric generalized inverse, for $\mathbf{A}^-$ and $(\mathbf{A}^-)^T$ are both generalized inverses of $\mathbf{A}$ hence so is the symmetric matrix $(1/2)[\mathbf{A}^- + (\mathbf{A}^-)^T]$.

Corollary 3.3.1.2 *For any matrix* $\mathbf{A}$, $[(\mathbf{A}^T\mathbf{A})^-]^T$ *is a generalized inverse of* $\mathbf{A}^T\mathbf{A}$.

Theorem 3.3.2 *For any matrix* $\mathbf{A}$ *and any matrices* $\mathbf{P}$ *and* $\mathbf{L}$ *of appropriate dimensions,* $\mathbf{A}^T\mathbf{A}\mathbf{P} = \mathbf{A}^T\mathbf{A}\mathbf{L}$ *if and only if* $\mathbf{A}\mathbf{P} = \mathbf{A}\mathbf{L}$.

Proof If $\mathbf{AP} = \mathbf{AL}$, then pre-multiplication of both sides by $\mathbf{A}^T$ yields $\mathbf{A}^T\mathbf{AP} = \mathbf{A}^T\mathbf{AL}$. Conversely, if $\mathbf{A}^T\mathbf{AP} = \mathbf{A}^T\mathbf{AL}$, then $\mathbf{A}^T\mathbf{A}(\mathbf{P} - \mathbf{L}) = \mathbf{0}$, which further implies that $(\mathbf{P} - \mathbf{L})^T\mathbf{A}^T\mathbf{A}(\mathbf{P} - \mathbf{L}) = \mathbf{0}$. Thus, $(\mathbf{AP} - \mathbf{AL})^T(\mathbf{AP} - \mathbf{AL}) = \mathbf{0}$, implying (by Corollary 2.10.4.1) that $\mathbf{AP} = \mathbf{AL}$. □

Theorem 3.3.3 *For any matrix* $\mathbf{A}$*:*

(a) $\mathbf{A}(\mathbf{A}^T\mathbf{A})^-\mathbf{A}^T\mathbf{A} = \mathbf{A}$*;*
(b) $\mathbf{A}[(\mathbf{A}^T\mathbf{A})^-]^T\mathbf{A}^T\mathbf{A} = \mathbf{A}$*;*
(c) $\mathbf{A}^T\mathbf{A}(\mathbf{A}^T\mathbf{A})^-\mathbf{A}^T = \mathbf{A}^T$*;*
(d) $\mathbf{A}^T\mathbf{A}[(\mathbf{A}^T\mathbf{A})^-]^T\mathbf{A}^T = \mathbf{A}^T$.

Proof Part (a) follows immediately from applying Theorem 3.3.2 to the identity $\mathbf{A}^T\mathbf{A}(\mathbf{A}^T\mathbf{A})^-\mathbf{A}^T\mathbf{A} = \mathbf{A}^T\mathbf{A}$ [with $\mathbf{P} = (\mathbf{A}^T\mathbf{A})^-\mathbf{A}^T\mathbf{A}$ and $\mathbf{L} = \mathbf{I}$]. The proof of part (b) is similar, starting with the identity $\mathbf{A}^T\mathbf{A}[(\mathbf{A}^T\mathbf{A})^-]^T\mathbf{A}^T\mathbf{A} = \mathbf{A}^T\mathbf{A}$. Parts (c) and (d) follow by matrix transposition of parts (b) and (a), respectively. □

Corollary 3.3.3.1 *For any matrix* $\mathbf{A}$, $\mathbf{A}(\mathbf{A}^T\mathbf{A})^-\mathbf{A}^T = \mathbf{A}[(\mathbf{A}^T\mathbf{A})^-]^T\mathbf{A}^T$, *i.e.,* $\mathbf{A}(\mathbf{A}^T\mathbf{A})^-\mathbf{A}^T$ *is symmetric.*

Proof By Theorem 3.3.3 c and d, $\mathbf{A}^T\mathbf{A}(\mathbf{A}^T\mathbf{A})^-\mathbf{A}^T = \mathbf{A}^T\mathbf{A}[(\mathbf{A}^T\mathbf{A})^-]^T\mathbf{A}^T$. The result follows immediately from Theorem 3.3.2 [with $\mathbf{P} = (\mathbf{A}^T\mathbf{A})^-\mathbf{A}^T$ and $\mathbf{L} = [(\mathbf{A}^T\mathbf{A})^-]^T\mathbf{A}^T$]. □

Theorem 3.3.4 *For any matrix* $\mathbf{A}$ *and any two generalized inverses* $\mathbf{G}_1$ *and* $\mathbf{G}_2$ *of* $\mathbf{A}^T\mathbf{A}$, $\mathbf{A}\mathbf{G}_1\mathbf{A}^T = \mathbf{A}\mathbf{G}_2\mathbf{A}^T$.

Proof Applying Theorem 3.3.2 to the identity $\mathbf{A}^T\mathbf{A}\mathbf{G}_1^T\mathbf{A}^T\mathbf{A} = \mathbf{A}^T\mathbf{A}\mathbf{G}_2^T\mathbf{A}^T\mathbf{A}$ (with $\mathbf{P} = \mathbf{G}_1^T\mathbf{A}^T\mathbf{A}$ and $\mathbf{L} = \mathbf{G}_2^T\mathbf{A}^T\mathbf{A}$) implies that $\mathbf{A}\mathbf{G}_1^T\mathbf{A}^T\mathbf{A} = \mathbf{A}\mathbf{G}_2^T\mathbf{A}^T\mathbf{A}$. Transposing both sides of this equality yields $\mathbf{A}^T\mathbf{A}\mathbf{G}_1\mathbf{A}^T = \mathbf{A}^T\mathbf{A}\mathbf{G}_2\mathbf{A}^T$. The result then follows upon applying Theorem 3.3.2 once again (this time with $\mathbf{P} = \mathbf{G}_1\mathbf{A}^T$ and $\mathbf{L} = \mathbf{G}_2\mathbf{A}^T$). □

Theorem 3.3.5 *For any $n \times m$ matrix* $\mathbf{A}$ *of rank r,* $\mathbf{AA}^-$ *and* $\mathbf{A}^-\mathbf{A}$ *are idempotent matrices of rank r, and* $\mathbf{I}_n - \mathbf{AA}^-$ *and* $\mathbf{I}_m - \mathbf{A}^-\mathbf{A}$ *are idempotent matrices of ranks $n - r$ and $m - r$, respectively.*

Proof Left as an exercise. □

Theorem 3.3.6 *Let* $\mathbf{M} = \begin{pmatrix} \mathbf{A} & \mathbf{B} \\ \mathbf{B}^T & \mathbf{D} \end{pmatrix}$ *represent a partitioned symmetric matrix of rank r. If* $\mathbf{A}$ *is $r \times r$ and nonsingular, then* $\begin{pmatrix} \mathbf{A}^{-1} & \mathbf{0} \\ \mathbf{0} & \mathbf{0} \end{pmatrix}$ *is a generalized inverse of* $\mathbf{M}$. *Similarly, if* $\mathbf{D}$ *is $r \times r$ and nonsingular, then* $\begin{pmatrix} \mathbf{0} & \mathbf{0} \\ \mathbf{0} & \mathbf{D}^{-1} \end{pmatrix}$ *is a generalized inverse of* $\mathbf{M}$.

Proof We prove only the first assertion; the proof of the second is similar. By Theorem 2.9.6, $\text{rank}(\mathbf{M}) = \text{rank}(\mathbf{A}) + \text{rank}(\mathbf{D} - \mathbf{B}^T\mathbf{A}^{-1}\mathbf{B})$, implying that $r = r + \text{rank}(\mathbf{D} - \mathbf{B}^T\mathbf{A}^{-1}\mathbf{B})$ or equivalently that $\text{rank}(\mathbf{D} - \mathbf{B}^T\mathbf{A}^{-1}\mathbf{B}) = 0$, implying still further that $\mathbf{D} = \mathbf{B}^T\mathbf{A}^{-1}\mathbf{B}$. Then,

$$\begin{pmatrix} \mathbf{A} & \mathbf{B} \\ \mathbf{B}^T & \mathbf{D} \end{pmatrix}\begin{pmatrix} \mathbf{A}^{-1} & \mathbf{0} \\ \mathbf{0} & \mathbf{0} \end{pmatrix}\begin{pmatrix} \mathbf{A} & \mathbf{B} \\ \mathbf{B}^T & \mathbf{D} \end{pmatrix} = \begin{pmatrix} \mathbf{I} & \mathbf{0} \\ \mathbf{B}^T\mathbf{A}^{-1} & \mathbf{0} \end{pmatrix}\begin{pmatrix} \mathbf{A} & \mathbf{B} \\ \mathbf{B}^T & \mathbf{D} \end{pmatrix} = \begin{pmatrix} \mathbf{A} & \mathbf{B} \\ \mathbf{B}^T & \mathbf{B}^T\mathbf{A}^{-1}\mathbf{B} \end{pmatrix}$$
$$= \begin{pmatrix} \mathbf{A} & \mathbf{B} \\ \mathbf{B}^T & \mathbf{D} \end{pmatrix}.$$

□

Theorem 3.3.7 *Let* $\mathbf{M} = \begin{pmatrix} \mathbf{A} & \mathbf{B} \\ \mathbf{C} & \mathbf{D} \end{pmatrix}$ *represent a partitioned matrix.*

(a) If $\mathcal{C}(\mathbf{B})$ *is a subspace of* $\mathcal{C}(\mathbf{A})$ *and* $\mathcal{R}(\mathbf{C})$ *is a subspace of* $\mathcal{R}(\mathbf{A})$ *(as would be the case, e.g., if* $\mathbf{A}$ *was nonsingular), then the partitioned matrix*

$$\begin{pmatrix} \mathbf{A}^- + \mathbf{A}^-\mathbf{BQ}^-\mathbf{CA}^- & -\mathbf{A}^-\mathbf{BQ}^- \\ -\mathbf{Q}^-\mathbf{CA}^- & \mathbf{Q}^- \end{pmatrix}$$

where $\mathbf{Q} = \mathbf{D} - \mathbf{CA}^-\mathbf{B}$, *is a generalized inverse of* $\mathbf{M}$.

(b) If $\mathcal{C}(\mathbf{C})$ *is a subspace of* $\mathcal{C}(\mathbf{D})$ *and* $\mathcal{R}(\mathbf{B})$ *is a subspace of* $\mathcal{R}(\mathbf{D})$ *(as would be the case, e.g., if* $\mathbf{D}$ *was nonsingular), then*

$$\begin{pmatrix} \mathbf{P}^- & -\mathbf{P}^-\mathbf{BD}^- \\ -\mathbf{D}^-\mathbf{CP}^- & \mathbf{D}^- + \mathbf{D}^-\mathbf{CP}^-\mathbf{BD}^- \end{pmatrix}$$

where $\mathbf{P} = \mathbf{A} - \mathbf{BD}^-\mathbf{C}$, *is a generalized inverse of* $\mathbf{M}$.

Proof Left as an exercise. □

Theorem 3.3.8 *Let* $\mathbf{A}$ *represent an* $n \times n$ *nonnegative definite matrix, let* $\mathbf{B}$ *represent an arbitrary* $n \times p$ *matrix, and let* $\mathbf{G} = \begin{pmatrix} \mathbf{G}_{11} & \mathbf{G}_{12} \\ \mathbf{G}_{21} & \mathbf{G}_{22} \end{pmatrix}$ *(where* $\mathbf{G}_{11}$ *is of dimensions* $n \times n$*) represent an arbitrary generalized inverse of the partitioned matrix* $\begin{pmatrix} \mathbf{A} & \mathbf{B} \\ \mathbf{B}^T & \mathbf{0} \end{pmatrix}$. *Then:*

(a) $\mathbf{B}^T\mathbf{G}_{12}\mathbf{B}^T = \mathbf{B}^T$ *and* $\mathbf{B}\mathbf{G}_{21}\mathbf{B} = \mathbf{B}$ *(i.e.,* $\mathbf{G}_{12}$ *is a generalized inverse of* $\mathbf{B}^T$ *and* $\mathbf{G}_{21}$ *is a generalized inverse of* $\mathbf{B}$*);*
(b) $\mathbf{A}\mathbf{G}_{12}\mathbf{B}^T = \mathbf{B}\mathbf{G}_{21}\mathbf{A} = -\mathbf{B}\mathbf{G}_{22}\mathbf{B}^T$*;*
(c) $\mathbf{A}\mathbf{G}_{11}\mathbf{B} = \mathbf{0}$, $\mathbf{B}^T\mathbf{G}_{11}\mathbf{A} = \mathbf{0}$, *and* $\mathbf{B}^T\mathbf{G}_{11}\mathbf{B} = \mathbf{0}$*;*
(d) $\mathbf{A} = \mathbf{A}\mathbf{G}_{11}\mathbf{A} - \mathbf{B}\mathbf{G}_{22}\mathbf{B}^T$*;*
(e) $tr(\mathbf{A}\mathbf{G}_{11}) = rank(\mathbf{A}, \mathbf{B}) - rank(\mathbf{B})$*;*
(f) $\mathbf{A}\mathbf{G}_{11}\mathbf{A}$, $\mathbf{A}\mathbf{G}_{12}\mathbf{B}^T$, $\mathbf{B}\mathbf{G}_{21}\mathbf{A}$, *and* $\mathbf{B}\mathbf{G}_{22}\mathbf{B}^T$ *are symmetric and invariant to the choice of generalize inverse* $\mathbf{G}$*; and*
(g) $tr(\mathbf{B}^T\mathbf{G}_{12}) = tr(\mathbf{B}\mathbf{G}_{21}) = rank(\mathbf{B})$.

Proof For proofs of parts (a)–(f), see Harville (1997, Section 19.4a). As for part (g), observe that part (a) and Theorem 3.3.5 imply that $\mathbf{B}^T\mathbf{G}_{12}$ and $\mathbf{B}\mathbf{G}_{21}$ are idempotent and that each has rank equal to rank($\mathbf{B}$). Part (g) follows immediately by Theorem 2.12.2. □

Theorem 3.3.9 *Let* $\mathbf{A}$ *represent an* $n \times n$ *nonnegative definite matrix, and let* $\mathbf{B}$ *represent an arbitrary* $n \times p$ *matrix. Then:*

(a) for any $n \times n$ *matrix* $\mathbf{U}$ *and any generalized inverse* $\begin{pmatrix} \mathbf{G}_{11} & \mathbf{G}_{12} \\ \mathbf{G}_{21} & \mathbf{G}_{22} \end{pmatrix}$ *(where* $\mathbf{G}_{11}$ *is of dimensions* $n \times n$*) of* $\begin{pmatrix} \mathbf{A} & \mathbf{B} \\ \mathbf{B}^T & \mathbf{0} \end{pmatrix}$, $\begin{pmatrix} \mathbf{G}_{11} & \mathbf{G}_{12} \\ \mathbf{G}_{21} & \mathbf{G}_{22} - \mathbf{U} \end{pmatrix}$ *is a generalized inverse of* $\begin{pmatrix} \mathbf{A} + \mathbf{B}\mathbf{U}\mathbf{B}^T & \mathbf{B} \\ \mathbf{B}^T & \mathbf{0} \end{pmatrix}$*;*
(b) conversely, if $\begin{pmatrix} \mathbf{H}_{11} & \mathbf{H}_{12} \\ \mathbf{H}_{21} & \mathbf{H}_{22} \end{pmatrix}$ *is a generalized inverse of* $\begin{pmatrix} \mathbf{A} + \mathbf{B}\mathbf{U}\mathbf{B}^T & \mathbf{B} \\ \mathbf{B}^T & \mathbf{0} \end{pmatrix}$, *then* $\begin{pmatrix} \mathbf{H}_{11} & \mathbf{H}_{12} \\ \mathbf{H}_{21} & \mathbf{H}_{22} + \mathbf{U} \end{pmatrix}$ *is a generalized inverse of* $\begin{pmatrix} \mathbf{A} & \mathbf{B} \\ \mathbf{B}^T & \mathbf{0} \end{pmatrix}$.

Proof See Harville (1997, Section 19.4a). □

Theorem 3.3.10 *Let* $\mathbf{M} = \mathbf{A} + \mathbf{H}$, *where* $\mathcal{C}(\mathbf{H})$ *and* $\mathcal{R}(\mathbf{H})$ *are subspaces of* $\mathcal{C}(\mathbf{A})$ *and* $\mathcal{R}(\mathbf{A})$, *respectively (as would be the case, e.g., if* $\mathbf{A}$ *was nonsingular), and take*

$\mathbf{B}$, $\mathbf{C}$, *and* $\mathbf{D}$ *to represent any three matrices such that* $\mathbf{H} = \mathbf{BCD}$. *Then, the matrix* $\mathbf{A}^- - \mathbf{A}^-\mathbf{BC}(\mathbf{C} + \mathbf{CDA}^-\mathbf{BC})^-\mathbf{CDA}^-$ *is a generalized inverse of* $\mathbf{M}$.

Proof Left as an exercise. □

The following corollary to Theorem 3.3.10 is the generalized-inverse analogue of the Sherman–Morrison–Woodbury formula (Corollary 2.9.7.1).

Corollary 3.3.10.1 *Let* $\mathbf{M} = \mathbf{A} + \mathbf{bd}^T$, *where* $\mathbf{b} \in \mathcal{C}(\mathbf{A})$ *and* $\mathbf{d}^T \in \mathcal{R}(\mathbf{A})$. *If* $\mathbf{d}^T\mathbf{A}^-\mathbf{b} \neq -1$, *then* $\mathbf{A}^- - (1 + \mathbf{d}^T\mathbf{A}^-\mathbf{b})^{-1}\mathbf{A}^-\mathbf{bd}^T\mathbf{A}^-$ *is a generalized inverse of* $\mathbf{M}$; *otherwise,* $\mathbf{A}^-$ *is a generalized inverse of* $\mathbf{M}$.

Proof By Theorem 3.3.10, under the given conditions on $\mathbf{b}$ and $\mathbf{d}^T$ one generalized inverse of $\mathbf{M}$ is $\mathbf{A}^- - \mathbf{A}^-\mathbf{b}(1 + \mathbf{d}^T\mathbf{A}^-\mathbf{b})^-\mathbf{d}^T\mathbf{A}^-$. The result for the case $\mathbf{d}^T\mathbf{A}^-\mathbf{b} \neq -1$ is then trivial; if $\mathbf{d}^T\mathbf{A}^-\mathbf{b} = -1$, then 0 is a generalized inverse of $1 + \mathbf{d}^T\mathbf{A}^-\mathbf{b}$, so that $\mathbf{A}^-$ is a generalized inverse of $\mathbf{M}$ in this case. □

Theorem 3.3.11 *Let* $\mathbf{A} = \begin{pmatrix} \mathbf{A}_{11} & \mathbf{A}_{12} \\ \mathbf{A}_{21} & \mathbf{A}_{22} \end{pmatrix}$, *where* $\mathbf{A}_{11}$ *is of dimensions* $n \times m$, *represent a partitioned matrix and let* $\mathbf{G} = \begin{pmatrix} \mathbf{G}_{11} & \mathbf{G}_{12} \\ \mathbf{G}_{21} & \mathbf{G}_{22} \end{pmatrix}$, *where* $\mathbf{G}_{11}$ *is of dimensions* $m \times n$, *represent any generalized inverse of* $\mathbf{A}$. *If* $\mathcal{R}(\mathbf{A}_{21}) \cap \mathcal{R}(\mathbf{A}_{11}) = \{\mathbf{0}\}$ *and* $\mathcal{C}(\mathbf{A}_{12}) \cap \mathcal{C}(\mathbf{A}_{11}) = \{\mathbf{0}\}$, *then* $\mathbf{G}_{11}$ *is a generalized inverse of* $\mathbf{A}_{11}$.

Proof Left as an exercise. □

Theorem 3.3.12 *Let* $\mathbf{A}$ *and* $\mathbf{B}$ *represent two* $m \times n$ *matrices. If* $\mathcal{C}(\mathbf{A}) \cap \mathcal{C}(\mathbf{B}) = \{\mathbf{0}\}$ *and* $\mathcal{R}(\mathbf{A}) \cap \mathcal{R}(\mathbf{B}) = \{\mathbf{0}\}$, *then any generalized inverse of* $\mathbf{A}+\mathbf{B}$ *is also a generalized inverse of* $\mathbf{A}$.

Proof Left as an exercise. □

3.4 The Moore–Penrose Inverse

Suppose that $\mathbf{A}$ is an arbitrary singular matrix. In the first section of this chapter, we noted that $\mathbf{A}$ has an infinite number of generalized inverses. For many purposes, including all the results presented so far, it makes no difference which generalized inverse of $\mathbf{A}$, among the infinitely many available, is used. However, for some other purposes a generalized inverse that has more properties in common with the ordinary inverse is more useful. For example, suppose $\mathbf{A}$ is not only singular but symmetric. Then a generalized inverse of $\mathbf{A}$ is not necessarily symmetric, whereas the inverse of a nonsingular symmetric matrix *is* symmetric (Corollary 2.9.2.1).

The Moore–Penrose inverse (defined below) is a particular type of generalized inverse that has more properties in common with the ordinary inverse.

Theorem 3.4.1 *For any matrix* $\mathbf{A}$, *a unique matrix* $\mathbf{G}$ *exists that satisfies the following four conditions:*

(i) $\mathbf{AGA} = \mathbf{A}$*;*
(ii) $\mathbf{GAG} = \mathbf{G}$*;*
(iii) $(\mathbf{GA})^T = \mathbf{GA}$*; and*
(iv) $(\mathbf{AG})^T = \mathbf{AG}$.

Furthermore, the unique matrix $\mathbf{G}$ *that satisfies these conditions is given by*

$$\mathbf{A}^T(\mathbf{AA}^T)^-\mathbf{A}(\mathbf{A}^T\mathbf{A})^-\mathbf{A}^T,$$

which is invariant to the choice of generalized inverses of $\mathbf{AA}^T$ *and* $\mathbf{A}^T\mathbf{A}$.

Proof For the specified $\mathbf{G}$, repeated use of Theorems 3.3.3 and 3.3.4 yields

$$\begin{aligned}
\mathbf{AGA} &= \mathbf{AA}^T(\mathbf{AA}^T)^-\mathbf{A}(\mathbf{A}^T\mathbf{A})^-\mathbf{A}^T\mathbf{A} = \mathbf{AA}^T(\mathbf{AA}^T)^-\mathbf{A} = \mathbf{A},\\
\mathbf{GAG} &= \mathbf{A}^T(\mathbf{AA}^T)^-\mathbf{A}(\mathbf{A}^T\mathbf{A})^-\mathbf{A}^T\mathbf{AA}^T(\mathbf{AA}^T)^-\mathbf{A}(\mathbf{A}^T\mathbf{A})^-\mathbf{A}^T\\
&= \mathbf{A}^T(\mathbf{AA}^T)^-\mathbf{AA}^T(\mathbf{AA}^T)^-\mathbf{A}(\mathbf{A}^T\mathbf{A})^-\mathbf{A}^T = \mathbf{A}^T(\mathbf{AA}^T)^-\mathbf{A}(\mathbf{A}^T\mathbf{A})^-\mathbf{A}^T = \mathbf{G},\\
(\mathbf{GA})^T &= \mathbf{A}^T\mathbf{A}[(\mathbf{A}^T\mathbf{A})^-]^T\mathbf{A}^T[(\mathbf{AA}^T)^-]^T\mathbf{A} = \mathbf{A}^T[(\mathbf{AA}^T)^-]^T\mathbf{A}\\
&= \mathbf{A}^T(\mathbf{AA}^T)^-\mathbf{A} = \mathbf{A}^T(\mathbf{AA}^T)^-\mathbf{A}(\mathbf{A}^T\mathbf{A})^-\mathbf{A}^T\mathbf{A} = \mathbf{GA}, \text{ and}\\
(\mathbf{AG})^T &= \mathbf{A}[(\mathbf{A}^T\mathbf{A})^-]^T\mathbf{A}^T[(\mathbf{AA}^T)^-]^T\mathbf{AA}^T = \mathbf{A}[(\mathbf{A}^T\mathbf{A})^-]^T\mathbf{A}^T\\
&= \mathbf{A}(\mathbf{A}^T\mathbf{A})^-\mathbf{A}^T = \mathbf{AA}^T(\mathbf{AA}^T)^-\mathbf{A}(\mathbf{A}^T\mathbf{A})^-\mathbf{A}^T = \mathbf{AG}.
\end{aligned}$$

Thus, $\mathbf{G}$ satisfies the four conditions, regardless of the choice of generalized inverses of $\mathbf{AA}^T$ and $\mathbf{A}^T\mathbf{A}$. To establish that the specified $\mathbf{G}$ is the only such matrix, let $\mathbf{G}_1$ and $\mathbf{G}_2$ represent any two matrices that satisfy the conditions. Then

$$\begin{aligned}
\mathbf{G}_1 &= \mathbf{G}_1\mathbf{AG}_1 = \mathbf{G}_1\mathbf{G}_1^T\mathbf{A}^T = \mathbf{G}_1\mathbf{G}_1^T\mathbf{A}^T\mathbf{G}_2^T\mathbf{A}^T = \mathbf{G}_1\mathbf{G}_1^T\mathbf{A}^T\mathbf{AG}_2\\
&= \mathbf{G}_1\mathbf{AG}_1\mathbf{AG}_2 = \mathbf{G}_1\mathbf{AG}_2 = \mathbf{G}_1\mathbf{AG}_2\mathbf{AG}_2 = \mathbf{G}_1\mathbf{AA}^T\mathbf{G}_2^T\mathbf{G}_2\\
&= \mathbf{A}^T\mathbf{G}_1^T\mathbf{A}^T\mathbf{G}_2^T\mathbf{G}_2 = \mathbf{A}^T\mathbf{G}_2^T\mathbf{G}_2 = \mathbf{G}_2\mathbf{AG}_2 = \mathbf{G}_2.
\end{aligned}$$

□

Definition 3.4.1 The four conditions of Theorem 3.4.1 are called the *Moore–Penrose conditions*, and the unique matrix $\mathbf{G}$ that satisfies those conditions is called the *Moore–Penrose inverse*.

Subsequently, we denote the Moore–Penrose inverse of an arbitrary matrix $\mathbf{A}$ by $\mathbf{A}^+$.

Example 3.4-1. Moore–Penrose Inverses of Some Special Matrices

Consider once again the six matrices for which a generalized inverse was given in Example 3.1-1: $\mathbf{I}_n$, $\mathbf{0}_{n\times m}$, $\mathbf{u}_i^{(n)}$, $\mathbf{1}_n$, $\mathbf{aa}^T$ where $\mathbf{a}$ is any nonnull vector, and $\mathbf{J}_2$. It may be easily verified that the generalized inverses given in that example for the first five of those matrices are, in fact, the Moore–Penrose inverses of those matrices. However, the generalized inverse of $\mathbf{J}_2$ given in the example, i.e., $\mathbf{G} = \begin{pmatrix} 1 & 0 \\ 0 & 0 \end{pmatrix}$, is not the Moore–Penrose inverse, as it fails to satisfy Moore–Penrose condition (iii):

$$\mathbf{GA} = \begin{pmatrix} 1 & 0 \\ 0 & 0 \end{pmatrix}\begin{pmatrix} 1 & 1 \\ 1 & 1 \end{pmatrix} = \begin{pmatrix} 1 & 1 \\ 0 & 0 \end{pmatrix},$$

which is not symmetric. However, it can be verified that $0.25\mathbf{J}_2$ satisfies all four Moore–Penrose conditions and is therefore the Moore–Penrose inverse of $\mathbf{J}_2$. ■

Theorem 3.4.2 *For any matrix* $\mathbf{A}$, $(\mathbf{A}^T)^+ = (\mathbf{A}^+)^T$.

Proof Left as an exercise. □

Corollary 3.4.2.1 *If* $\mathbf{A}$ *is symmetric, then so is* $\mathbf{A}^+$.

Theorem 3.4.3 *If* $\mathbf{A}$ *is nonnegative definite, then so is* $\mathbf{A}^+$.

Proof For any vector $\mathbf{x}$ having the same number of elements as the columns (and rows) of $\mathbf{A}$,

$$\mathbf{x}^T\mathbf{A}^+\mathbf{x} = \mathbf{x}^T\mathbf{A}^+\mathbf{A}\mathbf{A}^+\mathbf{x} = \mathbf{x}^T(\mathbf{A}^+)^T\mathbf{A}\mathbf{A}^+\mathbf{x} = (\mathbf{A}^+\mathbf{x})^T\mathbf{A}(\mathbf{A}^+\mathbf{x}) \geq 0$$

where we used Corollary 3.4.2.1 to obtain the third expression and the nonnegative definiteness of $\mathbf{A}$ to obtain the inequality. □

3.5 Exercises

1. Find a generalized inverse of each of the following matrices.
   (a) $c\mathbf{A}$, where $c \neq 0$ and $\mathbf{A}$ is an arbitrary matrix.
   (b) $\mathbf{ab}^T$, where $\mathbf{a}$ is a nonnull m-vector and $\mathbf{b}$ is a nonnull n-vector.

(c) $\mathbf{K}$, where $\mathbf{K}$ is nonnull and idempotent.
(d) $\mathbf{D} = \text{diag}(d_1, \ldots, d_n)$ where $d_1, \ldots, d_n$ are arbitrary real numbers.
(e) $\mathbf{C}$, where $\mathbf{C}$ is a square matrix whose elements are equal to zero except possibly on the cross-diagonal stretching from the lower left element to the upper right element.
(f) $\mathbf{PAQ}$, where $\mathbf{A}$ is any nonnull $m \times n$ matrix and $\mathbf{P}$ and $\mathbf{Q}$ are $m \times m$ and $n \times n$ orthogonal matrices.
(g) $\mathbf{P}$, where $\mathbf{P}$ is an $m \times n$ matrix such that $\mathbf{P}^T\mathbf{P} = \mathbf{I}_n$.
(h) $\mathbf{J}_{m\times n}$, and characterize the collection of *all* generalized inverses of $\mathbf{J}_{m\times n}$.
(i) $a\mathbf{I}_n + b\mathbf{J}_n$, where a and b are nonzero scalars. (Hint: Consider the nonsingular and singular cases separately, and use Corollary 2.9.7.1 for the former.)
(j) $\mathbf{A} + \mathbf{bd}^T$, where $\mathbf{A}$ is nonsingular and $\mathbf{A} + \mathbf{bd}^T$ is singular. (Hint: The singularity of $\mathbf{A} + \mathbf{bd}^T$ implies that $\mathbf{d}^T\mathbf{A}^{-1}\mathbf{b} = -1$ by Corollary 2.9.7.1.)
(k) $\mathbf{B}$, where $\mathbf{B} = \oplus_{i=1}^k \mathbf{B}_i$ and $\mathbf{B}_1, \ldots, \mathbf{B}_k$ are arbitrary matrices.
(l) $\mathbf{A} \otimes \mathbf{B}$, where $\mathbf{A}$ and $\mathbf{B}$ are arbitrary matrices.

2. Prove Theorem 3.1.1.
3. Let $\mathbf{A}$ represent any square matrix.
   (a) Prove that $\mathbf{A}$ has a nonsingular generalized inverse. (Hint: Use Theorem 3.1.1.)
   (b) Let $\mathbf{G}$ represent a nonsingular generalized inverse of $\mathbf{A}$. Show, by giving a counterexample, that $\mathbf{G}^{-1}$ need not equal $\mathbf{A}$.
4. Show, by giving a counterexample, that if the system of equations $\mathbf{Ax} = \mathbf{b}$ is not consistent, then $\mathbf{AA}^-\mathbf{b}$ need not equal $\mathbf{b}$.
5. Let $\mathbf{X}$ represent a matrix such that $\mathbf{X}^T\mathbf{X}$ is nonsingular, and let k represent an arbitrary real number. Consider generalized inverses of the matrix $\mathbf{I} + k\mathbf{X}(\mathbf{X}^T\mathbf{X})^{-1}\mathbf{X}^T$. For each $k \in \mathbb{R}$, determine S_k, where

$$S_k = \{c \in \mathbb{R} : \mathbf{I}+c\mathbf{X}(\mathbf{X}^T\mathbf{X})^{-1}\mathbf{X}^T \text{ is a generalized inverse of } \mathbf{I}+k\mathbf{X}(\mathbf{X}^T\mathbf{X})^{-1}\mathbf{X}^T\}.$$

6. Let $\mathbf{A}$ represent any $m \times n$ matrix, and let $\mathbf{B}$ represent any $n \times q$ matrix. Prove that for any choices of generalized inverses $\mathbf{A}^-$ and $\mathbf{B}^-$, $\mathbf{B}^-\mathbf{A}^-$ is a generalized inverse of $\mathbf{AB}$ if and only if $\mathbf{A}^-\mathbf{ABB}^-$ is idempotent.
7. Prove Theorem 3.3.5.
8. Prove Theorem 3.3.7.
9. Prove Theorem 3.3.10.
10. Prove Theorem 3.3.11.
11. Prove Theorem 3.3.12, and show that its converse is false by constructing a counterexample based on the matrices

$$\mathbf{A} = \begin{pmatrix} 1 & 0 \\ 0 & 0 \end{pmatrix} \quad \text{and} \quad \mathbf{B} = \begin{pmatrix} 0 & 0 \\ 0 & 1 \end{pmatrix}.$$

12. Determine Moore–Penrose inverses of each of the matrices listed in Exercise 3.1 except those in parts (c) and (j).
13. Prove Theorem 3.4.2.
14. Let **A** represent any matrix. Prove that $\text{rank}(\mathbf{A}^{+}) = \text{rank}(\mathbf{A})$. Does this result hold for an arbitrary generalized inverse of **A**? Prove or give a counterexample.

References

Harville, D. A. (1997). *Matrix algebra from a statistician's perspective*. New York, NY: Springer.

Rao, C. A. & Mitra, S. K. (1971). *Generalized inverse of matrices and its applications*. New York, NY: Wiley.

Moments of a Random Vector and of Linear and Quadratic Forms in a Random Vector

4

A considerable body of theory on point estimation of the parameters of linear models can be obtained using only some results pertaining to the first few moments of a random vector and of certain functions of that vector; it is not necessary to specify the random vector's probability distribution. This chapter presents these moment results. In the first section, moments up to fourth order (mean vector, variance–covariance matrix, skewness matrix, and kurtosis matrix) are defined. In the second section, moments (up to second order) of linear and quadratic forms in a random vector are derived in terms of the moments (up to fourth order) of the random vector.

4.1 Moments (up to Fourth-Order) of a Random Vector

Throughout this chapter, $\mathbf{x} = (x_i)$ represents a random n-vector, i.e., an n-vector of random variables. It is not (yet) required that this random vector follow a linear model.

Definition 4.1.1 The *mean*, or *expectation*, of a random n-vector $\mathbf{x} = (x_i)$, written as $\mathrm{E}(\mathbf{x})$, is the n-vector of means (first moments) of the elements of $\mathbf{x}$, provided that all these expectations exist; that is, $\mathrm{E}(\mathbf{x}) = [\mathrm{E}(x_1), \mathrm{E}(x_2), \ldots, \mathrm{E}(x_n)]^T$.

Definition 4.1.2 The *matrix of covariances* between a random n-vector $\mathbf{x}$ and a random m-vector $\mathbf{z}$, written as $\mathrm{cov}(\mathbf{x}, \mathbf{z})$, is the $n \times m$ matrix $\mathrm{E}\{[\mathbf{x}-\mathrm{E}(\mathbf{x})][\mathbf{z}-\mathrm{E}(\mathbf{z})]^T\}$, provided that all these expectations exist. In the special case in which $\mathbf{x} = \mathbf{z}$, the matrix of covariances is called the *variance–covariance matrix* of $\mathbf{x}$, and is written as $\mathrm{var}(\mathbf{x})$.

The (i, j)th element of $\mathrm{cov}(\mathbf{x}, \mathbf{z})$ is the covariance between x_i and z_j. The ith main diagonal element of $\mathrm{var}(\mathbf{x})$ is the variance (second central moment) of x_i, while

D. L. Zimmerman, *Linear Model Theory*,
https://doi.org/10.1007/978-3-030-52063-2_4

the (i, j)th element $(i \neq j)$ of var($\mathbf{x}$) is the covariance between x_i and x_j. Clearly, a variance–covariance matrix is symmetric. By the Cauchy–Schwartz inequality, var($\mathbf{x}$) exists if and only if var(x_i) exists for all i, and cov($\mathbf{x}$, $\mathbf{z}$) exists if and only if both var($\mathbf{x}$) and var($\mathbf{z}$) exist.

In nearly all of the linear models considered in this book, a variance–covariance matrix for the vector of observations will be specified, either completely or up to some unknown parameters. In this regard, it is important to recognize that not every symmetric matrix can serve as a variance–covariance matrix. The following theorem describes the conditions that a matrix must satisfy to be a variance–covariance matrix.

Theorem 4.1.1 *A variance–covariance matrix is nonnegative definite. Conversely, any nonnegative definite matrix is the variance–covariance matrix of some random vector.*

Proof Left as an exercise. □

The next theorem gives a sufficient condition for a variance–covariance matrix to be diagonal.

Theorem 4.1.2 *If the variances of all variables in a random vector* $\mathbf{x}$ *exist, and those variables are pairwise independent, then var(*$\mathbf{x}$*) is a diagonal matrix.*

Proof Left as an exercise. □

We now define two close relatives of a variance–covariance matrix.

Definition 4.1.3 The *precision matrix* of a random n-vector for which the variance–covariance matrix exists and is positive definite is the inverse of that variance–covariance matrix.

It follows from the Corollary 2.9.2.1 and Theorem 2.15.6 that a precision matrix is necessarily symmetric and positive definite, and from Theorems 2.9.3 and 4.1.2 that the precision matrix of pairwise independent random variables, when it exists, is diagonal.

Definition 4.1.4 The *correlation matrix* of a random n-vector $\mathbf{x} = (x_i)$ is the matrix whose (i, j)th element is equal to the correlation between x_i and x_j, which in turn is given by

$$\mathrm{corr}(x_i, x_j) = \frac{\mathrm{cov}(x_i, x_j)}{[\mathrm{var}(x_i)\mathrm{var}(x_j)]^{\frac{1}{2}}}$$

provided that the variances exist and are positive. If either var(x_i) or var(x_j) is equal to 0, then corr(x_i, x_j) is undefined.

A correlation matrix $\boldsymbol{\Upsilon}$ is a special case of a variance–covariance matrix, so (by Theorem 4.1.1) it must also be nonnegative definite. In addition, if all the variances are positive, then its main diagonal elements are all equal to 1 and it is related to the variance–covariance matrix $\boldsymbol{\Sigma} = (\sigma_{ij})$ via the matrix equation

$$\boldsymbol{\Upsilon} = \left(\oplus_{i=1}^{n}\sqrt{\sigma_{ii}}\right)^{-1}\boldsymbol{\Sigma}\left(\oplus_{i=1}^{n}\sqrt{\sigma_{ii}}\right)^{-1}.$$

Variance–covariance, precision, and correlation matrices that have special patterned structures are useful in several areas of statistics, including linear models. For example, a scalar multiple of $\mathbf{I}_n$ or some other diagonal matrix is the variance–covariance matrix of $\mathbf{e}$ in some linear models. Some other patterned variance–covariance matrices will be introduced in examples and exercises throughout the rest of the book. We give three such matrices now.

Example 4.1-1. Some Patterned Variance–Covariance Matrices

Consider an $n \times n$ matrix of the form $a\mathbf{I}_n + b\mathbf{J}_n$. In Example 2.15-1 we found conditions under which such a matrix is nonnegative definite, so by Theorem 4.1.1 it can be a variance–covariance matrix if (and only if) it satisfies those conditions. Moreover, its diagonal elements are all equal to $a + b$, and its off-diagonal elements are all equal to b, so if it is nonnegative definite, then it is a variance–covariance matrix with equal variances $a + b$ (*homoscedasticity*) and equal correlations $b/(a+b)$ (*equicorrelation*). It is customary to parameterize this variance–covariance matrix in such a way that the variances are represented by a parameter σ^2 and the correlations are represented by a parameter ρ, i.e., as $\sigma^2[(1-\rho)\mathbf{I}_n + \rho\mathbf{J}_n]$. A variance–covariance matrix of this form is called a *compound symmetric* variance covariance matrix. It is easily verified that the conditions for nonnegative definiteness derived in Example 2.15-1 for $a\mathbf{I}_n + b\mathbf{J}_n$ translate to $\sigma^2 \geq 0$ and $-\frac{1}{n-1} \leq \rho \leq 1$ for this parameterization, and the conditions for positive definiteness translate to $\sigma^2 > 0$ and $-\frac{1}{n-1} < \rho < 1$.

Next consider an $n \times n$ symmetric matrix of the form

$$\boldsymbol{\Sigma} = \sigma^2 \begin{pmatrix} 1 & 1 & \cdots & 1 & 1 \\ 1 & 2 & \cdots & 2 & 2 \\ \vdots & \vdots & \ddots & \vdots & \vdots \\ 1 & 2 & \cdots & n-1 & n-1 \\ 1 & 2 & \cdots & n-1 & n \end{pmatrix},$$

where $\sigma^2 > 0$. Observe that $(1/\sigma^2)\boldsymbol{\Sigma}$ may be written as

$$\mathbf{1}_n\mathbf{1}_n^T + \begin{pmatrix} 0 \\ \mathbf{1}_{n-1} \end{pmatrix}\begin{pmatrix} 0 \\ \mathbf{1}_{n-1} \end{pmatrix}^T + \begin{pmatrix} \mathbf{0}_2 \\ \mathbf{1}_{n-2} \end{pmatrix}\begin{pmatrix} \mathbf{0}_2 \\ \mathbf{1}_{n-2} \end{pmatrix}^T + \cdots + \begin{pmatrix} \mathbf{0}_{n-1} \\ 1 \end{pmatrix}\begin{pmatrix} \mathbf{0}_{n-1} \\ 1 \end{pmatrix}^T,$$

and that each summand is a nonnegative definite matrix by Theorem 2.15.9. Thus by Theorem 2.15.3, $\mathbf{\Sigma}$ is nonnegative definite and is therefore a variance–covariance matrix. In fact, it is not difficult to show that $\mathbf{\Sigma}$ is positive definite; see the exercises. Clearly, $\text{var}(x_i) = i\sigma^2$; it can also be shown that $\text{corr}(x_i, x_j) = \sqrt{i/j}$ for $i \leq j$.

Finally, consider an $n \times n$ symmetric matrix of the form

$$\mathbf{\Sigma} = \left(\oplus_{i=1}^n \sqrt{\sigma_{ii}}\right) \begin{pmatrix} 1 & \rho_1 & \rho_1\rho_2 & \rho_1\rho_2\rho_3 & \cdots & \prod_{i=1}^{n-2}\rho_i & \prod_{i=1}^{n-1}\rho_i \\ & 1 & \rho_2 & \rho_2\rho_3 & \cdots & \prod_{i=2}^{n-2}\rho_i & \prod_{i=2}^{n-1}\rho_i \\ & & 1 & \rho_3 & \cdots & \prod_{i=3}^{n-2}\rho_i & \prod_{i=3}^{n-1}\rho_i \\ & & & 1 & \cdots & \prod_{i=4}^{n-2}\rho_i & \prod_{i=4}^{n-1}\rho_i \\ & & & & \ddots & \vdots & \vdots \\ & & & & & 1 & \rho_{n-1} \\ & & & & & & 1 \end{pmatrix} \left(\oplus_{i=1}^n \sqrt{\sigma_{ii}}\right)$$

where $\sigma_{ii} > 0$ and $-1 < \rho_i < 1$ for all i. Using Theorem 2.15.11 and an induction argument, it can be shown that $\mathbf{\Sigma}$ is positive definite; see the exercises. Observe that $\text{var}(x_i) = \sigma_{ii}$ and $\text{corr}(x_i, x_j) = \prod_{k=i}^{j-1} \rho_k$ for $i < j$. This patterned form is called a *first-order antedependent variance–covariance matrix*. ■

Next, we consider matrices whose elements are higher-order moments of a random vector.

Definition 4.1.5 The *skewness matrix* of a random n-vector $\mathbf{x}$ is the $n \times n^2$ matrix $\mathbf{\Lambda}$ whose entry in the ith row and (j, k)th column is the third central moment $\lambda_{ijk} \equiv \text{E}\{[x_i - \text{E}(x_i)][x_j - \text{E}(x_j)][x_k - \text{E}(x_k)]\}$, for $i, j, k = 1, \ldots, n$, provided that all these expectations exist.

Two subscripts do not suffice for third-order central moments, so in order to lay out such moments in a two-dimensional array we have introduced a two-index scheme for labeling matrix columns in this definition. This scheme labels columns as (1,1), (1,2), ..., (1,n), (2,1), (2,2),..., (n, n). We will use a similar labeling scheme for both rows and columns for matrices holding the fourth-order central moments shortly. But first we state a theorem giving some special cases of skewness matrices.

Theorem 4.1.3 *Suppose that the skewness matrix* $\mathbf{\Lambda} = (\lambda_{ijk})$ *of a random* n*-vector* $\mathbf{x}$ *exists.*

(a) If the elements of $\mathbf{x}$ *are triple-wise independent (meaning that each trivariate marginal cdf of* $\mathbf{x}$ *factors into the product of its univariate marginal cdfs), then, for some constants* $\lambda_1, \ldots, \lambda_n$,

$$\lambda_{ijk} = \begin{cases} \lambda_i & \text{if } i = j = k, \\ 0 & \text{otherwise,} \end{cases}$$

or equivalently,

$$\boldsymbol{\Lambda} = (\lambda_1 \mathbf{u}_1 \mathbf{u}_1^T, \lambda_2 \mathbf{u}_2 \mathbf{u}_2^T, \ldots, \lambda_n \mathbf{u}_n \mathbf{u}_n^T)$$

where $\mathbf{u}_i$ *is the* i*th unit* n*-vector;*

(b) *If the distribution of* $\mathbf{x}$ *is symmetric about* $\mathbf{0}$ *(in the sense that the distributions of* $[\mathbf{x} - E(\mathbf{x})]$ *and* $-[\mathbf{x} - E(\mathbf{x})]$ *are identical), then* $\boldsymbol{\Lambda} = \mathbf{0}$.

Proof Left as an exercise. □

Definition 4.1.6 The *kurtosis matrix* of a random n-vector $\mathbf{x}$ is the $n^2 \times n^2$ matrix $\boldsymbol{\Gamma}$ whose entry in the (i, j)th row and (k, l)th column is the fourth central moment $\gamma_{ijkl} \equiv \mathrm{E}\{[x_i - \mathrm{E}(x_i)][x_j - \mathrm{E}(x_j)][x_k - \mathrm{E}(x_k)][x_l - \mathrm{E}(x_l)]\}$, for $i, j, k, l = 1, \ldots, n$, provided that all these expectations exist.

Theorem 4.1.4 *If the kurtosis matrix* $\boldsymbol{\Gamma} = (\gamma_{ijkl})$ *of a random* n*-vector* $\mathbf{x}$ *exists and the elements of* $\mathbf{x}$ *are quadruple-wise independent (meaning that each quadrivariate marginal cdf of* $\mathbf{x}$ *factors into the product of its univariate marginal cdfs) then for some constants* $\gamma_1, \ldots, \gamma_n$,

$$\gamma_{ijkl} = \begin{cases} \gamma_i & \text{if } i = j = k = l, \\ \sigma_{ii}\sigma_{kk} & \text{if } i = j \neq k = l, \\ \sigma_{ii}\sigma_{jj} & \text{if } i = k \neq j = l \text{ or } i = l \neq j = k, \\ 0 & \text{otherwise,} \end{cases}$$

where $var(\mathbf{x}) = (\sigma_{ij})$.

Proof Left as an exercise. □

Definition 4.1.7 The *excess kurtosis matrix* of a random n-vector $\mathbf{x}$ is the $n^2 \times n^2$ matrix $\boldsymbol{\Omega}$ whose entry in the (i, j)th row and (k, l)th column is $\gamma_{ijkl} - \sigma_{ij}\sigma_{kl} - \sigma_{ik}\sigma_{jl} - \sigma_{il}\sigma_{jk}$.

The adjective *excess* in Definition 4.1.7 refers to the additional kurtosis (element-wise) above and beyond that for a multivariate normal random vector $\mathbf{x}$. The kurtosis matrix for a multivariate normal vector will be derived in Chap. 14.

4.2 Means, Variances, and Covariances of Linear and Quadratic Forms in a Random Vector

Theorem 4.2.1 *Let* $\mathbf{x}$ *represent a random n-vector with mean* $\boldsymbol{\mu}$*, let* $\mathbf{A}$ *represent a* $t \times n$ *matrix of constants, and let* $\mathbf{a}$ *represent a t-vector of constants. Then*

$$E(\mathbf{Ax} + \mathbf{a}) = \mathbf{A}\boldsymbol{\mu} + \mathbf{a}.$$

Proof Left as an exercise. □

Theorem 4.2.2 *Let* $\mathbf{x}$ *and* $\mathbf{y}$ *represent random n-vectors, and let* $\mathbf{z}$ *and* $\mathbf{w}$ *represent random m-vectors. If* $var(\mathbf{x})$, $var(\mathbf{y})$, $var(\mathbf{z})$, *and* $var(\mathbf{w})$ *exist, then:*

(a) $cov(\mathbf{x} + \mathbf{y}, \mathbf{z} + \mathbf{w}) = cov(\mathbf{x}, \mathbf{z}) + cov(\mathbf{x}, \mathbf{w}) + cov(\mathbf{y}, \mathbf{z}) + cov(\mathbf{y}, \mathbf{w})$;
(b) $var(\mathbf{x} + \mathbf{y}) = var(\mathbf{x}) + var(\mathbf{y}) + cov(\mathbf{x}, \mathbf{y}) + [cov(\mathbf{x}, \mathbf{y})]^T$.

Proof Left as an exercise. □

Theorem 4.2.3 *Let* $\mathbf{x}$ *and* $\mathbf{z}$ *represent a random n-vector and a random m-vector, respectively, with matrix of covariances* $\boldsymbol{\Phi}$*; let* $\mathbf{A}$ *and* $\mathbf{B}$ *represent* $t \times n$ *and* $u \times m$ *matrices of constants, respectively; and let* $\mathbf{a}$ *and* $\mathbf{b}$ *represent a t-vector and u-vector of constants, respectively. Then*

$$cov(\mathbf{Ax} + \mathbf{a}, \mathbf{Bz} + \mathbf{b}) = \mathbf{A}\boldsymbol{\Phi}\mathbf{B}^T.$$

Proof Left as an exercise. □

Corollary 4.2.3.1 *Let* $\mathbf{x}$ *represent a random n-vector with variance–covariance matrix* $\boldsymbol{\Sigma}$*, and let* $\mathbf{A}$, $\mathbf{B}$, $\mathbf{a}$, *and* $\mathbf{b}$ *be defined as in Theorem 4.2.3 except that we take* $m = n$*. Then:*

(a) $cov(\mathbf{Ax} + \mathbf{a}, \mathbf{Bx} + \mathbf{b}) = \mathbf{A}\boldsymbol{\Sigma}\mathbf{B}^T$;
(b) $var(\mathbf{Ax} + \mathbf{a}) = \mathbf{A}\boldsymbol{\Sigma}\mathbf{A}^T$.

Theorem 4.2.4 *Let* $\mathbf{x}$ *represent a random n-vector with mean* $\boldsymbol{\mu}$ *and variance–covariance matrix* $\boldsymbol{\Sigma}$*, and let* $\mathbf{A}$ *represent an* $n \times n$ *matrix of constants. Then*

$$E(\mathbf{x}^T\mathbf{Ax}) = \boldsymbol{\mu}^T\mathbf{A}\boldsymbol{\mu} + tr(\mathbf{A}\boldsymbol{\Sigma}).$$

Proof Using the fact that $\mathbf{x}^T\mathbf{A}\mathbf{x}$ is a scalar (and is therefore equal to its trace), and using the cyclic and additive properties of the trace (Theorems 2.10.3 and 2.10.1b, respectively), we obtain

$$
\begin{aligned}
\mathrm{E}(\mathbf{x}^T\mathbf{A}\mathbf{x}) &= \mathrm{E}[\mathrm{tr}(\mathbf{x}^T\mathbf{A}\mathbf{x})] \\
&= \mathrm{E}[\mathrm{tr}(\mathbf{A}\mathbf{x}\mathbf{x}^T)] \\
&= \mathrm{tr}[\mathbf{A}\mathrm{E}(\mathbf{x}\mathbf{x}^T)] \\
&= \mathrm{tr}[\mathbf{A}(\boldsymbol{\mu}\boldsymbol{\mu}^T + \boldsymbol{\Sigma})] \\
&= \mathrm{tr}(\mathbf{A}\boldsymbol{\mu}\boldsymbol{\mu}^T) + \mathrm{tr}(\mathbf{A}\boldsymbol{\Sigma}) \\
&= \boldsymbol{\mu}^T\mathbf{A}\boldsymbol{\mu} + \mathrm{tr}(\mathbf{A}\boldsymbol{\Sigma}).
\end{aligned}
$$

□

Theorem 4.2.5 *Let* $\mathbf{x}$ *represent a random n-vector with mean vector* $\boldsymbol{\mu}$, *variance–covariance matrix* $\boldsymbol{\Sigma}$, *and skewness matrix* $\boldsymbol{\Lambda} = (\lambda_{ijk})$. *Let* $\mathbf{B} = (b_{ij})$ *and* $\mathbf{A} = (a_{ij})$ *represent a* $t \times n$ *matrix of constants and an* $n \times n$ *symmetric matrix of constants, respectively. Then,*

$$
cov(\mathbf{B}\mathbf{x}, \mathbf{x}^T\mathbf{A}\mathbf{x}) = \mathbf{B}\boldsymbol{\Lambda}vec(\mathbf{A}) + 2\mathbf{B}\boldsymbol{\Sigma}\mathbf{A}\boldsymbol{\mu}.
$$

Proof Let $\mathbf{z} = (z_i) = \mathbf{x} - \boldsymbol{\mu}$, so that $\mathbf{x} = \mathbf{z} + \boldsymbol{\mu}$. Also, let $\mathbf{b}_i^T$ represent the ith row of $\mathbf{B}$. Then

$$
\begin{aligned}
\mathrm{cov}(\mathbf{b}_i^T\mathbf{x}, \mathbf{x}^T\mathbf{A}\mathbf{x}) &= \mathrm{cov}[\mathbf{b}_i^T(\mathbf{z}+\boldsymbol{\mu}), (\mathbf{z}+\boldsymbol{\mu})^T\mathbf{A}(\mathbf{z}+\boldsymbol{\mu})] \\
&= \mathrm{cov}(\mathbf{b}_i^T\mathbf{z} + \mathbf{b}_i^T\boldsymbol{\mu}, \mathbf{z}^T\mathbf{A}\mathbf{z} + 2\boldsymbol{\mu}^T\mathbf{A}\mathbf{z} + \boldsymbol{\mu}^T\mathbf{A}\boldsymbol{\mu}) \\
&= \mathrm{cov}(\mathbf{b}_i^T\mathbf{z}, \mathbf{z}^T\mathbf{A}\mathbf{z}) + 2\mathrm{cov}(\mathbf{b}_i^T\mathbf{z}, \boldsymbol{\mu}^T\mathbf{A}\mathbf{z}) \\
&= \mathrm{E}(\mathbf{b}_i^T\mathbf{z}\mathbf{z}^T\mathbf{A}\mathbf{z}) + 2\mathbf{b}_i^T\boldsymbol{\Sigma}\mathbf{A}\boldsymbol{\mu}
\end{aligned}
$$

where we have used Theorem 4.2.2, Corollary 4.2.3.1, and the fact that $\mathrm{E}(\mathbf{b}_i^T\mathbf{z}) = \mathbf{0}$. Now

$$
\begin{aligned}
\mathrm{E}(\mathbf{b}_i^T\mathbf{z}\mathbf{z}^T\mathbf{A}\mathbf{z}) &= \mathrm{E}\left[\left(\sum_{j=1}^{n} b_{ij}z_j\right)\left(\sum_{k=1}^{n}\sum_{l=1}^{n} a_{kl}z_k z_l\right)\right] \\
&= \mathrm{E}\left(\sum_{j=1}^{n}\sum_{k=1}^{n}\sum_{l=1}^{n} b_{ij}a_{kl}z_j z_k z_l\right) \\
&= \sum_{j=1}^{n}\sum_{k=1}^{n}\sum_{l=1}^{n} b_{ij}a_{kl}\lambda_{jkl} \\
&= \mathbf{b}_i^T\boldsymbol{\Lambda}\mathrm{vec}(\mathbf{A}).
\end{aligned}
$$

Stacking these results for $i = 1, \ldots, n$ atop one another yields the desired result. □

Corollary 4.2.5.1 *If* $\boldsymbol{\Lambda} = \mathbf{0}$ *(as is the case, by Theorem 4.1.3b, when the distribution of* $\mathbf{x}$ *is symmetric in the sense described there), then*

$$cov(\mathbf{Bx}, \mathbf{x}^T\mathbf{Ax}) = 2\mathbf{B}\boldsymbol{\Sigma}\mathbf{A}\boldsymbol{\mu}.$$

Theorem 4.2.6 *Let* $\mathbf{x}$, $\boldsymbol{\mu}$, $\boldsymbol{\Sigma} = (\sigma_{ij})$, *and* $\boldsymbol{\Lambda}$ *be defined as in Theorem 4.2.5, and let* $\boldsymbol{\Gamma} = (\gamma_{ijkl})$ *and* $\boldsymbol{\Omega}$ *represent the kurtosis matrix and excess kurtosis matrix, respectively, of* $\mathbf{x}$. *Finally, let* $\mathbf{A}$ *and* $\mathbf{B}$ *represent symmetric* $n \times n$ *matrices of constants. Then,*

$$\begin{aligned} cov(\mathbf{x}^T\mathbf{Ax}, \mathbf{x}^T\mathbf{Bx}) &= [vec(\mathbf{A})]^T\boldsymbol{\Omega}vec(\mathbf{B}) + 2\boldsymbol{\mu}^T\mathbf{B}\boldsymbol{\Lambda}vec(\mathbf{A}) + 2\boldsymbol{\mu}^T\mathbf{A}\boldsymbol{\Lambda}vec(\mathbf{B}) \\ &\quad + 2tr(\mathbf{A}\boldsymbol{\Sigma}\mathbf{B}\boldsymbol{\Sigma}) + 4\boldsymbol{\mu}^T\mathbf{A}\boldsymbol{\Sigma}\mathbf{B}\boldsymbol{\mu}. \end{aligned}$$

Proof Let $\mathbf{z} = (z_i) = \mathbf{x} - \boldsymbol{\mu}$, in which case $\mathrm{E}(\mathbf{z}) = \mathbf{0}$ and $\mathbf{x} = \mathbf{z} + \boldsymbol{\mu}$. Then

$$\begin{aligned} \mathrm{cov}(\mathbf{x}^T\mathbf{Ax}, \mathbf{x}^T\mathbf{Bx}) &= \mathrm{cov}(\mathbf{z}^T\mathbf{Az} + 2\boldsymbol{\mu}^T\mathbf{Az} + \boldsymbol{\mu}^T\mathbf{A}\boldsymbol{\mu}, \mathbf{z}^T\mathbf{Bz} + 2\boldsymbol{\mu}^T\mathbf{Bz} + \boldsymbol{\mu}^T\mathbf{B}\boldsymbol{\mu}) \\ &= \mathrm{cov}(\mathbf{z}^T\mathbf{Az}, \mathbf{z}^T\mathbf{Bz}) + \mathrm{cov}(\mathbf{z}^T\mathbf{Az}, 2\boldsymbol{\mu}^T\mathbf{Bz}) \\ &\quad + \mathrm{cov}(2\boldsymbol{\mu}^T\mathbf{Az}, \mathbf{z}^T\mathbf{Bz}) + \mathrm{cov}(2\boldsymbol{\mu}^T\mathbf{Az}, 2\boldsymbol{\mu}^T\mathbf{Bz}) \\ &= \mathrm{cov}(\mathbf{z}^T\mathbf{Az}, \mathbf{z}^T\mathbf{Bz}) + 2\boldsymbol{\mu}^T\mathbf{A}\boldsymbol{\Lambda}\mathrm{vec}(\mathbf{B}) + 2\boldsymbol{\mu}^T\mathbf{B}\boldsymbol{\Lambda}\mathrm{vec}(\mathbf{A}) \\ &\quad + 4\boldsymbol{\mu}^T\mathbf{A}\boldsymbol{\Sigma}\mathbf{B}\boldsymbol{\mu} \end{aligned}$$

where we have used Theorems 4.2.2 and 4.2.5 (twice), and Corollary 4.2.3.1. Now,

$$\begin{aligned} \mathrm{cov}(\mathbf{z}^T\mathbf{Az}, \mathbf{z}^T\mathbf{Bz}) &= \mathrm{cov}\left(\sum_{i=1}^n\sum_{j=1}^n a_{ij}z_iz_j, \sum_{k=1}^n\sum_{l=1}^n b_{kl}z_kz_l\right) \\ &= \sum_{i=1}^n\sum_{j=1}^n\sum_{k=1}^n\sum_{l=1}^n a_{ij}b_{kl}\mathrm{cov}(z_iz_j, z_kz_l) \\ &= \sum_{i=1}^n\sum_{j=1}^n\sum_{k=1}^n\sum_{l=1}^n a_{ij}b_{kl}[\mathrm{E}(z_iz_jz_kz_l) - \mathrm{E}(z_iz_j)\mathrm{E}(z_kz_l)] \\ &= \sum_{i=1}^n\sum_{j=1}^n\sum_{k=1}^n\sum_{l=1}^n a_{ij}b_{kl}(\gamma_{ijkl} - \sigma_{ij}\sigma_{kl} - \sigma_{ik}\sigma_{jl} - \sigma_{il}\sigma_{jk}) \\ &\quad + \sum_{i=1}^n\sum_{j=1}^n\sum_{k=1}^n\sum_{l=1}^n a_{ij}(b_{lk}\sigma_{ik}\sigma_{jl} + b_{kl}\sigma_{il}\sigma_{jk}) \end{aligned}$$

$$
\begin{aligned}
&= [\mathrm{vec}(\mathbf{A})]^T \boldsymbol{\Omega} \mathrm{vec}(\mathbf{B}) + \sum_{i=1}^{n} \sum_{l=1}^{n} \left(\sum_{j=1}^{n} a_{ij}\sigma_{jl} \right) \left(\sum_{k=1}^{n} b_{lk}\sigma_{ki} \right) \\
&+ \sum_{i=1}^{n} \sum_{k=1}^{n} \left(\sum_{j=1}^{n} a_{ij}\sigma_{jk} \right) \left(\sum_{l=1}^{n} b_{kl}\sigma_{li} \right) \\
&= [\mathrm{vec}(\mathbf{A})]^T \boldsymbol{\Omega} \mathrm{vec}(\mathbf{B}) + 2\mathrm{tr}(\mathbf{A}\boldsymbol{\Sigma}\mathbf{B}\boldsymbol{\Sigma}),
\end{aligned}
$$

yielding the desired result. □

In many linear models, the distribution of the response vector is assumed to be such that either its skewness matrix or its excess kurtosis matrix (or both) is null; e.g., this is so if $\mathbf{y}$ is multivariate normal, as we show in Chap. 14. Consequently, we give three useful corollaries to Theorem 4.2.6.

Corollary 4.2.6.1 *If* $\boldsymbol{\Lambda} = \mathbf{0}$ *and* $\mathbf{A}$ *and* $\mathbf{B}$ *are symmetric* $n \times n$ *matrices of constants, then*

$$
cov(\mathbf{x}^T\mathbf{A}\mathbf{x}, \mathbf{x}^T\mathbf{B}\mathbf{x}) = [vec(\mathbf{A})]^T \boldsymbol{\Omega} vec(\mathbf{B}) + 2tr(\mathbf{A}\boldsymbol{\Sigma}\mathbf{B}\boldsymbol{\Sigma}) + 4\boldsymbol{\mu}^T\mathbf{A}\boldsymbol{\Sigma}\mathbf{B}\boldsymbol{\mu}.
$$

Corollary 4.2.6.2 *If* $\boldsymbol{\Omega} = \mathbf{0}$ *and* $\mathbf{A}$ *and* $\mathbf{B}$ *are symmetric* $n \times n$ *matrices of constants, then*

$$
cov(\mathbf{x}^T\mathbf{A}\mathbf{x}, \mathbf{x}^T\mathbf{B}\mathbf{x}) = 2\boldsymbol{\mu}^T\mathbf{B}\boldsymbol{\Lambda} vec(\mathbf{A}) + 2\boldsymbol{\mu}^T\mathbf{A}\boldsymbol{\Lambda} vec(\mathbf{B}) + 2tr(\mathbf{A}\boldsymbol{\Sigma}\mathbf{B}\boldsymbol{\Sigma}) + 4\boldsymbol{\mu}^T\mathbf{A}\boldsymbol{\Sigma}\mathbf{B}\boldsymbol{\mu}.
$$

Corollary 4.2.6.3 *If* $\boldsymbol{\Lambda} = \mathbf{0}$ *and* $\boldsymbol{\Omega} = \mathbf{0}$*, and* $\mathbf{A}$ *and* $\mathbf{B}$ *are symmetric* $n \times n$ *matrices of constants, then*

$$
cov(\mathbf{x}^T\mathbf{A}\mathbf{x}, \mathbf{x}^T\mathbf{B}\mathbf{x}) = 2tr(\mathbf{A}\boldsymbol{\Sigma}\mathbf{B}\boldsymbol{\Sigma}) + 4\boldsymbol{\mu}^T\mathbf{A}\boldsymbol{\Sigma}\mathbf{B}\boldsymbol{\mu}.
$$

Example 4.2-1. Expectations, Variances, and Covariance of the Mean and Variance of a Random Sample from a Population

Suppose that a random sample $x_1, x_2, \ldots, x_n$ is taken from a population that has mean μ, variance σ^2, skewness λ, and kurtosis γ. Put $\mathbf{x} = (x_1, \ldots, x_n)^T$. By the independence and identical distributions of the x_i's, and Theorems 4.1.3a and 4.1.4, the mean vector $\boldsymbol{\mu}$, variance–covariance matrix $\boldsymbol{\Sigma}$, skewness matrix $\boldsymbol{\Lambda}$, and kurtosis matrix $\boldsymbol{\Gamma}$ are as follows:

$$
\boldsymbol{\mu} = \mu\mathbf{1}_n, \quad \boldsymbol{\Sigma} = \sigma^2\mathbf{I}_n, \quad \boldsymbol{\Lambda} = \lambda(\mathbf{u}_1\mathbf{u}_1^T, \mathbf{u}_2\mathbf{u}_2^T, \ldots, \mathbf{u}_n\mathbf{u}_n^T), \quad \boldsymbol{\Gamma} = (\gamma_{ijkl}),
$$

where $\mathbf{u}_i$ is the ith unit n-vector and

$$\gamma_{ijkl} = \begin{cases} \gamma & \text{if } i = j = k = l \\ \sigma^4 & \text{if } i = j \neq k = l, i = k \neq j = l, \text{ or } i = l \neq j = k \\ 0, & \text{otherwise.} \end{cases}$$

The sample mean may be expressed as $\bar{x} = (1/n)\mathbf{1}^T\mathbf{x}$, and the sample variance may be expressed as $s^2 = \mathbf{x}^T[\mathbf{I} - (1/n)\mathbf{J}]\mathbf{x}/(n-1)$. (Here and throughout, a symbol with a bar placed over it, such as $\bar{x}$, will represent the arithmetic mean of a set of quantities which, prior to averaging, were represented by the same symbol but were indexed by one or more subscripts.) Using Theorems 4.2.1 and 4.2.4, we find that the expectations of $\bar{x}$ and s^2 are

$$\mathrm{E}(\bar{x}) = (1/n)\mathbf{1}^T(\mu\mathbf{1}) = \mu$$

and

$$\begin{aligned} \mathrm{E}(s^2) &= \mu\mathbf{1}^T\{[\mathbf{I} - (1/n)\mathbf{J}]/(n-1)\}\mu\mathbf{1} + \mathrm{tr}(\{[\mathbf{I} - (1/n)\mathbf{J}]/(n-1)\}\sigma^2\mathbf{I}) \\ &= 0 + \sigma^2\mathrm{tr}[\mathbf{I} - (1/n)\mathbf{J}]/(n-1) \\ &= \sigma^2, \end{aligned}$$

as is well known. Moreover, using Corollary 4.2.3.1 and Theorem 4.2.6, and noting for the latter that $\boldsymbol{\Omega} = (\omega_{ijkl})$ where

$$\omega_{ijkl} = \begin{cases} \gamma - 3\sigma^4 & \text{if } i = j = k = l \\ 0 & \text{otherwise,} \end{cases}$$

we obtain

$$\mathrm{var}(\bar{x}) = (1/n)\mathbf{1}^T(\sigma^2\mathbf{I})(1/n)\mathbf{1} = \sigma^2/n$$

and

$$\begin{aligned} \mathrm{var}(s^2) &= (\{\mathrm{vec}[\mathbf{I} - (1/n)\mathbf{J}]\}^T\boldsymbol{\Omega}\mathrm{vec}[\mathbf{I} - (1/n)\mathbf{J}] \\ &\quad + 2\mathrm{tr}\{[\mathbf{I} - (1/n)\mathbf{J}](\sigma^2\mathbf{I})[\mathbf{I} - (1/n)\mathbf{J}](\sigma^2\mathbf{I})\})/(n-1)^2 \\ &= [(\gamma - 3\sigma^4)/n] + 2\sigma^4/(n-1). \end{aligned}$$

Finally, using Theorem 4.2.5, we obtain

$$\begin{aligned}\text{cov}(\bar{x}, s^2) &= (1/n)\mathbf{1}^T[\lambda(\mathbf{u}_1\mathbf{u}_1^T, \mathbf{u}_2\mathbf{u}_2^T, \ldots, \mathbf{u}_n\mathbf{u}_n^T)]\text{vec}\{[\mathbf{I} - (1/n)\mathbf{J}]/(n-1)\} \\ &\quad +2(1/n)\mathbf{1}^T(\sigma^2\mathbf{I})\{[\mathbf{I} - (1/n)\mathbf{J}]/(n-1)\}\mu\mathbf{1} \\ &= \{\lambda/[n(n-1)]\}(\mathbf{u}_1^T, \mathbf{u}_2^T, \ldots, \mathbf{u}_n^T)\left[\begin{pmatrix}\mathbf{u}_1 \\ \mathbf{u}_2 \\ \vdots \\ \mathbf{u}_n\end{pmatrix} - (1/n)\mathbf{1}_{n^2}\right] + 0 \\ &= \lambda/n.\end{aligned}$$

■

4.3 Exercises

1. Prove Theorem 4.1.1.
2. Prove Theorem 4.1.2.
3. Prove that the second patterned variance–covariance matrix described in Example 4.1-1 is positive definite.
4. Prove that the third patterned variance–covariance matrix described in Example 4.1-1 is positive definite.
5. Prove Theorem 4.1.3.
6. Prove Theorem 4.1.4.
7. Prove Theorem 4.2.1.
8. Prove Theorem 4.2.2.
9. Prove Theorem 4.2.3.
10. Let $x_1, \ldots, x_n$ be uncorrelated random variables with common variance σ^2. Determine $\text{cov}(x_i - \bar{x}, x_j - \bar{x})$ for $i \leq j = 1, \ldots, n$.
11. Let $\mathbf{x}$ be a random n-vector with mean $\boldsymbol{\mu}$ and positive definite variance–covariance matrix $\boldsymbol{\Sigma}$. Define $\boldsymbol{\Sigma}^{-\frac{1}{2}} = (\boldsymbol{\Sigma}^{\frac{1}{2}})^{-1}$.
    (a) Show that $\boldsymbol{\Sigma}^{-\frac{1}{2}}(\mathbf{x} - \boldsymbol{\mu})$ has mean $\mathbf{0}$ and variance–covariance matrix $\mathbf{I}$.
    (b) Determine $\text{E}[(\mathbf{x} - \boldsymbol{\mu})^T\boldsymbol{\Sigma}^{-1}(\mathbf{x} - \boldsymbol{\mu})]$ and $\text{var}[(\mathbf{x} - \boldsymbol{\mu})^T\boldsymbol{\Sigma}^{-1}(\mathbf{x} - \boldsymbol{\mu})]$, assuming for the latter that the skewness matrix $\boldsymbol{\Lambda}$ and excess kurtosis matrix $\boldsymbol{\Omega}$ of $\mathbf{x}$ exist.
12. Let $\mathbf{x}$ be a random n-vector with variance–covariance matrix $\sigma^2\mathbf{I}$, where $\sigma^2 > 0$, and let $\mathbf{Q}$ represent an $n \times n$ orthogonal matrix. Determine the variance–covariance matrix of $\mathbf{Qx}$.
13. Let $\mathbf{x} = \begin{pmatrix}\mathbf{x}_1 \\ \mathbf{x}_2\end{pmatrix}$ have mean vector $\boldsymbol{\mu} = \begin{pmatrix}\boldsymbol{\mu}_1 \\ \boldsymbol{\mu}_2\end{pmatrix}$ and variance–covariance matrix $\boldsymbol{\Sigma} = \begin{pmatrix}\boldsymbol{\Sigma}_{11} & \boldsymbol{\Sigma}_{12} \\ \boldsymbol{\Sigma}_{21} & \boldsymbol{\Sigma}_{22}\end{pmatrix}$, where $\mathbf{x}_1$ and $\boldsymbol{\mu}_1$ are m-vectors and $\boldsymbol{\Sigma}_{11}$ is $m \times m$. Suppose

that $\boldsymbol{\Sigma}$ is positive definite, in which case both $\boldsymbol{\Sigma}_{22}$ and $\boldsymbol{\Sigma}_{11} - \boldsymbol{\Sigma}_{12}\boldsymbol{\Sigma}_{22}^{-1}\boldsymbol{\Sigma}_{21}$ are positive definite by Theorem 2.15.7a.

(a) Determine the mean vector and variance–covariance matrix of $\mathbf{x}_{1.2} = \mathbf{x}_1 - \boldsymbol{\mu}_1 - \boldsymbol{\Sigma}_{12}\boldsymbol{\Sigma}_{22}^{-1}(\mathbf{x}_2 - \boldsymbol{\mu}_2)$.

(b) Show that $\text{cov}(\mathbf{x}_1 - \boldsymbol{\Sigma}_{12}\boldsymbol{\Sigma}_{22}^{-1}\mathbf{x}_2, \mathbf{x}_2) = \mathbf{0}$.

(c) Determine $\text{E}\{\mathbf{x}_{1.2}^T [\text{var}(\mathbf{x}_{1.2})]^{-1} \mathbf{x}_{1.2}\}$.

14. Suppose that observations $x_1, \ldots, x_n$ have common mean μ, common variance σ^2, and common correlation ρ among pairs, so that $\boldsymbol{\mu} = \mu\mathbf{1}_n$ and $\boldsymbol{\Sigma} = \sigma^2[(1 - \rho)\mathbf{I}_n + \rho\mathbf{J}_n]$ for $\rho \in [-1/(n-1), 1]$. Determine $\text{E}(\bar{x})$, $\text{var}(\bar{x})$, and $\text{E}(s^2)$.
15. Prove that if $\mathbf{x}$ is a random vector for which $\boldsymbol{\mu} = \mathbf{0}$ and $\boldsymbol{\Lambda} = \mathbf{0}$, then any linear form $\mathbf{b}^T\mathbf{x}$ is uncorrelated with any quadratic form $\mathbf{x}^T\mathbf{A}\mathbf{x}$.
16. Determine how Theorems 4.2.1 and 4.2.4–4.2.6 specialize when $\boldsymbol{\mu}$ is orthogonal to every row of $\mathbf{A}$.
17. Let $\mathbf{x}$ be a random n-vector with mean $\boldsymbol{\mu}$ and variance–covariance matrix $\boldsymbol{\Sigma}$ of rank r. Find $\text{E}(\mathbf{x}^T\boldsymbol{\Sigma}^-\mathbf{x})$ and simplify it as much as possible.
18. Determine the covariance between the sample mean and sample variance of observations whose joint distribution is symmetric with common mean μ, variance–covariance matrix $\boldsymbol{\Sigma}$, and finite skewness matrix $\boldsymbol{\Lambda}$.

Types of Linear Models

5

The model equation for a linear model, i.e., $\mathbf{y} = \mathbf{X}\boldsymbol{\beta} + \mathbf{e}$, has two distinct terms on the right-hand side: $\mathbf{X}\boldsymbol{\beta}$ and $\mathbf{e}$. Because $\mathbf{X}\boldsymbol{\beta} = \mathrm{E}(\mathbf{y})$, we refer to any additional assumptions made about $\mathbf{X}\boldsymbol{\beta}$ (i.e., additional to $\mathbf{X}$ being nonrandom and known and $\boldsymbol{\beta}$ being nonrandom and unknown) as the model's mean structure, and to any additional assumptions made about $\mathbf{e}$ [i.e., additional to $\mathrm{E}(\mathbf{e}) = \mathbf{0}$] as the model's error structure. In this chapter, we give a brief overview of various linear models that are distinguishable from one another on the basis of these two types of structure, and we describe some terminology associated with each type. We also describe prediction-extended versions of these models, i.e., models for the mean structure and error structure of the joint distribution of $\mathbf{y}$ and a related, though unobservable, random vector $\mathbf{u}$. Throughout the remainder of the book, many aspects of linear model theory will be exemplified by the models introduced here.

5.1 Mean Structures

In this section, we identify three broad classes of mean structures for linear models: regression models, classificatory models (also known as ANOVA models), and analysis-of-covariance models. But first, we briefly distinguish mean structures on the basis of two other attributes: the rank of $\mathbf{X}$ and the presence or absence of constraints on $\boldsymbol{\beta}$.

5.1.1 Full-Rank Versus Less-Than-Full-Rank Models

Recall that the model matrix $\mathbf{X}$ has dimensions $n \times p$. Let p^* denote the rank of $\mathbf{X}$. Clearly, $p^* \leq p$. Throughout, we allow for the possibility that p^* is less than p. When $p^* = p$, the model is said to be a *full-rank linear model*; when $p^* < p$, the model is said to be a *less-than-full-rank linear model*.

D. L. Zimmerman, *Linear Model Theory*,
https://doi.org/10.1007/978-3-030-52063-2_5

Necessarily, $p^* \leq n$ also. In general, n can be less than p; classically it is not, but in the rapidly developing area within statistics known as *high-dimensional inference*, it is. For a full-rank linear model, $n \geq p$.

5.1.2 Constrained and Unconstrained Models

As noted in Chap. 1, $\boldsymbol{\beta}$ is often unrestricted, but in some cases $\boldsymbol{\beta}$ may be restricted, or *constrained*, to a specified subset of $\mathbb{R}^p$. In this book we consider only linear equality constraints on $\boldsymbol{\beta}$, i.e., constraints of the form $\mathbf{A}\boldsymbol{\beta} = \mathbf{h}$, where $\mathbf{A}$ is a specified $q \times p$ matrix and $\mathbf{h}$ is a specified q-vector. It is assumed that this system of constraint equations is consistent. A model with such constraints is said to be *constrained*, and a model with no such constraints is *unconstrained*.

Linear equality constraints in linear models may be labeled as either *real constraints* or *pseudo-constraints*. Real constraints are generally dictated by scientific or other subject-matter considerations; examples include requirements that $\mathrm{E}(y)$ pass through a specified point or that the derivative of $\mathrm{E}(y)$ at a given point is 0. Pseudo-constraints are constraints imposed for convenience, such as to obtain a unique estimator of $\boldsymbol{\beta}$ when the model is overparameterized. We shall encounter both types of constraints in this book.

5.1.3 Regression Models

In a *regression model*, the explanatory variables (with the possible exception of an overall constant) are measured on at least an ordinal numeric scale; very often, in fact, they are discrete or continuous. Such variables may be called *regression variables* or, more simply, *regressors*. Perhaps the simplest case of a regression model is the aptly named *simple linear regression model*, for which there is but one nonconstant regressor, labeled x. That is,

$$y_i = \beta_1 + \beta_2 x_i + e_i \qquad (i = 1, \ldots, n),$$

or equivalently

$$\mathbf{y} = (\mathbf{1}, \mathbf{x}) \begin{pmatrix} \beta_1 \\ \beta_2 \end{pmatrix} + \mathbf{e}$$

where $\mathbf{x} = (x_i)$. In this model, β_1 and β_2 are frequently called the *intercept* and *slope*, respectively; the intercept may be interpreted as the expectation of the response when $x = 0$, and the slope as the expected change in the response per unit change in x. Note that $\mathbf{X}$ has full column rank if and only if $\mathbf{1} \notin \mathcal{C}(\mathbf{x})$, i.e., if and only if at least one observed value of x is different than the others. Henceforth, whenever we refer to a simple linear regression model, it will be assumed implicitly that this condition is satisfied.

A re-expression (later called a reparameterization) of the same model, known as the *centered simple linear regression model*, has the form

$$y_i = \beta_1 + \beta_2(x_i - \bar{x}) + e_i \qquad (i = 1, \ldots, n).$$

Here, β_2 has the same interpretation as in the uncentered model, but β_1 is now interpreted as the expectation of the response when $x = \bar{x}$. The model matrix is $\mathbf{X} = (\mathbf{1}, \mathbf{x} - \bar{x}\mathbf{1})$. Furthermore, the columns of this matrix are orthogonal, unlike those of the model matrix for the uncentered simple linear regression model. This orthogonality has some advantages, as will be seen later.

The *no-intercept simple linear regression model* is a special case of the simple linear regression model in which the intercept is zero and the subscript on the slope is not needed hence often omitted, so that the general form $\mathbf{X}\boldsymbol{\beta}$ reduces to $\mathbf{x}\beta$.

The *multiple linear regression model* is an extension of the simple linear regression model to a situation with two or more regressors:

$$y_i = \beta_1 + \beta_2 x_{i2} + \beta_3 x_{i3} + \cdots + \beta_p x_{ip} + e_i \qquad (i = 1, \ldots, n),$$

or equivalently

$$\mathbf{y} = (\mathbf{1}, \mathbf{x}_2, \mathbf{x}_3, \ldots, \mathbf{x}_p) \begin{pmatrix} \beta_1 \\ \beta_2 \\ \beta_3 \\ \vdots \\ \beta_p \end{pmatrix} + \mathbf{e}$$

Centered and no-intercept versions of this model may also be defined.

A *polynomial regression model* in a single observed variable x is a multiple linear regression model in which each explanatory variable is a *monomial* in x, i.e., a nonnegative integer power of x. The simple linear regression model is a special case. A *complete* polynomial regression model of order $p - 1$ in a single variable is given by

$$y_i = \beta_1 + \beta_2 x_i + \beta_3 x_i^2 + \cdots + \beta_p x_i^{p-1} + e_i \qquad (i = 1, \ldots, n).$$

If any of the monomials of order less than $p - 1$ are not included as explanatory variables, the polynomial model is said to be *incomplete*.

Polynomial regression models can be extended to situations in which the explanatory variables are products of monomials of two or more distinct observed variables. For example, a complete polynomial regression model of order two in two variables is given by

$$y_i = \beta_1 + \beta_2 x_{i1} + \beta_3 x_{i2} + \beta_4 x_{i1}^2 + \beta_5 x_{i2}^2 + \beta_6 x_{i1} x_{i2} + e_i \qquad (i = 1, \ldots, n). \qquad (5.1)$$

The powers of each explanatory variable in (5.1) sum to two or less, which makes it a model of order two.

For reasons that will become clear later, it may be beneficial to re-express a polynomial regression model in terms of centered distinct observed variables. The centered version of model (5.1), for example, is

$$\begin{aligned} y_i &= \beta_1 + \beta_2(x_{i1} - \bar{x}_1) + \beta_3(x_{i2} - \bar{x}_2) + \beta_4(x_{i1} - \bar{x}_1)^2 + \beta_5(x_{i2} - \bar{x}_2)^2 \\ &\quad + \beta_6(x_{i1} - \bar{x}_1)(x_{i2} - \bar{x}_2) + e_i \\ &\quad (i = 1, \ldots, n). \end{aligned}$$

In this version, the parameters associated with the highest-order terms (which in this case are β_4, β_5, and β_6) have the same interpretations as in the uncentered model, but interpretations of parameters associated with all lower-order terms (in this case β_1, β_2, and β_3) are different.

5.1.4 Classificatory Models

In order to define a *classificatory model*, we must first define what is meant by a *factor of classification.*

Definition 5.1.1 A *factor of classification* is a criterion used to partition the set of observational units into exhaustive disjoint subsets. The subsets are called *levels.*

Examples of factors of classification abound. For instance, people may be classified by eye color (brown, blue, hazel, other) or educational level (no high school degree, high school degree, 2-year community college degree, college degree, Master's degree, Ph.D., postdoctoral).

In order to use a factor of classification in a linear model, we must somehow "quantify" that factor. This may be accomplished by forming a quantitative variable, called an *indicator variable*, for each of the levels of a factor. Thus, if a factor has q levels, then we associate q indicator variables $x_1, x_2, \ldots, x_q$ with that factor, where

$$x_j = \begin{cases} 1, & \text{for observational units classified to the } j\text{th level} \\ 0, & \text{for all other observational units.} \end{cases}$$

Definition 5.1.2 The model $\mathbf{y} = \mathbf{X}\boldsymbol{\beta} + \mathbf{e}$ is said to be a *classificatory* linear model if every explanatory variable, except possibly for an overall constant term, is an indicator variable.

The term "classificatory model" for models satisfying Definition 5.1.2 is by no means universally used. Searle (1971), Myers and Milton (1991), Hinkelmann and Kempthorne (1994), and Harville (2018) use the term, while Graybill (1976) refers to such models as "design models." Many other authors, e.g., Rencher

(2000), Ravishanker and Dey (2002), Khuri (2010), and Christensen (2011) call them "analysis-of-variance (ANOVA) models." We prefer "classificatory model" to "ANOVA model" because, as will be seen later, an analysis of variance may be constructed for any linear model with nonnull model matrix, not merely those that satisfy Definition 5.1.2. It is well, however, for the reader to be aware of the common usage of the latter term.

5.1.4.1 The One-Factor Model

The simplest nontrivial classificatory linear model is one in which the observational units are classified by a single factor, or the so-called *one-factor model*. For such a situation, let q denote the number of levels of the factor, and define variables $x_{i1} \equiv 1$ and

$$x_{i,j+1} = \begin{cases} 1, & \text{if unit } i \text{ is classified to level } j \text{ of the factor} \quad (j = 1, \dots, q) \\ 0, & \text{otherwise.} \end{cases}$$

The one-factor model may then be written as

$$y_i = \mu x_{i1} + \alpha_1 x_{i2} + \alpha_2 x_{i3} + \cdots + \alpha_q x_{i,q+1} + e_i \qquad (i = 1, \dots, n) \tag{5.2}$$

where μ is an overall constant common to all observations and α_j is the *effect* of the jth level of the factor on the response variable. A more common and certainly more convenient representation of the same model that we use in place of this one results from using two subscripts to identify observational units, i.e., from letting y_{ij} represent the observed response variable on the jth unit classified to the ith level of the factor and letting n_i denote the number of units classified to the ith level. This two-subscript representation is

$$y_{ij} = \mu + \alpha_i + e_{ij} \qquad (i = 1, \dots, q;\ j = 1, \dots, n_i). \tag{5.3}$$

Here i represents the level, and j the unit, which is the exact opposite of what these subscripts represented in (5.2). It is both helpful and customary to put the elements of $\mathbf{y}$ in lexicographic order, i.e., $\mathbf{y} = (y_{11}, y_{12}, \dots, y_{1n_1}, y_{21}, \dots, y_{qn_q})^T$, in which case

$$\mathbf{X} = (\mathbf{1}_n, \oplus_{i=1}^q \mathbf{1}_{n_i}) \tag{5.4}$$

where $n = \sum_{i=1}^q n_i$, and $\boldsymbol{\beta} = (\mu, \alpha_1, \alpha_2, \dots, \alpha_q)^T$. Clearly the columns of $\mathbf{X}$ are linearly dependent (the first column equals the sum of the remaining columns), so this model matrix is less than full rank. If $n_1 = n_2 = \cdots = n_q = r$ for some integer r, then the data are said to be *balanced* and $\mathbf{X} = (\mathbf{1}_n, \mathbf{I}_q \otimes \mathbf{1}_r)$ where $n = qr$.

Yet another representation of the one-factor model is the *cell means* representation

$$y_{ij} = \mu_i + e_{ij} \qquad (i = 1, \ldots, q;\ j = 1, \ldots, n_i), \tag{5.5}$$

so called because one can regard an observation classified to the ith level of the factor of classification as belonging to the ith cell (row-column combination) in a table consisting of q rows and one column. For this representation, $\mathbf{X} = \oplus_{i=1}^{q} \mathbf{1}_{n_i}$, which has linearly independent columns and therefore has full rank, and $\boldsymbol{\beta} = (\mu_1, \ldots, \mu_q)^T$.

5.1.4.2 Crossed and Nested Factors

When the observational units can be partitioned according to more than one factor of classification, certain relationships among those factors are important for determining an appropriate mean structure. For the following definitions, let A and B represent two factors of classification.

Definition 5.1.3 Factor B is said to be *nested* by (or in) Factor A if all observational units having the same level of B necessarily have the same level of A. If neither of the two factors are nested by the other, they are said to be *crossed.*

Examples: Take the observational units to be all people in the world who were born on January 1, 2000. Suppose that these people are classified by three factors: country of birth, city of birth, and gender. Then:

- City of birth is nested by country of birth because any two people born in the same city on January 1, 2000 were necessarily born in the same country.
- Country of birth and gender are crossed because any two people of the same gender were not necessarily born in the same country, and any two people born in the same country are not necessarily of the same gender.

Definition 5.1.4 Factors A and B are said to be *completely crossed* if it is possible for observational units to exist at every combination of the levels of Factor A and Factor B, regardless of whether observations are actually taken at every such combination. If Factors A and B are crossed, but not completely crossed, they are said to be *partially crossed.*

Examples:

- If the observational units are the people described in the examples of nested and crossed factors given previously, then country of birth and gender are completely crossed because it is possible for a person of any gender to have been born in any country on January 1, 2000, and vice versa.
- If the observational units are Big 10 conference basketball games in a given season, the two factors "home team" and "visiting team" are partially, rather

than completely, crossed because a Big 10 team cannot simultaneously be both a home team and a visiting team.

5.1.4.3 Two-Factor Crossed Models

Consider a situation in which there are two completely crossed factors of classification, A and B. Let q represent the number of levels of Factor A; m represent the number of levels of Factor B; $x_{i1} \equiv 1$; $x_{i2}, x_{i3}, \ldots, x_{i,q+1}$ represent indicator variables associated with Factor A; and $x_{i,q+2}, x_{i,q+3}, \ldots, x_{i,q+1+m}$ represent indicator variables associated with Factor B.

The first model one might consider in this context has effects corresponding to the levels of Factor A and effects corresponding to the levels of Factor B. Such a model, called the *two-way main effects model*, is given by

$$y_i = \mu x_{i1} + \sum_{j=1}^{q} \alpha_j x_{i,j+1} + \sum_{k=1}^{m} \gamma_k x_{i,q+1+k} + e_i \qquad (i = 1, \ldots, n)$$

where μ and α_j are defined as in the one-factor model, and γ_k is the effect on the response variable of the kth level of Factor B. A more common and more convenient representation of this same model results from using three subscripts to identify observational units, i.e., from letting y_{ijk} represent the kth of those units common to the ith level of Factor A and the jth level of Factor B. This alternative representation is

$$y_{ijk} = \mu + \alpha_i + \gamma_j + e_{ijk} \qquad (i = 1, \ldots, q;\ j = 1, \ldots, m;\ k = 1, \ldots, n_{ij}) \tag{5.6}$$

for those combinations of i and j for which $n_{ij} \geq 1$. Here, n_{ij} is the number of observational units common to the ith level of Factor A and the jth level of Factor B. Assuming that the elements of $\mathbf{y}$ are ordered lexicographically and that $n_{ij} \geq 1$ for every combination of i and j, a concise representation of the model matrix is

$$\mathbf{X} = \begin{pmatrix} \mathbf{1}_{n_{1\cdot}} & \mathbf{1}_{n_{1\cdot}} & \mathbf{0}_{n_{1\cdot}} & \cdots & \mathbf{0}_{n_{1\cdot}} & \oplus_{j=1}^{m} \mathbf{1}_{n_{1j}} \\ \mathbf{1}_{n_{2\cdot}} & \mathbf{0}_{n_{2\cdot}} & \mathbf{1}_{n_{2\cdot}} & \cdots & \mathbf{0}_{n_{2\cdot}} & \oplus_{j=1}^{m} \mathbf{1}_{n_{2j}} \\ \vdots & \vdots & \vdots & \ddots & \vdots & \vdots \\ \mathbf{1}_{n_{q\cdot}} & \mathbf{0}_{n_{q\cdot}} & \mathbf{0}_{n_{q\cdot}} & \cdots & \mathbf{1}_{n_{q\cdot}} & \oplus_{j=1}^{m} \mathbf{1}_{n_{qj}} \end{pmatrix}, \tag{5.7}$$

and $\boldsymbol{\beta} = (\mu, \alpha_1, \ldots, \alpha_q, \gamma_1, \ldots, \gamma_m)^T$. In (5.7) and subsequently, $n_{i\cdot} = \sum_{j=1}^{m} n_{ij}$. (Here and throughout, a dot "·" in place of a subscript on a quantity will indicate that the quantity has been summed over the range of that subscript.) If $n_{ij} = 0$ for one or more combinations of i and j, $\mathbf{X}$ is the matrix obtained by deleting the rows corresponding to those combinations from (5.7).

The two-factor main effects model implicitly assumes that the effect of the ith level of Factor A on the response variable of an observational unit does not depend on the level of Factor B to which that unit is classified. This *additivity* or *no-interaction* assumption is not always satisfied in practice. It can be relaxed by adding indicator variables to the model that allow for the presence of interaction. More specifically, an indicator variable is added for every combination (i, j) of the levels of the two factors, and this indicator variable is the product of the indicator variables corresponding to the ith level of Factor A and the jth level of Factor B. Denoting the corresponding interaction effect by ξ_{ij} and using a three-subscript representation, the *two-way model with interaction* is given by

$$y_{ijk} = \mu + \alpha_i + \gamma_j + \xi_{ij} + e_{ijk} \qquad (i = 1, \dots, q;\ j = 1, \dots, m;\ k = 1, \dots, n_{ij}), \tag{5.8}$$

where again this pertains only to those combinations of i and j for which $n_{ij} \geq 1$. Assuming again that the elements of $\mathbf{y}$ are ordered lexicographically and that there are no empty cells, the model matrix $\mathbf{X}$ may be represented concisely as follows:

$$\mathbf{X} = \begin{pmatrix} \mathbf{1}_{n_{1\cdot}} & \mathbf{1}_{n_{1\cdot}} & \mathbf{0}_{n_{1\cdot}} & \cdots & \mathbf{0}_{n_{1\cdot}} & \oplus_{j=1}^{m}\mathbf{1}_{n_{1j}} & \oplus_{j=1}^{m}\mathbf{1}_{n_{1j}} & \mathbf{0}_{n_{1\cdot}\times m} & \cdots & \mathbf{0}_{n_{1\cdot}\times m} \\ \mathbf{1}_{n_{2\cdot}} & \mathbf{0}_{n_{2\cdot}} & \mathbf{1}_{n_{2\cdot}} & \cdots & \mathbf{0}_{n_{2\cdot}} & \oplus_{j=1}^{m}\mathbf{1}_{n_{2j}} & \mathbf{0}_{n_{2\cdot}\times m} & \oplus_{j=1}^{m}\mathbf{1}_{n_{2j}} & \cdots & \mathbf{0}_{n_{2\cdot}\times m} \\ \vdots & \vdots & \vdots & \ddots & \vdots & \vdots & \vdots & \vdots & \ddots & \vdots \\ \mathbf{1}_{n_{q\cdot}} & \mathbf{0}_{n_{q\cdot}} & \mathbf{0}_{n_{q\cdot}} & \cdots & \mathbf{1}_{n_{q\cdot}} & \oplus_{j=1}^{m}\mathbf{1}_{n_{qj}} & \mathbf{0}_{n_{q\cdot}\times m} & \mathbf{0}_{n_{q\cdot}\times m} & \cdots & \oplus_{j=1}^{m}\mathbf{1}_{n_{qj}} \end{pmatrix}. \tag{5.9}$$

The corresponding $\boldsymbol{\beta}$ is $\boldsymbol{\beta} = (\mu, \alpha_1, \dots, \alpha_q, \gamma_1, \dots, \gamma_m, \xi_{11}, \xi_{12}, \dots, \xi_{qm})^T$.

The two-way model with interaction may be alternatively represented in cell-means model form as

$$y_{ijk} = \mu_{ij} + e_{ijk} \qquad (i = 1, \dots, q;\ j = 1, \dots, m;\ k = 1, \dots, n_{ij}). \tag{5.10}$$

Some additional terminology associated with two crossed factors is as follows. Because a two-factor crossed situation can be conceptualized as a two-way table with rows corresponding to the levels of Factor A, columns corresponding to the levels of Factor B, and the combination of any level of Factor A with any level of Factor B corresponding to a cell in the table, n_{ij} is called the (i, j)th *cell frequency*. Call $n_{i\cdot} \equiv \sum_j n_{ij}$ and $n_{\cdot j} \equiv \sum_i n_{ij}$ the ith *row total* and jth *column total*, respectively. If, for all i and j, $n_{ij} = r$ for some positive integer r, the data are said to be *balanced*. Otherwise, i.e., if one or more cell frequencies are different than others, the data are said to be *unbalanced*. A special case of unbalanced data, in which $n_{ij} = n_{i\cdot}n_{\cdot j}/n$ for all i and j, is referred to as *proportional frequencies*.

Example 5.1.4-1. A Two-Way Partially Crossed Model

Consider a particular two-way partially crossed model given by

$$y_{ijk} = \mu + \alpha_i - \alpha_j + e_{ijk} \quad (i \neq j = 1, \ldots, q;\ k = 1, \ldots, n_{ij})$$

for those combinations of i and j ($i \neq j$) for which $n_{ij} \geq 1$. Assuming that the elements of $\mathbf{y}$ are ordered lexicographically and that $n_{ij} \geq 1$ for all $i \neq j$, one representation of the model matrix is

$$\mathbf{X}_{n..\times(q+1)} = \left(\begin{array}{ccccccc}
\mathbf{1}_{n_{12}} & \mathbf{1}_{n_{12}} & -\mathbf{1}_{n_{12}} & \mathbf{0}_{n_{12}} & \cdots & \mathbf{0}_{n_{12}} & \mathbf{0}_{n_{12}} \\
\mathbf{1}_{n_{13}} & \mathbf{1}_{n_{13}} & \mathbf{0}_{n_{13}} & -\mathbf{1}_{n_{13}} & \cdots & \mathbf{0}_{n_{13}} & \mathbf{0}_{n_{13}} \\
\vdots & \vdots & \vdots & \vdots & \ddots & \vdots & \vdots \\
\mathbf{1}_{n_{1(q-1)}} & \mathbf{1}_{n_{1(q-1)}} & \mathbf{0}_{n_{1(q-1)}} & \mathbf{0}_{n_{1(q-1)}} & \cdots & -\mathbf{1}_{n_{1(q-1)}} & \mathbf{0}_{n_{1(q-1)}} \\
\mathbf{1}_{n_{1q}} & \mathbf{1}_{n_{1q}} & \mathbf{0}_{n_{1q}} & \mathbf{0}_{n_{1q}} & \cdots & \mathbf{0}_{n_{1q}} & -\mathbf{1}_{n_{1q}} \\
\hline
\mathbf{1}_{n_{21}} & -\mathbf{1}_{n_{21}} & \mathbf{1}_{n_{21}} & \mathbf{0}_{n_{21}} & \cdots & \mathbf{0}_{n_{21}} & \mathbf{0}_{n_{21}} \\
\mathbf{1}_{n_{23}} & \mathbf{0}_{n_{23}} & \mathbf{1}_{n_{23}} & -\mathbf{1}_{n_{23}} & \cdots & \mathbf{0}_{n_{23}} & \mathbf{0}_{n_{23}} \\
\vdots & \vdots & \vdots & \vdots & \ddots & \vdots & \vdots \\
\mathbf{1}_{n_{2(q-1)}} & \mathbf{0}_{n_{2(q-1)}} & \mathbf{1}_{n_{2(q-1)}} & \mathbf{0}_{n_{2(q-1)}} & \cdots & -\mathbf{1}_{n_{2(q-1)}} & \mathbf{0}_{n_{2(q-1)}} \\
\mathbf{1}_{n_{2q}} & \mathbf{0}_{n_{2q}} & \mathbf{1}_{n_{2q}} & \mathbf{0}_{n_{2q}} & \cdots & \mathbf{0}_{n_{2q}} & -\mathbf{1}_{n_{2q}} \\
\hline
\vdots & \vdots & \vdots & \vdots & \vdots & \vdots & \vdots \\
\hline
\mathbf{1}_{n_{q1}} & -\mathbf{1}_{n_{q1}} & \mathbf{0}_{n_{q1}} & \mathbf{0}_{n_{q1}} & \cdots & \mathbf{0}_{n_{q1}} & \mathbf{1}_{n_{q1}} \\
\mathbf{1}_{n_{q2}} & \mathbf{0}_{n_{q2}} & -\mathbf{1}_{n_{q2}} & \mathbf{0}_{n_{q2}} & \cdots & \mathbf{0}_{n_{q2}} & \mathbf{1}_{n_{q2}} \\
\vdots & \vdots & \vdots & \vdots & \ddots & \vdots & \vdots \\
\mathbf{1}_{n_{q(q-1)}} & \mathbf{0}_{n_{q(q-1)}} & \mathbf{0}_{n_{q(q-1)}} & \mathbf{0}_{n_{q(q-1)}} & \cdots & -\mathbf{1}_{n_{q(q-1)}} & \mathbf{1}_{n_{q(q-1)}}
\end{array} \right). \tag{5.11}$$

Again, if $n_{ij} = 0$ for one or more combinations of i and j, $\mathbf{X}$ is the matrix obtained by deleting the rows corresponding to those combinations from (5.11). In the case of balanced data (r observations in each cell (i, j), where $i \neq j$), $\mathbf{X}$ may be written more concisely as

$$\mathbf{X} = \left(\mathbf{1}_{rq(q-1)}, \ (\mathbf{1}_r \mathbf{v}_{ij}^T)_{i \neq j = 1,\ldots,q} \right),$$

where $\mathbf{v}_{ij} = \mathbf{u}_i^{(q)} - \mathbf{u}_j^{(q)}$. ■

5.1.4.4 The Two-Factor Nested Model

Consider a situation with two factors of classification, A and B, where B is nested by A. We give the corresponding classificatory linear model in a multiple-subscript form at the outset. Let q = # of levels of Factor A, m_i = # of levels of Factor B within the ith level of Factor A, and n_{ij} = # of observational units belonging to the jth level of Factor B within the ith level of Factor A. Then the *two-factor nested model* is given by

$$y_{ijk} = \mu + \alpha_i + \gamma_{ij} + e_{ijk} \quad (i = 1, \ldots, q;\ j = 1, \ldots, m_i;\ k = 1, \ldots, n_{ij})$$

where γ_{ij} is the effect of the jth level of Factor B within the ith level of Factor A. In this context, "balanced data" means that $m_i = m$ and $n_{ij} = r$ for all i and j, for some integers m and r.

Obtaining a concise representation of the model matrix corresponding to this model is left as an exercise.

5.1.4.5 Three-Factor Models

The structures that result from nesting and/or crossing among three or more factors of classification are more complicated than those that result from just two factors. For example, with three factors A, B, and C, there are essentially five different possibilities. These possibilities are as follows, with examples given in parentheses. The observational units in all these examples are all those people who were admitted to a hospital in the United States in a given year.

1. C (home address) nested in B (county) nested in A (state).
2. B (attending doctor) and C (attending nurse) nested in A (hospital), but B and C crossed.
3. B (county) nested in A (state), with A and B crossed with C (location of home address—rural or urban)
4. C (home address) nested in both A (state) and B (location—rural or urban), and B crossed with A.
5. A (state), B (home owned or rented), and C (home does or does not have cable TV) all crossed with each other.

The greater complexity of structures among three (or more) factors leads to more complexity in the models. A few of these are considered in the exercises.

5.1.5 Analysis-of-Covariance Models

An *analysis-of-covariance model* is a model for which the mean structure has both regression variables and indicator variables. The simplest case occurs when the observational units are classified according to a single factor, and a single regressor is observed on each unit. This gives rise to the one-factor single-regressor model

$$y_{ij} = \mu + \alpha_i + \gamma x_{ij} + e_{ij} \qquad (i = 1, \ldots, q;\ j = 1, \ldots, n_i),$$

where x_{ij} is the observed value of the regressor on the ijth unit. For this model,

$$\mathbf{X} = (\mathbf{1}_n, \oplus_{i=1}^{q} \mathbf{1}_{n_i}, \mathbf{x}), \quad \boldsymbol{\beta} = (\mu, \alpha_1, \ldots, \alpha_q, \gamma)^T,$$

and $\mathbf{x}$ is the vector of x_{ij}'s placed in lexicographic order. Additional examples of analysis-of-covariance models will be introduced in the exercises and in Chap. 9.

5.2 Error Structures

In this section, we give a brief overview of linear models that are distinguishable on the basis of what is assumed about $\mathbf{e}$, i.e., on the basis of their error structure. Recall from Chap. 1 that we assume throughout that the expectation of $\mathbf{e}$ is $\mathbf{0}$. The models presented in this section assume further that the variance–covariance matrix of $\mathbf{e}$ exists, but they differ in what is assumed about that variance–covariance matrix. Furthermore, for some of the models presented here, the variance–covariance matrix is the result of a decomposition of the error vector into a linear combination of other random vectors whose elements are called random effects.

5.2.1 Gauss–Markov Models

In a *Gauss–Markov model*, $\text{var}(\mathbf{e}) = \sigma^2\mathbf{I}$, where σ^2 is an unknown positive parameter. This variance–covariance matrix specifies that the errors are *homoscedastic* (i.e., they have common positive variance) and uncorrelated. The parameter space for the model is $\{\boldsymbol{\beta}, \sigma^2 : \boldsymbol{\beta} \in \mathbb{R}^p, \sigma^2 > 0\}$.

The Gauss–Markov model is named after Carl Friedrich Gauss (1777–1855) and Andrey Markov (1856–1922), two influential mathematicians who made contributions to the theory of estimation under the model. Gauss is generally credited with showing [in Gauss (1855)] that least squares estimators have smallest variance among all linear unbiased estimators, and Markov with clarifying some aspects of Gauss's arguments (Plackett, 1949).

Point estimation (more specifically, least squares estimation) of the parameters of a Gauss–Markov model (or functions thereof), and associated statistical methodologies not requiring distribution theory, are presented in Chaps. 7–9 of this book. Extension to the case of a constrained Gauss–Markov model is the topic of Chap. 10, and the effects that misspecifying the model may have on least squares estimation are considered in Chap. 12. Inference based on the response vector $\mathbf{y}$ having a multivariate normal distribution is presented in Chap. 15.

5.2.2 Aitken Models

In an *Aitken model*, var(**e**) $= \sigma^2\mathbf{W}$, where σ^2 is an unknown positive parameter and **W** is a specified nonnegative definite matrix. This variance–covariance matrix generalizes that of the Gauss–Markov model by allowing the errors to be *heteroscedastic* (have different variances) and/or correlated. The parameter space for the Aitken model is identical to that for the Gauss–Markov model. An important special case of an Aitken model occurs if **W** is positive definite; this case is called a *positive definite Aitken model.*

The Aitken model is named after Alexander Aitken (1895–1967), who extended Gauss's earlier work to show that generalized least squares estimators have smallest variance among all unbiased estimators under a model with a general positive definite variance–covariance matrix (Aitken, 1935). Aitken is also credited with being the first to extensively employ matrix notation in deriving results for linear models.

Point estimation (more specifically generalized least squares and best linear unbiased estimation) of the parameters of an Aitken model (or functions thereof), and associated statistical methodologies not requiring distribution theory, are presented in Chap. 11. Inferences that are possible under the additional assumption that **y** has a multivariate normal distribution are presented in Chap. 15.

Example 5.2.2-1. An Aitken Model for a Vector of Means

It is possible that among the n responses, some correspond to identical combinations of the explanatory variables. If so, then some of the rows of the model matrix **X** are identical to each other, and we may write **X** (by permuting rows, if necessary) as follows:

$$\mathbf{X} = \begin{pmatrix} \mathbf{1}_{n_1}\mathbf{x}_1^T \\ \mathbf{1}_{n_2}\mathbf{x}_2^T \\ \vdots \\ \mathbf{1}_{n_m}\mathbf{x}_m^T \end{pmatrix}$$

where $\mathbf{x}_1^T, \ldots, \mathbf{x}_m^T$ are the distinct rows of **X** (corresponding to the distinct combinations of the explanatory variables) and n_i is the number of replications of the ith combination. Of course, $\sum_{i=1}^m n_i = n$.

Let y_{ij} and e_{ij} denote the jth observed response and jth residual, respectively, associated with $\mathbf{x}_i^T$ ($j = 1, \ldots, n_i$; $i = 1, \ldots, m$). The reader will no doubt observe some similarity in notation and structure here to that of the one-factor model introduced in Sect. 5.1.4.1. Define $\bar{y}_{i\cdot} = n_i^{-1}\sum_{j=1}^{n_i} y_{ij}$ and $\bar{e}_{i\cdot} = n_i^{-1}\sum_{j=1}^{n_i} e_{ij}$. In words, $\bar{y}_{i\cdot}$ is the mean of the observations, and $\bar{e}_{i\cdot}$ is the mean of the residuals, corresponding to the ith combination of the explanatory variables.

Now suppose that the n observations follow the Gauss–Markov model, so that $\text{var}(\mathbf{y}) = \sigma^2\mathbf{I}$. Then it is easily shown that $\text{var}(\bar{y}_{i\cdot}) = \sigma^2/n_i$, and that the vector of means, $\bar{\mathbf{y}} = (\bar{y}_{1\cdot}, \ldots, \bar{y}_{m\cdot})^T$, satisfies the model

$$\bar{\mathbf{y}} = \breve{\mathbf{X}}\boldsymbol{\beta} + \bar{\mathbf{e}}$$

where

$$\breve{\mathbf{X}} = \begin{pmatrix} \mathbf{x}_1^T \\ \vdots \\ \mathbf{x}_m^T \end{pmatrix}, \qquad \bar{\mathbf{e}} = \begin{pmatrix} \bar{e}_{1\cdot} \\ \vdots \\ \bar{e}_{m\cdot} \end{pmatrix},$$

$\text{E}(\bar{\mathbf{e}}) = \mathbf{0}$, and $\text{var}(\bar{\mathbf{e}}) = \sigma^2[\oplus_{i=1}^m(1/n_i)]$. Plainly, this is a case of the positive definite Aitken model with $\mathbf{W} = \oplus_{i=1}^m(1/n_i)$. This model might be used, for example, if the original responses are lost but their means over each combination of the explanatory variables are known. ■

5.2.3 Random and Mixed Effects Models

In a *random effects model* or *mixed effects model*, the error vector $\mathbf{e}$ may be decomposed as follows:

$$\mathbf{e} = \mathbf{Z}\mathbf{b} + \mathbf{d}$$

where $\mathbf{Z}$ is a specified $n \times q$ matrix, $\mathbf{b}$ is a q-vector of zero-mean random variables called *random effects*, and $\mathbf{d}$ is an n-vector of zero-mean random variables. Furthermore, $\text{cov}(\mathbf{b}, \mathbf{d}) = \mathbf{0}$, $\text{var}(\mathbf{b}) = \sigma^2\mathbf{G}(\boldsymbol{\psi})$ and $\text{var}(\mathbf{d}) = \sigma^2\mathbf{R}(\boldsymbol{\psi})$, where the elements of $\mathbf{G}(\boldsymbol{\psi})$ and $\mathbf{R}(\boldsymbol{\psi})$ are known functions of an $(m-1)$-dimensional vector of unknown parameters $\boldsymbol{\psi} = (\psi_1, \ldots, \psi_{m-1})^T$. The parameter space for $\boldsymbol{\theta} = (\sigma^2, \boldsymbol{\psi}^T)^T$ is $\Theta = \{(\sigma^2, \boldsymbol{\psi}) : \sigma^2 > 0, \boldsymbol{\psi} \in \Psi\}$ where Ψ is the subset of $\mathbb{R}^{m-1}$ for which $\mathbf{G}(\boldsymbol{\psi})$ and $\mathbf{R}(\boldsymbol{\psi})$ are nonnegative definite or a specified subset of that set (such as the set of all $\boldsymbol{\psi}$ for which $\mathbf{G}(\boldsymbol{\psi})$ and $\mathbf{R}(\boldsymbol{\psi})$ are positive definite). Under this model,

$$\text{var}(\mathbf{e}) = \text{var}(\mathbf{Z}\mathbf{b} + \mathbf{d}) = \sigma^2[\mathbf{Z}\mathbf{G}(\boldsymbol{\psi})\mathbf{Z}^T + \mathbf{R}(\boldsymbol{\psi})], \tag{5.12}$$

which is nonnegative definite (by Corollary 2.15.12.1 and Theorem 2.15.3) but not a positive scalar multiple of a fully specified (i.e., fully known) matrix unless $\boldsymbol{\psi}$ is known.

A random effects model and a mixed effects model differ only with respect to the expectations of their responses. In a random effects model, there is at most one fixed effect, which is an overall constant, so the model equation is merely $\mathbf{y} = \mathbf{Z}\mathbf{b} + \mathbf{d}$ or $\mathbf{y} = \mathbf{1}\mu + \mathbf{Z}\mathbf{b} + \mathbf{d}$, and thus the expectations of the responses are equal to either zero or a common, unknown value. In a mixed effects model, the model equation is completely general with respect to its fixed effects: $\mathbf{y} = \mathbf{X}\boldsymbol{\beta} + \mathbf{Z}\mathbf{b} + \mathbf{d}$. The parameter space for the random effects model is either $\{(\sigma^2, \boldsymbol{\psi}) : \sigma^2 > 0, \boldsymbol{\psi} \in \Psi\}$ or $\{(\mu, \sigma^2, \boldsymbol{\psi}) : \mu \in \mathbb{R}, \sigma^2 > 0, \boldsymbol{\psi} \in \Psi\}$, according to whether μ is included in the model, while that for the (unconstrained) mixed effects model replaces $\mu \in \mathbb{R}$ with $\boldsymbol{\beta} \in \mathbb{R}^p$ but is otherwise the same.

A random or mixed effects model generalizes an Aitken model by allowing the variance–covariance matrix of $\mathbf{e}$ to be a function of unknown parameters. The nature of this functional dependence as given by (5.12), however, is highly structured. If $\boldsymbol{\psi}$ is known in (5.12), then this model reduces to an Aitken model.

The special case of a random or mixed effects model for which $\mathbf{G}(\boldsymbol{\psi})$ is a nonnegative definite diagonal matrix and $\mathbf{R} = \mathbf{I}$ is called a *components-of-variance model.*

Various types of inference for random and mixed effects models are considered in Chaps. 13, 15, and 16.

Example 5.2.3-1. Random and Mixed Coefficient Models

Simple and multiple regression models in which the intercept and slopes are regarded as random variables rather than fixed parameters are cases of random effects models known as *random coefficients models.* If some of these quantities are regarded as random variables and others as fixed parameters, the model is another case known as a *mixed coefficients model.* One of the simplest cases of the latter is the fixed-intercept, random-slope simple linear regression model

$$y_i = \beta_1 + (\beta_2 + b_2)x_i + d_i \qquad (i = 1, \ldots, n),$$

where

$$\mathrm{E}\begin{pmatrix} b_2 \\ \mathbf{d} \end{pmatrix} = \mathbf{0}, \qquad \mathrm{var}\begin{pmatrix} b_2 \\ \mathbf{d} \end{pmatrix} = \begin{pmatrix} \sigma_{22} & \mathbf{0}^T \\ \mathbf{0} & \sigma^2\mathbf{I} \end{pmatrix}.$$

The corresponding parameter space is $\{(\beta_1, \beta_2, \sigma^2, \sigma_{22}) : \beta_1 \in \mathbb{R},\ \beta_2 \in \mathbb{R},\ \sigma^2 > 0,\ \sigma_{22} > 0\}$. The resulting variance–covariance matrix for $\mathbf{y}$ is $\mathrm{var}(\mathbf{y}) = \sigma^2\mathbf{I} + \sigma_{22}\mathbf{x}\mathbf{x}^T$, where $\mathbf{x} = (x_1, \ldots, x_n)^T$.

One of the simplest cases of a random coefficient model is the random-intercept, random-slope model

$$y_i = (\beta_1 + b_1) + (\beta_2 + b_2)x_i + d_i \qquad (i = 1, \ldots, n),$$

where

$$\mathrm{E}\begin{pmatrix} b_1 \\ b_2 \\ \mathbf{d} \end{pmatrix} = \mathbf{0}, \qquad \mathrm{var}\begin{pmatrix} b_1 \\ b_2 \\ \mathbf{d} \end{pmatrix} = \begin{pmatrix} \sigma_{11} & \sigma_{12} & \mathbf{0}^T \\ \sigma_{12} & \sigma_{22} & \mathbf{0}^T \\ \mathbf{0} & \mathbf{0} & \sigma^2\mathbf{I} \end{pmatrix}.$$

The parameter space for this model is usually taken to be that for which the variance–covariance matrix of $(b_1, b_2, \mathbf{d})^T$ is positive definite, i.e., $\{\beta_1 \in \mathbb{R},\ \beta_2 \in \mathbb{R},\ (\sigma^2, \sigma_{11}, \sigma_{12}, \sigma_{22}) : \sigma^2 > 0,\ \sigma_{11} > 0,\ \sigma_{11}\sigma_{22} - \sigma_{12}^2 > 0\}$. The resulting variance–covariance matrix for $\mathbf{y}$ may be written as

$$\mathrm{var}(\mathbf{y}) = \sigma^2\mathbf{I} + (\mathbf{1}, \mathbf{x})\begin{pmatrix} \sigma_{11} & \sigma_{12} \\ \sigma_{12} & \sigma_{22} \end{pmatrix}\begin{pmatrix} \mathbf{1}^T \\ \mathbf{x}^T \end{pmatrix}.$$

■

Example 5.2.3-2. The One-Factor Random Effects Model

The one-factor random effects model is given by

$$y_{ij} = \mu + b_i + d_{ij} \qquad (i = 1, \ldots, q;\ j = 1, \ldots, n_i)$$

where $\mathrm{E}(b_i) = 0$ and $\mathrm{E}(d_{ij}) = 0$ for all i and j, the b_i's and d_{ij}'s are all uncorrelated, $\mathrm{var}(b_i) = \sigma_b^2$ for all i, and $\mathrm{var}(d_{ij}) = \sigma^2$ for all i and j. The parameter space is $\{\mu, \sigma_b^2, \sigma^2 : \mu \in \mathbb{R},\ \sigma_b^2 > 0,\ \sigma^2 > 0\}$. The variance–covariance matrix of $\mathbf{y}$ under this model may be written as

$$\begin{aligned} \mathrm{var}(\mathbf{y}) &= \left(\oplus_{i=1}^q \sigma_b^2 \mathbf{J}_{n_i}\right) + \sigma^2\mathbf{I}_n \\ &= (\sigma_b^2 + \sigma^2) \oplus_{i=1}^q [(1-\rho)\mathbf{I}_{n_i} + \rho\mathbf{J}_{n_i}] \end{aligned}$$

where $\rho = \sigma_b^2/(\sigma_b^2 + \sigma^2)$. Thus var($\mathbf{y}$) is block diagonal, where each block along its main diagonal of blocks is of the form $a\mathbf{I} + b\mathbf{J}$, a form that was discussed in Example 4.1-1. The quantity ρ is the common correlation coefficient within each of these blocks and is often called the *intraclass correlation coefficient*. Observe that $0 < \rho < 1$ for this model. ■

Example 5.2.3-3. A Split-Plot Model

"Split-plot" designs are common in applied experimentation. They typically consist of two types of "plots" (experimental units), whole plots and split plots, corresponding to two factors of classification. The split plots are nested within the whole plots. For example, several diets (Factor A) may be randomly assigned to human subjects (the whole plots), and two distinct exercise regimens (Factor B) may be randomly assigned to the subjects' right and left arms (the split plots). The response variable may be the change in muscle mass of each bicep from the beginning of the experiment to the end. More generally, if there are q levels of Factor A and m levels of Factor B, one type of split-plot model may be written as

$$y_{ijk} = \mu+\alpha_i+b_{ij}+\gamma_k+\xi_{ik}+d_{ijk} \qquad (i = 1,\ldots,q;\ j = 1,\ldots,r;\ k = 1,\ldots,m).$$

Here $\mathrm{E}(b_{ij}) = \mathrm{E}(d_{ijk}) = 0$ for all (i, j, k), the b_{ij}'s and d_{ijk}s are uncorrelated, $\mathrm{var}(b_{ij}) = \sigma_b^2$ for all i and j, and $\mathrm{var}(d_{ijk}) = \sigma^2$ for all (i, j, k). Furthermore, r is the number of replicates of each level of Factor A among the whole plots, while each level of Factor B is replicated exactly once within each whole plot. Fixed effect parameters μ, $\{\alpha_i\}$, $\{\gamma_k\}$, and $\{\xi_{ik}\}$ are unconstrained, and the parameter space for σ^2 and σ_b^2 is the same as it is for the one-factor random effects model of Example 5.2.3-2. The variance–covariance matrix of $\mathbf{y}$ is

$$\mathrm{var}(\mathbf{y}) = \sigma^2\mathbf{I}_{qrm} + \sigma_b^2(\mathbf{I}_{qr} \otimes \mathbf{J}_m).$$

■

5.2.4 General Mixed Linear Models

In a *general mixed linear model*, $\mathrm{var}(\mathbf{e}) = \mathbf{V}(\boldsymbol{\theta})$; that is, the elements of $\mathrm{var}(\mathbf{e})$ are known functions of an m-dimensional parameter vector $\boldsymbol{\theta}$. The parameter space for the model is $\{\boldsymbol{\beta}, \boldsymbol{\theta} : \boldsymbol{\beta} \in \mathbb{R}^p, \boldsymbol{\theta} \in \Theta\}$, where Θ is the set of vectors $\boldsymbol{\theta}$ for which $\mathbf{V}(\boldsymbol{\theta})$ is nonnegative definite or a given subset of that set. This is the most general type of linear model considered in this book; it subsumes all of the other models described previously. Inference for its parameters under a multivariate normal distributional assumption is considered in Chaps. 16 and 17.

Example 5.2.4-1. Time Series Models

Suppose that y_i is the value of a response variable of interest observed at time i, and that this variable is repeatedly observed at regularly spaced times (e.g., years) $1, 2, \ldots, n$. Such data are called *time series*. A model often used for time series is

$$y_i = \mu + u_i - \theta u_{i-1} \qquad (i = 1,\ldots,n)$$

where $u_0, u_1, \ldots, u_n$ are uncorrelated random variables with common mean 0 and common variance σ^2. This is a version of a model called the *moving average model of order one*. Under this model we find, using Theorem 4.2.2, that

$$\mathrm{var}(y_i) = \sigma^2(1+\theta^2), \qquad \mathrm{cov}(y_{i+1}, y_i) = \mathrm{cov}(u_{i+1} - \theta u_i, u_i - \theta u_{i-1}) = -\theta\sigma^2,$$

and $\mathrm{cov}(y_{i+k}, y_i) = 0$ for $k > 1$. The model is a special case of the general mixed linear model because we may write it as

$$\mathbf{y} = \mu \mathbf{1}_n + \mathbf{e}$$

where $\mathrm{E}(\mathbf{e}) = \mathbf{0}$ and

$$\mathrm{var}(\mathbf{e}) = \mathbf{V}(\sigma^2, \theta) = \sigma^2(1+\theta^2)\begin{pmatrix} 1 & \rho & 0 & 0 & \cdots & 0 \\ & 1 & \rho & 0 & \cdots & 0 \\ & & 1 & \rho & \cdots & 0 \\ & & & 1 & \cdots & 0 \\ & & & & \ddots & \vdots \\ & & & & & 1 \end{pmatrix},$$

where $\rho = -\theta/(1+\theta^2)$.

Another model often used for time series is

$$y_i = \mu + \rho(y_{i-1} - \mu) + u_i \qquad (i = 1, \ldots, n)$$

where $u_1, \ldots, u_n$ are uncorrelated random variables with common mean 0, $\mathrm{var}(u_1) = \sigma^2/(1-\rho^2)$, $\mathrm{var}(u_i) = \sigma^2$ for $i = 2, \ldots, n$, and $y_0 \equiv \mu$; the parameter space is $\{\mu, \rho, \sigma^2 : \mu \in \mathbb{R}, -1 < \rho < 1, \sigma^2 > 0\}$. This is a version of a model called the *stationary autoregressive model of order one*. It is left as an exercise to show that this model may be written as

$$\mathbf{y} = \mu \mathbf{1}_n + \mathbf{e},$$

where $\mathrm{E}(\mathbf{e}) = \mathbf{0}$ and

$$\mathrm{var}(\mathbf{e}) = \mathbf{V}(\sigma^2, \rho) = \frac{\sigma^2}{1-\rho^2}\begin{pmatrix} 1 & \rho & \rho^2 & \rho^3 & \cdots & \rho^{n-1} \\ & 1 & \rho & \rho^2 & \cdots & \rho^{n-2} \\ & & 1 & \rho & \cdots & \rho^{n-3} \\ & & & 1 & \cdots & \rho^{n-4} \\ & & & & \ddots & \vdots \\ & & & & & 1 \end{pmatrix}, \tag{5.13}$$

demonstrating that it too is a special case of the general mixed linear model. ■

5.3 Prediction-Extended Models

In any statistical linear model, the response vector $\mathbf{y}$ is a random vector, and it is observable. In a *prediction-extended* model, there is an additional random s-vector $\mathbf{u}$ that may be related to $\mathbf{y}$, but it is unobservable and we wish to predict its value or the value of some linear function of it and the unknown parameters. It is assumed that the mean of $\mathbf{u}$ is zero and that the variance–covariance matrix of $\mathbf{u}$ exists, in which case $\text{cov}(\mathbf{y}, \mathbf{u})$ also exists. Additional assumptions on these moments vary with the type of prediction-extended model. In the most general case we consider, the *prediction-extended general mixed linear model*, it is assumed that $\text{var}(\mathbf{u}) = \mathbf{V}_{uu}(\boldsymbol{\theta})$ and $\text{cov}(\mathbf{y}, \mathbf{u}) = \mathbf{V}_{yu}(\boldsymbol{\theta})$, where the elements of $\mathbf{V}_{uu}(\boldsymbol{\theta})$ and $\mathbf{V}_{yu}(\boldsymbol{\theta})$ are known functions of the elements of the same vector $\boldsymbol{\theta}$ of variance–covariance parameters that was defined for the general mixed linear model, and in this context the variance–covariance matrix of $\mathbf{y}$, formerly denoted by $\mathbf{V}(\boldsymbol{\theta})$, is denoted by $\mathbf{V}_{yy}(\boldsymbol{\theta})$. If the parameter space for $\boldsymbol{\theta}$ is the set of all $\boldsymbol{\theta}$ for which $\mathbf{V}_{yy}(\boldsymbol{\theta})$ is positive definite, the model is called the *prediction-extended positive definite general mixed linear model*. A very important special case is the *prediction-extended positive definite Aitken model*, for which it is assumed, in addition to $\text{var}(\mathbf{y}) = \sigma^2\mathbf{W}$ where $\mathbf{W}$ is a specified $n \times n$ positive definite matrix, that $\text{var}(\mathbf{u}) = \sigma^2\mathbf{H}$ and $\text{cov}(\mathbf{y}, \mathbf{u}) = \sigma^2\mathbf{K}$, where $\mathbf{H}$ and $\mathbf{K}$ are specified $s \times s$ and $n \times s$ matrices. A special case of this last model is the prediction-extended Gauss–Markov model, for which $\mathbf{W} = \mathbf{I}_n$, $\mathbf{H} = \mathbf{I}_s$, and $\mathbf{K} = \mathbf{0}$.

Point prediction for prediction-extended models is considered in Chap. 13, and interval prediction for such models under the assumption that the joint distribution of $\mathbf{y}$ and $\mathbf{u}$ is multivariate normal is taken up in Chap. 15. Some further results on inference for predictands under prediction-extended models are described in Chap. 17.

5.4 Shorthand Model Notation

Throughout the remainder of this book, we use a (modified) shorthand notational convention common to the literature on linear model theory. This convention is to represent a linear model

$$\mathbf{y} = \mathbf{X}\boldsymbol{\beta} + \mathbf{e}$$

for which the variance–covariance matrix is unspecified by the duplet $\{\mathbf{y}, \mathbf{X}\boldsymbol{\beta}\}$, and to add the variance–covariance matrix as a third component when it is specified (either partially or fully). Thus the Gauss–Markov model with response vector $\mathbf{y}$, model matrix $\mathbf{X}$, and variance–covariance matrix $\sigma^2\mathbf{I}$ is represented by the triplet $\{\mathbf{y}, \mathbf{X}\boldsymbol{\beta}, \sigma^2\mathbf{I}\}$, and the Aitken model with the same response vector and model matrix is represented by $\{\mathbf{y}, \mathbf{X}\boldsymbol{\beta}, \sigma^2\mathbf{W}\}$. The general mixed linear model may be represented by $\{\mathbf{y}, \mathbf{X}\boldsymbol{\beta}, \mathbf{V}(\boldsymbol{\theta})\}$. Prediction-extended models are represented by a string of quantities of which the first is the vector of responses $\mathbf{y}$ and unobserved **u**, the second is the vector of expectations, and the third is the variance–covariance matrix. Thus a prediction-extended Gauss–Markov model is represented by $\left\{\begin{pmatrix}\mathbf{y}\\ \mathbf{u}\end{pmatrix}, \begin{pmatrix}\mathbf{X}\boldsymbol{\beta}\\ \mathbf{0}\end{pmatrix}, \sigma^2\begin{pmatrix}\mathbf{I} & \mathbf{0}\\ \mathbf{0} & \mathbf{I}\end{pmatrix}\right\}$; a prediction-extended Aitken model is represented by $\left\{\begin{pmatrix}\mathbf{y}\\ \mathbf{u}\end{pmatrix}, \begin{pmatrix}\mathbf{X}\boldsymbol{\beta}\\ \mathbf{0}\end{pmatrix}, \sigma^2\begin{pmatrix}\mathbf{W} & \mathbf{K}\\ \mathbf{K}^T & \mathbf{H}\end{pmatrix}\right\}$; and a prediction-extended general mixed linear model is represented by $\left\{\begin{pmatrix}\mathbf{y}\\ \mathbf{u}\end{pmatrix}, \begin{pmatrix}\mathbf{X}\boldsymbol{\beta}\\ \mathbf{0}\end{pmatrix}, \begin{pmatrix}\mathbf{V}_{yy}(\boldsymbol{\theta}) & \mathbf{V}_{yu}(\boldsymbol{\theta})\\ \mathbf{V}_{yu}(\boldsymbol{\theta})^T & \mathbf{V}_{uu}(\boldsymbol{\theta})\end{pmatrix}\right\}$. All of the preceding are representations of unconstrained models. If the model is constrained, a colon is placed at the end of the list of other symbols and the constraints are listed after the colon. Thus, for example, the Gauss–Markov model $\{\mathbf{y}, \mathbf{X}\boldsymbol{\beta}, \sigma^2\mathbf{I}\}$ with constraints $\mathbf{A}\boldsymbol{\beta} = \mathbf{h}$ is represented by $\{\mathbf{y}, \mathbf{X}\boldsymbol{\beta}, \sigma^2\mathbf{I} : \mathbf{A}\boldsymbol{\beta} = \mathbf{h}\}$.

The shorthand notation does not fully describe some features of a linear model that, depending on the context, may be important, such as multivariate normality of the response vector or positive definiteness of the variance–covariance matrix. In such cases, the additional features are specified using words. For example, we may refer to a constrained Aitken model for which the distribution of **y** is multivariate normal and **W** is positive definite as the normal positive definite constrained Aitken model $\{\mathbf{y}, \mathbf{X}\boldsymbol{\beta}, \sigma^2\mathbf{W} : \mathbf{A}\boldsymbol{\beta} = \mathbf{h}\}$.

5.5 Exercises

1. Give as concise a representation of the model matrix **X** as possible for a two-way main effects model when the data are balanced, and specialize further to the case $r = 1$.
2. Give a concise representation of the model matrix **X** for the cell-means representation of the two-way model with interaction. Is this **X** full-rank? Specialize to the case of balanced data.
3. Give the model matrix **X** for a three-way main effects model in which each factor has two levels and each cell of the $2 \times 2 \times 2$ layout has exactly one observation, i.e., for the model

$$y_{ijk} = \mu + \alpha_i + \gamma_j + \delta_k + e_{ijk} \qquad (i = 1, 2;\ j = 1, 2;\ k = 1, 2).$$

4. Give a concise representation of the model matrix **X** for the two-factor nested model, and specialize it to the case of balanced data.
5. Give a concise representation of the model matrix **X** for the balanced two-factor partially crossed model

$$y_{ijk} = \mu + \alpha_i - \gamma_j + e_{ijk} \quad (i \neq j = 1, \ldots, q;\ k = 1, \ldots, r).$$

6. Give a concise representation of the model matrix $\mathbf{X}$ for a one-factor, factor-specific-slope, analysis-of-covariance model

$$y_{ij} = \mu + \alpha_i + \gamma_i x_{ij} + e_{ij} \qquad (i = 1, \ldots, q;\ j = 1, \ldots, n_i).$$

7. Determine the variance–covariance matrix for each of the following two-factor crossed mixed effects models:
   (a) $y_{ij} = \mu + \alpha_i + b_j + d_{ijk} \quad (i = 1, \ldots, q;\ j = 1, \ldots, m;\ k = 1, \ldots, n_{ij})$, where $\mathrm{E}(b_j) = 0$ and $\mathrm{E}(d_{ijk}) = 0$ for all i, j, k, the b_j's and d_{ijk}'s are all uncorrelated, $\mathrm{var}(b_j) = \sigma_b^2$ for all j, and $\mathrm{var}(d_{ijk}) = \sigma^2$ for all i, j, k; with parameter space $\{\mu, \sigma_b^2, \sigma^2 : \mu \in \mathbb{R},\ \sigma_b^2 > 0,\ \sigma^2 > 0\}$.
   (b) $y_{ij} = \mu + a_i + \gamma_j + d_{ijk} \quad (i = 1, \ldots, q;\ j = 1, \ldots, m;\ k = 1, \ldots, n_{ij})$, where $\mathrm{E}(a_i) = 0$ and $\mathrm{E}(d_{ijk}) = 0$ for all i, j, k, the a_i's and d_{ijk}'s are all uncorrelated, $\mathrm{var}(a_i) = \sigma_a^2$ for all i, and $\mathrm{var}(d_{ijk}) = \sigma^2$ for all i, j, k; with parameter space $\{\mu, \sigma_a^2, \sigma^2 : \mu \in \mathbb{R},\ \sigma_a^2 > 0,\ \sigma^2 > 0\}$.
   (c) $y_{ij} = \mu + \alpha_i + b_j + c_{ij} + d_{ijk} \quad (i = 1, \ldots, q;\ j = 1, \ldots, m;\ k = 1, \ldots, n_{ij})$, where $\mathrm{E}(b_j) = \mathrm{E}(c_{ij}) = 0$ and $\mathrm{E}(d_{ijk}) = 0$ for all i, j, k, the b_j's, c_{ij}'s, and d_{ijk}'s are all uncorrelated, $\mathrm{var}(b_j) = \sigma_b^2$ for all j, $\mathrm{var}(c_{ij}) = \sigma_c^2$ for all i and j, and $\mathrm{var}(d_{ijk}) = \sigma^2$ for all i, j, k; with parameter space $\{\mu, \sigma_b^2, \sigma_c^2, \sigma^2 : \mu \in \mathbb{R},\ \sigma_b^2 > 0,\ \sigma_c^2 > 0,\ \sigma^2 > 0\}$.
   (d) $y_{ij} = \mu + a_i + \gamma_j + c_{ij} + d_{ijk} \quad (i = 1, \ldots, q;\ j = 1, \ldots, m;\ k = 1, \ldots, n_{ij})$, where $\mathrm{E}(a_i) = \mathrm{E}(c_{ij}) = 0$ and $\mathrm{E}(d_{ijk}) = 0$ for all i, j, k, the a_i's, c_{ij}'s, and d_{ijk}'s are all uncorrelated, $\mathrm{var}(a_i) = \sigma_a^2$ for all i, $\mathrm{var}(c_{ij}) = \sigma_c^2$ for all i and j, and $\mathrm{var}(d_{ijk}) = \sigma^2$ for all i, j, k with parameter space $\{\mu, \sigma_a^2, \sigma_c^2, \sigma^2 : \mu \in \mathbb{R},\ \sigma_a^2 > 0,\ \sigma_c^2 > 0,\ \sigma^2 > 0\}$.
   (e) $y_{ij} = \mu + a_i + b_j + c_{ij} + d_{ijk} \quad (i = 1, \ldots, q;\ j = 1, \ldots, m;\ k = 1, \ldots, n_{ij})$, where $\mathrm{E}(a_i) = \mathrm{E}(b_j) = \mathrm{E}(c_{ij}) = 0$ and $\mathrm{E}(d_{ijk}) = 0$ for all i, j, k, the a_i's, b_j's, c_{ij}'s, and d_{ijk}'s are all uncorrelated, $\mathrm{var}(a_i) = \sigma_a^2$ for all i, $\mathrm{var}(b_j) = \sigma_b^2$ for all j, $\mathrm{var}(c_{ij}) = \sigma_c^2$ for all i and j, and $\mathrm{var}(d_{ijk}) = \sigma^2$ for all i, j, k; with parameter space $\{\mu, \sigma_a^2, \sigma_b^2, \sigma_c^2, \sigma^2 : \mu \in \mathbb{R},\ \sigma_a^2 > 0,\ \sigma_b^2 > 0,\ \sigma_c^2 > 0,\ \sigma^2 > 0\}$.
8. Verify that the stationary autoregressive model of order one has the variance–covariance matrix given by (5.13).
9. Determine the variance–covariance matrix for $\mathbf{y} = (y_1, y_2, \ldots, y_n)^T$ where

$$y_i = \mu + \sum_{j=1}^{i} u_j \qquad (i = 1, \ldots, n),$$

and the u_j's are uncorrelated random variables with mean 0 and variance $\sigma^2 > 0$.
10. Determine the variance–covariance matrix for $\mathbf{y} = (y_1, y_2, y_3)^T$ where

$$y_i = \mu_i + \phi_{i-1}(y_{i-1} - \mu_{i-1}) + u_i \qquad (i = 1, 2, 3),$$

$y_0 \equiv \mu_0$, and u_1, u_2, u_3 are uncorrelated random variables with common mean 0 and $\text{var}(u_i) = \sigma_i^2$ for $i = 1, 2, 3$; the parameter space is $\{\mu_0, \mu_1, \mu_2, \mu_3, \phi_0, \phi_1, \phi_2, \sigma_1^2, \sigma_2^2, \sigma_3^2 : \mu_i \in \mathbb{R}$ for all i, $\phi_i \in \mathbb{R}$ for all i, and $\sigma_i^2 > 0$ for all $i\}$. This model is an extension of the first-order stationary autoregressive model (for three observations) called the *first-order antedependence model*.

References

Aitken, A. C. (1935). On least squares and linear combination of observations. *Proceedings of the Royal Society of Edinburgh, 55*, 42–48.

Christensen, R. (2011). *Plane answers to complex questions* (4th ed.) New York: Springer.

Gauss, C. F. (1855). *Methode des moindres Carres*. Paris: Mallet-Bachelier.

Graybill, F. A. (1976). *Theory and application of the linear model*. Belmont, CA: Wadsworth.

Harville, D. A. (2018). *Linear models and the relevant distributions and matrix algebra*. Boca Raton, FL: CRC.

Hinkelmann, K., & Kempthorne, O. (1994). *Design and analysis of experiments*. New York: Wiley.

Khuri, A. I. (2010). *Linear model methodology*. Boca Raton, FL: Chapman & Hall/CRC Press.

Myers, R. H. & Milton, J. S. (1991). *A first course in the theory of linear statistical models*. Boston: PWS-KENT.

Plackett, R. L. (1949). A historical note on the method of least squares. *Biometrika, 36*, 458–460.

Ravishanker, N. & Dey, D. K. (2002). *A first course in linear model theory*. Boca Raton, FL: Chapman & Hall/CRC Press.

Rencher, A. C. (2000). *Linear models in statistics*. New York: Wiley.

Searle, S. R. (1971). *Linear models*. New York: Wiley.

Estimability

6

In this chapter, it is not yet our objective to learn how to estimate the parameters of a linear model (or functions thereof) that are of interest, but rather to learn whether it is even *possible* to "sensibly estimate" those parameters or functions. Precisely what is meant by "sensibly estimate"? Several definitions are possible, but in this book the phrase is synonymous with "estimate unbiasedly." Parameters, or functions thereof, that can be estimated unbiasedly from the available data (**X** and **y**) are said to be *estimable*, with the remainder designated as *nonestimable*. In this chapter we focus exclusively on the estimability of linear functions of $\boldsymbol{\beta}$, i.e., functions of the form $\mathbf{c}^T\boldsymbol{\beta}$ where $\mathbf{c}$ is a specified, nonrandom p-vector. Later it will be shown that the estimability of a linear function of $\boldsymbol{\beta}$ is necessary and sufficient for a "best" (minimum variance) linear unbiased estimator of that function to exist, and we will eventually derive that estimator. We defer the issue of estimability for variance–covariance parameters (or functions thereof) to Chap. 16.

Loosely speaking, linear functions of $\boldsymbol{\beta}$ can be nonestimable under a model $\{\mathbf{y}, \mathbf{X}\boldsymbol{\beta}\}$ for two reasons: (1) the data in $\mathbf{y}$ are too limited to support the estimation of those functions; (2) the model's mean structure, $\mathbf{X}\boldsymbol{\beta}$, is overparameterized, i.e., more parameters than necessary are used to describe the means of the observations. A toy example will make the point. Consider a main effects model for data in a 2×2 layout in which only the upper left and lower right cells are occupied by two and three observations respectively (as depicted in the figure below, where each observation is represented by an asterisk, $*$). Intuitively, it seems that the data would support the estimation of the parametric functions $\mu + \alpha_1 + \gamma_1$ and $\mu + \alpha_2 + \gamma_2$ by the obvious estimators $\bar{y}_{11\cdot}$ and $\bar{y}_{22\cdot}$ (the sample means of the observations in the upper left and lower right cells), respectively; but a sensible estimator of $\mu + \alpha_1 + \gamma_2$ or $\mu + \alpha_2 + \gamma_1$ does not come to mind. Furthermore, even if additional observations were taken in the two empty cells, so that each of the four functions $\mu + \alpha_1 + \gamma_1$, $\mu + \alpha_2 + \gamma_2$, $\mu + \alpha_1 + \gamma_2$ and $\mu + \alpha_2 + \gamma_1$ could be estimated sensibly by some function of the observations in the corresponding

D. L. Zimmerman, *Linear Model Theory*,
https://doi.org/10.1007/978-3-030-52063-2_6

cell, $\boldsymbol{\beta}$ consists of five individual parameters ($\mu, \alpha_1, \alpha_2, \gamma_1$, and γ_2) and it is not immediately obvious whether any of them can be estimated sensibly.

		Factor B	
		$j = 1$	$j = 2$
Factor A	$i = 1$	$**$	
	$i = 2$		$***$

The estimability of a linear function of $\boldsymbol{\beta}$ may be affected by whether the model is constrained or unconstrained. This chapter is divided into two sections according to this dichotomy.

6.1 Estimability Under an Unconstrained Model

We begin with two definitions.

Definition 6.1.1 An estimator $\text{t}(\mathbf{y})$ of a linear function $\mathbf{c}^T\boldsymbol{\beta}$ associated with the model $\{\mathbf{y}, \mathbf{X}\boldsymbol{\beta}\}$ is said to be *linear* if $\text{t}(\mathbf{y}) \equiv t_0 + \mathbf{t}^T\mathbf{y}$ for some constant t_0 and some n-vector $\mathbf{t}$ of constants.

Definition 6.1.2 An estimator $\text{t}(\mathbf{y})$ of a linear function $\mathbf{c}^T\boldsymbol{\beta}$ associated with the model $\{\mathbf{y}, \mathbf{X}\boldsymbol{\beta}\}$ is said to be *unbiased* under that model if $\text{E}[\text{t}(\mathbf{y})] = \mathbf{c}^T\boldsymbol{\beta}$ for all $\boldsymbol{\beta} \in \mathbb{R}^p$.

Observe that, under the model $\{\mathbf{y}, \mathbf{X}\boldsymbol{\beta}\}$,

$$\text{E}(t_0 + \mathbf{t}^T\mathbf{y}) = t_0 + \mathbf{t}^T\mathbf{X}\boldsymbol{\beta} \quad \text{for all } \boldsymbol{\beta} \in \mathbb{R}^p, \tag{6.1}$$

regardless of the (unspecified) values of the elements of the variance–covariance matrix.

Theorem 6.1.1 *A linear estimator $t_0+\mathbf{t}^T\mathbf{y}$ of a linear function $\mathbf{c}^T\boldsymbol{\beta}$ associated with the model $\{\mathbf{y}, \mathbf{X}\boldsymbol{\beta}\}$ is unbiased under any unconstrained model with model matrix $\mathbf{X}$ if and only if $t_0 = 0$ and $\mathbf{t}^T\mathbf{X} = \mathbf{c}^T$.*

Proof The sufficiency of the given conditions on t_0 and $\mathbf{t}$ follows directly upon applying them to (6.1). For the necessity of those conditions, suppose that $t_0 + \mathbf{t}^T\mathbf{y}$ is unbiased for $\mathbf{c}^T\boldsymbol{\beta}$ under the specified model. Then it follows from (6.1) that $t_0 + \mathbf{t}^T\mathbf{X}\boldsymbol{\beta} = \mathbf{c}^T\boldsymbol{\beta}$ for all $\boldsymbol{\beta}$, and in particular for $\boldsymbol{\beta} = \mathbf{0}$ (implying that $t_0 = 0$), and for $\boldsymbol{\beta} = \mathbf{u}_1^{(p)}, \ldots, \boldsymbol{\beta} = \mathbf{u}_p^{(p)}$ (implying that $\mathbf{t}^T\mathbf{X} = \mathbf{c}^T$). □

Definition 6.1.3 A function $\mathbf{c}^T\boldsymbol{\beta}$ is said to be *estimable* under the model $\{\mathbf{y}, \mathbf{X}\boldsymbol{\beta}\}$ if a linear estimator exists that estimates it unbiasedly under that model. Otherwise, the function is said to be *nonestimable* (under that model).

Definition 6.1.3 of estimability applies to functions associated with constrained models as well as unconstrained models. It should be noted, however, that regardless of whether the model is constrained or unconstrained, this definition is slightly more restrictive than the definition of parameter estimability given in some mathematical statistics books and courses. In those, estimability is usually defined without a restriction to linear estimators and the definition here typically would be called *linear estimability*. Furthermore, some authors (e.g., Lehmann and Casella, 1998) refer to a parametric function that has an unbiased estimator as "U-estimable" (where "U" is short for "Unbiased") in order to avoid giving the false impression that functions that cannot be estimated unbiasedly cannot be estimated sensibly in some other sense (e.g., consistently).

Theorem 6.1.2 *A function* $\mathbf{c}^T\boldsymbol{\beta}$ *is estimable under the model* $\{\mathbf{y}, \mathbf{X}\boldsymbol{\beta}\}$ *if and only if* $\mathbf{c}^T \in \mathcal{R}(\mathbf{X})$.

Proof If $\mathbf{c}^T \in \mathcal{R}(\mathbf{X})$, then an n-vector $\mathbf{a}$ exists such that $\mathbf{a}^T\mathbf{X} = \mathbf{c}^T$, implying further that $\mathrm{E}(\mathbf{a}^T\mathbf{y}) = \mathbf{a}^T\mathbf{X}\boldsymbol{\beta} = \mathbf{c}^T\boldsymbol{\beta}$ for all $\boldsymbol{\beta}$. Thus, $\mathbf{c}^T\boldsymbol{\beta}$ is estimable under the specified model. Conversely, if $\mathbf{c}^T\boldsymbol{\beta}$ is estimable under the specified model, then by definition an unbiased estimator of $\mathbf{c}^T\boldsymbol{\beta}$ of the form $t_0+\mathbf{t}^T\mathbf{y}$ exists, where $\mathbf{t}^T\mathbf{X} = \mathbf{c}^T$. Thus, $\mathbf{c}^T \in \mathcal{R}(\mathbf{X})$. □

Theorem 6.1.2 has several corollaries, listed next, that can be more helpful than the theorem itself for actually determining which functions are estimable under a given model. Short proofs of the corollaries may be constructed easily, but are saved for the exercises. One of the corollaries refers to functions that are linearly independent as well as estimable, so we define those first.

Definition 6.1.4 Functions $\mathbf{c}_1^T\boldsymbol{\beta},\ \mathbf{c}_2^T\boldsymbol{\beta},\ \ldots,\ \mathbf{c}_k^T\boldsymbol{\beta}$ are said to be *linearly independent* if $\{\mathbf{c}_1, \mathbf{c}_2, \ldots, \mathbf{c}_k\}$ are linearly independent vectors.

Corollary 6.1.2.1 *If* $\mathbf{c}_1^T\boldsymbol{\beta},\ \mathbf{c}_2^T\boldsymbol{\beta},\ \ldots,\ \mathbf{c}_k^T\boldsymbol{\beta}$ *are estimable under the model* $\{\mathbf{y}, \mathbf{X}\boldsymbol{\beta}\}$*, then so is any linear combination of them; that is, the set of estimable functions under a given unconstrained model is a linear space.*

Corollary 6.1.2.2 *The elements of* $\mathbf{X}\boldsymbol{\beta}$ *are estimable under the model* $\{\mathbf{y}, \mathbf{X}\boldsymbol{\beta}\}$*; in fact, those elements span the space of estimable functions for that model.*

Corollary 6.1.2.3 *A set of* p^* *[*$=$ *rank*$(\mathbf{X})$*] linearly independent estimable functions under the model* $\{\mathbf{y}, \mathbf{X}\boldsymbol{\beta}\}$ *exists such that any estimable function under that model can be written as a linear combination of functions in this set.*

The set of p^* functions defined by Corollary 6.1.2.3 is called a *basis* for the space of estimable functions for the model $\{\mathbf{y}, \mathbf{X}\boldsymbol{\beta}\}$. Corollary 6.1.2.3 implies that if we can find such a basis, its dimension will equal rank$(\mathbf{X})$. This is sometimes a useful way to determine the rank of $\mathbf{X}$.

Corollary 6.1.2.4 $\mathbf{c}^T\boldsymbol{\beta}$ *is estimable for* every *vector* $\mathbf{c}$ *under the model* $\{\mathbf{y}, \mathbf{X}\boldsymbol{\beta}\}$ *if and only if* $p^* = p$.

Corollary 6.1.2.5 $\mathbf{c}^T\boldsymbol{\beta}$ *is estimable under the model* $\{\mathbf{y}, \mathbf{X}\boldsymbol{\beta}\}$ *if and only if* $\mathbf{c} \perp \mathcal{N}(\mathbf{X})$.

Corollary 6.1.2.6 $\mathbf{c}^T\boldsymbol{\beta}$ *is estimable under the model* $\{\mathbf{y}, \mathbf{X}\boldsymbol{\beta}\}$ *if and only if* $\mathbf{c}^T \in \mathcal{R}(\mathbf{X}^T\mathbf{X})$.

A proof of Corollary 6.1.2.6 can be constructed using the following theorem, which is of interest in its own right.

Theorem 6.1.3 *For any matrix* $\mathbf{X}$, $\mathcal{R}(\mathbf{X}^T\mathbf{X}) = \mathcal{R}(\mathbf{X})$.

Proof $\mathbf{X}^T\mathbf{X} = (\mathbf{X}^T)\mathbf{X}$, and by Theorem 3.3.3a $\mathbf{X} = [\mathbf{X}(\mathbf{X}^T\mathbf{X})^-]\mathbf{X}^T\mathbf{X}$. Thus, by Theorem 2.6.1 each of $\mathcal{R}(\mathbf{X}^T\mathbf{X})$ and $\mathcal{R}(\mathbf{X})$ is contained in the other; hence, they are equal. □

We now give several examples of estimable functions and unbiased estimators of those functions, in the context of various unconstrained regression and classificatory models. Because the error structures of the models are unimportant insofar as estimability of linear functions of $\boldsymbol{\beta}$ is concerned, the error structures are left unspecified.

Example 6.1-1. Estimability Under the Polynomial Regression Model in a Single Variable

Consider the complete polynomial regression model in a single variable,

$$y_i = \beta_1 + \beta_2 x_i + \beta_3 x_i{}^2 + \cdots + \beta_p x_i{}^{p-1} + e_i \quad (i = 1, \ldots, n).$$

Under what conditions are all of the parameters $\beta_1, \ldots, \beta_p$ estimable? By Corollary 6.1.2.4, an equivalent question is, under what conditions does $\text{rank}(\mathbf{X}) = p$? To answer this question, let q denote the number of distinct values of x among $x_1, \ldots, x_n$. If $q < p$, then clearly $\text{rank}(\mathbf{X}) < p$, so a necessary condition for all the parameters to be estimable is $q \geq p$. But is this condition sufficient as well?

Suppose that $q \geq p$, in which case we can write $\mathbf{X}$, possibly after rearranging its rows, as

$$\mathbf{X} = \begin{bmatrix} 1 & x_{\boxed{1}} & \cdots & x_{\boxed{1}}^{p-1} \\ 1 & x_{\boxed{2}} & \cdots & x_{\boxed{2}}^{p-1} \\ \vdots & \vdots & & \vdots \\ 1 & x_{\boxed{p}} & \cdots & x_{\boxed{p}}^{p-1} \\ 1 & x_{\boxed{p+1}} & \cdots & x_{\boxed{p+1}}^{p-1} \\ \vdots & \vdots & & \vdots \\ 1 & x_{\boxed{q}} & \cdots & x_{\boxed{q}}^{p-1} \\ 1 & x_{q+1} & \cdots & x_{q+1}^{p-1} \\ \vdots & \vdots & & \vdots \\ 1 & x_n & \cdots & x_n^{p-1} \end{bmatrix} = \begin{bmatrix} \mathbf{A} \\ \mathbf{B} \end{bmatrix},$$

where $x_{\boxed{1}}, x_{\boxed{2}}, \ldots, x_{\boxed{q}}$ represent the q distinct values of x among the n observational units and $\mathbf{A}$ is the submatrix of $\mathbf{X}$ consisting of its first p rows. In fact, $\mathbf{A}$ is a special type of matrix called a *Vandermonde matrix*, for which the determinant is

$$|\mathbf{A}| = \prod_{i>j, i=1,\ldots,p} (x_{\boxed{i}} - x_{\boxed{j}})$$

[see Harville (1977, Section 13.6) for details]. Because $x_{\boxed{1}}, x_{\boxed{2}}, \ldots, x_{\boxed{p}}$ are distinct, $|\mathbf{A}| \neq 0$, which implies (by Theorem 2.11.2) that $\mathbf{A}$ is nonsingular, so that $\text{rank}(\mathbf{X}) = \text{rank}(\mathbf{A}) = p$. Thus, a necessary and sufficient condition for all of $\beta_1, \ldots, \beta_p$ to be estimable is that the number of distinct values of x is larger than the order of the polynomial. For the simple linear regression model, for example, at least two distinct values of x are required, as was determined previously from other considerations (Sect. 5.2). For a quadratic regression model, at least three distinct values of x are required, and so on. ■

Example 6.1-2. Estimability Under the One-Factor Model

Which functions are estimable under the one-factor model? To answer this question, we use Theorem 6.1.2 and attempt to find a simple characterization of $\mathcal{R}(\mathbf{X})$. From (5.4), we see that the matrix of distinct rows of $\mathbf{X}$ is $(\mathbf{1}_q, \mathbf{I}_q)$, which implies that $\mathcal{R}(\mathbf{X}) = \mathcal{R}(\mathbf{1}_q, \mathbf{I}_q)$. An arbitrary element of $\mathcal{R}(\mathbf{X})$ may therefore be expressed as $\mathbf{a}^T(\mathbf{1}_q, \mathbf{I}_q)$, or equivalently as $(a_., a_1, a_2, \ldots, a_q)$ where $\mathbf{a}^T = (a_1, a_2, \ldots, a_q)$ is a vector in $\mathbb{R}^q$. Observe that although $\mathbf{a}$ is unrestricted, $\mathbf{a}^T(\mathbf{1}_q, \mathbf{I}_q)$ is not: its first element must equal the sum of its remaining elements. Thus, by Theorem 6.1.2, $\mathbf{c}^T\boldsymbol{\beta}$ is estimable if and only if the coefficient on μ (the first element of $\mathbf{c}^T$) is equal to the sum of the coefficients on the α_i's (the remaining elements of $\mathbf{c}^T$). Thus, for example, $\mu + \alpha_i$ and $\alpha_i - \alpha_{i'}$ are estimable for all i and i', but μ and α_i are nonestimable (for any i). Furthermore, any contrast $\sum_{i=1}^q d_i\alpha_i$, where $\sum_{i=1}^q d_i = 0$, is estimable. In the following table we list these estimable functions and two (among many other possible) unbiased estimators of each one; the reader should verify that the estimators are indeed unbiased, either by directly taking expectations or by verifying the conditions of Theorem 6.1.1.

A basis for the space of estimable functions is $\{\mu+\alpha_1, \alpha_2-\alpha_1, \alpha_3-\alpha_1, \ldots, \alpha_q-\alpha_1\}$. How do we know this is such a basis? Each function in the list is estimable; the functions are linearly independent (because each includes a parameter not present in any other function); and each element of $\mathbf{X}\boldsymbol{\beta}$ may be obtained as a linear combination of functions in the list (implying that these functions span the space of estimable functions). Because there are q functions in this basis, $\text{rank}(\mathbf{X}) = q$ by Corollary 6.1.2.3. Of course, in this case it is easy to see directly that $\text{rank}(\mathbf{X}) = \text{rank}(\mathbf{1}_q, \mathbf{I}_q) = q$, so it is not necessary to find $\text{rank}(\mathbf{X})$ in this way.

Function	Unbiased estimators
$\mu + \alpha_i$	y_{i1}, $\bar{y}_{i\cdot}$
$\sum_{i=1}^q d_i\alpha_i$, where $\sum_{i=1}^q d_i = 0$	$\sum_{i=1}^q d_i y_{i1}$, $\sum_{i=1}^q d_i \bar{y}_{i\cdot}$

■

Example 6.1-3. Estimability Under the Two-Way Main Effects Model with No Empty Cells

Consider the two-way main effects model, and suppose that no cells are empty. From (5.7), the matrix of distinct rows of $\mathbf{X}$ is $(\mathbf{1}_{qm}, \mathbf{I}_q \otimes \mathbf{1}_m, \mathbf{1}_q \otimes \mathbf{I}_m)$, a matrix whose rows correspond, one by one, to the cells of the two-way layout. Any element of $\mathcal{R}(\mathbf{X})$ may therefore be expressed as $\mathbf{a}^T(\mathbf{1}_{qm}, \mathbf{I}_q \otimes \mathbf{1}_m, \mathbf{1}_q \otimes \mathbf{I}_m)$, or equivalently as $(a_{\cdot\cdot}, a_{1\cdot}, a_{2\cdot}, \ldots, a_{q\cdot}, a_{\cdot 1}, a_{\cdot 2}, \ldots, a_{\cdot m})$, where $\mathbf{a}^T = (a_{11}, a_{12}, \ldots, a_{1m}, a_{21}, \ldots, a_{qm})$. (We have used double subscripts to index the elements of $\mathbf{a}$ here so that each such element can be identified with a particular cell of the two-way layout.) Thus, $\mathbf{c}^T\boldsymbol{\beta}$ is estimable if and only if the coefficients on the α_i's sum to the coefficient on μ and the coefficients on the γ_j's do likewise. For example, $\mu + \alpha_i + \gamma_j$ is estimable for every i and j, as is any Factor A contrast $\sum_{i=1}^q d_i\alpha_i$ (where $\sum_{i=1}^q d_i = 0$) and any Factor B contrast $\sum_{j=1}^m g_j\gamma_j$ (where $\sum_{j=1}^m g_j = 0$). The following table lists these estimable functions and two (among many other possible) unbiased estimators of each one; the reader should verify that the estimators are indeed unbiased. A basis for the space of estimable functions is $\{\mu+\alpha_1+\gamma_1, \alpha_2-\alpha_1, \alpha_3-\alpha_1, \ldots, \alpha_q-\alpha_1, \gamma_2-\gamma_1, \gamma_3-\gamma_1, \ldots, \gamma_m-\gamma_1\}$. None of the individual parameters μ, α_i, or γ_j are estimable, and neither are any of their pairwise sums $\mu+\alpha_i$, $\mu+\gamma_j$, or $\alpha_i+\gamma_j$. By Corollary 6.1.2.3, $\text{rank}(\mathbf{X}) = q+m-1$.

Function	Unbiased estimators
$\mu + \alpha_i + \gamma_j$	y_{ij1}, $\bar{y}_{ij\cdot}$
$\sum_{i=1}^q d_i \alpha_i$, where $\sum_{i=1}^q d_i = 0$	$\sum_{i=1}^q d_i y_{i11}$, $\sum_{i=1}^q d_i \bar{y}_{i\cdot\cdot}$
$\sum_{j=1}^m g_j \gamma_j$, where $\sum_{j=1}^m g_j = 0$	$\sum_{j=1}^m g_j y_{1j1}$, $\sum_{j=1}^m g_j \bar{y}_{\cdot j\cdot}$

■

Example 6.1-4. Estimability Under the Two-Way Main Effects Model with One or More Empty Cells

In this scenario, which functions are estimable depends on which cells are empty. Furthermore, it turns out that using Corollaries 6.1.2.1 and 6.1.2.2 in tandem to ascertain whether certain functions of interest are estimable is usually easier than trying to directly characterize $\mathcal{R}(\mathbf{X})$. To illustrate, consider the following two 3×3 layouts, both of which have five observations.

*	*	*
*		
		*

*	*	
*	*	
		*

In the layout on the left, $\mu+\alpha_i+\gamma_j$ is estimable for $(i, j) \in \{(1, 1), (1, 2), (1, 3), (2, 1), (3, 3)\}$ by Corollary 6.1.2.2, implying further (by Corollary 6.1.2.1) that $\mu + \alpha_2 + \gamma_2 = (\mu + \alpha_1 + \gamma_2) + (\mu + \alpha_2 + \gamma_1) - (\mu + \alpha_1 + \gamma_1)$ is estimable, despite cell (2,2) being empty. In a similar fashion it can be demonstrated that $\mu + \alpha_i + \gamma_j$ is estimable for all i and j. From this it follows that the space of all estimable functions for this layout is identical to what it is when no cells are empty.

The situation is considerably different in the layout on the right, however. Corollary 6.1.2.2 tells us here that $\mu + \alpha_i + \gamma_j$ is estimable for $(i, j) \in \{(1, 1), (1, 2), (2, 1), (2, 2), (3, 3)\}$. But unlike before, Corollary 6.1.2.1 does not establish the estimability of additional functions of the form $\mu + \alpha_i + \gamma_j$. Instead, because $\mu + \alpha_2 + \gamma_2$ is a linear combination of $\{\mu + \alpha_i + \gamma_j : (i, j) = (1, 1), (1, 2), (2, 1)\}$, a basis for the space of estimable functions is $\{\mu+\alpha_1+\gamma_1, \mu+\alpha_1+\gamma_2, \mu+\alpha_2+\gamma_1, \mu+\alpha_3+\gamma_3\}$. Note that $\text{rank}(\mathbf{X}) = 4$ for this layout, compared to $\text{rank}(\mathbf{X}) = 5$ for the layout on the left.

Similar ideas apply to the more general situation of q rows and m columns. Let us define

$$\delta_{ij} = \begin{cases} 1, & \text{if } n_{ij} > 0, \\ 0, & \text{otherwise.} \end{cases}$$

If $\delta_{ij} = 1$, then $\mu + \alpha_i + \gamma_j$ is estimable by Corollary 6.1.2.2. Furthermore, because $\mu + \alpha_{i'} + \gamma_{j'} = (\mu + \alpha_{i'} + \gamma_j) + (\mu + \alpha_i + \gamma_{j'}) - (\mu + \alpha_i + \gamma_j)$, it follows from Corollary 6.1.2.1 that a sufficient condition for $\mu + \alpha_{i'} + \gamma_{j'}$ to be estimable is $\delta_{ij} = \delta_{i'j} = \delta_{ij'} = 1$; it is not necessary that $\delta_{i'j'} = 1$.

This suggests the following procedure, which is henceforth called the "3 + e" procedure, for determining which functions of the form $\mu + \alpha_i + \gamma_j$ are estimable. First, draw the two-way table of cells and put an asterisk in each cell for which $n_{ij} > 0$. Then examine the 2×2 subarray formed as the Cartesian product of any two rows and any two columns, and put an "e" (for "estimable") in the unoccupied cell of the subarray if three of the four cells are occupied by an asterisk. Repeat this for every 2×2 subarray, putting an "e" in the unoccupied cell if three of the four cells are occupied by either an asterisk or an "e." Continue the process until no 2×2 subarray has only three of its cells so occupied. Upon completion of this procedure, those cell means $\mu + \alpha_i + \gamma_j$ for which the corresponding cell (i, j) is occupied by an asterisk or an "e" are estimable. Furthermore, the collection of all such cell means will span the space of estimable functions.

The three examples that follow can be used to illustrate the $3 + e$ procedure (the reader should attempt to carry out the procedure on these):

*		*	
	*		*
	*		
*	*	*	
		*	*

*	*		
*		*	
*			*
	*	*	
	*		*
		*	*

*			*		
	*			*	
*		*	*		
				*	*
		*			

It turns out that the procedure puts an asterisk or an "e" in every cell of the first two layouts displayed above, but not in the third layout. Those readers familiar with experimental design may recognize the second layout as a balanced incomplete block design (BIBD); specifically, it is a BIBD corresponding to a situation in which four treatments are applied to units that are grouped into blocks of size two. The estimability of all cell means (and hence of all Factor A and Factor B contrasts) is one nice property of a BIBD, but not the only one; another is revealed in an exercise in Chap. 7. Furthermore, the third layout can be rearranged, by interchanging the second and fourth columns and then interchanging the second and fifth rows, to

*	*	e			
e	e	*			
*	*	*			
			e	*	*
			*	*	e

which reveals that the original two-way table consists of two distinct two-way subtables (located at the upper left and lower right) such that all cell means within each subtable are estimable, but cell means corresponding to cells in the original, larger table that do not belong to one of the two subtables are nonestimable. Thus, a basis for the space of estimable functions is

$$\{\mu+\alpha_1+\gamma_1, \alpha_5-\alpha_1, \alpha_3-\alpha_1, \gamma_4-\gamma_1, \gamma_3-\gamma_1\}\cup\{\mu+\alpha_4+\gamma_2, , \alpha_2-\alpha_4, \gamma_5-\gamma_2, \gamma_6-\gamma_2\},$$

showing that the rank of the corresponding model matrix is 9. In general, if the procedure yields an array that can be written as two distinct two-way subtables, the rank of the corresponding model matrix is $q + m - 2$; moreover, if the procedure yields an array that can be written as s distinct two-way subtables (possibly after interchanging entire rows and/or entire columns), the rank of the corresponding model matrix is $q + m - s$.

Some terminology has been developed to describe two-way layouts in which the $3 + e$ procedure results in every cell filled with an asterisk or an "e." Before leaving this example, we present some of this terminology and a theorem.

Definition 6.1.5 If $\alpha_i - \alpha_{i'}$ is estimable for all i and i' under the main effects model, the two-way layout is said to be *row-connected*, and if $\gamma_j - \gamma_{j'}$ is estimable for all j and j' under the main effects model, the two-way layout is said to be *column-connected.*

Using Corollary 6.1.2.1, it is easy to show that if a two-way layout is row-connected it is column-connected and vice versa; therefore we refer to such a layout merely as *connected.* Using the same corollary in tandem with Corollary 6.1.2.3, one can prove the following theorem; the proof is left as an exercise.

Theorem 6.1.4 *Under the two-way main effects model for a $q \times m$ layout, each of the following four conditions implies the other three:*

(a) the layout is connected;
(b) rank($\mathbf{X}$) $= q + m - 1$;
(c) $\mu + \alpha_i + \gamma_j$ is estimable for all i and j;
(d) all α-contrasts and γ-contrasts are estimable. ■

Example 6.1-5. Estimability Under the Two-Way Model with Interaction and No Empty Cells

It follows from (5.9) that the matrix of distinct rows of $\mathbf{X}$ is

$$[\mathbf{1}_{qm}, \mathbf{I}_q \otimes \mathbf{1}_m, \mathbf{1}_q \otimes \mathbf{I}_m, \mathbf{I}_q \otimes \mathbf{I}_m].$$

Any linear combination of the rows of $\mathbf{X}$, i.e., $\mathbf{a}^T\mathbf{X}$, may therefore be expressed as $(a_{\cdot\cdot}, a_{1\cdot}, a_{2\cdot}, \ldots, a_{q\cdot}, a_{\cdot 1}, a_{\cdot 2}, \ldots, a_{\cdot m}, a_{11}, a_{12}, \ldots, a_{1m}, a_{21}, \ldots, a_{qm})$, where $\mathbf{a}^T = (a_{11}, a_{12}, \ldots, a_{1m}, a_{21}, \ldots, a_{qm})$. Thus, $\mathbf{c}^T\boldsymbol{\beta}$ is estimable if and only if the following three conditions hold:

(i) for each fixed i, the coefficients on the ξ_{ij}'s sum (over j) to the coefficient on α_i;
(ii) for each fixed j, the coefficients on the ξ_{ij}'s sum (over i) to the coefficient on γ_j;
(iii) the coefficients on the α_i's sum to the coefficient on μ.

Note that (i)–(iii) imply that the coefficients on the γ_j's, like those of the α_i's, sum to the coefficient on μ. A basis for the space of estimable functions is simply the set of qm cell means $\{\mu + \alpha_i + \gamma_j + \xi_{ij} : i = 1, \ldots, q;\ j = 1, \ldots, m\}$; hence $\text{rank}(\mathbf{X}) = qm$. No α-contrast or γ-contrast is estimable, but contrasts among the interaction effects of the form $\xi_{ij} - \xi_{ij'} - \xi_{i'j} + \xi_{i'j'}$ are estimable.

If some cells are empty, at least some interaction contrasts of the form just mentioned are no longer estimable. Further investigation into this is the topic of an exercise. ■

Example 6.1-6. Estimability Under the Two-Factor Nested Model

The solution to Exercise 5.4 informs us that the matrix of distinct rows of $\mathbf{X}$ is

$$(\mathbf{1}_{\sum_{i=1}^q m_i}, \oplus_{i=1}^q \mathbf{1}_{m_i}, \oplus_{i=1}^q \mathbf{I}_{m_i}).$$

Any linear combination of the rows of $\mathbf{X}$, i.e., $\mathbf{a}^T\mathbf{X}$, may therefore be expressed as $(a_{\cdot\cdot}, a_{1\cdot}, a_{2\cdot}, \ldots, a_{q\cdot},\ a_{11}, a_{12}, \ldots, a_{1m_1}, a_{21}, \ldots, a_{qm_q})$, where $\mathbf{a}^T = (a_{11}, a_{12}, \ldots, a_{1m_1}, a_{21}, \ldots, a_{qm_q})$. Thus, $\mathbf{c}^T\boldsymbol{\beta}$ is estimable if and only if the coefficients on the α_i's sum to the coefficient on μ and the coefficients on the γ_{ij}'s sum (over j) to the coefficient on α_i for each fixed i. Some examples of estimable functions and unbiased estimators of them are as follows:

Function	Unbiased estimators
$\mu + \alpha_i + \gamma_{ij}$	$y_{ij1}, \ \bar{y}_{ij\cdot}$
$\sum_{j=1}^{m_i} d_j \gamma_{ij}$ where $\sum_{j=1}^{m_i} d_j = 0$	$\sum_{j=1}^{m_i} d_j y_{ij1}, \ \sum_{j=1}^{m_i} d_j \bar{y}_{ij\cdot}$

A basis for the space of estimable functions is $\{\mu + \alpha_i + \gamma_{i1}, \gamma_{i2} - \gamma_{i1}, \gamma_{i3} - \gamma_{i1}, \ldots, \gamma_{im_i} - \gamma_{i1} : i = 1, \ldots, q\}$. Thus rank($\mathbf{X}$) $= \sum_{i=1}^{q} m_i$, and no α-contrast is estimable. ■

Example 6.1-7. Confounding of Treatment and Block Effects in Incomplete Block Designs

Consider a situation in which there are two crossed treatment factors, say A and B, each with two levels, and a third factor, C, representing blocks, that also has two levels and is crossed with both A and B. Suppose that the blocks are of size two, so that only two of the four combinations of levels of A and B can occur in any one block, i.e., we must use an incomplete block design. Consider the following three-way main effects model for these observations:

$$y_{ijk} = \mu + \alpha_i + \gamma_j + \delta_k + e_{ijk},$$

with i, j, k denoting the levels of A, B, and C respectively. Suppose that the treatment combinations assigned to the first block correspond to $(i, j) = (1, 1)$ and $(i, j) = (2, 2)$, and the other two combinations are assigned to the second block. Which of the functions $\alpha_2 - \alpha_1$, $\gamma_2 - \gamma_1$, and $\delta_2 - \delta_1$ are estimable? It turns out, by Corollaries 6.1.2.1 and 6.1.2.2, that all of them are estimable because:

- $-\frac{1}{2}(\mu + \alpha_1 + \gamma_1 + \delta_1) + \frac{1}{2}(\mu + \alpha_2 + \gamma_2 + \delta_1) - \frac{1}{2}(\mu + \alpha_1 + \gamma_2 + \delta_2) + \frac{1}{2}(\mu + \alpha_2 + \gamma_1 + \delta_2) = \alpha_2 - \alpha_1$;
- $-\frac{1}{2}(\mu + \alpha_1 + \gamma_1 + \delta_1) + \frac{1}{2}(\mu + \alpha_2 + \gamma_2 + \delta_1) + \frac{1}{2}(\mu + \alpha_1 + \gamma_2 + \delta_2) - \frac{1}{2}(\mu + \alpha_2 + \gamma_1 + \delta_2) = \gamma_2 - \gamma_1$;
- $-\frac{1}{2}(\mu + \alpha_1 + \gamma_1 + \delta_1) - \frac{1}{2}(\mu + \alpha_2 + \gamma_2 + \delta_1) + \frac{1}{2}(\mu + \alpha_1 + \gamma_2 + \delta_2) + \frac{1}{2}(\mu + \alpha_2 + \gamma_1 + \delta_2) = \delta_2 - \delta_1$.

Because the Factor A difference and Factor B difference are estimable, we say that Factors A and B are *not confounded with blocks*. In this case, a basis for the space of estimable functions is $\{\mu + \alpha_1 + \gamma_1 + \delta_1, \alpha_2 - \alpha_1, \gamma_2 - \gamma_1, \delta_2 - \delta_1\}$, hence rank($\mathbf{X}$) $= 4$.

Now, suppose that the assignment of treatment combinations to blocks is changed, so that those assigned to the first block correspond to $(i, j) = (1, 1)$ and $(i, j) = (2, 1)$ (and again the other two combinations are assigned to the second block). Which of the functions $\alpha_2 - \alpha_1$, $\gamma_2 - \gamma_1$, and $\delta_2 - \delta_1$ are estimable now? In this case $\alpha_2 - \alpha_1$ is estimable because

$$-\frac{1}{2}(\mu + \alpha_1 + \gamma_1 + \delta_1) + \frac{1}{2}(\mu + \alpha_2 + \gamma_1 + \delta_1) - \frac{1}{2}(\mu + \alpha_1 + \gamma_2 + \delta_2) + \frac{1}{2}(\mu + \alpha_2 + \gamma_2 + \delta_2) = \alpha_2 - \alpha_1.$$

However, neither $\gamma_2 - \gamma_1$ nor $\delta_2 - \delta_1$ is estimable, as is easily verified. Thus, in this case we say that Factor B is *confounded with blocks*. A basis for the space of estimable functions in this case is $\{\mu + \alpha_1 + \gamma_1 + \delta_1, \alpha_2 - \alpha_1, \gamma_2 - \gamma_1 + \delta_2 - \delta_1\}$, showing that rank$(\mathbf{X}) = 3$, one less than it was in the first case. ■

6.2 Estimability Under a Constrained Model

Definition 6.2.1 An estimator $t(\mathbf{y})$ of a linear function $\mathbf{c}^T\boldsymbol{\beta}$ associated with the model $\{\mathbf{y}, \mathbf{X}\boldsymbol{\beta} : \mathbf{A}\boldsymbol{\beta} = \mathbf{h}\}$ is said to be *unbiased* under that model if $\mathrm{E}[t(\mathbf{y})] = \mathbf{c}^T\boldsymbol{\beta}$ for all $\boldsymbol{\beta}$ satisfying $\mathbf{A}\boldsymbol{\beta} = \mathbf{h}$.

Comparing Definitions 6.1.2 and 6.2.1, it may be seen that unbiasedness under a constrained model is *less* restrictive than unbiasedness under the corresponding unconstrained model. The latter definition requires the estimator's expectation to equal $\mathbf{c}^T\boldsymbol{\beta}$ merely for those $\boldsymbol{\beta}$ satisfying $\mathbf{A}\boldsymbol{\beta} = \mathbf{h}$, rather than for all $\boldsymbol{\beta} \in \mathbb{R}^p$. Thus, an estimator of $\mathbf{c}^T\boldsymbol{\beta}$ that is unbiased under an unconstrained model is also unbiased under any corresponding constrained model (any constrained model with the same model matrix), but an estimator of $\mathbf{c}^T\boldsymbol{\beta}$ that is unbiased under a constrained model is not necessarily unbiased under the corresponding unconstrained model.

The first theorem in this section gives a necessary and sufficient condition for a linear estimator to be unbiased for $\mathbf{c}^T\boldsymbol{\beta}$ under a constrained model.

Theorem 6.2.1 *A linear estimator $t_0 + \mathbf{t}^T\mathbf{y}$ of a linear function $\mathbf{c}^T\boldsymbol{\beta}$ associated with the model $\{\mathbf{y}, \mathbf{X}\boldsymbol{\beta} : \mathbf{A}\boldsymbol{\beta} = \mathbf{h}\}$ is unbiased under that model if and only if a vector $\mathbf{g}$ exists such that $t_0 = \mathbf{g}^T\mathbf{h}$ and $\mathbf{t}^T\mathbf{X} + \mathbf{g}^T\mathbf{A} = \mathbf{c}^T$.*

Proof

$$\begin{aligned}
&\mathrm{E}(t_0 + \mathbf{t}^T\mathbf{y}) = \mathbf{c}^T\boldsymbol{\beta} \text{ for all } \boldsymbol{\beta} \text{ such that } \mathbf{A}\boldsymbol{\beta} = \mathbf{h} \\
\Leftrightarrow\ & t_0 + (\mathbf{t}^T\mathbf{X} - \mathbf{c}^T)\boldsymbol{\beta} = 0 \text{ for all } \boldsymbol{\beta} \text{ such that } \mathbf{A}\boldsymbol{\beta} = \mathbf{h} \\
\Leftrightarrow\ & t_0 + (\mathbf{t}^T\mathbf{X} - \mathbf{c}^T)[\mathbf{A}^-\mathbf{h} + (\mathbf{I} - \mathbf{A}^-\mathbf{A})\mathbf{z}] = 0 \text{ for all } \mathbf{z} \\
\Leftrightarrow\ & t_0 = (\mathbf{c}^T - \mathbf{t}^T\mathbf{X})\mathbf{A}^-\mathbf{h} \text{ and } (\mathbf{t}^T\mathbf{X} - \mathbf{c}^T)(\mathbf{I} - \mathbf{A}^-\mathbf{A}) = \mathbf{0}^T \\
\Leftrightarrow\ & t_0 = (\mathbf{c}^T - \mathbf{t}^T\mathbf{X})\mathbf{A}^-\mathbf{h} \text{ and } (\mathbf{t}^T\mathbf{X} - \mathbf{c}^T) \in \mathcal{R}(\mathbf{A})
\end{aligned}$$

$$\Leftrightarrow t_0 = (\mathbf{c}^T - \mathbf{t}^T\mathbf{X})\mathbf{A}^-\mathbf{h} \text{ and a vector } \mathbf{g} \text{ exists such that } \mathbf{t}^T\mathbf{X} - \mathbf{c}^T = -\mathbf{g}^T\mathbf{A}$$

$$\Leftrightarrow \text{a vector } \mathbf{g} \text{ exists such that such that } t_0 = \mathbf{g}^T\mathbf{h} \text{ and } \mathbf{t}^T\mathbf{X} + \mathbf{g}^T\mathbf{A} = \mathbf{c}^T.$$

Here the second two-way implication is obtained by a direct application of Theorem 3.2.3, and the fourth two-way implication holds because if $(\mathbf{t}^T\mathbf{X} - \mathbf{c}^T)(\mathbf{I} - \mathbf{A}^-\mathbf{A}) = \mathbf{0}^T$, then $(\mathbf{t}^T\mathbf{X} - \mathbf{c}^T) = (\mathbf{t}^T\mathbf{X} - \mathbf{c}^T)\mathbf{A}^-\mathbf{A}$, which is an element of $\mathcal{R}(\mathbf{A})$; conversely if $(\mathbf{t}^T\mathbf{X} - \mathbf{c}^T) \in \mathcal{R}(\mathbf{A})$, then $(\mathbf{t}^T\mathbf{X} - \mathbf{c}^T) = \boldsymbol{\ell}^T\mathbf{A}$ for some $\boldsymbol{\ell}$, in which case $(\mathbf{t}^T\mathbf{X} - \mathbf{c}^T)(\mathbf{I} - \mathbf{A}^-\mathbf{A}) = \boldsymbol{\ell}^T\mathbf{A}(\mathbf{I} - \mathbf{A}^-\mathbf{A}) = \boldsymbol{\ell}^T\mathbf{A} - \boldsymbol{\ell}^T\mathbf{A}\mathbf{A}^-\mathbf{A} = \mathbf{0}^T$. For the last implication we have used the consistency of the system of constraint equations, i.e., the existence of a p-vector $\boldsymbol{\beta}^*$ such that $\mathbf{A}\boldsymbol{\beta}^* = \mathbf{h}$, hence $\mathbf{g}^T\mathbf{A}\mathbf{A}^-\mathbf{h} = \mathbf{g}^T\mathbf{A}\mathbf{A}^-\mathbf{A}\boldsymbol{\beta}^* = \mathbf{g}^T\mathbf{A}\boldsymbol{\beta}^* = \mathbf{g}^T\mathbf{h}$. The remaining implications are transparent. □

Observe that the necessary and sufficient condition for unbiasedness of a linear estimator under a constrained model given in Theorem 6.2.1 reduces to that given in Theorem 6.1.1 if $\mathbf{g} = \mathbf{0}$. This is consistent with the fact, noted previously, that a linear estimator that is unbiased under an unconstrained model is also unbiased under any corresponding constrained model.

Theorem 6.2.2 *A linear function* $\mathbf{c}^T\boldsymbol{\beta}$ *is estimable under the model* $\{\mathbf{y}, \mathbf{X}\boldsymbol{\beta} : \mathbf{A}\boldsymbol{\beta} = \mathbf{h}\}$ *if and only if* $\mathbf{c}^T \in \mathcal{R}\begin{pmatrix}\mathbf{X}\\ \mathbf{A}\end{pmatrix}$.

Proof Suppose that $\mathbf{c}^T \in \mathcal{R}\begin{pmatrix}\mathbf{X}\\ \mathbf{A}\end{pmatrix}$. Then, for some row vector $(\mathbf{a}_1^T, \mathbf{a}_2^T)$,

$$\mathbf{c}^T = (\mathbf{a}_1^T, \mathbf{a}_2^T)\begin{pmatrix}\mathbf{X}\\ \mathbf{A}\end{pmatrix},$$

implying that

$$\mathbf{c}^T\boldsymbol{\beta} = (\mathbf{a}_1^T, \mathbf{a}_2^T)\begin{pmatrix}\mathbf{X}\\ \mathbf{A}\end{pmatrix}\boldsymbol{\beta} \quad \text{for all } \boldsymbol{\beta} \text{ such that } \mathbf{A}\boldsymbol{\beta} = \mathbf{h}.$$

Because the right-hand side of this equality may be expressed as $\mathrm{E}(\mathbf{a}_1^T\mathbf{y} + \mathbf{a}_2^T\mathbf{h})$, it is apparent that a linear unbiased estimator $t_0 + \mathbf{t}^T\mathbf{y}$ of $\mathbf{c}^T\boldsymbol{\beta}$ exists; specifically, the estimator of this form with $t_0 = \mathbf{a}_2^T\mathbf{h}$ and $\mathbf{t} = \mathbf{a}_1$ is unbiased. This establishes that $\mathbf{c}^T\boldsymbol{\beta}$ is estimable under the specified model.

Conversely, suppose that $\mathbf{c}^T\boldsymbol{\beta}$ is estimable under the specified model. Then by definition, a linear unbiased estimator $t_0 + \mathbf{t}^T\mathbf{y}$ of it exists. By Theorem 6.2.1, a vector $\mathbf{g}$ exists such that $\mathbf{t}^T\mathbf{X} + \mathbf{g}^T\mathbf{A} = \mathbf{c}^T$, i.e., a vector $\mathbf{g}$ exists such that

$$\mathbf{c}^T = (\mathbf{t}^T, \mathbf{g}^T)\begin{pmatrix}\mathbf{X}\\ \mathbf{A}\end{pmatrix},$$

i.e.,

$$\mathbf{c}^T \in \mathcal{R}\begin{pmatrix}\mathbf{X}\\ \mathbf{A}\end{pmatrix}.$$

□

Recall that a necessary and sufficient condition for $\mathbf{c}^T\boldsymbol{\beta}$ to be estimable under the unconstrained model $\{\mathbf{y}, \mathbf{X}\boldsymbol{\beta}\}$ is $\mathbf{c}^T \in \mathcal{R}(\mathbf{X})$ (Theorem 6.1.2). We see from Theorem 6.2.2, therefore, that any function $\mathbf{c}^T\boldsymbol{\beta}$ that is estimable under the model $\{\mathbf{y}, \mathbf{X}\boldsymbol{\beta}\}$ is estimable under a constrained model with model matrix $\mathbf{X}$, but not necessarily vice versa.

Next, we give five corollaries to Theorem 6.2.2 that are analogous to the first five corollaries to Theorem 6.1.2. Proofs of the corollaries are left as exercises.

Corollary 6.2.2.1 *If* $\mathbf{c}_1^T\boldsymbol{\beta},\ \mathbf{c}_2^T\boldsymbol{\beta},\ \ldots,\ \mathbf{c}_k^T\boldsymbol{\beta}$ *are estimable under the model* $\{\mathbf{y}, \mathbf{X}\boldsymbol{\beta} : \mathbf{A}\boldsymbol{\beta} = \mathbf{h}\}$, *then so is any linear combination of them; that is, the set of estimable functions under a given constrained model is a linear space.*

Corollary 6.2.2.2 *The elements of* $\mathbf{X}\boldsymbol{\beta}$ *are estimable under the model* $\{\mathbf{y}, \mathbf{X}\boldsymbol{\beta} : \mathbf{A}\boldsymbol{\beta} = \mathbf{h}\}$.

Corollary 6.2.2.3 *A set of* $\text{rank}\begin{pmatrix}\mathbf{X}\\ \mathbf{A}\end{pmatrix}$ *linearly independent estimable functions under the model* $\{\mathbf{y}, \mathbf{X}\boldsymbol{\beta} : \mathbf{A}\boldsymbol{\beta} = \mathbf{h}\}$ *exists such that any estimable function under that model can be written as a linear combination of functions in this set.*

Corollary 6.2.2.4 $\mathbf{c}^T\boldsymbol{\beta}$ *is estimable for* every *vector* $\mathbf{c}$ *under the model* $\{\mathbf{y}, \mathbf{X}\boldsymbol{\beta} : \mathbf{A}\boldsymbol{\beta} = \mathbf{h}\}$ *if and only if* $\text{rank}\begin{pmatrix}\mathbf{X}\\ \mathbf{A}\end{pmatrix} = p$.

Corollary 6.2.2.5 $\mathbf{c}^T\boldsymbol{\beta}$ *is estimable under the model* $\{\mathbf{y}, \mathbf{X}\boldsymbol{\beta} : \mathbf{A}\boldsymbol{\beta} = \mathbf{h}\}$ *if and only if* $\mathbf{c}^T \in \mathcal{R}\begin{pmatrix}\mathbf{X}^T\mathbf{X}\\ \mathbf{A}\end{pmatrix}$.

Corollary 6.2.2.5 is an immediate consequence of the following theorem, which is of interest in its own right.

Theorem 6.2.3 $\mathcal{R}\begin{pmatrix}\mathbf{X}^T\mathbf{X}\\ \mathbf{A}\end{pmatrix} = \mathcal{R}\begin{pmatrix}\mathbf{X}\\ \mathbf{A}\end{pmatrix}$.

Proof

$$\begin{pmatrix} \mathbf{X}^T\mathbf{X} \\ \mathbf{A} \end{pmatrix} = \begin{pmatrix} \mathbf{X}^T & \mathbf{0} \\ \mathbf{0} & \mathbf{I} \end{pmatrix} \begin{pmatrix} \mathbf{X} \\ \mathbf{A} \end{pmatrix}$$

and

$$\begin{pmatrix} \mathbf{X} \\ \mathbf{A} \end{pmatrix} = \begin{pmatrix} \mathbf{X}(\mathbf{X}^T\mathbf{X})^- & \mathbf{0} \\ \mathbf{0} & \mathbf{I} \end{pmatrix} \begin{pmatrix} \mathbf{X}^T\mathbf{X} \\ \mathbf{A} \end{pmatrix},$$

so by Theorem 2.6.1 each row space is contained in the other. The desired result follows immediately. □

Example 6.2-1. Estimability Under Constrained One-Factor Models

Consider the one-factor model

$$y_{ij} = \mu + \alpha_i + e_{ij}, \quad (i = 1, \ldots, q;\ j = 1, \ldots, n_i),$$

and let $\boldsymbol{\beta} = (\mu, \alpha_1, \ldots, \alpha_q)^T$. Recall from Example 6.1-2 that if there are no constraints, a linear function $\mathbf{c}^T\boldsymbol{\beta}$ is estimable if and only if $\mathbf{c}^T = (c_1, c_2, \ldots, c_{q+1}) \in \mathcal{R}(\mathbf{X}) = \{\mathbf{c}^T : c_1 = \sum_{i=2}^{q+1} c_i\}$, and that $\text{rank}(\mathbf{X}) = q$. Let us first consider applying the single constraint $\alpha_q = 0$ to this model. For this constraint, $\mathbf{A} = (\mathbf{0}_q^T, 1) \notin \mathcal{R}(\mathbf{X})$, and $\text{rank}(\mathbf{A}) = 1$. Thus $\mathcal{R}\begin{pmatrix} \mathbf{X} \\ \mathbf{A} \end{pmatrix} = \mathbb{R}^{q+1}$, implying (by Corollary 6.2.2.4) that every linear function $\mathbf{c}^T\boldsymbol{\beta}$ is estimable under the one-factor model with this constraint. The same would be true if the constraint was changed to $\alpha_q = h$ for any real number h, or to any other single constraint for which the corresponding matrix $\mathbf{A}$ is not an element of $\mathcal{R}(\mathbf{X})$, such as $\mu = h$ or $\sum_{i=1}^{q} \alpha_i = h$.

On the other hand, if the constraint is $\mu + \alpha_1 = h$, then $\mathbf{A} = (1, 1, \mathbf{0}_{q-1}^T) \in \mathcal{R}(\mathbf{X})$, in which case $\mathcal{R}\begin{pmatrix} \mathbf{X} \\ \mathbf{A} \end{pmatrix} = \mathcal{R}(\mathbf{X})$, implying that a function is estimable under a model with this constraint if and only if it is estimable under the unconstrained model. ■

6.3 Exercises

1. A linear estimator $t(\mathbf{y}) = t_0 + \mathbf{t}^T\mathbf{y}$ of $\mathbf{c}^T\boldsymbol{\beta}$ associated with the model $\{\mathbf{y}, \mathbf{X}\boldsymbol{\beta}\}$ is said to be *location equivariant* if $t(\mathbf{y} + \mathbf{X}\mathbf{d}) = t(\mathbf{y}) + \mathbf{c}^T\mathbf{d}$ for all $\mathbf{d}$.
   (a) Prove that a linear estimator $t_0 + \mathbf{t}^T\mathbf{y}$ of $\mathbf{c}^T\boldsymbol{\beta}$ is location equivariant if and only if $\mathbf{t}^T\mathbf{X} = \mathbf{c}^T$.

(b) Is a location equivariant estimator of $\mathbf{c}^T\boldsymbol{\beta}$ necessarily unbiased, or vice versa? Explain.

2. Prove Corollary 6.1.2.1.
3. Prove Corollary 6.1.2.2.
4. Prove Corollary 6.1.2.3.
5. Prove Corollary 6.1.2.4.
6. Prove Corollary 6.1.2.6.
7. Prove Corollary 6.1.2.5.
8. For each of the following two questions, answer "yes" or "no." If the answer is "yes," give an example; if the answer is "no," prove it.
   (a) Can the sum of an estimable function and a nonestimable function be estimable?
   (b) Can the sum of two nonestimable functions be estimable?
   (c) Can the sum of two linearly independent nonestimable functions be nonestimable?
9. Determine $\mathcal{N}(\mathbf{X})$ for the one-factor model and for the two-way main effects model with no empty cells, and then use Corollary 6.1.2.5 to obtain the same characterizations of estimable functions given in Examples 6.1-2 and 6.1-3 for these two models.
10. Prove that a two-way layout that is row-connected is also column-connected, and vice versa.
11. Prove Theorem 6.1.4.
12. For a two-way main effects model with q levels of Factor A and m levels of Factor B, what is the smallest possible value of rank($\mathbf{X}$)? Explain.
13. Consider the following two-way layout, where the number in each cell indicates how many observations are in that cell:

	Levels of B									
Levels of A	1	2	3	4	5	6	7	8	9	10
1			1				2			
2									1	
3				4		1				1
4					2					2
5							1	1		
6					1					
7				1	2					
8		1							3	
9	2									

Consider the main effects model

$$y_{ijk} = \mu + \alpha_i + \gamma_j + e_{ijk} \qquad (i = 1, \ldots, 9;\ j = 1, \ldots, 10)$$

for these observations (where k indexes the observations, if any, within cell (i, j) of the table).
(a) Which α-contrasts and γ-contrasts are estimable?
(b) Give a basis for the set of all estimable functions.
(c) What is the rank of the associated model matrix $\mathbf{X}$?

14. Consider the following two-way layout, where the number in each cell indicates how many observations are in that cell:

	Levels of B							
Levels of A	1	2	3	4	5	6	7	8
1	1				4	1		
2	3					1		
3								1
4		1					2	
5		1						
6			2	1				1
7						3		
8		2					1	
9	1				1			

Consider the main effects model

$$y_{ijk} = \mu + \alpha_i + \gamma_j + e_{ijk} \qquad (i = 1, \ldots, 9;\ j = 1, \ldots, 8)$$

for these observations (where k indexes the observations, if any, within cell (i, j) of the table).
(a) Which α-contrasts and γ-contrasts are estimable?
(b) Give a basis for the set of all estimable functions.
(c) What is the rank of the associated model matrix $\mathbf{X}$?
(d) Suppose that it was possible to take more observations in any cells of the table, but that each additional observation in cell (i, j) of the table will cost the investigator

$$(\$20 \times i) + [\$25 \times (8 - j)].$$

How many more observations are necessary for all α-contrasts and all γ-contrasts to be estimable, and in which cells should those observations be taken to minimize the additional cost to the investigator?

15. Consider the two-way partially crossed model introduced in Example 5.1.4-1, i.e.,

$$y_{ijk} = \mu + \alpha_i - \alpha_j + e_{ijk} \quad (i \neq j = 1, \ldots, q;\ j = 1, \ldots, n_i)$$

for those combinations of i and j $(i \neq j)$ for which $n_{ij} \geq 1$.

(a) For the case of no empty cells, what conditions must the elements of $\mathbf{c}$ satisfy for $\mathbf{c}^T\boldsymbol{\beta}$ to be estimable? In particular, are μ and all differences $\alpha_i - \alpha_j$ estimable? Determine the rank of the model matrix $\mathbf{X}$ for this case and find a basis for the set of all estimable functions.
(b) Now consider the case where all cells below the main diagonal of the two-way layout are empty, but no cells above the main diagonal are empty. How, if at all, would your answers to the same questions in part (a) change?
(c) Finally, consider the case where there are empty cells in arbitrary locations. Describe a method for determining which cell means are estimable. Under what conditions on the locations of the empty cells are all functions estimable that were estimable in part (a)?

16. For a 3×3 layout, consider a two-way model with interaction, i.e.,

$$y_{ijk} = \mu + \alpha_i + \gamma_j + \xi_{ij} + e_{ijk} \qquad (i = 1, 2, 3;\ j = 1, 2, 3;\ k = 1, \ldots, n_{ij}),$$

where some of the nine cells may be empty.
(a) Give a basis for the interaction contrasts, $\psi_{iji'j'} \equiv \xi_{ij} - \xi_{ij'} - \xi_{i'j} + \xi_{i'j'}$, that are estimable if no cells are empty.
(b) If only one of the nine cells is empty, how many interaction contrasts are there in a basis for the estimable interaction contrasts? What is rank($\mathbf{X}$)? Does the answer depend on which cell is empty?
(c) If exactly two of the nine cells are empty, how many interaction contrasts are there in a basis for the estimable interaction contrasts? What is rank($\mathbf{X}$)? Does the answer depend on which cells are empty?
(d) What is the maximum number of empty cells possible for a 3×3 layout that has at least one estimable interaction contrast? What is rank($\mathbf{X}$) in this case?

17. Consider the three-way main effects model

$$y_{ijkl} = \mu + \alpha_i + \gamma_j + \delta_k + e_{ijkl}$$
$$(i = 1, \ldots, q;\ j = 1, \ldots, m;\ k = 1, \ldots, s;\ l = 1, \ldots, n_{ijk}),$$

and suppose that there are no empty cells (i.e., $n_{ijk} > 0$ for all i, j, k).
(a) What conditions must the elements of $\mathbf{c}$ satisfy for $\mathbf{c}^T\boldsymbol{\beta}$ to be estimable?
(b) Find a basis for the set of all estimable functions.
(c) What is the rank of the model matrix $\mathbf{X}$?

18. Consider a three-factor crossed classification in which Factors A, B, and C each have two levels, and suppose that there is exactly one observation in the following four (of the eight) cells: 111, 122, 212, 221. The other cells are empty. Consider the following main effects model for these observations:

$$y_{ijk} = \mu + \alpha_i + \gamma_j + \delta_k + e_{ijk}$$

where $(i, j, k) \in \{(1, 1, 1),\ (1, 2, 2),\ (2, 1, 2),\ (2, 2, 1)\}$.

(a) Which of the functions $\alpha_2 - \alpha_1$, $\gamma_2 - \gamma_1$, and $\delta_2 - \delta_1$ are estimable?
(b) Find a basis for the set of all estimable functions.
(c) What is the rank of the model matrix $\mathbf{X}$?
Note: This exercise and the next one are relevant to the issue of "confounding" in a 2^2 factorial design in blocks of size two.

19. Consider a three-factor crossed classification in which Factors A, B, and C each have two levels, and suppose that there is exactly one observation in the following four (of the eight) cells: 111, 211, 122, 222. The other cells are empty. Consider the following main effects model for these observations:

$$y_{ijk} = \mu + \alpha_i + \gamma_j + \delta_k + e_{ijk}$$

where $(i, j, k) \in \{(1, 1, 1), (2, 1, 1), (1, 2, 2), (2, 2, 2)\}$.
(a) Which of the functions $\alpha_2 - \alpha_1$, $\gamma_2 - \gamma_1$, and $\delta_2 - \delta_1$ are estimable?
(b) Find a basis for the set of all estimable functions.
(c) What is the rank of the model matrix $\mathbf{X}$?

20. Consider a three-factor crossed classification in which Factors A and B have two levels and Factor C has three levels (i.e., a $2 \times 2 \times 3$ layout), but some cells may be empty. Consider the following main effects model for this situation:

$$y_{ijkl} = \mu + \alpha_i + \gamma_j + \delta_k + e_{ijkl}$$

where $i = 1, 2$; $j = 1, 2$; $k = 1, 2, 3$; and $l = 1, \ldots, n_{ijk}$ for the nonempty cells.
(a) What is the fewest number of observations that, if they are placed in appropriate cells, will make all functions of the form $\mu + \alpha_i + \gamma_j + \delta_k$ estimable?
(b) If all functions of the form $\mu + \alpha_i + \gamma_j + \delta_k$ are estimable, what is the rank of the model matrix?
(c) Let a represent the correct answer to part (a). If a observations are placed in *any* cells (one observation per cell), will all functions of the form $\mu + \alpha_i + \gamma_j + \delta_k$ necessarily be estimable? Explain.
(d) If $a + 1$ observations are placed in *any* cells (one observation per cell), will all functions of the form $\mu + \alpha_i + \gamma_j + \delta_k$ necessarily be estimable? Explain.
(e) Suppose there is exactly one observation in the following cells: 211, 122, 123, 221. The other cells are empty. Determine p^* and find a basis for the set of all estimable functions.

21. A "Latin square" design is an experimental design with three partially crossed factors, called "Rows," "Columns," and "Treatments." The number of levels of each factor is common across factors, i.e., if there are q levels of Row, then there must be q levels of Column and q levels of Treatment. The first two factors are completely crossed, so a two-way table can be used to represent their combinations; but the Treatment factor is only partially crossed with the other two factors because there is only one observational unit for each Row$\times$Column combination. The levels of Treatment are assigned to

Row×Column combinations in such a way that each level occurs exactly once in each row and column (rather than once in each combination, as would be the case if Treatment was completely crossed with Row and Column). An example of a 3 × 3 "Latin square" design is depicted below:

2	1	3
3	2	1
1	3	2

The number displayed in each of the nine cells of this layout is the Treatment level (i.e., the Treatment label) which the observation in that cell received.

For observations in a Latin square design with q treatments, consider the following model:

$$y_{ijk} = \mu + \alpha_i + \gamma_j + \tau_k + e_{ijk} \qquad (i = 1, \ldots, q;\ j = 1, \ldots, q;\ k = 1, \ldots, q)$$

where y_{ijk} is the observation in row i and column j (and k is the treatment assigned to that cell); μ is an overall effect; α_i is the effect of Row i; γ_j is the effect of Column j; τ_k is the effect of Treatment k; and e_{ijk} is the error corresponding to y_{ijk}. These errors are uncorrelated with mean 0 and variance σ^2.

(a) Show that $\tau_k - \tau_{k'}$ is estimable for all k and k', and give an unbiased estimator for it.
(b) What condition(s) must the elements of $\mathbf{c}^T\boldsymbol{\beta}$ satisfy for $\mathbf{c}^T\boldsymbol{\beta}$ to be estimable?
(c) Find a basis for the set of all estimable functions.
(d) What is the rank of the model matrix $\mathbf{X}$?

22. An experiment is being planned that involves two continuous explanatory variables, x_1 and x_2, and a response variable y. The model relating y to the explanatory variables is

$$y_i = \beta_1 + \beta_2 x_{i1} + \beta_3 x_{i2} + e_i \qquad (i = 1, \ldots, n).$$

Suppose that only three observations can be taken (i.e., $n = 3$), and that these three observations must be taken at distinct values of (x_{i1}, x_{i2}) such that $x_{i1} = 1, 2,$ or 3, and $x_{i2} = 1, 2,$ or 3 (see Fig. 6.1 for a depiction of the nine allowable (x_{i1}, x_{i2})-pairs, which are labeled as A, B, …, I for easier reference).

A three-observation design can thus be represented by three letters: for example, ACI is the design consisting of the three points A, C, and I. For some of the three-observation designs, all parameters in the model's mean structure (β_1, β_2, and β_3) are estimable. However, there are some three-observation designs for which not all of these parameters are estimable. List 8 designs of the latter type, and justify your list.

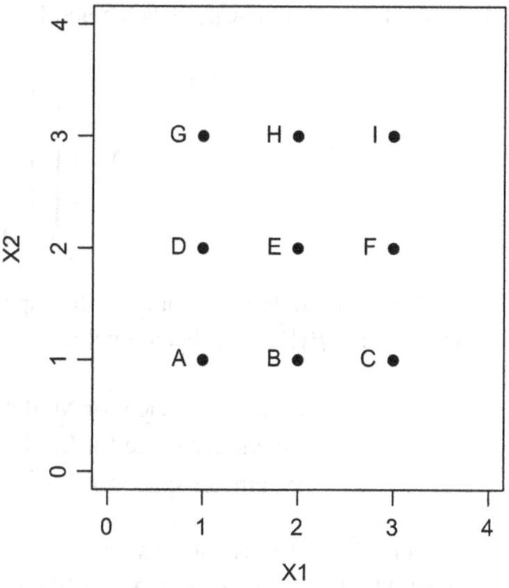

Fig. 6.1 Experimental layout for Exercise 6.22

23. Suppose that **y** follows a Gauss–Markov model in which

$$\mathbf{X} = \begin{pmatrix} 1 & 1 & t \\ 1 & 2 & 2t \\ \vdots & \vdots & \vdots \\ 1 & t & t^2 \\ 1 & 1 & s \\ 1 & 2 & 2s \\ \vdots & \vdots & \vdots \\ 1 & s & s^2 \end{pmatrix}, \quad \boldsymbol{\beta} = \begin{pmatrix} \beta_1 \\ \beta_2 \\ \beta_3 \end{pmatrix}.$$

Here, t and s are integers such that $t \geq 2$ and $s \geq 2$. [Note: The corresponding model is called the "conditional linear model" and has been used for data that are informatively right-censored; see, e.g., Wu and Bailey (1989).]

(a) Determine which of the following linear functions of the elements of $\boldsymbol{\beta}$ are always estimable, with no further assumptions on t and s.

(i) β_1

(ii) $\beta_1 + \beta_2 t + \beta_3 s$

(iii) $\beta_2 + \beta_3(s + t)/2$

(b) Determine necessary and sufficient conditions on t and s for all linear functions of the elements of $\boldsymbol{\beta}$ to be estimable.

24. Consider three linear models labeled I, II, and III, which all have model matrix

$$\mathbf{X} = \begin{pmatrix} 1 & 1 & 1 & 1 \\ 0 & 1 & 0 & 1 \\ 1 & 0 & 1 & 0 \\ 1 & 1 & 1 & 1 \\ 1 & 0 & 1 & 0 \end{pmatrix}.$$

The three models differ only with respect to their parameter spaces for $\boldsymbol{\beta} = [\beta_1, \beta_2, \beta_3, \beta_4]^T$, as indicated below:

$$\begin{aligned} &\text{Parameter space for Model I} : \{\boldsymbol{\beta} : \boldsymbol{\beta} \in \mathbb{R}^4\}, \\ &\text{Parameter space for Model II} : \{\boldsymbol{\beta} : \beta_1 + \beta_4 = 5\}, \\ &\text{Parameter space for Model III} : \{\boldsymbol{\beta} : \beta_1 + \beta_4 = 0\}. \end{aligned}$$

Under which, if any, of these models does an unbiased estimator of $\beta_2 + \beta_3$ exist which is of the form $\mathbf{t}^T\mathbf{y}$? Justify your answer.

25. Consider two linear models of the form

$$\mathbf{y} = \mathbf{X}\boldsymbol{\beta} + \mathbf{e}$$

where $\mathbf{y}$ is a 4-vector, $\boldsymbol{\beta}$ is the 3-vector $(\beta_1, \beta_2, \beta_3)^T$, and $\mathbf{X}$ is the 4×3 matrix

$$\begin{pmatrix} 1 & 3 & 5 \\ 0 & 4 & 4 \\ -2 & 6 & 2 \\ 4 & -3 & 5 \end{pmatrix}.$$

(Observe that the third column of $\mathbf{X}$ is equal to the second column plus twice the first column, but the first and the second columns are linearly independent, so the rank of $\mathbf{X}$ is 2). In the first model, $\boldsymbol{\beta}$ is unconstrained in $\mathbb{R}^3$. In the second model, $\boldsymbol{\beta}$ is constrained to the subset of $\mathbb{R}^3$ for which $\beta_1 = \beta_2$.

(a) Under which of the two models is the function $\beta_2 + \beta_3$ estimable? (Your answer may be neither model, the unconstrained model only, the constrained model only, or both models.)

(b) Under which of the two models is the function $\beta_1 + \beta_3$ estimable? (Your answer may be neither model, the unconstrained model only, the constrained model only, or both models.)

26. Prove Corollary 6.2.2.1.
27. Prove Corollary 6.2.2.2.
28. Prove Corollary 6.2.2.3.
29. Prove Corollary 6.2.2.4.
30. Prove Corollary 6.2.2.5.
31. Consider the constrained model $\{\mathbf{y}, \mathbf{X}\boldsymbol{\beta} : \mathbf{A}\boldsymbol{\beta} = \mathbf{h}\}$.

(a) Suppose that a function $\mathbf{c}^T\boldsymbol{\beta}$ exists that is estimable under this model but is nonestimable under the unconstrained model $\{\mathbf{y}, \mathbf{X}\boldsymbol{\beta}\}$. Under these circumstances, how do $\mathcal{R}(\mathbf{X}) \cap \mathcal{R}(\mathbf{A})$, $\mathcal{R}\begin{pmatrix}\mathbf{X}\\ \mathbf{A}\end{pmatrix}$, and $\mathcal{R}(\mathbf{X})$ compare to each other? Indicate which subset inclusions, if any, are strict.

(b) Again suppose that a function $\mathbf{c}^T\boldsymbol{\beta}$ exists that is estimable under this model but is nonestimable under the model $\{\mathbf{y}, \mathbf{X}\boldsymbol{\beta}\}$. Under these circumstances, if a vector $\mathbf{t}$ exists such that $\mathbf{t}^T\mathbf{y}$ is an unbiased estimator of $\mathbf{c}^T\boldsymbol{\beta}$ under the constrained model, what can be said about $\mathbf{A}$ and $\mathbf{h}$? Be as specific as possible.

References

Harville, D. A. (1977). Maximum likelihood approaches to variance component estimation and to related problems. *Journal of the American Statistical Association,72*, 320–338.

Lehmann, E. L., & Casella, G. (1998). *Theory of point estimation* (2nd ed.). New York: Springer.

Wu, M. C., & Bailey, K. R. (1989). Estimation and comparison of changes in the presence of informative right censoring: Conditional linear model. *Biometrics, 45*, 939–955.

Least Squares Estimation for the Gauss–Markov Model

7

Now that we have determined which functions $\mathbf{c}^T\boldsymbol{\beta}$ in a linear model $\{\mathbf{y}, \mathbf{X}\boldsymbol{\beta}\}$ can be estimated unbiasedly, we can consider how we might actually estimate them. This chapter presents the estimation method known as *least squares*. Least squares estimation involves finding a value of $\boldsymbol{\beta}$ that minimizes the distance between $\mathbf{y}$ and $\mathbf{X}\boldsymbol{\beta}$, as measured by the squared length of the vector $\mathbf{y} - \mathbf{X}\boldsymbol{\beta}$. Although this seems like it should lead to reasonable estimators of the elements of $\mathbf{X}\boldsymbol{\beta}$, it is not obvious that it will lead to estimators of all estimable functions $\mathbf{c}^T\boldsymbol{\beta}$ that are optimal in any sense. It is shown in this chapter that the least squares estimator of any estimable function $\mathbf{c}^T\boldsymbol{\beta}$ associated with the model $\{\mathbf{y}, \mathbf{X}\boldsymbol{\beta}\}$ is linear and unbiased under that model, and that it is, in fact, the "best" (minimum variance) linear unbiased estimator of $\mathbf{c}^T\boldsymbol{\beta}$ under the Gauss–Markov model $\{\mathbf{y}, \mathbf{X}\boldsymbol{\beta}, \sigma^2\mathbf{I}\}$. It is also shown that if the mean structure of the model is reparameterized, the least squares estimators of estimable functions are materially unaffected.

7.1 Least Squares Estimators

Definition 7.1.1 The *residual sum of squares function* for the model $\{\mathbf{y}, \mathbf{X}\boldsymbol{\beta}\}$ is the function

$$Q(\boldsymbol{\beta}) = (\mathbf{y} - \mathbf{X}\boldsymbol{\beta})^T(\mathbf{y} - \mathbf{X}\boldsymbol{\beta}),$$

defined for all $\boldsymbol{\beta} \in \mathbb{R}^p$.

Definition 7.1.2 (Initial Definition of a Least Squares Estimator) A *least squares estimator* of an estimable function $\mathbf{c}^T\boldsymbol{\beta}$ associated with the model $\{\mathbf{y}, \mathbf{X}\boldsymbol{\beta}\}$ is $\mathbf{c}^T\hat{\boldsymbol{\beta}}$, where $\hat{\boldsymbol{\beta}}$ is any vector that minimizes the residual sum of squares function for that model.

D. L. Zimmerman, *Linear Model Theory*,
https://doi.org/10.1007/978-3-030-52063-2_7

Because the domain of the residual sum of squares function has no boundary, all points $\boldsymbol{\beta} \in \mathbb{R}^p$ are interior points. Thus, as noted in Sect. 2.19, a necessary condition for $Q(\boldsymbol{\beta})$ to assume a global minimum at $\boldsymbol{\beta} = \hat{\boldsymbol{\beta}}$ is that

$$\nabla Q(\hat{\boldsymbol{\beta}}) = \left.\frac{\partial Q}{\partial \boldsymbol{\beta}}\right|_{\boldsymbol{\beta}=\hat{\boldsymbol{\beta}}} = \mathbf{0}.$$

Although this condition is necessary, it may not be sufficient; without further analysis that demonstrates otherwise, the point at which the partial derivatives equal 0 could be a local minimum, a local maximum, or even a saddle point. Nevertheless, we proceed with partial differentiation of $Q(\boldsymbol{\beta})$ and see where it leads. Using Theorem 2.19.1, we obtain

$$\frac{\partial Q}{\partial \boldsymbol{\beta}} = \frac{\partial(\mathbf{y}^T\mathbf{y})}{\partial \boldsymbol{\beta}} - \frac{\partial(2\mathbf{y}^T\mathbf{X}\boldsymbol{\beta})}{\partial \boldsymbol{\beta}} + \frac{\partial(\boldsymbol{\beta}^T\mathbf{X}^T\mathbf{X}\boldsymbol{\beta})}{\partial \boldsymbol{\beta}} = -2\mathbf{X}^T\mathbf{y} + 2\mathbf{X}^T\mathbf{X}\boldsymbol{\beta}.$$

Setting this equal to $\mathbf{0}$, we see that a necessary condition that $Q(\boldsymbol{\beta})$ assume a global minimum at $\boldsymbol{\beta} = \hat{\boldsymbol{\beta}}$ is that $\hat{\boldsymbol{\beta}}$ satisfy the system of equations

$$\mathbf{X}^T\mathbf{X}\boldsymbol{\beta} = \mathbf{X}^T\mathbf{y}.$$

Definition 7.1.3 The system of equations $\mathbf{X}^T\mathbf{X}\boldsymbol{\beta} = \mathbf{X}^T\mathbf{y}$ associated with the linear model $\{\mathbf{y}, \mathbf{X}\boldsymbol{\beta}\}$ are called the *normal equations* for that model.

Theorem 7.1.1 *The normal equations are consistent.*

Proof Theorem 3.3.3c implies that

$$\mathbf{X}^T\mathbf{X}(\mathbf{X}^T\mathbf{X})^-\mathbf{X}^T\mathbf{y} = \mathbf{X}^T\mathbf{y},$$

demonstrating that $(\mathbf{X}^T\mathbf{X})^-\mathbf{X}^T\mathbf{y}$ is a solution to the normal equations. □

As noted above, the fact that a global minimizer of $Q(\boldsymbol{\beta})$ necessarily satisfies the normal equations does not guarantee that an arbitrary solution to the normal equations minimizes $Q(\boldsymbol{\beta})$. But it suggests that it might. The following theorem establishes that indeed it does.

Theorem 7.1.2 *$Q(\boldsymbol{\beta})$ for the model $\{\mathbf{y}, \mathbf{X}\boldsymbol{\beta}\}$ assumes a global minimum at $\boldsymbol{\beta} = \hat{\boldsymbol{\beta}}$ if and only if $\hat{\boldsymbol{\beta}}$ is a solution to the normal equations for that model.*

Proof Let $\hat{\boldsymbol{\beta}}$ represent any solution to the normal equations for the specified model. Then,

$$\begin{aligned} Q(\boldsymbol{\beta}) &= (\mathbf{y} - \mathbf{X}\boldsymbol{\beta})^T(\mathbf{y} - \mathbf{X}\boldsymbol{\beta}) \\ &= [(\mathbf{y} - \mathbf{X}\hat{\boldsymbol{\beta}}) + \mathbf{X}(\hat{\boldsymbol{\beta}} - \boldsymbol{\beta})]^T[(\mathbf{y} - \mathbf{X}\hat{\boldsymbol{\beta}}) + \mathbf{X}(\hat{\boldsymbol{\beta}} - \boldsymbol{\beta})] \\ &= (\mathbf{y} - \mathbf{X}\hat{\boldsymbol{\beta}})^T(\mathbf{y} - \mathbf{X}\hat{\boldsymbol{\beta}}) + [\mathbf{X}(\hat{\boldsymbol{\beta}} - \boldsymbol{\beta})]^T[\mathbf{X}(\hat{\boldsymbol{\beta}} - \boldsymbol{\beta})], \end{aligned}$$

because $\mathbf{X}^T(\mathbf{y}-\mathbf{X}\hat{\boldsymbol{\beta}}) = \mathbf{X}^T\mathbf{y}-\mathbf{X}^T\mathbf{X}\hat{\boldsymbol{\beta}} = \mathbf{0}$. Thus, $Q(\boldsymbol{\beta}) \geq Q(\hat{\boldsymbol{\beta}})$, which establishes the sufficiency. For the necessity, let $\bar{\boldsymbol{\beta}}$ be any other global minimizer of $Q(\boldsymbol{\beta})$. Then by Corollary 2.10.4.1,

$$Q(\bar{\boldsymbol{\beta}}) = Q(\hat{\boldsymbol{\beta}}) \Rightarrow [\mathbf{X}(\hat{\boldsymbol{\beta}}-\bar{\boldsymbol{\beta}})]^T[\mathbf{X}(\hat{\boldsymbol{\beta}}-\bar{\boldsymbol{\beta}})] = 0 \Rightarrow \mathbf{X}\bar{\boldsymbol{\beta}} = \mathbf{X}\hat{\boldsymbol{\beta}} \Rightarrow \mathbf{X}^T\mathbf{X}\bar{\boldsymbol{\beta}} = \mathbf{X}^T\mathbf{X}\hat{\boldsymbol{\beta}} = \mathbf{X}^T\mathbf{y},$$

demonstrating that $\bar{\boldsymbol{\beta}}$ is also a solution to the normal equations for the specified model. □

Using Theorem 7.1.2, we may update the definition of a least squares estimator of an estimable function as follows.

Definition 7.1.4 (Updated Definition of a Least Squares Estimator) A *least squares estimator* of an estimable function $\mathbf{c}^T\boldsymbol{\beta}$ associated with the model $\{\mathbf{y}, \mathbf{X}\boldsymbol{\beta}\}$ is $\mathbf{c}^T\hat{\boldsymbol{\beta}}$, where $\hat{\boldsymbol{\beta}}$ is any solution to the normal equations for that model.

In the proof of Theorem 7.1.2, a secondary result emerged that we state as a corollary, as it is important in its own right.

Corollary 7.1.2.1 $\mathbf{X}\hat{\boldsymbol{\beta}} = \mathbf{X}\tilde{\boldsymbol{\beta}}$ *for every two solutions* $\hat{\boldsymbol{\beta}}$ *and* $\tilde{\boldsymbol{\beta}}$ *to the normal equations for the model* $\{\mathbf{y}, \mathbf{X}\boldsymbol{\beta}\}$.

Definition 7.1.5 The unique vector $\hat{\mathbf{y}} = \mathbf{X}\hat{\boldsymbol{\beta}}$, where $\hat{\boldsymbol{\beta}}$ is any solution to the normal equations for the model $\{\mathbf{y}, \mathbf{X}\boldsymbol{\beta}\}$, is called the *(least squares) fitted values vector* for that model. The unique vector $\hat{\mathbf{e}} = \mathbf{y} - \hat{\mathbf{y}}$ is called the *(least squares) fitted residuals vector* for that model.

If $\hat{\boldsymbol{\beta}}$ is a solution to the normal equations for the model $\{\mathbf{y}, \mathbf{X}\boldsymbol{\beta}\}$, then $\mathbf{X}^T\mathbf{X}\hat{\boldsymbol{\beta}} = \mathbf{X}^T\mathbf{y}$, which implies that $\mathbf{X}^T(\mathbf{y} - \mathbf{X}\hat{\boldsymbol{\beta}}) = \mathbf{0}$, or equivalently that $\mathbf{X}^T\hat{\mathbf{e}} = \mathbf{0}$, i.e., $\hat{\mathbf{e}} \in \mathcal{N}(\mathbf{X}^T)$. That is, the least squares fitted residuals vector is orthogonal to the vector of observed values of each explanatory variable. If, as is often the case, $\mathbf{1}_n$ is one of the columns of $\mathbf{X}$, then the least squares fitted residuals sum to 0.

Because (by Corollary 7.1.2.1) each element of the estimable vector $\mathbf{X}\boldsymbol{\beta}$ has a unique least squares estimator, the reader may be motivated to ask, "Does every linear function $\mathbf{c}^T\boldsymbol{\beta}$ have a unique least squares estimator?" The following theorem reveals that the answer to this question is "no"; however, the answer to the modified question "Does every *estimable* linear function $\mathbf{c}^T\boldsymbol{\beta}$ have a unique least squares estimator?" is yes.

Theorem 7.1.3 *Let* $\mathbf{c}^T\boldsymbol{\beta}$ *be a linear function. Then,* $\mathbf{c}^T\hat{\boldsymbol{\beta}} = \mathbf{c}^T\tilde{\boldsymbol{\beta}}$ *for every two solutions* $\hat{\boldsymbol{\beta}}$ *and* $\tilde{\boldsymbol{\beta}}$ *to the normal equations for the model* $\{\mathbf{y}, \mathbf{X}\boldsymbol{\beta}\}$ *if and only if* $\mathbf{c}^T\boldsymbol{\beta}$ *is estimable under that model.*

Proof First let us prove the sufficiency of the estimability condition. If $\mathbf{c}^T\boldsymbol{\beta}$ is estimable under the specified model, then $\mathbf{c}^T = \mathbf{a}^T\mathbf{X}$ for some n-vector $\mathbf{a}$ by Theorem 6.1.2. Using Corollary 7.1.2.1, we have

$$\mathbf{c}^T\hat{\boldsymbol{\beta}} = \mathbf{a}^T\mathbf{X}\hat{\boldsymbol{\beta}} = \mathbf{a}^T\mathbf{X}\tilde{\boldsymbol{\beta}} = \mathbf{c}^T\tilde{\boldsymbol{\beta}}$$

for every two solutions $\hat{\boldsymbol{\beta}}$ and $\tilde{\boldsymbol{\beta}}$ to the normal equations for that model.

Next we prove the necessity. Let $(\mathbf{X}^T\mathbf{X})^-$ be an arbitrary generalized inverse of $\mathbf{X}^T\mathbf{X}$. In light of Theorems 3.2.3 and 7.1.1, all solutions to the normal equations for the specified model are obtained by forming

$$(\mathbf{X}^T\mathbf{X})^-\mathbf{X}^T\mathbf{y} + [\mathbf{I} - (\mathbf{X}^T\mathbf{X})^-\mathbf{X}^T\mathbf{X}]\mathbf{z}$$

and letting $\mathbf{z}$ range over the space of all possible p-vectors. Thus, if $\mathbf{c}^T\hat{\boldsymbol{\beta}} = \mathbf{c}^T\tilde{\boldsymbol{\beta}}$ for every two solutions to the normal equations, then

$$\mathbf{c}^T\{(\mathbf{X}^T\mathbf{X})^-\mathbf{X}^T\mathbf{y}+[\mathbf{I}-(\mathbf{X}^T\mathbf{X})^-\mathbf{X}^T\mathbf{X}]\mathbf{z}_1\} = \mathbf{c}^T\{(\mathbf{X}^T\mathbf{X})^-\mathbf{X}^T\mathbf{y}+[\mathbf{I}-(\mathbf{X}^T\mathbf{X})^-\mathbf{X}^T\mathbf{X}]\mathbf{z}_2\}$$

for all $\mathbf{z}_1$ and $\mathbf{z}_2$, i.e.,

$$\mathbf{c}^T[\mathbf{I} - (\mathbf{X}^T\mathbf{X})^-\mathbf{X}^T\mathbf{X}]\mathbf{z} = 0$$

for all $\mathbf{z}$, i.e.,

$$\mathbf{c}^T[\mathbf{I} - (\mathbf{X}^T\mathbf{X})^-\mathbf{X}^T\mathbf{X}] = \mathbf{0}^T,$$

i.e.,

$$\mathbf{c}^T = \mathbf{c}^T(\mathbf{X}^T\mathbf{X})^-\mathbf{X}^T\mathbf{X},$$

i.e.,

$$\mathbf{c}^T \in \mathcal{R}(\mathbf{X}).$$

The estimability of $\mathbf{c}^T\boldsymbol{\beta}$ under the specified model follows from Theorem 6.1.2. □

Theorem 7.1.3 allows us to finalize the definition of "least squares estimator," replacing the indefinite article "a" with the definite article "the." In the finalized definition, we also extend the term to vectors of estimable functions.

Definition 7.1.6 (Final Definition of Least Squares Estimator) The *least squares estimator* of an estimable function $\mathbf{c}^T\boldsymbol{\beta}$ associated with the model $\{\mathbf{y}, \mathbf{X}\boldsymbol{\beta}\}$ is $\mathbf{c}^T\hat{\boldsymbol{\beta}}$, where $\hat{\boldsymbol{\beta}}$ is any solution to the normal equations for that model. The *least squares estimator* of a vector $\mathbf{C}^T\boldsymbol{\beta}$ of estimable functions associated with the model $\{\mathbf{y}, \mathbf{X}\boldsymbol{\beta}\}$ is the vector $\mathbf{C}^T\hat{\boldsymbol{\beta}}$ of least squares estimators of those functions.

Theorem 7.1.4 *The least squares estimator of a vector of estimable functions* $\mathbf{C}^T\boldsymbol{\beta}$ *associated with the model* $\{\mathbf{y}, \mathbf{X}\boldsymbol{\beta}\}$ *is* $\mathbf{C}^T(\mathbf{X}^T\mathbf{X})^-\mathbf{X}^T\mathbf{y}$, *and this expression is invariant to the choice of generalized inverse of* $\mathbf{X}^T\mathbf{X}$.

Proof Because $\mathbf{C}^T\boldsymbol{\beta}$ is estimable under the model $\{\mathbf{y}, \mathbf{X}\boldsymbol{\beta}\}$, $\mathbf{C}^T = \mathbf{A}^T\mathbf{X}$ for some matrix $\mathbf{A}$. Let $\hat{\boldsymbol{\beta}}$ represent any solution to the normal equations for that model. Then for any generalized inverse of $\mathbf{X}^T\mathbf{X}$, there exists (by Theorem 3.2.3) a p-vector $\mathbf{z}$ such that $\hat{\boldsymbol{\beta}}$ is expressible as

$$\hat{\boldsymbol{\beta}} = (\mathbf{X}^T\mathbf{X})^-\mathbf{X}^T\mathbf{y} + [\mathbf{I} - (\mathbf{X}^T\mathbf{X})^-\mathbf{X}^T\mathbf{X}]\mathbf{z}.$$

Consequently,

$$\begin{aligned}\mathbf{C}^T\hat{\boldsymbol{\beta}} &= \mathbf{C}^T(\mathbf{X}^T\mathbf{X})^-\mathbf{X}^T\mathbf{y} + \mathbf{A}^T\mathbf{X}[\mathbf{I} - (\mathbf{X}^T\mathbf{X})^-\mathbf{X}^T\mathbf{X}]\mathbf{z} \\ &= \mathbf{C}^T(\mathbf{X}^T\mathbf{X})^-\mathbf{X}^T\mathbf{y}.\end{aligned}$$

Because the generalized inverse in Theorem 3.2.3 is arbitrary, this expression has the claimed invariance. □

Corollary 7.1.4.1 *If* $\mathbf{C}^T\boldsymbol{\beta}$ *is estimable under the model* $\{\mathbf{y}, \mathbf{X}\boldsymbol{\beta}\}$ *and* $\mathbf{L}^T$ *is a matrix of constants having the same number of columns as* $\mathbf{C}^T$ *has rows, then* $\mathbf{L}^T\mathbf{C}^T\boldsymbol{\beta}$ *is estimable and its least squares estimator (associated with the specified model) is* $\mathbf{L}^T(\mathbf{C}^T\hat{\boldsymbol{\beta}})$.

Proof Left as an exercise. □

Example 7.1-1. Least Squares Estimators of Intercept and Slope in the Simple Linear Regression Model

Consider the simple linear regression model, for which $\mathbf{X} = (\mathbf{1}_n, \mathbf{x})$ where $\mathbf{1}_n \notin \mathcal{C}(\mathbf{x})$. By Corollary 6.1.2.4, β_1 and β_2 are estimable, and their least squares estimators are the elements of

$$\begin{aligned}
\begin{pmatrix} \hat{\beta}_1 \\ \hat{\beta}_2 \end{pmatrix} &= [(\mathbf{1}_n, \mathbf{x})^T(\mathbf{1}_n, \mathbf{x})]^{-1}(\mathbf{1}_n, \mathbf{x})^T\mathbf{y} \\
&= \begin{pmatrix} n & n\bar{x} \\ n\bar{x} & \sum_{i=1}^n x_i^2 \end{pmatrix}^{-1} \begin{pmatrix} n\bar{y} \\ \sum_{i=1}^n x_i y_i \end{pmatrix} \\
&= (1/SXX) \begin{pmatrix} \sum_{i=1}^n x_i^2/n & -\bar{x} \\ -\bar{x} & 1 \end{pmatrix} \begin{pmatrix} n\bar{y} \\ \sum_{i=1}^n x_i y_i \end{pmatrix} \\
&= (1/SXX) \begin{pmatrix} \bar{y}(\sum_{i=1}^n x_i^2 - n\bar{x}^2) - \bar{x}(\sum_{i=1}^n x_i y_i - n\bar{x}\bar{y}) \\ \sum_{i=1}^n x_i y_i - n\bar{x}\bar{y} \end{pmatrix} \\
&= \begin{pmatrix} \bar{y} - \hat{\beta}_2\bar{x} \\ SXY/SXX \end{pmatrix}
\end{aligned}$$

where here and throughout the rest of the book, $SXX = \sum_{i=1}^n x_i^2 - n\bar{x}^2 = \sum_{i=1}^n (x_i - \bar{x})^2$ and $SXY = \sum_{i=1}^n x_i y_i - n\bar{x}\bar{y} = \sum_{i=1}^n (x_i - \bar{x})(y_i - \bar{y})$. ■

Example 7.1-2. Least Squares Estimators of Level Means and Contrasts in the One-Factor Model

Consider the one-factor model, for which $\mathbf{X} = (\mathbf{1}_n, \oplus_{i=1}^q \mathbf{1}_{n_i})$. It may be verified that

$$\mathbf{X}^T\mathbf{X} = \begin{pmatrix} n & \mathbf{n}^T \\ \mathbf{n} & \oplus_{i=1}^q n_i \end{pmatrix}$$

where $\mathbf{n} = (n_1, n_2, \ldots, n_q)^T$. Now, $\mathbf{X}^T\mathbf{X}$ is a $(q+1) \times (q+1)$ symmetric matrix, and by Theorem 2.8.8 and the results of Example 6.1-2, $\text{rank}(\mathbf{X}^T\mathbf{X}) = \text{rank}(\mathbf{X}) = q$. Because $\oplus_{i=1}^q n_i$ is nonsingular,

$$\begin{pmatrix} 0 & \mathbf{0}^T \\ \mathbf{0} & \oplus_{i=1}^q n_i^{-1} \end{pmatrix}$$

is a generalized inverse of $\mathbf{X}^T\mathbf{X}$ (by Theorem 3.3.6). Consider first the estimable functions $\{\mu + \alpha_i : i = 1, \ldots, q\}$, which are the elements of $\mathbf{C}^T\boldsymbol{\beta}$ where $\mathbf{C}^T = (\mathbf{1}_q, \mathbf{I}_q)$. By Theorem 7.1.4, the least squares estimator of this vector is

$$(\mathbf{1}_q, \mathbf{I}_q)\begin{pmatrix} 0 & \mathbf{0}^T \\ \mathbf{0} & \oplus_{i=1}^q n_i^{-1} \end{pmatrix}(\mathbf{1}_n, \oplus_{i=1}^q \mathbf{1}_{n_i})^T\mathbf{y} = (\oplus_{i=1}^q n_i^{-1}\mathbf{1}_{n_i})^T\mathbf{y}$$

$$= \begin{pmatrix} \bar{y}_{1\cdot} \\ \bar{y}_{2\cdot} \\ \vdots \\ \bar{y}_{q\cdot} \end{pmatrix} \equiv \bar{\mathbf{y}},$$

say. Because $\alpha_i - \alpha_{i'} = (\mathbf{u}_i - \mathbf{u}_{i'})^T\mathbf{C}^T\boldsymbol{\beta}$ where $\mathbf{u}_i$ is the ith unit q-vector, the least squares estimator of any difference $\alpha_i - \alpha_{i'}$ is (by Corollary 7.1.4.1)

$$(\mathbf{u}_i - \mathbf{u}_{i'})^T\bar{\mathbf{y}} = \bar{y}_{i\cdot} - \bar{y}_{i'\cdot},$$

and more generally, the least squares estimator of any contrast $\sum_{i=1}^q d_i\alpha_i$ is $\sum_{i=1}^q d_i\bar{y}_{i\cdot}$. Observe that this estimator coincides with the second estimator listed in the second row of the table in Example 6.1-2. ■

Example 7.1-3. Least Squares Estimators of Cell Means and Factor Differences in the Two-Way Main Effects Model for Some 2 × 2 Layouts

In this example, we obtain expressions for the least squares estimators of some estimable functions when the data follow a two-way main effects model with two levels of each factor and are laid out according to the 2 × 2 layouts depicted below. Unlike layouts of this type that were displayed in Chap. 6, here we denote the observations by y_{ijk} rather than by an asterisk:

y_{111}		y_{111}	y_{121}	y_{111}	y_{121}	y_{111}, y_{112}	y_{121}
	y_{221}	y_{211}		y_{211}	y_{221}	y_{211}	

From left to right, the first layout is not connected; the second layout has an empty cell but is connected; the third layout has equal frequencies (of size 1); and the fourth layout has unequal frequencies and one empty cell. For reasons that will become clear later, let us modify the usual ordering of elements in $\boldsymbol{\beta}$ by writing $\boldsymbol{\beta} = (\mu, \alpha_1, \gamma_1, \alpha_2, \gamma_2)^T$.

For the first layout, we know (from Example 6.1-4) that the only cell means that are estimable are those corresponding to the occupied cells, namely $\mu + \alpha_1 + \gamma_1$ and

$\mu + \alpha_2 + \gamma_2$, and that neither the Factor A difference nor the Factor B difference is estimable. Furthermore,

$$\mathbf{X} = \begin{pmatrix} 1 & 1 & 1 & 0 & 0 \\ 1 & 0 & 0 & 1 & 1 \end{pmatrix}, \quad \mathbf{X}^T\mathbf{X} = \begin{pmatrix} 2 & 1 & 1 & 1 & 1 \\ 1 & 1 & 1 & 0 & 0 \\ 1 & 1 & 1 & 0 & 0 \\ 1 & 0 & 0 & 1 & 1 \\ 1 & 0 & 0 & 1 & 1 \end{pmatrix},$$

and because $\text{rank}(\mathbf{X}^T\mathbf{X}) = 2$, one generalized inverse of $\mathbf{X}^T\mathbf{X}$ is

$$(\mathbf{X}^T\mathbf{X})^- = \begin{pmatrix} 1 & -1 & 0 & 0 & 0 \\ -1 & 2 & 0 & 0 & 0 \\ 0 & 0 & 0 & 0 & 0 \\ 0 & 0 & 0 & 0 & 0 \\ 0 & 0 & 0 & 0 & 0 \end{pmatrix}$$

(by Theorem 3.3.6). So, by Theorem 7.1.4 the least squares estimators of the estimable cell means are

$$\begin{pmatrix} \widehat{\mu + \alpha_1 + \gamma_1} \\ \widehat{\mu + \alpha_2 + \gamma_2} \end{pmatrix} = \begin{pmatrix} 1 & 1 & 1 & 0 & 0 \\ 1 & 0 & 0 & 1 & 1 \end{pmatrix} \begin{pmatrix} 1 & -1 & 0 & 0 & 0 \\ -1 & 2 & 0 & 0 & 0 \\ 0 & 0 & 0 & 0 & 0 \\ 0 & 0 & 0 & 0 & 0 \\ 0 & 0 & 0 & 0 & 0 \end{pmatrix} \begin{pmatrix} 1 & 1 \\ 1 & 0 \\ 1 & 0 \\ 0 & 1 \\ 0 & 1 \end{pmatrix} \begin{pmatrix} y_{111} \\ y_{221} \end{pmatrix} = \begin{pmatrix} y_{111} \\ y_{221} \end{pmatrix},$$

as might be expected.

Because the second layout is connected, all cell means and the Factor A and Factor B differences are estimable (Theorem 6.1.4). Moreover,

$$\mathbf{X} = \begin{pmatrix} 1 & 1 & 1 & 0 & 0 \\ 1 & 1 & 0 & 0 & 1 \\ 1 & 0 & 1 & 1 & 0 \end{pmatrix}, \quad \mathbf{X}^T\mathbf{X} = \begin{pmatrix} 3 & 2 & 2 & 1 & 1 \\ 2 & 2 & 1 & 0 & 1 \\ 2 & 1 & 2 & 1 & 0 \\ 1 & 0 & 1 & 1 & 0 \\ 1 & 1 & 0 & 0 & 1 \end{pmatrix},$$

and because $\text{rank}(\mathbf{X}^T\mathbf{X}) = 3$, one generalized inverse of $\mathbf{X}^T\mathbf{X}$ is

$$(\mathbf{X}^T\mathbf{X})^- = \begin{pmatrix} 0 & 0 & 0 & 0 & 0 \\ 0 & 0 & 0 & 0 & 0 \\ 0 & 0 & 1 & -1 & 0 \\ 0 & 0 & -1 & 2 & 0 \\ 0 & 0 & 0 & 0 & 1 \end{pmatrix}.$$

(Here, we obtained this generalized inverse "by hand," using the second part of Theorem 3.3.6 with

$$\mathbf{D} = \begin{pmatrix} 2 & 1 & 0 \\ 1 & 1 & 0 \\ 0 & 0 & 1 \end{pmatrix}$$

and Corollary 2.9.5.1 with $\mathbf{B} = 1$.) Using Theorem 7.1.4, it may then be verified easily that the least squares estimators of the cell means and Factor A and B differences are

$$\begin{pmatrix} \widehat{\mu + \alpha_1 + \gamma_1} \\ \widehat{\mu + \alpha_1 + \gamma_2} \\ \widehat{\mu + \alpha_2 + \gamma_1} \\ \widehat{\mu + \alpha_2 + \gamma_2} \end{pmatrix} = \begin{pmatrix} y_{111} \\ y_{121} \\ y_{211} \\ y_{121} + y_{211} - y_{111} \end{pmatrix} \quad \text{and} \quad \begin{pmatrix} \widehat{\alpha_1 - \alpha_2} \\ \widehat{\gamma_1 - \gamma_2} \end{pmatrix} = \begin{pmatrix} y_{111} - y_{211} \\ y_{111} - y_{121} \end{pmatrix},$$

respectively. Thus, the least squares estimators of the cell means corresponding to occupied cells are merely the observations in those cells, but the least squares estimator of the cell mean corresponding to the empty cell is a linear combination of all three observations.

All cell means and Factor A and Factor B differences are estimable in the third layout also, but the least squares estimators of some of these functions differ from those in the second layout, as we now show. In this case,

$$\mathbf{X} = \begin{pmatrix} 1 & 1 & 1 & 0 & 0 \\ 1 & 1 & 0 & 0 & 1 \\ 1 & 0 & 1 & 1 & 0 \\ 1 & 0 & 0 & 1 & 1 \end{pmatrix}, \quad \mathbf{X}^T\mathbf{X} = \begin{pmatrix} 4 & 2 & 2 & 2 & 2 \\ 2 & 2 & 1 & 0 & 1 \\ 2 & 1 & 2 & 1 & 0 \\ 2 & 0 & 1 & 2 & 1 \\ 2 & 1 & 0 & 1 & 2 \end{pmatrix},$$

and one generalized inverse of $\mathbf{X}^T\mathbf{X}$ is

$$(\mathbf{X}^T\mathbf{X})^- = \begin{pmatrix} 3/4 & -1/2 & -1/2 & 0 & 0 \\ -1/2 & 1 & 0 & 0 & 0 \\ -1/2 & 0 & 1 & 0 & 0 \\ 0 & 0 & 0 & 0 & 0 \\ 0 & 0 & 0 & 0 & 0 \end{pmatrix}.$$

(We obtained this generalized inverse by: (1) noting from Theorem 3.3.6 that the 5×5 matrix whose upper left 3×3 submatrix is the ordinary inverse of

$$\begin{pmatrix} 4 & 2 & 2 \\ 2 & 2 & 1 \\ 2 & 1 & 2 \end{pmatrix}$$

and remaining elements are zeros is a generalized inverse of $\mathbf{X}^T\mathbf{X}$, and (2) obtaining the ordinary inverse of that 3×3 matrix using a computer.) It can be verified that the least squares estimators of the cell means and Factor A and B differences are

$$\begin{pmatrix} \widehat{\mu + \alpha_1 + \gamma_1} \\ \widehat{\mu + \alpha_1 + \gamma_2} \\ \widehat{\mu + \alpha_2 + \gamma_1} \\ \widehat{\mu + \alpha_2 + \gamma_2} \end{pmatrix} = \begin{pmatrix} (3y_{111}/4) + (y_{121} + y_{211} - y_{221})/4 \\ (3y_{121}/4) + (y_{111} + y_{221} - y_{211})/4 \\ (3y_{211}/4) + (y_{111} + y_{221} - y_{121})/4 \\ (3y_{221}/4) + (y_{121} + y_{211} - y_{111})/4 \end{pmatrix} \quad \text{and}$$

$$\begin{pmatrix} \widehat{\alpha_1 - \alpha_2} \\ \widehat{\gamma_1 - \gamma_2} \end{pmatrix} = \begin{pmatrix} \bar{y}_{1\cdot\cdot} - \bar{y}_{2\cdot\cdot} \\ \bar{y}_{\cdot 1\cdot} - \bar{y}_{\cdot 2\cdot} \end{pmatrix},$$

respectively. Thus, the least squares estimator of each cell mean is not merely the observation in the corresponding cell but is a nontrivial linear combination of all the observations, though more weight is given to the observation in the corresponding cell than to the others. Least squares estimators of the Factor A and Factor B differences are merely the differences in the level means for Factor A and Factor B, respectively. The extension of these results to a balanced two-way layout of arbitrary size is the topic of an exercise.

Finally, for the last layout we have

$$\mathbf{X} = \begin{pmatrix} 1 & 1 & 1 & 0 & 0 \\ 1 & 1 & 1 & 0 & 0 \\ 1 & 1 & 0 & 0 & 1 \\ 1 & 0 & 1 & 1 & 0 \end{pmatrix}, \quad \mathbf{X}^T\mathbf{X} = \begin{pmatrix} 4 & 3 & 3 & 1 & 1 \\ 3 & 3 & 2 & 0 & 1 \\ 3 & 2 & 3 & 1 & 0 \\ 1 & 0 & 1 & 1 & 0 \\ 1 & 1 & 0 & 0 & 1 \end{pmatrix},$$

and one generalized inverse of $\mathbf{X}^T\mathbf{X}$ is

$$(\mathbf{X}^T\mathbf{X})^- = \frac{1}{2}\begin{pmatrix} 0 & 0 & 0 & 0 & 0 \\ 0 & 0 & 0 & 0 & 0 \\ 0 & 0 & 1 & -1 & 0 \\ 0 & 0 & -1 & 3 & 0 \\ 0 & 0 & 0 & 0 & 2 \end{pmatrix}.$$

(Here again, we obtained this generalized inverse "by hand," using Theorem 3.3.6 and Corollary 2.9.5.1.) After the appropriate calculations, the least squares estimators of the cell means and Factor A and B differences are

$$\begin{pmatrix} \widehat{\mu + \alpha_1 + \gamma_1} \\ \widehat{\mu + \alpha_1 + \gamma_2} \\ \widehat{\mu + \alpha_2 + \gamma_1} \\ \widehat{\mu + \alpha_2 + \gamma_2} \end{pmatrix} = \begin{pmatrix} \bar{y}_{11\cdot} \\ y_{121} \\ y_{211} \\ y_{121} + y_{211} - \bar{y}_{11\cdot} \end{pmatrix} \quad \text{and} \quad \begin{pmatrix} \widehat{\alpha_1 - \alpha_2} \\ \widehat{\gamma_1 - \gamma_2} \end{pmatrix} = \begin{pmatrix} \bar{y}_{11\cdot} - y_{211} \\ \bar{y}_{11\cdot} - y_{121} \end{pmatrix}.$$

■

7.2 Properties of Least Squares Estimators

The least squares estimators of estimable functions have several important properties. We establish some of these properties in this section, via three theorems.

Theorem 7.2.1 *The least squares estimator of an estimable function* $\mathbf{c}^T\boldsymbol{\beta}$ *associated with the model* $\{\mathbf{y}, \mathbf{X}\boldsymbol{\beta}\}$ *is linear and unbiased under any unconstrained model with model matrix* $\mathbf{X}$.

Proof Because $\mathbf{c}^T\boldsymbol{\beta}$ is estimable under the specified model, $\mathbf{c}^T = \mathbf{a}^T\mathbf{X}$ for some n-vector $\mathbf{a}$, and the least squares estimator of $\mathbf{c}^T\boldsymbol{\beta}$ may be written as $\mathbf{c}^T(\mathbf{X}^T\mathbf{X})^-\mathbf{X}^T\mathbf{y}$, invariant to the choice of generalized inverse of $\mathbf{X}^T\mathbf{X}$ (Theorem 7.1.4). Thus, the least squares estimator is of linear-estimator form

$$t_0 + \mathbf{t}^T\mathbf{y},$$

where $t_0 = 0$ and $\mathbf{t}^T = \mathbf{c}^T(\mathbf{X}^T\mathbf{X})^-\mathbf{X}^T$. The unbiasedness under any unconstrained model with model matrix $\mathbf{X}$ follows from Theorems 6.1.1 and 3.3.3c upon observing that

$$\mathbf{t}^T\mathbf{X} = \mathbf{c}^T(\mathbf{X}^T\mathbf{X})^-\mathbf{X}^T\mathbf{X} = \mathbf{a}^T\mathbf{X}(\mathbf{X}^T\mathbf{X})^-\mathbf{X}^T\mathbf{X} = \mathbf{a}^T\mathbf{X} = \mathbf{c}^T.$$

□

The next theorem gives expressions for the matrix of covariances of two vectors of least squares estimators, and for the variance–covariance matrix of a vector of least squares estimators, under the Gauss–Markov model $\{\mathbf{y}, \mathbf{X}\boldsymbol{\beta}, \sigma^2\mathbf{I}\}$.

Theorem 7.2.2 *Let* $\mathbf{C}_1^T\boldsymbol{\beta}$ *and* $\mathbf{C}_2^T\boldsymbol{\beta}$ *be two vectors of estimable functions under the model* $\{\mathbf{y}, \mathbf{X}\boldsymbol{\beta}\}$, *and let* $\hat{\boldsymbol{\beta}}$ *represent any solution to the normal equations for that model, so that* $\mathbf{C}_1^T\hat{\boldsymbol{\beta}}$ *and* $\mathbf{C}_2^T\hat{\boldsymbol{\beta}}$ *are the least squares estimators of* $\mathbf{C}_1^T\boldsymbol{\beta}$ *and* $\mathbf{C}_2^T\boldsymbol{\beta}$, *respectively, associated with that model. Then under the Gauss–Markov model* $\{\mathbf{y}, \mathbf{X}\boldsymbol{\beta}, \sigma^2\mathbf{I}\}$,

$$cov(\mathbf{C}_1^T\hat{\boldsymbol{\beta}}, \mathbf{C}_2^T\hat{\boldsymbol{\beta}}) = \sigma^2\mathbf{C}_1^T(\mathbf{X}^T\mathbf{X})^-\mathbf{C}_2 \quad and \quad var(\mathbf{C}_1^T\hat{\boldsymbol{\beta}}) = \sigma^2\mathbf{C}_1^T(\mathbf{X}^T\mathbf{X})^-\mathbf{C}_1,$$

and these expressions are invariant to the choice of generalized inverse of $\mathbf{X}^T\mathbf{X}$. *Furthermore, if the functions in* $\mathbf{C}_1^T\boldsymbol{\beta}$ *are linearly independent, then* $var(\mathbf{C}_1^T\hat{\boldsymbol{\beta}})$ *is positive definite (hence nonsingular).*

Proof Because $\mathbf{C}_1^T\boldsymbol{\beta}$ and $\mathbf{C}_2^T\boldsymbol{\beta}$ are estimable under the specified model, there exist (by Theorem 6.1.2) matrices $\mathbf{A}_1$ and $\mathbf{A}_2$ such that

$$\mathbf{C}_1^T = \mathbf{A}_1^T\mathbf{X} \quad \text{and} \quad \mathbf{C}_2^T = \mathbf{A}_2^T\mathbf{X};$$

furthermore,

$$\mathbf{C}_1^T\hat{\boldsymbol{\beta}} = \mathbf{C}_1^T(\mathbf{X}^T\mathbf{X})^-\mathbf{X}^T\mathbf{y} \quad \text{and} \quad \mathbf{C}_2^T\hat{\boldsymbol{\beta}} = \mathbf{C}_2^T(\mathbf{X}^T\mathbf{X})^-\mathbf{X}^T\mathbf{y},$$

invariant to the choice of generalized inverse of $\mathbf{X}^T\mathbf{X}$ (by Theorem 7.1.4). Then using Corollary 4.2.3.1 and Theorem 3.3.3d, we obtain

$$\begin{aligned}
\text{cov}(\mathbf{C}_1^T\hat{\boldsymbol{\beta}}, \mathbf{C}_2^T\hat{\boldsymbol{\beta}}) &= \text{cov}[\mathbf{A}_1^T\mathbf{X}(\mathbf{X}^T\mathbf{X})^-\mathbf{X}^T\mathbf{y},\ \mathbf{A}_2^T\mathbf{X}(\mathbf{X}^T\mathbf{X})^-\mathbf{X}^T\mathbf{y}] \\
&= \mathbf{A}_1^T\mathbf{X}(\mathbf{X}^T\mathbf{X})^-\mathbf{X}^T(\sigma^2\mathbf{I})\mathbf{X}[(\mathbf{X}^T\mathbf{X})^-]^T\mathbf{X}^T\mathbf{A}_2 \\
&= \sigma^2\mathbf{A}_1^T\mathbf{X}(\mathbf{X}^T\mathbf{X})^-\mathbf{X}^T\mathbf{A}_2 \\
&= \sigma^2\mathbf{C}_1^T(\mathbf{X}^T\mathbf{X})^-\mathbf{C}_2.
\end{aligned}$$

The invariance of this quantity with respect to the choice of generalized inverse of $\mathbf{X}^T\mathbf{X}$ can be seen by applying Theorem 3.3.4 to the expression after the next-to-last equality. The expression for $\text{var}(\mathbf{C}_1^T\hat{\boldsymbol{\beta}})$ may be obtained merely by substituting $\mathbf{C}_1$ for $\mathbf{C}_2$ in the corresponding expression for the matrix of covariances. Now, if the functions in $\mathbf{C}_1^T\boldsymbol{\beta}$ are linearly independent, then letting s denote the number of rows of $\mathbf{C}_1^T$ and using Theorems 3.3.3a and 2.8.4, we have

$$s = \text{rank}(\mathbf{C}_1^T) = \text{rank}(\mathbf{A}_1^T\mathbf{X}) = \text{rank}[\mathbf{A}_1^T\mathbf{X}(\mathbf{X}^T\mathbf{X})^-\mathbf{X}^T\mathbf{X}] \le \text{rank}[\mathbf{A}_1^T\mathbf{X}(\mathbf{X}^T\mathbf{X})^-\mathbf{X}^T] \le s,$$

where the last inequality holds because the dimensions of $\mathbf{A}_1^T\mathbf{X}(\mathbf{X}^T\mathbf{X})^-\mathbf{X}^T$ are $s \times n$. Thus

$$\text{rank}[\mathbf{A}_1^T\mathbf{X}(\mathbf{X}^T\mathbf{X})^-\mathbf{X}^T] = s.$$

Furthermore, using Theorem 3.3.3d again,

$$\begin{aligned}
\mathbf{C}_1^T(\mathbf{X}^T\mathbf{X})^-\mathbf{C}_1 &= \mathbf{A}_1^T\mathbf{X}(\mathbf{X}^T\mathbf{X})^-\mathbf{X}^T\mathbf{A}_1 \\
&= \mathbf{A}_1^T\mathbf{X}(\mathbf{X}^T\mathbf{X})^-\mathbf{X}^T\mathbf{X}[(\mathbf{X}^T\mathbf{X})^-]^T\mathbf{X}^T\mathbf{A}_1 \\
&= [\mathbf{A}_1^T\mathbf{X}(\mathbf{X}^T\mathbf{X})^-\mathbf{X}^T][\mathbf{A}_1^T\mathbf{X}(\mathbf{X}^T\mathbf{X})^-\mathbf{X}^T]^T.
\end{aligned}$$

It follows immediately from Theorem 2.15.9 that $\mathbf{C}_1^T(\mathbf{X}^T\mathbf{X})^-\mathbf{C}_1$ is positive definite. □

The last theorem in this section, commonly known as the Gauss–Markov theorem, informs us that the least squares estimator of an estimable function associated with the model $\{\mathbf{y}, \mathbf{X}\boldsymbol{\beta}\}$ is not merely linear and unbiased under any unconstrained model with model matrix $\mathbf{X}$, but under the Gauss–Markov model $\{\mathbf{y}, \mathbf{X}\boldsymbol{\beta}, \sigma^2\mathbf{I}\}$ it is the unique *best* estimator (in the sense of having smallest variance) among all linear unbiased estimators, or more succinctly, it is the best linear unbiased estimator (BLUE).

Theorem 7.2.3 *Under the Gauss–Markov model* $\{\mathbf{y}, \mathbf{X}\boldsymbol{\beta}, \sigma^2\mathbf{I}\}$, *the variance of the least squares estimator of an estimable function* $\mathbf{c}^T\boldsymbol{\beta}$ *associated with the model* $\{\mathbf{y}, \mathbf{X}\boldsymbol{\beta}\}$ *is uniformly (in* $\boldsymbol{\beta}$ *and* σ^2*) less than that of any other linear unbiased estimator of* $\mathbf{c}^T\boldsymbol{\beta}$.

Proof Because $\mathbf{c}^T\boldsymbol{\beta}$ is estimable under the specified model, $\mathbf{c}^T = \mathbf{a}^T\mathbf{X}$ for some n-vector $\mathbf{a}$. Let $\mathbf{c}^T\hat{\boldsymbol{\beta}}$ and $t_0+\mathbf{t}^T\mathbf{y}$ be the least squares estimator and any linear unbiased estimator, respectively, of $\mathbf{c}^T\boldsymbol{\beta}$. Then (by Theorem 6.1.1) $t_0 = 0$ and $\mathbf{t}^T\mathbf{X} = \mathbf{c}^T$. Consequently, by Theorem 4.2.2b,

$$\begin{aligned}
\text{var}(t_0 + \mathbf{t}^T\mathbf{y}) &= \text{var}(\mathbf{t}^T\mathbf{y}) \\
&= \text{var}[\mathbf{c}^T\hat{\boldsymbol{\beta}} + (\mathbf{t}^T\mathbf{y} - \mathbf{c}^T\hat{\boldsymbol{\beta}})] \\
&= \text{var}(\mathbf{c}^T\hat{\boldsymbol{\beta}}) + \text{var}(\mathbf{t}^T\mathbf{y} - \mathbf{c}^T\hat{\boldsymbol{\beta}}) + 2\,\text{cov}(\mathbf{c}^T\hat{\boldsymbol{\beta}}, \mathbf{t}^T\mathbf{y} - \mathbf{c}^T\hat{\boldsymbol{\beta}}),
\end{aligned}$$

where, by Corollary 4.2.3.1,

$$\begin{aligned}
\text{cov}(\mathbf{c}^T\hat{\boldsymbol{\beta}}, \mathbf{t}^T\mathbf{y} - \mathbf{c}^T\hat{\boldsymbol{\beta}}) &= \text{cov}\{\mathbf{c}^T(\mathbf{X}^T\mathbf{X})^-\mathbf{X}^T\mathbf{y}, [\mathbf{t}^T - \mathbf{c}^T(\mathbf{X}^T\mathbf{X})^-\mathbf{X}^T]\mathbf{y}\} \\
&= \mathbf{c}^T(\mathbf{X}^T\mathbf{X})^-\mathbf{X}^T(\sigma^2\mathbf{I})\{\mathbf{t} - \mathbf{X}[(\mathbf{X}^T\mathbf{X})^-]^T\mathbf{c}\} \\
&= \sigma^2\mathbf{c}^T(\mathbf{X}^T\mathbf{X})^-\{\mathbf{X}^T\mathbf{t} - \mathbf{X}^T\mathbf{X}[(\mathbf{X}^T\mathbf{X})^-]^T\mathbf{X}^T\mathbf{a}\} \\
&= \sigma^2\mathbf{c}^T(\mathbf{X}^T\mathbf{X})^-(\mathbf{c} - \mathbf{c}) \\
&= 0.
\end{aligned}$$

Thus,

$$\begin{aligned}\text{var}(t_0 + \mathbf{t}^T\mathbf{y}) &= \text{var}(\mathbf{c}^T\hat{\boldsymbol{\beta}}) + \text{var}(\mathbf{t}^T\mathbf{y} - \mathbf{c}^T\hat{\boldsymbol{\beta}}) \\ &\geq \text{var}(\mathbf{c}^T\hat{\boldsymbol{\beta}}),\end{aligned}$$

and equality holds if and only if $\text{var}(\mathbf{t}^T\mathbf{y} - \mathbf{c}^T\hat{\boldsymbol{\beta}}) = 0$, i.e., if and only if $\sigma^2\{\mathbf{t}^T - \mathbf{c}^T(\mathbf{X}^T\mathbf{X})^-\mathbf{X}^T\}\{\mathbf{t} - \mathbf{X}[(\mathbf{X}^T\mathbf{X})^-]^T\mathbf{c}\} = 0$, i.e., if and only if $\mathbf{t}^T = \mathbf{c}^T(\mathbf{X}^T\mathbf{X})^-\mathbf{X}^T$, i.e., if and only if $\mathbf{t}^T\mathbf{y} = \mathbf{c}^T(\mathbf{X}^T\mathbf{X})^-\mathbf{X}^T\mathbf{y}$ for all $\mathbf{y}$, i.e., if and only if $\mathbf{t}^T\mathbf{y} = \mathbf{c}^T\hat{\boldsymbol{\beta}}$ for all $\mathbf{y}$. □

Example 7.2-1. Variance–Covariance Matrix of the Least Squares Estimators of Slope and Intercept in Simple Linear Regression

Recall from Example 7.1-1 that the least squares estimators of intercept β_1 and slope β_2 in the simple linear regression model are, respectively,

$$\hat{\beta}_1 = \bar{y} - \hat{\beta}_2\bar{x}, \quad \text{and} \quad \hat{\beta}_2 = \frac{SXY}{SXX},$$

where $SXY = \sum_{i=1}^n (x_i - \bar{x})(y_i - \bar{y})$ and $SXX = \sum_{i=1}^n (x_i - \bar{x})^2$. Also recall that for such a model,

$$(\mathbf{X}^T\mathbf{X})^{-1} = \frac{1}{SXX}\begin{pmatrix} \sum_{i=1}^n x_i^2/n & -\bar{x} \\ -\bar{x} & 1 \end{pmatrix}.$$

By Theorem 7.2.2, the variance–covariance matrix of these estimators under the Gauss–Markov model is

$$\frac{\sigma^2}{SXX}\begin{pmatrix} \sum_{i=1}^n x_i^2/n & -\bar{x} \\ -\bar{x} & 1 \end{pmatrix} = \begin{pmatrix} \sigma^2[(1/n) + (\bar{x}^2/SXX)] & -\sigma^2\bar{x}/SXX \\ -\sigma^2\bar{x}/SXX & \sigma^2/SXX \end{pmatrix}.$$

Both $\text{var}(\hat{\beta}_1)$ and $\text{var}(\hat{\beta}_2)$ decrease as the dispersion of the x_i's, as measured by *SXX*, increases. The variance of $\hat{\beta}_1$ also decreases as the center of the x_i's, as measured by $\bar{x}$, gets closer to 0. Also,

$$\text{corr}(\hat{\beta}_1, \hat{\beta}_2) = \frac{-\bar{x}}{[(SXX/n) + \bar{x}^2]^{\frac{1}{2}}},$$

which has sign opposite to that of $\bar{x}$, and equals 0 if and only if $\bar{x} = 0$. ■

Example 7.2-2. Variance–Covariance Matrices of the Least Squares Estimators of Level Means and Differences in the One-Factor Model

Recall from Example 7.1-2 that the least squares estimator of the vector of functions $(\mu+\alpha_1, \ldots, \mu+\alpha_q)^T$ is $(\oplus_{i=1}^q n_i^{-1}\mathbf{1}_{n_i})^T\mathbf{y} = (\bar{y}_{1\cdot}, \ldots, \bar{y}_{q\cdot})^T \equiv \bar{\mathbf{y}}$, say. Using results from that example, we obtain, under the Gauss–Markov model,

$$\begin{aligned}\text{var}(\bar{\mathbf{y}}) &= (\oplus_{i=1}^q n_i^{-1}\mathbf{1}_{n_i})^T(\sigma^2\mathbf{I})(\oplus_{i=1}^q n_i^{-1}\mathbf{1}_{n_i}) \\ &= \sigma^2 \oplus_{i=1}^q n_i^{-1}.\end{aligned}$$

Equivalently,

$$\text{cov}(\bar{y}_{i\cdot}, \bar{y}_{j\cdot}) = \begin{cases} \sigma^2/n_i & \text{if } i = j, \\ 0 & \text{otherwise.} \end{cases} \tag{7.1}$$

Also recall that the least squares estimator of the $[q(q-1)/2]$-vector of all pairwise differences $(\alpha_1 - \alpha_2, \alpha_1 - \alpha_3, \ldots, \alpha_{q-1} - \alpha_q)^T$ is

$$(\bar{y}_{1\cdot} - \bar{y}_{2\cdot}, \bar{y}_{1\cdot} - \bar{y}_{3\cdot}, \ldots, \bar{y}_{q-1,\cdot} - \bar{y}_{q\cdot})^T.$$

The variance–covariance matrix of this last vector could be obtained by writing the vector as $\mathbf{C}^T\hat{\boldsymbol{\beta}}$ for some $\mathbf{C}^T$ and then using Theorem 7.2.2, but it is easier in this case to obtain it more directly using Theorem 4.2.2a and (7.1). Taking the latter route, we obtain, for $i \leq j,\ i < i'$, and $j < j'$,

$$\begin{aligned}\text{cov}(\bar{y}_{i\cdot} - \bar{y}_{i'\cdot}, \bar{y}_{j\cdot} - \bar{y}_{j'\cdot}) &= \text{cov}(\bar{y}_{i\cdot}, \bar{y}_{j\cdot}) - \text{cov}(\bar{y}_{i'\cdot}, \bar{y}_{j\cdot}) - \text{cov}(\bar{y}_{i\cdot}, \bar{y}_{j'\cdot}) + \text{cov}(\bar{y}_{i'\cdot}, \bar{y}_{j'\cdot}) \\ &= \begin{cases} \sigma^2\left(\frac{1}{n_i} + \frac{1}{n_{i'}}\right) & \text{if } i = j \text{ and } i' = j', \\ \sigma^2\left(\frac{1}{n_i}\right) & \text{if } i = j \text{ and } i' \neq j', \\ \sigma^2\left(\frac{1}{n_{i'}}\right) & \text{if } i \neq j \text{ and } i' = j', \\ -\sigma^2\left(\frac{1}{n_{i'}}\right) & \text{if } i' = j, \\ 0, & \text{otherwise.} \end{cases}\end{aligned}$$

■

Example 7.2-3. Variance–Covariance Matrix of the Least Squares Estimators of Estimable Functions in a Two-Way Main Effects Model for Some 2 × 2 Layouts

Consider again the two-way main effects model for the four two-way layouts considered in Example 7.1-3, and take the model to be the Gauss–Markov model. For the first layout, the least squares estimators of the estimable cell means were

$$\begin{pmatrix} \widehat{\mu + \alpha_1 + \gamma_1} \\ \widehat{\mu + \alpha_2 + \gamma_2} \end{pmatrix} = \begin{pmatrix} y_{111} \\ y_{221} \end{pmatrix},$$

so it follows easily that

$$\text{var}\begin{pmatrix} \widehat{\mu + \alpha_1 + \gamma_1} \\ \widehat{\mu + \alpha_2 + \gamma_2} \end{pmatrix} = \sigma^2 \mathbf{I}.$$

Thus, the least squares estimators of the cell means corresponding to the occupied cells are homoscedastic and uncorrelated.

For the second layout, the least squares estimators of the cell means and the Factor A and B differences are

$$\begin{pmatrix} \widehat{\mu + \alpha_1 + \gamma_1} \\ \widehat{\mu + \alpha_1 + \gamma_2} \\ \widehat{\mu + \alpha_2 + \gamma_1} \\ \widehat{\mu + \alpha_2 + \gamma_2} \end{pmatrix} = \begin{pmatrix} y_{111} \\ y_{121} \\ y_{211} \\ y_{121} + y_{211} - y_{111} \end{pmatrix} \quad \text{and} \quad \begin{pmatrix} \widehat{\alpha_1 - \alpha_2} \\ \widehat{\gamma_1 - \gamma_2} \end{pmatrix} = \begin{pmatrix} y_{111} - y_{211} \\ y_{111} - y_{121} \end{pmatrix},$$

respectively. It is easy to show that the variance–covariance matrix of the first of these vectors of least squares estimators is

$$\text{var}\begin{pmatrix} \widehat{\mu + \alpha_1 + \gamma_1} \\ \widehat{\mu + \alpha_1 + \gamma_2} \\ \widehat{\mu + \alpha_2 + \gamma_1} \\ \widehat{\mu + \alpha_2 + \gamma_2} \end{pmatrix} = \sigma^2 \begin{pmatrix} 1 & 0 & 0 & -1 \\ 0 & 1 & 0 & 1 \\ 0 & 0 & 1 & 1 \\ -1 & 1 & 1 & 3 \end{pmatrix},$$

while that of the second is

$$\text{var}\begin{pmatrix} \widehat{\alpha_1 - \alpha_2} \\ \widehat{\gamma_1 - \gamma_2} \end{pmatrix} = \sigma^2 \begin{pmatrix} 2 & 1 \\ 1 & 2 \end{pmatrix}.$$

Thus, the least squares estimator of the cell mean corresponding to the empty cell is correlated with, and has a variance three times larger than, the least squares estimators of the other cell means. Furthermore, the least squares estimators of the Factor A and Factor B differences are (positively) correlated.

For the third layout, the variance–covariance matrices of the least squares estimators of the same quantities are

$$\text{var}\begin{pmatrix} \widehat{\mu+\alpha_1+\gamma_1} \\ \widehat{\mu+\alpha_1+\gamma_2} \\ \widehat{\mu+\alpha_2+\gamma_1} \\ \widehat{\mu+\alpha_2+\gamma_2} \end{pmatrix} = \text{var}\begin{pmatrix} (3y_{111}/4)+(y_{121}+y_{211}-y_{221})/4 \\ (3y_{121}/4)+(y_{111}+y_{221}-y_{211})/4 \\ (3y_{211}/4)+(y_{111}+y_{221}-y_{121})/4 \\ (3y_{221}/4)+(y_{121}+y_{211}-y_{111})/4 \end{pmatrix}$$

$$= (\sigma^2/4)\begin{pmatrix} 3 & 1 & 1 & -1 \\ 1 & 3 & -1 & 1 \\ 1 & -1 & 3 & 1 \\ -1 & 1 & 1 & 3 \end{pmatrix}$$

and

$$\text{var}\begin{pmatrix} \widehat{\alpha_1-\alpha_2} \\ \widehat{\gamma_1-\gamma_2} \end{pmatrix} = \text{var}\begin{pmatrix} \bar{y}_{1\cdot\cdot}-\bar{y}_{2\cdot\cdot} \\ \bar{y}_{\cdot1\cdot}-\bar{y}_{\cdot2\cdot} \end{pmatrix} = \sigma^2\mathbf{I}.$$

In this case the least squares estimators of the cell means are homoscedastic but correlated, while the least squares estimators of the Factor A and Factor B differences are homoscedastic and uncorrelated—a property that most would consider desirable. Furthermore, upon comparison of the variance–covariance matrix of the least squares estimators of the Factor A and B differences for this layout with that for the second layout, we see that the addition of only one observation in a suitably chosen cell cuts the variances of both estimated differences in half.

Determination of the variance–covariance matrices of the same two sets of least squares estimators for the fourth layout is left as an exercise. ■

Methodological Interlude #1: Variance Inflation Factors

Consider the case of a full-rank linear model that includes an intercept, and without loss of generality suppose that the model has been centered so that we can write the model matrix $\mathbf{X}$ as follows:

$$\mathbf{X} = (\mathbf{1}_n, \mathbf{x}_2 - \bar{x}_2\mathbf{1}_n, \mathbf{x}_3 - \bar{x}_3\mathbf{1}_n, \ldots, \mathbf{x}_p - \bar{x}_p\mathbf{1}_n) = (\mathbf{1}_n, \mathbf{X}_{\bar{2}}).$$

For such a model, β_i is estimable for every i; the interpretation of each β_i except β_1 is the same after centering as before; and, by Theorem 7.2.2, the variance of the least squares estimator of β_i under the Gauss–Markov model $\{\mathbf{y}, \mathbf{X}\boldsymbol{\beta}, \sigma^2\mathbf{I}\}$ is the ith diagonal element of the matrix $\sigma^2(\mathbf{X}^T\mathbf{X})^{-1}$. Because the model is centered,

$$\mathbf{X}^T\mathbf{X} = \text{diag}(n, \mathbf{X}_{\bar{2}}^T\mathbf{X}_{\bar{2}}),$$

implying further that

$$(\mathbf{X}^T\mathbf{X})^{-1} = \text{diag}[1/n, (\mathbf{X}_{\bar{2}}^T\mathbf{X}_{\bar{2}})^{-1}].$$

Note also that $\mathbf{X}^T\mathbf{X}$ is positive definite (by Theorem 2.15.9), hence so is $\mathbf{X}_{\bar{2}}^T\mathbf{X}_{\bar{2}}$ (by Theorem 2.15.4). Now consider in particular the variance of $\hat{\beta}_2$, which is equal to σ^2 times the first main diagonal element of $(\mathbf{X}_{\bar{2}}^T\mathbf{X}_{\bar{2}})^{-1}$. Partitioning the latter matrix before inversion as

$$\mathbf{X}_{\bar{2}}^T\mathbf{X}_{\bar{2}} = \begin{pmatrix} \sum_{i=1}^n (x_{i2} - \bar{x}_2)^2 & \mathbf{a}^T \\ \mathbf{a} & \mathbf{A} \end{pmatrix},$$

where the $(k-2)$th element of $\mathbf{a}$ is $\sum_{i=1}^n (x_{i2} - \bar{x}_2)(x_{ik} - \bar{x}_k)$ $(k = 3, \ldots, p)$ and the $(k-2, l-2)$th element of $\mathbf{A}$ is $\sum_{i=1}^n (x_{ik} - \bar{x}_k)(x_{il} - \bar{x}_l)$ $(k, l = 3, \ldots, p)$, and applying Theorems 2.15.7, 2.15.5, and 2.9.5b, we obtain

$$\begin{aligned} \text{var}(\hat{\beta}_2) &= \sigma^2 \left[\sum_{i=1}^n (x_{i2} - \bar{x}_2)^2 - \mathbf{a}^T\mathbf{A}^{-1}\mathbf{a} \right]^{-1} \\ &= \sigma^2 \left\{ \sum_{i=1}^n (x_{i2} - \bar{x}_2)^2 \left[1 - \frac{\mathbf{a}^T\mathbf{A}^{-1}\mathbf{a}}{\sum_{i=1}^n (x_{i2} - \bar{x}_2)^2} \right] \right\}^{-1} \\ &= \frac{\sigma^2}{\sum_{i=1}^n (x_{i2} - \bar{x}_2)^2} \times \left[1 - \frac{\mathbf{a}^T\mathbf{A}^{-1}\mathbf{a}}{\sum_{i=1}^n (x_{i2} - \bar{x}_2)^2} \right]^{-1}. \end{aligned} \tag{7.2}$$

The variance of $\hat{\beta}_2$ equals the first term of the product in (7.2) if and only if either the centered x_2 is the only explanatory variable in the model or $\mathbf{a} = \mathbf{0}$, i.e., the vector of centered x_2-values is orthogonal to the vector of centered values of every other explanatory variable. The second term indicates how the variance of $\hat{\beta}_2$ is affected by the presence of other explanatory variables not orthogonal to x_2. Because this second term is greater than or equal to one, it is often called the *variance inflation factor* corresponding to x_2. Furthermore, because the columns of $\mathbf{X}_{\bar{2}}$ may be rearranged to put any explanatory variable in the second column, there is a variance inflation factor corresponding to each explanatory variable (apart from the overall constant term), i.e.,

$$\text{var}(\hat{\beta}_j) = \frac{\sigma^2}{\sum_{i=1}^n (x_{ij} - \bar{x}_j)^2} \times VIF_j \qquad (j = 2, \ldots, p).$$

Variance inflation factors are often introduced in applied regression courses. An alternative expression for VIF_j that has a useful interpretation will be derived in Chap. 9. ■

7.3 Least Squares Estimation for Reparameterized Models

There may be more than one way to represent essentially the same mean structure for a linear model. This section considers these alternative representations, which are known as model reparameterizations, and examines how least squares estimation for one representation relates to that for another.

Definition 7.3.1 Two unconstrained linear models, one with mean structure $\mathbf{X}\boldsymbol{\beta}$ and the other with mean structure $\breve{\mathbf{X}}\boldsymbol{\tau}$ but otherwise the same, are said to be *reparameterizations* of each other if $\mathcal{C}(\mathbf{X}) = \mathcal{C}(\breve{\mathbf{X}})$.

Throughout this section we assume that $\mathcal{C}(\mathbf{X}) = \mathcal{C}(\breve{\mathbf{X}})$, and we refer to the model with mean structure $\mathbf{X}\boldsymbol{\beta}$ as the *original model* and to the model with mean structure $\breve{\mathbf{X}}\boldsymbol{\tau}$ as the *reparameterized model.* Let t denote the number of columns of $\breve{\mathbf{X}}$. Because $\mathcal{C}(\mathbf{X}) = \mathcal{C}(\breve{\mathbf{X}})$, $p \times t$ matrices $\mathbf{T}$ and $\mathbf{S}$ exist such that

$$\breve{\mathbf{X}} = \mathbf{X}\mathbf{T} \quad \text{and} \quad \mathbf{X} = \breve{\mathbf{X}}\mathbf{S}^T = \mathbf{X}\mathbf{T}\mathbf{S}^T.$$

(The converse of this statement is also true.) In what follows, let $\mathbf{T}$ and $\mathbf{S}$ represent such matrices. Observe that $\mathbf{X}\boldsymbol{\beta} = \breve{\mathbf{X}}\mathbf{S}^T\boldsymbol{\beta} = \breve{\mathbf{X}}\boldsymbol{\tau}$ say, where $\boldsymbol{\tau} = \mathbf{S}^T\boldsymbol{\beta}$. Thus, to relate the parameters of the reparameterized model to those of the original model in practice, we must determine a suitable $\mathbf{S}^T$.

Example 7.3-1. Some Reparameterizations of a Two-Way Main Effects Model for a 2 × 2 Layout

Consider the following three matrices:

$$\mathbf{X} = \begin{pmatrix} 1 & 1 & 0 & 1 & 0 \\ 1 & 1 & 0 & 0 & 1 \\ 1 & 0 & 1 & 1 & 0 \\ 1 & 0 & 1 & 0 & 1 \end{pmatrix}, \quad \breve{\mathbf{X}}_1 = \begin{pmatrix} 1 & 1 & 0 & 2 \\ 1 & 1 & 0 & 1 \\ 1 & 0 & 1 & 2 \\ 1 & 0 & 1 & 1 \end{pmatrix}, \quad \breve{\mathbf{X}}_2 = \begin{pmatrix} 1 & 1 & 1 \\ 1 & 1 & 0 \\ 1 & 0 & 1 \\ 1 & 0 & 0 \end{pmatrix}.$$

The reader will recognize $\mathbf{X}$ as the model matrix corresponding to a two-way main effects model in which there are two levels of each factor and exactly one observation in each cell, i.e.,

$$y_{ij} = \mu + \alpha_i + \gamma_j + e_{ij} \quad (i = 1, 2;\ j = 1, 2).$$

Plainly, $\mathcal{C}(\breve{\mathbf{X}}_1) \subseteq \mathcal{C}(\mathbf{X})$ (the first three columns of the matrices are identical, and the fourth column of $\breve{\mathbf{X}}_1$ is the sum of the first and fourth columns of $\mathbf{X}$), and $\mathcal{C}(\mathbf{X}) \subseteq \mathcal{C}(\breve{\mathbf{X}}_1)$ (the fourth column of $\mathbf{X}$ is the difference of the fourth and first columns of $\breve{\mathbf{X}}_1$, and the fifth column of $\mathbf{X}$ is twice the first column of $\breve{\mathbf{X}}_1$ minus the fourth column of $\breve{\mathbf{X}}_1$). Thus, $\mathcal{C}(\breve{\mathbf{X}}_1) = \mathcal{C}(\mathbf{X})$, so the models with mean structures $\mathbf{X}\boldsymbol{\beta}$ and $\breve{\mathbf{X}}_1\boldsymbol{\tau}$ are reparameterizations of each other. Suitable matrices $\mathbf{T}$ and $\mathbf{S}^T$ such that $\breve{\mathbf{X}}_1 = \mathbf{XT}$ and $\mathbf{X} = \breve{\mathbf{X}}_1\mathbf{S}^T$ are

$$\mathbf{T} = \begin{pmatrix} 1&0&0&0 \\ 0&1&0&0 \\ 0&0&1&0 \\ 0&0&0&2 \\ 0&0&0&1 \end{pmatrix}, \qquad \mathbf{S}^T = \begin{pmatrix} 1&0&0&-1&2 \\ 0&1&0&0&0 \\ 0&0&1&0&0 \\ 0&0&0&1&-1 \end{pmatrix},$$

but these are not unique; alternatives are

$$\mathbf{T} = \begin{pmatrix} 1&0&0&1 \\ 0&1&0&0 \\ 0&0&1&0 \\ 0&0&0&1 \\ 0&0&0&0 \end{pmatrix}, \qquad \mathbf{S}^T = \begin{pmatrix} 1&0&0&-2&2 \\ 0&1&0&1&0 \\ 0&0&1&1&0 \\ 0&0&0&1&-1 \end{pmatrix}.$$

Similarly, $\mathcal{C}(\breve{\mathbf{X}}_2) \subseteq \mathcal{C}(\mathbf{X})$ (obviously) and $\mathcal{C}(\mathbf{X}) \subseteq \mathcal{C}(\breve{\mathbf{X}}_2)$ (the third column of $\mathbf{X}$ is the difference of the first two columns of $\breve{\mathbf{X}}_2$, and the fifth column of $\mathbf{X}$ is the difference of the first and third columns of $\breve{\mathbf{X}}_2$), so the models with mean structures $\mathbf{X}\boldsymbol{\beta}$ and $\breve{\mathbf{X}}_2\boldsymbol{\tau}$ likewise are reparameterizations of each other. Suitable matrices $\mathbf{T}$ and $\mathbf{S}^T$ for this case are

$$\mathbf{T} = \begin{pmatrix} 1&0&0 \\ 0&1&0 \\ 0&0&0 \\ 0&0&1 \\ 0&0&0 \end{pmatrix}, \qquad \mathbf{S}^T = \begin{pmatrix} 1&0&1&0&1 \\ 0&1&-1&0&0 \\ 0&0&0&1&-1 \end{pmatrix},$$

and although $\mathbf{T}$ is not unique, $\mathbf{S}^T$ is unique in this case because the columns of $\breve{\mathbf{X}}_2$ constitute a basis for $\mathcal{C}(\mathbf{X})$. The elements of $\boldsymbol{\tau} = \mathbf{S}^T\boldsymbol{\beta}$ corresponding to the three cases of $\mathbf{S}^T$ listed above are as follows:

$$\begin{pmatrix} \mu - \gamma_1 + 2\gamma_2 \\ \alpha_1 \\ \alpha_2 \\ \gamma_1 - \gamma_2 \end{pmatrix}, \qquad \begin{pmatrix} \mu - 2\gamma_1 + 2\gamma_2 \\ \alpha_1 + \gamma_1 \\ \alpha_2 + \gamma_1 \\ \gamma_1 - \gamma_2 \end{pmatrix}, \qquad \begin{pmatrix} \mu + \alpha_2 + \gamma_2 \\ \alpha_1 - \alpha_2 \\ \gamma_1 - \gamma_2 \end{pmatrix}.$$

■

Because $\mathbf{T}\boldsymbol{\tau}$ is a solution (for $\boldsymbol{\beta}$) to the system of equations $\mathbf{X}\boldsymbol{\beta} = \breve{\mathbf{X}}\boldsymbol{\tau}$ for every value of $\boldsymbol{\tau}$, we might guess that the estimation of a function $\mathbf{c}^T\boldsymbol{\beta}$ that is estimable under the original model should be essentially the same as the estimation of $\mathbf{c}^T\mathbf{T}\boldsymbol{\tau}$ under the reparameterized model. Likewise, because $\mathbf{S}^T\boldsymbol{\beta}$ is a solution (for $\boldsymbol{\tau}$) to the system of equations $\breve{\mathbf{X}}\boldsymbol{\tau} = \mathbf{X}\boldsymbol{\beta}$ for every value of $\boldsymbol{\beta}$, we might guess that, if $\mathbf{b}^T\boldsymbol{\tau}$ is estimable under the reparameterized model, then its estimation under that model should be essentially the same as the estimation of $\mathbf{b}^T\mathbf{S}^T\boldsymbol{\beta}$ under the original model. As indicated by the following theorem, both guesses are correct.

Theorem 7.3.1

(a) *If $\mathbf{c}^T\boldsymbol{\beta}$ is estimable under the original model, then $\mathbf{c}^T\mathbf{T}\boldsymbol{\tau}$ is estimable under the reparameterized model, and its least squares estimator may be expressed as either $\mathbf{c}^T\hat{\boldsymbol{\beta}}$ or $\mathbf{c}^T\mathbf{T}\hat{\boldsymbol{\tau}}$, where $\hat{\boldsymbol{\beta}}$ is any solution to the normal equations $\mathbf{X}^T\mathbf{X}\boldsymbol{\beta} = \mathbf{X}^T\mathbf{y}$ associated with the original model and $\hat{\boldsymbol{\tau}}$ is any solution to the normal equations $\breve{\mathbf{X}}^T\breve{\mathbf{X}}\boldsymbol{\tau} = \breve{\mathbf{X}}^T\mathbf{y}$ associated with the reparameterized model.*

(b) *If $\mathbf{b}^T\boldsymbol{\tau}$ is estimable under the reparameterized model, then $\mathbf{b}^T\mathbf{S}^T\boldsymbol{\beta}$ is estimable under the original model and its least squares estimator may be expressed as either $\mathbf{b}^T\mathbf{S}^T\hat{\boldsymbol{\beta}}$ or $\mathbf{b}^T\hat{\boldsymbol{\tau}}$, where $\hat{\boldsymbol{\beta}}$ and $\hat{\boldsymbol{\tau}}$ are defined as in part (a).*

Proof We prove only part (a); the proof of part (b) is very similar. Suppose that $\mathbf{c}^T\boldsymbol{\beta}$ is estimable under the original model. Then, by Theorem 6.1.2, $\mathbf{c}^T = \mathbf{a}^T\mathbf{X}$ for some $\mathbf{a}$, implying that $\mathbf{c}^T\mathbf{T} = \mathbf{a}^T\mathbf{X}\mathbf{T} = \mathbf{a}^T\breve{\mathbf{X}}$ for that $\mathbf{a}$. This establishes that $\mathbf{c}^T\mathbf{T}\boldsymbol{\tau}$ is estimable under the reparameterized model. Moreover, if $\hat{\boldsymbol{\tau}}$ is any solution to the normal equations $\breve{\mathbf{X}}^T\breve{\mathbf{X}}\boldsymbol{\tau} = \breve{\mathbf{X}}^T\mathbf{y}$ associated with the reparameterized model, then

$$\mathbf{X}^T\mathbf{X}\mathbf{T}\hat{\boldsymbol{\tau}} = \mathbf{X}^T\breve{\mathbf{X}}\hat{\boldsymbol{\tau}} = \mathbf{S}\breve{\mathbf{X}}^T\breve{\mathbf{X}}\hat{\boldsymbol{\tau}} = \mathbf{S}\breve{\mathbf{X}}^T\mathbf{y} = \mathbf{X}^T\mathbf{y},$$

which shows that $\mathbf{T}\hat{\boldsymbol{\tau}}$ is a solution to the normal equations for the original model. That $\mathbf{c}^T\mathbf{T}\hat{\boldsymbol{\tau}}$ is the least squares estimator of $\mathbf{c}^T\boldsymbol{\beta}$ (and therefore coincides with $\mathbf{c}^T\hat{\boldsymbol{\beta}}$) follows from Definition 7.1.6 of the least squares estimator. □

The converses of the estimability claims in Theorem 7.3.1 are not generally true. That is, it is possible, for example, for $\mathbf{c}^T\mathbf{T}\boldsymbol{\tau}$ to be estimable under the reparameterized model while $\mathbf{c}^T\boldsymbol{\beta}$ is not estimable under the original model; see the exercises for an example. However, such possibilities do not take anything away from the important fact that any function that is estimable under one of the models may be estimated (by least squares) using a solution to the normal equations corresponding to the other model (in conjunction with the appropriate matrix $\mathbf{T}$ or $\mathbf{S}^T$).

Why would we ever want to consider a model reparameterization? One reason might be computational simplicity. For certain reparameterizations, the computations associated with least squares estimation are considerably less burdensome to perform "by hand." This is particularly so when some or all of the columns of the model matrix $\breve{\mathbf{X}}$ of the reparameterized model are orthogonal, as in the following two examples.

Example 7.3-2. Mean-Centered Models

Consider the multiple linear regression model with intercept,

$$y_i = \beta_1 + \beta_2 x_{i2} + \beta_3 x_{i3} + \cdots + \beta_p x_{ip} + e_i \qquad (i = 1, \ldots, n),$$

or equivalently $\mathbf{y} = \mathbf{X}\boldsymbol{\beta} + \mathbf{e}$ where

$$\mathbf{X} = (\mathbf{1}_n, \mathbf{x}_2, \mathbf{x}_3, \ldots, \mathbf{x}_p).$$

By subtracting the mean (over the n observations) of each explanatory variable from all values of that variable, we obtain the model

$$y_i = \tau_1 + \tau_2(x_{i2} - \bar{x}_2) + \tau_3(x_{i3} - \bar{x}_3) + \cdots + \tau_p(x_{ip} - \bar{x}_p) + e_i \qquad (i = 1, \ldots, n),$$

or equivalently, $\mathbf{y} = \breve{\mathbf{X}}\boldsymbol{\tau} + \mathbf{e}$ where

$$\breve{\mathbf{X}} = (\mathbf{1}_n, \mathbf{x}_2 - \bar{x}_2\mathbf{1}_n, \mathbf{x}_3 - \bar{x}_3\mathbf{1}_n, \ldots, \mathbf{x}_p - \bar{x}_p\mathbf{1}_n).$$

Clearly $\mathcal{C}(\breve{\mathbf{X}}) = \mathcal{C}(\mathbf{X})$, so the second model is a reparameterization of the first. Furthermore, the first column of $\breve{\mathbf{X}}$ is orthogonal to all other columns of $\breve{\mathbf{X}}$. This property was put to advantage in Methodological Interlude #1 and will be useful later as well. ■

Example 7.3-3. Orthogonal Polynomial Regression Models

Consider the cubic polynomial regression model in one variable and four observations, i.e.,

$$y_i = \beta_1 + \beta_2 x_i + \beta_3 x_i^2 + \beta_4 x_i^3 + e_i \qquad (i = 1, \ldots, 4),$$

and suppose that $x_i = i$. Then

$$\mathbf{X} = (\mathbf{1}_4, \mathbf{x}_2, \mathbf{x}_3, \mathbf{x}_4) = \begin{pmatrix} 1 & 1 & 1 & 1 \\ 1 & 2 & 4 & 8 \\ 1 & 3 & 9 & 27 \\ 1 & 4 & 16 & 64 \end{pmatrix}.$$

Now define

$$\breve{\mathbf{X}} = (\mathbf{1}_4, 2\mathbf{x}_2 - 5\mathbf{1}_4, \mathbf{x}_3 - 5\mathbf{x}_2 + 5\mathbf{1}_4, \frac{10}{3}\mathbf{x}_4 - 25\mathbf{x}_3 + \frac{167}{3}\mathbf{x}_2 - 35\mathbf{1}_4) = \begin{pmatrix} 1 & -3 & 1 & -1 \\ 1 & -1 & -1 & 3 \\ 1 & 1 & -1 & -3 \\ 1 & 3 & 1 & 1 \end{pmatrix}.$$

Clearly $\mathcal{C}(\breve{\mathbf{X}}) = \mathcal{C}(\mathbf{X})$, implying that the model $\mathbf{y} = \breve{\mathbf{X}}\boldsymbol{\tau} + \mathbf{e}$ is a reparameterization of the original model; furthermore, it is easily verified that the columns of $\breve{\mathbf{X}}$ are orthogonal, so $\breve{\mathbf{X}}^T\breve{\mathbf{X}}$ is a diagonal matrix. The columns of this $\breve{\mathbf{X}}$ are known as *orthogonal polynomial contrast coefficients*. Such coefficients are listed, for various orders of the polynomial and various numbers of equally spaced observations, in many statistical methods books; see, for example, Table A19 in Snedecor and Cochran (1980). ■

Another reason why one might want to reparameterize is to create a model matrix that has full column rank. If $\breve{\mathbf{X}}$ has exactly p^* columns, and those columns are linearly independent, then the reparameterization of the model is called a *full-rank reparametrization*. A full-rank reparametrization always exists because $\breve{\mathbf{X}}$ may be formed from any basis for $\mathcal{C}(\mathbf{X})$. Furthermore, a full-rank reparameterization has two properties that might be regarded as desirable:

1. All (linear) functions of $\boldsymbol{\tau}$ are estimable (by Corollary 6.1.2.4);
2. The $p^* \times p^*$ coefficient matrix $\breve{\mathbf{X}}^T\breve{\mathbf{X}}$ of the normal equations associated with a full-rank reparameterization has rank p^* (by Theorem 2.8.8) and is thereby nonsingular, implying that the normal equations for a full-rank reparameterization can be solved using an ordinary matrix inverse and that there is a unique solution, which is given by $\hat{\boldsymbol{\tau}} = (\breve{\mathbf{X}}^T\breve{\mathbf{X}})^{-1}\breve{\mathbf{X}}^T\mathbf{y}$.

In light of these two properties and Theorem 7.3.1, it might seem, in retrospect, that Chap. 6 (on estimability) was "much ado about nothing." It might also seem that the presentation of least squares estimation in the current chapter was unnecessarily complicated, based as it was on solving the normal equations for a less-than-full-rank model using a generalized inverse of the coefficient matrix of those equations, when it could have been based on solving the normal equations for a full-rank reparameterized model using the ordinary inverse of the coefficient matrix. For some linear models, the author would agree with this assessment. For example, suppose that $\mathbf{X}$ is such that the mere deletion of easily identified columns results in a full-rank matrix $\breve{\mathbf{X}}$ with the same column space as $\mathbf{X}$. Such a case is provided by Example 7.3-1, where it is easy to see that the third and fifth columns of $\mathbf{X}$ may be deleted to yield the three-column matrix $\breve{\mathbf{X}}_2$ of rank three, which has the same column space as $\mathbf{X}$. In any such situation we have $\mathbf{X} = \breve{\mathbf{X}}\mathbf{A}$ for some $p^* \times (p - p^*)$ matrix $\mathbf{A}$, and we may take

$$\mathbf{S}^T = (\breve{\mathbf{X}}^T\breve{\mathbf{X}})^{-1}\breve{\mathbf{X}}^T\mathbf{X} \tag{7.3}$$

because for this $\mathbf{S}^T$ we have

$$\breve{\mathbf{X}}\mathbf{S}^T = \breve{\mathbf{X}}(\breve{\mathbf{X}}^T\breve{\mathbf{X}})^{-1}\breve{\mathbf{X}}^T\mathbf{X} = \breve{\mathbf{X}}(\breve{\mathbf{X}}^T\breve{\mathbf{X}})^{-1}\breve{\mathbf{X}}^T\breve{\mathbf{X}}\mathbf{A} = \breve{\mathbf{X}}\mathbf{A} = \mathbf{X}.$$

Furthermore, we may take

$$\mathbf{T} = \mathbf{S}(\mathbf{S}^T\mathbf{S})^{-1} \tag{7.4}$$

because for such a $\mathbf{T}$,

$$\mathbf{X}\mathbf{T} = \breve{\mathbf{X}}\mathbf{S}^T\mathbf{T} = \breve{\mathbf{X}}\mathbf{S}^T\mathbf{S}(\mathbf{S}^T\mathbf{S})^{-1} = \breve{\mathbf{X}}.$$

(The nonsingularity of $\mathbf{S}^T\mathbf{S}$ for a full-rank reparameterization is guaranteed by Theorem 7.3.2a, which is given later in this section). Indeed, the matrices $\mathbf{S}^T$ and $\mathbf{T}$ corresponding to the full-rank reparameterization with model matrix $\breve{\mathbf{X}}_2$ in Example 7.3-1 were obtained in exactly this manner. Two more such cases are given in the following example.

Example 7.3-4. The Cell-Means Reparameterizations of the One-Factor Model and the Two-Way Model with Interaction

In Sect. 5.1.4, it was claimed that the one-factor model could be represented as either the "effects model"

$$y_{ij} = \mu + \alpha_i + e_{ij} \quad (i = 1, \ldots, q;\ j = 1, \ldots, n_i)$$

with model matrix $\mathbf{X} = (\mathbf{1}_n, \oplus_{i=1}^q \mathbf{1}_{n_i})$ and parameter vector $\boldsymbol{\beta} = (\mu, \alpha_1, \ldots, \alpha_q)^T$, or the cell-means model

$$y_{ij} = \mu_i + e_{ij} \quad (i = 1, \ldots, q;\ j = 1, \ldots, n_i)$$

with model matrix $\breve{\mathbf{X}} = \oplus_{i=1}^q \mathbf{1}_{n_i}$ and parameter vector $\boldsymbol{\tau} = (\mu_1, \ldots, \mu_q)^T = (\mu + \alpha_1, \ldots, \mu + \alpha_q)^T$. Clearly, because the columns of $\oplus_{i=1}^q \mathbf{1}_{n_i}$ sum to $\mathbf{1}_n$, the cell-means model is a reparameterization of the effects model; in fact, the cell-means model is a full-rank reparameterization. Moreover, the columns of $\breve{\mathbf{X}}$ are orthogonal, implying that $\breve{\mathbf{X}}^T\breve{\mathbf{X}}$ is diagonal. Thus, the least squares estimators of $\mu_1, \ldots, \mu_q$ are given by the elements of

$$\hat{\boldsymbol{\tau}} = (\breve{\mathbf{X}}^T\breve{\mathbf{X}})^{-1}\breve{\mathbf{X}}^T\mathbf{y} = [(\oplus_{i=1}^q \mathbf{1}_{n_i})^T(\oplus_{i=1}^q \mathbf{1}_{n_i})]^{-1}(\oplus_{i=1}^q \mathbf{1}_{n_i})^T\mathbf{y} = \begin{pmatrix} \bar{y}_{1\cdot} \\ \bar{y}_{2\cdot} \\ \vdots \\ \bar{y}_{q\cdot} \end{pmatrix},$$

which agrees with a result obtained previously in Example 7.1-2 using the effects model parameterization.

Also in Sect. 5.1.4, it was claimed that the two-way model with interaction could be represented as either

$$y_{ijk} = \mu + \alpha_i + \gamma_j + \xi_{ij} + e_{ijk} \quad (i = 1, \ldots, q;\ j = 1, \ldots, m;\ k = 1, \ldots, n_{ij})$$

or as

$$y_{ijk} = \mu_{ij} + e_{ijk} \quad (i = 1, \ldots, q;\ j = 1, \ldots, m;\ k = 1, \ldots, n_{ij}),$$

where the latter was again referred to as a cell-means representation. Suppose that there are no empty cells. Then the model matrix for the first of these representations was given by (5.9) and is repeated below:

$$\mathbf{X} = \begin{pmatrix} \mathbf{1}_{n_{1\cdot}} & \mathbf{1}_{n_{1\cdot}} & \mathbf{0}_{n_{1\cdot}} & \cdots & \mathbf{0}_{n_{1\cdot}} & \oplus_{j=1}^m \mathbf{1}_{n_{1j}} & \oplus_{j=1}^m \mathbf{1}_{n_{1j}} & \mathbf{0}_{n_{1\cdot}\times m} & \cdots & \mathbf{0}_{n_{1\cdot}\times m} \\ \mathbf{1}_{n_{2\cdot}} & \mathbf{0}_{n_{2\cdot}} & \mathbf{1}_{n_{2\cdot}} & \cdots & \mathbf{0}_{n_{2\cdot}} & \oplus_{j=1}^m \mathbf{1}_{n_{2j}} & \mathbf{0}_{n_{2\cdot}\times m} & \oplus_{j=1}^m \mathbf{1}_{n_{2j}} & \cdots & \mathbf{0}_{n_{2\cdot}\times m} \\ \vdots & \vdots & \vdots & \ddots & \vdots & \vdots & \vdots & \vdots & \ddots & \vdots \\ \mathbf{1}_{n_{q\cdot}} & \mathbf{0}_{n_{q\cdot}} & \mathbf{0}_{n_{q\cdot}} & \cdots & \mathbf{1}_{n_{q\cdot}} & \oplus_{j=1}^m \mathbf{1}_{n_{qj}} & \mathbf{0}_{n_{q\cdot}\times m} & \mathbf{0}_{n_{q\cdot}\times m} & \cdots & \oplus_{j=1}^m \mathbf{1}_{n_{qj}} \end{pmatrix}.$$

It is obvious that the first $1 + q + m$ columns of $\mathbf{X}$ are trivial linear combinations of the other columns, and that those other columns are linearly independent. Thus, the cell-means representation, which has model matrix

$$\breve{\mathbf{X}} = \begin{pmatrix} \oplus_{j=1}^m \mathbf{1}_{n_{1j}} & \mathbf{0}_{n_{1\cdot}\times m} & \cdots & \mathbf{0}_{n_{1\cdot}\times m} \\ \mathbf{0}_{n_{2\cdot}\times m} & \oplus_{j=1}^m \mathbf{1}_{n_{2j}} & \cdots & \mathbf{0}_{n_{2\cdot}\times m} \\ \vdots & \vdots & \ddots & \vdots \\ \mathbf{0}_{n_{q\cdot}\times m} & \mathbf{0}_{n_{q\cdot}\times m} & \cdots & \oplus_{j=1}^m \mathbf{1}_{n_{qj}} \end{pmatrix},$$

is a full-rank reparameterization of the original model. The corresponding parameter vector is

$$\boldsymbol{\tau} = \begin{pmatrix} \mu_{11} \\ \mu_{12} \\ \vdots \\ \mu_{qm} \end{pmatrix} = \begin{pmatrix} \mu + \alpha_1 + \gamma_1 + \xi_{11} \\ \mu + \alpha_1 + \gamma_2 + \xi_{12} \\ \vdots \\ \mu + \alpha_q + \gamma_m + \xi_{qm} \end{pmatrix}.$$

Now, $\breve{\mathbf{X}}^T\breve{\mathbf{X}} = \oplus_{i=1}^{q} \oplus_{j=1}^{m} n_{ij}$, which is nonsingular and diagonal. Thus, the least squares estimators of the μ_{ij}'s are given by the elements of

$$(\breve{\mathbf{X}}^T\breve{\mathbf{X}})^{-1}\breve{\mathbf{X}}^T\mathbf{y} = \left[\oplus_{i=1}^{q} \oplus_{j=1}^{m} (1/n_{ij})\right] \begin{pmatrix} y_{11\cdot} \\ y_{12\cdot} \\ \vdots \\ y_{qm\cdot} \end{pmatrix} = \begin{pmatrix} \bar{y}_{11\cdot} \\ \bar{y}_{12\cdot} \\ \vdots \\ \bar{y}_{qm\cdot} \end{pmatrix}.$$

In Example 6.1-5 it was shown that contrasts among the interaction effects in the original parameterization that are of the form $\xi_{ij} - \xi_{ij'} - \xi_{i'j} + \xi_{i'j'}$ are estimable. It is easily verified that, in terms of the reparameterized model, any such contrast is of the form $\mu_{ij} - \mu_{ij'} - \mu_{i'j} + \mu_{i'j'}$. By Corollary 7.1.4.1, the least squares estimator of such a contrast is $\bar{y}_{ij} - \bar{y}_{ij'} - \bar{y}_{i'j} + \bar{y}_{i'j'}$. ■

In general, it is not as easy to obtain a full-rank reparameterization as it was for the cases considered in Example 7.3-4. In fact, the computations required to obtain a full-rank reparameterization and to relate the parameters of that reparameterization to those in the original model generally are not that different from those required to determine estimability in the original model and to obtain a generalized inverse of the normal equations for the less-than-full-rank model. Two main strategies to perform these computations exist. The first is to find a basis for $\mathcal{C}(\mathbf{X})$ and then form $\breve{\mathbf{X}}$ from the column vectors in that basis. A basis can be found in various ways; one way that may be familiar to many readers is to reduce $\mathbf{X}$ to "column echelon form" via a series of elementary column operations. Another method for finding a basis is to include the first column of $\mathbf{X}$ in the basis if it is nonnull, then include the second column in the basis if it is not expressible as a scalar multiple of the first column, then include the third column if it is not expressible as a linear combination of the first two columns, etc., until all columns of $\mathbf{X}$ have been considered for inclusion.

The second strategy for obtaining a full-rank reparameterization is motivated by the following theorem.

Theorem 7.3.2 *For a full-rank reparameterization of the model* $\{\mathbf{y}, \mathbf{X}\boldsymbol{\beta}\}$, $\mathcal{R}(\mathbf{S}^T) = \mathcal{R}(\mathbf{X})$ *and hence:*

(a) rank$(\mathbf{S}^T) = p^*$ *and*
(b) a function $\mathbf{c}^T\boldsymbol{\beta}$ *is estimable if and only if* $\mathbf{c}^T \in \mathcal{R}(\mathbf{S}^T)$.

Proof Left as an exercise. □

As a consequence of Theorem 7.3.2b, a full-rank reparameterization may be obtained for any linear model by finding a basis for the space of estimable functions [rather than a basis for $\mathcal{C}(\mathbf{X})$], and forming $\mathbf{S}^T$ from the row vectors in that basis. Then, $\mathbf{T}$ may be obtained via (7.4), and $\breve{\mathbf{X}}$ may be set equal to $\mathbf{XT}$.

Example 7.3-5. Some Full-Rank Reparameterizations of a Two-Way Main Effects Model for a 2 × 2 Layout

Consider once again the two-way main effects model for a 2×2 layout featured in Example 7.3-1, and observe that $\breve{\mathbf{X}}_2$ in that example was a full-rank reparameterization. Here we consider two more full-rank reparameterizations, which are obtained by implementing the two strategies just described. The first such reparameterization corresponds to the basis for $\mathcal{C}(\mathbf{X})$ obtained by reducing $\mathbf{X}$ to column echelon form. It may be easily verified that the reduction of $\mathbf{X}$ to column echelon form yields

$$\breve{\mathbf{X}}_3 = \begin{pmatrix} 1 & 0 & 0 \\ 1 & 1 & 0 \\ 1 & 0 & 1 \\ 1 & 1 & 1 \end{pmatrix}.$$

Following the prescription described by (7.3) and (7.4), we obtain

$$\mathbf{S}^T = (\breve{\mathbf{X}}_3^T \breve{\mathbf{X}}_3)^{-1} \breve{\mathbf{X}}_3^T \mathbf{X} = \begin{pmatrix} 1 & 1 & 0 & 1 & 0 \\ 0 & 0 & 0 & -1 & 1 \\ 0 & -1 & 1 & 0 & 0 \end{pmatrix}$$

and

$$\mathbf{T} = \mathbf{S}(\mathbf{S}^T \mathbf{S})^{-1} = \begin{pmatrix} \frac{1}{2} & \frac{1}{4} & \frac{1}{4} \\ \frac{1}{4} & \frac{1}{8} & -\frac{3}{8} \\ \frac{1}{4} & \frac{1}{8} & \frac{5}{8} \\ \frac{1}{4} & -\frac{3}{8} & \frac{1}{8} \\ \frac{1}{4} & \frac{5}{8} & \frac{1}{8} \end{pmatrix}.$$

The form of $\mathbf{S}^T$ reveals that $\boldsymbol{\tau} = (\mu + \alpha_1 + \gamma_1, \gamma_2 - \gamma_1, \alpha_2 - \alpha_1)^T$.

Next consider the full-rank reparameterization corresponding to the basis $\{\mu + \alpha_1 + \gamma_1, \mu + \alpha_1 + \gamma_2, \mu + \alpha_2 + \gamma_1\}$ for the space of estimable functions. For this basis,

$$\mathbf{S}^T = \begin{pmatrix} 1 & 1 & 0 & 1 & 0 \\ 1 & 1 & 0 & 0 & 1 \\ 1 & 0 & 1 & 1 & 0 \end{pmatrix},$$

$$\mathbf{T} = \mathbf{S}(\mathbf{S}^T\mathbf{S})^{-1} = \begin{pmatrix} 0 & \frac{1}{4} & \frac{1}{4} \\ \frac{1}{2} & \frac{1}{8} & -\frac{3}{8} \\ -\frac{1}{2} & \frac{1}{8} & \frac{5}{8} \\ \frac{1}{2} & -\frac{3}{8} & \frac{1}{8} \\ -\frac{1}{2} & \frac{5}{8} & \frac{1}{8} \end{pmatrix}, \quad \text{and}$$

$$\breve{\mathbf{X}}_4 = \mathbf{X}\mathbf{T} = \begin{pmatrix} 1\,0\,0 \\ 0\,1\,0 \\ 0\,0\,1 \\ -1\,1\,1 \end{pmatrix}.$$

■

In light of the foregoing results and discussion, in this book a pragmatic approach is taken to the use of full-rank reparameterizations. When a full-rank reparameterization of the model $\{\mathbf{y}, \mathbf{X}\boldsymbol{\beta}\}$ is "staring us in the face," as in Example 7.3-4, we often use it to carry out the computations associated with least squares estimation. For most models, however, we continue to use generalized inverses for this purpose, as well as for presentation of theory.

7.4 Exercises

1. The proof of the consistency of the normal equations (Theorem 7.1.1) relies upon an identity (Theorem 3.3.3c) involving a generalized inverse of $\mathbf{X}^T\mathbf{X}$. Provide an alternate proof of the consistency of the normal equations that does not rely on this identity, by showing that the normal equations are compatible.
2. For observations following a certain linear model, the normal equations are

$$\begin{pmatrix} 7 & -2 & -5 \\ -2 & 3 & -1 \\ -5 & -1 & 6 \end{pmatrix} \begin{pmatrix} \beta_1 \\ \beta_2 \\ \beta_3 \end{pmatrix} = \begin{pmatrix} 17 \\ 34 \\ -51 \end{pmatrix}.$$

Observe that the three rows of the coefficient matrix sum to $\mathbf{0}_3^T$.

(a) Obtain two distinct solutions, say $\hat{\boldsymbol{\beta}}_1$ and $\hat{\boldsymbol{\beta}}_2$, to the normal equations.

(b) Characterize, as simply as possible, the collection of linear functions $\mathbf{c}^T\boldsymbol{\beta} = c_1\beta_1 + c_2\beta_2 + c_3\beta_3$ that are estimable.

(c) List one nonzero estimable function, $\mathbf{c}_E^T\boldsymbol{\beta}$, and one nonestimable function, $\mathbf{c}_{NE}^T\boldsymbol{\beta}$. (Give numerical entries for $\mathbf{c}_E^T$ and $\mathbf{c}_{NE}^T$.)

(d) Determine numerically whether the least squares estimator of your estimable function from part (c) is the same for both of your solutions from part (a), i.e. determine whether $\mathbf{c}_E^T\hat{\boldsymbol{\beta}}_1 = \mathbf{c}_E^T\hat{\boldsymbol{\beta}}_2$. Similarly, determine whether $\mathbf{c}_{NE}^T\hat{\boldsymbol{\beta}}_1 = \mathbf{c}_{NE}^T\hat{\boldsymbol{\beta}}_2$. Which theorem does this exemplify?

3. Consider a linear model for which

$$\mathbf{y} = \begin{pmatrix} y_1 \\ y_2 \\ y_3 \\ y_4 \\ y_5 \\ y_6 \\ y_7 \\ y_8 \end{pmatrix}, \quad \mathbf{X} = \begin{pmatrix} 1 & 1 & 1 & -1 \\ 1 & 1 & 1 & -1 \\ 1 & 1 & -1 & 1 \\ 1 & 1 & -1 & 1 \\ 1 & -1 & 1 & 1 \\ 1 & -1 & 1 & 1 \\ -1 & 1 & 1 & 1 \\ -1 & 1 & 1 & 1 \end{pmatrix}, \quad \boldsymbol{\beta} = \begin{pmatrix} \beta_1 \\ \beta_2 \\ \beta_3 \\ \beta_4 \end{pmatrix}.$$

(a) Obtain the normal equations for this model and solve them.
(b) Are all functions $\mathbf{c}^T\boldsymbol{\beta}$ estimable? Justify your answer.
(c) Obtain the least squares estimator of $\beta_1 + \beta_2 + \beta_3 + \beta_4$.

4. Prove Corollary 7.1.4.1.
5. Prove that, under the Gauss–Markov model $\{\mathbf{y}, \mathbf{X}\boldsymbol{\beta}, \sigma^2\mathbf{I}\}$, the least squares estimator of an estimable function $\mathbf{c}^T\boldsymbol{\beta}$ associated with that model has uniformly (in $\boldsymbol{\beta}$ and σ^2) smaller mean squared error than any other linear location equivariant estimator of $\mathbf{c}^T\boldsymbol{\beta}$. (Refer back to Exercise 6.1 for the definition of a location equivariant estimator.)
6. Consider the model $\{\mathbf{y}, \mathbf{X}\boldsymbol{\beta}\}$ with observations partitioned into two groups, so that the model may be written alternatively as $\left\{\begin{pmatrix} \mathbf{y}_1 \\ \mathbf{y}_2 \end{pmatrix}, \begin{pmatrix} \mathbf{X}_1 \\ \mathbf{X}_2 \end{pmatrix}\boldsymbol{\beta}\right\}$, where $\mathbf{y}_1$ is an n_1-vector. Suppose that $\mathbf{c}^T \in \mathcal{R}(\mathbf{X}_1)$ and $\mathcal{R}(\mathbf{X}_1) \cap \mathcal{R}(\mathbf{X}_2) = \{\mathbf{0}\}$. Prove that the least squares estimator of $\mathbf{c}^T\boldsymbol{\beta}$ associated with the model $\{\mathbf{y}_1, \mathbf{X}_1\boldsymbol{\beta}\}$ for the first n_1 observations is identical to the least squares estimator of $\mathbf{c}^T\boldsymbol{\beta}$ associated with the model for all the observations; i.e., the additional observations do not affect the least squares estimator.
7. For an arbitrary model $\{\mathbf{y}, \mathbf{X}\boldsymbol{\beta}\}$:
(a) Prove that $\bar{x}_1\beta_1 + \bar{x}_2\beta_2 + \cdots + \bar{x}_p\beta_p$ is estimable.
(b) Prove that if one of the columns of $\mathbf{X}$ is a column of ones, then the least squares estimator of the function in part (a) is $\bar{y}$.
8. For the centered simple linear regression model

$$y_i = \beta_1 + \beta_2(x_i - \bar{x}) + e_i \quad (i = 1, \ldots, n),$$

obtain the least squares estimators of β_1 and β_2 and their variance–covariance matrix (under Gauss–Markov assumptions).
9. For the no-intercept simple linear regression model, obtain the least squares estimator of the slope and its variance (under Gauss–Markov assumptions).
10. For the two-way main effects model with cell frequencies specified by the last 2×2 layout of Example 7.1-3, obtain the variance–covariance matrix (under

Gauss–Markov assumptions) of the least squares estimators of:
(a) the cell means;
(b) the Factor A and Factor B differences.

11. For the two-way main effects model with equal cell frequencies r, obtain the least squares estimators, and their variance–covariance matrix (under Gauss–Markov assumptions), of:
(a) the cell means;
(b) the Factor A and Factor B differences.

12. Consider the connected 6×4 layout displayed in Example 6.1-4. Suppose that the model for the "data" in the occupied cells is the Gauss–Markov two-way main effects model

$$y_{ij} = \mu + \alpha_i + \gamma_j + e_{ij}$$

where $(i, j) \in \{(1,1), (1,2), (2,1), (2,3), (3,1), (3,4), (4,2), (4,3), (5,2), (5,4), (6,3), (6,4)\}$ (the occupied cells). Obtain specialized expressions for the least squares estimators of $\gamma_j - \gamma_{j'}$ $(j' > j = 1, 2, 3, 4)$, and show that their variances are equal. The homoscedasticity of the variances of these differences is a nice property of balanced incomplete block designs such as this one.

13. For the two-factor nested model, obtain the least squares estimators and their variance–covariance matrix (under Gauss–Markov assumptions), of:
(a) the cell means;
(b) the Factor B differences (within levels of Factor A).

14. Consider the two-way partially crossed model introduced in Example 5.1.4-1, with one observation per cell, i.e.,

$$y_{ij} = \mu + \alpha_i - \alpha_j + e_{ij} \qquad (i \neq j = 1, \ldots, q),$$

where the e_{ij}'s satisfy Gauss–Markov assumptions.
(a) Obtain expressions for the least squares estimators of those functions in the set $\{\mu, \alpha_i - \alpha_j : i \neq j\}$ that are estimable.
(b) Obtain expressions for the variances and covariances (under Gauss–Markov assumptions) of the least squares estimators obtained in part (a).

15. For the Latin square design with q treatments described in Exercise 6.21:
(a) Obtain the least squares estimators of the treatment differences $\tau_k - \tau_{k'}$, where $k \neq k' = 1, \ldots, q$.
(b) Obtain expressions for the variances and covariances (under Gauss–Markov assumptions) of those least squares estimators.

16. Consider a simple linear regression model (with Gauss–Markov assumptions) for (x, y)-observations $\{(i, y_i) : i = 1, 2, 3, 4, 5\}$.
(a) Give a nonmatrix expression for the least squares estimator, $\hat{\beta}_2$, of the slope, in terms of the y_i's only.
(b) Give the variance of $\hat{\beta}_2$ as a multiple of σ^2.

(c) Show that each of the linear estimators

$$\hat{\gamma} = y_2 - y_1 \quad \text{and} \quad \hat{\eta} = (y_5 - y_2)/3$$

is unbiased for the slope, and determine their variances. Are these variances larger, smaller, or equal to the variance of $\hat{\beta}_2$?

(d) Determine the bias and variance of

$$c\hat{\gamma} + (1 - c)\hat{\eta}$$

(where $0 \leq c \leq 1$) as an estimator of the slope.

(e) Find the choice of c that minimizes the variance in part (d), and determine this minimized variance. Is this variance larger, smaller, or equal to the variance of $\hat{\beta}_2$?

17. Consider the cell-means parameterization of the two-way model with interaction, as given by (5.10). Suppose that none of the cells are empty. Under Gauss–Markov assumptions, obtain expressions for the variance of the least squares estimator of $\mu_{ij} - \mu_{ij'} - \mu_{i'j} + \mu_{i'j'}$ and the covariance between the least squares estimators of every pair $\mu_{ij} - \mu_{ij'} - \mu_{i'j} + \mu_{i'j'}$ and $\mu_{st} - \mu_{st'} - \mu_{s't} + \mu_{s't'}$ of such functions.

18. This exercise generalizes Theorem 7.2.3 (the Gauss–Markov theorem) to a situation in which one desires to estimate several estimable functions and say something about their joint optimality. Let $\mathbf{C}^T\boldsymbol{\beta}$ be a k-vector of estimable functions under the Gauss–Markov model $\{\mathbf{y}, \mathbf{X}\boldsymbol{\beta}, \sigma^2\mathbf{I}\}$, and let $\mathbf{C}^T\hat{\boldsymbol{\beta}}$ be the k-vector of least squares estimators of those functions associated with that model. Furthermore, let $\mathbf{k} + \mathbf{B}^T\mathbf{y}$ be any vector of linear unbiased estimators of $\mathbf{C}^T\boldsymbol{\beta}$. Prove the following results (the last two results follow from the first). For the last result, use the following lemma, which is essentially the same as Theorem 18.1.6 of Harville (1997): If $\mathbf{M}$ is a positive definite matrix, $\mathbf{Q}$ is a nonnegative definite matrix of the same dimensions as $\mathbf{M}$, and $\mathbf{M}-\mathbf{Q}$ is nonnegative definite, then $|\mathbf{M} - \mathbf{Q}| \leq |\mathbf{M}|$.

(a) $\text{var}(\mathbf{k} + \mathbf{B}^T\mathbf{y}) - \text{var}(\mathbf{C}^T\hat{\boldsymbol{\beta}})$ is nonnegative definite;

(b) $\text{tr}[\text{var}(\mathbf{k} + \mathbf{B}^T\mathbf{y})] \geq \text{tr}[\text{var}(\mathbf{C}^T\hat{\boldsymbol{\beta}})]$;

(c) $|\text{var}(\mathbf{k} + \mathbf{B}^T\mathbf{y})| \geq |\text{var}(\mathbf{C}^T\hat{\boldsymbol{\beta}})|$.

19. For a balanced two-factor crossed classification with two levels of each factor and r observations per cell:

(a) Show that the model

$$y_{ijk} = \theta + (-1)^i\tau_1 + (-1)^j\tau_2 + e_{ijk} \quad (i = 1, 2;\ j = 1, 2;\ k = 1, \ldots, r)$$

is a reparameterization of the two-way main effects model given by (5.6).

(b) Determine how θ, τ_1, and τ_2 are related to the parameters μ, α_1, α_2, γ_1, and γ_2 for parameterization (5.6) of the two-way main effects model.

(c) What "nice" property does the model matrix for this reparameterization have? Does it still have this property if the data are unbalanced?
(d) Indicate how you might reparameterize a three-way main effects model with two levels for each factor and balanced data in a similar way, and give the corresponding model matrix.

20. Consider the second-order polynomial regression model

$$y_i = \beta_1 + \beta_2 x_i + \beta_3 x_i^2 + e_i \quad (i = 1, \ldots, n)$$

and the reparameterization

$$y_i = \tau_1 + \tau_2(x_i - \bar{x}) + \tau_3(x_i - \bar{x})^2 + e_i \quad (i = 1, \ldots, n).$$

The second model is obtained from the first by centering the x_i's. Suppose that at least three of the x_i's are distinct and that the errors satisfy Gauss–Markov assumptions.
(a) Let $\mathbf{X}$ and $\breve{\mathbf{X}}$ represent the model matrices corresponding to these two models. Determine the columns of $\mathbf{X}$ and $\breve{\mathbf{X}}$ and verify that the second model is a reparameterization of the first model.
(b) Let $\hat{\tau}_j$ denote the least squares estimators of τ_j $(j = 1, 2, 3)$ in the model $\mathbf{y} = \breve{\mathbf{X}}\boldsymbol{\tau} + \mathbf{e}$. Suppose that $x_i = i$ $(i = 1 \ldots, n)$. Determine $\text{cov}(\hat{\tau}_1, \hat{\tau}_2)$ and $\text{cov}(\hat{\tau}_2, \hat{\tau}_3)$.
(c) Determine the variance inflation factors for the regressors in the second model.

21. Consider the Gauss–Markov model $\{\mathbf{y}, \mathbf{X}\boldsymbol{\beta}, \sigma^2\mathbf{I}\}$, and suppose that $\mathbf{X} = (x_{ij})$ $(i = 1, \ldots, n;\ j = 1, \ldots, p)$ has full column rank. Let $\hat{\beta}_j$ represent the least squares estimator of β_j $(j = 1, \ldots, p)$ associated with this model.
(a) Prove that

$$\text{var}(\hat{\beta}_j) \geq \frac{\sigma^2}{\sum_{i=1}^n x_{ij}^2} \quad \text{for all } j = 1, \ldots, p,$$

with equality if and only if $\sum_{i=1}^n x_{ij}x_{ik} = 0$ for all $j \neq k = 1, \ldots, p$. (Hint: Without loss of generality take $j = 1$, and use Theorem 2.9.5.)
(b) Suppose that $n = 8$ and $p = 4$. Let $\mathcal{X}_c$ represent the set of all 8×4 model matrices $\mathbf{X} = (x_{ij})$ for which $\sum_{i=1}^8 x_{ij}^2 \leq c$ for all $j = 1, \ldots, 4$. From part (a), it follows that if an $\mathbf{X} \in \mathcal{X}_c$ exists for which $\sum_{i=1}^8 x_{ij}x_{ik} = 0$ for all $j \neq k = 1, \ldots, 4$ and $\sum_{i=1}^8 x_{ij}^2 = c$ for $j = 1, \ldots, 4$, then $\mathbf{X}$ minimizes

var($\hat{\beta}_j$), for all $j = 1, \ldots, 4$, over $\mathcal{X}_c$. Use this fact to show that

$$\mathbf{X} = \begin{pmatrix} 1 & -1 & -1 & -1 \\ 1 & -1 & -1 & 1 \\ 1 & -1 & 1 & -1 \\ 1 & -1 & 1 & 1 \\ 1 & 1 & -1 & -1 \\ 1 & 1 & -1 & 1 \\ 1 & 1 & 1 & -1 \\ 1 & 1 & 1 & 1 \end{pmatrix}$$

minimizes var($\hat{\beta}_j$) for all $j = 1, \ldots, 4$ among all 8×4 model matrices for which $-1 \leq x_{ij} \leq 1$ ($i = 1, \ldots, 8$ and $j = 1, \ldots, 4$).

(c) The model matrix $\mathbf{X}$ displayed in part (b) corresponds to what is called a 2^3 *factorial design*, i.e., an experimental design for which there are three completely crossed experimental factors, each having two levels (coded as -1 and 1), with one observation per combination of factor levels. According to part (b), a 2^3 factorial design minimizes var($\hat{\beta}_j$) (for all $j = 1, \ldots, 4$) among all 8×4 model matrices for which $x_{i1} \equiv 1$ and $-1 \leq x_{ij} \leq 1$ ($i = 1, \ldots, 8$ and $j = 2, 3, 4$), under the "first-order" model

$$y_i = \beta_1 + \beta_2 x_{i2} + \beta_3 x_{i3} + \beta_4 x_{i4} + e_i, \quad i = 1, \ldots, 8.$$

Now consider adding seven additional combinations of the three quantitative explanatory variables to the eight combinations in a 2^3 factorial design. These seven additional combinations are as listed in rows 9 through 15 of the following new, larger model matrix:

$$\mathbf{X} = \begin{pmatrix} 1 & -1 & -1 & -1 \\ 1 & -1 & -1 & 1 \\ 1 & -1 & 1 & -1 \\ 1 & -1 & 1 & 1 \\ 1 & 1 & -1 & -1 \\ 1 & 1 & -1 & 1 \\ 1 & 1 & 1 & -1 \\ 1 & 1 & 1 & 1 \\ 1 & 0 & 0 & 0 \\ 1 & -\alpha & 0 & 0 \\ 1 & \alpha & 0 & 0 \\ 1 & 0 & -\alpha & 0 \\ 1 & 0 & \alpha & 0 \\ 1 & 0 & 0 & -\alpha \\ 1 & 0 & 0 & \alpha \end{pmatrix}.$$

Here α is any positive number. The experimental design associated with this model matrix is known as a 2^3 *factorial central composite design*. Use the results of parts (a) and (b) to show that this design minimizes $\mathrm{var}(\hat{\beta}_j)$ $(j = 2, 3, 4)$ among all 15×4 model matrices for which $x_{i1} \equiv 1$ and $\sum_{i=1}^{15} x_{ij}^2 \leq 8 + 2\alpha^2$ $(j = 2, 3, 4)$, under the "first-order" model

$$y_i = \beta_1 + \beta_2 x_{i2} + \beta_3 x_{i3} + \beta_4 x_{i4} + e_i, \quad i = 1, \ldots, 15.$$

22. Consider the model $\{\mathbf{y}, \mathbf{X}\boldsymbol{\beta}\}$ and a reparameterization $\{\mathbf{y}, \breve{\mathbf{X}}\boldsymbol{\tau}\}$ of it. Let matrices $\mathbf{T}$ and $\mathbf{S}^T$ be defined such that $\breve{\mathbf{X}} = \mathbf{X}\mathbf{T}$ and $\mathbf{X} = \breve{\mathbf{X}}\mathbf{S}^T$. Show that the converse of Theorem 7.3.1 is false by letting

$$\mathbf{X} = \begin{pmatrix} 1 & 2 & 1 & 2 \\ 1 & 2 & 0 & 0 \\ 1 & 2 & 0 & 0 \end{pmatrix}, \quad \breve{\mathbf{X}} = \begin{pmatrix} 1 & 1 & 2 \\ 1 & 0 & 0 \\ 1 & 0 & 0 \end{pmatrix}, \quad \mathbf{S}^T = \begin{pmatrix} 1 & 2 & 0 & 0 \\ 0 & 0 & 1 & 2 \\ 0 & 0 & 0 & 0 \end{pmatrix},$$

and finding a vector $\mathbf{b}$ such that $\mathbf{b}^T\mathbf{S}^T\boldsymbol{\beta}$ is estimable under the original model but $\mathbf{b}^T\boldsymbol{\tau}$ is not estimable under the reparameterized model.

23. Prove Theorem 7.3.2.

References

Harville, D. A. (1997). *Matrix algebra from a statistician's perspective*. New York: Springer.

Snedecor, G. W., & Cochran, W. G. (1980). *Statistical methods* (7th ed.). Iowa: Iowa State University Press.

Least Squares Geometry and the Overall ANOVA

8

In Chap. 7, we defined least squares estimators associated with a given model *algebraically*, i.e., in terms of the problem of minimizing the residual sum of squares function for the model. But least squares also have a geometric interpretation, which is demonstrated in this chapter. This additional interpretation of least squares yields new insights and suggests a decomposition of the total variability of the response vector into components attributable to the model's mean structure and error structure, which is known as the overall analysis of variance (or overall ANOVA). It also leads to an estimator of the residual variance in the Gauss–Markov model $\{\mathbf{y}, \mathbf{X}\boldsymbol{\beta}, \sigma^2\mathbf{I}\}$ that is unbiased under that model.

8.1 Least Squares as an Orthogonal Projection

Let $\hat{\boldsymbol{\beta}}$ represent any solution to the normal equations for the model $\{\mathbf{y}, \mathbf{X}\boldsymbol{\beta}\}$, and as in Sect. 7.1 put

$$\hat{\mathbf{y}} = \mathbf{X}\hat{\boldsymbol{\beta}} \quad \text{and} \quad \hat{\mathbf{e}} = \mathbf{y} - \hat{\mathbf{y}} = \mathbf{y} - \mathbf{X}\hat{\boldsymbol{\beta}},$$

so that the vector $\mathbf{y}$ of observations admits the decomposition

$$\mathbf{y} = \hat{\mathbf{y}} + \hat{\mathbf{e}}. \tag{8.1}$$

The elements of $\hat{\mathbf{y}}$, of course, comprise the least squares estimators of the corresponding elements of $\mathbf{X}\boldsymbol{\beta}$. The elements of $\hat{\mathbf{e}}$ are not estimators of any function of $\boldsymbol{\beta}$, but might be regarded as "predictors" (estimators of random variables) of the corresponding elements of the random vector $\mathbf{e}$. Clearly $\hat{\mathbf{y}} \in \mathcal{C}(\mathbf{X})$, and recall from Sect. 7.1 that $\hat{\mathbf{e}} \in \mathcal{N}(\mathbf{X}^T)$. Thus, by Theorem 2.6.2, $\hat{\mathbf{y}}$ and $\hat{\mathbf{e}}$ are orthogonal. This can

D. L. Zimmerman, *Linear Model Theory*,
https://doi.org/10.1007/978-3-030-52063-2_8

also be verified directly:

$$\hat{\mathbf{y}}^T\hat{\mathbf{e}} = \hat{\boldsymbol{\beta}}^T\mathbf{X}^T(\mathbf{y} - \mathbf{X}\hat{\boldsymbol{\beta}}) = \hat{\boldsymbol{\beta}}^T(\mathbf{X}^T\mathbf{y} - \mathbf{X}^T\mathbf{X}\hat{\boldsymbol{\beta}}) = 0.$$

The following theorem makes a much stronger claim regarding decomposition (8.1).

Theorem 8.1.1 *For the model* $\{\mathbf{y}, \mathbf{X}\boldsymbol{\beta}\}$, *the response vector* $\mathbf{y}$ *admits a unique decomposition*

$$\mathbf{y} = \mathbf{y}_1 + \mathbf{y}_2$$

where $\mathbf{y}_1 \in \mathcal{C}(\mathbf{X})$ *and* $\mathbf{y}_2 \in \mathcal{N}(\mathbf{X}^T)$, *and this decomposition is given by* $\mathbf{y}_1 = \hat{\mathbf{y}} = \mathbf{X}\hat{\boldsymbol{\beta}}$ *and* $\mathbf{y}_2 = \hat{\mathbf{e}} = \mathbf{y} - \mathbf{X}\hat{\boldsymbol{\beta}}$, *where* $\hat{\boldsymbol{\beta}}$ *is any solution to the normal equations for the model.*

Proof As noted above, $\mathbf{y} = \hat{\mathbf{y}} + \hat{\mathbf{e}}$, where $\hat{\mathbf{y}} \in \mathcal{C}(\mathbf{X})$ and $\hat{\mathbf{e}} \in \mathcal{N}(\mathbf{X}^T)$. It remains to verify the uniqueness of the decomposition. Suppose that $\mathbf{y}_1 \in \mathcal{C}(\mathbf{X})$, $\mathbf{y}_2 \in \mathcal{N}(\mathbf{X}^T)$, and $\mathbf{y} = \mathbf{y}_1 + \mathbf{y}_2$. Then, $\hat{\mathbf{y}} - \mathbf{y}_1 \in \mathcal{C}(\mathbf{X})$ and

$$\hat{\mathbf{y}} - \mathbf{y}_1 = (\mathbf{y} - \mathbf{y}_1) - (\mathbf{y} - \hat{\mathbf{y}}) = \mathbf{y}_2 - \hat{\mathbf{e}} \in \mathcal{N}(\mathbf{X}^T),$$

implying that $(\hat{\mathbf{y}} - \mathbf{y}_1)^T(\hat{\mathbf{y}} - \mathbf{y}_1) = 0$. This implies further (by Corollary 2.10.4.1) that $\hat{\mathbf{y}} - \mathbf{y}_1 = \mathbf{0}$, i.e., that $\mathbf{y}_1 = \hat{\mathbf{y}}$. It follows immediately that $\mathbf{y}_2 = \hat{\mathbf{e}}$ also. □

The upshot of Theorem 8.1.1, when considered in conjunction with Theorem 7.1.2, is that $\hat{\mathbf{y}}$, the vector of least squares fitted values, is the unique point in $\mathcal{C}(\mathbf{X})$ that is closest (in terms of Euclidean distance) to $\mathbf{y}$.

Definition 8.1.1 The vector $\hat{\mathbf{y}}$ of least squares fitted values is also called *the orthogonal projection of* $\mathbf{y}$ *onto* $\mathcal{C}(\mathbf{X})$.

We may write $\hat{\mathbf{y}} = \mathbf{X}\hat{\boldsymbol{\beta}} = \mathbf{X}(\mathbf{X}^T\mathbf{X})^-\mathbf{X}^T\mathbf{y} = \mathbf{P_X}\mathbf{y}$, say, where $\mathbf{P_X} = \mathbf{X}(\mathbf{X}^T\mathbf{X})^-\mathbf{X}^T$. By Theorem 3.3.4, $\mathbf{P_X}$ is invariant to the choice of generalized inverse of $\mathbf{X}^T\mathbf{X}$. Furthermore, if $\mathbf{A}$ is any matrix for which $\mathbf{Ay}$ is the orthogonal projection of $\mathbf{y}$ onto $\mathcal{C}(\mathbf{X})$, then $\mathbf{Ay} = \mathbf{P_X}\mathbf{y}$ for all $\mathbf{y}$, implying (by Theorem 2.1.1) that $\mathbf{A} = \mathbf{P_X}$. Hence $\mathbf{P_X}$ is the *unique* matrix that projects $\mathbf{y}$ orthogonally onto $\mathcal{C}(\mathbf{X})$.

Definition 8.1.2 The matrix

$$\mathbf{P_X} = \mathbf{X}(\mathbf{X}^T\mathbf{X})^-\mathbf{X}^T$$

is called *the orthogonal projection matrix onto the column space of* $\mathbf{X}$.

In Chap. 7 we noted that the least squares fitted residuals vector is orthogonal to the vector of observed values of each explanatory variable, and that, as a special case, if $\mathbf{1}_n$ is one of the columns of $\mathbf{X}$, then the least squares fitted residuals sum to 0. Theorem 8.1.1 allows this result to be strengthened, viz., the least squares fitted residuals vector is orthogonal to every vector in $\mathcal{C}(\mathbf{X})$. Thus, for example, if $\mathbf{1}_n$ is not one of the columns of $\mathbf{X}$ but $\mathbf{1}_n \in \mathcal{C}(\mathbf{X})$, the least squares fitted residuals still sum to 0.

The following corollary, which is an immediate consequence of Theorem 8.1.1, indicates that the orthogonal projection of $\mathbf{y}$ onto the column space of the model matrix is invariant to model reparameterization.

Corollary 8.1.1.1 *If* $\breve{\mathbf{X}}$ *is any matrix for which* $\mathcal{C}(\breve{\mathbf{X}}) = \mathcal{C}(\mathbf{X})$, *then* $\mathbf{P}_{\breve{\mathbf{X}}} = \mathbf{P}_{\mathbf{X}}$ *and* $\breve{\mathbf{X}}\hat{\boldsymbol{\tau}} = \mathbf{X}\hat{\boldsymbol{\beta}}$, *where* $\hat{\boldsymbol{\beta}}$ *is any solution to* $\mathbf{X}^T\mathbf{X}\boldsymbol{\beta} = \mathbf{X}^T\mathbf{y}$ *and* $\hat{\boldsymbol{\tau}}$ *is any solution to* $\breve{\mathbf{X}}^T\breve{\mathbf{X}}\boldsymbol{\tau} = \breve{\mathbf{X}}^T\mathbf{y}$.

Example 8.1-1. Toy Example of an Orthogonal Projection onto $\mathcal{C}(\mathbf{X})$

A toy example will illustrate Theorem 8.1.1 and Corollary 8.1.1.1. Suppose that $n = 2$, $p = 1$, $\mathbf{y} = \begin{pmatrix} 1 \\ 3 \end{pmatrix}$, and $\mathbf{X} = \mathbf{1}_2$. Then, the normal "equations" (there is only one equation in this case) are $2\boldsymbol{\beta} = 4$, which have the unique solution $\hat{\boldsymbol{\beta}} = 2$. Figure 8.1 depicts the situation. The orthogonal projection of $\mathbf{y}$ onto $\mathcal{C}(\mathbf{X})$ is $2\mathbf{1}_2$; furthermore, $\hat{\mathbf{y}} = 2\mathbf{1}_2$, $\hat{\mathbf{e}} = \mathbf{y} - \hat{\mathbf{y}} = \begin{pmatrix} -1 \\ 1 \end{pmatrix}$, and $\hat{\mathbf{y}}$ and $\hat{\mathbf{e}}$ are orthogonal. Each circle in the figure represents a locus of points equidistant (in terms of Euclidean distance) from $\mathbf{y}$.

Now suppose we define $\breve{\mathbf{X}} = -7\mathbf{1}_2$ [in which case $\mathcal{C}(\breve{\mathbf{X}}) = \mathcal{C}(\mathbf{X})$]. It can be verified easily that the orthogonal projection of $\mathbf{y}$ onto $\mathcal{C}(\breve{\mathbf{X}})$ is also $2\mathbf{1}_2$, and that

$$\mathbf{P}_{\mathbf{X}} = \mathbf{P}_{\breve{\mathbf{X}}} = \begin{pmatrix} 0.5 & 0.5 \\ 0.5 & 0.5 \end{pmatrix}. \qquad \blacksquare$$

Besides the term "orthogonal projection matrix onto $\mathcal{C}(\mathbf{X})$," $\mathbf{P}_{\mathbf{X}}$ is known by a few other names. In applied regression courses, $\mathbf{P}_{\mathbf{X}}$ is often called the *hat matrix*, which derives from the fact that multiplication of $\mathbf{y}$ by $\mathbf{P}_{\mathbf{X}}$ produces $\hat{\mathbf{y}}$. $\mathbf{P}_{\mathbf{X}}$ has also been called the *influence matrix*, for if a "small" number δ_j is added to y_j, then the effect on the ith element of $\mathbf{X}\hat{\boldsymbol{\beta}}$ is to perturb (translate) it by the amount $\delta_j p_{ij}$, where p_{ij} is the ijth element of $\mathbf{P}_{\mathbf{X}}$. Thus, the elements of $\mathbf{P}_{\mathbf{X}}$ determine how much influence a perturbation in $\mathbf{y}$ will have on the least squares fitted values.

$\mathbf{P}_{\mathbf{X}}$ and its "complement" $\mathbf{I} - \mathbf{P}_{\mathbf{X}}$ have several more important properties, as described by the next five theorems.

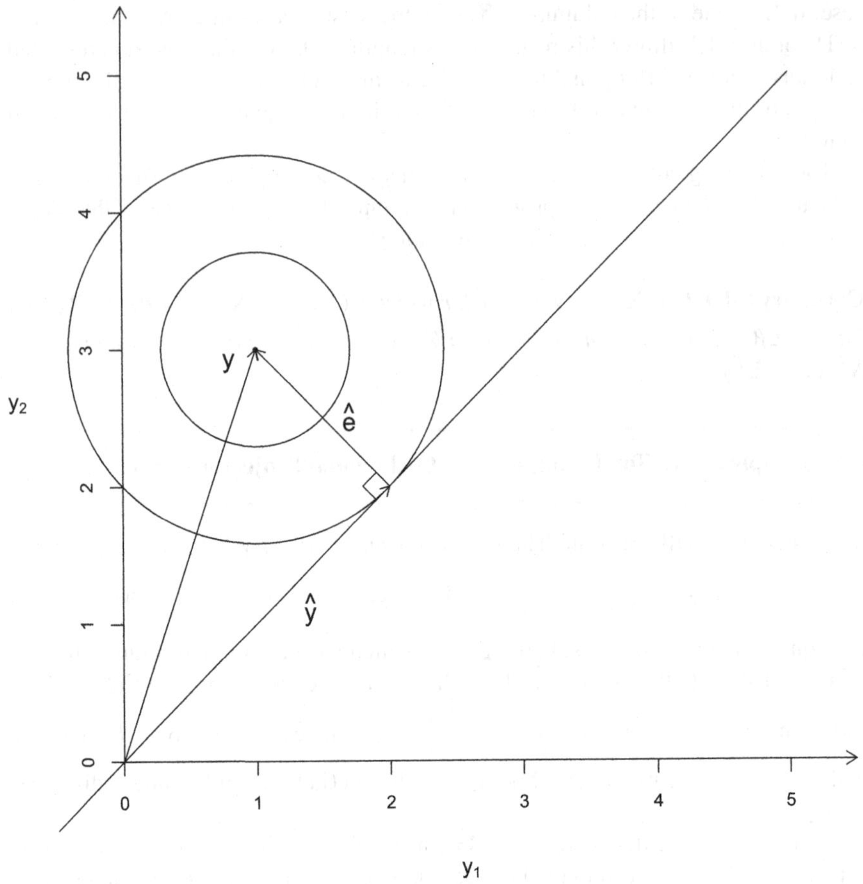

Fig. 8.1 Orthogonal projection of $\mathbf{y} = (1, 3)^T$ onto $\mathcal{C}(\mathbf{1}_2)$

Theorem 8.1.2

(a) $\mathbf{P_X}$*, the orthogonal projection matrix onto* $\mathcal{C}(\mathbf{X})$*, has the following properties:*

(i) $\mathbf{P_X X} = \mathbf{X}$*;*

(ii) $\mathbf{P_X P_X} = \mathbf{P_X}$*, i.e.,* $\mathbf{P_X}$ *is idempotent;*

(iii) $\mathbf{P_X^T} = \mathbf{P_X}$*, i.e.,* $\mathbf{P_X}$ *is symmetric;*

(iv) $\mathcal{C}(\mathbf{P_X}) = \mathcal{C}(\mathbf{X})$*; and*

(v) $rank(\mathbf{P_X}) = rank(\mathbf{X})$*.*

(b) Conversely, suppose that a matrix $\mathbf{P}$ *has the following properties:*

(i) $\mathbf{PX} = \mathbf{X}$*;*

(ii) $\mathbf{PP} = \mathbf{P}$*;*

(iii) $\mathbf{P}^T = \mathbf{P}$*; and*

(iv) $rank(\mathbf{P}) = rank(\mathbf{X})$*.*

Then $\mathbf{P} = \mathbf{P_X}$*.*

Proof First we prove part (a). Observe that $\mathbf{P_X X} = \mathbf{X(X^T X)^- X^T X} = \mathbf{X}$ directly by Theorem 3.3.3a, which establishes (i). Then

$$\mathbf{P_X P_X} = \mathbf{P_X X(X^T X)^- X^T} = \mathbf{X(X^T X)^- X^T} = \mathbf{P_X},$$

which establishes (ii). Part (iii), the symmetry of $\mathbf{P_X}$, follows immediately from Corollary 3.3.3.1. For part (iv), note that $\mathcal{C}(\mathbf{X}) \subseteq \mathcal{C}(\mathbf{P_X})$ by part (i) and Theorem 2.6.1, while $\mathcal{C}(\mathbf{P_X}) \subseteq \mathcal{C}(\mathbf{X})$ by the definition of $\mathbf{P_X}$ and Theorem 2.6.1. Part (v) follows easily from part (iv).

Next we prove part (b). Recall that $\mathbf{P_X} = \mathbf{XB}$ for some matrix $\mathbf{B}$. So by (i), $\mathbf{PP_X} = \mathbf{PXB} = \mathbf{XB} = \mathbf{P_X}$. By (iii) and (a)(iii), $\mathbf{P_X P} = \mathbf{P_X}$ and $\mathbf{P} - \mathbf{P_X}$ is symmetric. By these results and (ii),

$$(\mathbf{P} - \mathbf{P_X})(\mathbf{P} - \mathbf{P_X}) = \mathbf{P} - \mathbf{P_X P} - \mathbf{PP_X} + \mathbf{P_X} = \mathbf{P} - \mathbf{P_X},$$

i.e., $\mathbf{P} - \mathbf{P_X}$ is idempotent. Thus $\mathbf{P} - \mathbf{P_X}$ is symmetric and idempotent, so by Theorem 2.12.2 we obtain

$$\text{rank}(\mathbf{P} - \mathbf{P_X}) = \text{tr}(\mathbf{P} - \mathbf{P_X}) = \text{tr}(\mathbf{P}) - \text{tr}(\mathbf{P_X}) = \text{rank}(\mathbf{P}) - \text{rank}(\mathbf{P_X}) = 0$$

where we used (iv) and (a)(v) for the last equality. Thus $\mathbf{P} - \mathbf{P_X} = \mathbf{0}$, i.e., $\mathbf{P} = \mathbf{P_X}$. □

Theorem 8.1.3 *The matrix* $\mathbf{I} - \mathbf{P_X}$ *has the following properties:*

(a) $(\mathbf{I} - \mathbf{P_X})\mathbf{X} = \mathbf{0}$;
(b) $(\mathbf{I} - \mathbf{P_X})(\mathbf{I} - \mathbf{P_X}) = \mathbf{I} - \mathbf{P_X}$, *i.e.,* $\mathbf{I} - \mathbf{P_X}$ *is idempotent;*
(c) $(\mathbf{I} - \mathbf{P_X})^T = \mathbf{I} - \mathbf{P_X}$, *i.e.,* $\mathbf{I} - \mathbf{P_X}$ *is symmetric;*
(d) *rank*$(\mathbf{I} - \mathbf{P_X}) = n -$ *rank*$(\mathbf{X})$.

Proof $(\mathbf{I} - \mathbf{P_X})\mathbf{X} = \mathbf{X} - \mathbf{P_X X} = \mathbf{X} - \mathbf{X} = \mathbf{0}$ by Theorem 8.1.2a(i), proving part (a). The idempotency of $\mathbf{I} - \mathbf{P_X}$ [part (b)] follows immediately from that of $\mathbf{P_X}$ by Theorem 2.12.1. Furthermore, $(\mathbf{I} - \mathbf{P_X})^T = \mathbf{I}^T - \mathbf{P_X}^T = \mathbf{I} - \mathbf{P_X}$ by Theorem 8.1.2a(iii), which establishes part (c). Finally, for part (d), by part (b) and Theorems 2.12.2 and 8.1.2a(ii), (v),

$$\text{rank}(\mathbf{I} - \mathbf{P_X}) = \text{tr}(\mathbf{I} - \mathbf{P_X}) = \text{tr}(\mathbf{I}) - \text{tr}(\mathbf{P_X}) = n - \text{rank}(\mathbf{P_X}) = n - \text{rank}(\mathbf{X}).$$

□

Theorem 8.1.4 $\mathbf{P_X}(\mathbf{I} - \mathbf{P_X}) = \mathbf{0}$.

Proof By Theorem 8.1.2a(ii),

$$\mathbf{P_X}(\mathbf{I} - \mathbf{P_X}) = \mathbf{P_X} - \mathbf{P_X P_X} = \mathbf{P_X} - \mathbf{P_X} = \mathbf{0}.$$

□

Theorem 8.1.5 *$\mathcal{C}(\mathbf{I} - \mathbf{P_X})$ is the orthogonal complement of $\mathcal{C}(\mathbf{X})$ (relative to $\mathbb{R}^n$).*

Proof If $\mathbf{y} \in \mathcal{C}(\mathbf{I}-\mathbf{P_X})$, then $\mathbf{y} = (\mathbf{I}-\mathbf{P_X})\mathbf{a}$ for some $\mathbf{a}$. Thus $\mathbf{X}^T\mathbf{y} = \mathbf{X}^T(\mathbf{I}-\mathbf{P_X})\mathbf{a} = \mathbf{0}_{p\times n}\mathbf{a} = \mathbf{0}$ by the transpose of the result in Theorem 8.1.3a, which establishes that $\mathbf{y}$ is orthogonal to $\mathcal{C}(\mathbf{X})$. Conversely, if $\mathbf{y}$ is orthogonal to $\mathcal{C}(\mathbf{X})$ then $\mathbf{X}^T\mathbf{y} = \mathbf{0}$, implying that $\mathbf{P_X}\mathbf{y} = \mathbf{0}$, implying further that $\mathbf{y} - \mathbf{P_X}\mathbf{y} = \mathbf{y}$, i.e., $\mathbf{y} = (\mathbf{I} - \mathbf{P_X})\mathbf{y}$. Thus $\mathbf{y} \in \mathcal{C}(\mathbf{I} - \mathbf{P_X})$. □

Theorem 8.1.6 *Let p_{ij} denote the ijth element of $\mathbf{P_X}$. Then:*

(a) $\sum_{i=1}^n p_{ii} = \mathrm{rank}(\mathbf{X})$;
(b) $p_{ii} \geq 0$;
(c) $p_{ii} \leq 1/r_i$, *where r_i is the number of rows of $\mathbf{X}$ that are identical to the ith row.*

If, in addition, $\mathbf{1}_n$ is one of the columns of $\mathbf{X}$, then:

(d) $p_{ii} \geq 1/n$;
(e) $\sum_{i=1}^n p_{ij} = \sum_{j=1}^n p_{ij} = 1$.

Proof By Theorems 8.1.2a(ii) and a(v) and Theorem 2.12.2,

$$\mathrm{rank}(\mathbf{X}) = \mathrm{rank}(\mathbf{P_X}) = \mathrm{tr}(\mathbf{P_X}) = \sum_{i=1}^{n} p_{ii},$$

which proves part (a). Also by Theorem 8.1.2a(ii), $\mathbf{P_X} = \mathbf{P_X}\mathbf{P_X}$, implying (by Theorem 8.1.2a(iii)) that

$$p_{ii} = \sum_{j=1}^{n} p_{ij}p_{ji} = \sum_{j=1}^{n} p_{ij}^2 \geq 0.$$

This establishes part (b). Proofs of the remaining parts are left as exercises. □

We conclude this section with a theorem giving the means, variance–covariance matrices, and matrix of covariances of the two components of the orthogonal decomposition (8.1) under the Gauss–Markov model.

Theorem 8.1.7 *Let $\hat{\mathbf{y}}$ and $\hat{\mathbf{e}}$ be defined as in Theorem 8.1.1. Then, under the Gauss–Markov model $\{\mathbf{y}, \mathbf{X}\boldsymbol{\beta}, \sigma^2\mathbf{I}\}$:*

(a) $E(\hat{\mathbf{y}}) = \mathbf{X}\boldsymbol{\beta}$;
(b) $E(\hat{\mathbf{e}}) = \mathbf{0}$;
(c) $\mathrm{var}(\hat{\mathbf{y}}) = \sigma^2\mathbf{P_X}$;
(d) $\mathrm{var}(\hat{\mathbf{e}}) = \sigma^2(\mathbf{I} - \mathbf{P_X})$;
(e) $\mathrm{cov}(\hat{\mathbf{y}}, \hat{\mathbf{e}}) = \mathbf{0}$.

Proof $\mathrm{E}(\hat{\mathbf{y}}) = \mathrm{E}(\mathbf{P_X y}) = \mathbf{P_X X}\boldsymbol{\beta} = \mathbf{X}\boldsymbol{\beta}$ by Theorem 8.1.2a(i), proving part (a). Part (b) follows easily from part (a) and Theorem 8.1.3a, as $\mathrm{E}(\hat{\mathbf{e}}) = \mathrm{E}[(\mathbf{I} - \mathbf{P_X})\mathbf{y}] = (\mathbf{I} - \mathbf{P_X})\mathbf{X}\boldsymbol{\beta} = \mathbf{0}$. Furthermore, by Theorems 8.1.2a(ii) and a(iii) and 8.1.3b, c,

$$\mathrm{var}(\hat{\mathbf{y}}) = \mathrm{var}(\mathbf{P_X y}) = \mathbf{P_X}(\sigma^2\mathbf{I})\mathbf{P_X}^T = \sigma^2\mathbf{P_X}$$

and

$$\mathrm{var}(\hat{\mathbf{e}}) = \mathrm{var}[(\mathbf{I} - \mathbf{P_X})\mathbf{y}] = (\mathbf{I} - \mathbf{P_X})(\sigma^2\mathbf{I})(\mathbf{I} - \mathbf{P_X})^T = \sigma^2(\mathbf{I} - \mathbf{P_X}),$$

and by Theorems 8.1.3c and 8.1.4,

$$\mathrm{cov}(\hat{\mathbf{y}}, \hat{\mathbf{e}}) = \mathrm{cov}[\mathbf{P_X y}, (\mathbf{I} - \mathbf{P_X})\mathbf{y}] = \mathbf{P_X}(\sigma^2\mathbf{I})(\mathbf{I} - \mathbf{P_X})^T = \mathbf{0}.$$

□

Example 8.1-2. $\mathbf{P_X}$, $\mathrm{var}(\hat{\mathbf{y}})$, and $\mathrm{var}(\hat{\mathbf{e}})$ for the Simple Linear Regression Model

For the (Gauss–Markov) simple linear regression model,

$$\begin{aligned}\mathbf{P_X} &= (1/SXX)(\mathbf{1}_n, \mathbf{x}) \begin{pmatrix} \sum_{i=1}^n x_i^2/n & -\bar{x} \\ -\bar{x} & 1 \end{pmatrix} \begin{pmatrix} \mathbf{1}_n^T \\ \mathbf{x}^T \end{pmatrix} \\ &= (1/SXX)\left[\left(\sum_{i=1}^n x_i^2/n\right)\mathbf{1}\mathbf{1}^T - \bar{x}(\mathbf{x}\mathbf{1}^T + \mathbf{1}\mathbf{x}^T) + \mathbf{x}\mathbf{x}^T\right],\end{aligned}$$

which has (i, j)th element

$$\begin{aligned}(1/SXX)\left[\left(\sum_{k=1}^n x_k^2/n\right) - \bar{x}(x_i + x_j) + x_i x_j\right] &= \frac{1}{n} + [\bar{x}^2 - \bar{x}(x_i + x_j) + x_i x_j]/SXX \\ &= \frac{1}{n} + \frac{(x_i - \bar{x})(x_j - \bar{x})}{SXX}.\end{aligned}$$

Consequently, by Theorem 8.1.7, $\mathrm{var}(\hat{\mathbf{y}})$ is the $n \times n$ matrix with (i, j)th element $\sigma^2[(1/n) + (x_i - \bar{x})(x_j - \bar{x})/SXX]$ and $\mathrm{var}(\hat{\mathbf{e}})$ is the $n \times n$ matrix with (i, j)th element equal to $\sigma^2[1 - (1/n) - (x_i - \bar{x})^2/SXX]$ if $i = j$, and $\sigma^2[-(1/n) - (x_i - \bar{x})(x_j - \bar{x})/SXX]$ otherwise. ■

Example 8.1-3. $\mathbf{P_X}$, var($\hat{\mathbf{y}}$), and var($\hat{\mathbf{e}}$) for the One-Factor Model

For the (Gauss–Markov) one-factor model,

$$\mathbf{P_X} = (\mathbf{1}_n, \oplus_{i=1}^q \mathbf{1}_{n_i}) \begin{pmatrix} 0 & \mathbf{0}^T \\ \mathbf{0} & \oplus_{i=1}^q (1/n_i) \end{pmatrix} \begin{pmatrix} \mathbf{1}_n^T \\ \oplus_{i=1}^q \mathbf{1}_{n_i}^T \end{pmatrix}$$
$$= \oplus_{i=1}^q (1/n_i)\mathbf{J}_{n_i}.$$

Consequently, by Theorem 8.1.7, var($\hat{\mathbf{y}}$) and var($\hat{\mathbf{e}}$) are the block diagonal matrices

$$\sigma^2 \oplus_{i=1}^q (1/n_i)\mathbf{J}_{n_i} \qquad \text{and} \qquad \sigma^2[\mathbf{I} - \oplus_{i=1}^q (1/n_i)\mathbf{J}_{n_i}],$$

respectively. ■

Methodological Interlude #2: Residual Plots

Recall the identity $\mathbf{y} = \mathbf{X}\hat{\boldsymbol{\beta}} + \hat{\mathbf{e}}$ given by Theorem 8.1.1. Because $\mathbf{X}\hat{\boldsymbol{\beta}}$ is an unbiased estimator of $\mathbf{X}\boldsymbol{\beta}$, it appears that $\hat{\mathbf{e}} = (\hat{e}_i)$ could be considered an unbiased "estimator" (later we call it a "predictor") of $\mathbf{e} = (e_i)$. For the same reason, we might expect that the sampling distribution of $\hat{\mathbf{e}}$ will emulate, to some extent, that of $\mathbf{e}$. These notions have led users of linear models to construct "residual plots," i.e., plots of the fitted residuals $\hat{e}_i$ versus either i, x_{ij} for one or more js, or $\hat{y}_i$ $(i = 1, \ldots, n)$, with a goal of assessing the reasonableness of the assumed model. Later in this and subsequent chapters, we describe some additional diagnostic plots based on the fitted residuals from various models; here, we merely document some properties of the fitted residuals that hold under the Gauss–Markov model $\{\mathbf{y}, \mathbf{X}\boldsymbol{\beta}, \sigma^2\mathbf{I}\}$.

First, as noted previously, the fitted residuals sum to 0 when $\mathbf{1}_n \in \mathcal{C}(\mathbf{X})$; this, in fact, is true whether or not the assumed model is correct. Second, by Theorem 8.1.7b, E$(\hat{e}_i) = 0$ for all i, implying that the existence of a discernible trend in the residual plot is highly unlikely (under the assumed model). Third, by Theorem 8.1.7d, var$(\hat{e}_i) = \sigma^2(1 - p_{ii})$, revealing (by Theorem 8.1.6b) that var$(\hat{e}_i) \leq$ var(e_i) and that generally the $\hat{e}_i$'s, unlike the e_i's, are heteroscedastic. Last, but also by Theorem 8.1.7d,

$$\text{corr}(\hat{e}_i, \hat{e}_j) = \frac{-p_{ij}}{\sqrt{(1 - p_{ii})(1 - p_{jj})}}, \qquad i \neq j.$$

Because p_{ij} need not equal 0, in general the $\hat{e}_i$'s, unlike the e_is, are correlated.

Although the fitted residuals generally are heteroscedastic and correlated, if the next two examples provide any indication, the extent of the residuals' departure from homoscedasticity and uncorrelatedness often is not large in practice (assuming that the fitted model is correct). Thus, if the fitted residuals in a residual plot exhibit a substantial trend in variability, or excessive persistence or rapid change in sign, the model is likely to be misspecified. ■

Example 8.1-4. Variances and Correlations of Residuals Under a Simple Linear Regression Model with Equally Spaced Values of the Regressor

Using the expression for p_{ij} under a simple linear regression model, given in Example 8.1-2, we obtain (assuming a Gauss–Markov model)

$$\text{var}(\hat{e}_i) = \sigma^2[1 - (1/n) - (x_i - \bar{x})^2/SXX]$$

and

$$\text{corr}(\hat{e}_i, \hat{e}_j) = \frac{-(1/n) - (x_i - \bar{x})(x_j - \bar{x})/SXX}{[1 - (1/n) - (x_i - \bar{x})^2/SXX]^{\frac{1}{2}}[1 - (1/n) - (x_j - \bar{x})^2/SXX]^{\frac{1}{2}}}.$$

Suppose that the values of the regressor are equally spaced, or equivalently that $x_i = i$ $(i = 1, \ldots, n)$. Then $\bar{x} = (n + 1)/2$ and $SXX = \sum_{i=1}^{n} x_i^2 - n\bar{x}^2 = n(n + 1)(n - 1)/12$ (after some algebra). Thus, in this case

$$\text{var}(\hat{e}_i) = \sigma^2\{1 - (1/n) - (i - \bar{x})^2/[n(n + 1)(n - 1)/12]\},$$

which ranges from approximately $\sigma^2[1 - (4/n)]$ when $i = 1$ or $i = n$ to approximately $\sigma^2[1 - (1/n)]$ when i is equal or close to $\bar{x}$. Moreover,

$$\text{corr}(\hat{e}_i, \hat{e}_j) = \frac{-(1/n) - (i - \bar{x})(j - \bar{x})/SXX}{[1 - (1/n) - (i - \bar{x})^2/SXX]^{\frac{1}{2}}[1 - (1/n) - (j - \bar{x})^2/SXX]^{\frac{1}{2}}}.$$

The absolute value of the numerator of this last quantity is bounded above by $4/n$ and the denominator is bounded below by $1-(4/n)$, with the result that $|\text{corr}(\hat{e}_i, \hat{e}_j)|$ is bounded above by $\min[1, 4/(n - 4)]$. As n increases, this bound decreases and gets arbitrarily close to 0. ■

Example 8.1-5. Variances and Correlations of Residuals Under the One-Factor Model

According to the expression for $\text{var}(\hat{\mathbf{e}})$ under the (Gauss–Markov) one-factor model given in Example 8.1-3, the residuals corresponding to observations within the ith level of the factor have common variance $\sigma^2[1 - (1/n_i)]$. Thus, the residuals are homoscedastic if and only if the data are balanced. Furthermore, the correlation between $\hat{e}_{ij}$ and $\hat{e}_{i'j'}$ is equal to $\frac{-1/n_i}{1-(1/n_i)} = \frac{-1}{n_i-1}$ if $i = i'$, and equal to 0 otherwise. ■

8.2 The Overall ANOVA and the Estimation of Residual Variance

In the previous section, it was shown that any response vector $\mathbf{y}$ following a linear model $\{\mathbf{y}, \mathbf{X}\boldsymbol{\beta}\}$ can be decomposed uniquely as the sum of two orthogonal vectors $\mathbf{y}_1$ and $\mathbf{y}_2$, where $\mathbf{y}_1 \in \mathcal{C}(\mathbf{X})$ and $\mathbf{y}_2 \in \mathcal{N}(\mathbf{X}^T)$. It was also shown that this unique decomposition is given by

$$\mathbf{y} = \mathbf{P_X}\mathbf{y} + (\mathbf{I} - \mathbf{P_X})\mathbf{y}. \tag{8.2}$$

In this section we consider the corresponding decomposition of the squared length (sum of squares) of $\mathbf{y}$ and we describe some properties and by-products of this decomposition. One important by-product is an unbiased estimator (under the Gauss–Markov model) of the residual variance, σ^2.

Pre-multiplication of both sides of (8.2) by $\mathbf{y}^T$ yields

$$\mathbf{y}^T\mathbf{y} = \mathbf{y}^T[\mathbf{P_X}\mathbf{y} + (\mathbf{I} - \mathbf{P_X})\mathbf{y}] = \mathbf{y}^T\mathbf{P_X}\mathbf{y} + \mathbf{y}^T(\mathbf{I} - \mathbf{P_X})\mathbf{y},$$

or equivalently (because of the symmetry and idempotency of $\mathbf{P_X}$ and $\mathbf{I} - \mathbf{P_X}$),

$$\|\mathbf{y}\|^2 = \|\mathbf{P_X}\mathbf{y}\|^2 + \|(\mathbf{I} - \mathbf{P_X})\mathbf{y}\|^2.$$

That is, the squared length of the vector $\mathbf{y}$ is equal to the sum of the squared lengths of vectors $\mathbf{P_X}\mathbf{y}$ and $(\mathbf{I} - \mathbf{P_X})\mathbf{y}$. So we see that not only may $\mathbf{y}$ be decomposed into $\mathbf{P_X}\mathbf{y}$ and $(\mathbf{I} - \mathbf{P_X})\mathbf{y}$, but the squared length (or sum of squares of the elements) of $\mathbf{y}$ may also be decomposed into the squared lengths (sums of squares of the elements) of $\mathbf{P_X}\mathbf{y}$ and $(\mathbf{I}-\mathbf{P_X})\mathbf{y}$. Customarily, the sum of squares of the elements of $\mathbf{y}$ is called the *total sum of squares*; the sum of squares of the elements of $\mathbf{P_X}\mathbf{y}$ is called the *model sum of squares* (the term we use) or *regression sum of squares*; and the sum of squares of the elements of $(\mathbf{I} - \mathbf{P_X})\mathbf{y}$ is called the *residual sum of squares* (the term we use) or *error sum of squares*.

Definition 8.2.1 An *analysis of variance*, or *ANOVA*, of a linear model $\{\mathbf{y}, \mathbf{X}\boldsymbol{\beta}\}$ is an additive decomposition of the sum of squares of $\mathbf{y}$ (or of the sum of squares of some linear transformation of $\mathbf{y}$) into two or more sums of squares, each one being the squared length of a vector in an orthogonal decomposition of $\mathbf{y}$ (or of the linear transformation of $\mathbf{y}$), accompanied by the dimensionalities (ranks) of the linear spaces to which the vectors belong.

The decomposition of the total sum of squares into the model sum of squares and residual sum of squares, accompanied by a tally of the ranks of $\mathbf{P_X}$ and $\mathbf{I} - \mathbf{P_X}$ [the dimensionalities of the linear spaces to which $\mathbf{P_X y}$ and $(\mathbf{I} - \mathbf{P_X})\mathbf{y}$ belong] satisfies the definition of an ANOVA, and has come to be known as the *overall ANOVA*. The overall ANOVA is often laid out in tabular form, as follows:

Source	Rank	Sum of squares
Model	$\text{rank}(\mathbf{P_X})$	$\mathbf{y}^T\mathbf{P_X y}$
Residual	$\text{rank}(\mathbf{I} - \mathbf{P_X})$	$\mathbf{y}^T(\mathbf{I} - \mathbf{P_X})\mathbf{y}$
Total	$\text{rank}(\mathbf{I})$	$\mathbf{y}^T\mathbf{y}$

Using Theorems 8.1.2a(v) and 8.1.3d, the expressions in the "Rank" column may be given alternatively as $\text{rank}(\mathbf{X})$, $n-\text{rank}(\mathbf{X})$, and n. Furthermore, a column labeled as "Mean square," each entry of which is the ratio of the corresponding entries in the "Sum of squares" and "Rank" columns, is often appended to the right, with the entry for the "Total" row generally omitted. For both of these ratios to be well defined, we must assume that $p^* > 0$ and $n - p^* > 0$, or equivalently that $\mathbf{X} \neq \mathbf{0}$ and $\mathcal{C}(\mathbf{X}) \neq \mathbb{R}^n$. The resulting overall ANOVA table is then as follows:

Source	Rank	Sum of squares	Mean square
Model	$\text{rank}(\mathbf{X})$	$\mathbf{y}^T\mathbf{P_X y}$	$\mathbf{y}^T\mathbf{P_X y}/\text{rank}(\mathbf{X})$
Residual	$n - \text{rank}(\mathbf{X})$	$\mathbf{y}^T(\mathbf{I} - \mathbf{P_X})\mathbf{y}$	$[\mathbf{y}^T(\mathbf{I} - \mathbf{P_X})\mathbf{y}]/[n - \text{rank}(\mathbf{X})]$
Total	n	$\mathbf{y}^T\mathbf{y}$	

In practice, some shortcuts are possible to complete the overall ANOVA table. For example, the model and total sums of squares may be computed and then the residual sum of squares obtained merely by subtraction of the former from the latter. Also, to compute the model sum of squares, it is not actually necessary to determine $\mathbf{P_X}$. This is evident upon observing that if $\hat{\boldsymbol{\beta}}$ is any solution to the normal equations, then

$$\mathbf{y}^T\mathbf{P_X y} = \mathbf{y}^T\mathbf{P_X}^T\mathbf{P_X y} = (\mathbf{P_X y})^T(\mathbf{P_X y}) = \hat{\boldsymbol{\beta}}^T\mathbf{X}^T\mathbf{X}\hat{\boldsymbol{\beta}} = \hat{\boldsymbol{\beta}}^T\mathbf{X}^T\mathbf{y},$$

which shows that the model sum of squares can be computed alternatively as either $\hat{\boldsymbol{\beta}}^T\mathbf{X}^T\mathbf{X}\hat{\boldsymbol{\beta}}$ or $\hat{\boldsymbol{\beta}}^T\mathbf{X}^T\mathbf{y}$. Using these alternative expressions, specialized formulas for the model sum of squares may be obtained that are free of matrix/vector notation for many linear models. Some such specialized formulas will be illustrated in the examples and exercises of this chapter.

The relevance of the overall ANOVA to hypothesis testing under the Gauss–Markov model is described in Chap. 15 in detail. For now, we merely observe that the model mean square may be interpreted as the portion of the total sum of squares explained by the model matrix $\mathbf{X}$ per unit of model space dimension, while the residual mean square may be interpreted as the portion of the total sum of squares not explained by $\mathbf{X}$ per unit of residual space dimension. Although both of these mean squares are random variables, the following theorem and its first corollary suggest that it is reasonable to expect their ratio to be large if the squared length of $\mathbf{X}\boldsymbol{\beta}$ is large relative to $p^*\sigma^2$.

Theorem 8.2.1 *Under the Gauss–Markov model* $\{\mathbf{y}, \mathbf{X}\boldsymbol{\beta}, \sigma^2\mathbf{I}\}$*:*

(a) $E(\mathbf{y}^T\mathbf{P}_\mathbf{X}\mathbf{y}) = \boldsymbol{\beta}^T\mathbf{X}^T\mathbf{X}\boldsymbol{\beta} + p^*\sigma^2$ *and*
(b) $E[\mathbf{y}^T(\mathbf{I} - \mathbf{P}_\mathbf{X})\mathbf{y}] = (n - p^*)\sigma^2$.

Proof Using Theorem 4.2.4 and various parts of Theorems 8.1.2 and 8.1.3, we obtain

$$\begin{aligned} E(\mathbf{y}^T\mathbf{P}_\mathbf{X}\mathbf{y}) &= (\mathbf{X}\boldsymbol{\beta})^T\mathbf{P}_\mathbf{X}(\mathbf{X}\boldsymbol{\beta}) + \mathrm{tr}[\mathbf{P}_\mathbf{X}(\sigma^2\mathbf{I})] \\ &= \boldsymbol{\beta}^T\mathbf{X}^T\mathbf{X}\boldsymbol{\beta} + p^*\sigma^2 \end{aligned}$$

and

$$\begin{aligned} E[\mathbf{y}^T(\mathbf{I} - \mathbf{P}_\mathbf{X})\mathbf{y}] &= (\mathbf{X}\boldsymbol{\beta})^T(\mathbf{I} - \mathbf{P}_\mathbf{X})(\mathbf{X}\boldsymbol{\beta}) + \mathrm{tr}[(\mathbf{I} - \mathbf{P}_\mathbf{X})(\sigma^2\mathbf{I})] \\ &= 0 + \sigma^2[\mathrm{rank}(\mathbf{I} - \mathbf{P}_\mathbf{X})] \\ &= (n - p^*)\sigma^2. \end{aligned}$$

□

Corollary 8.2.1.1 *Under the Gauss–Markov model* $\{\mathbf{y}, \mathbf{X}\boldsymbol{\beta}, \sigma^2\mathbf{I}\}$, *if* $0 < p^* < n$, *then the ratio of the expectation of the model mean square to that of the residual mean square is equal to* $1 + \|\mathbf{X}\boldsymbol{\beta}\|^2/(p^*\sigma^2)$.

Subsequently, we represent the residual mean square for the model $\{\mathbf{y}, \mathbf{X}\boldsymbol{\beta}\}$ by $\hat{\sigma}^2$. One expression for $\hat{\sigma}^2$ is

$$\hat{\sigma}^2 = \frac{\mathbf{y}^T(\mathbf{I} - \mathbf{P}_\mathbf{X})\mathbf{y}}{n - p^*},$$

and others may be obtained by replacing $\mathbf{y}^T(\mathbf{I} - \mathbf{P}_\mathbf{X}\mathbf{y}$ in the numerator with $Q(\hat{\boldsymbol{\beta}})$, $\mathbf{y}^T\mathbf{y} - \hat{\boldsymbol{\beta}}^T\mathbf{X}^T\mathbf{y}$, or $\hat{\mathbf{e}}^T\hat{\mathbf{e}}$. Another corollary to Theorem 8.2.1 states that $\hat{\sigma}^2$ is an unbiased estimator of the Gauss–Markov model's residual variance under that model.

Corollary 8.2.1.2 *Under the Gauss–Markov model* $\{\mathbf{y}, \mathbf{X}\boldsymbol{\beta}, \sigma^2\mathbf{I}\}$, *if* $n > p^*$ *then* $\hat{\sigma}^2$ *is an unbiased estimator of* σ^2.

It follows immediately from Theorem 7.2.2 and Corollary 8.2.1.2 that, under the Gauss–Markov model $\{\mathbf{y}, \mathbf{X}\boldsymbol{\beta}, \sigma^2\mathbf{I}\}$, $\hat{\sigma}^2\mathbf{C}_1^T(\mathbf{X}^T\mathbf{X})^-\mathbf{C}_2$ is an unbiased estimator of the matrix of covariances between vectors of least squares estimators of estimable vectors $\mathbf{C}_1^T\boldsymbol{\beta}$ and $\mathbf{C}_2^T\boldsymbol{\beta}$ and $\hat{\sigma}^2\mathbf{C}^T(\mathbf{X}^T\mathbf{X})^-\mathbf{C}$ is an unbiased estimator of the variance–covariance matrix of the least squares estimator of an estimable vector $\mathbf{C}^T\boldsymbol{\beta}$.

According to Corollary 8.2.1.2, $\mathrm{E}(\hat{\sigma}^2) = \sigma^2$ under the Gauss–Markov model $\{\mathbf{y}, \mathbf{X}\boldsymbol{\beta}, \sigma^2\mathbf{I}\}$. The next two theorems give additional moment properties of the residual mean square under certain conditions.

Theorem 8.2.2 *Let $\mathbf{c}^T\hat{\boldsymbol{\beta}}$ be the least squares estimator of an estimable function $\mathbf{c}^T\boldsymbol{\beta}$ associated with the model $\{\mathbf{y}, \mathbf{X}\boldsymbol{\beta}\}$. Suppose further that $n > p^*$, and let $\hat{\sigma}^2$ be the residual mean square for that model. Under a Gauss–Markov model $\{\mathbf{y}, \mathbf{X}\boldsymbol{\beta}, \sigma^2\mathbf{I}\}$ with skewness matrix $\mathbf{0}$, $cov(\mathbf{c}^T\hat{\boldsymbol{\beta}}, \hat{\sigma}^2) = 0$.*

Proof Because $\mathbf{c}^T\boldsymbol{\beta}$ is estimable, $\mathbf{c}^T = \mathbf{a}^T\mathbf{X}$ for some $\mathbf{a}$. By Corollary 4.2.5.1 and Theorem 8.1.3a,

$$\begin{aligned}
\mathrm{cov}(\mathbf{c}^T\hat{\boldsymbol{\beta}}, \hat{\sigma}^2) &= \mathrm{cov}[\mathbf{a}^T\mathbf{P_X}\mathbf{y}, \mathbf{y}^T(\mathbf{I}-\mathbf{P_X})\mathbf{y}/(n-p^*)] \\
&= 2\mathbf{a}^T\mathbf{P_X}(\sigma^2\mathbf{I})(\mathbf{I}-\mathbf{P_X})\mathbf{X}\boldsymbol{\beta}/(n-p^*) \\
&= 2\sigma^2\mathbf{a}^T\mathbf{P_X}(\mathbf{X}-\mathbf{X})\boldsymbol{\beta}/(n-p^*) \\
&= 0.
\end{aligned}$$

□

Theorem 8.2.3 *Under a Gauss–Markov model with $n > p^*$ and excess kurtosis matrix $\mathbf{0}$, $var(\hat{\sigma}^2) = 2\sigma^4/(n-p^*)$.*

Proof Let $\boldsymbol{\Lambda}$ represent the skewness matrix of $\mathbf{y}$. By Corollary 4.2.6.2 and Theorem 8.1.3a, b, d,

$$\begin{aligned}
\mathrm{var}(\hat{\sigma}^2) &= \mathrm{var}[\mathbf{y}^T(\mathbf{I}-\mathbf{P_X})\mathbf{y}/(n-p^*)] \\
&= \{4(\mathbf{X}\boldsymbol{\beta})^T(\mathbf{I}-\mathbf{P_X})\boldsymbol{\Lambda}\mathrm{vec}(\mathbf{I}-\mathbf{P_X}) + 2\mathrm{tr}[(\mathbf{I}-\mathbf{P_X})(\sigma^2\mathbf{I})(\mathbf{I}-\mathbf{P_X})(\sigma^2\mathbf{I})] \\
&\quad +4(\mathbf{X}\boldsymbol{\beta})^T(\mathbf{I}-\mathbf{P_X})(\sigma^2\mathbf{I})(\mathbf{I}-\mathbf{P_X})\mathbf{X}\boldsymbol{\beta})\}/(n-p^*)^2 \\
&= 0 + 2\sigma^4\mathrm{tr}(\mathbf{I}-\mathbf{P_X})/(n-p^*)^2 + 0 \\
&= 2\sigma^4/(n-p^*).
\end{aligned}$$

□

Example 8.2-1. The Overall ANOVA for Simple Linear Regression

To obtain the overall ANOVA for simple linear regression, it is most convenient to consider the centered version of the model, i.e.,

$$y_i = \beta_1 + \beta_2(x_i - \bar{x}) + e_i \qquad (i = 1, \ldots, n).$$

For this version,

$$\mathbf{X}^T\mathbf{X} = \begin{pmatrix} n & 0 \\ 0 & SXX \end{pmatrix},$$

hence

$$\hat{\boldsymbol{\beta}} = \begin{pmatrix} n & 0 \\ 0 & SXX \end{pmatrix}^{-1} \begin{pmatrix} \sum_{i=1}^n y_i \\ \sum_{i=1}^n (x_i - \bar{x})y_i \end{pmatrix} = \begin{pmatrix} \bar{y} \\ \hat{\beta}_2 \end{pmatrix},$$

where $\hat{\beta}_2 = SXY/SXX$. Then

$$\mathbf{y}^T\mathbf{P_X}\mathbf{y} = \hat{\boldsymbol{\beta}}^T\mathbf{X}^T\mathbf{X}\hat{\boldsymbol{\beta}} = \begin{pmatrix} \bar{y} \\ \hat{\beta}_2 \end{pmatrix}^T \begin{pmatrix} n & 0 \\ 0 & SXX \end{pmatrix} \begin{pmatrix} \bar{y} \\ \hat{\beta}_2 \end{pmatrix} = n\bar{y}^2 + \hat{\beta}_2^2 SXX$$

and

$$\mathbf{y}^T\mathbf{y} - \mathbf{y}^T\mathbf{P_X}\mathbf{y} = \sum_{i=1}^n y_i^2 - [n\bar{y}^2 + \hat{\beta}_2^2 SXX] = \sum_{i=1}^n (y_i - \bar{y})^2 - \hat{\beta}_2^2 SXX.$$

Because rank($\mathbf{X}$) = 2, the overall ANOVA may be laid out as follows:

Source	Rank	Sum of squares
Model	2	$n\bar{y}^2 + \hat{\beta}_2^2 SXX$
Residual	$n-2$	$\sum_{i=1}^n (y_i - \bar{y})^2 - \hat{\beta}_2^2 SXX$
Total	n	$\sum_{i=1}^n y_i^2$

In this case

$$\hat{\sigma}^2 = \frac{\sum_{i=1}^n (y_i - \bar{y})^2 - \hat{\beta}_2^2 SXX}{n-2}.$$

■

Example 8.2-2. The Overall ANOVA for the One-Factor Model

To obtain the overall ANOVA for the one-factor model, note that

$$\mathbf{y}^T\mathbf{P_X y} = (\mathbf{X}\hat{\boldsymbol{\beta}})^T\mathbf{X}\hat{\boldsymbol{\beta}} = \begin{pmatrix} \bar{y}_{1\cdot}\mathbf{1}_{n_1} \\ \bar{y}_{2\cdot}\mathbf{1}_{n_2} \\ \vdots \\ \bar{y}_{q\cdot}\mathbf{1}_{n_q} \end{pmatrix}^T \begin{pmatrix} \bar{y}_{1\cdot}\mathbf{1}_{n_1} \\ \bar{y}_{2\cdot}\mathbf{1}_{n_2} \\ \vdots \\ \bar{y}_{q\cdot}\mathbf{1}_{n_q} \end{pmatrix} = \sum_{i=1}^{q} n_i \bar{y}_{i\cdot}^2$$

and

$$\mathbf{y}^T\mathbf{y} - \mathbf{y}^T\mathbf{P_X y} = \sum_{i=1}^{q}\sum_{j=1}^{n_i} y_{ij}^2 - \sum_{i=1}^{q} n_i \bar{y}_{i\cdot}^2 = \sum_{i=1}^{q}\sum_{j=1}^{n_i}(y_{ij} - \bar{y}_{i\cdot})^2.$$

From Example 6.1-2, rank($\mathbf{X}$) $= q$; hence the overall ANOVA may be laid out as follows:

Source	Rank	Sum of squares
Model	q	$\sum_{i=1}^{q} n_i \bar{y}_{i\cdot}^2$
Residual	$n - q$	$\sum_{i=1}^{q}\sum_{j=1}^{n_i}(y_{ij} - \bar{y}_{i\cdot})^2$
Total	n	$\sum_{i=1}^{q}\sum_{j=1}^{n_i} y_{ij}^2$

Also, in this case

$$\hat{\sigma}^2 = \frac{\sum_{i=1}^{q}\sum_{j=1}^{n_i}(y_{ij} - \bar{y}_{i\cdot})^2}{n - q}.$$

■

Methodological Interlude #3: Influence Diagnostics

In applied data analysis it is generally regarded as undesirable for one or very few observations to have a much larger effect on the conclusions than the others. Instead, it is preferable for all of the data to "tell the same story." Accordingly, users of linear models have developed several measures of the influence that an observation has on the results of the analysis. These *influence diagnostics* typically measure how much a selected quantity (such as a parameter estimate or a fitted value) changes when computed with and without each observation. To be more specific, let $\mathbf{y}_{-i}$ and $\mathbf{X}_{-i}$ represent the response vector and model matrix without the ith observation for the model $\{\mathbf{y}, \mathbf{X}\boldsymbol{\beta}, \sigma^2\mathbf{I}\}$, and suppose that rank($\mathbf{X}_{-i}$) $=$ rank($\mathbf{X}$) $= p$. Further, let

$\hat{\boldsymbol{\beta}}_{-i} = (\hat{\beta}_{j,-i})$, $\hat{\mathbf{y}}_{-i} = (\hat{y}_{j,-i})$, $\hat{\mathbf{e}}_{-i} = (\hat{e}_{j,-i})$, and $\hat{\sigma}^2_{-i}$ for $i = 1, \ldots, n$ denote respectively the least squares estimator of $\boldsymbol{\beta}$, fitted values vector, fitted residuals vector, and residual mean square without the ith observation. The following are some commonly used influence diagnostics:

$$i\text{th PRESS residual} = \hat{e}_{i,-i} \quad (i - 1, \ldots, n),$$

$$\text{DFBETAS}_{j,i} = \frac{\hat{\beta}_j - \hat{\beta}_{j,-i}}{\hat{\sigma}_{-i}\sqrt{[(\mathbf{X}^T\mathbf{X})^{-1}]_{jj}}} \quad (j = 1, \ldots, p;\ i = 1, \ldots, n),$$

$$\text{DFFITS}_i = \frac{\hat{y}_i - \hat{y}_{i,-i}}{\hat{\sigma}_{-i}\sqrt{p_{ii}}} \quad (i = 1, \ldots, n),$$

$$\text{Cook's } D_i = \frac{(\hat{\boldsymbol{\beta}} - \hat{\boldsymbol{\beta}}_{-i})^T\mathbf{X}^T\mathbf{X}(\hat{\boldsymbol{\beta}} - \hat{\boldsymbol{\beta}}_{-i})}{p\hat{\sigma}^2} \quad (i = 1, \ldots, n),$$

$$\text{COVRATIO}_i = \frac{|(\mathbf{X}_{-i}^T\mathbf{X}_{-i})^{-1}\hat{\sigma}^2_{-i}|}{|(\mathbf{X}^T\mathbf{X})^{-1}\hat{\sigma}^2|} \quad (i = 1, \ldots, n).$$

Original sources for these influence diagnostics are Allen (1974), Cook (1977), Welsch and Kuh (1977), and Belsley et al. (1980). The ith PRESS residual, when compared to the ith fitted residual $\hat{e}_i$, measures the influence of the ith observation on the ith fitted residual. Also, $\text{DFBETAS}_{j,i}$ measures the influence of the ith observation on the jth estimated regression coefficient; DFFITS_i measures the influence of the ith observation on the ith fitted value; Cook's D_i may be viewed as a measure of the influence of the ith observation on either the entire p-vector of estimated regression coefficients or the entire n-vector of fitted values; and COVRATIO_i measures the influence of the ith observation on the usual unbiased estimate of the variance–covariance matrix of estimated regression coefficients. The reader is referred to textbooks on applied regression analysis for guidance regarding the magnitudes of these diagnostics that are noteworthy and/or cause for concern.

The expressions for the influence diagnostics given above would seem to suggest that their calculation would be an exceedingly tedious exercise, requiring the setting aside of each observation one-by-one for recalculation of the various quantities (least squares estimator, fitted values, and residual mean square) without that observation. However, it turns out that equivalent expressions may be given that involve quantities computed once, and only once, from the full collection of observations. Writing the model matrix in terms of its rows as $\mathbf{X} = \begin{pmatrix} \mathbf{x}_1^T \\ \vdots \\ \mathbf{x}_n^T \end{pmatrix}$, we see that

$$\mathbf{X}^T\mathbf{X} = \sum_{i=1}^{n} \mathbf{x}_i\mathbf{x}_i^T, \quad \mathbf{X}_{-i}^T\mathbf{X}_{-i} = \sum_{j \neq i} \mathbf{x}_i\mathbf{x}_i^T = \mathbf{X}^T\mathbf{X} - \mathbf{x}_i\mathbf{x}_i^T, \quad \text{and} \quad \mathbf{X}_{-i}^T\mathbf{y}_{-i} = \mathbf{X}^T\mathbf{y} - \mathbf{x}_i y_i.$$

By Corollary 2.9.7.1 (with $\mathbf{X}^T\mathbf{X}$, $-\mathbf{x}_i$, and $\mathbf{x}_i$ here playing the roles of **A**, **b**, and **d** in the corollary),

$$(\mathbf{X}_{-i}^T\mathbf{X}_{-i})^{-1} = (\mathbf{X}^T\mathbf{X})^{-1} + \left(\frac{1}{1-p_{ii}}\right)(\mathbf{X}^T\mathbf{X})^{-1}\mathbf{x}_i\mathbf{x}_i^T(\mathbf{X}^T\mathbf{X})^{-1}.$$

Hence

$$\begin{aligned}
\hat{\boldsymbol{\beta}}_{-i} &= \left[(\mathbf{X}^T\mathbf{X})^{-1} + \left(\frac{1}{1-p_{ii}}\right)(\mathbf{X}^T\mathbf{X})^{-1}\mathbf{x}_i\mathbf{x}_i^T(\mathbf{X}^T\mathbf{X})^{-1}\right](\mathbf{X}^T\mathbf{y} - \mathbf{x}_i y_i) \\
&= \hat{\boldsymbol{\beta}} - (\mathbf{X}^T\mathbf{X})^{-1}\mathbf{x}_i y_i + \left(\frac{1}{1-p_{ii}}\right)(\mathbf{X}^T\mathbf{X})^{-1}\mathbf{x}_i(\hat{y}_i - p_{ii}y_i) \\
&= \hat{\boldsymbol{\beta}} - \left(\frac{1}{1-p_{ii}}\right)(\mathbf{X}^T\mathbf{X})^{-1}\mathbf{x}_i(y_i - p_{ii}y_i - \hat{y}_i + p_{ii}y_i) \\
&= \hat{\boldsymbol{\beta}} - \left(\frac{\hat{e}_i}{1-p_{ii}}\right)(\mathbf{X}^T\mathbf{X})^{-1}\mathbf{x}_i,
\end{aligned}$$

implying that

$$\hat{\boldsymbol{\beta}} - \hat{\boldsymbol{\beta}}_{-i} = \left(\frac{\hat{e}_i}{1-p_{ii}}\right)(\mathbf{X}^T\mathbf{X})^{-1}\mathbf{x}_i. \tag{8.3}$$

Using (8.3), it can be shown (see the exercises) that alternative expressions for the ith PRESS residual, $\text{DFBETAS}_{j,i}$, DFFITS_i, and Cook's D_i are as follows:

$$\begin{aligned}
\hat{e}_{i,-i} &= \frac{\hat{e}_i}{1-p_{ii}}, \\
\text{DFBETAS}_{j,i} &= \frac{\hat{e}_i[(\mathbf{X}^T\mathbf{X})^{-1}\mathbf{x}_i]_j}{\hat{\sigma}_{-i}(1-p_{ii})\sqrt{[(\mathbf{X}^T\mathbf{X})^{-1}]_{jj}}}, \\
\text{DFFITS}_i &= \frac{\hat{e}_i\sqrt{p_{ii}}}{\hat{\sigma}_{-i}(1-p_{ii})}, \\
\text{Cook's } D_i &= \frac{\hat{e}_i^2 p_{ii}}{p\hat{\sigma}^2(1-p_{ii})^2}.
\end{aligned}$$

Using Theorems 2.11.4 and 2.11.8 it may also be shown that an alternative expression for COVRATIO_i is

$$\text{COVRATIO}_i = \frac{1}{1-p_{ii}}\left(\frac{\hat{\sigma}_{-i}^2}{\hat{\sigma}^2}\right)^p.$$

Furthermore, again using (8.3) and letting $\mathbf{p}_i$ denote the ith column of $\mathbf{P_X}$, we obtain

$$\begin{aligned}
(n-p-1)\hat{\sigma}^2_{-i} &= (\mathbf{y}_{-i} - \mathbf{X}_{-i}\hat{\boldsymbol{\beta}}_{-i})^T(\mathbf{y}_{-i} - \mathbf{X}_{-i}\hat{\boldsymbol{\beta}}_{-i}) \\
&= (\mathbf{y} - \mathbf{X}\hat{\boldsymbol{\beta}}_{-i})^T(\mathbf{y} - \mathbf{X}\hat{\boldsymbol{\beta}}_{-i}) - (y_i - \mathbf{x}_i^T\hat{\boldsymbol{\beta}}_{-i})^2 \\
&= \left[\mathbf{y} - \mathbf{X}\left(\hat{\boldsymbol{\beta}} - \left(\frac{\hat{e}_i}{1-p_{ii}}\right)(\mathbf{X}^T\mathbf{X})^{-1}\mathbf{x}_i\right)\right]^T \\
&\quad \times \left[\mathbf{y} - \mathbf{X}\left(\hat{\boldsymbol{\beta}} - \left(\frac{\hat{e}_i}{1-p_{ii}}\right)(\mathbf{X}^T\mathbf{X})^{-1}\mathbf{x}_i\right)\right] \\
&\quad - \left[y_i - \mathbf{x}_i^T\left(\hat{\boldsymbol{\beta}} - \left(\frac{\hat{e}_i}{1-p_{ii}}\right)(\mathbf{X}^T\mathbf{X})^{-1}\mathbf{x}_i\right)\right]^2 \\
&= \left[\hat{\mathbf{e}} - \left(\frac{\hat{e}_i}{1-p_{ii}}\right)\mathbf{p}_i\right]^T\left[\hat{\mathbf{e}} - \left(\frac{\hat{e}_i}{1-p_{ii}}\right)\mathbf{p}_i\right] - \left[\hat{e}_i + \left(\frac{p_{ii}\hat{e}_i}{1-p_{ii}}\right)\right]^2 \\
&= \hat{\mathbf{e}}^T\hat{\mathbf{e}} + \left(\frac{\hat{e}_i}{1-p_{ii}}\right)^2 p_{ii} - \frac{\hat{e}_i^2}{(1-p_{ii})^2} \\
&= (n-p)\hat{\sigma}^2 - \frac{\hat{e}_i^2}{1-p_{ii}},
\end{aligned}$$

where we used Theorem 8.1.4 for the fifth equality. Thus,

$$\hat{\sigma}_{-i} = \sqrt{\frac{(n-p)\hat{\sigma}^2 - [\hat{e}_i^2/(1-p_{ii})]}{n-p-1}},$$

which together with the alternative expressions for influence diagnostics displayed immediately after (8.3), indicates that all of them may be computed from one least squares fit using all the observations. ■

8.3 Exercises

1. Prove Corollary 8.1.1.1.
2. Prove Theorem 8.1.6c–e. [Hint: To prove part (c), use Theorem Ginv14 to show (without loss of generality) that the result holds for p_{11} by representing $\mathbf{X}$ as $\begin{pmatrix} \mathbf{1}_{r_1-1}\mathbf{x}_1^T \\ \mathbf{X}_{\boxed{2}} \end{pmatrix}$ where $\mathbf{x}_1^T$ is the first distinct row of $\mathbf{X}$ and r_1 is the number of replicates of that row, and $\mathbf{X}_{\boxed{2}}$ consists of the remaining rows of $\mathbf{X}$ including, as its first row, the last replicate of $\mathbf{x}_1^T$. To prove part (d), consider properties of $\mathbf{P_X} - (1/n)\mathbf{J}_n$.]
3. Find var($\hat{\mathbf{e}} - \mathbf{e}$) and cov($\hat{\mathbf{e}} - \mathbf{e}, \mathbf{e}$) under the Gauss–Markov model $\{\mathbf{y}, \mathbf{X}\boldsymbol{\beta}, \sigma^2\mathbf{I}\}$.

4. Determine $\mathbf{P_X}$ for each of the following models:
   (a) the no-intercept simple linear regression model;
   (b) the two-way main effects model with balanced data;
   (c) the two-way model with interaction and balanced data;
   (d) the two-way partially crossed model introduced in Example 5.1.4-1, with one observation per cell;
   (e) the two-factor nested model.
5. Determine the overall ANOVA, including a nonmatrix expression for the model sum of squares, for each of the following models:
   (a) the no-intercept simple linear regression model;
   (b) the two-way main effects model with balanced data;
   (c) the two-way model with interaction and balanced data;
   (d) the two-way partially crossed model introduced in Example 5.1.4-1, with one observation per cell;
   (e) the two-factor nested model.
6. For the special case of a (Gauss–Markov) simple linear regression model with even sample size n and $n/2$ observations at each of $x = -n$ and $x = n$, obtain the variances of the fitted residuals and the correlations among them.
7. For the (Gauss–Markov) no-intercept simple linear regression model, obtain $\text{var}(\hat{e}_i)$ and $\text{corr}(\hat{e}_i, \hat{e}_j)$.
8. Consider the Gauss–Markov model $\{\mathbf{y}, \mathbf{X}\boldsymbol{\beta}, \sigma^2\mathbf{I}\}$. Prove that if $\mathbf{X}^T\mathbf{X} = k\mathbf{I}$ for some $k > 0$, and rows i and j of $\mathbf{X}$ are orthogonal, then $\text{corr}(\hat{e}_i, \hat{e}_j) = 0$.
9. Under the Gauss–Markov model $\{\mathbf{y}, \mathbf{X}\boldsymbol{\beta}, \sigma^2\mathbf{I}\}$ with $n > p^*$ and excess kurtosis matrix $\mathbf{0}$, find the estimator of σ^2 that minimizes the mean squared error within the class of estimators of the form $\mathbf{y}^T(\mathbf{I} - \mathbf{P_X})\mathbf{y}/k$, where $k > 0$. How does the minimized mean square error compare to the mean squared error of $\hat{\sigma}^2$ specified in Theorem 8.2.3?
10. Define a *quadratic estimator* of σ^2 associated with the Gauss–Markov model $\{\mathbf{y}, \mathbf{X}\boldsymbol{\beta}, \sigma^2\mathbf{I}\}$ to be any quadratic form $\mathbf{y}^T\mathbf{A}\mathbf{y}$, where $\mathbf{A}$ is a positive definite matrix.
    (a) Prove that a quadratic estimator $\mathbf{y}^T\mathbf{A}\mathbf{y}$ of σ^2 is unbiased under the specified Gauss–Markov model if and only if $\mathbf{A}\mathbf{X} = \mathbf{0}$ and $\text{tr}(\mathbf{A}) = 1$. (An estimator $t(\mathbf{y})$ is said to be unbiased for σ^2 under a Gauss–Markov model if $\text{E}[t(\mathbf{y})] = \sigma^2$ for all $\boldsymbol{\beta} \in \mathbb{R}^p$ and all $\sigma^2 > 0$ under that model.)
    (b) Prove the following extension of Theorem 8.2.2: Let $\mathbf{c}^T\hat{\boldsymbol{\beta}}$ be the least squares estimator of an estimable function $\mathbf{c}^T\boldsymbol{\beta}$ associated with the model $\{\mathbf{y}, \mathbf{X}\boldsymbol{\beta}\}$, suppose that $n > p^*$, and let $\tilde{\sigma}^2$ be a quadratic unbiased estimator of σ^2 under the Gauss–Markov model $\{\mathbf{y}, \mathbf{X}\boldsymbol{\beta}, \sigma^2\mathbf{I}\}$. If $\mathbf{e}$ has skewness matrix $\mathbf{0}$, then $\text{cov}(\mathbf{c}^T\hat{\boldsymbol{\beta}}, \tilde{\sigma}^2) = 0$.
    (c) Determine as simple an expression as possible for the variance of a quadratic unbiased estimator of σ^2 under a Gauss–Markov model $\{\mathbf{y}, \mathbf{X}\boldsymbol{\beta}, \sigma^2\mathbf{I}\}$ for which the skewness matrix of $\mathbf{e}$ equals $\mathbf{0}$ and the excess kurtosis matrix of $\mathbf{e}$ equals $\mathbf{0}$.

11. Verify the alternative expressions for the five influence diagnostics described in Methodological Interlude #3.
12. Find $\mathrm{E}(\hat{e}_{i,-i})$, $\mathrm{var}(\hat{e}_{i,-i})$, and $\mathrm{corr}(\hat{e}_{i,-i}, \hat{e}_{j,-j})$ (for $i \neq j$) under the Gauss–Markov model $\{\mathbf{y}, \mathbf{X}\boldsymbol{\beta}, \sigma^2\mathbf{I}\}$.

References

Allen, D. M. (1974). The relationship between variable selection and data augmentation and a method for prediction. *Technometrics, 16*, 125–127.

Belsley, D. A., Kuh, E., & Welsch, R. E. (1980). *Regression diagnostics; identifying influential data and sources of collinearity*. New York: Wiley.

Cook, R. D. (1977). Detection of influential observations in linear regression. *Technometrics, 19*, 15–18.

Welsch, R. E., & Kuh, E. (1977). *Linear regression diagnostics*. Working Paper No. 173. National Bureau of Economic Research.

Least Squares Estimation and ANOVA for Partitioned Models

9

In the previous chapter, we alluded to the fact that we will eventually use the overall ANOVA to judge whether the variables comprising the model matrix $\mathbf{X}$ are, as a whole, important in "explaining" the variation in the elements of $\mathbf{y}$. Often we want to also make this same kind of judgment about groups of explanatory variables or even individual variables. For this and other reasons, we consider in this chapter how the model sum of squares in the overall ANOVA may be partitioned into sums of squares that correspond to the explanatory value of groups of, or individual, variables. We also consider a partition of the residual sum of squares that is possible when multiple observations are taken at one or more combinations of values of the explanatory variables.

9.1 Partitioned Models and Associated Projection Matrices

We begin by defining a linear model in which the explanatory variables are grouped into k subsets. Some or all of these subsets could be individual variables.

Definition 9.1.1 The model

$$\mathbf{y} = \mathbf{X}_1\boldsymbol{\beta}_1 + \mathbf{X}_2\boldsymbol{\beta}_2 + \cdots + \mathbf{X}_k\boldsymbol{\beta}_k + \mathbf{e},$$

represented alternatively by $\{\mathbf{y}, \mathbf{X}_1\boldsymbol{\beta}_1 + \mathbf{X}_2\boldsymbol{\beta}_2 + \cdots + \mathbf{X}_k\boldsymbol{\beta}_k\}$, is called an *ordered k-part linear model.*

Of course, an ordered k-part linear model is just another way of writing the unpartitioned model $\{\mathbf{y}, \mathbf{X}\boldsymbol{\beta}\}$, where $\mathbf{X} = (\mathbf{X}_1, \ldots, \mathbf{X}_k)$ and $\boldsymbol{\beta} = (\boldsymbol{\beta}_1^T, \ldots, \boldsymbol{\beta}_k^T)^T$. From any k-part ordered model, we can form $k! - 1$ other ordered k-part models by permuting the k terms.

D. L. Zimmerman, *Linear Model Theory*,
https://doi.org/10.1007/978-3-030-52063-2_9

The simplest case of an ordered k-part model to consider is that of $k = 2$, for which $\mathbf{X} = (\mathbf{X}_1, \mathbf{X}_2)$ and there are two ordered two-part models:

$$\{\mathbf{y}, \mathbf{X}_1\boldsymbol{\beta}_1 + \mathbf{X}_2\boldsymbol{\beta}_2\} \qquad \text{and} \qquad \{\mathbf{y}, \mathbf{X}_2\boldsymbol{\beta}_2 + \mathbf{X}_1\boldsymbol{\beta}_1\}. \tag{9.1}$$

By Theorem 3.3.3c, a matrix $\dot{\mathbf{B}}$ exists such that $\mathbf{X}^T\mathbf{X}\dot{\mathbf{B}} = \mathbf{X}^T$; for example, we can take $\dot{\mathbf{B}} = (\mathbf{X}^T\mathbf{X})^-\mathbf{X}^T$ using any generalized inverse of $\mathbf{X}^T\mathbf{X}$. For such a $\dot{\mathbf{B}}$, $\mathbf{X}\dot{\mathbf{B}}$ ($= \mathbf{P}_\mathbf{X}$) is the orthogonal projection matrix onto $\mathcal{C}(\mathbf{X})$. Or, in the equivalent two-part form, matrices $\dot{\mathbf{B}}_1$ and $\dot{\mathbf{B}}_2$ exist such that

$$\begin{pmatrix} \mathbf{X}_1^T\mathbf{X}_1 & \mathbf{X}_1^T\mathbf{X}_2 \\ \mathbf{X}_2^T\mathbf{X}_1 & \mathbf{X}_2^T\mathbf{X}_2 \end{pmatrix} \begin{pmatrix} \dot{\mathbf{B}}_1 \\ \dot{\mathbf{B}}_2 \end{pmatrix} = \begin{pmatrix} \mathbf{X}_1^T \\ \mathbf{X}_2^T \end{pmatrix}. \tag{9.2}$$

Furthermore,

$$\mathbf{P}_\mathbf{X} = (\mathbf{X}_1, \mathbf{X}_2)\begin{pmatrix} \dot{\mathbf{B}}_1 \\ \dot{\mathbf{B}}_2 \end{pmatrix} = \mathbf{X}_1\dot{\mathbf{B}}_1 + \mathbf{X}_2\dot{\mathbf{B}}_2.$$

In this context we also denote $\mathbf{P}_\mathbf{X}$ by $\mathbf{P}_{12}$. Thus, $\mathbf{P}_{12}$ is symmetric and idempotent, $\mathbf{P}_{12}\mathbf{X}_1 = \mathbf{X}_1$ and $\mathbf{P}_{12}\mathbf{X}_2 = \mathbf{X}_2$, and $\text{rank}(\mathbf{P}_{12}) = \text{rank}(\mathbf{X}_1, \mathbf{X}_2)$.

Now consider the "submodel" $\{\mathbf{y}, \mathbf{X}_1\boldsymbol{\beta}_1\}$. Again by Theorem 3.3.3c, a matrix $\mathbf{B}_1$ exists such that $\mathbf{X}_1^T\mathbf{X}_1\mathbf{B}_1 = \mathbf{X}_1^T$. In general, however, $\mathbf{B}_1 \neq \dot{\mathbf{B}}_1$. Of course, $\mathbf{P}_1 \equiv \mathbf{X}_1\mathbf{B}_1$ is the orthogonal projection matrix onto $\mathcal{C}(\mathbf{X}_1)$. Thus, $\mathbf{P}_1$ is symmetric and idempotent, $\mathbf{P}_1\mathbf{X}_1 = \mathbf{X}_1$, and $\text{rank}(\mathbf{P}_1) = \text{rank}(\mathbf{X}_1)$. Several additional results pertaining to $\mathbf{P}_{12}$, $\mathbf{P}_1$ and other related matrices are assembled into the following theorem.

Theorem 9.1.1 *For the orthogonal projection matrices $\mathbf{P}_{12}$ and $\mathbf{P}_1$ corresponding to the ordered two-part model $\{\mathbf{y}, \mathbf{X}_1\boldsymbol{\beta}_1 + \mathbf{X}_2\boldsymbol{\beta}_2\}$ and its submodel $\{\mathbf{y}, \mathbf{X}_1\boldsymbol{\beta}_1\}$, respectively, the following results hold:*

(a) $\mathbf{P}_{12}\mathbf{P}_1 = \mathbf{P}_1$ *and* $\mathbf{P}_1\mathbf{P}_{12} = \mathbf{P}_1$*;*
(b) $\mathbf{P}_1(\mathbf{P}_{12} - \mathbf{P}_1) = \mathbf{0}$*;*
(c) $\mathbf{P}_1(\mathbf{I} - \mathbf{P}_{12}) = \mathbf{0}$*;*
(d) $(\mathbf{P}_{12} - \mathbf{P}_1)(\mathbf{I} - \mathbf{P}_{12}) = \mathbf{0}$*;*
(e) $(\mathbf{P}_{12} - \mathbf{P}_1)(\mathbf{P}_{12} - \mathbf{P}_1) = \mathbf{P}_{12} - \mathbf{P}_1$*;*
(f) $\mathbf{P}_{12} - \mathbf{P}_1$ *is symmetric and rank*$(\mathbf{P}_{12} - \mathbf{P}_1) =$ *rank*$(\mathbf{X}_1, \mathbf{X}_2) -$ *rank*$(\mathbf{X}_1)$.

Proof Left as an exercise. □

Consider the projection matrices appearing in Theorem 9.1.1. As already noted, $\mathbf{P}_{12}$ and $\mathbf{P}_1$ are the orthogonal projection matrices onto $\mathcal{C}(\mathbf{X}_1, \mathbf{X}_2)$ and $\mathcal{C}(\mathbf{X}_1)$, respectively. Furthermore, recall from Theorem 8.1.5 that $\mathbf{I} - \mathbf{P}_{12}$ is the orthogonal projection matrix onto the orthogonal complement of $\mathcal{C}(\mathbf{X}_1, \mathbf{X}_2)$. There is another matrix appearing in Theorem 9.1.1, namely $\mathbf{P}_{12} - \mathbf{P}_1$. Is this matrix, like those already mentioned, an orthogonal projection matrix onto some space? Because $\mathbf{P}_{12}-\mathbf{P}_1$ is symmetric and idempotent (Theorem 9.1.1e, f), by Theorem 8.1.2b (with $\mathbf{P}_{12}-\mathbf{P}_1$ here taken to be both $\mathbf{P}$ and $\mathbf{X}$ in the theorem) $\mathbf{P}_{12}-\mathbf{P}_1$ is the orthogonal projection matrix onto $\mathcal{C}(\mathbf{P}_{12}-\mathbf{P}_1)$. The following theorem gives another name for $\mathcal{C}(\mathbf{P}_{12}-\mathbf{P}_1)$, and clarifies its nature.

Theorem 9.1.2 $\mathbf{P}_{12}-\mathbf{P}_1$ *is the orthogonal projection matrix onto* $\mathcal{C}[(\mathbf{I}-\mathbf{P}_1)\mathbf{X}_2]$, *which is the orthogonal complement of* $\mathcal{C}(\mathbf{X}_1)$ *relative to* $\mathcal{C}(\mathbf{X}_1, \mathbf{X}_2)$.

Proof Let $\mathbf{X}_1\mathbf{a}_1+\mathbf{X}_2\mathbf{a}_2$ be any vector in $\mathcal{C}(\mathbf{X}_1, \mathbf{X}_2)$. Then

$$\mathbf{X}_1\mathbf{a}_1+\mathbf{X}_2\mathbf{a}_2 = \mathbf{X}_1\mathbf{a}_1+\mathbf{P}_1\mathbf{X}_2\mathbf{a}_2+(\mathbf{I}-\mathbf{P}_1)\mathbf{X}_2\mathbf{a}_2 = \mathbf{X}_1(\mathbf{a}_1+\mathbf{B}_1\mathbf{X}_2\mathbf{a}_2)+(\mathbf{I}-\mathbf{P}_1)\mathbf{X}_2\mathbf{a}_2,$$

showing that $\mathcal{C}(\mathbf{X}_1, \mathbf{X}_2) \subseteq \mathcal{C}[\mathbf{X}_1, (\mathbf{I}-\mathbf{P}_1)\mathbf{X}_2]$. Also,

$$\mathbf{X}_1\mathbf{a}_1+(\mathbf{I}-\mathbf{P}_1)\mathbf{X}_2\mathbf{a}_2 = \mathbf{X}_1\mathbf{a}_1-\mathbf{P}_1\mathbf{X}_2\mathbf{a}_2+\mathbf{X}_2\mathbf{a}_2 = \mathbf{X}_1(\mathbf{a}_1-\mathbf{B}_1\mathbf{X}_2\mathbf{a}_2)+\mathbf{X}_2\mathbf{a}_2,$$

which, together with the previous result, shows that $\mathcal{C}(\mathbf{X}_1, \mathbf{X}_2) = \mathcal{C}[\mathbf{X}_1, (\mathbf{I}-\mathbf{P}_1)\mathbf{X}_2]$. Moreover, for any vector $\boldsymbol{v}$ with as many elements as $\mathbf{X}_1$ has columns,

$$(\mathbf{X}_1\boldsymbol{v})^T(\mathbf{I}-\mathbf{P}_1)\mathbf{X}_2\mathbf{a}_2 = 0,$$

which implies (by Theorem 2.4.1) that $\mathcal{C}(\mathbf{X}_1) \cap \mathcal{C}[(\mathbf{I}-\mathbf{P}_1)\mathbf{X}_2] = \{\mathbf{0}\}$. By Theorem 2.8.6, $\text{rank}(\mathbf{X}_1, \mathbf{X}_2) = \text{rank}(\mathbf{X}_1) + \text{rank}[(\mathbf{I}-\mathbf{P}_1)\mathbf{X}_2]$, or equivalently

$$\text{rank}[(\mathbf{I}-\mathbf{P}_1)\mathbf{X}_2] = \text{rank}(\mathbf{X}_1, \mathbf{X}_2) - \text{rank}(\mathbf{X}_1) = \text{rank}(\mathbf{P}_{12}-\mathbf{P}_1)$$

where we used Theorem 9.1.1f for the last equality. Now

$$\begin{aligned}(\mathbf{P}_{12}-\mathbf{P}_1)(\mathbf{I}-\mathbf{P}_1)\mathbf{X}_2 &= (\mathbf{P}_{12}-\mathbf{P}_1)\mathbf{X}_2\\ &= \mathbf{X}_2-\mathbf{P}_1\mathbf{X}_2\\ &= (\mathbf{I}-\mathbf{P}_1)\mathbf{X}_2,\end{aligned}$$

implying that $\mathcal{C}[(\mathbf{I}-\mathbf{P}_1)\mathbf{X}_2] \subseteq \mathcal{C}(\mathbf{P}_{12}-\mathbf{P}_1)$. The result then follows by Theorem 2.8.5. □

Corresponding to the other submodel of the ordered two-part model, i.e., $\{\mathbf{y}, \mathbf{X}_2\boldsymbol{\beta}_2\}$, are projection matrices that are analogous to those appearing in Theorems 9.1.1 and 9.1.2. In particular, let $\mathbf{P}_2 = \mathbf{X}_2\mathbf{B}_2$, where $\mathbf{B}_2$ satisfies $\mathbf{X}_2^T\mathbf{X}_2\mathbf{B}_2 = \mathbf{X}_2^T$; in general, $\mathbf{B}_2 \neq \dot{\mathbf{B}}_2$. Of course, $\mathbf{P}_2$ is the orthogonal projection matrix onto $\mathcal{C}(\mathbf{X}_2)$. All the results in Theorems 9.1.1 and 9.1.2 involving $\mathbf{P}_1$ hold for $\mathbf{P}_2$ also, but with $\mathbf{X}_2$ appearing in place of $\mathbf{X}_1$. In particular, $\mathbf{P}_{12}-\mathbf{P}_2$ is the orthogonal projection matrix onto $\mathcal{C}[(\mathbf{I}-\mathbf{P}_2)\mathbf{X}_1]$, which is the orthogonal complement of $\mathcal{C}(\mathbf{X}_2)$ relative to $\mathcal{C}(\mathbf{X}_1, \mathbf{X}_2)$.

Moreover, all of the results presented here for projection matrices associated with the ordered two-part model extend easily to projection matrices relevant to the general case of a k-part model. For the extension, we partition the model matrix and the parameter vector of the model $\{\mathbf{y}, \mathbf{X}\boldsymbol{\beta}\}$ as follows:

$$\mathbf{X}\boldsymbol{\beta} = (\mathbf{X}_1, \ldots, \mathbf{X}_k)\begin{pmatrix}\boldsymbol{\beta}_1 \\ \vdots \\ \boldsymbol{\beta}_k\end{pmatrix}, \tag{9.3}$$

where the dimensions of $\mathbf{X}_j$ and $\boldsymbol{\beta}_j$ are $n \times p_j$ and $p_j \times 1$, respectively, and $\sum_{j=1}^{k} p_j = p$. Consider fitting (by least squares) each of the k ordered j-part models

$$\mathbf{y} = \mathbf{X}_1\boldsymbol{\beta}_1 + \cdots + \mathbf{X}_j\boldsymbol{\beta}_j + \mathbf{e} \quad (j = 1, \ldots, k).$$

Define orthogonal projection matrices

$$\mathbf{P}_{12\cdots j} = (\mathbf{X}_1, \ldots, \mathbf{X}_j)[(\mathbf{X}_1, \ldots, \mathbf{X}_j)^T(\mathbf{X}_1, \ldots, \mathbf{X}_j)]^-(\mathbf{X}_1, \ldots, \mathbf{X}_j)^T \qquad (j = 1, \ldots, k) \tag{9.4}$$

onto the column spaces of the model matrices of those models. These orthogonal projection matrices and their "successive differences" $\mathbf{P}_{12} - \mathbf{P}_1$, $\mathbf{P}_{123} - \mathbf{P}_{12}$, ..., $\mathbf{P}_{12\cdots k} - \mathbf{P}_{12\cdots k-1}$ satisfy properties like those listed in Theorem 9.1.1, as described by the following theorem.

Theorem 9.1.3 *For the matrices $\mathbf{P}_{12\cdots j}$ $(j = 1, \ldots, k)$ given by (9.4) and their successive differences, the following properties hold:*

(a) $\mathbf{P}_{12\cdots j}$ *is symmetric and idempotent, and rank*$(\mathbf{P}_{12\cdots j}) =$ *rank*$(\mathbf{X}_1, \ldots, \mathbf{X}_j)$ *for* $j = 1, \ldots, k$;
(b) $\mathbf{P}_{12\cdots j}\mathbf{X}_{j'} = \mathbf{X}_{j'}$ *for* $1 \leq j' \leq j$, $j = 1, \ldots, k$;
(c) $\mathbf{P}_{12\cdots j}\mathbf{P}_{12\cdots j'} = \mathbf{P}_{12\cdots j'}$ *and* $\mathbf{P}_{12\cdots j'}\mathbf{P}_{12\cdots j} = \mathbf{P}_{12\cdots j'}$ *for* $j' < j = 2, \ldots, k$;
(d) $\mathbf{P}_{12\cdots j}(\mathbf{I} - \mathbf{P}_{12\cdots k}) = \mathbf{0}$ *for* $j = 1, \ldots, k$ *and* $(\mathbf{P}_{12\cdots j} - \mathbf{P}_{12\cdots j-1})(\mathbf{I} - \mathbf{P}_{12\cdots k}) = \mathbf{0}$ *for* $j = 2, \ldots, k$;
(e) $\mathbf{P}_{12\cdots j-1}(\mathbf{P}_{12\cdots j} - \mathbf{P}_{12\cdots j-1}) = \mathbf{0}$ *for* $j = 2, \ldots, k$ *and* $(\mathbf{P}_{12\cdots j} - \mathbf{P}_{12\cdots j-1})(\mathbf{P}_{12\cdots j'} - \mathbf{P}_{12\cdots j'-1}) = \mathbf{0}$ *for all* $j \neq j'$;
(f) $(\mathbf{P}_{12\cdots j} - \mathbf{P}_{12\cdots j-1})(\mathbf{P}_{12\cdots j} - \mathbf{P}_{12\cdots j-1}) = \mathbf{P}_{12\cdots j} - \mathbf{P}_{12\cdots j-1}$ *for* $j = 2, \ldots, k$;
(g) $\mathbf{P}_{12\cdots j} - \mathbf{P}_{12\cdots j-1}$ *is symmetric and rank*$(\mathbf{P}_{12\cdots j} - \mathbf{P}_{12\cdots j-1}) =$ *rank*$(\mathbf{X}_1, \ldots, \mathbf{X}_j)$ $-$ *rank*$(\mathbf{X}_1, \ldots, \mathbf{X}_{j-1})$ *for* $j = 2, \ldots, k$;
(h) $\mathbf{P}_{12\cdots j} - \mathbf{P}_{12\cdots j-1}$ *is the orthogonal projection matrix onto* $\mathcal{C}[(\mathbf{I} - \mathbf{P}_{12\cdots j-1})\mathbf{X}_{12\cdots j}]$, *which is the orthogonal complement of* $\mathcal{C}(\mathbf{X}_{12\cdots j-1})$ *relative to* $\mathcal{C}(\mathbf{X}_{12\cdots j})$ *for* $j = 2, \ldots, k$.

Proof Left as an exercise. □

9.2 Least Squares Estimation for Partitioned Models

Because an ordered k-part model is merely another way of writing an unpartitioned model, there is nothing fundamentally different about least squares estimation for a partitioned model from what has been presented in previous chapters. Consider, for example, the least squares estimation of a linear function under the first of the two-part ordered models in (9.1). By Theorem 6.1.2, a linear function $\mathbf{c}^T\boldsymbol{\beta} = (\mathbf{c}_1^T, \mathbf{c}_2^T)\begin{pmatrix}\boldsymbol{\beta}_1\\ \boldsymbol{\beta}_2\end{pmatrix} = \mathbf{c}_1^T\boldsymbol{\beta}_1 + \mathbf{c}_2^T\boldsymbol{\beta}_2$ is estimable under this model if and only if $(\mathbf{c}_1^T, \mathbf{c}_2^T) = \mathbf{a}^T(\mathbf{X}_1, \mathbf{X}_2)$ for some $\mathbf{a}$, i.e., if and only if $\mathbf{c}_1^T = \mathbf{a}^T\mathbf{X}_1$ and $\mathbf{c}_2^T = \mathbf{a}^T\mathbf{X}_2$ for some $\mathbf{a}$. Furthermore, by Theorem 7.1.4 the least squares estimator of such a function may be written as $\mathbf{c}_1^T\hat{\boldsymbol{\beta}}_1 + \mathbf{c}_2^T\hat{\boldsymbol{\beta}}_2$, where $\hat{\boldsymbol{\beta}}_1$ and $\hat{\boldsymbol{\beta}}_2$ comprise a solution to the "two-part normal equations" (the normal equations written in two-part form)

$$\begin{pmatrix}\mathbf{X}_1^T\mathbf{X}_1 & \mathbf{X}_1^T\mathbf{X}_2\\ \mathbf{X}_2^T\mathbf{X}_1 & \mathbf{X}_2^T\mathbf{X}_2\end{pmatrix}\begin{pmatrix}\boldsymbol{\beta}_1\\ \boldsymbol{\beta}_2\end{pmatrix} = \begin{pmatrix}\mathbf{X}_1^T\mathbf{y}\\ \mathbf{X}_2^T\mathbf{y}\end{pmatrix}. \tag{9.5}$$

Still further, by Theorem 7.2.2 the variance–covariance matrix of the least squares estimators of a vector $\mathbf{C}^T\boldsymbol{\beta} = (\mathbf{C}_1^T, \mathbf{C}_2^T)\boldsymbol{\beta} = \mathbf{C}_1^T\boldsymbol{\beta}_1 + \mathbf{C}_2^T\boldsymbol{\beta}_2$ of estimable functions under Gauss–Markov assumptions is

$$\sigma^2(\mathbf{C}_1^T, \mathbf{C}_2^T)\begin{pmatrix}\mathbf{X}_1^T\mathbf{X}_1 & \mathbf{X}_1^T\mathbf{X}_2\\ \mathbf{X}_2^T\mathbf{X}_1 & \mathbf{X}_2^T\mathbf{X}_2\end{pmatrix}^-\begin{pmatrix}\mathbf{C}_1\\ \mathbf{C}_2\end{pmatrix},$$

invariant to the choice of generalized inverse.

However, the partitioned form of the normal equations may provide an opportunity to obtain a solution in a manner that is more computationally efficient than might otherwise be considered, using a technique known as *absorption* that we now describe. We continue to consider the two-part case only; extension to the general k-part model is straightforward. Suppose that we pre-multiply the "top" subset of equations in (9.5) by $\mathbf{B}_1^T$, where $\mathbf{B}_1$ is, as in the previous section, any matrix such that $\mathbf{X}_1^T\mathbf{X}_1\mathbf{B}_1 = \mathbf{X}_1^T$. Then

$$\mathbf{B}_1^T\mathbf{X}_1^T\mathbf{X}_1\boldsymbol{\beta}_1 + \mathbf{B}_1^T\mathbf{X}_1^T\mathbf{X}_2\boldsymbol{\beta}_2 = \mathbf{B}_1^T\mathbf{X}_1^T\mathbf{y},$$

implying that

$$\mathbf{X}_1\boldsymbol{\beta}_1 + \mathbf{P}_1\mathbf{X}_2\boldsymbol{\beta}_2 = \mathbf{P}_1\mathbf{y},$$

or equivalently that

$$\mathbf{X}_1\boldsymbol{\beta}_1 = \mathbf{P}_1\mathbf{y} - \mathbf{P}_1\mathbf{X}_2\boldsymbol{\beta}_2. \tag{9.6}$$

Substituting this expression for $\mathbf{X}_1\boldsymbol{\beta}_1$ into the second, or "bottom," subset of equations in (9.5) yields

$$\mathbf{X}_2^T(\mathbf{P}_1\mathbf{y} - \mathbf{P}_1\mathbf{X}_2\boldsymbol{\beta}_2) + \mathbf{X}_2^T\mathbf{X}_2\boldsymbol{\beta}_2 = \mathbf{X}_2^T\mathbf{y},$$

or upon rearrangement,

$$\mathbf{X}_2^T(\mathbf{I} - \mathbf{P}_1)\mathbf{X}_2\boldsymbol{\beta}_2 = \mathbf{X}_2^T(\mathbf{I} - \mathbf{P}_1)\mathbf{y}.$$

This last system of equations has the form of normal equations, with $(\mathbf{I} - \mathbf{P}_1)\mathbf{X}_2$ playing the role usually played by the model matrix $\mathbf{X}$.

Definition 9.2.1 Under the two-part model $\mathbf{y} = \mathbf{X}_1\boldsymbol{\beta}_1 + \mathbf{X}_2\boldsymbol{\beta}_2 + \mathbf{e}$, the system of equations

$$\mathbf{X}_2^T(\mathbf{I} - \mathbf{P}_1)\mathbf{X}_2\boldsymbol{\beta}_2 = \mathbf{X}_2^T(\mathbf{I} - \mathbf{P}_1)\mathbf{y}$$

are called the *reduced normal equations for* $\boldsymbol{\beta}_2$. Similarly, the system of equations $\mathbf{X}_1^T(\mathbf{I} - \mathbf{P}_2)\mathbf{X}_1\boldsymbol{\beta}_1 = \mathbf{X}_1^T(\mathbf{I} - \mathbf{P}_2)\mathbf{y}$ are called the *reduced normal equations for* $\boldsymbol{\beta}_1$.

The series of algebraic steps used to arrive at either set of reduced normal equations from the normal equations for the two-part model is what is commonly referred to as absorption.

How might absorption reduce the computational burden of solving the normal equations? Suppose, for example, that it is very easy to determine a generalized inverse of $\mathbf{X}_1^T\mathbf{X}_1$, but not so easy to determine a generalized inverse of $\mathbf{X}^T\mathbf{X}$. Then we could obtain $\mathbf{P}_1$ easily and solve the reduced normal equations for $\boldsymbol{\beta}_2$ (which could be a much smaller system of equations than the full set of normal equations) to obtain a $\hat{\boldsymbol{\beta}}_2$, and finally "backsolve" for a $\hat{\boldsymbol{\beta}}_1$ using the "top" subset of equations in (9.5) (rearranged slightly), i.e.,

$$\mathbf{X}_1^T\mathbf{X}_1\boldsymbol{\beta}_1 = \mathbf{X}_1^T\mathbf{y} - \mathbf{X}_1^T\mathbf{X}_2\hat{\boldsymbol{\beta}}_2.$$

This approach to solving the normal equations would require the (generalized) inversion of two matrices, one of dimensions $p_1 \times p_1$ and the other of dimensions $p_2 \times p_2$, rather than a single $p \times p$ matrix, which would reduce the computational burden. (Inversion of a $p \times p$ matrix in general requires $O(p^3)$ operations.) Alternatively, instead of absorbing the "top" subset of normal equations into the "bottom" subset, we could absorb the "bottom" subset into the "top" subset. This would be advantageous computationally if it is $\mathbf{X}_2^T\mathbf{X}_2$ (rather than $\mathbf{X}_1^T\mathbf{X}_1$) for which a generalized inverse can be determined easily. If it is equally easy to obtain a generalized inverse of $\mathbf{X}_1^T\mathbf{X}_1$ and $\mathbf{X}_2^T\mathbf{X}_2$, greater computational efficiency usually results from absorbing the "top" subset of normal equations into the "bottom" subset if the dimensions of $\mathbf{X}_2^T\mathbf{X}_2$ are smaller than the dimensions of $\mathbf{X}_1^T\mathbf{X}_1$ (and vice versa otherwise).

Example 9.2-1. Solving the Normal Equations for the Two-Way Main Effects Model Using Absorption

Consider the normal equations for the two-way main effects model. Using (5.7), the model matrix may be written in two parts as $\mathbf{X} = (\mathbf{X}_1, \mathbf{X}_2)$ where $\mathbf{X}_1$ and $\mathbf{X}_2$ are matrices obtained from the larger matrices

$$\begin{pmatrix} \mathbf{1}_{n_{1\cdot}} & \mathbf{1}_{n_{1\cdot}} & \mathbf{0}_{n_{1\cdot}} & \cdots & \mathbf{0}_{n_{1\cdot}} \\ \mathbf{1}_{n_{2\cdot}} & \mathbf{0}_{n_{2\cdot}} & \mathbf{1}_{n_{2\cdot}} & \cdots & \mathbf{0}_{n_{2\cdot}} \\ \vdots & \vdots & \vdots & \ddots & \vdots \\ \mathbf{1}_{n_{q\cdot}} & \mathbf{0}_{n_{q\cdot}} & \mathbf{0}_{n_{q\cdot}} & \cdots & \mathbf{1}_{n_{q\cdot}} \end{pmatrix} \quad \text{and} \quad \begin{pmatrix} \oplus_{j=1}^m \mathbf{1}_{n_{1j}} \\ \oplus_{j=1}^m \mathbf{1}_{n_{2j}} \\ \vdots \\ \oplus_{j=1}^m \mathbf{1}_{n_{qj}} \end{pmatrix},$$

respectively, by deleting rows corresponding to empty cells. The parameter vector $\boldsymbol{\beta}$ may be partitioned accordingly as $\boldsymbol{\beta} = (\boldsymbol{\beta}_1^T, \boldsymbol{\beta}_2^T)^T$ where $\boldsymbol{\beta}_1 = (\mu, \alpha_1, \ldots, \alpha_q)^T$ and $\boldsymbol{\beta}_2 = (\gamma_1, \ldots, \gamma_m)^T$. The two-part normal equations corresponding to this partition are

$$\begin{pmatrix} n & \mathbf{n}_{*\cdot}^T & \mathbf{n}_{\cdot *}^T \\ \mathbf{n}_{*\cdot} & \oplus_{i=1}^q n_{i\cdot} & \mathbf{N} \\ \mathbf{n}_{\cdot *} & \mathbf{N}^T & \oplus_{j=1}^m n_{\cdot j} \end{pmatrix} \begin{pmatrix} \boldsymbol{\beta}_1 \\ \boldsymbol{\beta}_2 \end{pmatrix} = \begin{pmatrix} y_{\cdots} \\ \mathbf{y}_{*\cdot} \\ \mathbf{y}_{\cdot *} \end{pmatrix},$$

where

$$\mathbf{n}_{*\cdot} = \begin{pmatrix} n_{1\cdot} \\ n_{2\cdot} \\ \vdots \\ n_{q\cdot} \end{pmatrix}, \quad \mathbf{n}_{\cdot *} = \begin{pmatrix} n_{\cdot 1} \\ n_{\cdot 2} \\ \vdots \\ n_{\cdot m} \end{pmatrix}, \quad \mathbf{N} = (n_{ij}), \quad \mathbf{y}_{*\cdot} = \begin{pmatrix} y_{1\cdot\cdot} \\ y_{2\cdot\cdot} \\ \vdots \\ y_{q\cdot\cdot} \end{pmatrix}, \quad \mathbf{y}_{\cdot *} = \begin{pmatrix} y_{\cdot 1\cdot} \\ y_{\cdot 2\cdot} \\ \vdots \\ y_{\cdot m\cdot} \end{pmatrix}.$$

It is easily verified (as noted previously in Example 7.1-2) that one generalized inverse of $\mathbf{X}_1^T\mathbf{X}_1$ is

$$\begin{pmatrix} 0 & \mathbf{0}^T \\ \mathbf{0} & \oplus_{i=1}^q n_{i\cdot}^{-1} \end{pmatrix}.$$

Consequently,

$$\begin{aligned} \mathbf{X}_2^T(\mathbf{I} - \mathbf{P}_1)\mathbf{X}_2 &= \mathbf{X}_2^T\mathbf{X}_2 - \mathbf{X}_2^T\mathbf{X}_1(\mathbf{X}_1^T\mathbf{X}_1)^-\mathbf{X}_1^T\mathbf{X}_2 \\ &= \oplus_{j=1}^m n_{\cdot j} - (\mathbf{n}_{\cdot *}, \mathbf{N}^T) \begin{pmatrix} 0 & \mathbf{0}^T \\ \mathbf{0} & \oplus_{i=1}^q n_{i\cdot}^{-1} \end{pmatrix} \begin{pmatrix} \mathbf{n}_{\cdot *}^T \\ \mathbf{N} \end{pmatrix}, \end{aligned}$$

which has jkth element

$$\begin{cases} n_{\cdot j} - \sum_i \frac{n_{ij}^2}{n_{i\cdot}} & \text{if } k = j, \\ -\sum_i \frac{n_{ij}n_{ik}}{n_{i\cdot}} & \text{if } k \neq j. \end{cases}$$

Furthermore,

$$\mathbf{X}_2^T(\mathbf{I}-\mathbf{P}_1)\mathbf{y} = \mathbf{X}_2^T\mathbf{y} - \mathbf{X}_2^T\mathbf{X}_1(\mathbf{X}_1^T\mathbf{X}_1)^-\mathbf{X}_1^T\mathbf{y} = \mathbf{y}_{\cdot *} - (\mathbf{n}_{\cdot *}, \mathbf{N}^T)\begin{pmatrix} 0 & \mathbf{0}^T \\ \mathbf{0} & \oplus_{i=1}^q n_{i\cdot}^{-1} \end{pmatrix}\begin{pmatrix} y_{\cdots} \\ \mathbf{y}_{*\cdot}^T \end{pmatrix},$$

which has jth element

$$y_{\cdot j \cdot} - \sum_i \left(\frac{n_{ij}}{n_{i\cdot}}\right) y_{i\cdots}.$$

These results could be used to solve (numerically) the reduced normal equations for $\boldsymbol{\beta}_2$, obtaining a solution $\hat{\boldsymbol{\beta}}_2 = (\hat{\gamma}_1, \hat{\gamma}_2, \ldots, \hat{\gamma}_m)^T$. A solution $\hat{\boldsymbol{\beta}}_1$ can then be obtained by back-substituting, i.e.,

$$\hat{\boldsymbol{\beta}}_1 = (\mathbf{X}_1^T\mathbf{X}_1)^-\mathbf{X}_1^T(\mathbf{y} - \mathbf{X}_2\hat{\boldsymbol{\beta}}_2) = \begin{pmatrix} 0 & \mathbf{0}^T \\ \mathbf{0} & \oplus_{i=1}^q n_{i\cdot}^{-1} \end{pmatrix}\begin{pmatrix} y_{\cdots} \\ \mathbf{y}_{*\cdot}^T \end{pmatrix} - \begin{pmatrix} 0 & \mathbf{0}^T \\ \mathbf{0} & \oplus_{i=1}^q n_{i\cdot}^{-1} \end{pmatrix}\begin{pmatrix} \mathbf{n}_{\cdot *}^T \\ \mathbf{N} \end{pmatrix}\hat{\boldsymbol{\beta}}_2.$$

Thus, $\hat{\boldsymbol{\beta}}_1 = (\hat{\mu}, \hat{\alpha}_1, \ldots, \hat{\alpha}_q)^T$ is a solution, where

$$\hat{\mu} = 0, \qquad \hat{\alpha}_i = \bar{y}_{i\cdot\cdot} - \sum_j \left(\frac{n_{ij}}{n_{i\cdot}}\right)\hat{\gamma}_j.$$

■

Methodological Interlude #4: The "Added Variable" Plot

Suppose that the model matrix $\mathbf{X}$ has $\mathbf{1}_n$ as one of its columns, in which case we may write $\mathbf{X}$, without loss of generality, as $(\mathbf{1}_n, \mathbf{x}_2, \mathbf{x}_3, \ldots, \mathbf{x}_p)$. For $j = 2, \ldots, p$, let $\mathbf{X}_{(-j)}$ denote the $n \times (p-1)$ matrix obtained by deleting $\mathbf{x}_j$ from $\mathbf{X}$, and let $\boldsymbol{\beta}_{-j}$ denote the corresponding vector of parameters obtained by deleting β_j from $\boldsymbol{\beta}$. Define

$$\mathbf{P}_{-j} = \mathbf{X}_{(-j)}(\mathbf{X}_{(-j)}^T\mathbf{X}_{(-j)})^-\mathbf{X}_{(-j)}^T, \quad \hat{\mathbf{e}}_{y|-j} = \mathbf{y} - \mathbf{P}_{-j}\mathbf{y}, \quad \hat{\mathbf{e}}_{x_j|-j} = \mathbf{x}_j - \mathbf{P}_{-j}\mathbf{x}_j.$$

Note that $\hat{\mathbf{e}}_{y|-j}$ is the vector of least squares fitted residuals obtained by regressing $\mathbf{y}$ on all explanatory variables except $\mathbf{x}_j$, and $\hat{\mathbf{e}}_{x_j|-j}$ is the vector of least squares fitted residuals obtained by regressing $\mathbf{x}_j$ on all other explanatory variables.

Consider the following simple linear regression model (with an intercept) for regressing $\hat{\mathbf{e}}_{y|-j}$ on $\hat{\mathbf{e}}_{x_j|-j}$:

$$(\mathbf{I}-\mathbf{P}_{-j})\mathbf{y} = \mathbf{1}_n\gamma_{0j} + (\mathbf{I}-\mathbf{P}_{-j})\mathbf{x}_j\gamma_{1j} + \mathbf{e}_j^*$$

where $\mathbf{e}_j^*$ represents a generic zero-mean random n-vector. The two-part normal equations for this model are

$$\begin{pmatrix} \mathbf{1}_n^T\mathbf{1}_n & \mathbf{1}_n^T(\mathbf{I}-\mathbf{P}_{-j})\mathbf{x}_j \\ \mathbf{x}_j^T(\mathbf{I}-\mathbf{P}_{-j})\mathbf{1}_n & \mathbf{x}_j^T(\mathbf{I}-\mathbf{P}_{-j})\mathbf{x}_j \end{pmatrix}\begin{pmatrix} \gamma_{0j} \\ \gamma_{1j} \end{pmatrix} = \begin{pmatrix} \mathbf{1}_n^T(\mathbf{I}-\mathbf{P}_{-j})\mathbf{y} \\ \mathbf{x}_j^T(\mathbf{I}-\mathbf{P}_{-j})\mathbf{y} \end{pmatrix}$$

or equivalently, because $\mathbf{1}_n \in \mathcal{C}(\mathbf{P}_{-j})$ and thus $\mathbf{P}_{-j}\mathbf{1}_n = \mathbf{1}_n$,

$$\begin{pmatrix} n & 0 \\ 0 & \mathbf{x}_j^T(\mathbf{I}-\mathbf{P}_{-j})\mathbf{x}_j \end{pmatrix}\begin{pmatrix} \gamma_{0j} \\ \gamma_{1j} \end{pmatrix} = \begin{pmatrix} 0 \\ \mathbf{x}_j^T(\mathbf{I}-\mathbf{P}_{-j})\mathbf{y} \end{pmatrix}. \tag{9.7}$$

Assuming that $\mathbf{x}_j \notin \mathcal{C}(\mathbf{X}_{(-j)})$, the unique solution to this system of equations is

$$\hat{\gamma}_{0j} = 0, \qquad \hat{\gamma}_{1j} = [\mathbf{x}_j^T(\mathbf{I}-\mathbf{P}_{-j})\mathbf{x}_j]^{-1}\mathbf{x}_j^T(\mathbf{I}-\mathbf{P}_{-j})\mathbf{y}.$$

The reduced normal equation for the scalar parameter β_j for the model $\{\mathbf{y}, \mathbf{X}\boldsymbol{\beta}\} = \{\mathbf{y}, \mathbf{x}_j\beta_j + \mathbf{X}_{(-j)}\boldsymbol{\beta}_{-j}\}$ coincides with the second equation in (9.7). Thus, the least squares estimator of the slope parameter in the simple linear regression of $\hat{\mathbf{e}}_{y|-j}$ on $\hat{\mathbf{e}}_{x_j|-j}$ is identical to the least squares estimator of β_j associated with the complete model $\{\mathbf{y}, \mathbf{X}\boldsymbol{\beta}\}$. That is, we can gain some understanding of the effect that the explanatory variable x_j has on y, in the context of a model with all the other explanatory variables, via a plot of the elements of $\hat{\mathbf{e}}_{y|-j}$ versus the corresponding elements of $\hat{\mathbf{e}}_{x_j|-j}$ and the fit of a simple linear regression model to these points. Such a plot is called an *added variable plot* or *partial regression plot* and is frequently presented in applied regression courses. ■

9.3 Partitioning the Model Sum of Squares

9.3.1 The Sequential ANOVA

Consider once again the ordered two-part model $\{\mathbf{y}, \mathbf{X}_1\boldsymbol{\beta}_1 + \mathbf{X}_2\boldsymbol{\beta}_2\}$. For this model, consider expanding upon the decomposition (8.2) of $\mathbf{y}$, as follows:

$$\begin{aligned} \mathbf{y} &= \mathbf{P}_{12}\mathbf{y} + (\mathbf{I}-\mathbf{P}_{12})\mathbf{y} \\ &= \mathbf{P}_1\mathbf{y} + (\mathbf{P}_{12}-\mathbf{P}_1)\mathbf{y} + (\mathbf{I}-\mathbf{P}_{12})\mathbf{y}. \end{aligned} \tag{9.8}$$

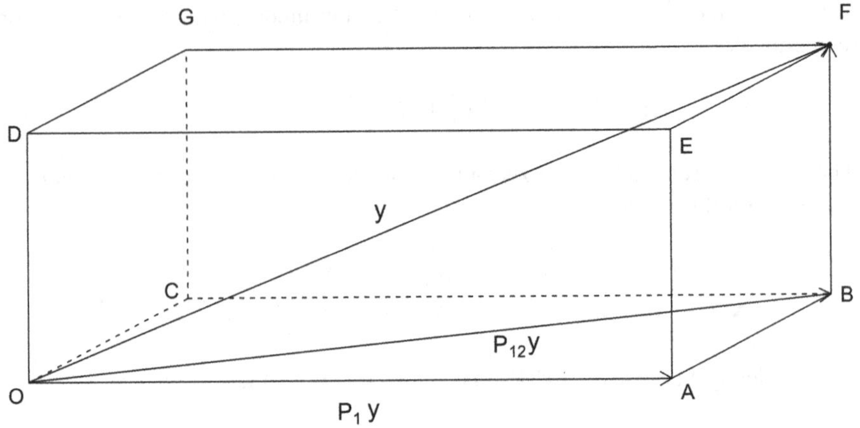

Fig. 9.1 Orthogonal projections of $\mathbf{y}$ onto $\mathcal{C}(\mathbf{X}_1)$ and $\mathcal{C}(\mathbf{X}_1, \mathbf{X}_2)$

The last term in the expanded decomposition is identical to the last term in (8.2). The first term in the expanded decomposition is the orthogonal projection of $\mathbf{y}$ onto $\mathcal{C}(\mathbf{X}_1)$, and the second term is the difference between the orthogonal projection of $\mathbf{y}$ onto $\mathcal{C}(\mathbf{X})$ and the first term; equivalently, by Theorem 9.1.2, the second term is the orthogonal projection of $\mathbf{y}$ onto $\mathcal{C}[(\mathbf{I} - \mathbf{P}_1)\mathbf{X}_2]$. By Theorem 9.1.1b–d, all three terms in the decomposition are orthogonal to each other.

The "geometry" of decomposition (9.8) for the special case of $n = 3$ is displayed in Fig. 9.1. The figure displays a rectangular cuboid, of which some sides and edges are portions of relevant column spaces. The origin is at O and $\mathbf{y}$ is the vector OF. Rectangle $OABC$ represents part of $\mathcal{C}(\mathbf{X}_1, \mathbf{X}_2)$, and line segment OA represents part of $\mathcal{C}(\mathbf{X}_1)$. OA also represents $\mathbf{P}_1\mathbf{y}$; OG represents $(\mathbf{I} - \mathbf{P}_1)\mathbf{y}$ (and AF is a translation of it); OB represents $\mathbf{P}_{12}\mathbf{y}$; OD represents $(\mathbf{I} - \mathbf{P}_{12})\mathbf{y}$ (and BF is a translation of it); and OC represents $(\mathbf{P}_{12} - \mathbf{P}_1)\mathbf{y}$ (and AB is a translation of it).

Pre-multiplication of both sides of (9.8) by $\mathbf{y}^T$ yields the corresponding decomposition of the total sum of squares:

$$\mathbf{y}^T\mathbf{y} = \mathbf{y}^T\mathbf{P}_1\mathbf{y} + \mathbf{y}^T(\mathbf{P}_{12} - \mathbf{P}_1)\mathbf{y} + \mathbf{y}^T(\mathbf{I} - \mathbf{P}_{12})\mathbf{y}.$$

Like the decomposition of the total sum of squares introduced in Chap. 8, this sum of squares decomposition can be laid out in tabular form, as follows:

Source	Rank	Sum of squares
$\mathbf{X}_1$	$\text{rank}(\mathbf{X}_1)$	$\mathbf{y}^T\mathbf{P}_1\mathbf{y}$
$\mathbf{X}_2\|\mathbf{X}_1$	$\text{rank}(\mathbf{X}_1, \mathbf{X}_2) - \text{rank}(\mathbf{X}_1)$	$\mathbf{y}^T(\mathbf{P}_{12} - \mathbf{P}_1)\mathbf{y}$
Residual	$n - \text{rank}(\mathbf{X}_1, \mathbf{X}_2)$	$\mathbf{y}^T(\mathbf{I} - \mathbf{P}_{12})\mathbf{y}$
Total	n	$\mathbf{y}^T\mathbf{y}$

This is called the *sequential ANOVA of the ordered two-part model* $\{\mathbf{y}, \mathbf{X}_1\boldsymbol{\beta}_1 + \mathbf{X}_2\boldsymbol{\beta}_2\}$. Alternative expressions for $\mathbf{y}^T\mathbf{P}_1\mathbf{y}$ include $\tilde{\boldsymbol{\beta}}_1^T\mathbf{X}_1^T\mathbf{y}$ and $\tilde{\boldsymbol{\beta}}_1^T\mathbf{X}_1^T\mathbf{X}_1\tilde{\boldsymbol{\beta}}_1$, where $\tilde{\boldsymbol{\beta}}_1$ is any solution to $\mathbf{X}_1^T\mathbf{X}_1\boldsymbol{\beta}_1 = \mathbf{X}_1^T\mathbf{y}$. An alternative expression for $\mathbf{y}^T(\mathbf{P}_{12} - \mathbf{P}_1)\mathbf{y}$ is $\hat{\boldsymbol{\beta}}_1^T\mathbf{X}_1^T\mathbf{y} + \hat{\boldsymbol{\beta}}_2^T\mathbf{X}_2^T\mathbf{y} - \tilde{\boldsymbol{\beta}}_1^T\mathbf{X}_1^T\mathbf{y}$, where $(\hat{\boldsymbol{\beta}}_1^T, \hat{\boldsymbol{\beta}}_2^T)^T$ is any solution to the normal equations corresponding to the aforementioned two-part model. In general, $\mathbf{X}_1\tilde{\boldsymbol{\beta}}_1 \neq \mathbf{X}_1\hat{\boldsymbol{\beta}}_1$.

Corresponding to the other two-part model in (9.1), we could also consider the following decomposition of $\mathbf{y}$:

$$\mathbf{y} = \mathbf{P}_2\mathbf{y} + (\mathbf{P}_{12} - \mathbf{P}_2)\mathbf{y} + (\mathbf{I} - \mathbf{P}_{12})\mathbf{y}. \tag{9.9}$$

Results for $\mathbf{P}_2$ analogous to those for $\mathbf{P}_1$ in Theorem 9.1.1 establish that (9.9) is also an orthogonal decomposition and that the total sum of squares can be decomposed as shown in the following table:

Source	Rank	Sum of squares
$\mathbf{X}_2$	$\text{rank}(\mathbf{X}_2)$	$\mathbf{y}^T\mathbf{P}_2\mathbf{y}$
$\mathbf{X}_1\vert\mathbf{X}_2$	$\text{rank}(\mathbf{X}_1, \mathbf{X}_2) - \text{rank}(\mathbf{X}_2)$	$\mathbf{y}^T(\mathbf{P}_{12} - \mathbf{P}_2)\mathbf{y}$
Residual	$n - \text{rank}(\mathbf{X}_1, \mathbf{X}_2)$	$\mathbf{y}^T(\mathbf{I} - \mathbf{P}_{12})\mathbf{y}$
Total	n	$\mathbf{y}^T\mathbf{y}$

This is the sequential ANOVA of the ordered two-part model $\{\mathbf{y}, \mathbf{X}_2\boldsymbol{\beta}_2 + \mathbf{X}_1\boldsymbol{\beta}_1\}$.

The method of absorption introduced in the previous section can reduce the computational burden of not only obtaining a solution to the normal equations, but also of performing the sequential ANOVA. In fact, the sum of squares for $\mathbf{X}_2|\mathbf{X}_1$ can be expressed in terms of a solution vector and the vector on the right-hand side of the reduced normal equations for $\boldsymbol{\beta}_2$, i.e., $\text{SS}(\mathbf{X}_2|\mathbf{X}_1) = \hat{\boldsymbol{\beta}}_2^T\mathbf{X}_2^T(\mathbf{I} - \mathbf{P}_1)\mathbf{y}$. This can be verified as follows:

$$\begin{aligned}
\text{SS}(\mathbf{X}_2|\mathbf{X}_1) &= \mathbf{y}^T(\mathbf{P}_{12} - \mathbf{P}_1)\mathbf{y} \\
&= \hat{\boldsymbol{\beta}}_1^T\mathbf{X}_1^T\mathbf{y} + \hat{\boldsymbol{\beta}}_2^T\mathbf{X}_2^T\mathbf{y} - \mathbf{y}^T\mathbf{P}_1\mathbf{y} \\
&= (\mathbf{y}^T\mathbf{P}_1 - \hat{\boldsymbol{\beta}}_2^T\mathbf{X}_2^T\mathbf{P}_1)\mathbf{y} + \hat{\boldsymbol{\beta}}_2^T\mathbf{X}_2^T\mathbf{y} - \mathbf{y}^T\mathbf{P}_1\mathbf{y} \\
&= \hat{\boldsymbol{\beta}}_2^T\mathbf{X}_2^T\mathbf{y} - \hat{\boldsymbol{\beta}}_2^T\mathbf{X}_2^T\mathbf{P}_1\mathbf{y} \\
&= \hat{\boldsymbol{\beta}}_2^T\mathbf{X}_2^T(\mathbf{I} - \mathbf{P}_1)\mathbf{y},
\end{aligned} \tag{9.10}$$

where (9.6) was used to obtain the third equality. A similar expression holds for $\text{SS}(\mathbf{X}_1|\mathbf{X}_2)$ in terms of a solution vector and the vector on the right-hand side of the reduced normal equations for $\boldsymbol{\beta}_1$.

The results established in this section for ordered two-part models can be extended without difficulty to the general ordered k-part model. As a consequence of the properties listed in Theorem 9.1.3, the decomposition

$$\mathbf{y} = \mathbf{P}_1\mathbf{y} + (\mathbf{P}_{12} - \mathbf{P}_1)\mathbf{y} + (\mathbf{P}_{123} - \mathbf{P}_{12})\mathbf{y} + \cdots + (\mathbf{P}_{12\cdots k} - \mathbf{P}_{12\cdots k-1})\mathbf{y} + (\mathbf{I} - \mathbf{P}_{12\cdots k})\mathbf{y}$$

is an orthogonal decomposition and the total sum of squares can be partitioned and laid out in tabular form as follows:

Source	Rank	Sum of squares
$\mathbf{X}_1$	$\text{rank}(\mathbf{X}_1)$	$\mathbf{y}^T\mathbf{P}_1\mathbf{y}$
$\mathbf{X}_2\|\mathbf{X}_1$	$\text{rank}(\mathbf{X}_1, \mathbf{X}_2) - \text{rank}(\mathbf{X}_1)$	$\mathbf{y}^T(\mathbf{P}_{12} - \mathbf{P}_1)\mathbf{y}$
$\mathbf{X}_3\|(\mathbf{X}_1, \mathbf{X}_2)$	$\text{rank}(\mathbf{X}_1, \mathbf{X}_2, \mathbf{X}_3) - \text{rank}(\mathbf{X}_1, \mathbf{X}_2)$	$\mathbf{y}^T(\mathbf{P}_{123} - \mathbf{P}_{12})\mathbf{y}$
⋮	⋮	⋮
$\mathbf{X}_k\|(\mathbf{X}_1, \ldots, \mathbf{X}_{k-1})$	$\text{rank}(\mathbf{X}_1, \ldots, \mathbf{X}_k) - \text{rank}(\mathbf{X}_1, \ldots, \mathbf{X}_{k-1})$	$\mathbf{y}^T(\mathbf{P}_{12\cdots k} - \mathbf{P}_{12\cdots k-1})\mathbf{y}$
Residual	$n - \text{rank}(\mathbf{X}_1, \ldots, \mathbf{X}_k)$	$\mathbf{y}^T(\mathbf{I} - \mathbf{P}_{12\cdots k})\mathbf{y}$
Total	n	$\mathbf{y}^T\mathbf{y}$

This is called *the sequential ANOVA of the ordered k-part model* (9.3). Because $k!$ ordered k-part models may be constructed by permuting terms, for a given k-part partition of $\mathbf{X}$ there are $k!$ distinct k-part sequential ANOVAs.

9.3.2 Expected Mean Squares

In many statistical methods textbooks, displayed ANOVA tables include a column (usually located at the far right) called *Expected Mean Square*, or EMS. Each entry in this column is, as advertised, the expectation of the mean square in the corresponding row, under a Gauss–Markov model. An expression for each EMS under the Gauss–Markov model $\{\mathbf{y}, \mathbf{X}\boldsymbol{\beta}, \sigma^2\mathbf{I}\}$ may be obtained by observing that for any quadratic form $\mathbf{y}^T\mathbf{A}_i\mathbf{y}$ for which $\mathbf{A}_i$ is nonnull and idempotent, Theorem 4.2.4 yields

$$\begin{aligned} \text{E}(\mathbf{y}^T\mathbf{A}_i\mathbf{y}) &= (\mathbf{X}\boldsymbol{\beta})^T\mathbf{A}_i(\mathbf{X}\boldsymbol{\beta}) + \text{tr}[\mathbf{A}_i(\sigma^2\mathbf{I})] \\ &= (\mathbf{X}\boldsymbol{\beta})^T\mathbf{A}_i(\mathbf{X}\boldsymbol{\beta}) + \sigma^2\text{rank}(\mathbf{A}_i), \end{aligned}$$

and thus

$$\text{E}\left(\frac{\mathbf{y}^T\mathbf{A}_i\mathbf{y}}{\text{rank}(\mathbf{A}_i)}\right) = \sigma^2 + \frac{(\mathbf{X}\boldsymbol{\beta})^T\mathbf{A}_i(\mathbf{X}\boldsymbol{\beta})}{\text{rank}(\mathbf{A}_i)}.$$

As a special case of this expression, the expected residual mean square is, by Theorem 8.1.3a, σ^2; of course, we already knew this (Corollary 8.2.1.2). More interestingly, the EMS on any given line of an ANOVA table above the Residual line is equal to σ^2 plus a scaled quadratic form in $\mathbf{X}\boldsymbol{\beta}$ of the same form as the quadratic form (in $\mathbf{y}$) involved in the sum of squares on that line. This means that a nonmatrix expression for the EMS in any row of an ANOVA table may be derived by first replacing the elements of $\mathbf{y}$ in a nonmatrix expression for the sum of squares in that row with the elements of its expectation $\mathbf{X}\boldsymbol{\beta}$, then dividing the result by the rank of the matrix of the quadratic form, and finally adding that result to σ^2. Several examples in the remainder of this chapter will illustrate this procedure.

9.3.3 The "Corrected for the Mean" Overall ANOVA

Many linear models used in practice have a column of ones as one of the columns of $\mathbf{X}$, i.e., $\mathbf{X} = (\mathbf{1}_n, \mathbf{X}_2)$ where

$$\mathbf{X}_2 = \begin{pmatrix} x_{12} & x_{13} & \cdots & x_{1p} \\ \vdots & \vdots & & \vdots \\ x_{n2} & x_{n3} & \cdots & x_{np} \end{pmatrix}.$$

We may write such a model as an ordered two-part model

$$\mathbf{y} = \mathbf{1}\mu + \mathbf{X}_2\boldsymbol{\beta}_2 + \mathbf{e}. \tag{9.11}$$

The parameter μ that corresponds to the column $\mathbf{1}$ is variously called the overall mean, the constant term, or the intercept. The two-part normal equations are

$$n\mu + \mathbf{1}^T\mathbf{X}_2\boldsymbol{\beta}_2 = \sum_{i=1}^{n} y_i$$

$$\mathbf{X}_2^T\mathbf{1}\mu + \mathbf{X}_2^T\mathbf{X}_2\boldsymbol{\beta}_2 = \mathbf{X}_2^T\mathbf{y}.$$

Furthermore, $\mathbf{P}_1 = (1/n)\mathbf{J}_n$ so the reduced normal equations for $\boldsymbol{\beta}_2$ are

$$\mathbf{X}_2^T[\mathbf{I} - (1/n)\mathbf{J}]\mathbf{X}_2\boldsymbol{\beta}_2 = \mathbf{X}_2^T[\mathbf{I} - (1/n)\mathbf{J}]\mathbf{y}.$$

Now

$$[\mathbf{I} - (1/n)\mathbf{J}]\mathbf{X}_2 = \begin{pmatrix} x_{12} - \bar{x}_2 & x_{13} - \bar{x}_3 & \cdots & x_{1p} - \bar{x}_p \\ x_{22} - \bar{x}_2 & x_{23} - \bar{x}_3 & \cdots & x_{2p} - \bar{x}_p \\ \vdots & \vdots & & \vdots \\ x_{n2} - \bar{x}_2 & x_{n3} - \bar{x}_3 & \cdots & x_{np} - \bar{x}_p \end{pmatrix}$$

and

$$[\mathbf{I} - (1/n)\mathbf{J}]\mathbf{y} = \begin{pmatrix} y_1 - \bar{y} \\ y_2 - \bar{y} \\ \vdots \\ y_n - \bar{y} \end{pmatrix}.$$

Accordingly, the coefficient matrix of the reduced normal equations for $\boldsymbol{\beta}_2$, i.e., $\mathbf{X}_2^T[\mathbf{I} - (1/n)\mathbf{J}_n]\mathbf{X}_2$, is often called the "corrected" sum of squares and cross products matrix, and the n-vector with ith element $y_i - \bar{y}$ is called the "corrected" response vector. The modifier "corrected" does not imply that any error was made previously; it merely indicates that the values of each variable have been replaced with the differences of those values from their sample mean. Letting $\hat{\boldsymbol{\beta}}_2 = (\hat{\beta}_{2j})$ be a solution to the reduced normal equations for $\boldsymbol{\beta}_2$ and back-solving to obtain $\hat{\mu} = \bar{y} - \sum_{j=2}^{p} \bar{x}_j \hat{\beta}_{2j}$, we obtain a solution to the (unreduced) normal equations.

Now, if $\mathbf{1} \notin \mathcal{C}(\mathbf{X}_2)$, then the sequential ANOVA corresponding to the ordered two-part model (9.11) is as follows:

Source	Rank	Sum of squares
$\mathbf{1}$	1	$\mathbf{y}^T[(1/n)\mathbf{J}]\mathbf{y}$
$\mathbf{X}_2\|\mathbf{1}$	$\text{rank}(\mathbf{X}_2)$	$\mathbf{y}^T[\mathbf{P}_\mathbf{X} - (1/n)\mathbf{J}]\mathbf{y} = \hat{\boldsymbol{\beta}}_2^T\mathbf{X}_2^T[\mathbf{I} - (1/n)\mathbf{J}]\mathbf{y}$
Residual	$n - 1 - \text{rank}(\mathbf{X}_2)$	$\mathbf{y}^T(\mathbf{I} - \mathbf{P}_\mathbf{X})\mathbf{y}$
Total	n	$\mathbf{y}^T\mathbf{y}$

If $\mathbf{1} \in \mathcal{C}(\mathbf{X}_2)$ instead, the sequential ANOVA is identical to that just displayed, except that the rank of $\mathbf{X}_2|\mathbf{1}$ is decreased by one and the rank of Residual is increased by one.

Most statistical methods textbooks that display ANOVA tables for models that have an overall mean do not include a line in the table for that term; instead, they subtract that line from the "Total" line and either relabel the resulting line as "Corrected Total" or merely leave the label as is, with the implicit understanding that the correction for the mean has been applied to either the response variable or all variables (it makes no difference which because $\mathbf{X}_2^T[\mathbf{I} - (1/n)\mathbf{J}]\mathbf{y} = \{[\mathbf{I} - (1/n)\mathbf{J}]\mathbf{X}_2\}^T][\mathbf{I} - (1/n)\mathbf{J}]\mathbf{y})$. In addition, the label $\mathbf{X}_2|\mathbf{1}$ is invariably shortened to $\mathbf{X}_2$, although the ranks and sum of squares in this line are not modified. The resulting ANOVA table, assuming that $\mathbf{1} \notin \mathcal{C}(\mathbf{X}_2)$, is as follows:

Source	Rank	Sum of squares
$\mathbf{X}_2$	$\text{rank}(\mathbf{X}_2)$	$\mathbf{y}^T[\mathbf{P}_\mathbf{X} - (1/n)\mathbf{J}]\mathbf{y}$
Residual	$n - 1 - \text{rank}(\mathbf{X}_2)$	$\mathbf{y}^T(\mathbf{I} - \mathbf{P}_\mathbf{X})\mathbf{y}$
Corrected Total	$n - 1$	$\mathbf{y}^T[\mathbf{I} - (1/n)\mathbf{J}]\mathbf{y}$

This ANOVA is called the *corrected (for the mean) overall ANOVA*. Non-matrix expressions may be given for the sums of squares in this table. Specifically,

$$\mathbf{y}^T[\mathbf{P_X} - (1/n)\mathbf{J}]\mathbf{y} = \mathbf{y}^T[\mathbf{P_X} - (1/n)\mathbf{J}]^T[\mathbf{P_X} - (1/n)\mathbf{J}]\mathbf{y}$$
$$= \sum_{i=1}^{n}(\hat{y}_i - \bar{y})^2,$$

$$\mathbf{y}^T(\mathbf{I} - \mathbf{P_X})\mathbf{y} = \mathbf{y}^T(\mathbf{I} - \mathbf{P_X})^T(\mathbf{I} - \mathbf{P_X})\mathbf{y}$$
$$= \sum_{i=1}^{n}(y_i - \hat{y}_i)^2,$$

and

$$\mathbf{y}^T[\mathbf{I} - (1/n)\mathbf{J}]\mathbf{y} = \mathbf{y}^T[\mathbf{I} - (1/n)\mathbf{J}]^T[\mathbf{I} - (1/n)\mathbf{J}]\mathbf{y}$$
$$= \sum_{i=1}^{n}(y_i - \bar{y})^2.$$

Thus, when $\mathbf{1} \notin \mathcal{C}(\mathbf{X}_2)$, we may rewrite the corrected overall ANOVA table as follows:

Source	Rank	Sum of squares
$\mathbf{X}_2$	$\text{rank}(\mathbf{X}_2)$	$\sum_{i=1}^{n}(\hat{y}_i - \bar{y})^2$
Residual	$n - 1 - \text{rank}(\mathbf{X}_2)$	$\sum_{i=1}^{n}(y_i - \hat{y}_i)^2$
Corrected Total	$n - 1$	$\sum_{i=1}^{n}(y_i - \bar{y})^2$

When $\mathbf{1} \in \mathcal{C}(\mathbf{X}_2)$, the corrected overall ANOVA table is as follows:

Source	Rank	Sum of squares
$\mathbf{X}_2$	$\text{rank}(\mathbf{X}_2) - 1$	$\sum_{i=1}^{n}(\hat{y}_i - \bar{y})^2$
Residual	$n - \text{rank}(\mathbf{X}_2)$	$\sum_{i=1}^{n}(y_i - \hat{y}_i)^2$
Corrected Total	$n - 1$	$\sum_{i=1}^{n}(y_i - \bar{y})^2$

In either case, the following quantity is often computed from two of the sums of squares in the corrected overall ANOVA:

$$R^2 = \frac{\text{SS}(\mathbf{X}_2)}{\text{Corrected Total SS}} = \frac{\sum_{i=1}^{n}(\hat{y}_i - \bar{y})^2}{\sum_{i=1}^{n}(y_i - \bar{y})^2}.$$

This quantity, known as the *coefficient of determination*, measures the proportion of the total variation of the observations of y about their sample mean that can be "explained" by or attributed to the explanatory variables in the model. Obviously, $0 \leq R^2 \leq 1$.

Example 9.3.3-1. The Corrected Overall ANOVA for Simple Linear Regression

In the case of simple linear regression, $\mathbf{1} \notin \mathcal{C}(\mathbf{X}_2)$ so the corrected overall ANOVA table is as follows:

Source	Rank	Sum of squares	Expected mean square
Model	1	$\hat{\beta}_2^2 SXX$	$\sigma^2 + \beta_2^2 SXX$
Residual	$n-2$	$\sum_{i=1}^n (y_i - \bar{y})^2 - \hat{\beta}_2^2 SXX$	σ^2
Corrected Total	$n-1$	$\sum_{i=1}^n (y_i - \bar{y})^2$	

The reader should compare this table to the "uncorrected" overall ANOVA table for simple linear regression given previously in Example 8.2-1. The expected model mean square was determined by adding

$$[\mathrm{E}(\hat{\beta}_2)]^2 SXX/1 = \beta_2^2 SXX$$

to σ^2. ■

Example 9.3.3-2. The Corrected One-Factor Overall ANOVA

In the case of the one-factor model, $\mathbf{1} \in \mathcal{C}(\mathbf{X}_2)$ so the corrected overall ANOVA table is as follows:

Source	Rank	Sum of squares	Expected mean square
Model	$q-1$	$\sum_{i=1}^q n_i(\bar{y}_{i\cdot} - \bar{y}_{\cdot\cdot})^2$	$\sigma^2 + \frac{1}{q-1}\sum_{i=1}^q n_i \left(\alpha_i - \frac{\sum_{k=1}^q n_k\alpha_k}{n}\right)^2$
Residual	$n-q$	$\sum_{i=1}^q \sum_{j=1}^{n_i} (y_{ij} - \bar{y}_{i\cdot})^2$	σ^2
Corrected Total	$n-1$	$\sum_{i=1}^q \sum_{j=1}^{n_i} (y_{ij} - \bar{y}_{\cdot\cdot})^2$	

The reader should compare this table to the "uncorrected" overall ANOVA table for the one-factor model given previously in Example 8.2-2. Here, the expected model mean square was determined by replacing y_{ij} in the expression for the model sum of squares with its expectation $\mu + \alpha_i$, yielding

$$\sum_{i=1}^{q} n_i \left(\frac{\sum_{j=1}^{n_i}(\mu + \alpha_i)}{n_i} - \frac{\sum_{k=1}^{q}\sum_{m=1}^{n_k}(\mu + \alpha_k)}{n} \right)^2$$

$$= \sum_{i=1}^{q} n_i \left(\frac{n_i\mu + n_i\alpha_i}{n_i} - \frac{n\mu + \sum_{k=1}^{q} n_k\alpha_k}{n} \right)^2$$

$$= \sum_{i=1}^{q} n_i \left(\alpha_i - \frac{\sum_{k=1}^{q} n_k\alpha_k}{n} \right)^2,$$

then dividing this quantity by $q - 1$ and adding the result to σ^2. ■

The same correction for the mean can be applied to the sequential ANOVA of a three-part (or higher) model also, yielding the ANOVA (shown for the case $\mathbf{1} \notin \mathcal{C}(\mathbf{X})$ only)

Source	Rank	Sum of squares
$\mathbf{X}_2$	$\text{rank}(\mathbf{X}_2)$	$\mathbf{y}^T[\mathbf{P}_{12} - (1/n)\mathbf{J}]\mathbf{y}$
$\mathbf{X}_3\|\mathbf{X}_2$	$\text{rank}(\mathbf{X}_2, \mathbf{X}_3) - \text{rank}(\mathbf{X}_2)$	$\mathbf{y}^T(\mathbf{P}_{123} - \mathbf{P}_{12})\mathbf{y}$
$\vdots$	$\vdots$	$\vdots$
$\mathbf{X}_k\|(\mathbf{X}_2, \ldots, \mathbf{X}_{k-1})$	$\text{rank}(\mathbf{X}_2, \ldots, \mathbf{X}_k) - \text{rank}(\mathbf{X}_2, \ldots, \mathbf{X}_{k-1})$	$\mathbf{y}^T(\mathbf{P}_{12\cdots k} - \mathbf{P}_{12\cdots k-1})\mathbf{y}$
Residual	$n - \text{rank}(\mathbf{X}_2, \ldots, \mathbf{X}_k) - 1$	$\mathbf{y}^T(\mathbf{I} - \mathbf{P}_{12\cdots k})\mathbf{y}$
Total	$n - 1$	$\mathbf{y}^T[\mathbf{I} - (1/n)\mathbf{J}]\mathbf{y}$

Methodological Interlude #5: Variance Inflation Factors, Revisited

Consider the case of a full-rank, centered linear model that includes an intercept, with model matrix

$$\mathbf{X} = (\mathbf{1}_n, \mathbf{x}_2 - \bar{x}_2\mathbf{1}_n, \mathbf{x}_3 - \bar{x}_3\mathbf{1}_n, \ldots, \mathbf{x}_p - \bar{x}_p\mathbf{1}_n) = (\mathbf{1}_n, \mathbf{X}_{\bar{2}}),$$

say. Recall from Methodological Interlude #1 that under a Gauss–Markov version of such a model,

$$\text{var}(\hat{\beta}_j) = \frac{\sigma^2}{\sum_{i=1}^{n}(x_{ij} - \bar{x}_j)^2} \times VIF_j \qquad (j = 2, \ldots, p),$$

where VIF_j is the variance inflation factor corresponding to x_j. Also recall, from expression (7.2), that

$$VIF_2 = \left[1 - \frac{\mathbf{a}^T \mathbf{A}^{-1} \mathbf{a}}{\sum_{i=1}^n (x_{i2} - \bar{x}_2)^2}\right]^{-1},$$

where the $(k-2)$th element of $\mathbf{a}$ is $\sum_{i=1}^n (x_{i2} - \bar{x}_2)(x_{ik} - \bar{x}_k)$ $(k = 3, \ldots, p)$ and the $(k-2, l-2)$th element of $\mathbf{A}$ is $\sum_{i=1}^n (x_{ik} - \bar{x}_k)(x_{il} - \bar{x}_l)$ $(k, l = 3, \ldots, p)$, with analogous expressions for the variance inflation factors corresponding to the other explanatory variables.

Now define

$$\mathbf{X}_{\bar{3}} = (\mathbf{x}_3 - \bar{x}_3 \mathbf{1}_n, \ldots, \mathbf{x}_p - \bar{x}_p \mathbf{1}_n),$$

and observe that

$$\mathbf{a}^T = \mathbf{x}_2^T [\mathbf{I} - (1/n)\mathbf{J}] \mathbf{X}_{\bar{3}}$$

and

$$\mathbf{A} = \mathbf{X}_{\bar{3}}^T [\mathbf{I} - (1/n)\mathbf{J}] \mathbf{X}_{\bar{3}}.$$

Therefore,

$$\begin{aligned} \mathbf{a}^T \mathbf{A}^{-1} \mathbf{a} &= \mathbf{x}_2^T [\mathbf{I} - (1/n)\mathbf{J}] \mathbf{X}_{\bar{3}} \{\mathbf{X}_{\bar{3}}^T [(\mathbf{I} - (1/n)\mathbf{J}] \mathbf{X}_{\bar{3}}\}^{-1} \mathbf{X}_{\bar{3}}^T [\mathbf{I} - (1/n)\mathbf{J}] \mathbf{x}_2 \\ &\equiv \mathbf{x}_2^T \mathbf{B} \mathbf{x}_2, \end{aligned}$$

where $\mathbf{B}$ is the orthogonal projection matrix onto $\mathcal{C}\{[\mathbf{I} - (1/n)\mathbf{J}]\mathbf{X}_{\bar{3}}\}$. But by Theorem 9.1.2, this orthogonal projection matrix may be expressed as $\mathbf{P}_{(\mathbf{1}, \mathbf{x}_3, \ldots, \mathbf{x}_p)} - (1/n)\mathbf{J}$. Thus,

$$\frac{\mathbf{a}^T \mathbf{A}^{-1} \mathbf{a}}{\sum_{i=1}^n (x_{i2} - \bar{x}_2)^2} = \frac{\mathbf{x}_2^T [\mathbf{P}_{(\mathbf{1}, \mathbf{x}_3, \ldots, \mathbf{x}_p)} - (1/n)\mathbf{J}] \mathbf{x}_2}{\mathbf{x}_2^T [\mathbf{I} - (1/n)\mathbf{J}] \mathbf{x}_2} = R_2^2,$$

where R_2^2 is the coefficient of determination from a least squares regression of x_2 on the other explanatory variables (including the intercept). Thus,

$$VIF_2 = \frac{1}{1 - R_2^2}.$$

More generally, the jth variance inflation factor may be written as $1/(1 - R_j^2)$, where R_j^2 is the coefficient of determination from a least squares regression of x_j on the other explanatory variables. Thus, the higher the coefficient of determination in this regression, or equivalently the more strongly that x_j is correlated (in this sense) with the other explanatory variables, the more that the variance of $\hat{\beta}_j$ is inflated by the inclusion of those variables in the model for $\mathbf{y}$. ■

9.3.4 Conditions for Order-Invariance of a Sequential ANOVA

In general, the numerical values of the sums of squares in sequential ANOVAs corresponding to two ordered k-part models are not identical. For example, the sum of squares for $\mathbf{X}_1$ in the sequential ANOVA of the ordered two-part model $\{\mathbf{y}, \mathbf{X}_1\boldsymbol{\beta}_1 + \mathbf{X}_2\boldsymbol{\beta}_2\}$ is generally not equal to the sum of squares for $\mathbf{X}_1|\mathbf{X}_2$ in the sequential ANOVA of the ordered two-part model $\{\mathbf{y}, \mathbf{X}_2\boldsymbol{\beta}_2 + \mathbf{X}_1\boldsymbol{\beta}_1\}$. However, conditions exist under which these sums of squares are identical, implying that under those conditions it does not matter which ordered model (and which sequential ANOVA) one uses. This section obtains such conditions. We consider the two-part case first.

Inspection of the two sequential ANOVAs corresponding to the two ordered two-part models reveals that the sums of squares from these ANOVAs are identical (apart from order of listing) if and only if either (a) $\mathbf{y}^T\mathbf{P}_1\mathbf{y} = \mathbf{y}^T\mathbf{P}_2\mathbf{y}$ for all $\mathbf{y}$ or (b) $\mathbf{y}^T\mathbf{P}_1\mathbf{y} = \mathbf{y}^T(\mathbf{P}_{12} - \mathbf{P}_2)\mathbf{y}$ for all $\mathbf{y}$. By Theorem 2.14.1, Condition (a) occurs if and only $\mathbf{P}_1 = \mathbf{P}_2$. Clearly, $\mathbf{P}_1 = \mathbf{P}_2$ (in which case $\mathbf{P}_{12} - \mathbf{P}_1 = \mathbf{P}_{12} - \mathbf{P}_2 = \mathbf{0}$) if and only if $\mathcal{C}(\mathbf{X}_1) = \mathcal{C}(\mathbf{X}_2)$, i.e., if and only if each explanatory variable in $\mathbf{X}_2$ is a linear combination of explanatory variables in $\mathbf{X}_1$ and vice versa. In this case there is no reason to use the two-part model; the model $\{\mathbf{y}, \mathbf{X}_1\boldsymbol{\beta}_1\}$ will suffice. Condition (b) occurs if and only if $\mathbf{P}_1 = \mathbf{P}_{12} - \mathbf{P}_2$, or equivalently if and only if $\mathbf{P}_1 + \mathbf{P}_2 = \mathbf{P}_{12}$. This case is more interesting and is addressed by the following theorem.

Theorem 9.3.1 *Consider the ordered two-part models $\{\mathbf{y}, \mathbf{X}_1\boldsymbol{\beta}_1 + \mathbf{X}_2\boldsymbol{\beta}_2\}$ and $\{\mathbf{y}, \mathbf{X}_2\boldsymbol{\beta}_2 + \mathbf{X}_1\boldsymbol{\beta}_1\}$, and suppose that $\mathcal{C}(\mathbf{X}_1) \neq \mathcal{C}(\mathbf{X}_2)$. The sequential ANOVAs corresponding to these models are identical (apart from order of listing) if and only if $\mathbf{X}_1^T\mathbf{X}_2 = \mathbf{0}$, i.e., if and only if the columns of $\mathbf{X}_1$ are orthogonal to the columns of $\mathbf{X}_2$.*

Proof Let $\mathbf{B}_1$ and $\mathbf{B}_2$ represent matrices that satisfy the systems of equations $\mathbf{X}_1^T\mathbf{X}_1\mathbf{B}_1 = \mathbf{X}_1^T$ and $\mathbf{X}_2^T\mathbf{X}_2\mathbf{B}_2 = \mathbf{X}_2^T$, respectively. Equivalently, $\mathbf{B}_1$ and $\mathbf{B}_2$ satisfy

$$\begin{pmatrix} \mathbf{X}_1^T\mathbf{X}_1 & \mathbf{0} \\ \mathbf{0} & \mathbf{X}_2^T\mathbf{X}_2 \end{pmatrix} \begin{pmatrix} \mathbf{B}_1 \\ \mathbf{B}_2 \end{pmatrix} = \begin{pmatrix} \mathbf{X}_1^T \\ \mathbf{X}_2^T \end{pmatrix}. \tag{9.12}$$

But if $\mathbf{X}_1^T\mathbf{X}_2 = \mathbf{0}$, then Eqs. (9.2) and (9.12) are identical, so $\begin{pmatrix} \mathbf{B}_1 \\ \mathbf{B}_2 \end{pmatrix}$ also satisfies (9.2). Thus, $\mathbf{P}_{12} = \begin{pmatrix} \mathbf{X}_1 & \mathbf{X}_2 \end{pmatrix} \begin{pmatrix} \mathbf{B}_1 \\ \mathbf{B}_2 \end{pmatrix} = \mathbf{X}_1\mathbf{B}_1 + \mathbf{X}_2\mathbf{B}_2 = \mathbf{P}_1 + \mathbf{P}_2$, which proves the sufficiency. Conversely, if $\mathbf{P}_1 + \mathbf{P}_2 = \mathbf{P}_{12}$, then pre- and post-multiplying by $\mathbf{X}_1^T$ and $\mathbf{X}_2$, respectively, yields $\mathbf{X}_1^T\mathbf{X}_2 + \mathbf{X}_1^T\mathbf{X}_2 = \mathbf{X}_1^T\mathbf{X}_2$, implying that $\mathbf{X}_1^T\mathbf{X}_2 = \mathbf{0}$. □

Theorem 9.3.1 may be extended to the general case of a k-part model; doing so is left as an exercise.

Example 9.3.4-1. The Sequential ANOVA for Polynomial Regression

Consider the complete polynomial regression model in a single observed variable,

$$y_i = \beta_1 + \beta_2 x_i + \beta_3 x_i^2 + \cdots + \beta_p x_i^{p-1} + e_i \qquad (i = 1, \ldots, n).$$

Regarding this model as an ordered k-part model [with $k = p$, $\mathbf{X}_1 = \mathbf{1}_n$, and $\mathbf{X}_j = (x_1^{j-1}, \ldots, x_n^{j-1})^T$ for $j = 2, \ldots, p$] is natural because the permissible shapes of $\mathrm{E}(y)$ as a function of x get progressively "rougher," or more wiggly, as the order of the polynomial increases. Then each successive sum of squares in the sequential ANOVA has rank one and measures the extent to which the jth term in the model explains the variation in y beyond that which has been explained by the terms of order up to $j-1$. These sums of squares generally are not invariant to order of listing in the ANOVA; however, by the result of Exercise 9.8 they do have this invariance property if the model is reparameterized to the *orthogonal* polynomial regression model described in Example 7.3-3. ■

For three-part and higher-order models with an overall intercept term $\mathbf{1}\mu$, another interesting question is, under what conditions (if any) are the corrected sequential ANOVAs corresponding to the subset of ordered models in which $\mathbf{1}\mu$ is listed (and fitted) first identical (apart possibly from order of listing)? For instance, in a three-part model with $\mathbf{X}_1 = \mathbf{1}$ and $\mathcal{C}[(\mathbf{I} - \frac{1}{n}\mathbf{J})\mathbf{X}_2] \neq \mathcal{C}[(\mathbf{I} - \frac{1}{n}\mathbf{J})\mathbf{X}_3]$, the question is, when does $\mathrm{SS}(\mathbf{X}_2|\mathbf{1})$ equal $\mathrm{SS}(\mathbf{X}_2|\mathbf{1}, \mathbf{X}_3)$? A somewhat more general question is, when does $\mathrm{SS}(\mathbf{X}_2|\mathbf{X}_1)$ equal $\mathrm{SS}(\mathbf{X}_2|\mathbf{X}_1, \mathbf{X}_3)$ in a three-part model for which $\mathcal{C}[(\mathbf{I} - \mathbf{P}_1)\mathbf{X}_2] \neq \mathcal{C}[(\mathbf{I} - \mathbf{P}_1)\mathbf{X}_3]$? The answer to this question is provided by the following theorem; its proof is left as an exercise.

Theorem 9.3.2 *Consider the two ordered three-part models*

$$\mathbf{y} = \mathbf{X}_1\boldsymbol{\beta}_1 + \mathbf{X}_2\boldsymbol{\beta}_2 + \mathbf{X}_3\boldsymbol{\beta}_3 + \mathbf{e}$$

and

$$\mathbf{y} = \mathbf{X}_1\boldsymbol{\beta}_1 + \mathbf{X}_3\boldsymbol{\beta}_3 + \mathbf{X}_2\boldsymbol{\beta}_2 + \mathbf{e},$$

and suppose that $\mathcal{C}[(\mathbf{I} - \mathbf{P}_1)\mathbf{X}_2] \neq \mathcal{C}[(\mathbf{I} - \mathbf{P}_1)\mathbf{X}_3]$. *A necessary and sufficient condition for the sequential ANOVAs of these models to be identical (apart from order of listing) is*

$$\mathbf{X}_2^T(\mathbf{I} - \mathbf{P}_1)\mathbf{X}_3 = \mathbf{0}. \tag{9.13}$$

Example 9.3.4-2. Conditions for Order-Invariance of the Corrected Sequential ANOVA for the Two-Way Main Effects Model

Let us consider the two-way main effects model as an unordered three-part model with

$$\boldsymbol{\beta}_1 = \mu, \qquad \boldsymbol{\beta}_2 = (\alpha_1, \ldots, \alpha_q)^T, \qquad \text{and} \qquad \boldsymbol{\beta}_3 = (\gamma_1, \ldots, \gamma_m)^T,$$

and

$$\mathbf{X}_1 = \mathbf{1}_n, \quad \mathbf{X}_2 = \begin{pmatrix} \mathbf{1}_{n_{1\cdot}} & \mathbf{0}_{n_{1\cdot}} & \cdots & \mathbf{0}_{n_{1\cdot}} \\ \mathbf{0}_{n_{2\cdot}} & \mathbf{1}_{n_{2\cdot}} & \cdots & \mathbf{0}_{n_{2\cdot}} \\ \vdots & \vdots & \ddots & \vdots \\ \mathbf{0}_{n_{q\cdot}} & \mathbf{0}_{n_{q\cdot}} & \cdots & \mathbf{1}_{n_{q\cdot}} \end{pmatrix}, \quad \mathbf{X}_3 = \begin{pmatrix} \oplus_{j=1}^m \mathbf{1}_{n_{1j}} \\ \oplus_{j=1}^m \mathbf{1}_{n_{2j}} \\ \vdots \\ \oplus_{j=1}^m \mathbf{1}_{n_{qj}} \end{pmatrix},$$

where it is understood that there are no rows in $\mathbf{X}$ corresponding to cells for which $n_{ij} = 0$. By Theorem 9.3.2, the corrected sequential ANOVA corresponding to fitting the Factor A effects first followed by the Factor B effects is identical, apart from order of listing, to the corrected sequential ANOVA corresponding to fitting the Factor B effects first followed by the Factor A effects, if and only if

$$\mathbf{X}_2^T(\mathbf{I} - (1/n)\mathbf{J})\mathbf{X}_3 = \mathbf{0},$$

i.e., if and only if

$$\mathbf{X}_2^T\mathbf{X}_3 = (1/n)\mathbf{X}_2^T\mathbf{1}_n\mathbf{1}_n^T\mathbf{X}_3,$$

i.e., if and only if

$$\begin{pmatrix} n_{11} & n_{12} & \cdots & n_{1m} \\ n_{21} & n_{22} & \cdots & n_{2m} \\ \vdots & \vdots & \vdots & \vdots \\ n_{q1} & n_{21} & \cdots & n_{qm} \end{pmatrix} = (1/n) \begin{pmatrix} n_{1\cdot} \\ n_{2\cdot} \\ \vdots \\ n_{q\cdot} \end{pmatrix} \begin{pmatrix} n_{\cdot 1} & n_{\cdot 2} & \cdots & n_{\cdot m} \end{pmatrix},$$

i.e., if and only if

$$n_{ij} = \frac{n_{i\cdot}n_{\cdot j}}{n} \qquad (i = 1, \ldots, q;\ j = 1, \ldots, m),$$

or in words, if and only if the cell frequencies are proportional. Equal cell frequencies are not required for the desired invariance, but they are sufficient. ■

Example 9.3.4-3. The Corrected Sequential ANOVA for the Two-Way Main Effects Model with One Observation per Cell

Consider the two-way main effects model with one observation per cell, for which only two subscripts are needed:

$$y_{ij} = \mu + \alpha_i + \gamma_j + e_{ij} \qquad (i = 1, \ldots, q;\ j = 1, \ldots, m).$$

By Example 9.3.3-2, the sum of squares for the Factor A effects in the corrected sequential ANOVA corresponding to fitting those effects first is $m \sum_{i=1}^{q} (\bar{y}_{i\cdot} - \bar{y}_{\cdot\cdot})^2$. By the same result, the sum of squares for the Factor B effects in the corrected sequential ANOVA corresponding to fitting those effects first is $q \sum_{j=1}^{m} (\bar{y}_{\cdot j} - \bar{y}_{\cdot\cdot})^2$. By the invariance result established in Example 9.3.4-2, the sums of squares corresponding to these two factors are invariant to the order in which the factors are fitted. Thus, the corrected sequential ANOVA for the model with Factor A effects fitted first is as follows:

Source	Rank	Sum of squares	Expected mean square
Factor A	$q - 1$	$m \sum_{i=1}^{q} (\bar{y}_{i\cdot} - \bar{y}_{\cdot\cdot})^2$	$\sigma^2 + \frac{m}{q-1} \sum_{i=1}^{q} (\alpha_i - \bar{\alpha})^2$
Factor B	$m - 1$	$q \sum_{j=1}^{m} (\bar{y}_{\cdot j} - \bar{y}_{\cdot\cdot})^2$	$\sigma^2 + \frac{q}{m-1} \sum_{j=1}^{m} (\gamma_j - \bar{\gamma})^2$
Residual	$(q - 1)(m - 1)$	By subtraction	σ^2
Corrected Total	$qm - 1$	$\sum_{i=1}^{q} \sum_{j=1}^{m} (y_{ij} - \bar{y}_{\cdot\cdot})^2$	

where $\bar{\alpha} = \frac{1}{q} \sum_{i=1}^{q} \alpha_i$ and $\bar{\gamma} = \frac{1}{m} \sum_{j=1}^{m} \gamma_j$. The corrected sequential ANOVA for the model with Factor B effects fitted first is identical, apart from interchanging the rows labeled "Factor A" and "Factor B." Verifying the given expressions for the expected mean squares associated with Factor A and Factor B is left as an exercise. ■

9.4 The Analysis of Covariance

Recall from Sect. 5.1.5 that an analysis-of-covariance model is a model that includes quantitative explanatory variables and classificatory explanatory variables. Such a model may be regarded as a two-part model, where $\mathbf{X}_1$ comprises the values of the classificatory variables and $\mathbf{X}_2$ the values of the quantitative variables, or vice versa. An analysis of covariance is merely a sequential analysis of variance of such a model. Here we consider a specific case of an analysis of covariance in which there is a single factor of classification and a single quantitative variable; some other cases are considered in the exercises. Thus the model considered here is as follows:

$$y_{ij} = \mu + \alpha_i + \gamma z_{ij} + e_{ij} \quad (i = 1, \ldots, q;\ j = 1, \ldots, n_i).$$

The z_{ij}'s represent values of the single quantitative variable (z) and γ is the "slope parameter" associated with z. This model can be formulated as a special case of the two-part model

$$\mathbf{y} = \mathbf{X}_1\boldsymbol{\beta}_1 + \mathbf{z}\gamma + \mathbf{e},$$

where $\mathbf{X}_1 = (\mathbf{1}_n, \oplus_{i=1}^q \mathbf{1}_{n_i})$, $\boldsymbol{\beta}_1 = (\mu, \alpha_1, \ldots, \alpha_q)^T$, $\mathbf{z}$ (which is $\mathbf{X}_2$ by another name) is the n-vector of z_{ij}'s, and γ is $\boldsymbol{\beta}_2$ by another name. The corresponding two-part normal equations simplify to

$$\begin{pmatrix} n & \mathbf{n}^T \\ \mathbf{n} & \oplus_{i=1}^q n_i \end{pmatrix} \boldsymbol{\beta}_1 + \begin{pmatrix} z_{\cdot\cdot} \\ z_{1\cdot} \\ \vdots \\ z_{q\cdot} \end{pmatrix} \gamma = \begin{pmatrix} y_{\cdot\cdot} \\ y_{1\cdot} \\ \vdots \\ y_{q\cdot} \end{pmatrix},$$

$$(z_{\cdot\cdot}, z_{1\cdot}, \ldots, z_{q\cdot})\boldsymbol{\beta}_1 + \sum_{i=1}^q \sum_{j=1}^{n_i} z_{ij}^2 \gamma = \sum_{i=1}^q \sum_{j=1}^{n_i} z_{ij} y_{ij},$$

where $\mathbf{n}^T = (n_1, \ldots, n_q)$. Because γ is a scalar, there is only one reduced normal equation for γ, which is

$$\mathbf{z}^T[\mathbf{I} - \oplus_{i=1}^q (1/n_i)\mathbf{J}_{n_i}]\mathbf{z}\gamma = \mathbf{z}^T[\mathbf{I} - \oplus_{i=1}^q (1/n_i)\mathbf{J}_{n_i}]\mathbf{y}.$$

(Here we have used the expression for the orthogonal projection matrix onto $\mathcal{C}(\mathbf{X}_1)$ that was derived in Example 8.1-3). Equivalently, in nonmatrix notation, the reduced normal equation for γ is

$$\sum_{i=1}^q \sum_{j=1}^{n_i} (z_{ij} - \bar{z}_{i\cdot})^2 \gamma = \sum_{i=1}^q \sum_{j=1}^{n_i} (z_{ij} - \bar{z}_{i\cdot})(y_{ij} - \bar{y}_{i\cdot}).$$

Assuming that the inverse of the coefficient matrix of this equation exists (which is almost always the case in practice), the unique solution to the reduced normal equation is

$$\hat{\gamma} = \frac{\sum_{i=1}^q \sum_{j=1}^{n_i} (z_{ij} - \bar{z}_{i\cdot})(y_{ij} - \bar{y}_{i\cdot})}{\sum_{i=1}^q \sum_{j=1}^{n_i} (z_{ij} - \bar{z}_{i\cdot})^2}. \tag{9.14}$$

Back-solving for $\boldsymbol{\beta}_1$ yields a solution

$$\hat{\boldsymbol{\beta}}_1 = \begin{pmatrix} \hat{\mu} \\ \hat{\alpha}_1 \\ \vdots \\ \hat{\alpha}_q \end{pmatrix} = \begin{pmatrix} 0 & \mathbf{0}^T \\ \mathbf{0} & \oplus_{i=1}^q (1/n_i) \end{pmatrix} \begin{pmatrix} y_{\cdot\cdot} - \hat{\gamma} z_{\cdot\cdot} \\ y_{1\cdot} - \hat{\gamma} z_{1\cdot} \\ \vdots \\ y_{q\cdot} - \hat{\gamma} z_{q\cdot} \end{pmatrix} = \begin{pmatrix} 0 \\ \bar{y}_{1\cdot} - \hat{\gamma} \bar{z}_{1\cdot} \\ \vdots \\ \bar{y}_{q\cdot} - \hat{\gamma} \bar{z}_{q\cdot} \end{pmatrix}.$$

There are certain similarities between the form of these least squares estimators and the least squares estimators corresponding to a simple linear regression model. In particular, the numerator of $\hat{\gamma}$ is the sum of the numerators of the least squares estimators of slope under a simple linear regression model for the data at each factor level, and the denominator is the sum of the denominators of those same slope estimators. Thus, $\hat{\gamma}$ is a "pooled" (over the factor levels) slope estimator. Also, the least squares estimators $\hat{\mu} + \hat{\alpha}_i = \bar{y}_{i\cdot} - \hat{\gamma}\bar{z}_{i\cdot}$ of the level means may be viewed as estimators of the intercepts of simple linear regression models for the data at each factor level. These latter estimators are sometimes called "adjusted means" or "least squares means," as they have the form of a sample mean of the y-observations at level i adjusted by subtracting the product of the corresponding slope estimator and sample mean of the z-observations.

In order to obtain the corrected sequential ANOVAs corresponding to this model, it is helpful to think of it now as a three-part model in which $\mathbf{X}_1 = \mathbf{1}_n$, $\mathbf{X}_2 = \oplus_{i=1}^q \mathbf{1}_{n_i}$, and $\mathbf{X}_3 = \mathbf{z}$. The corrected sequential ANOVAs corresponding to the two ordered three-part models in which $\mathbf{1}\mu$ is fitted first may be obtained as follows, where we note that in this instance $\mathbf{1} \in \mathcal{C}(\mathbf{X}_2)$:

Source	Rank	Sum of squares
Classes	$q-1$	$\sum_{i=1}^q n_i(\bar{y}_{i\cdot} - \bar{y}_{\cdot\cdot})^2$
$\mathbf{z}$|Classes	1	$\hat{\gamma}\sum_{i=1}^q \sum_{j=1}^{n_i} (z_{ij} - \bar{z}_{i\cdot})(y_{ij} - \bar{y}_{i\cdot})$
Residual	$n-q-1$	By subtraction
Corrected Total	$n-1$	$\sum_{i=1}^q \sum_{j=1}^{n_i} y_{ij}^2 - n\bar{y}_{\cdot\cdot}^2$

Source	Rank	Sum of squares
$\mathbf{z}$	1	$\frac{\left[\sum_{i=1}^q \sum_{j=1}^{n_i} (y_{ij}-\bar{y}_{\cdot\cdot})(z_{ij}-\bar{z}_{\cdot\cdot})\right]^2}{\sum_{i=1}^q \sum_{j=1}^{n_i} (z_{ij}-\bar{z}_{\cdot\cdot})^2}$
Classes|$\mathbf{z}$	$q-1$	$\sum_{i=1}^q n_i(\bar{y}_{i\cdot} - \bar{y}_{\cdot\cdot})^2 + \hat{\gamma}\sum_{i=1}^q \sum_{j=1}^{n_i} (z_{ij} - \bar{z}_{i\cdot})(y_{ij} - \bar{y}_{i\cdot}) - \frac{\left[\sum_{i=1}^q \sum_{j=1}^{n_i} (y_{ij}-\bar{y}_{\cdot\cdot})(z_{ij}-\bar{z}_{\cdot\cdot})\right]^2}{\sum_{i=1}^q \sum_{j=1}^{n_i} (z_{ij}-\bar{z}_{\cdot\cdot})^2}$
Residual	$n-q-1$	By subtraction
Corrected Total	$n-1$	$\sum_{i=1}^q \sum_{j=1}^{n_i} y_{ij}^2 - n\bar{y}_{\cdot\cdot}^2$

In the first of these sequential ANOVAs, the expression for the classes sum of squares was copied from the corrected overall ANOVA table for the one-factor model given in Example 9.3.3-2, while the expression for SS($\mathbf{z}$|Classes) was obtained using (9.10). An alternative expression for SS($\mathbf{z}$|Classes), obtained using (9.14), is $\hat{\gamma}^2 \sum_{i=1}^{q} \sum_{j=1}^{n_i} (z_{ij} - \bar{z}_{i\cdot})^2$. In the second sequential ANOVA, the expression for SS($\mathbf{z}$) was obtained from the corrected overall ANOVA table for simple linear regression given in Example 9.3.3-1, while the expression for SS(Classes|$\mathbf{z}$) is merely the difference between the complete model sum of squares [obtained by summing SS(Classes) and SS($\mathbf{z}$|Classes) from the first sequential ANOVA] and SS($\mathbf{z}$). Of the two sequential ANOVAs, the second is usually of greater practical interest, as it allows for testing for differences among the factor level means in a way that adjusts for the values of the covariate observed at those levels.

9.5 Partitioning the Residual Sum of Squares

As noted in Example 5.2.2-1, it is possible that among the n y-observations, some correspond to identical combinations of the explanatory variables. If so, then some of the rows of the model matrix $\mathbf{X}$ are identical to each other, and $\mathbf{X}$ may be written (by permuting rows, if necessary) as follows:

$$\mathbf{X} = \begin{pmatrix} \mathbf{1}_{n_1}\mathbf{x}_1^T \\ \mathbf{1}_{n_2}\mathbf{x}_2^T \\ \vdots \\ \mathbf{1}_{n_m}\mathbf{x}_m^T \end{pmatrix},$$

where $\mathbf{x}_1^T, \ldots, \mathbf{x}_m^T$ are the distinct rows of $\mathbf{X}$ (i.e., the distinct combinations of the explanatory variables), n_i is the number of replications of the ith combination, and $\sum_{i=1}^{m} n_i = n$. Assume that $p^* < m < n$.

Let y_{ij} denote the jth observed y-value associated with $\mathbf{x}_i^T$ ($j = 1, \ldots, n_i$; $i = 1, \ldots, m$). The vector of least squares fitted residuals, $(\mathbf{I} - \mathbf{P_X})\mathbf{y}$, may be decomposed in this case as follows:

$$(\mathbf{I} - \mathbf{P_X})\mathbf{y} = (\mathbf{I} - \bar{\mathbf{J}})\mathbf{y} + (\bar{\mathbf{J}} - \mathbf{P_X})\mathbf{y}, \tag{9.15}$$

where $\bar{\mathbf{J}} = \oplus_{i=1}^{m} (1/n_i)\mathbf{J}_{n_i}$. The following theorem pertains to this decomposition.

Theorem 9.5.1 *For the $n \times n$ matrix $\bar{\mathbf{J}}$ just defined, the following results hold:*

(a) $\bar{\mathbf{J}}$ is symmetric and idempotent, and rank($\bar{\mathbf{J}}$) $= m$;
(b) $\mathbf{I} - \bar{\mathbf{J}}$ is symmetric and idempotent, and rank($\mathbf{I} - \bar{\mathbf{J}}$) $= n - m$;
(c) $\bar{\mathbf{J}}\mathbf{X} = \mathbf{X}$;
(d) $\bar{\mathbf{J}}\mathbf{P_X} = \mathbf{P_X}$;

(e) $(\mathbf{I} - \bar{\mathbf{J}})(\bar{\mathbf{J}} - \mathbf{P_X}) = \mathbf{0}$;
(f) $\bar{\mathbf{J}} - \mathbf{P_X}$ *is symmetric and idempotent, and rank*$(\bar{\mathbf{J}} - \mathbf{P_X}) = m - p^*$.

Proof Left as an exercise. □

By Theorem 9.5.1e, (9.15) is an orthogonal decomposition of $(\mathbf{I} - \mathbf{P_X})\mathbf{y}$. Furthermore, pre-multiplication of (9.15) by $\mathbf{y}^T$ yields the decomposition

$$\mathbf{y}^T(\mathbf{I} - \mathbf{P_X})\mathbf{y} = \mathbf{y}^T(\mathbf{I} - \bar{\mathbf{J}})\mathbf{y} + \mathbf{y}^T(\bar{\mathbf{J}} - \mathbf{P_X})\mathbf{y} \tag{9.16}$$

of the residual sum of squares. By Theorem 9.5.1b, f, the two terms on the right-hand side of (9.16) are the sums of squares, respectively, of vectors $(\mathbf{I} - \bar{\mathbf{J}})\mathbf{y}$ and $(\bar{\mathbf{J}} - \mathbf{P_X})\mathbf{y}$, and the corresponding ranks are $n - m$ and $m - p^*$, respectively. Within the context of the overall ANOVA, this orthogonal decomposition may be laid out in tabular form as follows:

Source	Rank	Sum of squares
$\mathbf{X}$	p^*	$\mathbf{y}^T\mathbf{P_X}\mathbf{y}$
Pure error	$n - m$	$\mathbf{y}^T(\mathbf{I} - \bar{\mathbf{J}})\mathbf{y}$
Lack of fit	$m - p^*$	$\mathbf{y}^T(\bar{\mathbf{J}} - \mathbf{P_X})\mathbf{y}$
Total	n	$\mathbf{y}^T\mathbf{y}$

Equivalently, using nonmatrix notation, the corrected ANOVA [assuming that $\mathbf{1} \notin \mathcal{C}(\mathbf{X}_2)$] is as follows:

Source	Rank	Sum of squares
$\mathbf{X}_2$	$p^* - 1$	$\sum_{i=1}^m n_i(\hat{y}_i - \bar{y}_{\cdot\cdot})^2$
Pure error	$n - m$	$\sum_{i=1}^m \sum_{j=1}^{n_i}(y_{ij} - \bar{y}_{i\cdot})^2$
Lack of fit	$m - p^*$	$\sum_{i=1}^m n_i(\bar{y}_{i\cdot} - \hat{y}_i)^2$
Corrected Total	$n - 1$	$\sum_{i=1}^m \sum_{j=1}^{n_i}(y_{ij} - \bar{y}_{\cdot\cdot})^2$

The labels "Pure error" and "Lack of fit" for the second and third lines of these tables are motivated by the results of the following theorem, which considers the expectation of the sums of squares in those lines under two scenarios. Under Scenario I, the expectation of y_{ij} is exactly what the linear model $\mathbf{y} = \mathbf{X}\boldsymbol{\beta} + \mathbf{e}$ purports it to be, i.e., $\mathrm{E}_I(y_{ij}) = \mathbf{x}_i^T\boldsymbol{\beta}$. Under Scenario II, the expectation of y_{ij} is $f(\mathbf{x}_i)$, where $f(\cdot)$ is an arbitrary (possibly even nonlinear) function of p variables. As a consequence, $\mathrm{E}_{II}(y_{ij}) = \mathrm{E}_{II}(y_{i1}) = f(\mathbf{x}_i)$ for all i and $j = 1, \ldots, n_i$.

Theorem 9.5.2 *Under Gauss–Markov assumptions:*

(a) $E_I[\mathbf{y}^T(\mathbf{I}-\bar{\mathbf{J}})\mathbf{y}]/(n-m)=\sigma^2$ *and* $E_I[\mathbf{y}^T(\bar{\mathbf{J}}-\mathbf{P_X})\mathbf{y}]/(m-p^*)=\sigma^2$;
(b) $E_{II}[\mathbf{y}^T(\mathbf{I}-\bar{\mathbf{J}})\mathbf{y}]/(n-m)=\sigma^2$ *and* $E_{II}[\mathbf{y}^T(\bar{\mathbf{J}}-\mathbf{P_X})\mathbf{y}]/(m-p^*)\geq\sigma^2$.

Proof Under Gauss–Markov assumptions, and using Theorems 4.2.4 and 9.5.1b, c, f,

$$\begin{aligned}\mathrm{E}_I[\mathbf{y}^T(\mathbf{I}-\bar{\mathbf{J}})\mathbf{y}] &= (\mathbf{X}\boldsymbol{\beta})^T(\mathbf{I}-\bar{\mathbf{J}})\mathbf{X}\boldsymbol{\beta}+\mathrm{tr}[(\mathbf{I}-\bar{\mathbf{J}})(\sigma^2\mathbf{I})]\\ &= 0+\sigma^2\mathrm{rank}(\mathbf{I}-\bar{\mathbf{J}})\\ &= \sigma^2(n-m)\end{aligned}$$

and

$$\begin{aligned}\mathrm{E}_I[\mathbf{y}^T(\bar{\mathbf{J}}-\mathbf{P_X})\mathbf{y}] &= (\mathbf{X}\boldsymbol{\beta})^T(\bar{\mathbf{J}}-\mathbf{P_X})\mathbf{X}\boldsymbol{\beta}+\mathrm{tr}[(\bar{\mathbf{J}}-\mathbf{P_X})(\sigma^2\mathbf{I})]\\ &= 0+\sigma^2\mathrm{rank}(\bar{\mathbf{J}}-\mathbf{P_X})\\ &= \sigma^2(m-p^*).\end{aligned}$$

Similarly,

$$\begin{aligned}\mathrm{E}_{II}[\mathbf{y}^T(\mathbf{I}-\bar{\mathbf{J}})\mathbf{y}] &= \begin{pmatrix}\mathbf{1}_{n_1}f(\mathbf{x}_1)\\ \mathbf{1}_{n_2}f(\mathbf{x}_2)\\ \vdots\\ \mathbf{1}_{n_m}f(\mathbf{x}_m)\end{pmatrix}^T(\mathbf{I}-\bar{\mathbf{J}})\begin{pmatrix}\mathbf{1}_{n_1}f(\mathbf{x}_1)\\ \mathbf{1}_{n_2}f(\mathbf{x}_2)\\ \vdots\\ \mathbf{1}_{n_m}f(\mathbf{x}_m)\end{pmatrix}+\mathrm{tr}[(\mathbf{I}-\bar{\mathbf{J}})(\sigma^2\mathbf{I})]\\ &= 0+\sigma^2\mathrm{rank}(\mathbf{I}-\bar{\mathbf{J}})\\ &= \sigma^2(n-m),\end{aligned}$$

but

$$\begin{aligned}&\mathrm{E}_{II}[\mathbf{y}^T(\bar{\mathbf{J}}-\mathbf{P_X})\mathbf{y}]\\ &= \begin{pmatrix}\mathbf{1}_{n_1}f(\mathbf{x}_1)\\ \mathbf{1}_{n_2}f(\mathbf{x}_2)\\ \vdots\\ \mathbf{1}_{n_m}f(\mathbf{x}_m)\end{pmatrix}^T(\bar{\mathbf{J}}-\mathbf{P_X})\begin{pmatrix}\mathbf{1}_{n_1}f(\mathbf{x}_1)\\ \mathbf{1}_{n_2}f(\mathbf{x}_2)\\ \vdots\\ \mathbf{1}_{n_m}f(\mathbf{x}_m)\end{pmatrix}+\mathrm{tr}[(\bar{\mathbf{J}}-\mathbf{P_X})(\sigma^2\mathbf{I})]\end{aligned}$$

$$
= \begin{pmatrix} \mathbf{1}_{n_1}\mathbf{x}_1^T\boldsymbol{\beta} + \mathbf{1}_{n_1}[f(\mathbf{x}_1) - \mathbf{x}_1^T\boldsymbol{\beta}] \\ \mathbf{1}_{n_2}\mathbf{x}_2^T\boldsymbol{\beta} + \mathbf{1}_{n_2}[f(\mathbf{x}_2) - \mathbf{x}_2^T\boldsymbol{\beta}] \\ \vdots \\ \mathbf{1}_{n_m}\mathbf{x}_m^T\boldsymbol{\beta} + \mathbf{1}_{n_m}[f(\mathbf{x}_m) - \mathbf{x}_m^T\boldsymbol{\beta}] \end{pmatrix}^T (\bar{\mathbf{J}} - \mathbf{P_X}) \begin{pmatrix} \mathbf{1}_{n_1}\mathbf{x}_1^T\boldsymbol{\beta} + \mathbf{1}_{n_1}[f(\mathbf{x}_1) - \mathbf{x}_1^T\boldsymbol{\beta}] \\ \mathbf{1}_{n_2}\mathbf{x}_2^T\boldsymbol{\beta} + \mathbf{1}_{n_2}[f(\mathbf{x}_2) - \mathbf{x}_2^T\boldsymbol{\beta}] \\ \vdots \\ \mathbf{1}_{n_m}\mathbf{x}_m^T\boldsymbol{\beta} + \mathbf{1}_{n_m}[f(\mathbf{x}_m) - \mathbf{x}_m^T\boldsymbol{\beta}] \end{pmatrix}
$$

$$
+\sigma^2\text{rank}(\bar{\mathbf{J}} - \mathbf{P_X})
$$

$$
= \begin{pmatrix} \mathbf{1}_{n_1}[f(\mathbf{x}_1) - \mathbf{x}_1^T\boldsymbol{\beta}] \\ \mathbf{1}_{n_2}[f(\mathbf{x}_2) - \mathbf{x}_2^T\boldsymbol{\beta}] \\ \vdots \\ \mathbf{1}_{n_m}[f(\mathbf{x}_m) - \mathbf{x}_m^T\boldsymbol{\beta}] \end{pmatrix}^T (\bar{\mathbf{J}} - \mathbf{P_X}) \begin{pmatrix} \mathbf{1}_{n_1}[f(\mathbf{x}_1) - \mathbf{x}_1^T\boldsymbol{\beta}] \\ \mathbf{1}_{n_2}[f(\mathbf{x}_2) - \mathbf{x}_2^T\boldsymbol{\beta}] \\ \vdots \\ \mathbf{1}_{n_m}[f(\mathbf{x}_m) - \mathbf{x}_m^T\boldsymbol{\beta}] \end{pmatrix} + \sigma^2(m - p^*)
$$

$$
= K + \sigma^2(m - p^*),
$$

where $K \geq 0$ by Corollary 2.15.9.1. □

By Theorem 9.5.2, $[\mathbf{y}^T(\mathbf{I} - \bar{\mathbf{J}})\mathbf{y}]/(n - m)$ is an unbiased estimator of σ^2 under each of the two scenarios, but the same cannot be said of $[\mathbf{y}^T(\bar{\mathbf{J}} - \mathbf{P_X})\mathbf{y}]/(m - p^*)$. Accordingly, the first term on the right-hand side of (9.16) is frequently called the *pure error sum of squares*. The second term is called the *lack-of-fit sum of squares* because the farther that the $\mathbf{x}_i^T\boldsymbol{\beta}$'s are from the $f(\mathbf{x}_i)$'s, the larger this term tends to be. In Chap. 15 we will see how these two sums of squares may be used to assess the adequacy of the assumed functional form for $\text{E}(y_{ij})$.

9.6 Exercises

1. Prove Theorem 9.1.1.
2. Prove Theorem 9.1.3.
3. For the ordered k-part Gauss–Markov model $\{\mathbf{y}, \sum_{l=1}^k \mathbf{X}_l\boldsymbol{\beta}_l, \sigma^2\mathbf{I}\}$, show that $\text{E}[\mathbf{y}^T(\mathbf{P}_{12\cdots j} - \mathbf{P}_{12\cdots j-1})\mathbf{y}] = (\sum_{l=j}^k \mathbf{X}_l\boldsymbol{\beta}_l)^T(\mathbf{P}_{12\cdots j} - \mathbf{P}_{12\cdots j-1})(\sum_{l=j}^k \mathbf{X}_l\boldsymbol{\beta}_l) + \sigma^2[\text{rank}(\mathbf{X}_1, \ldots, \mathbf{X}_j) - \text{rank}(\mathbf{X}_1, \ldots, \mathbf{X}_{j-1})]$.
4. Verify the expressions for the expected mean squares in the corrected sequential ANOVA for the two-way main effects model with one observation per cell given in Example 9.3.4-3.
5. Obtain the corrected sequential ANOVA for a balanced (r observations per cell) two-way main effects model, with Factor A fitted first (but after the overall mean, of course). Give nonmatrix expressions for the sums of squares in this ANOVA table and for the corresponding expected mean squares (under a Gauss–Markov version of the model).

6. For the balanced ($r \geq 2$ observations per cell) two-way model with interaction:
   (a) Obtain the corrected sequential ANOVA with Factor A fitted first (but after the overall mean, of course), then Factor B, then the interaction effects. Give nonmatrix expressions for the sums of squares in this ANOVA table and for the corresponding expected mean squares (under a Gauss–Markov version of the model).
   (b) Would the corrected sequential ANOVA corresponding to the ordered model in which the overall mean is fitted first, then Factor B, then Factor A, then the interaction be the same (apart from order) as the sequential ANOVA of part (a)? Justify your answer.
7. Obtain the corrected sequential ANOVA for a two-factor nested model in which the Factor A effects are fitted first (but after the overall mean, of course). Give nonmatrix expressions for the sums of squares in this ANOVA table and for the corresponding expected mean squares (under a Gauss–Markov version of the model).
8. Suppose, in a k-part model, that $\mathcal{C}(\mathbf{X}_j) \neq \mathcal{C}(\mathbf{X}_{j'})$ for all $j \neq j'$. Prove that the sequential ANOVAs corresponding to all ordered k-part models are identical (apart from order of listing) if and only if $\mathbf{X}_j^T \mathbf{X}_{j'} = \mathbf{0}$ for all $j \neq j'$.
9. Prove Theorem 9.3.2.
10. Consider the linear model specified in Exercise 7.3.
    (a) Obtain the overall ANOVA (Source, Rank, and Sum of squares) for this model. Give nonmatrix expressions for the model and total sums of squares (the residual sum of squares may be obtained by subtraction).
    (b) Obtain the sequential ANOVA (Source, Rank, and Sum of squares) for the ordered two-part model $\{\mathbf{y}, \mathbf{X}_1\boldsymbol{\beta}_1 + \mathbf{X}_2\boldsymbol{\beta}_2\}$, where $\mathbf{X}_1$ is the submatrix of $\mathbf{X}$ consisting of its first two columns and $\mathbf{X}_2$ is the submatrix of $\mathbf{X}$ consisting of its last two columns. (Again, give nonmatrix expressions for all sums of squares except the residual sum of squares).
    (c) Would the sequential ANOVA for the ordered two-part model $\{\mathbf{y}, \mathbf{X}_2\boldsymbol{\beta}_2 + \mathbf{X}_1\boldsymbol{\beta}_1\}$ be identical to the ANOVA you obtained in part (b), apart from order of listing? Justify your answer.
11. Consider the two-way partially crossed model introduced in Example 5.1.4-1, with one observation per cell, i.e.,

$$y_{ij} = \mu + \alpha_i - \alpha_j + e_{ij} \quad (i \neq j = 1, \ldots, q).$$

    (a) Would the sequential ANOVAs corresponding to the ordered two-part models in which the overall mean is fitted first and last be identical (apart from order of listing)? Justify your answer.
    (b) Obtain the sequential ANOVA (Source, Rank, and Sum of squares) corresponding to the ordered two-part version of this model in which the overall mean is fitted first. Give nonmatrix expressions for the sums of squares and the corresponding expected mean squares (under a Gauss–Markov version of the model).

12. Consider the two corrected sequential ANOVAs for the analysis-of-covariance model having a single factor of classification and a single quantitative variable, which was described in Sect. 9.4.
    (a) Obtain nonmatrix expressions for the expected mean squares (under a Gauss–Markov version of the model) in both ANOVAs.
    (b) Obtain a necessary and sufficient condition for the two ANOVAs to be identical (apart from order of listing).
13. Consider the analysis-of-covariance model having a single factor of classification with q levels and a single quantitative variable, i.e.,

$$y_{ij} = \mu + \alpha_i + \gamma z_{ij} + e_{ij} \qquad (i = 1, \ldots, q;\ j = 1, \ldots, n_i),$$

where the e_{ij}'s satisfy Gauss–Markov assumptions. Assume that $n_i \geq 2$ and $z_{i1} \neq z_{i2}$ for all $i = 1, \ldots, q$. In Sect. 9.4 it was shown that one solution to the normal equations for this model is $\hat{\mu} = 0$, $\hat{\alpha}_i = \bar{y}_{i\cdot} - \hat{\gamma}\bar{z}_{i\cdot}$ $(i = 1, \ldots, q)$, and

$$\hat{\gamma} = \frac{\sum_{i=1}^{q} \sum_{j=1}^{n_i} (z_{ij} - \bar{z}_{i\cdot}) y_{ij}}{\sum_{i=1}^{q} \sum_{j=1}^{n_i} (z_{ij} - \bar{z}_{i\cdot})^2}.$$

Thus, the least squares estimators of $\mu + \alpha_i$ are $\bar{y}_{i\cdot} - \hat{\gamma}\bar{z}_{i\cdot}$ $(i = 1, \ldots, q)$. Obtain nonmatrix expressions for:
    (a) $\text{var}(\hat{\gamma})$
    (b) $\text{var}(\bar{y}_{i\cdot} - \hat{\gamma}\bar{z}_{i\cdot})$ $\quad (i = 1, \ldots, q)$
    (c) $\text{cov}(\bar{y}_{i\cdot} - \hat{\gamma}\bar{z}_{i\cdot}, \bar{y}_{i'\cdot} - \hat{\gamma}\bar{z}_{i'\cdot})$ $(i' > i = 1, \ldots, q)$
14. Consider the one-factor, factor-specific-slope analysis-of-covariance model

$$y_{ij} = \mu + \alpha_i + \gamma_i (z_{ij} - \bar{z}_{i\cdot}) + e_{ij} \quad (i = 1, \ldots, q;\ j = 1, \ldots, n_i).$$

and assume that $z_{i1} \neq z_{i2}$ for all i.
    (a) Obtain the least squares estimators of $\mu + \alpha_i$ and γ_i $(i = 1, \ldots, q)$.
    (b) Obtain the corrected sequential ANOVA corresponding to the ordered version of this model in which $\mathbf{Z}\boldsymbol{\gamma}$ appears first (but after the overall mean, of course), where $\mathbf{Z} = \oplus_{i=1}^{q}(\mathbf{z}_i - \bar{z}_{i\cdot}\mathbf{1}_{n_i})$ and $\mathbf{z}_i = (z_{i1}, z_{i2}, \ldots, z_{in_i})^T$. Assume that $\mathbf{z}_i$ has at least two distinct elements for each i. Give nonmatrix expressions for the sums of squares in the ANOVA table.
15. Consider the balanced two-way main effects analysis-of-covariance model with one observation per cell, i.e.,

$$y_{ij} = \mu + \alpha_i + \gamma_j + \xi z_{ij} + e_{ij} \quad (i = 1, \ldots, q;\ j = 1, \ldots, m).$$

    (a) Obtain expressions for the least squares estimators of $\alpha_i - \alpha_{i'}$, $\gamma_j - \gamma_{j'}$, and ξ.
    (b) Obtain the corrected sequential ANOVAs corresponding to the two ordered four-part models in which $\mathbf{1}\mu$ appears first and $\mathbf{z}\xi$ appears second.

16. Consider an analysis-of-covariance extension of the model in Exercise 9.6, written in cell-means form, as

$$y_{ijk} = \mu_{ij} + \gamma z_{ijk} + e_{ijk} \quad (i = 1, \ldots, q;\ j = 1, \ldots, m;\ k = 1, \ldots, r).$$

   (a) Give the reduced normal equation for γ (in either matrix or nonmatrix form), and assuming that the inverse of its coefficient "matrix" exists, obtain the unique solution to the reduced normal equation in nonmatrix form.
   (b) Back-solve to obtain a solution for the μ_{ij}'s.
   (c) Using the overall ANOVA from Exercise 9.15 as a starting point, give the sequential ANOVA (sums of squares and ranks of the corresponding matrices), uncorrected for the mean, for the ordered two-part model $\{\mathbf{y}, \mathbf{X}_1\boldsymbol{\beta}_1 + \mathbf{z}\gamma\}$, where $\mathbf{X}_1$ is the model matrix without the covariate and $\mathbf{z}$ is the vector of covariates.
17. Prove Theorem 9.5.1.

Constrained Least Squares Estimation and ANOVA

10

In our consideration of least squares estimation up to this point, $\boldsymbol{\beta}$ was unrestricted, i.e., $\boldsymbol{\beta}$ could assume any value in $\mathbb{R}^p$. We now consider least squares estimation for models in which $\boldsymbol{\beta}$ is restricted to the subset of $\mathbb{R}^p$ consisting of all those $\boldsymbol{\beta}$-values that satisfy the consistent system of linear equations

$$\mathbf{A}\boldsymbol{\beta} = \mathbf{h}, \tag{10.1}$$

where $\mathbf{A}$ is a specified $q \times p$ matrix of rank q^* and $\mathbf{h}$ is a specified q-vector. We introduced such *constrained linear models* in Chap. 5 and considered the estimability of functions $\mathbf{c}^T\boldsymbol{\beta}$ under them in Chap. 6. This chapter presents constrained least squares estimation and ANOVA via a theoretical development similar to that by which (unconstrained) least squares estimation and ANOVA was presented in Chaps. 7 and 8. In fact, the results obtained here specialize to results obtained in earlier chapters for unconstrained models by setting $\mathbf{A} = \mathbf{0}_p^T$ and $\mathbf{h} = 0$.

There are at least three reasons why equality-constrained linear models are important. First, some linear models that might be useful in applications truly are constrained. An example that will be featured in this chapter considers a model that consists of two simple linear regression models, each defined over one of two intervals that partition the real line, which are required to join at the point separating the two intervals. This requirement imposes a linear constraint on the parameters of the model's mean structure. Second, one important class of inferential problems in linear models involves testing a null hypothesis that a vector of estimable linear functions of $\boldsymbol{\beta}$ is equal to a specified vector of constants (which is often $\mathbf{0}$ in practice), versus an alternative hypothesis that those functions of $\boldsymbol{\beta}$ are not so constrained. The classical approach to deriving a test for these hypotheses (to be considered in detail in Chap. 15) requires estimating the vector of functions under each hypothesis. Third, as will be seen in this chapter, it turns out that imposing linear equality constraints of a certain number and type on $\boldsymbol{\beta}$ can make the normal

D. L. Zimmerman, *Linear Model Theory*,
https://doi.org/10.1007/978-3-030-52063-2_10

equations for the constrained model full rank, so that generalized inverses may not be needed to deal with a model having an overparameterized mean structure.

10.1 The General Case

Consider a constrained linear model $\{\mathbf{y}, \mathbf{X}\boldsymbol{\beta} : \mathbf{A}\boldsymbol{\beta} = \mathbf{h}\}$, for which, until further notice, the variance–covariance matrix is left unspecified. Recall from Theorem 6.2.2 that under this model, a linear function $\mathbf{c}^T\boldsymbol{\beta}$ is estimable if and only if $\mathbf{c}^T \in \mathcal{R}\begin{pmatrix}\mathbf{X}\\ \mathbf{A}\end{pmatrix}$. We begin with a definition of a constrained least squares estimator of such a function.

Definition 10.1.1 A *constrained least squares estimator* of an estimable function $\mathbf{c}^T\boldsymbol{\beta}$ associated with the model $\{\mathbf{y}, \mathbf{X}\boldsymbol{\beta} : \mathbf{A}\boldsymbol{\beta} = \mathbf{h}\}$ is $\mathbf{c}^T\breve{\boldsymbol{\beta}}$, where $\breve{\boldsymbol{\beta}}$ is any value of $\boldsymbol{\beta}$ that minimizes the residual sum of squares function $Q(\boldsymbol{\beta}) = (\mathbf{y} - \mathbf{X}\boldsymbol{\beta})^T(\mathbf{y} - \mathbf{X}\boldsymbol{\beta})$ among all $\boldsymbol{\beta}$ satisfying the constraints $\mathbf{A}\boldsymbol{\beta} = \mathbf{h}$.

How do we obtain a constrained least squares estimator of an estimable function $\mathbf{c}^T\boldsymbol{\beta}$? Consider the Lagrangian function for the problem of minimizing $Q(\boldsymbol{\beta})$ for the model $\{\mathbf{y}, \mathbf{X}\boldsymbol{\beta}\}$ with respect to $\boldsymbol{\beta}$, subject to the constraints $\mathbf{A}\boldsymbol{\beta} = \mathbf{h}$. This function is

$$L(\boldsymbol{\beta}, \boldsymbol{\lambda}) = Q(\boldsymbol{\beta}) + 2\boldsymbol{\lambda}^T(\mathbf{A}\boldsymbol{\beta} - \mathbf{h}),$$

where $2\boldsymbol{\lambda}$ is a q-vector of Lagrange multipliers. (We have written the vector of Lagrange multipliers as $2\boldsymbol{\lambda}$ rather than as $\boldsymbol{\lambda}$ so that $\boldsymbol{\lambda}$, rather than the slightly more cumbersome quantity $(1/2)\boldsymbol{\lambda}$, will appear in the system of equations that eventually must be solved.) Taking partial derivatives, we have

$$\frac{\partial L}{\partial \boldsymbol{\beta}} = -2\mathbf{X}^T\mathbf{y} + 2\mathbf{X}^T\mathbf{X}\boldsymbol{\beta} + 2\mathbf{A}^T\boldsymbol{\lambda},$$

$$\frac{\partial L}{\partial \boldsymbol{\lambda}} = 2(\mathbf{A}\boldsymbol{\beta} - \mathbf{h}).$$

Equating these partial derivatives to zero yields the system

$$\mathbf{X}^T\mathbf{X}\boldsymbol{\beta} + \mathbf{A}^T\boldsymbol{\lambda} = \mathbf{X}^T\mathbf{y},$$

$$\mathbf{A}\boldsymbol{\beta} = \mathbf{h}.$$

Thus, a necessary condition for $Q(\boldsymbol{\beta})$ to assume a minimum at $\boldsymbol{\beta} = \breve{\boldsymbol{\beta}}$, for $\boldsymbol{\beta}$ satisfying $\mathbf{A}\boldsymbol{\beta} = \mathbf{h}$, is that the system of equations

$$\begin{pmatrix} \mathbf{X}^T\mathbf{X} & \mathbf{A}^T \\ \mathbf{A} & \mathbf{0} \end{pmatrix} \begin{pmatrix} \boldsymbol{\beta} \\ \boldsymbol{\lambda} \end{pmatrix} = \begin{pmatrix} \mathbf{X}^T\mathbf{y} \\ \mathbf{h} \end{pmatrix}$$

have a solution whose first p-component subvector is $\breve{\boldsymbol{\beta}}$. We refer to this system as the *constrained normal equations* for the model $\{\mathbf{y}, \mathbf{X}\boldsymbol{\beta} : \mathbf{A}\boldsymbol{\beta} = \mathbf{h}\}$.

Theorem 10.1.1 *The constrained normal equations are consistent.*

Proof We show that the constrained normal equations are compatible; the consistency will follow because compatibility implies consistency (Theorem 3.2.1). Let $\mathbf{a}^T = (\mathbf{a}_1^T, \mathbf{a}_2^T)$ be any $(p+q)$-vector such that

$$\mathbf{a}^T \begin{pmatrix} \mathbf{X}^T\mathbf{X} & \mathbf{A}^T \\ \mathbf{A} & \mathbf{0} \end{pmatrix} = (\mathbf{a}_1^T\mathbf{X}^T\mathbf{X} + \mathbf{a}_2^T\mathbf{A}, \mathbf{a}_1^T\mathbf{A}^T) = (\mathbf{0}^T, \mathbf{0}^T).$$

Post-multiplication of the "left" subset of equations by $\mathbf{a}_1$ yields the equation

$$\mathbf{a}_1^T\mathbf{X}^T\mathbf{X}\mathbf{a}_1 + \mathbf{a}_2^T\mathbf{A}\mathbf{a}_1 = 0,$$

implying in turn that $\mathbf{a}_1^T\mathbf{X}^T\mathbf{X}\mathbf{a}_1 = 0$ (because $\mathbf{A}\mathbf{a}_1 = \mathbf{0}$ by the "right" subset of equations), $\mathbf{a}_1^T\mathbf{X}^T = \mathbf{0}^T$, and $\mathbf{a}_1^T\mathbf{X}^T\mathbf{y} = 0$. Also, because the system of constraint equations $\mathbf{A}\boldsymbol{\beta} = \mathbf{h}$ is consistent by assumption, a vector $\boldsymbol{\ell}$ exists such that $\mathbf{A}\boldsymbol{\ell} = \mathbf{h}$. Consequently, using the "left" subset of equations once more,

$$\mathbf{a}_2^T\mathbf{h} = \mathbf{a}_2^T\mathbf{A}\boldsymbol{\ell} = -\mathbf{a}_1^T\mathbf{X}^T\mathbf{X}\boldsymbol{\ell} = -(\mathbf{0}^T)\mathbf{X}\boldsymbol{\ell} = 0.$$

□

The next theorem establishes that the aforementioned condition for $Q(\boldsymbol{\beta})$ to assume a minimum at $\boldsymbol{\beta} = \breve{\boldsymbol{\beta}}$, among those $\boldsymbol{\beta}$ that satisfy the constraints $\mathbf{A}\boldsymbol{\beta} = \mathbf{h}$, is not merely necessary, but sufficient as well.

Theorem 10.1.2 *$Q(\boldsymbol{\beta})$ for the model $\{\mathbf{y}, \mathbf{X}\boldsymbol{\beta}\}$ assumes a global minimum at $\boldsymbol{\beta} = \breve{\boldsymbol{\beta}}$, among those $\boldsymbol{\beta}$ that satisfy $\mathbf{A}\boldsymbol{\beta} = \mathbf{h}$, if and only if the constrained normal equations for the model $\{\mathbf{y}, \mathbf{X}\boldsymbol{\beta} : \mathbf{A}\boldsymbol{\beta} = \mathbf{h}\}$ have a solution whose first p-component subvector is $\breve{\boldsymbol{\beta}}$.*

Proof Let $(\breve{\boldsymbol{\beta}}^T, \breve{\boldsymbol{\lambda}}^T)^T$ represent any solution to the constrained normal equations for the specified model, and let $\boldsymbol{\beta}^*$ represent any value of $\boldsymbol{\beta}$ that satisfies the constraints $\mathbf{A}\boldsymbol{\beta} = \mathbf{h}$. Then,

$$\begin{aligned}
Q(\boldsymbol{\beta}^*) &= (\mathbf{y} - \mathbf{X}\boldsymbol{\beta}^*)^T(\mathbf{y} - \mathbf{X}\boldsymbol{\beta}^*) \\
&= [(\mathbf{y} - \mathbf{X}\breve{\boldsymbol{\beta}}) + \mathbf{X}(\breve{\boldsymbol{\beta}} - \boldsymbol{\beta}^*)]^T[(\mathbf{y} - \mathbf{X}\breve{\boldsymbol{\beta}}) + \mathbf{X}(\breve{\boldsymbol{\beta}} - \boldsymbol{\beta}^*)] \\
&= Q(\breve{\boldsymbol{\beta}}) + [\mathbf{X}(\breve{\boldsymbol{\beta}} - \boldsymbol{\beta}^*)]^T[\mathbf{X}(\breve{\boldsymbol{\beta}} - \boldsymbol{\beta}^*)]
\end{aligned}$$

because

$$(\breve{\boldsymbol{\beta}} - \boldsymbol{\beta}^*)^T\mathbf{X}^T(\mathbf{y} - \mathbf{X}\breve{\boldsymbol{\beta}}) = (\breve{\boldsymbol{\beta}} - \boldsymbol{\beta}^*)^T\mathbf{A}^T\breve{\boldsymbol{\lambda}} = (\mathbf{h} - \mathbf{h})^T\breve{\boldsymbol{\lambda}} = 0.$$

Thus, among those $\boldsymbol{\beta}$ that satisfy the constraints $\mathbf{A}\boldsymbol{\beta} = \mathbf{h}$, $Q(\boldsymbol{\beta})$ assumes a global minimum at $\boldsymbol{\beta} = \breve{\boldsymbol{\beta}}$. Moreover, if $\bar{\boldsymbol{\beta}}$ is any other global minimizer of $Q(\boldsymbol{\beta})$ among those $\boldsymbol{\beta}$ that satisfy the constraints, then $Q(\bar{\boldsymbol{\beta}}) = Q(\breve{\boldsymbol{\beta}})$, implying that $[\mathbf{X}(\breve{\boldsymbol{\beta}} - \bar{\boldsymbol{\beta}})]^T[\mathbf{X}(\breve{\boldsymbol{\beta}} - \bar{\boldsymbol{\beta}})] = 0$, i.e., that $\mathbf{X}\bar{\boldsymbol{\beta}} = \mathbf{X}\breve{\boldsymbol{\beta}}$. This implies further that

$$\mathbf{X}^T\mathbf{X}\bar{\boldsymbol{\beta}} + \mathbf{A}^T\breve{\boldsymbol{\lambda}} = \mathbf{X}^T\mathbf{X}\breve{\boldsymbol{\beta}} + \mathbf{A}^T\breve{\boldsymbol{\lambda}} = \mathbf{X}^T\mathbf{y},$$

which shows that $(\bar{\boldsymbol{\beta}}^T, \breve{\boldsymbol{\lambda}}^T)^T$ is also a solution to the constrained normal equations for the specified model. □

As a consequence of Theorem 10.1.2, we may update the definition of a constrained least squares estimator of an estimable function as follows.

Definition 10.1.2 (Updated Version of Definition 10.1.1) A *constrained least squares estimator* of an estimable function $\mathbf{c}^T\boldsymbol{\beta}$ associated with the model $\{\mathbf{y}, \mathbf{X}\boldsymbol{\beta} : \mathbf{A}\boldsymbol{\beta} = \mathbf{h}\}$ is $\mathbf{c}^T\breve{\boldsymbol{\beta}}$, where $\breve{\boldsymbol{\beta}}$ is the first p-component subvector of any solution to the constrained normal equations for that model.

Theorem 10.1.3 $\mathbf{c}^T\breve{\boldsymbol{\beta}} = \mathbf{c}^T\bar{\boldsymbol{\beta}}$ *for the first p-component subvectors* $\breve{\boldsymbol{\beta}}$ *and* $\bar{\boldsymbol{\beta}}$ *of every two solutions to the constrained normal equations for the model* $\{\mathbf{y}, \mathbf{X}\boldsymbol{\beta} : \mathbf{A}\boldsymbol{\beta} = \mathbf{h}\}$ *if and only if* $\mathbf{c}^T\boldsymbol{\beta}$ *is estimable under that model.*

Proof Let $\mathbf{G} = \begin{pmatrix} \mathbf{G}_{11} & \mathbf{G}_{12} \\ \mathbf{G}_{21} & \mathbf{G}_{22} \end{pmatrix}$, where $\mathbf{G}_{11}$ is $p \times p$, be any generalized inverse of $\begin{pmatrix} \mathbf{X}^T\mathbf{X} & \mathbf{A}^T \\ \mathbf{A} & \mathbf{0} \end{pmatrix}$. By Theorem 3.2.3, all solutions to the constrained normal equations for the specified model are given by

$$\mathbf{G}\begin{pmatrix} \mathbf{X}^T\mathbf{y} \\ \mathbf{h} \end{pmatrix} + \left[\mathbf{I}_{p+q} - \begin{pmatrix} \mathbf{G}_{11}\mathbf{X}^T\mathbf{X} + \mathbf{G}_{12}\mathbf{A} & \mathbf{G}_{11}\mathbf{A}^T \\ \mathbf{G}_{21}\mathbf{X}^T\mathbf{X} + \mathbf{G}_{22}\mathbf{A} & \mathbf{G}_{21}\mathbf{A}^T \end{pmatrix}\right]\mathbf{z}$$

as $\mathbf{z}$ ranges over $\mathbb{R}^{p+q}$. From the "top" subset of these equations, we see that $\mathbf{c}^T\breve{\boldsymbol{\beta}} = \mathbf{c}^T\bar{\boldsymbol{\beta}}$ for every two solutions $(\breve{\boldsymbol{\beta}}^T, \breve{\boldsymbol{\lambda}}^T)^T$ and $(\bar{\boldsymbol{\beta}}^T, \bar{\boldsymbol{\lambda}}^T)^T$ to the constrained normal equations if and only if

$$\mathbf{c}^T[\mathbf{I}_p - \mathbf{G}_{11}\mathbf{X}^T\mathbf{X} - \mathbf{G}_{12}\mathbf{A}, -\mathbf{G}_{11}\mathbf{A}^T]\mathbf{z} = 0 \quad \text{for all } \mathbf{z},$$

i.e., if and only if

$$\mathbf{c}^T[\mathbf{I}_p - \mathbf{G}_{11}\mathbf{X}^T\mathbf{X} - \mathbf{G}_{12}\mathbf{A}, -\mathbf{G}_{11}\mathbf{A}^T] = \mathbf{0}^T,$$

i.e., if and only if

$$\mathbf{c}^T = \mathbf{c}^T\mathbf{G}_{11}\mathbf{X}^T\mathbf{X} + \mathbf{c}^T\mathbf{G}_{12}\mathbf{A} \quad \text{and} \quad \mathbf{c}^T\mathbf{G}_{11}\mathbf{A}^T = \mathbf{0}^T,$$

i.e., if and only if vectors $\mathbf{r}$ and $\mathbf{s}$ exist such that

$$\mathbf{c}^T = \mathbf{r}^T\mathbf{X}^T\mathbf{X} + \mathbf{s}^T\mathbf{A}.$$

For the reverse implication of this last result, we rely on Theorem 3.3.8a–d (with $\mathbf{X}^T\mathbf{X}$ and $\mathbf{A}^T$ here playing the roles of $\mathbf{A}$ and $\mathbf{B}$, respectively, in that theorem) to establish that

$$\begin{aligned}(\mathbf{r}^T\mathbf{X}^T\mathbf{X} + \mathbf{s}^T\mathbf{A})(\mathbf{G}_{11}\mathbf{X}^T\mathbf{X} + \mathbf{G}_{12}\mathbf{A}) &= \mathbf{r}^T\mathbf{X}^T\mathbf{X}\mathbf{G}_{11}\mathbf{X}^T\mathbf{X} + \mathbf{s}^T\mathbf{A}\mathbf{G}_{11}\mathbf{X}^T\mathbf{X} \\ &\quad + \mathbf{r}^T\mathbf{X}^T\mathbf{X}\mathbf{G}_{12}\mathbf{A} + \mathbf{s}^T\mathbf{A}\mathbf{G}_{12}\mathbf{A} \\ &= \mathbf{r}^T(\mathbf{X}^T\mathbf{X}\mathbf{G}_{11}\mathbf{X}^T\mathbf{X} - \mathbf{A}^T\mathbf{G}_{22}\mathbf{A}) + \mathbf{s}^T(\mathbf{0} + \mathbf{A}) \\ &= \mathbf{r}^T\mathbf{X}^T\mathbf{X} + \mathbf{s}^T\mathbf{A}\end{aligned}$$

and

$$\begin{aligned}(\mathbf{r}^T\mathbf{X}^T\mathbf{X} + \mathbf{s}^T\mathbf{A})\mathbf{G}_{11}\mathbf{A}^T &= \mathbf{r}^T\mathbf{X}^T\mathbf{X}\mathbf{G}_{11}\mathbf{A}^T + \mathbf{s}^T\mathbf{A}\mathbf{G}_{11}\mathbf{A}^T \\ &= \mathbf{r}^T\mathbf{0} + \mathbf{s}^T\mathbf{0} \\ &= \mathbf{0}^T.\end{aligned}$$

Now, if vectors $\mathbf{r}$ and $\mathbf{s}$ exist such that $\mathbf{c}^T = \mathbf{r}^T\mathbf{X}^T\mathbf{X} + \mathbf{s}^T\mathbf{A}$, then

$$\mathbf{c}^T = (\mathbf{r}^T, \mathbf{s}^T)\begin{pmatrix}\mathbf{X}^T\mathbf{X} \\ \mathbf{A}\end{pmatrix},$$

i.e.,

$$\mathbf{c}^T \in \mathcal{R}\begin{pmatrix}\mathbf{X}^T\mathbf{X} \\ \mathbf{A}\end{pmatrix}$$

and vice versa, which is equivalent (by Corollary 6.2.2.5) to the estimability of $\mathbf{c}^T\boldsymbol{\beta}$ under the specified model. □

Corollary 10.1.3.1 $\mathbf{X}\breve{\boldsymbol{\beta}} = \mathbf{X}\bar{\boldsymbol{\beta}}$ *for the first p-component subvectors* $\breve{\boldsymbol{\beta}}$ *and* $\bar{\boldsymbol{\beta}}$ *of any two solutions to the constrained normal equations for the model* $\{\mathbf{y}, \mathbf{X}\boldsymbol{\beta} : \mathbf{A}\boldsymbol{\beta} = \mathbf{h}\}$.

Theorem 10.1.3 establishes that a constrained least squares estimator of a linear function $\mathbf{c}^T\boldsymbol{\beta}$ associated with the model $\{\mathbf{y}, \mathbf{X}\boldsymbol{\beta} : \mathbf{A}\boldsymbol{\beta} = \mathbf{h}\}$ is unique if and only if the function is estimable under that model. This allows us to finalize the definition of constrained least squares estimator, replacing the indefinite article "a" with the definite article "the."

Definition 10.1.3 (Final Version of Definition 10.1.1) The *constrained least squares estimator* of an estimable function $\mathbf{c}^T\boldsymbol{\beta}$ associated with the constrained linear model $\{\mathbf{y}, \mathbf{X}\boldsymbol{\beta} : \mathbf{A}\boldsymbol{\beta} = \mathbf{h}\}$ is $\mathbf{c}^T\breve{\boldsymbol{\beta}}$, where $\breve{\boldsymbol{\beta}}$ is the first p-component subvector of any solution to the constrained normal equations for that model. The *constrained least squares estimator* of a vector $\mathbf{C}^T\boldsymbol{\beta}$ of estimable functions associated with that constrained model is the vector $\mathbf{C}^T\breve{\boldsymbol{\beta}}$ of constrained least squares estimators of those functions.

The next three theorems establish some properties of constrained least squares estimators of estimable functions.

Theorem 10.1.4 *The constrained least squares estimator of an estimable function* $\mathbf{c}^T\boldsymbol{\beta}$ *associated with the model* $\{\mathbf{y}, \mathbf{X}\boldsymbol{\beta} : \mathbf{A}\boldsymbol{\beta} = \mathbf{h}\}$ *may be expressed alternatively as* $\mathbf{c}^T(\mathbf{G}_{11}\mathbf{X}^T\mathbf{y}+\mathbf{G}_{12}\mathbf{h})$, *where* $\begin{pmatrix} \mathbf{G}_{11} & \mathbf{G}_{12} \\ \mathbf{G}_{21} & \mathbf{G}_{22} \end{pmatrix}$ *is any generalized inverse of* $\begin{pmatrix} \mathbf{X}^T\mathbf{X} & \mathbf{A}^T \\ \mathbf{A} & \mathbf{0} \end{pmatrix}$ *such that* $\mathbf{G}_{11}$ *has dimensions* $p \times p$. *Furthermore, this estimator is linear and unbiased under the specified constrained model.*

Proof Let $\breve{\boldsymbol{\beta}}$ represent the first p-component subvector of any solution to the constrained normal equations for the specified model. Then,

$$\mathbf{c}^T\breve{\boldsymbol{\beta}} = (\mathbf{c}^T, \mathbf{0}^T)\begin{pmatrix} \mathbf{G}_{11} & \mathbf{G}_{12} \\ \mathbf{G}_{21} & \mathbf{G}_{22} \end{pmatrix}\begin{pmatrix} \mathbf{X}^T\mathbf{y} \\ \mathbf{h} \end{pmatrix} = \mathbf{c}^T(\mathbf{G}_{11}\mathbf{X}^T\mathbf{y} + \mathbf{G}_{12}\mathbf{h}).$$

Now because $\mathbf{c}^T\boldsymbol{\beta}$ is estimable, by Corollary 6.2.2.5 vectors $\mathbf{r}$ and $\mathbf{s}$ exist such that $\mathbf{c}^T = \mathbf{r}^T\mathbf{X}^T\mathbf{X} + \mathbf{s}^T\mathbf{A}$. Thus,

$$\begin{aligned} \mathbf{c}^T\breve{\boldsymbol{\beta}} &= (\mathbf{r}^T\mathbf{X}^T\mathbf{X} + \mathbf{s}^T\mathbf{A})(\mathbf{G}_{11}\mathbf{X}^T\mathbf{y} + \mathbf{G}_{12}\mathbf{h}) \\ &= \mathbf{r}^T\mathbf{X}^T\mathbf{X}\mathbf{G}_{12}\mathbf{h} + \mathbf{s}^T\mathbf{A}\mathbf{G}_{12}\mathbf{h} + (\mathbf{r}^T\mathbf{X}^T\mathbf{X}\mathbf{G}_{11}\mathbf{X}^T + \mathbf{s}^T\mathbf{A}\mathbf{G}_{11}\mathbf{X}^T)\mathbf{y} \\ &= t_0 + \mathbf{t}^T\mathbf{y}, \end{aligned}$$

say, showing that $\mathbf{c}^T\breve{\boldsymbol{\beta}}$ is a linear estimator. Observe that $t_0 = \mathbf{g}^T\mathbf{h}$ where $\mathbf{g}^T = \mathbf{r}^T\mathbf{X}^T\mathbf{X}\mathbf{G}_{12} + \mathbf{s}^T\mathbf{A}\mathbf{G}_{12}$. Now, using Theorem 3.3.8a–d (with $\mathbf{X}^T\mathbf{X}$ and $\mathbf{A}^T$ here playing the roles of $\mathbf{A}$ and $\mathbf{B}$, respectively, in that theorem),

$$\begin{aligned}
\mathbf{t}^T\mathbf{X} + \mathbf{g}^T\mathbf{A} &= \mathbf{r}^T\mathbf{X}^T\mathbf{X}\mathbf{G}_{11}\mathbf{X}^T\mathbf{X} + \mathbf{s}^T\mathbf{A}\mathbf{G}_{11}\mathbf{X}^T\mathbf{X} + \mathbf{r}^T\mathbf{X}^T\mathbf{X}\mathbf{G}_{12}\mathbf{A} + \mathbf{s}^T\mathbf{A}\mathbf{G}_{12}\mathbf{A} \\
&= \mathbf{r}^T\mathbf{X}^T\mathbf{X}\mathbf{G}_{11}\mathbf{X}^T\mathbf{X} + \mathbf{s}^T\mathbf{0} - \mathbf{r}^T\mathbf{A}^T\mathbf{G}_{22}\mathbf{A} + \mathbf{s}^T\mathbf{A} \\
&= \mathbf{r}^T(\mathbf{X}^T\mathbf{X}\mathbf{G}_{11}\mathbf{X}^T\mathbf{X} - \mathbf{A}^T\mathbf{G}_{22}\mathbf{A}) + \mathbf{s}^T\mathbf{A} \\
&= \mathbf{r}^T\mathbf{X}^T\mathbf{X} + \mathbf{s}^T\mathbf{A} \\
&= \mathbf{c}^T.
\end{aligned}$$

The unbiasedness of $\mathbf{c}^T\breve{\boldsymbol{\beta}}$ under the specified model follows immediately from Theorem 6.2.1. □

Theorem 10.1.5 *Let $\mathbf{C}_1^T\boldsymbol{\beta}$ and $\mathbf{C}_2^T\boldsymbol{\beta}$ be two vectors of estimable functions under the model $\{\mathbf{y}, \mathbf{X}\boldsymbol{\beta} : \mathbf{A}\boldsymbol{\beta} = \mathbf{h}\}$, and let $\breve{\boldsymbol{\beta}}$ represent the first p-component subvector of any solution to the constrained normal equations for that model, so that $\mathbf{C}_1^T\breve{\boldsymbol{\beta}}$ and $\mathbf{C}_2^T\breve{\boldsymbol{\beta}}$ are the vectors of constrained least squares estimators of $\mathbf{C}_1^T\boldsymbol{\beta}$ and $\mathbf{C}_2^T\boldsymbol{\beta}$, respectively, associated with that model. Finally let $\begin{pmatrix} \mathbf{G}_{11} & \mathbf{G}_{12} \\ \mathbf{G}_{21} & \mathbf{G}_{22} \end{pmatrix}$ (where $\mathbf{G}_{11}$ is $p \times p$) represent any generalized inverse of $\begin{pmatrix} \mathbf{X}^T\mathbf{X} & \mathbf{A}^T \\ \mathbf{A} & \mathbf{0} \end{pmatrix}$. Then under the constrained Gauss–Markov model $\{\mathbf{y}, \mathbf{X}\boldsymbol{\beta}, \sigma^2\mathbf{I} : \mathbf{A}\boldsymbol{\beta} = \mathbf{h}\}$,*

$$cov(\mathbf{C}_1^T\breve{\boldsymbol{\beta}}, \mathbf{C}_2^T\breve{\boldsymbol{\beta}}) = \sigma^2\mathbf{C}_1^T\mathbf{G}_{11}\mathbf{C}_2 \quad and \quad var(\mathbf{C}_1^T\breve{\boldsymbol{\beta}}) = \sigma^2\mathbf{C}_1^T\mathbf{G}_{11}\mathbf{C}_1$$

and these expressions are invariant to the choice of the generalized inverse.

Proof Because $\mathbf{C}_1^T\boldsymbol{\beta}$ and $\mathbf{C}_2^T\boldsymbol{\beta}$ are estimable under the specified model, by Corollary 6.2.2.5 matrices $\mathbf{R}_1, \mathbf{S}_1, \mathbf{R}_2$, and $\mathbf{S}_2$ exist such that

$$\mathbf{C}_1^T = \mathbf{R}_1^T\mathbf{X}^T\mathbf{X} + \mathbf{S}_1^T\mathbf{A} \quad \text{and} \quad \mathbf{C}_2^T = \mathbf{R}_2^T\mathbf{X}^T\mathbf{X} + \mathbf{S}_2^T\mathbf{A}.$$

Using Corollary 4.2.3.1 and the representations of $\mathbf{C}_1^T\breve{\boldsymbol{\beta}}$ and $\mathbf{C}_2^T\breve{\boldsymbol{\beta}}$ given by Theorem 10.1.4, we have

$$\begin{aligned}
\text{cov}(\mathbf{C}_1^T\breve{\boldsymbol{\beta}}, \mathbf{C}_2^T\breve{\boldsymbol{\beta}}) &= \text{cov}[(\mathbf{R}_1^T\mathbf{X}^T\mathbf{X} + \mathbf{S}_1^T\mathbf{A})(\mathbf{G}_{11}\mathbf{X}^T\mathbf{y} + \mathbf{G}_{12}\mathbf{h}), \\
&\qquad (\mathbf{R}_2^T\mathbf{X}^T\mathbf{X} + \mathbf{S}_2^T\mathbf{A})(\mathbf{G}_{11}\mathbf{X}^T\mathbf{y} + \mathbf{G}_{12}\mathbf{h})] \\
&= (\mathbf{R}_1^T\mathbf{X}^T\mathbf{X} + \mathbf{S}_1^T\mathbf{A})[\text{var}(\mathbf{G}_{11}\mathbf{X}^T\mathbf{y})](\mathbf{X}^T\mathbf{X}\mathbf{R}_2 + \mathbf{A}^T\mathbf{S}_2) \\
&= \sigma^2(\mathbf{R}_1^T\mathbf{X}^T\mathbf{X} + \mathbf{S}_1^T\mathbf{A})\mathbf{G}_{11}\mathbf{X}^T\mathbf{X}\mathbf{G}_{11}^T(\mathbf{X}^T\mathbf{X}\mathbf{R}_2 + \mathbf{A}^T\mathbf{S}_2)
\end{aligned}$$

$$= \sigma^2 \mathbf{R}_1^T \mathbf{X}^T \mathbf{X} \mathbf{G}_{11} \mathbf{X}^T \mathbf{X} \mathbf{G}_{11}^T \mathbf{X}^T \mathbf{X} \mathbf{R}_2$$
$$= \sigma^2 \mathbf{R}_1^T (\mathbf{X}^T \mathbf{X} + \mathbf{A}^T \mathbf{G}_{22} \mathbf{A}) \mathbf{G}_{11}^T \mathbf{X}^T \mathbf{X} \mathbf{R}_2$$
$$= \sigma^2 \mathbf{R}_1^T \mathbf{X}^T \mathbf{X} \mathbf{G}_{11} \mathbf{X}^T \mathbf{X} \mathbf{R}_2$$
$$= \sigma^2 (\mathbf{R}_1^T \mathbf{X}^T \mathbf{X} + \mathbf{S}_1^T \mathbf{A}) \mathbf{G}_{11} (\mathbf{X}^T \mathbf{X} \mathbf{R}_2 + \mathbf{A}^T \mathbf{S}_2)$$
$$= \sigma^2 \mathbf{C}_1^T \mathbf{G}_{11} \mathbf{C}_2,$$

where for the fourth, fifth, sixth, and seventh equalities we used Theorem 3.3.8a–d (with $\mathbf{X}^T\mathbf{X}$ and $\mathbf{A}^T$ here playing the roles of $\mathbf{A}$ and $\mathbf{B}$, respectively, in that theorem). Furthermore, Theorem 3.3.8f establishes that the expression on the right-hand side of the sixth equality, and hence the final expression, is invariant to the choice of generalized inverse. The expression for $\text{var}(\mathbf{C}_1^T \breve{\boldsymbol{\beta}})$ may be obtained merely by substituting $\mathbf{C}_1$ for $\mathbf{C}_2$ in the corresponding expression for the matrix of covariances. □

Upon comparing Theorem 7.2.2 with Theorem 10.1.5, we see that while the former claimed that the variance–covariance matrix of the vector of least squares estimators of linearly independent estimable functions is positive definite under a Gauss–Markov model, the latter does not make an analogous claim about the variance–covariance matrix of the vector of constrained least squares estimators of linearly independent estimable functions under a constrained Gauss–Markov model. Indeed, it will be demonstrated in forthcoming examples that $\text{var}(\mathbf{C}_1^T \breve{\boldsymbol{\beta}})$ may not be positive definite, even when the estimable functions in $\mathbf{C}_1^T \boldsymbol{\beta}$ are linearly independent.

The next theorem claims that the constrained least squares estimator of an estimable function associated with a constrained model is best (in a certain sense) among all linear unbiased estimators of that function under the corresponding constrained Gauss–Markov model. As such, the theorem may be considered an extension of the Gauss–Markov theorem to constrained linear models. Its proof, which is left as an exercise, parallels that of the "unconstrained" Gauss–Markov theorem (Theorem 7.2.3).

Theorem 10.1.6 *Under the constrained Gauss–Markov model* $\{\mathbf{y}, \mathbf{X}\boldsymbol{\beta}, \sigma^2\mathbf{I} : \mathbf{A}\boldsymbol{\beta} = \mathbf{h}\}$, *the variance of the constrained least squares estimator of an estimable function* $\mathbf{c}^T\boldsymbol{\beta}$ *associated with the model* $\{\mathbf{y}, \mathbf{X}\boldsymbol{\beta} : \mathbf{A}\boldsymbol{\beta} = \mathbf{h}\}$ *is uniformly (in* $\boldsymbol{\beta}$ *and* σ^2*) less than that of any other linear unbiased estimator of* $\mathbf{c}^T\boldsymbol{\beta}$.

Next we establish the rank of the coefficient matrix of the constrained normal equations, which will help us to obtain simplified expressions for the constrained least squares estimators and will be useful for other purposes as well.

Theorem 10.1.7

$$rank\begin{pmatrix}\mathbf{X}^T\mathbf{X} & \mathbf{A}^T \\ \mathbf{A} & \mathbf{0}\end{pmatrix} = rank\begin{pmatrix}\mathbf{X}^T\mathbf{X} \\ \mathbf{A}\end{pmatrix} + rank(\mathbf{A}).$$

Proof For any two vectors $\boldsymbol{\ell}$ and $\mathbf{m}$ such that

$$\begin{pmatrix}\mathbf{X}^T\mathbf{X} \\ \mathbf{A}\end{pmatrix}\boldsymbol{\ell} = \begin{pmatrix}\mathbf{A}^T \\ \mathbf{0}\end{pmatrix}\mathbf{m},$$

we have $\boldsymbol{\ell}^T\mathbf{X}^T\mathbf{X}\boldsymbol{\ell} = \boldsymbol{\ell}^T\mathbf{A}^T\mathbf{m} = (\mathbf{A}\boldsymbol{\ell})^T\mathbf{m} = 0$, implying that $\mathbf{X}\boldsymbol{\ell} = \mathbf{0}$ and thus that $\begin{pmatrix}\mathbf{X}^T\mathbf{X} \\ \mathbf{A}\end{pmatrix}\boldsymbol{\ell} = \mathbf{0}$ and $\begin{pmatrix}\mathbf{A}^T \\ \mathbf{0}\end{pmatrix}\mathbf{m} = \mathbf{0}$. This establishes that $\mathcal{C}\begin{pmatrix}\mathbf{X}^T\mathbf{X} \\ \mathbf{A}\end{pmatrix}$ and $\mathcal{C}\begin{pmatrix}\mathbf{A}^T \\ \mathbf{0}\end{pmatrix}$ are essentially disjoint. So, by Theorem 2.8.6,

$$\text{rank}\begin{pmatrix}\mathbf{X}^T\mathbf{X} & \mathbf{A}^T \\ \mathbf{A} & \mathbf{0}\end{pmatrix} = \text{rank}\begin{pmatrix}\mathbf{X}^T\mathbf{X} \\ \mathbf{A}\end{pmatrix} + \text{rank}(\mathbf{A}).$$

□

Corollary 10.1.7.1 $rank\begin{pmatrix}\mathbf{X}^T\mathbf{X} & \mathbf{A}^T \\ \mathbf{A} & \mathbf{0}\end{pmatrix} = rank\begin{pmatrix}\mathbf{X} \\ \mathbf{A}\end{pmatrix} + q^*.$

Proof rank$\begin{pmatrix}\mathbf{X}^T\mathbf{X} \\ \mathbf{A}\end{pmatrix}$ = rank$\begin{pmatrix}\mathbf{X} \\ \mathbf{A}\end{pmatrix}$ by Theorem 6.2.3, and rank($\mathbf{A}$) $= q^*$ by the model specification. □

Now we are ready for an example illustrating some of the results presented thus far.

Example 10.1-1. Constrained Least Squares Estimators and Their Variance–Covariance Matrices Under Some Constrained One-Factor Models

Consider once again the one-factor model with each of the constraints considered in Example 6.2-1. Recall that when the constraint is $\alpha_q = h$, all linear functions $\mathbf{c}^T\boldsymbol{\beta}$ are estimable. The constrained normal equations for the model with this constraint may be written as

$$\begin{pmatrix} n & \mathbf{n}_{-q}^T & n_q & 0 \\ \mathbf{n}_{-q} & \oplus_{i=1}^{q-1} n_i & \mathbf{0}_{q-1} & \mathbf{0}_{q-1} \\ n_q & \mathbf{0}_{q-1}^T & n_q & 1 \\ 0 & \mathbf{0}_{q-1}^T & 1 & 0 \end{pmatrix}\begin{pmatrix}\mu \\ \boldsymbol{\alpha}_{-q} \\ \alpha_q \\ \lambda\end{pmatrix} = \begin{pmatrix} y_{\cdot\cdot} \\ \mathbf{y}_{-q\cdot} \\ y_{q\cdot} \\ h\end{pmatrix},$$

where $\mathbf{n}_{-q} = (n_1, \ldots, n_{q-1})^T$, $\boldsymbol{\alpha}_{-q} = (\alpha_1, \ldots, \alpha_{q-1})^T$, and $\mathbf{y}_{-q\cdot} = (y_{1\cdot}, \ldots, y_{q-1\cdot})^T$. The coefficient matrix has dimensions $(q+2) \times (q+2)$, and by Theorem 10.1.7 its rank is equal to $(q+1)+1 = q+2$. Thus the coefficient matrix is nonsingular, and by inspection so are its upper left $q \times q$ block $\begin{pmatrix} n & \mathbf{n}_{-q}^T \\ \mathbf{n}_{-q} & \oplus_{i=1}^{q-1} n_i \end{pmatrix}$ and the Schur complement of that block,

$$\begin{pmatrix} n_q & 1 \\ 1 & 0 \end{pmatrix} - \begin{pmatrix} n_q & \mathbf{0}_{q-1}^T \\ 0 & \mathbf{0}_{q-1}^T \end{pmatrix} \begin{pmatrix} n & \mathbf{n}_{-q}^T \\ \mathbf{n}_{-q} & \oplus_{i=1}^{q-1} n_i \end{pmatrix}^{-1} \begin{pmatrix} n_q & 0 \\ \mathbf{0}_{q-1} & \mathbf{0}_{q-1} \end{pmatrix} = \begin{pmatrix} 0 & 1 \\ 1 & 0 \end{pmatrix}.$$

Theorem 2.9.5 may therefore be used to obtain the inverse of the coefficient matrix, from which $\mathbf{G}_{11}$ and $\mathbf{G}_{12}$ may be extracted, yielding (by Theorem 10.1.4) the following general expression for the constrained least squares estimator of any function $\mathbf{c}^T\boldsymbol{\beta}$ (see the exercises for details):

$$\mathbf{c}^T\breve{\boldsymbol{\beta}} = \mathbf{c}^T \begin{pmatrix} \bar{y}_{q\cdot} - h \\ \bar{y}_{1\cdot} - \bar{y}_{q\cdot} + h \\ \bar{y}_{2\cdot} - \bar{y}_{q\cdot} + h \\ \vdots \\ \bar{y}_{q-1\cdot} - \bar{y}_{q\cdot} + h \\ h \end{pmatrix}. \tag{10.2}$$

Some functions and their constrained least squares estimators are listed in the following table:

Function	Constrained least squares estimator
μ	$\bar{y}_{q\cdot} - h$
$\alpha_i \ (i = 1, \ldots, q-1)$	$\bar{y}_{i\cdot} - \bar{y}_{q\cdot} + h$
α_q	h
$\mu + \alpha_i$	$\bar{y}_{i\cdot}$
$\alpha_i - \alpha_{i'}$	$\bar{y}_{i\cdot} - \bar{y}_{i'\cdot}$

Furthermore, by Theorem 10.1.5 we have

$$\text{var}(\breve{\boldsymbol{\beta}}) = \sigma^2\mathbf{G}_{11} = \sigma^2 \begin{pmatrix} n_q^{-1} & -n_q^{-1}\mathbf{1}_{q-1}^T & 0 \\ -n_q^{-1}\mathbf{1}_{q-1} & \oplus_{i=1}^{q-1} n_i^{-1} + n_q^{-1}\mathbf{J}_{q-1} & \mathbf{0}_{q-1} \\ 0 & \mathbf{0}_{q-1}^T & 0 \end{pmatrix},$$

and thus, for example,

$$\begin{aligned}\mathrm{var}(\breve{\mu}+\breve{\alpha}_i) &= \sigma^2 \begin{pmatrix} 1 & \mathbf{u}_i^T \end{pmatrix} \begin{pmatrix} n_q^{-1} & -n_q^{-1}\mathbf{1}_{q-1}^T & 0 \\ -n_q^{-1}\mathbf{1}_{q-1} & \oplus_{i=1}^{q-1} n_i^{-1} + n_q^{-1}\mathbf{J}_{q-1} & \mathbf{0}_{q-1} \\ 0 & \mathbf{0}_{q-1}^T & 0 \end{pmatrix} \begin{pmatrix} 1 \\ \mathbf{u}_i \end{pmatrix} \\ &= \sigma^2/n_i \quad (i = 1, \ldots, q).\end{aligned}$$

Observe that $\mathrm{var}(\breve{\boldsymbol{\beta}})$ is not positive definite, and in particular that $\mathrm{var}(\breve{\alpha}_q) = 0$.

For any other single constraint for which the corresponding matrix $\mathbf{A}$ (which is actually a row vector in this example) is not an element of $\mathcal{R}(\mathbf{X})$, again all functions $\mathbf{c}^T\boldsymbol{\beta}$ will be estimable and the coefficient matrix of the constrained normal equations will be nonsingular, so a similar approach can be used to obtain a general expression for the constrained least squares estimator of any function. It turns out that although the constrained least squares estimators of the individual parameters μ and α_i may differ from their counterparts in the table above, the constrained least squares estimators of the level means $\mu+\alpha_i$ and the level mean differences $\alpha_i - \alpha_{i'}$ coincide with their counterparts in the table; see the exercises. Furthermore, the latter estimators coincide with the least squares estimators under the unconstrained one-factor model. More will be said about these coincidences later in this chapter.

However, now consider the constraint $\mu + \alpha_1 = h$, which differs from the previously considered constraints in an important way: the matrix $\mathbf{A}$ corresponding to this constraint is an element of $\mathcal{R}(\mathbf{X})$. Recall from Example 6.2-1 that in this case $\mathbf{c}^T\boldsymbol{\beta}$ is estimable if and only if it is estimable under the unconstrained model. Thus the functions μ and α_i $(i = 1, \ldots, q)$, which were estimable under the first constrained model, are no longer so. The coefficient matrix of the constrained normal equations may be written as

$$\begin{pmatrix} n & \mathbf{n}^T & 1 \\ \mathbf{n} & \oplus_{i=1}^q n_i & \mathbf{u}_1^{(q)} \\ 1 & \mathbf{u}_1^{(q)T} & 0 \end{pmatrix}.$$

By Theorem 10.1.7 this matrix has rank $q+1$, so it is singular and the method used for the previous case to obtain a general expression for constrained least squares estimators must be modified. Noting as we did in Example 7.1-2 that $\begin{pmatrix} 0 & \mathbf{0}^T \\ \mathbf{0} & \oplus_{i=1}^q n_i^{-1} \end{pmatrix}$ is a generalized inverse of the upper left $(q+1) \times (q+1)$ block $\begin{pmatrix} n & \mathbf{n}^T \\ \mathbf{n} & \oplus_{i=1}^q n_i \end{pmatrix}$, Theorem 3.3.7 may be used to obtain a generalized inverse of the coefficient matrix, from which $\mathbf{G}_{11}$ and $\mathbf{G}_{12}$ may be extracted, yielding (by Theorem 10.1.4) the following general expression for the constrained least squares estimator of any estimable function:

$$\mathbf{c}^T\breve{\boldsymbol{\beta}} = \mathbf{c}^T \begin{pmatrix} 0 \\ h \\ \bar{y}_{2\cdot} \\ \vdots \\ \bar{y}_{q\cdot} \end{pmatrix}. \tag{10.3}$$

The following table lists the constrained least squares estimators for those functions included in the previous table that are estimable under this constrained model:

Estimable function	Constrained least squares estimator
$\mu + \alpha_1$	h
$\mu + \alpha_i \ (i = 2, \ldots, q)$	$\bar{y}_{i\cdot}$
$\alpha_i - \alpha_1 \ (i = 2, \ldots, q)$	$\bar{y}_{i\cdot} - h$
$\alpha_i - \alpha_{i'} \ (i > i' = 2, \ldots, q)$	$\bar{y}_{i\cdot} - \bar{y}_{i'\cdot}$

Note that constrained least squares estimators of some of these functions coincide with their counterparts under the unconstrained one-factor model, but others do not. Letting $\mathbf{C}^T = (\mathbf{1}_q, \mathbf{I}_q)$ so that $\mathbf{C}^T\boldsymbol{\beta} = (\mu + \alpha_1, \ldots, \mu + \alpha_q)^T$, we have

$$\text{var}(\mathbf{C}^T\breve{\boldsymbol{\beta}}) = \sigma^2 \begin{pmatrix} 0 & \mathbf{0}_{q-1}^T \\ \mathbf{0}_{q-1} & \oplus_{i=2}^q n_i^{-1} \end{pmatrix},$$

and thus

$$\text{var}(\breve{\mu} + \breve{\alpha}_i) = \begin{cases} 0 & \text{if } i = 1, \\ \sigma^2/n_i & \text{if } i = 2, \ldots, q. \end{cases}$$

Verifying all of these expressions is left as an exercise. ■

Now we obtain some results that will help us understand the geometry of constrained least squares estimation. Again, let $\breve{\boldsymbol{\beta}}$ represent the first p-component subvector of a solution to the constrained normal equations for the model $\{\mathbf{y}, \mathbf{X}\boldsymbol{\beta} : \mathbf{A}\boldsymbol{\beta} = \mathbf{h}\}$, and let $\begin{pmatrix} \mathbf{G}_{11} & \mathbf{G}_{12} \\ \mathbf{G}_{21} & \mathbf{G}_{22} \end{pmatrix}$ (where $\mathbf{G}_{11}$ is $p \times p$) represent any generalized inverse of $\begin{pmatrix} \mathbf{X}^T\mathbf{X} & \mathbf{A}^T \\ \mathbf{A} & \mathbf{0} \end{pmatrix}$. Then because $\mathbf{X}\boldsymbol{\beta}$ is estimable under the specified model, by Theorem 10.1.4 the constrained least squares estimator of $\mathbf{X}\boldsymbol{\beta}$ is

$$\mathbf{X}\breve{\boldsymbol{\beta}} = \mathbf{X}\,(\mathbf{G}_{11},\ \mathbf{G}_{12}) \begin{pmatrix} \mathbf{X}^T\mathbf{y} \\ \mathbf{h} \end{pmatrix} = \mathbf{X}\mathbf{G}_{11}\mathbf{X}^T\mathbf{y} + \mathbf{X}\mathbf{G}_{12}\mathbf{h}. \tag{10.4}$$

This estimator is the sum of two vectors: the first vector is a linear transformation of $\mathbf{y}$, and the second is a vector that is not functionally dependent on $\mathbf{y}$. The next theorem identifies some properties of the linear transformation matrix and the second vector.

Theorem 10.1.8 *The matrix* $\mathbf{X}\mathbf{G}_{11}\mathbf{X}^T$ *appearing in (10.4):*

(a) is symmetric;
(b) is idempotent;
(c) is invariant to the choice of upper left $p \times p$ *block* $\mathbf{G}_{11}$ *of a generalized inverse of* $\begin{pmatrix} \mathbf{X}^T\mathbf{X} & \mathbf{A}^T \\ \mathbf{A} & \mathbf{0} \end{pmatrix}$*; and*
(d) has rank equal to $p^*_{\mathbf{A}} \equiv \text{rank}\begin{pmatrix} \mathbf{X} \\ \mathbf{A} \end{pmatrix} - \text{rank}(\mathbf{A})$.
(e) Furthermore, the additive term $\mathbf{X}\mathbf{G}_{12}\mathbf{h}$ *appearing in (10.4) is invariant to the choice of upper right* $p \times q$ *block* $\mathbf{G}_{12}$ *of the aforementioned generalized inverse.*

Proof By Theorem 3.3.8f (with $\mathbf{X}^T\mathbf{X}$ here playing the role of $\mathbf{A}$ in the theorem), $\mathbf{X}^T\mathbf{X}\mathbf{G}_{11}\mathbf{X}^T\mathbf{X}$ is symmetric. Hence $\mathbf{X}^T\mathbf{X}\mathbf{G}_{11}\mathbf{X}^T\mathbf{X} = (\mathbf{X}^T\mathbf{X}\mathbf{G}_{11}\mathbf{X}^T\mathbf{X})^T = \mathbf{X}^T\mathbf{X}\mathbf{G}_{11}^T\mathbf{X}^T\mathbf{X}$. Applying Theorem 3.3.2 to this matrix equation twice (once "in front" and once "in back"), we obtain the relation $\mathbf{X}\mathbf{G}_{11}\mathbf{X}^T = \mathbf{X}\mathbf{G}_{11}^T\mathbf{X}^T = (\mathbf{X}\mathbf{G}_{11}\mathbf{X})^T$, which establishes part (a).

To establish part (b), note by Theorem 3.3.8d (with $\mathbf{X}^T\mathbf{X}$ and $\mathbf{A}^T$ here playing the roles of $\mathbf{A}$ and $\mathbf{B}$ in the theorem) that

$$\mathbf{X}^T\mathbf{X}\mathbf{G}_{11}\mathbf{X}^T\mathbf{X} - \mathbf{A}^T\mathbf{G}_{22}\mathbf{A} = \mathbf{X}^T\mathbf{X}.$$

Post-multiplication of both sides of this matrix equation by $\mathbf{G}_{11}\mathbf{X}^T$ yields

$$\mathbf{X}^T\mathbf{X}\mathbf{G}_{11}\mathbf{X}^T\mathbf{X}\mathbf{G}_{11}\mathbf{X}^T = \mathbf{X}^T\mathbf{X}\mathbf{G}_{11}\mathbf{X}^T$$

because $\mathbf{A}\mathbf{G}_{11}\mathbf{X}^T = \mathbf{0}$ as a consequence of Theorem 3.3.8c (with $\mathbf{X}^T\mathbf{X}$ and $\mathbf{A}^T$ again playing the roles of $\mathbf{A}$ and $\mathbf{B}$ in the theorem). Then by Theorem 3.3.2 again,

$$\mathbf{X}\mathbf{G}_{11}\mathbf{X}^T\mathbf{X}\mathbf{G}_{11}\mathbf{X}^T = \mathbf{X}\mathbf{G}_{11}\mathbf{X}^T,$$

i.e., $\mathbf{X}\mathbf{G}_{11}\mathbf{X}^T$ is idempotent.

Next, let $\mathbf{G}_{11}$ and $\mathbf{G}_{11}^*$ denote the upper left $p \times p$ blocks of any two generalized inverses of $\begin{pmatrix} \mathbf{X}^T\mathbf{X} & \mathbf{A}^T \\ \mathbf{A} & \mathbf{0} \end{pmatrix}$. Then by Theorem 3.3.8f,

$$\mathbf{X}^T\mathbf{X}\mathbf{G}_{11}\mathbf{X}^T\mathbf{X} = \mathbf{X}^T\mathbf{X}\mathbf{G}_{11}^*\mathbf{X}^T\mathbf{X},$$

and using Theorem 3.3.2 twice more we obtain

$$\mathbf{X}\mathbf{G}_{11}\mathbf{X}^T = \mathbf{X}\mathbf{G}_{11}^*\mathbf{X}^T.$$

This establishes part (c).

To prove part (d), note that

$$\begin{aligned}
\text{rank}\begin{pmatrix} \mathbf{X} \\ \mathbf{A} \end{pmatrix} + \text{rank}(\mathbf{A}) &= \text{rank}\begin{pmatrix} \mathbf{X}^T\mathbf{X} & \mathbf{A}^T \\ \mathbf{A} & \mathbf{0} \end{pmatrix} \\
&= \text{rank}\left[\begin{pmatrix} \mathbf{X}^T\mathbf{X} & \mathbf{A}^T \\ \mathbf{A} & \mathbf{0} \end{pmatrix}\begin{pmatrix} \mathbf{G}_{11} & \mathbf{G}_{12} \\ \mathbf{G}_{21} & \mathbf{G}_{22} \end{pmatrix}\right] \\
&= \text{rank}\begin{pmatrix} \mathbf{X}^T\mathbf{X}\mathbf{G}_{11} + \mathbf{A}^T\mathbf{G}_{21} & \mathbf{X}^T\mathbf{X}\mathbf{G}_{12} + \mathbf{A}^T\mathbf{G}_{22} \\ \mathbf{A}\mathbf{G}_{11} & \mathbf{A}\mathbf{G}_{12} \end{pmatrix} \\
&= \text{tr}\begin{pmatrix} \mathbf{X}^T\mathbf{X}\mathbf{G}_{11} + \mathbf{A}^T\mathbf{G}_{21} & \mathbf{X}^T\mathbf{X}\mathbf{G}_{12} + \mathbf{A}^T\mathbf{G}_{22} \\ \mathbf{A}\mathbf{G}_{11} & \mathbf{A}\mathbf{G}_{12} \end{pmatrix} \\
&= \text{tr}(\mathbf{X}^T\mathbf{X}\mathbf{G}_{11} + \mathbf{A}^T\mathbf{G}_{21}) + \text{tr}(\mathbf{A}\mathbf{G}_{12}) \\
&= \text{tr}(\mathbf{X}\mathbf{G}_{11}\mathbf{X}^T) + \text{rank}(\mathbf{A}) + \text{rank}(\mathbf{A}) \\
&= \text{rank}(\mathbf{X}\mathbf{G}_{11}\mathbf{X}^T) + 2 \cdot \text{rank}(\mathbf{A}),
\end{aligned}$$

where the first equality holds by Corollary 10.1.7.1, the second by Theorem 3.3.5, the fourth by Theorems 3.3.5 and 2.12.2, the sixth by Theorems 2.10.2 and 3.3.8g, and the last by part (b) of this theorem and Theorem 2.12.2. Part (d) follows immediately.

Finally, let $\mathbf{G}_{12}$ and $\mathbf{G}_{12}^*$ denote the upper right $p \times q$ blocks of any two generalized inverses of $\begin{pmatrix} \mathbf{X}^T\mathbf{X} & \mathbf{A}^T \\ \mathbf{A} & \mathbf{0} \end{pmatrix}$. Then by Theorem 3.3.8f,

$$\mathbf{X}^T\mathbf{X}\mathbf{G}_{12}\mathbf{A} = \mathbf{X}^T\mathbf{X}\mathbf{G}_{12}^*\mathbf{A},$$

and using Theorem 3.3.2 again we obtain

$$\mathbf{X}\mathbf{G}_{12}\mathbf{A} = \mathbf{X}\mathbf{G}_{12}^*\mathbf{A},$$

which, upon post-multiplying both sides by $\boldsymbol{\beta}$, establishes part (e). □

Theorem 10.1.8 implies that $\mathbf{XG}_{11}\mathbf{X}^T$ is an orthogonal projection matrix. Onto what space does it project $\mathbf{y}$? One answer, of course, is $\mathcal{C}(\mathbf{XG}_{11}\mathbf{X}^T)$, which is a subspace of $\mathcal{C}(\mathbf{X})$ of dimension $p^*_{\mathbf{A}} \leq p^*$ by Theorems 2.6.1 and 10.1.8d. The next theorem provides an equivalent but somewhat more illuminating answer to the same question.

Theorem 10.1.9 $\mathbf{XG}_{11}\mathbf{X}^T$ *is the orthogonal projection matrix onto* $\mathcal{C}[\mathbf{X}(\mathbf{I}-\mathbf{A}^-\mathbf{A})]$, *where* $\mathbf{A}^-$ *is an arbitrary generalized inverse of* $\mathbf{A}$.

Proof We begin by showing that $\mathcal{C}[\mathbf{X}(\mathbf{I}-\mathbf{A}^-\mathbf{A})]$ is well defined, i.e., invariant to the choice of generalized inverse of $\mathbf{A}$. Let $\mathbf{G}_1$ and $\mathbf{G}_2$ represent two distinct generalized inverses of $\mathbf{A}$. Then

$$\mathbf{X}(\mathbf{I}-\mathbf{G}_1\mathbf{A})(\mathbf{I}-\mathbf{G}_2\mathbf{A}) = \mathbf{X}(\mathbf{I}-\mathbf{G}_1\mathbf{A}-\mathbf{G}_2\mathbf{A}+\mathbf{G}_1\mathbf{A}\mathbf{G}_2\mathbf{A}) = \mathbf{X}(\mathbf{I}-\mathbf{G}_2\mathbf{A}).$$

Likewise, upon reversing the roles of $\mathbf{G}_1$ and $\mathbf{G}_2$, we obtain

$$\mathbf{X}(\mathbf{I}-\mathbf{G}_2\mathbf{A})(\mathbf{I}-\mathbf{G}_1\mathbf{A}) = \mathbf{X}(\mathbf{I}-\mathbf{G}_1\mathbf{A}).$$

This establishes (by Theorem 2.6.1) that $\mathcal{C}[\mathbf{X}(\mathbf{I}-\mathbf{G}_2\mathbf{A})] \subseteq \mathcal{C}(\mathbf{X}(\mathbf{I}-\mathbf{G}_1\mathbf{A})]$ and vice versa, and hence that $\mathcal{C}[\mathbf{X}(\mathbf{I}-\mathbf{G}_2\mathbf{A})] = \mathcal{C}[\mathbf{X}(\mathbf{I}-\mathbf{G}_1\mathbf{A})]$. Next, note that $\mathbf{XG}_{11}\mathbf{X}^T$ is symmetric and idempotent by Theorem 10.1.8a, b. By Theorem 3.3.8b, d (with $\mathbf{X}^T\mathbf{X}$ and $\mathbf{A}^T$ here playing the roles of $\mathbf{A}$ and $\mathbf{B}$ in the theorem),

$$\mathbf{X}^T\mathbf{X} = \mathbf{X}^T\mathbf{XG}_{11}\mathbf{X}^T\mathbf{X} - \mathbf{A}^T\mathbf{G}_{22}\mathbf{A} = \mathbf{X}^T(\mathbf{XG}_{11}\mathbf{X}^T\mathbf{X}+\mathbf{XG}_{12}\mathbf{A}),$$

so using Theorem 3.3.2, $\mathbf{XG}_{11}\mathbf{X}^T\mathbf{X} = \mathbf{X}-\mathbf{XG}_{12}\mathbf{A}$. Thus

$$\begin{aligned}\mathbf{XG}_{11}\mathbf{X}^T\mathbf{X}(\mathbf{I}-\mathbf{A}^-\mathbf{A}) &= (\mathbf{X}-\mathbf{XG}_{12}\mathbf{A})(\mathbf{I}-\mathbf{A}^-\mathbf{A})\\ &= \mathbf{X}(\mathbf{I}-\mathbf{A}^-\mathbf{A}).\end{aligned}$$

By Theorem 8.1.2, it remains to show that $\text{rank}[\mathbf{X}(\mathbf{I}-\mathbf{A}^-\mathbf{A})] = \text{rank}(\mathbf{XG}_{11}\mathbf{X}^T)$. To that end, observe that

$$\begin{pmatrix}\mathbf{X}(\mathbf{I}-\mathbf{A}^-\mathbf{A})\\ \mathbf{A}\end{pmatrix} = \begin{pmatrix}\mathbf{I} & -\mathbf{XA}^-\\ \mathbf{0} & \mathbf{I}\end{pmatrix}\begin{pmatrix}\mathbf{X}\\ \mathbf{A}\end{pmatrix}$$

and

$$\begin{pmatrix}\mathbf{X}\\ \mathbf{A}\end{pmatrix} = \begin{pmatrix}\mathbf{I} & \mathbf{XA}^-\\ \mathbf{0} & \mathbf{I}\end{pmatrix}\begin{pmatrix}\mathbf{X}(\mathbf{I}-\mathbf{A}^-\mathbf{A})\\ \mathbf{A}\end{pmatrix},$$

which shows that

$$\mathcal{R}\begin{pmatrix} \mathbf{X}(\mathbf{I}-\mathbf{A}^-\mathbf{A}) \\ \mathbf{A} \end{pmatrix} = \mathcal{R}\begin{pmatrix} \mathbf{X} \\ \mathbf{A} \end{pmatrix}. \tag{10.5}$$

Furthermore, suppose that vectors $\boldsymbol{\ell}$ and $\mathbf{m}$ exist such that $\boldsymbol{\ell}^T\mathbf{X}(\mathbf{I}-\mathbf{A}^-\mathbf{A}) = \mathbf{m}^T\mathbf{A}$. Then, after post-multiplication by $\mathbf{I}-\mathbf{A}^-\mathbf{A}$ we have

$$\boldsymbol{\ell}^T\mathbf{X}(\mathbf{I}-\mathbf{A}^-\mathbf{A})^2 = \mathbf{m}^T\mathbf{A}(\mathbf{I}-\mathbf{A}^-\mathbf{A}) = \mathbf{0}^T,$$

or equivalently $\boldsymbol{\ell}^T\mathbf{X}(\mathbf{I}-\mathbf{A}^-\mathbf{A}) = \mathbf{0}^T$, which implies that $\mathcal{R}[\mathbf{X}(\mathbf{I}-\mathbf{A}^-\mathbf{A})] \cap \mathcal{R}(\mathbf{A}) = \mathbf{0}$. This and (10.5) imply that

$$\text{rank}[\mathbf{X}(\mathbf{I}-\mathbf{A}^-\mathbf{A})] = \text{rank}\begin{pmatrix} \mathbf{X} \\ \mathbf{A} \end{pmatrix} - \text{rank}(\mathbf{A}),$$

which by Theorem 10.1.8 is the same as $\text{rank}(\mathbf{X}\mathbf{G}_{11}\mathbf{X}^T)$. □

The following result, which emerged within the proof of Theorem 10.1.9, is worth documenting so we state it as a corollary.

Corollary 10.1.9.1 *rank*$(\mathbf{P}_{\mathbf{X}(\mathbf{I}-\mathbf{A}^-\mathbf{A})}) = p^*_{\mathbf{A}}$.

Expression (10.4) together with Theorems 10.1.8 and 10.1.9 establish that

$$\mathbf{X}\breve{\boldsymbol{\beta}} = \mathbf{P}_{\mathbf{X}(\mathbf{I}-\mathbf{A}^-\mathbf{A})}\mathbf{y} + \mathbf{X}\mathbf{G}_{12}\mathbf{h}, \tag{10.6}$$

which reveals that the vector of fitted values obtained by constrained least squares estimation is the sum of two vectors, one of which is the orthogonal projection of $\mathbf{y}$ onto $\mathcal{C}[\mathbf{X}(\mathbf{I}-\mathbf{A}^-\mathbf{A})]$. Thus $\mathbf{X}\breve{\boldsymbol{\beta}}$ is the sum of an n-vector in a $p^*_{\mathbf{A}}$-dimensional subspace of $\mathcal{C}(\mathbf{X})$ and a fixed (not functionally dependent on $\mathbf{y}$) n-vector, the latter generally belonging to a different subspace of $\mathcal{C}(\mathbf{X})$. The mathematical term for such a geometric structure (the set of all vectors equal to the sum of a vector in a given linear subspace and another fixed vector of the same dimensions) is an *affine space*.

Example 10.1-2. Toy Example of the Geometry of Constrained Least Squares Estimation

A toy example will serve to illustrate the geometry of constrained least squares estimation just described. Suppose that $n = 3$, $p = p^* = 2$,

$$\mathbf{y} = \begin{pmatrix} 1 \\ 2 \\ 3 \end{pmatrix}, \quad \mathbf{X} = \begin{pmatrix} 1 & 0 \\ 0 & 1 \\ 0 & 0 \end{pmatrix}, \quad \mathbf{A} = \begin{pmatrix} 1 & 1 \end{pmatrix},$$

and $\mathbf{h} = 1$. Here, $\mathcal{C}(\mathbf{X}) = \mathbb{R}^2$ and $p^*_{\mathbf{A}} = 2 - 1 = 1$. One generalized inverse of $\mathbf{A}$ is $\mathbf{A}^- = (0.5, 0.5)^T$, for which

$$\mathbf{I} - \mathbf{A}^-\mathbf{A} = \begin{pmatrix} 0.5 & -0.5 \\ -0.5 & 0.5 \end{pmatrix}.$$

It is easily verified that

$$\mathbf{X}(\mathbf{I} - \mathbf{A}^-\mathbf{A}) = \begin{pmatrix} 0.5 & -0.5 \\ -0.5 & 0.5 \\ 0 & 0 \end{pmatrix}$$

and hence that $\mathcal{C}[\mathbf{X}(\mathbf{I} - \mathbf{A}^-\mathbf{A})] = \{(1, -1, 0)^T a : a \in \mathbb{R}\}$. Furthermore, the coefficient matrix of the constrained normal equations (which is nonsingular) and its inverse are

$$\begin{pmatrix} 1 & 0 & 1 \\ 0 & 1 & 1 \\ 1 & 1 & 0 \end{pmatrix} \quad \text{and} \quad \begin{pmatrix} 0.5 & -0.5 & 0.5 \\ -0.5 & 0.5 & 0.5 \\ 0.5 & 0.5 & -0.5 \end{pmatrix},$$

respectively, where we used Theorem 2.9.5 to find the inverse. Extracting $\mathbf{G}_{12}$ as the upper right 2×1 matrix from the inverse, we find that $\mathbf{X}\mathbf{G}_{12} = (0.5, 0.5, 0)^T$. Figure 10.1 depicts the geometry of the situation. The orthogonal projection of $\mathbf{y}$ onto $\mathcal{C}[\mathbf{X}(\mathbf{I} - \mathbf{A}^-\mathbf{A})]$ is $(-0.5, 0.5, 0)^T$, and the vector $\mathbf{X}\mathbf{G}_{12}\mathbf{h}$, when added to it, to yields $\mathbf{X}\breve{\boldsymbol{\beta}} = (0, 1, 0)^T$. ∎

The next theorem gives some results on the first two moments of the vector of fitted values and vector of residuals from constrained least squares estimation. Its proof is left as an exercise.

Theorem 10.1.10 *Define $\breve{\mathbf{y}} = \mathbf{X}\breve{\boldsymbol{\beta}}$ and $\breve{\mathbf{e}} = \mathbf{y} - \mathbf{X}\breve{\boldsymbol{\beta}}$, where $\mathbf{X}\breve{\boldsymbol{\beta}}$ is given by (10.6). Then, under the constrained Gauss–Markov model $\{\mathbf{y}, \mathbf{X}\boldsymbol{\beta}, \sigma^2\mathbf{I} : \mathbf{A}\boldsymbol{\beta} = \mathbf{h}\}$:*

(a) $E(\breve{\mathbf{y}}) = \mathbf{X}\boldsymbol{\beta}$;
(b) $E(\breve{\mathbf{e}}) = \mathbf{0}$;
(c) $var(\breve{\mathbf{y}}) = \sigma^2\mathbf{P}_{\mathbf{X}(\mathbf{I}-\mathbf{A}^-\mathbf{A})}$;
(d) $var(\breve{\mathbf{e}}) = \sigma^2(\mathbf{I} - \mathbf{P}_{\mathbf{X}(\mathbf{I}-\mathbf{A}^-\mathbf{A})})$; and
(e) $cov(\breve{\mathbf{y}}, \breve{\mathbf{e}}) = \mathbf{0}$.

Next, we consider the estimation of σ^2 under a constrained Gauss–Markov model. To this end, we obtain an expression for the constrained residual sum of squares, i.e., $Q(\breve{\boldsymbol{\beta}})$, in terms of a solution to the constrained normal equations and known or observable quantities in the model.

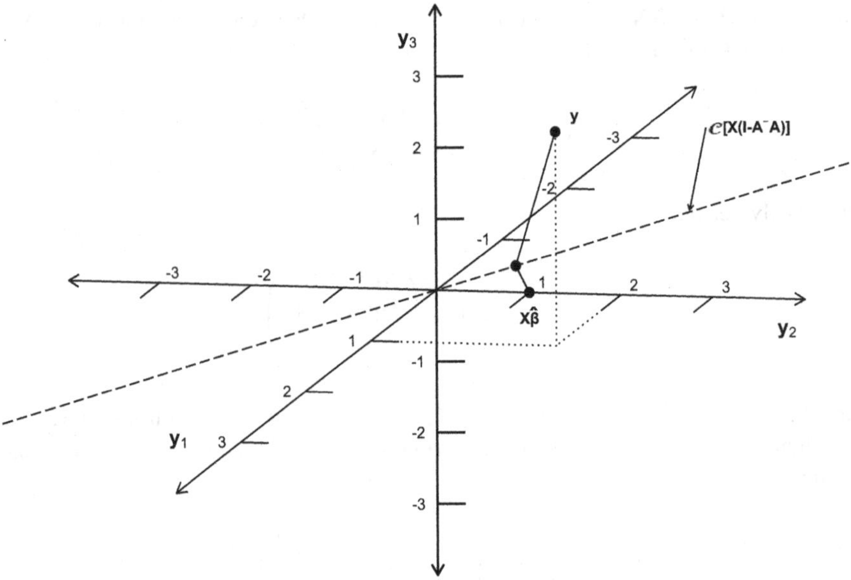

Fig. 10.1 A toy example of the geometry of constrained least squares estimation

Theorem 10.1.11 *Let* $(\breve{\boldsymbol{\beta}}^T, \breve{\boldsymbol{\lambda}}^T)^T$ *represent any solution to the constrained normal equations for the model* $\{\mathbf{y}, \mathbf{X}\boldsymbol{\beta} : \mathbf{A}\boldsymbol{\beta} = \mathbf{h}\}$. *Then,*

$$(\mathbf{X}\breve{\boldsymbol{\beta}})^T(\mathbf{X}\breve{\boldsymbol{\beta}}) = \breve{\boldsymbol{\beta}}^T\mathbf{X}^T\mathbf{y} - \breve{\boldsymbol{\lambda}}^T\mathbf{h} \quad \textit{and} \quad Q(\breve{\boldsymbol{\beta}}) = \mathbf{y}^T\mathbf{y} - \breve{\boldsymbol{\beta}}^T\mathbf{X}^T\mathbf{y} - \breve{\boldsymbol{\lambda}}^T\mathbf{h}.$$

Proof Using the constrained normal equations, we find that

$$(\mathbf{X}\breve{\boldsymbol{\beta}})^T(\mathbf{X}\breve{\boldsymbol{\beta}}) = \breve{\boldsymbol{\beta}}^T\mathbf{X}^T\mathbf{X}\breve{\boldsymbol{\beta}} = \breve{\boldsymbol{\beta}}^T(\mathbf{X}^T\mathbf{y} - \mathbf{A}^T\breve{\boldsymbol{\lambda}}) = \breve{\boldsymbol{\beta}}^T\mathbf{X}^T\mathbf{y} - \breve{\boldsymbol{\lambda}}^T\mathbf{h}$$

and

$$\begin{aligned} Q(\breve{\boldsymbol{\beta}}) &= (\mathbf{y} - \mathbf{X}\breve{\boldsymbol{\beta}})^T(\mathbf{y} - \mathbf{X}\breve{\boldsymbol{\beta}}) \\ &= \mathbf{y}^T\mathbf{y} - \mathbf{y}^T\mathbf{X}\breve{\boldsymbol{\beta}} - \breve{\boldsymbol{\beta}}^T(\mathbf{X}^T\mathbf{y} - \mathbf{X}^T\mathbf{X}\breve{\boldsymbol{\beta}}) \\ &= \mathbf{y}^T\mathbf{y} - \breve{\boldsymbol{\beta}}^T\mathbf{X}^T\mathbf{y} - \breve{\boldsymbol{\beta}}^T\mathbf{A}^T\breve{\boldsymbol{\lambda}} \\ &= \mathbf{y}^T\mathbf{y} - \breve{\boldsymbol{\beta}}^T\mathbf{X}^T\mathbf{y} - \breve{\boldsymbol{\lambda}}^T\mathbf{h}. \end{aligned}$$

□

Now we may use the result of the previous theorem to obtain an unbiased estimator of the residual variance σ^2 under a constrained Gauss–Markov model.

Theorem 10.1.12 *Let* $(\breve{\boldsymbol{\beta}}^T, \breve{\boldsymbol{\lambda}}^T)^T$ *be any solution to the constrained normal equations for the model* $\{\mathbf{y}, \mathbf{X}\boldsymbol{\beta} : \mathbf{A}\boldsymbol{\beta} = \mathbf{h}\}$, *and suppose that* $n > p^*_{\mathbf{A}}$. *Then, under the constrained Gauss–Markov model* $\{\mathbf{y}, \mathbf{X}\boldsymbol{\beta}, \sigma^2\mathbf{I} : \mathbf{A}\boldsymbol{\beta} = \mathbf{h}\}$,

$$\breve{\sigma}^2 = (\mathbf{y}^T\mathbf{y} - \breve{\boldsymbol{\beta}}^T\mathbf{X}^T\mathbf{y} - \breve{\boldsymbol{\lambda}}^T\mathbf{h})/(n - p^*_{\mathbf{A}})$$

is an unbiased estimator of σ^2.

Proof Let $\begin{pmatrix} \mathbf{G}_{11} & \mathbf{G}_{12} \\ \mathbf{G}_{21} & \mathbf{G}_{22} \end{pmatrix}$ represent any generalized inverse of $\begin{pmatrix} \mathbf{X}^T\mathbf{X} & \mathbf{A}^T \\ \mathbf{A} & \mathbf{0} \end{pmatrix}$, where $\mathbf{G}_{11}$ is $p \times p$. Then

$$\begin{aligned}
&\mathrm{E}(\mathbf{y}^T\mathbf{y} - \breve{\boldsymbol{\beta}}^T\mathbf{X}^T\mathbf{y} - \breve{\boldsymbol{\lambda}}^T\mathbf{h}) \\
&= \mathrm{E}[\mathbf{y}^T\mathbf{y} - (\mathbf{G}_{11}\mathbf{X}^T\mathbf{y} + \mathbf{G}_{12}\mathbf{h})^T\mathbf{X}^T\mathbf{y} - (\mathbf{G}_{21}\mathbf{X}^T\mathbf{y} + \mathbf{G}_{22}\mathbf{h})^T\mathbf{h}] \\
&= \mathrm{tr}(\sigma^2\mathbf{I}) + \boldsymbol{\beta}^T\mathbf{X}^T\mathbf{X}\boldsymbol{\beta} - \mathrm{tr}(\sigma^2\mathbf{X}\mathbf{G}_{11}^T\mathbf{X}^T) - \boldsymbol{\beta}^T\mathbf{X}^T\mathbf{X}\mathbf{G}_{11}^T\mathbf{X}^T\mathbf{X}\boldsymbol{\beta} \\
&\quad - \mathbf{h}^T\mathbf{G}_{12}^T\mathbf{X}^T\mathbf{X}\boldsymbol{\beta} - \boldsymbol{\beta}^T\mathbf{X}^T\mathbf{X}\mathbf{G}_{21}^T\mathbf{h} - \mathbf{h}^T\mathbf{G}_{22}^T\mathbf{h} \\
&= n\sigma^2 - \sigma^2\mathrm{tr}(\mathbf{X}^T\mathbf{X}\mathbf{G}_{11}^T) + \boldsymbol{\beta}^T(\mathbf{X}^T\mathbf{X} - \mathbf{X}^T\mathbf{X}\mathbf{G}_{11}^T\mathbf{X}^T\mathbf{X} - \mathbf{A}^T\mathbf{G}_{12}^T\mathbf{X}^T\mathbf{X} \\
&\quad - \mathbf{X}^T\mathbf{X}\mathbf{G}_{21}^T\mathbf{A} - \mathbf{A}^T\mathbf{G}_{22}^T\mathbf{A})\boldsymbol{\beta} \\
&= n\sigma^2 - \sigma^2 p^*_{\mathbf{A}} \\
&\quad + \boldsymbol{\beta}^T(-\mathbf{A}^T\mathbf{G}_{22}\mathbf{A} - \mathbf{A}^T\mathbf{G}_{12}^T\mathbf{X}^T\mathbf{X} - \mathbf{X}^T\mathbf{X}\mathbf{G}_{21}^T\mathbf{A} - \mathbf{A}^T\mathbf{G}_{22}\mathbf{A})\boldsymbol{\beta} \\
&= \sigma^2(n - p^*_{\mathbf{A}})
\end{aligned}$$

where we used Theorem 4.2.4 for the second equality and Theorem 3.3.8d, e and b (with $\mathbf{X}^T\mathbf{X}$ and $\mathbf{A}^T$ here playing the roles of **A** and **B**, respectively, in the theorem) for the fourth and fifth equalities. The unbiasedness of $\breve{\sigma}^2$ follows immediately. □

Henceforth, we refer to $\breve{\sigma}^2$, the unbiased estimator of σ^2 defined in Theorem 10.1.12, as the *constrained residual mean square*.

Example 10.1-3. Constrained Least Squares Estimation for a Piecewise Continuous Simple Linear Regression Model

Consider the model

$$y_i = \begin{cases} \beta_1 + \beta_2(x_i - \bar{x}_-) + e_i, & \text{if } x_i \le c \\ \gamma_1 + \gamma_2(x_i - \bar{x}_+) + e_i, & \text{if } x_i > c \end{cases} \qquad (i = 1, \ldots, n)$$

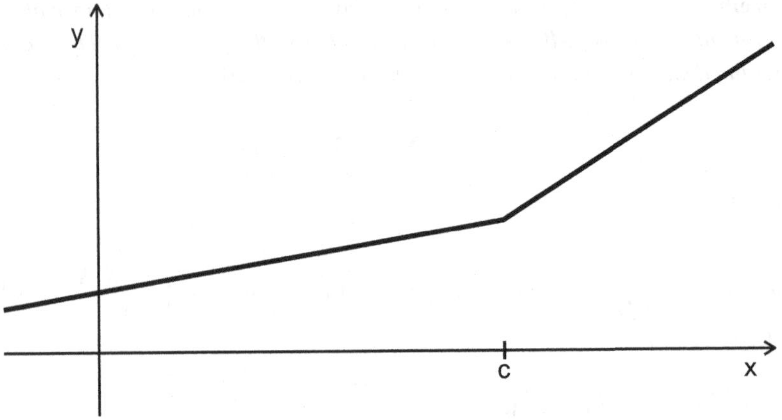

Fig. 10.2 Depiction of E(y) for the piecewise continuous simple linear regression model

with the restriction that E(y) be continuous at $x = c$ (see Fig. 10.2). Here c is a specified real number, $\bar{x}_-$ is the average of those x_i's that are less than or equal to c, $\bar{x}_+$ is the average of the other x_i's, and the e_i's are uncorrelated and have common variance $\sigma^2 > 0$. This model is known as the (Gauss–Markov) piecewise continuous simple linear regression model. The model may be written as

$$\mathbf{y} = \mathbf{X}\boldsymbol{\beta} + \mathbf{e}, \qquad \mathbf{A}\boldsymbol{\beta} = \mathbf{h},$$

where $\mathbf{A} = (1, c - \bar{x}_-, -1, -(c - \bar{x}_+))$, $\boldsymbol{\beta} = (\beta_1, \beta_2, \gamma_1, \gamma_2)^T$, $\mathbf{h} = 0$, E($\mathbf{e}$) $= \mathbf{0}$, and var($\mathbf{e}$) $= \sigma^2\mathbf{I}$. Thus, it is a constrained linear model of the type described in this chapter.

Assume that there are at least two distinct x_i's less than c and at least two distinct x_i's greater than c. Then $\text{rank}(\mathbf{X}) = \text{rank}\begin{pmatrix}\mathbf{X}\\ \mathbf{A}\end{pmatrix} = 4$, implying that every function $\mathbf{c}^T\boldsymbol{\beta}$ is estimable and that $p^*_{\mathbf{A}} = 4 - 1 = 3$. Let

$$n_- = \#\{x_i \leq c\}, \quad n_+ = n - n_-, \quad S_- = \{i : x_i \leq c\}, \quad S_+ = \{i : x_i > c\},$$

$$SXX_- = \sum_{S_-}(x_i - \bar{x}_-)^2, \quad SXX_+ = \sum_{S_+}(x_i - \bar{x}_+)^2,$$

$$SXY_- = \sum_{S_-}(x_i - \bar{x}_-)y_i, \quad SXY_+ = \sum_{S_+}(x_i - \bar{x}_+)y_i.$$

Then by Corollary 10.1.7.1, the rank of the coefficient matrix of the constrained normal equations is equal to $4 + 1 = 5$, so in this case there is a unique solution to the constrained normal equations

$$\begin{pmatrix} n_- & 0 & 0 & 0 & 1 \\ 0 & SXX_- & 0 & 0 & c - \bar{x}_- \\ 0 & 0 & n_+ & 0 & -1 \\ 0 & 0 & 0 & SXX_+ & -(c - \bar{x}_+) \\ 1 & c - \bar{x}_- & -1 & -(c - \bar{x}_+) & 0 \end{pmatrix} \begin{pmatrix} \beta_1 \\ \beta_2 \\ \gamma_1 \\ \gamma_2 \\ \lambda \end{pmatrix} = \begin{pmatrix} \sum_{S_-} y_i \\ SXY_- \\ \sum_{S_+} y_i \\ SXY_+ \\ 0 \end{pmatrix}.$$

Because $\mathbf{h} = 0$ here, by Theorem 10.1.4 the constrained least squares estimator of $\boldsymbol{\beta}$ is $\breve{\boldsymbol{\beta}} = \mathbf{G}_{11}\mathbf{X}^T\mathbf{y}$. Using Theorem 2.9.5, the inverse of the coefficient matrix of the constrained normal equations may be obtained easily and the upper left 4×4 matrix $\mathbf{G}_{11}$ extracted from it, yielding

$$\begin{pmatrix} \breve{\beta}_1 \\ \breve{\beta}_2 \\ \breve{\gamma}_1 \\ \breve{\gamma}_2 \end{pmatrix} = \begin{pmatrix} \frac{1}{n_-}(1 + \frac{1}{an_-}) & \frac{c-\bar{x}_-}{an_-SXX_-} & -\frac{1}{an_-n_+} & -\frac{c-\bar{x}_+}{an_-SXX_+} \\ & \frac{1}{SXX_-} + \frac{(c-\bar{x}_-)^2}{aSXX_-^2} & -\frac{(c-\bar{x}_-)}{aSXX_-n_+} & -\frac{(c-\bar{x}_-)(c-\bar{x}_+)}{aSXX_-SXX_+} \\ & & \frac{1}{n_+}(1 + \frac{1}{an_+}) & \frac{c-\bar{x}_+}{an_+SXX_+} \\ & & & \frac{1}{SXX_+} + \frac{(c-\bar{x}_+)^2}{aSXX_+^2} \end{pmatrix} \begin{pmatrix} \sum_{S_-} y_i \\ SXY_- \\ \sum_{S_+} y_i \\ SXY_+ \end{pmatrix}.$$

where

$$a = -\frac{1}{n_-} - \frac{(c - \bar{x}_-)^2}{SXX_-} - \frac{1}{n_+} - \frac{(c - \bar{x}_+)^2}{SXX_+}.$$

By Theorem 10.1.5, the variance–covariance matrix of $\breve{\boldsymbol{\beta}}$ is $\sigma^2\mathbf{G}_{11}$. Furthermore, the constrained residual mean square is

$$\breve{\sigma}^2 = \left(\sum_{S_-} y_i^2 + \sum_{S_+} y_i^2 - \breve{\beta}_1 \sum_{S_-} y_i - \breve{\beta}_2 SXY_- - \breve{\gamma}_1 \sum_{S_+} y_i - \breve{\gamma}_2 SXY_+ \right) \Big/ (n-3).$$

■

Recall that under an unconstrained Gauss–Markov model, if $\hat{\boldsymbol{\beta}}$ is a solution to the normal equations for that model then $Q(\hat{\boldsymbol{\beta}})/(n - p^*)$ is not only unbiased for σ^2 but is also the residual mean square from an ANOVA (the overall ANOVA of the unconstrained model). This, and the unbiasedness (just established by Theorem 10.1.12) of $\breve{\sigma}^2 = Q(\breve{\boldsymbol{\beta}})/(n - p^*_{\mathbf{A}})$ for σ^2 under a constrained Gauss–Markov model, lead naturally to the following question: Is there an ANOVA associated with the constrained model? More precisely, is there an orthogonal decomposition of $\mathbf{y}$ for which the corresponding sums of squares include $Q(\breve{\boldsymbol{\beta}})$ and sum to $\mathbf{y}^T\mathbf{y}$? In general, the answer is no. To see this, let $\begin{pmatrix} \mathbf{G}_{11} & \mathbf{G}_{12} \\ \mathbf{G}_{21} & \mathbf{G}_{22} \end{pmatrix}$ (where

$\mathbf{G}_{11}$ is $p \times p$) represent an arbitrary generalized inverse of the coefficient matrix $\begin{pmatrix} \mathbf{X}^T\mathbf{X} & \mathbf{A}^T \\ \mathbf{A} & \mathbf{0} \end{pmatrix}$ of the constrained normal equations for the model $\{\mathbf{y}, \mathbf{X}\boldsymbol{\beta} : \mathbf{A}\boldsymbol{\beta} = \mathbf{h}\}$ and note, by Corollary 3.3.1.1, that $\begin{pmatrix} \mathbf{G}_{11}^T & \mathbf{G}_{21}^T \\ \mathbf{G}_{12}^T & \mathbf{G}_{22}^T \end{pmatrix}$ is also a generalized inverse of $\begin{pmatrix} \mathbf{X}^T\mathbf{X} & \mathbf{A}^T \\ \mathbf{A} & \mathbf{0} \end{pmatrix}$. Then

$$\begin{aligned} Q(\breve{\boldsymbol{\beta}}) &= \mathbf{y}^T\mathbf{y} - \breve{\boldsymbol{\beta}}^T\mathbf{X}^T\mathbf{y} - \breve{\boldsymbol{\lambda}}^T\mathbf{h} \\ &= \mathbf{y}^T\mathbf{y} - (\mathbf{G}_{11}\mathbf{X}^T\mathbf{y} + \mathbf{G}_{12}\mathbf{h})^T\mathbf{X}^T\mathbf{y} - (\mathbf{G}_{21}\mathbf{X}^T\mathbf{y} + \mathbf{G}_{22}\mathbf{h})^T\mathbf{h} \\ &= \mathbf{y}^T\mathbf{y} - \mathbf{y}^T\mathbf{X}\mathbf{G}_{11}^T\mathbf{X}^T\mathbf{y} - \mathbf{h}^T\mathbf{G}_{12}^T\mathbf{X}^T\mathbf{y} - \mathbf{y}^T\mathbf{X}\mathbf{G}_{21}^T\mathbf{h} - \mathbf{h}^T\mathbf{G}_{22}^T\mathbf{h} \\ &= \mathbf{y}^T\mathbf{y} - \mathbf{y}^T\mathbf{X}\mathbf{G}_{11}\mathbf{X}^T\mathbf{y} - \mathbf{h}^T\mathbf{G}_{21}\mathbf{X}^T\mathbf{y} - \mathbf{h}^T\mathbf{G}_{21}\mathbf{X}^T\mathbf{y} - \mathbf{h}^T\mathbf{G}_{22}\mathbf{h} \\ &= \mathbf{y}^T\mathbf{y} - \mathbf{y}^T\mathbf{X}\mathbf{G}_{11}\mathbf{X}^T\mathbf{y} - 2\mathbf{h}^T\mathbf{G}_{21}\mathbf{X}^T\mathbf{y} - \mathbf{h}^T\mathbf{G}_{22}\mathbf{h}, \end{aligned}$$

where we used Theorem 10.1.11 for the first equality, Theorem 10.1.4 for the second equality, and symmetry of scalars plus Theorem 10.1.8e for the fourth equality. Thus, the total sum of squares may be decomposed as follows:

$$\mathbf{y}^T\mathbf{y} = (\mathbf{y}^T\mathbf{X}\mathbf{G}_{11}\mathbf{X}^T\mathbf{y} + 2\mathbf{h}^T\mathbf{G}_{21}\mathbf{X}^T\mathbf{y} + \mathbf{h}^T\mathbf{G}_{22}\mathbf{h}) + Q(\breve{\boldsymbol{\beta}}). \tag{10.7}$$

In general, neither $\mathbf{y}^T\mathbf{X}\mathbf{G}_{11}\mathbf{X}^T\mathbf{y} + 2\mathbf{h}^T\mathbf{G}_{21}\mathbf{X}^T\mathbf{y} + \mathbf{h}^T\mathbf{G}_{22}\mathbf{h}$ nor $Q(\breve{\boldsymbol{\beta}})$ is a sum of squares of $\mathbf{y}$ (or of a linear transformation of $\mathbf{y}$); equivalently, neither term is a nonnegative definite quadratic form of $\mathbf{y}$. Therefore, this decomposition does not satisfy the definition of an ANOVA given in Chap. 8. Instead, both terms in this decomposition are, in general, second-order polynomial functions of $\mathbf{y}$. (A second-order polynomial function of $\mathbf{y}$ is the sum of a quadratic form in $\mathbf{y}$, a linear form in $\mathbf{y}$, and a constant.) In some important cases of the constrained linear model, however, these second-order polynomial functions reduce to sums of squares, and furthermore, those sums of squares correspond to an orthogonal decomposition of $\mathbf{y}$. In those cases, the decomposition just described is a *bona fide* ANOVA. And in another important case, even though (10.7) is not an ANOVA, it still is a useful decomposition for some inferential purposes. These cases are considered in the remaining sections of this chapter. In any case, we may write (10.7) in tabular form as follows:

Source	Rank	Second-order polynomial
Constrained Model	$p_{\mathbf{A}}^*$	$\mathbf{y}^T\mathbf{X}\mathbf{G}_{11}\mathbf{X}^T\mathbf{y} + 2\mathbf{h}^T\mathbf{G}_{21}\mathbf{X}^T\mathbf{y} + \mathbf{h}^T\mathbf{G}_{22}\mathbf{h}$
Constrained Residual	$n - p_{\mathbf{A}}^*$	$Q(\breve{\boldsymbol{\beta}})$
Total	n	$\mathbf{y}^T\mathbf{y}$

Here, by analogy with an ANOVA, the "Rank" column gives the rank of the matrix of the quadratic form involved in each second-order polynomial function.

That the ranks of these two matrices are $p^*_{\mathbf{A}}$ and $n - p^*_{\mathbf{A}}$ is a direct consequence of Theorems 10.1.8 and 10.1.12.

10.2 The Case of Homogeneous Constraints

The first special case we consider is that of homogeneous constraints, i.e., constraints of the form $\mathbf{A}\boldsymbol{\beta} = \mathbf{0}$. In this case, Theorem 6.2.1 specializes, telling us that a linear estimator $t_0 + \mathbf{t}^T\mathbf{y}$ is unbiased for a linear parametric function $\mathbf{c}^T\boldsymbol{\beta}$ if and only if $t_0 = 0$ and $\mathbf{t}^T\mathbf{X} + \mathbf{g}^T\mathbf{A} = \mathbf{c}^T$ for some q-vector $\mathbf{g}$. However, the necessary and sufficient condition for estimability of a function $\mathbf{c}^T\boldsymbol{\beta}$ specified by Theorem 6.2.2 remains the same: $\mathbf{c}^T \in \mathcal{R}\begin{pmatrix}\mathbf{X}\\ \mathbf{A}\end{pmatrix}$. Likewise, although the right-hand side of the constrained normal equations specializes to $\begin{pmatrix}\mathbf{X}^T\mathbf{y}\\ \mathbf{0}\end{pmatrix}$, the rank of the coefficient matrix of those equations is unaffected. The constrained least squares estimator of an estimable function $\mathbf{c}^T\boldsymbol{\beta}$ may be expressed either as $\mathbf{c}^T\breve{\boldsymbol{\beta}}$, where $\breve{\boldsymbol{\beta}}$ is the first p-component subvector of any solution to the constrained normal equations, or (by specializing Theorem 10.1.4) as $\mathbf{c}^T\mathbf{G}_{11}\mathbf{X}^T\mathbf{y}$. However, it can be seen from Theorem 10.1.5 that the fact that the constraints are homogeneous has no impact on the variance–covariance matrix of constrained least squares estimators of a vector of estimable functions.

One specialized feature that arises when the constraints are homogeneous, however, manifests in the vector of fitted values. It can be seen from (10.6) that in this case $\mathbf{X}\breve{\boldsymbol{\beta}}$ is an orthogonal projection of $\mathbf{y}$; more specifically,

$$\mathbf{X}\breve{\boldsymbol{\beta}} = \mathbf{X}\mathbf{G}_{11}\mathbf{X}^T\mathbf{y} = \mathbf{P}_{\mathbf{X}(\mathbf{I}-\mathbf{A}^-\mathbf{A})}\mathbf{y}. \tag{10.8}$$

Thus in this case $\mathbf{y}$ admits the orthogonal decomposition

$$\mathbf{y} = \mathbf{P}_{\mathbf{X}(\mathbf{I}-\mathbf{A}^-\mathbf{A})}\mathbf{y} + (\mathbf{I} - \mathbf{P}_{\mathbf{X}(\mathbf{I}-\mathbf{A}^-\mathbf{A})})\mathbf{y}, \tag{10.9}$$

which upon pre-multiplication by $\mathbf{y}^T$ yields the following decomposition of the total sum of squares:

$$\mathbf{y}^T\mathbf{y} = \mathbf{y}^T\mathbf{P}_{\mathbf{X}(\mathbf{I}-\mathbf{A}^-\mathbf{A})}\mathbf{y} + \mathbf{y}^T(\mathbf{I} - \mathbf{P}_{\mathbf{X}(\mathbf{I}-\mathbf{A}^-\mathbf{A})})\mathbf{y}.$$

We call this decomposition, which is a specialization of (10.7), the *constrained overall ANOVA*. Furthermore, by Theorem 10.1.8d,

$$\text{rank}(\mathbf{P}_{\mathbf{X}(\mathbf{I}-\mathbf{A}^-\mathbf{A})}) = p^*_{\mathbf{A}}$$

and

$$\text{rank}(\mathbf{I} - \mathbf{P}_{\mathbf{X}(\mathbf{I}-\mathbf{A}^{-}\mathbf{A})}) = \text{tr}(\mathbf{I} - \mathbf{P}_{\mathbf{X}(\mathbf{I}-\mathbf{A}^{-}\mathbf{A})}) = n - p_{\mathbf{A}}^{*}.$$

Therefore, we may write the constrained overall ANOVA as follows:

Source	Rank	Sum of squares	Mean square
Constrained Model	$p_{\mathbf{A}}^{*}$	$\mathbf{y}^T\mathbf{P}_{\mathbf{X}(\mathbf{I}-\mathbf{A}^{-}\mathbf{A})}\mathbf{y}$	$\mathbf{y}^T\mathbf{P}_{\mathbf{X}(\mathbf{I}-\mathbf{A}^{-}\mathbf{A})}\mathbf{y}/p_{\mathbf{A}}^{*}$
Constrained Residual	$n - p_{\mathbf{A}}^{*}$	$\mathbf{y}^T(\mathbf{I} - \mathbf{P}_{\mathbf{X}(\mathbf{I}-\mathbf{A}^{-}\mathbf{A})})\mathbf{y}$	$\mathbf{y}^T(\mathbf{I} - \mathbf{P}_{\mathbf{X}(\mathbf{I}-\mathbf{A}^{-}\mathbf{A})})\mathbf{y}/(n - p_{\mathbf{A}}^{*})$
Total	n	$\mathbf{y}^T\mathbf{y}$	

By (10.8) and the idempotency of $\mathbf{P}_{\mathbf{X}(\mathbf{I}-\mathbf{A}^{-}\mathbf{A})}$, the sum of squares in the "Constrained Model" line of the ANOVA table may be written alternatively as $\breve{\boldsymbol{\beta}}\mathbf{X}^T\mathbf{X}\breve{\boldsymbol{\beta}}$ or (by putting $\mathbf{h} = \mathbf{0}$ into Theorem 10.1.11) as $\breve{\boldsymbol{\beta}}^T\mathbf{X}^T\mathbf{y}$. Furthermore, the sum of squares in the "Constrained Residual" line is equivalent to $\mathbf{y}^T(\mathbf{I} - \mathbf{P}_{\mathbf{X}(\mathbf{I}-\mathbf{A}^{-}\mathbf{A})})^T(\mathbf{I} - \mathbf{P}_{\mathbf{X}(\mathbf{I}-\mathbf{A}^{-}\mathbf{A})})\mathbf{y}$, which is also equivalent to the expression for the constrained residual sum of squares $Q(\breve{\boldsymbol{\beta}})$ obtained upon setting $\mathbf{h} = \mathbf{0}$ in Theorem 10.1.11: $\mathbf{y}^T\mathbf{y} - \breve{\boldsymbol{\beta}}^T\mathbf{X}^T\mathbf{y}$. Thus, when the constraints are homogeneous, the residual mean square from the constrained overall ANOVA coincides with the constrained residual mean square, $\breve{\sigma}^2$.

10.3 The Case of Estimable Constraints

In this section we suppose that the constraints for the model $\{\mathbf{y}, \mathbf{X}\boldsymbol{\beta} : \mathbf{A}\boldsymbol{\beta} = \mathbf{h}\}$ are estimable under the corresponding unconstrained model $\{\mathbf{y}, \mathbf{X}\boldsymbol{\beta}\}$, i.e., that $\mathcal{R}(\mathbf{A}) \subseteq \mathcal{R}(\mathbf{X})$ or equivalently that $\mathbf{A} = \mathbf{M}\mathbf{X}$ for some $q \times n$ matrix $\mathbf{M}$. In this case, the necessary and sufficient condition for estimability of a function $\mathbf{c}^T\boldsymbol{\beta}$ specified by Theorem 6.2.2 specializes to $\mathbf{c}^T \in \mathcal{R}(\mathbf{X})$, i.e., $\mathbf{c}^T\boldsymbol{\beta}$ is estimable under the constrained model if and only if it is estimable under the unconstrained model. The coefficient matrix of the constrained normal equations specializes to

$$\begin{pmatrix} \mathbf{X}^T\mathbf{X} & \mathbf{X}^T\mathbf{M}^T \\ \mathbf{M}\mathbf{X} & \mathbf{0} \end{pmatrix}.$$

Because $\mathcal{R}(\mathbf{M}\mathbf{X}) \subseteq \mathcal{R}(\mathbf{X}) = \mathcal{R}(\mathbf{X}^T\mathbf{X})$ and $\mathcal{C}(\mathbf{X}^T\mathbf{M}^T) \subseteq \mathcal{C}(\mathbf{X}^T) = \mathcal{C}(\mathbf{X}^T\mathbf{X})$, by Theorem 3.3.7 one generalized inverse of the coefficient matrix is

$$\begin{pmatrix} \mathbf{G}_{11} & \mathbf{G}_{12} \\ \mathbf{G}_{21} & \mathbf{G}_{22} \end{pmatrix},$$

where

$$\begin{aligned}\mathbf{G}_{11} &= (\mathbf{X}^T\mathbf{X})^- + (\mathbf{X}^T\mathbf{X})^-\mathbf{X}^T\mathbf{M}^T[-\mathbf{M}\mathbf{X}(\mathbf{X}^T\mathbf{X})^-\mathbf{X}^T\mathbf{M}^T]^-\mathbf{M}\mathbf{X}(\mathbf{X}^T\mathbf{X})^- \\ &= (\mathbf{X}^T\mathbf{X})^- - (\mathbf{X}^T\mathbf{X})^-\mathbf{A}^T[\mathbf{A}(\mathbf{X}^T\mathbf{X})^-\mathbf{A}^T]^-\mathbf{A}(\mathbf{X}^T\mathbf{X})^-, \qquad (10.10)\\ \mathbf{G}_{12} &= (\mathbf{X}^T\mathbf{X})^-\mathbf{A}^T[\mathbf{A}(\mathbf{X}^T\mathbf{X})^-\mathbf{A}^T]^-, \qquad (10.11)\\ \mathbf{G}_{21} &= [\mathbf{A}(\mathbf{X}^T\mathbf{X})^-\mathbf{A}^T]^-\mathbf{A}(\mathbf{X}^T\mathbf{X})^-, \qquad (10.12)\end{aligned}$$

and

$$\mathbf{G}_{22} = -[\mathbf{A}(\mathbf{X}^T\mathbf{X})^-\mathbf{A}^T]^-. \qquad (10.13)$$

Using (10.10) and (10.11) and Theorem 10.1.4, it can be shown (see the exercises) that the constrained least squares estimator $\mathbf{c}^T\breve{\boldsymbol{\beta}}$ of an estimable function $\mathbf{c}^T\boldsymbol{\beta}$ is related to the least squares estimator $\mathbf{c}^T\hat{\boldsymbol{\beta}}$ of that function associated with the corresponding unconstrained model as follows:

$$\mathbf{c}^T\breve{\boldsymbol{\beta}} = \mathbf{c}^T\hat{\boldsymbol{\beta}} - \mathbf{c}^T(\mathbf{X}^T\mathbf{X})^-\mathbf{A}^T[\mathbf{A}(\mathbf{X}^T\mathbf{X})^-\mathbf{A}^T]^-(\mathbf{A}\hat{\boldsymbol{\beta}} - \mathbf{h}). \qquad (10.14)$$

Furthermore, it may be shown (see the exercises) that in this case

$$\mathbf{P}_{\mathbf{X}(\mathbf{I}-\mathbf{A}^-\mathbf{A})} = \mathbf{X}\mathbf{G}_{11}\mathbf{X}^T = \mathbf{P}_{\mathbf{X}} - \mathbf{P}_{\mathbf{X}(\mathbf{X}^T\mathbf{X})^-\mathbf{A}^T}$$

and that

$$\begin{aligned}(\mathbf{P}_{\mathbf{X}} - \mathbf{P}_{\mathbf{X}(\mathbf{X}^T\mathbf{X})^-\mathbf{A}^T})\mathbf{P}_{\mathbf{X}(\mathbf{X}^T\mathbf{X})^-\mathbf{A}^T} &= \mathbf{0},\\ (\mathbf{P}_{\mathbf{X}} - \mathbf{P}_{\mathbf{X}(\mathbf{X}^T\mathbf{X})^-\mathbf{A}^T})(\mathbf{I} - \mathbf{P}_{\mathbf{X}}) &= \mathbf{0}, \quad \text{and}\\ \mathbf{P}_{\mathbf{X}(\mathbf{X}^T\mathbf{X})^-\mathbf{A}^T}(\mathbf{I} - \mathbf{P}_{\mathbf{X}}) &= \mathbf{0}.\end{aligned}$$

As a consequence, the following is an orthogonal decomposition of $\mathbf{y}$:

$$\mathbf{y} = (\mathbf{P}_{\mathbf{X}} - \mathbf{P}_{\mathbf{X}(\mathbf{X}^T\mathbf{X})^-\mathbf{A}^T})\mathbf{y} + \mathbf{P}_{\mathbf{X}(\mathbf{X}^T\mathbf{X})^-\mathbf{A}^T}\mathbf{y} + (\mathbf{I} - \mathbf{P}_{\mathbf{X}})\mathbf{y}. \qquad (10.15)$$

Pre-multiplication of this decomposition by $\mathbf{y}^T$ leads to an ANOVA that is distinct from the constrained overall ANOVA introduced in the previous section and that can be viewed as a certain partitioned ANOVA of the unconstrained model, which we call the *constraint-partitioned ANOVA*. In this ANOVA the unconstrained model sum of squares, $\mathbf{y}^T\mathbf{P}_{\mathbf{X}}\mathbf{y}$, is partitioned into two parts, corresponding to the first two terms on the right-hand side of (10.15). The constraint-partitioned ANOVA may be displayed in tabular form as follows:

Source	Rank	Sum of squares	Mean square
Constrained Model	$p^* - q^*$	$\mathbf{y}^T(\mathbf{P_X} - \mathbf{P}_{\mathbf{X}(\mathbf{X}^T\mathbf{X})^-\mathbf{A}^T})\mathbf{y}$	$\mathbf{y}^T(\mathbf{P_X} - \mathbf{P}_{\mathbf{X}(\mathbf{X}^T\mathbf{X})^-\mathbf{A}^T})\mathbf{y}/p^*_{\mathbf{A}}$
Unconstrained|Constrained	q^*	$\mathbf{y}^T\mathbf{P}_{\mathbf{X}(\mathbf{X}^T\mathbf{X})^-\mathbf{A}^T}\mathbf{y}$	$\mathbf{y}^T\mathbf{P}_{\mathbf{X}(\mathbf{X}^T\mathbf{X})^-\mathbf{A}^T}\mathbf{y}/(p^* - p^*_{\mathbf{A}})$
Residual	$n - p^*$	$\mathbf{y}^T(\mathbf{I} - \mathbf{P_X})\mathbf{y}$	$\mathbf{y}^T(\mathbf{I} - \mathbf{P_X})\mathbf{y}/(n - p^*)$
Total	n	$\mathbf{y}^T\mathbf{y}$	

The first line of this ANOVA table coincides with the first line of the constrained overall ANOVA table, so it is given the same label, "Constrained Model." The sum of squares in the second line, however, is the additional variation in $\mathbf{y}$ explained by removing the constraints; therefore, in partial accordance with the labeling introduced in Chap. 9 for partitioned ANOVAs, we label the second line "Unconstrained|Constrained." The third line is simply the residual sum of squares for the unconstrained model so it retains the same label it has in an overall or sequential ANOVA for such a model. Alternative expressions for the sums of squares in the "Constrained Model" and "Unconstrained|Constrained" lines of the ANOVA are

$$\hat{\boldsymbol{\beta}}^T\mathbf{X}^T\mathbf{y} - (\mathbf{A}\hat{\boldsymbol{\beta}})^T[\mathbf{A}(\mathbf{X}^T\mathbf{X})^-\mathbf{A}^T]^-(\mathbf{A}\hat{\boldsymbol{\beta}})$$

and

$$(\mathbf{A}\hat{\boldsymbol{\beta}})^T[\mathbf{A}(\mathbf{X}^T\mathbf{X})^-\mathbf{A}^T]^-(\mathbf{A}\hat{\boldsymbol{\beta}}),$$

respectively, as may be easily verified. The entries in the "Rank" column have been simplified using the fact that $\text{rank}\begin{pmatrix}\mathbf{X}\\ \mathbf{A}\end{pmatrix} = \text{rank}\begin{pmatrix}\mathbf{X}\\ \mathbf{MX}\end{pmatrix} = \text{rank}(\mathbf{X})$.

How decomposition (10.7) of $\mathbf{y}^T\mathbf{y}$ into second-order polynomial functions specializes in this case is also of interest, and turns out to be related (but not identical) to the constraint-partitioned ANOVA just introduced. Using the expressions for $\mathbf{G}_{11}$, $\mathbf{G}_{12}$, $\mathbf{G}_{21}$, and $\mathbf{G}_{22}$ given by (10.10)–(10.13), it may be shown (see the exercises) that the "Constrained Model" and "Constrained Residual" second-order polynomials in decomposition (10.7) specialize to

$$\mathbf{y}^T\mathbf{P_X}\mathbf{y} - (\mathbf{A}\hat{\boldsymbol{\beta}} - \mathbf{h})^T[\mathbf{A}(\mathbf{X}^T\mathbf{X})^-\mathbf{A}^T]^-(\mathbf{A}\hat{\boldsymbol{\beta}} - \mathbf{h}) \tag{10.16}$$

and

$$\mathbf{y}^T(\mathbf{I} - \mathbf{P_X})\mathbf{y} + (\mathbf{A}\hat{\boldsymbol{\beta}} - \mathbf{h})^T[\mathbf{A}(\mathbf{X}^T\mathbf{X})^-\mathbf{A}^T]^-(\mathbf{A}\hat{\boldsymbol{\beta}} - \mathbf{h}), \tag{10.17}$$

respectively. Thus, in this case (10.7) may be written in tabular form as follows:

Source	Rank	Second-order polynomial
Constrained Model	$p^* - q^*$	$\mathbf{y}^T\mathbf{P_X}\mathbf{y} - (\mathbf{A}\hat{\boldsymbol{\beta}} - \mathbf{h})^T[\mathbf{A}(\mathbf{X}^T\mathbf{X})^-\mathbf{A}^T]^-(\mathbf{A}\hat{\boldsymbol{\beta}} - \mathbf{h})$
Constrained Residual	$n - p^* + q^*$	$\mathbf{y}^T(\mathbf{I} - \mathbf{P_X})\mathbf{y} + (\mathbf{A}\hat{\boldsymbol{\beta}} - \mathbf{h})^T[\mathbf{A}(\mathbf{X}^T\mathbf{X})^-\mathbf{A}^T]^-(\mathbf{A}\hat{\boldsymbol{\beta}} - \mathbf{h})$
Total	n	$\mathbf{y}^T\mathbf{y}$

Furthermore, the Constrained Residual line may be partitioned to yield the following:

Source	Rank	Second-order polynomial
Constrained Model	$p^* - q^*$	$\mathbf{y}^T\mathbf{P_X}\mathbf{y} - (\mathbf{A}\hat{\boldsymbol{\beta}} - \mathbf{h})^T[\mathbf{A}(\mathbf{X}^T\mathbf{X})^-\mathbf{A}^T]^-(\mathbf{A}\hat{\boldsymbol{\beta}} - \mathbf{h})$
Unconstrained|Constrained	q^*	$(\mathbf{A}\hat{\boldsymbol{\beta}} - \mathbf{h})^T[\mathbf{A}(\mathbf{X}^T\mathbf{X})^-\mathbf{A}^T]^-(\mathbf{A}\hat{\boldsymbol{\beta}} - \mathbf{h})$
Residual	$n - p^*$	$\mathbf{y}^T(\mathbf{I} - \mathbf{P_X})\mathbf{y}$
Total	n	$\mathbf{y}^T\mathbf{y}$

Here we have used the same labels for the lines comprising the two parts of the Constrained Residual second-order polynomial that were used for two of the sums of squares in the constraint-partitioned ANOVA. Indeed, this last decomposition strongly resembles the constraint-partitioned ANOVA, but in general it is different: the Constrained Model second-order polynomial differs from the Constrained Model sum of squares by $2(\mathbf{A}\hat{\boldsymbol{\beta}})^T[\mathbf{A}(\mathbf{X}^T\mathbf{X})^-\mathbf{A}^T]^-\mathbf{h} - \mathbf{h}^T[\mathbf{A}(\mathbf{X}^T\mathbf{X})^-\mathbf{A}^T]^-\mathbf{h}$, and the Unconstrained|Constrained second-order polynomial differs from the Unconstrained|Constrained sum of squares by the negative of the same quantity. The two decompositions will coincide if the constraints are homogeneous.

In Chap. 15, it will be shown that this last representation of the second-order polynomial decomposition for the case of homogeneous constraints is highly relevant to testing certain linear hypotheses about $\boldsymbol{\beta}$. With that application in mind, we give one more representation of it—this one in terms of residual sums of squares under the constrained and unconstrained models:

Source	Rank	Second-order polynomial
Constrained Model	$p^* - q^*$	$\mathbf{y}^T\mathbf{y} - Q(\breve{\boldsymbol{\beta}})$
Unconstrained|Constrained	q^*	$Q(\breve{\boldsymbol{\beta}}) - Q(\hat{\boldsymbol{\beta}})$
Residual	$n - p^*$	$Q(\hat{\boldsymbol{\beta}})$
Total	n	$\mathbf{y}^T\mathbf{y}$

Example 10.3-1. The Constraint-Partitioned ANOVA for a Piecewise Continuous Simple Linear Regression Model

For the piecewise continuous simple linear regression model considered previously in Example 10.1-3, the constraints are homogeneous and estimable. Thus the constraint-partitioned ANOVA specializes as follows:

Source	Rank	Sum of squares	Mean square
Constrained Model	3	$\mathbf{y}^T(\mathbf{P_X} - \mathbf{P}_{\mathbf{X}(\mathbf{X}^T\mathbf{X})^-\mathbf{A}^T})\mathbf{y}$	$\mathbf{y}^T(\mathbf{P_X} - \mathbf{P}_{\mathbf{X}(\mathbf{X}^T\mathbf{X})^-\mathbf{A}^T})\mathbf{y}/p_{\mathbf{A}}^*$
Unconstrained|Constrained	1	$\mathbf{y}^T\mathbf{P}_{\mathbf{X}(\mathbf{X}^T\mathbf{X})^-\mathbf{A}^T}\mathbf{y}$	$\mathbf{y}^T\mathbf{P}_{\mathbf{X}(\mathbf{X}^T\mathbf{X})^-\mathbf{A}^T}\mathbf{y}/(p^* - p_{\mathbf{A}}^*)$
Residual	$n-4$	$\mathbf{y}^T(\mathbf{I} - \mathbf{P_X})\mathbf{y}$	$\mathbf{y}^T(\mathbf{I} - \mathbf{P_X})\mathbf{y}/(n - p^*)$
Total	n	$\mathbf{y}^T\mathbf{y}$	

where the "Unconstrained|Constrained" sum of squares is

$$(\mathbf{A}\hat{\boldsymbol{\beta}})^T[\mathbf{A}(\mathbf{X}^T\mathbf{X})^{-1}\mathbf{A}]^{-1}(\mathbf{A}\hat{\boldsymbol{\beta}})$$
$$= \frac{[\bar{y}_- + (c - \bar{x}_-)(SXY_-/SXX_-) - \bar{y}_+ - (c - \bar{x}_+)(SXY_+/SXX_+)]^2}{(1/n_-) + (c - \bar{x}_-)^2/SXX_- + (1/n_+) + (c - \bar{x}_+)^2/SXX_+},$$

and the "Constrained Model" sum of squares is $\hat{\boldsymbol{\beta}}^T\mathbf{X}^T\mathbf{y}$ minus the Unconstrained| Constrained sum of squares where

$$\hat{\boldsymbol{\beta}}^T\mathbf{X}^T\mathbf{y} = n_-\bar{y}_-^2 + (SXY_-^2/SXX_-) + n_+\bar{y}_+^2 + (SXY_+^2/SXX_+).$$

Because the constraints are homogeneous, this ANOVA coincides with the decomposition of $\mathbf{y}^T\mathbf{y}$ into second-order polynomial functions presented in this section and, upon summing the lines labeled "Constrained Model" and "Unconstrained|Constrained," with the constrained overall ANOVA described in Sect. 10.2. ■

10.4 The Case of Jointly Nonestimable Constraints

Finally, let us consider the case in which the constraints are jointly nonestimable, as defined immediately below. For a model with such constraints, some very important relationships with least squares estimation associated with the corresponding unconstrained model emerge.

Definition 10.4.1 The constraints $\mathbf{A}\boldsymbol{\beta} = \mathbf{h}$ are said to be *jointly nonestimable* under the model $\{\mathbf{y}, \mathbf{X}\boldsymbol{\beta}\}$ if no nonzero linear combination $\boldsymbol{\ell}^T\mathbf{A}\boldsymbol{\beta}$ is estimable under that model.

Observe that the right-hand side vector of the system of constraint equations, i.e., $\mathbf{h}$, has no effect on whether or not the constraints are jointly nonestimable.

The following theorem gives a somewhat more direct way to characterize jointly nonestimable constraints.

Theorem 10.4.1 *Constraints* $\mathbf{A}\boldsymbol{\beta} = \mathbf{h}$ *are jointly nonestimable under the model* $\{\mathbf{y}, \mathbf{X}\boldsymbol{\beta}\}$ *if and only if* $\mathcal{R}(\mathbf{X}) \cap \mathcal{R}(\mathbf{A}) = \{\mathbf{0}\}$, *i.e., if and only if* $\mathcal{R}(\mathbf{X})$ *and* $\mathcal{R}(\mathbf{A})$ *are essentially disjoint.*

Proof By Theorem 6.1.2, $\boldsymbol{\ell}^T\mathbf{A}\boldsymbol{\beta}$ is nonestimable under the specified model if and only if $\boldsymbol{\ell}^T\mathbf{A} \notin \mathcal{R}(\mathbf{X})$. The desired result follows because $\boldsymbol{\ell}^T\mathbf{A} \in \mathcal{R}(\mathbf{A})$. □

The next theorem establishes some relationships between constrained least squares estimation and (unconstrained) least squares estimation when the constraints are jointly nonestimable under the unconstrained model.

Theorem 10.4.2 *Consider a constrained linear model* $\{\mathbf{y}, \mathbf{X}\boldsymbol{\beta} : \mathbf{A}\boldsymbol{\beta} = \mathbf{h}\}$ *for which the constraints are jointly nonestimable under the corresponding unconstrained model, and let* $(\breve{\boldsymbol{\beta}}^T, \breve{\boldsymbol{\lambda}}^T)^T$ *be any solution to the constrained normal equations for the constrained model. Then:*

(a) $\breve{\boldsymbol{\beta}}$ *satisfies the normal equations* $\mathbf{X}^T\mathbf{X}\boldsymbol{\beta} = \mathbf{X}^T\mathbf{y}$ *for the unconstrained model;*
(b) $\mathbf{X}\mathbf{G}_{11}\mathbf{X}^T = \mathbf{P}_\mathbf{X}$, $\mathbf{X}\mathbf{G}_{12}\mathbf{h} = \mathbf{0}$, *and* $\mathbf{h}^T\mathbf{G}_{22}\mathbf{h} = 0$;
(c) $p^*_\mathbf{A} = rank(\mathbf{X})$;
(d) $\mathbf{A}^T\breve{\boldsymbol{\lambda}} = \mathbf{0}$ *and* $\breve{\boldsymbol{\lambda}}^T\mathbf{h} = 0$.

Proof The "top subset" of constrained normal equations are

$$\mathbf{X}^T\mathbf{X}\breve{\boldsymbol{\beta}} + \mathbf{A}^T\breve{\boldsymbol{\lambda}} = \mathbf{X}^T\mathbf{y},$$

or equivalently,

$$\mathbf{X}^T(\mathbf{X}\breve{\boldsymbol{\beta}} - \mathbf{y}) = -\mathbf{A}^T\breve{\boldsymbol{\lambda}},$$

or equivalently,

$$(\mathbf{X}\breve{\boldsymbol{\beta}} - \mathbf{y})^T\mathbf{X} = -\breve{\boldsymbol{\lambda}}^T\mathbf{A}.$$

By Theorem 10.4.1, both sides of this last system of equations are null vectors, from which part (a) and the first equality in part (d) follow immediately. And, because $\breve{\boldsymbol{\lambda}}^T \mathbf{h} = \breve{\boldsymbol{\lambda}}^T \mathbf{A}\breve{\boldsymbol{\beta}} = \mathbf{0}^T \breve{\boldsymbol{\beta}} = 0$, the second equality in part (d) holds as well. Furthermore, part (c) follows immediately from Theorem 2.8.6. The proof of part (b) is left as an exercise. □

Corollary 10.4.2.1 *If the constraints* $\mathbf{A}\boldsymbol{\beta} = \mathbf{h}$ *are jointly nonestimable under the corresponding unconstrained model, then:*

(a) the least squares estimators of functions that are estimable under the unconstrained model are identical to the constrained least squares estimators of those functions;
(b) the variance–covariance matrix of the vector of least squares estimators described in part (a) under the unconstrained Gauss–Markov model coincides with the variance–covariance matrix of the corresponding vector of constrained least squares estimators under the constrained Gauss–Markov model;
(c) the residual mean squares for the constrained and corresponding unconstrained models are identical; and
(d) the constrained overall ANOVA exists even if the constraints are not homogeneous, and it coincides with the overall ANOVA for the corresponding unconstrained model.

Proof That the least squares estimators of functions that are estimable under the unconstrained model are identical to the constrained least squares estimators of those functions follows immediately from part (a) of the theorem. According to Theorem 10.1.5, under the constrained Gauss–Markov model the variance–covariance matrix of the constrained least squares estimator of a vector $\mathbf{C}^T\boldsymbol{\beta}$ of functions that are estimable under the constrained model is $\sigma^2\mathbf{C}^T\mathbf{G}_{11}\mathbf{C}$, where $\mathbf{G}_{11}$ is the upper left $p \times p$ block of a generalized inverse of the coefficient matrix of the constrained normal equations. But because these functions are also estimable under the corresponding unconstrained model, $\mathbf{C}^T = \mathbf{B}^T\mathbf{X}$ for some matrix $\mathbf{B}$, whereupon the aforementioned variance–covariance matrix may be written, using part (b) of Theorem 10.4.2, as

$$\sigma^2\mathbf{B}^T\mathbf{X}\mathbf{G}_{11}\mathbf{X}^T\mathbf{B} = \sigma^2\mathbf{B}^T\mathbf{P_X}\mathbf{B} = \sigma^2\mathbf{C}^T(\mathbf{X}^T\mathbf{X})^-\mathbf{C}.$$

Part (b) of the corollary then follows immediately from Theorem 7.2.2. The residual mean square under the constrained Gauss–Markov model is, by Theorem 10.1.12, $(\mathbf{y}^T\mathbf{y} - \breve{\boldsymbol{\beta}}^T\mathbf{X}^T\mathbf{y} - \breve{\boldsymbol{\lambda}}^T\mathbf{h})/(n - p_{\mathbf{A}}^*)$. Letting $\hat{\boldsymbol{\beta}}$ represent any solution to the normal equations for the unconstrained model, the numerator of this estimator is equal to $\mathbf{y}^T\mathbf{y} - \hat{\boldsymbol{\beta}}^T\mathbf{X}^T\mathbf{y}$ by parts (a) and (d), and the denominator is equal to $n - p^*$ by part (c). Thus, the residual sums of squares (and the residual mean squares) of the constrained and corresponding unconstrained Gauss–Markov models are equal. It follows immediately that the constrained overall ANOVA coincides with the overall ANOVA for the corresponding unconstrained model. □

A cautionary note is in order. Corollary 10.4.2.1 does *not* say that if $\mathcal{R}(\mathbf{X}) \cap \mathcal{R}(\mathbf{A}) = \{\mathbf{0}\}$, then the constrained least squares estimators of functions that are estimable *under the constrained model* are identical to the least squares estimators of those functions under the corresponding unconstrained model. In fact, those functions may not even be estimable under the unconstrained model. We see, however, that because the elements of $\mathbf{X}\boldsymbol{\beta}$ are estimable under both the constrained and unconstrained models, the fitted values (and hence also the residuals) for the constrained and unconstrained models are identical in this case.

As a consequence of Corollary 10.4.2.1, if the model adopted for the data actually is unconstrained, we may augment it with jointly nonestimable "pseudo-constraints," if we wish, and use methods associated with constrained least squares estimation to perform inferences for functions that are estimable under the original unconstrained model, and those inferences will be identical to the inferences that would be obtained using the unconstrained-model methodology presented in Chaps. 7 through 9. Why would we ever want to do such a thing? Two reasons are as follows:

1. Augmenting the model with an appropriate number of such constraints can make the coefficient matrix of the constrained normal equations for that model, i.e., $\begin{pmatrix} \mathbf{X}^T\mathbf{X} & \mathbf{A}^T \\ \mathbf{A} & \mathbf{0} \end{pmatrix}$, nonsingular, even when the coefficient matrix of the normal equations for the unconstrained model (i.e., $\mathbf{X}^T\mathbf{X}$) is singular. This would allow one to dispense with generalized inverses for the purpose of obtaining least squares estimators, using instead the ordinary inverse of the coefficient matrix of the constrained normal equations to obtain a unique solution to those equations. This unique solution would be given by the first p-component subvector of
$$\begin{pmatrix} \mathbf{X}^T\mathbf{X} & \mathbf{A}^T \\ \mathbf{A} & \mathbf{0} \end{pmatrix}^{-1} \begin{pmatrix} \mathbf{X}^T\mathbf{y} \\ \mathbf{h} \end{pmatrix}.$$
In fact, this is precisely what occurred for the first two constraints considered in Example 10.1-1.
2. The unique solution mentioned in the previous point, which (by Theorem 10.4.2) is also a solution to the normal equations for the unconstrained model, can be given a meaningful interpretation as the least squares estimator of a certain vector of estimable (under the unconstrained model) functions.

To establish the first claim above, we give one final theorem. The second claim will be illustrated via some examples.

Theorem 10.4.3 *If $q = p - p^*$ constraints $\mathbf{A}\boldsymbol{\beta} = \mathbf{h}$ are jointly nonestimable under the model $\{\mathbf{y}, \mathbf{X}\boldsymbol{\beta}\}$, and they are linearly independent (so that rank($\mathbf{A}$) $= q$), then*

$$rank\begin{pmatrix} \mathbf{X}^T\mathbf{X} & \mathbf{A}^T \\ \mathbf{A} & \mathbf{0} \end{pmatrix} = p + q,$$

i.e., the coefficient matrix of the constrained normal equations for the model $\{\mathbf{y}, \mathbf{X}\boldsymbol{\beta} : \mathbf{A}\boldsymbol{\beta} = \mathbf{h}\}$ has full rank.

Proof By Corollary 10.1.7.1, the joint nonestimability condition, and the linear independence of the constraints, respectively, we find that

$$\begin{aligned} \text{rank}\begin{pmatrix} \mathbf{X}^T\mathbf{X} & \mathbf{A}^T \\ \mathbf{A} & \mathbf{0} \end{pmatrix} &= \text{rank}\begin{pmatrix} \mathbf{X} \\ \mathbf{A} \end{pmatrix} + q^* \\ &= \text{rank}(\mathbf{X}) + \text{rank}(\mathbf{A}) + q^* \\ &= p^* + 2q, \end{aligned}$$

which equals $p + q$ if q is chosen to equal $p - p^*$. □

It follows from Theorem 10.4.3 that it is always possible to obtain least squares estimators of functions that are estimable under an unconstrained model by working with full-rank versions of certain constrained normal equations. We need only augment the unconstrained model with $p - p^*$ linearly independent, jointly nonestimable pseudo-constraints and then proceed with all of the usual inference goals using constrained least squares methods.

For less-than-full-rank models that have "real" constraints, a similar approach of model augmentation with independent, jointly nonestimable pseudo-constraints (under the model constrained only by the real constraints) may be used to obtain full-rank versions of constrained normal equations whose unique solution is also a solution to the constrained normal equations for the less-than-full-rank constrained model; this is left as an exercise.

Example 10.4-1. Constrained Least Squares Estimation for the One-Factor Model with Jointly Nonestimable Constraints

Consider the one-factor model

$$y_{ij} = \mu + \alpha_i + e_{ij} \quad (i = 1, \ldots, q;\ j = 1, \ldots, n_i),$$

which is unconstrained but less than full rank. The normal equations, in matrix notation, are

$$\mathbf{X}^T\mathbf{X}\boldsymbol{\beta} = \mathbf{X}^T\mathbf{y},$$

where $\boldsymbol{\beta} = (\mu, \alpha_1, \alpha_2, \ldots, \alpha_q)^T$; in nonmatrix notation, these same equations are

$$n\mu + \sum_{i=1}^{q} n_i\alpha_i = y_{\cdot\cdot},$$

$$n_i\mu + n_i\alpha_i = y_{i\cdot} \quad (i = 1, \ldots, q).$$

In Example 6.1-2, it was shown that μ and α_i (for any i) are (individually) nonestimable under this model. So are, e.g., $\sum_{i=1}^q \alpha_i$ and $\sum_{i=1}^q n_i\alpha_i$. It was also shown that the rank of this model is q, i.e., $p^* = q$, while $p = q+1$. Consequently, to obtain constrained normal equations that are full rank, we must augment the unconstrained model with $p - p^* = (q+1) - q = 1$ jointly nonestimable pseudo-constraint. (Actually, because $p - p^* = 1$ we could omit the word "jointly" in this case). Consider the following possibilities for this constraint:

Pseudo-constraint	Unique solution to constrained normal equations
$\mu = 0$	$\breve{\boldsymbol{\beta}} = (\breve{\mu}, \breve{\alpha}_1, \ldots, \breve{\alpha}_q)$ where $\breve{\mu} = 0$, $\breve{\alpha}_i = \bar{y}_{i\cdot}$ $(i = 1, \ldots, q)$
$\sum_{i=1}^q n_i\alpha_i = 0$	$\breve{\boldsymbol{\beta}}^* = (\breve{\mu}^*, \breve{\alpha}_1^*, \ldots, \breve{\alpha}_q^*)$ where $\breve{\mu}^* = \bar{y}_{\cdot\cdot}$, $\breve{\alpha}_i^* = \bar{y}_{i\cdot} - \bar{y}_{\cdot\cdot}$ $(i = 1, \ldots, q)$
$\sum_{i=1}^q \alpha_i = 0$	$\breve{\boldsymbol{\beta}}^{**} = (\breve{\mu}^{**}, \breve{\alpha}_1^{**}, \ldots, \breve{\alpha}_q^{**})$ where $\breve{\mu}^{**} = \frac{1}{q}\sum_{i=1}^q \bar{y}_{i\cdot}$, $\breve{\alpha}_i^{**} = \bar{y}_{i\cdot} - \frac{1}{q}\sum_{j=1}^q \bar{y}_{j\cdot}$

If $\mathbf{c}^T\boldsymbol{\beta}$ is any function that is estimable under the unconstrained model, then its least squares estimator $\mathbf{c}^T\hat{\boldsymbol{\beta}}$ coincides with the constrained least squares estimators $\mathbf{c}^T\breve{\boldsymbol{\beta}}$, $\mathbf{c}^T\breve{\boldsymbol{\beta}}^*$, and $\mathbf{c}^T\breve{\boldsymbol{\beta}}^{**}$, as may be easily verified. The constrained overall ANOVA is the same under all three constrained models, and coincides with the (unconstrained) overall ANOVA. Furthermore, the following interpretations may be given to the unique solutions. If the constraint is $\mu = 0$, then $\breve{\mu}$ is the least squares and constrained least squares estimator of 0 and $\breve{\alpha}_i$ is the constrained least squares estimator of $\mathrm{E}(\bar{y}_{i\cdot}) = \mu + \alpha_i$ $(i = 1, \ldots, q)$ (the ith level mean). If the constraint is $\sum_{i=1}^q n_i\alpha_i = 0$, then $\breve{\mu}^*$ is the least squares and constrained least squares estimator of $\mathrm{E}(\bar{y}_{\cdot\cdot}) = \mu + \sum_{i=1}^q (n_i/n)\alpha_i$ (the mean of all the observations) and $\breve{\alpha}_i^*$ is the least squares and constrained least squares estimator of $\mathrm{E}(\bar{y}_{i\cdot}) - \mathrm{E}(\bar{y}_{\cdot\cdot}) = \alpha_i - \sum_{j=1}^q (n_j/n)\alpha_j$ $(i = 1, \ldots, q)$. If the constraint is $\sum_{i=1}^q \alpha_i = 0$, then $\breve{\mu}^{**}$ is the least squares and constrained least squares estimator of $\mathrm{E}(\frac{1}{q}\sum_{i=1}^q \bar{y}_{i\cdot}) = \mu + \frac{1}{q}\sum_{i=1}^q \alpha_i$ and $\breve{\alpha}_i^{**}$ is the least squares and constrained least squares estimator of $\mathrm{E}(\bar{y}_{i\cdot}) - \mathrm{E}(\frac{1}{q}\sum_{j=1}^q \bar{y}_{j\cdot}) = \alpha_i - \frac{1}{q}\sum_{j=1}^q \alpha_j$ $(i = 1, \ldots, q)$. ■

10.5 Exercises

1. Consider the constrained one-factor models featured in Example 10.1-1.
   (a) Under the model with constraint $\alpha_q = h$, verify that (10.2) is the constrained least squares estimator of an arbitrary estimable function.
   (b) Under the model with constraint $\sum_{i=1}^{q} \alpha_i = h$, obtain a general expression for the constrained least squares estimator of an arbitrary estimable function, and verify that the estimators of level means and differences coincide with their counterparts under the unconstrained model.
   (c) Under the model with constraint $\mu + \alpha_1 = h$, verify that (10.3) is the constrained least squares estimator of an arbitrary estimable function. Also verify the expressions for the constrained least squares estimators of estimable functions and for $\mathrm{var}(\mathbf{C}^T\breve{\boldsymbol{\beta}})$ that immediately follow (10.3).
2. Consider the one-factor model. For this model, the normal equations may be written without using matrix and vector notation as

$$n\mu + \sum_{i=1}^{q} n_i\alpha_i = y_{\cdot\cdot},$$

$$n_i\mu + n_i\alpha_i = y_{i\cdot} \quad (i = 1, \ldots, q).$$

   Using the approach described in Example 7.1-2, it may be shown that one solution to these equations is $\hat{\boldsymbol{\beta}} = (0, \bar{y}_{1\cdot}, \bar{y}_{2\cdot}, \ldots, \bar{y}_{q\cdot})^T$. The following five functions might be of interest: μ, α_1, $\alpha_1 - \alpha_2$, $\mu + \frac{1}{q}\sum \alpha_i$, $\mu + \frac{1}{n}\sum n_i\alpha_i$.
   (a) For a model without constraints, which of these five functions are estimable and which are not? Give the BLUEs for those that are estimable.
   (b) For each function that is nonestimable under a model without constraints, show that it is estimable under the model with the constraint $\alpha_q = 0$ and give its BLUE under this model.
   (c) For each function that is estimable under the model without constraints, show that its least squares estimator in terms of the elements of $\hat{\boldsymbol{\beta}}$, and its constrained least squares estimator under the model with constraint $\alpha_q = 0$, are identical.
3. Consider the two-way main effects model, with exactly one observation per cell.
   (a) Write out the normal equations for this model without using matrix and vector notation.
   (b) Show that one solution to the normal equations is $\hat{\mu} = \bar{y}_{\cdot\cdot}$, $\hat{\alpha}_i = \bar{y}_{i\cdot} - \bar{y}_{\cdot\cdot}$ $(i = 1, \ldots, q)$, $\hat{\gamma}_j = \bar{y}_{\cdot j} - \bar{y}_{\cdot\cdot}$ $(j = 1, \ldots, m)$.
   (c) Solve the constrained normal equations obtained by augmenting the model with an appropriate number of jointly nonestimable pseudo-constraints. List your pseudo-constraints explicitly and explain why they are jointly nonestimable.
   (d) The following five functions might be of interest: μ, α_1, $\gamma_1 - \gamma_2$, $\alpha_1 - \gamma_2$, $\mu + \frac{1}{q}\sum_{i=1}^{q} \alpha_i + \frac{1}{m}\sum_{j=1}^{m} \gamma_j$.

(i) For a model without constraints, which of these five functions are estimable and which are not? Give the BLUEs for those that are estimable.

(ii) For each function that is nonestimable under the model without constraints, show that it is estimable under the model with the pseudo-constraints you gave in part (c) and give its BLUE under this constrained model.

(iii) For each function that is estimable under the model without constraints, show that its least squares estimator in terms of the elements of the solution to the normal equations given in part (b) is identical to its constrained least squares estimator under the model with the constraints you gave in part (c).

4. Consider the two-factor nested model.

(a) Write out the normal equations for this model without using matrix and vector notation.

(b) Show that one solution to the normal equations is $\hat{\mu} = \bar{y}_{\cdots}$, $\hat{\alpha}_i = \bar{y}_{i\cdot\cdot} - \bar{y}_{\cdots}$ $(i = 1, \ldots, q)$, $\hat{\gamma}_{ij} = \bar{y}_{ij\cdot} - \bar{y}_{i\cdot\cdot}$ $(i = 1, \ldots, q;\ j = 1, \ldots, m_i)$.

(c) Solve the constrained normal equations obtained by augmenting the model with an appropriate number of jointly nonestimable pseudo-constraints. List your pseudo-constraints explicitly and explain why they are jointly nonestimable.

(d) The following six functions might be of interest: $\mu, \alpha_1, \mu+\alpha_1, \gamma_{11}-\gamma_{12}, \mu+\alpha_1+(1/m_1)\sum_{j=1}^{m_1}\gamma_{1j}, \mu+(1/m.)\sum_{i=1}^{q} m_i\alpha_i+(1/m.)\sum_{i=1}^{q}\sum_{j=1}^{m_i}\gamma_{ij}$.

(i) For a model without constraints, which of these six functions are estimable and which are not? Give the BLUEs for those that are estimable.

(ii) For each function that is nonestimable under a model without constraints, show that it is estimable under the model with the pseudo-constraints you gave in part (c) and give its BLUE under this constrained model.

(iii) For each function that is estimable under a model without constraints, show that its least squares estimator in terms of the elements of the solution to the normal equations given in part (b) is identical to its constrained least squares estimator under the model with the constraints you gave in part (c).

5. Consider the one-factor model

$$y_{ij} = \mu + \alpha_i + e_{ij} \quad (i = 1, 2; \quad j = 1, \ldots, r)$$

with the constraint

$$\alpha_1 + \alpha_2 = 24.$$

(a) Write out the constrained normal equations without using matrix and vector notation.

(b) Obtain the unique solution $\breve{\mu}$, $\breve{\alpha}_1$, and $\breve{\alpha}_2$ to the constrained normal equations.
(c) Is $\breve{\alpha}_1 - \breve{\alpha}_2$ identical to the least squares estimator of $\alpha_1 - \alpha_2$ under the corresponding unconstrained model? If you answer yes, verify your answer. If you answer no, give a reason why not.

6. Prove Theorem 10.1.6.
7. Prove Theorem 10.1.10.
8. Under the constrained Gauss–Markov linear model

$$\mathbf{y} = \mathbf{X}\boldsymbol{\beta} + \mathbf{e}, \quad \mathbf{A}\boldsymbol{\beta} = \mathbf{h},$$

consider the estimation of an estimable function $\mathbf{c}^T\boldsymbol{\beta}$.
(a) Show that the least squares estimator of $\mathbf{c}^T\boldsymbol{\beta}$ associated with the corresponding unconstrained model is unbiased under the constrained model.
(b) Theorem 10.1.6, in tandem with the unbiasedness of the least squares estimator shown in part (a), reveals that the variance of the least squares estimator of $\mathbf{c}^T\boldsymbol{\beta}$ under the constrained Gauss–Markov model is at least as large as the variance of the constrained least squares estimator of $\mathbf{c}^T\boldsymbol{\beta}$ under that model. For the case in which the constraints are estimable under the corresponding unconstrained model, obtain an expression for the amount by which the former exceeds the latter, and verify that this exceedance is nonnegative.
(c) Suppose that $n > \text{rank}(\mathbf{X})$, and let $\hat{\sigma}^2$ denote the residual mean square associated with the corresponding unconstrained model, i.e., $\hat{\sigma}^2 = \mathbf{y}^T(\mathbf{I} - \mathbf{P_X})\mathbf{y}/[n - \text{rank}(\mathbf{X})]$. Is $\hat{\sigma}^2$ unbiased under the constrained model?

9. Prove Theorem 10.4.2b. (Hint: Use Theorems 3.3.11 and 3.3.8b.)
10. For the case of estimable constraints:
(a) Obtain expression (10.14) for the constrained least squares estimator of an estimable function $\mathbf{c}^T\boldsymbol{\beta}$.
(b) Show that $\mathbf{P}_{\mathbf{X}(\mathbf{I}-\mathbf{A}^-\mathbf{A})} = \mathbf{P_X} - \mathbf{P}_{\mathbf{X}(\mathbf{X}^T\mathbf{X})^-\mathbf{A}^T}$.
(c) Show that $(\mathbf{P_X} - \mathbf{P}_{\mathbf{X}(\mathbf{X}^T\mathbf{X})^-\mathbf{A}^T})\mathbf{P}_{\mathbf{X}(\mathbf{X}^T\mathbf{X})^-\mathbf{A}^T} = \mathbf{0}$, $(\mathbf{P_X} - \mathbf{P}_{\mathbf{X}(\mathbf{X}^T\mathbf{X})^-\mathbf{A}^T})(\mathbf{I} - \mathbf{P_X}) = \mathbf{0}$, and $\mathbf{P}_{\mathbf{X}(\mathbf{X}^T\mathbf{X})^-\mathbf{A}^T}(\mathbf{I} - \mathbf{P_X}) = \mathbf{0}$.
(d) Obtain expressions (10.16) and (10.17) for the Constrained Model and Constrained Residual second-order polynomial functions, respectively.

11. Consider a constrained version of the simple linear regression model

$$y_i = \beta_1 + \beta_2 x_i + e_i \quad (i = 1, \ldots, n)$$

where the constraint is that the line $\beta_1 + \beta_2 x$ passes through a known point (a, b).
(a) Give the constrained normal equations for this model in matrix form, with numbers given, where possible, for the elements of the matrices and vectors involved.
(b) Give simplified expressions for the constrained least squares estimators of β_1 and β_2.

(c) Give an expression for the constrained residual mean square in terms of a solution to the constrained normal equations (and other quantities).

12. Consider a constrained version of the mean-centered second-order polynomial regression model

$$y_i = \beta_1 + \beta_2(x_i - \bar{x}) + \beta_3(x_i - \bar{x})^2 + e_i \quad (i = 1, \ldots, 5)$$

where $x_i = i$ and the constraint is that the first derivative of the second-order polynomial is equal to 0 at $x = a$ where a is known.

(a) Give the constrained normal equations for this model in matrix form, with numbers given, where possible, for the elements of the matrices and vectors involved.

(b) Give nonmatrix expressions for the constrained least squares estimators of β_1, β_2, and β_3.

(c) Give a nonmatrix expression for the constrained residual mean square in terms of a solution to the constrained normal equations (and other quantities).

13. Recall from Sect. 10.4 that one can augment an unconstrained model of rank p^* with $p - p^*$ jointly nonestimable "pseudo-constraints" so as to obtain a unique solution to the constrained normal equations, from which the least squares estimators of all functions that are estimable under an unconstrained model may be obtained. Extend this idea to a constrained linear model $\mathbf{y} = \mathbf{X}\boldsymbol{\beta} + \mathbf{e}$, where $\boldsymbol{\beta}$ satisfies q_1 real, linearly independent, consistent constraints $\mathbf{A}_1\boldsymbol{\beta} = \mathbf{h}_1$, so as to obtain a unique solution to constrained normal equations corresponding to this constrained model augmented by pseudo-constraints $\mathbf{A}_2\boldsymbol{\beta}_2 = \mathbf{h}_2$, from which constrained least squares estimators of all functions that are estimable under the original constrained model may be obtained.

Best Linear Unbiased Estimation for the Aitken Model

11

Recall from Chap. 7 that the least squares estimators of estimable functions are best linear unbiased estimators (BLUEs) of those functions under the Gauss–Markov model. But it turns out that this is not necessarily so under linear models having a more general variance–covariance structure, such as the Aitken model. In this chapter, we consider estimators that are best linear unbiased under the Aitken model. The first section considers the special case of an Aitken model in which the variance–covariance matrix is positive definite; BLUE in this case is also called generalized least squares estimation. The second section considers the general case. The third section characterizes those Aitken models for which the least squares estimators of estimable functions are BLUEs of those functions. A final section briefly considers an attempt to extend BLUE to the general mixed linear model.

11.1 Generalized Least Squares Estimation

Consider the positive definite Aitken model $\{\mathbf{y}, \mathbf{X}\boldsymbol{\beta}, \sigma^2\mathbf{W}\}$, where σ^2 is an unknown positive parameter and $\mathbf{W}$ is a specified positive definite (hence nonsingular) matrix. By Theorem 2.15.12, a unique positive definite (hence nonsingular) matrix $\mathbf{W}^{\frac{1}{2}}$ exists such that $\mathbf{W} = \mathbf{W}^{\frac{1}{2}}\mathbf{W}^{\frac{1}{2}}$. Furthermore, by Theorem 2.9.4, $\mathbf{W}^{-1} = \mathbf{W}^{-\frac{1}{2}}\mathbf{W}^{-\frac{1}{2}}$ [where we write $\mathbf{W}^{-\frac{1}{2}}$ for $(\mathbf{W}^{\frac{1}{2}})^{-1}$]. Let $\mathbf{z} = \mathbf{W}^{-\frac{1}{2}}\mathbf{y}$ and $\mathbf{M} = \mathbf{W}^{-\frac{1}{2}}\mathbf{X}$. Pre-multiplying the left-hand and right-hand sides of the model equation by $\mathbf{W}^{-\frac{1}{2}}$, we obtain

$$\mathbf{z} = \mathbf{M}\boldsymbol{\beta} + \mathbf{W}^{-\frac{1}{2}}\mathbf{e}.$$

Observe that

$$\mathrm{E}(\mathbf{W}^{-\frac{1}{2}}\mathbf{e}) = \mathbf{0} \quad \text{and} \quad \mathrm{var}(\mathbf{W}^{-\frac{1}{2}}\mathbf{e}) = \mathbf{W}^{-\frac{1}{2}}(\sigma^2\mathbf{W})\mathbf{W}^{-\frac{1}{2}} = \sigma^2\mathbf{I},$$

D. L. Zimmerman, *Linear Model Theory*,
https://doi.org/10.1007/978-3-030-52063-2_11

indicating that a positive definite Aitken model for $\mathbf{y}$ implies a Gauss–Markov model for $\mathbf{z}$ (and vice versa). We refer to the Gauss–Markov model

$$\mathbf{z} = \mathbf{M}\boldsymbol{\beta} + \mathbf{d},$$

where the distribution of $\mathbf{d}$ is assumed to be identical to that of $\mathbf{W}^{-\frac{1}{2}}\mathbf{e}$ under the positive definite Aitken model just described, and where the parameter space is taken to be $\{\boldsymbol{\beta}, \sigma^2 : \boldsymbol{\beta} \in \mathbb{R}^p, \sigma^2 > 0\}$, as the *transformed model*.

Because the transformed model is an unconstrained Gauss–Markov model, we know from Theorems 7.1.2 and 7.1.3 that the ordinary least squares estimator of an estimable function $\mathbf{c}^T\boldsymbol{\beta}$ under the transformed model is given uniquely by $\mathbf{c}^T\tilde{\boldsymbol{\beta}}$, where $\tilde{\boldsymbol{\beta}}$ is any value at which the residual sum of squares function $(\mathbf{z} - \mathbf{M}\boldsymbol{\beta})^T(\mathbf{z} - \mathbf{M}\boldsymbol{\beta})$ assumes a minimum, or equivalently, where $\tilde{\boldsymbol{\beta}}$ is any solution to the consistent system of equations $\mathbf{M}^T\mathbf{M}\boldsymbol{\beta} = \mathbf{M}^T\mathbf{z}$. We can re-express this in terms of quantities in the positive definite Aitken model by noting that

$$\begin{aligned}(\mathbf{z} - \mathbf{M}\boldsymbol{\beta})^T(\mathbf{z} - \mathbf{M}\boldsymbol{\beta}) &= (\mathbf{W}^{-\frac{1}{2}}\mathbf{y} - \mathbf{W}^{-\frac{1}{2}}\mathbf{X}\boldsymbol{\beta})^T(\mathbf{W}^{-\frac{1}{2}}\mathbf{y} - \mathbf{W}^{-\frac{1}{2}}\mathbf{X}\boldsymbol{\beta}) \\ &= (\mathbf{y} - \mathbf{X}\boldsymbol{\beta})^T\mathbf{W}^{-1}(\mathbf{y} - \mathbf{X}\boldsymbol{\beta}),\end{aligned}$$

$$\mathbf{M}^T\mathbf{M} = (\mathbf{W}^{-\frac{1}{2}}\mathbf{X})^T(\mathbf{W}^{-\frac{1}{2}}\mathbf{X}) = \mathbf{X}^T\mathbf{W}^{-1}\mathbf{X},$$

and

$$\mathbf{M}^T\mathbf{z} = (\mathbf{W}^{-\frac{1}{2}}\mathbf{X})^T\mathbf{W}^{-\frac{1}{2}}\mathbf{y} = \mathbf{X}^T\mathbf{W}^{-1}\mathbf{y}.$$

Thus, the ordinary least squares estimator under the transformed model is $\mathbf{c}^T\tilde{\boldsymbol{\beta}}$, where $\tilde{\boldsymbol{\beta}}$ is any value at which the *generalized residual sum of squares function* $Q_{\mathbf{W}}(\boldsymbol{\beta}) = (\mathbf{y} - \mathbf{X}\boldsymbol{\beta})^T\mathbf{W}^{-1}(\mathbf{y} - \mathbf{X}\boldsymbol{\beta})$ assumes a minimum, or equivalently, where $\tilde{\boldsymbol{\beta}}$ is any solution to the *Aitken equations*

$$\mathbf{X}^T\mathbf{W}^{-1}\mathbf{X}\boldsymbol{\beta} = \mathbf{X}^T\mathbf{W}^{-1}\mathbf{y}$$

for the positive definite Aitken model $\{\mathbf{y}, \mathbf{X}\boldsymbol{\beta}, \sigma^2\mathbf{W}\}$. Because the Aitken equations are normal equations for the appropriate transformed model, their consistency is ensured by Theorem 7.1.1.

Definition 11.1.1 The *generalized least squares estimator* of an estimable function $\mathbf{c}^T\boldsymbol{\beta}$ associated with the positive definite Aitken model $\{\mathbf{y}, \mathbf{X}\boldsymbol{\beta}, \sigma^2\mathbf{W}\}$ is $\mathbf{c}^T\tilde{\boldsymbol{\beta}}$, where $\tilde{\boldsymbol{\beta}}$ is any solution to the Aitken equations for that model. The generalized least squares estimator of a vector $\mathbf{C}^T\boldsymbol{\beta}$ of estimable functions associated with the positive definite Aitken model $\{\mathbf{y}, \mathbf{X}\boldsymbol{\beta}, \sigma^2\mathbf{W}\}$ is the vector $\mathbf{C}^T\tilde{\boldsymbol{\beta}}$ of generalized least squares estimators of those functions.

The next five theorems, which generalize results on ordinary least squares estimation given in Chap. 7 to the present context of generalized least squares estimation, follow immediately from applying Theorems 7.1.4, 7.2.1–7.2.3 and 7.3.1 to the transformed model and then, if appropriate, re-expressing results in terms of quantities in the Aitken model.

Theorem 11.1.1 *The generalized least squares estimator of a vector of estimable functions* $\mathbf{C}^T\boldsymbol{\beta}$ *associated with the positive definite Aitken model* $\{\mathbf{y}, \mathbf{X}\boldsymbol{\beta}, \sigma^2\mathbf{W}\}$ *is* $\mathbf{C}^T(\mathbf{X}^T\mathbf{W}^{-1}\mathbf{X})^-\mathbf{X}^T\mathbf{W}^{-1}\mathbf{y}$, *and this expression is invariant to the choice of generalized inverse of* $\mathbf{X}^T\mathbf{W}^{-1}\mathbf{X}$.

Theorem 11.1.2 *Under any unconstrained linear model with model matrix* $\mathbf{X}$, *the generalized least squares estimator of an estimable function* $\mathbf{c}^T\boldsymbol{\beta}$ *associated with the positive definite Aitken model* $\{\mathbf{y}, \mathbf{X}\boldsymbol{\beta}, \sigma^2\mathbf{W}\}$ *is linear and unbiased.*

Theorem 11.1.3 *Let* $\mathbf{C}_1^T\boldsymbol{\beta}$ *and* $\mathbf{C}_2^T\boldsymbol{\beta}$ *be two vectors of functions that are estimable under any unconstrained linear model with model matrix* $\mathbf{X}$, *and let* $\tilde{\boldsymbol{\beta}}$ *represent any solution to the Aitken equations for the positive definite Aitken model* $\{\mathbf{y}, \mathbf{X}\boldsymbol{\beta}, \sigma^2\mathbf{W}\}$, *so that* $\mathbf{C}_1^T\tilde{\boldsymbol{\beta}}$ *and* $\mathbf{C}_2^T\tilde{\boldsymbol{\beta}}$ *are the vectors of generalized least squares estimators of* $\mathbf{C}_1^T\boldsymbol{\beta}$ *and* $\mathbf{C}_2^T\boldsymbol{\beta}$, *respectively, associated with that model. Then*

$$cov(\mathbf{C}_1^T\tilde{\boldsymbol{\beta}}, \mathbf{C}_2^T\tilde{\boldsymbol{\beta}}) = \sigma^2\mathbf{C}_1^T(\mathbf{X}^T\mathbf{W}^{-1}\mathbf{X})^-\mathbf{C}_2 \quad \text{and} \quad var(\mathbf{C}_1^T\tilde{\boldsymbol{\beta}}) = \sigma^2\mathbf{C}_1^T(\mathbf{X}^T\mathbf{W}^{-1}\mathbf{X})^-\mathbf{C}_1,$$

and these expressions are invariant to the choice of generalized inverse of $\mathbf{X}^T\mathbf{W}^{-1}\mathbf{X}$. *Furthermore, if the functions in* $\mathbf{C}_1^T\boldsymbol{\beta}$ *are linearly independent, then* $var(\mathbf{C}_1^T\tilde{\boldsymbol{\beta}})$ *is positive definite.*

Theorem 11.1.4 *Under the positive definite Aitken model* $\{\mathbf{y}, \mathbf{X}\boldsymbol{\beta}, \sigma^2\mathbf{W}\}$, *the variance of the generalized least squares estimator* $\mathbf{c}^T\tilde{\boldsymbol{\beta}}$ *of an estimable linear function* $\mathbf{c}^T\boldsymbol{\beta}$ *associated with that model is uniformly (in* $\boldsymbol{\beta}$ *and* σ^2*) less than that of any other linear unbiased estimator of* $\mathbf{c}^T\boldsymbol{\beta}$; *that is, the generalized least squares estimator is the BLUE.*

Theorem 11.1.5 *Consider the reparameterized positive definite Aitken model* $\{\mathbf{y}, \breve{\mathbf{X}}\boldsymbol{\tau}, \sigma^2\mathbf{W}\}$, *where* $C(\breve{\mathbf{X}}) = C(\mathbf{X})$. *Let* $\mathbf{S}$ *and* $\mathbf{T}$ *represent* $p \times t$ *matrices such that* $\breve{\mathbf{X}} = \mathbf{X}\mathbf{T}$ *and* $\mathbf{X} = \breve{\mathbf{X}}\mathbf{S}^T$.

(a) If $\mathbf{c}^T\boldsymbol{\beta}$ *is estimable under the positive definite Aitken model* $\{\mathbf{y}, \mathbf{X}\boldsymbol{\beta}, \sigma^2\mathbf{W}\}$, *then* $\mathbf{c}^T\mathbf{T}\boldsymbol{\tau}$ *is estimable under the reparameterized Aitken model and its generalized least squares estimator may be expressed as either* $\mathbf{c}^T\tilde{\boldsymbol{\beta}}$ *or* $\mathbf{c}^T\mathbf{T}\tilde{\boldsymbol{\tau}}$, *where* $\tilde{\boldsymbol{\beta}}$ *is any solution to the Aitken equations* $\mathbf{X}^T\mathbf{W}^{-1}\mathbf{X}\boldsymbol{\beta} = \mathbf{X}^T\mathbf{W}^{-1}\mathbf{y}$ *for the original model and* $\tilde{\boldsymbol{\tau}}$ *is any solution to the Aitken equations* $\breve{\mathbf{X}}^T\mathbf{W}^{-1}\breve{\mathbf{X}}\boldsymbol{\tau} = \breve{\mathbf{X}}^T\mathbf{W}^{-1}\mathbf{y}$ *for the reparameterized model.*

(b) *If* $\mathbf{b}^T\boldsymbol{\tau}$ *is estimable under the reparameterized Aitken model, then* $\mathbf{b}^T\mathbf{S}^T\boldsymbol{\beta}$ *is estimable under the positive definite Aitken model* $\{\mathbf{y}, \mathbf{X}\boldsymbol{\beta}, \sigma^2\mathbf{W}\}$ *and its generalized least squares estimator may be expressed as either* $\mathbf{b}^T\mathbf{S}^T\tilde{\boldsymbol{\beta}}$ *or* $\mathbf{b}^T\tilde{\boldsymbol{\tau}}$, *where* $\tilde{\boldsymbol{\beta}}$ *and* $\tilde{\boldsymbol{\tau}}$ *are defined as in part (a).*

The next theorem facilitates the consideration of the geometry of generalized least squares and is useful for other purposes as well.

Theorem 11.1.6 *For any* $n \times p$ *matrix* $\mathbf{X}$ *of rank* p^* *and any* $n \times n$ *positive definite matrix* $\mathbf{W}$*:*

(a) $\mathbf{X}(\mathbf{X}^T\mathbf{W}^{-1}\mathbf{X})^-\mathbf{X}^T\mathbf{W}^{-1}\mathbf{X} = \mathbf{X}$ *for any generalized inverse of* $\mathbf{X}^T\mathbf{W}^{-1}\mathbf{X}$*;*
(b) $\mathbf{X}(\mathbf{X}^T\mathbf{W}^{-1}\mathbf{X})^-\mathbf{X}^T$ *is invariant to the choice of generalized inverse of* $\mathbf{X}^T\mathbf{W}^{-1}\mathbf{X}$*;*
(c) *if* $\breve{\mathbf{X}}$ *is any* $n \times p^*$ *matrix whose columns are linearly independent columns of* $\mathbf{X}$*, then*

$$\breve{\mathbf{X}}(\breve{\mathbf{X}}^T\mathbf{W}^{-1}\breve{\mathbf{X}})^{-1}\breve{\mathbf{X}}^T = \mathbf{X}(\mathbf{X}^T\mathbf{W}^{-1}\mathbf{X})^-\mathbf{X}^T$$

for any generalized inverse of $\mathbf{X}^T\mathbf{W}^{-1}\mathbf{X}$*;*
(d) $\mathbf{E} \equiv \mathbf{W}^{-1} - \mathbf{W}^{-1}\mathbf{X}(\mathbf{X}^T\mathbf{W}^{-1}\mathbf{X})^-\mathbf{X}^T\mathbf{W}^{-1}$ *is symmetric, and it satisfies* $\mathbf{EX} = \mathbf{0}$ *and* $\mathbf{EWE} = \mathbf{E}$.

Proof Applying Theorem 3.3.3a to the matrix $\mathbf{W}^{-\frac{1}{2}}\mathbf{X}$ yields

$$\mathbf{W}^{-\frac{1}{2}}\mathbf{X}(\mathbf{X}^T\mathbf{W}^{-1}\mathbf{X})^-\mathbf{X}^T\mathbf{W}^{-1}\mathbf{X} = \mathbf{W}^{-\frac{1}{2}}\mathbf{X},$$

where the generalized inverse of $\mathbf{X}^T\mathbf{W}^{-1}\mathbf{X}$ is arbitrary. Pre-multiplication of both sides by $\mathbf{W}^{\frac{1}{2}}$ yields part (a). Next, let $\mathbf{G}_1$ and $\mathbf{G}_2$ represent two generalized inverses of $\mathbf{X}^T\mathbf{W}^{-1}\mathbf{X}$. Applying Theorem 3.3.4 to the matrix $\mathbf{W}^{-\frac{1}{2}}\mathbf{X}$, we obtain

$$\mathbf{W}^{-\frac{1}{2}}\mathbf{X}\mathbf{G}_1\mathbf{X}^T\mathbf{W}^{-\frac{1}{2}} = \mathbf{W}^{-\frac{1}{2}}\mathbf{X}\mathbf{G}_2\mathbf{X}^T\mathbf{W}^{-\frac{1}{2}}.$$

Pre- and post-multiplying both sides by $\mathbf{W}^{\frac{1}{2}}$ establishes part (b). For part (c), note that if $\mathbf{c} \in \mathcal{C}(\mathbf{W}^{-\frac{1}{2}}\mathbf{X})$ and $\mathbf{S}$ is any $p \times p^*$ matrix such that $\mathbf{X} = \breve{\mathbf{X}}\mathbf{S}^T$, then for some p-vector $\mathbf{a}$, $\mathbf{c} = \mathbf{W}^{-\frac{1}{2}}\mathbf{X}\mathbf{a} = \mathbf{W}^{-\frac{1}{2}}\breve{\mathbf{X}}\mathbf{S}^T\mathbf{a}$, showing that $\mathbf{c} \in \mathcal{C}(\mathbf{W}^{-\frac{1}{2}}\breve{\mathbf{X}})$. Similarly, if $\mathbf{c}^* \in \mathcal{C}(\mathbf{W}^{-\frac{1}{2}}\breve{\mathbf{X}})$ and $\mathbf{T}$ is any $p \times p^*$ matrix such that $\breve{\mathbf{X}} = \mathbf{XT}$, then for some p^*-vector $\mathbf{a}^*$, $\mathbf{c}^* = \mathbf{W}^{-\frac{1}{2}}\breve{\mathbf{X}}\mathbf{a}^* = \mathbf{W}^{-\frac{1}{2}}\mathbf{XTa}^*$, showing that $\mathbf{c}^* \in \mathcal{C}(\mathbf{W}^{-\frac{1}{2}}\mathbf{X})$. Thus $\mathcal{C}(\mathbf{W}^{-\frac{1}{2}}\mathbf{X}) = \mathcal{C}(\mathbf{W}^{-\frac{1}{2}}\breve{\mathbf{X}})$. So by Corollary 8.1.1.1,

$$\mathbf{W}^{-\frac{1}{2}}\breve{\mathbf{X}}(\breve{\mathbf{X}}^T\mathbf{W}^{-1}\breve{\mathbf{X}})^-\breve{\mathbf{X}}^T\mathbf{W}^{-\frac{1}{2}} = \mathbf{W}^{-\frac{1}{2}}\mathbf{X}(\mathbf{X}^T\mathbf{W}^{-1}\mathbf{X})^-\mathbf{X}^T\mathbf{W}^{-\frac{1}{2}}.$$

Part (c) follows upon pre- and post-multiplying both sides by $\mathbf{W}^{\frac{1}{2}}$ and noting that $\breve{\mathbf{X}}^T\mathbf{W}^{-1}\breve{\mathbf{X}}$ is nonsingular. The symmetry claim of part (d) follows directly from part (b) and Corollary 3.3.1.1, and using part (a) we obtain

$$\mathbf{EX} = [\mathbf{W}^{-1} - \mathbf{W}^{-1}\mathbf{X}(\mathbf{X}^T\mathbf{W}^{-1}\mathbf{X})^-\mathbf{X}^T\mathbf{W}^{-1}]\mathbf{X} = \mathbf{W}^{-1}\mathbf{X} - \mathbf{W}^{-1}\mathbf{X} = \mathbf{0}$$

and

$$\begin{aligned}\mathbf{EWE} &= [\mathbf{W}^{-1} - \mathbf{W}^{-1}\mathbf{X}(\mathbf{X}^T\mathbf{W}^{-1}\mathbf{X})^-\mathbf{X}^T\mathbf{W}^{-1}]\mathbf{W}[\mathbf{W}^{-1} - \mathbf{W}^{-1}\mathbf{X}(\mathbf{X}^T\mathbf{W}^{-1}\mathbf{X})^-\mathbf{X}^T\mathbf{W}^{-1}]\\ &= \mathbf{W}^{-1} - 2\mathbf{W}^{-1}\mathbf{X}(\mathbf{X}^T\mathbf{W}^{-1}\mathbf{X})^-\mathbf{X}^T\mathbf{W}^{-1}\\ &\quad +\mathbf{W}^{-1}\mathbf{X}(\mathbf{X}^T\mathbf{W}^{-1}\mathbf{X})^-\mathbf{X}^T\mathbf{W}^{-1}\mathbf{X}(\mathbf{X}^T\mathbf{W}^{-1}\mathbf{X})^-\mathbf{X}^T\mathbf{W}^{-1}\\ &= \mathbf{W}^{-1} - 2\mathbf{W}^{-1}\mathbf{X}(\mathbf{X}^T\mathbf{W}^{-1}\mathbf{X})^-\mathbf{X}^T\mathbf{W}^{-1} + \mathbf{W}^{-1}\mathbf{X}(\mathbf{X}^T\mathbf{W}^{-1}\mathbf{X})^-\mathbf{X}^T\mathbf{W}^{-1}\\ &= \mathbf{W}^{-1} - \mathbf{W}^{-1}\mathbf{X}(\mathbf{X}^T\mathbf{W}^{-1}\mathbf{X})^-\mathbf{X}^T\mathbf{W}^{-1}.\end{aligned}$$

This establishes part (d). □

By Theorem 11.1.1, the generalized least squares estimator of $\mathbf{X}\boldsymbol{\beta}$ corresponding to the positive definite Aitken model $\{\mathbf{y}, \mathbf{X}\boldsymbol{\beta}, \sigma^2\mathbf{W}\}$ is

$$\begin{aligned}\mathbf{X}\tilde{\boldsymbol{\beta}} &= \mathbf{X}(\mathbf{X}^T\mathbf{W}^{-1}\mathbf{X})^-\mathbf{X}^T\mathbf{W}^{-1}\mathbf{y}\\ &\equiv \tilde{\mathbf{P}}_\mathbf{X}\mathbf{y},\end{aligned}$$

say. Now recall that $\mathbf{X}\hat{\boldsymbol{\beta}}$, where $\hat{\boldsymbol{\beta}}$ is any solution to the normal equations $\mathbf{X}^T\mathbf{X}\boldsymbol{\beta} = \mathbf{X}^T\mathbf{y}$, is the orthogonal projection of $\mathbf{y}$ onto the column space of $\mathbf{X}$, i.e., $\mathbf{X}\hat{\boldsymbol{\beta}} = \mathbf{P}_\mathbf{X}\mathbf{y}$. Can we say something similar about $\mathbf{X}\tilde{\boldsymbol{\beta}}$? Clearly $\tilde{\mathbf{P}}_\mathbf{X}$ is a generalization of $\mathbf{P}_\mathbf{X}$, in the sense that $\tilde{\mathbf{P}}_\mathbf{X} = \mathbf{P}_\mathbf{X}$ when $\mathbf{W} = \mathbf{I}$. But does $\tilde{\mathbf{P}}_\mathbf{X}$ retain all the "nice" properties of $\mathbf{P}_\mathbf{X}$ delineated in Chap. 8 (and in Theorem 8.1.2 in particular)? The following theorem and subsequent discussion answer this question.

Theorem 11.1.7 *The matrix* $\tilde{\mathbf{P}}_\mathbf{X}$*:*

(a) is invariant to the choice of generalized inverse of $\mathbf{X}^T\mathbf{W}^{-1}\mathbf{X}$*;*
(b) satisfies $\tilde{\mathbf{P}}_\mathbf{X}\mathbf{X} = \mathbf{X}$*;*
(c) is idempotent;
(d) satisfies $\mathcal{C}(\tilde{\mathbf{P}}_\mathbf{X}) = \mathcal{C}(\mathbf{X})$*;*
(e) satisfies $rank(\tilde{\mathbf{P}}_\mathbf{X}) = rank(\mathbf{X})$*.*

Proof Part (a) follows immediately from Theorem 11.1.6b, and part (b) is just a restatement of Theorem 11.1.6a. For part (c), observe that

$$\tilde{\mathbf{P}}_\mathbf{X}\tilde{\mathbf{P}}_\mathbf{X} = \tilde{\mathbf{P}}_\mathbf{X}\mathbf{X}(\mathbf{X}^T\mathbf{W}^{-1}\mathbf{X})^-\mathbf{X}^T\mathbf{W}^{-1} = \mathbf{X}(\mathbf{X}^T\mathbf{W}^{-1}\mathbf{X})^-\mathbf{X}^T\mathbf{W}^{-1} = \tilde{\mathbf{P}}_\mathbf{X}$$

where the second equality holds by part (b). Part (d) is established by noting that the definition of $\tilde{\mathbf{P}}_{\mathbf{X}}$ implies that $\mathcal{C}(\tilde{\mathbf{P}}_{\mathbf{X}}) \subseteq \mathcal{C}(\mathbf{X})$, while part (b) implies that $\mathcal{C}(\mathbf{X}) \subseteq \mathcal{C}(\tilde{\mathbf{P}}_{\mathbf{X}})$. Part (e) follows directly from part (d). □

Upon comparing Theorem 11.1.7 to Theorem 8.1.2, we see that one property that holds for $\mathbf{P}_{\mathbf{X}}$ is not claimed to hold for $\tilde{\mathbf{P}}_{\mathbf{X}}$: symmetry. In fact, in general $\tilde{\mathbf{P}}_{\mathbf{X}}^T$ does not equal $\tilde{\mathbf{P}}_{\mathbf{X}}$. Consequently, $\tilde{\mathbf{P}}_{\mathbf{X}}$ is generally not the orthogonal projection matrix onto $\mathcal{C}(\mathbf{X})$; in fact, we already know that $\mathbf{P}_{\mathbf{X}}$ is that matrix. Rather, $\tilde{\mathbf{P}}_{\mathbf{X}}$ is what is called a *skew projection matrix onto* $\mathcal{C}(\mathbf{X})$. Specifically, it is the skew projection matrix onto $\mathcal{C}(\mathbf{X})$ associated with $\mathbf{W}$. $\tilde{\mathbf{P}}_{\mathbf{X}}$ projects $\mathbf{y}$ onto $\mathcal{C}(\mathbf{X})$, but not orthogonally (in general).

Example 11.1-1. Toy Example of a Skew Projection onto $\mathcal{C}(\mathbf{X})$

Consider a toy example similar to Example 8.1-1, with $n = 2$, $\mathbf{y} = (1, 3)^T$, and $\mathbf{X} = \mathbf{1}_2$. Skew projections of $\mathbf{y}$ onto $\mathcal{C}(\mathbf{X})$ corresponding to several choices of $\mathbf{W}$ are illustrated in Fig. 11.1. Each ellipse in the figure represents a locus of points $\mathbf{v}$ equidistant, in terms of the generalized distance $[(\mathbf{y}-\mathbf{v})^T\mathbf{W}^{-1}(\mathbf{y}-\mathbf{v})]^{\frac{1}{2}}$, from $\mathbf{y}$. ■

In Chap. 8, we constructed an overall ANOVA for the Gauss–Markov model based upon the orthogonal projection of $\mathbf{y}$ onto $\mathcal{C}(\mathbf{X})$. Might we do something similar for the positive definite Aitken model? We may write

$$\mathbf{y} = \tilde{\mathbf{P}}_{\mathbf{X}}\mathbf{y} + (\mathbf{I} - \tilde{\mathbf{P}}_{\mathbf{X}})\mathbf{y},$$

or equivalently, upon letting $\tilde{\mathbf{y}} = \tilde{\mathbf{P}}_{\mathbf{X}}\mathbf{y}$ and $\tilde{\mathbf{e}} = (\mathbf{I} - \tilde{\mathbf{P}}_{\mathbf{X}})\mathbf{y}$,

$$\mathbf{y} = \tilde{\mathbf{y}} + \tilde{\mathbf{e}}. \tag{11.1}$$

However, $\tilde{\mathbf{y}}$ and $\tilde{\mathbf{e}}$ are not orthogonal, so unlike (8.1), (11.1) is not an orthogonal decomposition of $\mathbf{y}$. Nevertheless, it is worthwhile to give names to the terms on the right-hand side of (11.1) and to document some of their properties.

Definition 11.1.2 Under the positive definite Aitken model $\{\mathbf{y}, \mathbf{X}\boldsymbol{\beta}, \sigma^2\mathbf{W}\}$, the unique vector $\tilde{\mathbf{y}} = \mathbf{X}\tilde{\boldsymbol{\beta}}$, where $\tilde{\boldsymbol{\beta}}$ is any solution to the Aitken equations corresponding to that model, is called the *generalized least squares fitted values vector* for that model. The unique vector $\tilde{\mathbf{e}} = \mathbf{y} - \tilde{\mathbf{y}}$ is called the *generalized least squares fitted residuals vector* for that model.

The next theorem gives means and variances of the generalized least squares fitted values and residuals vectors. Its proof is left as an exercise.

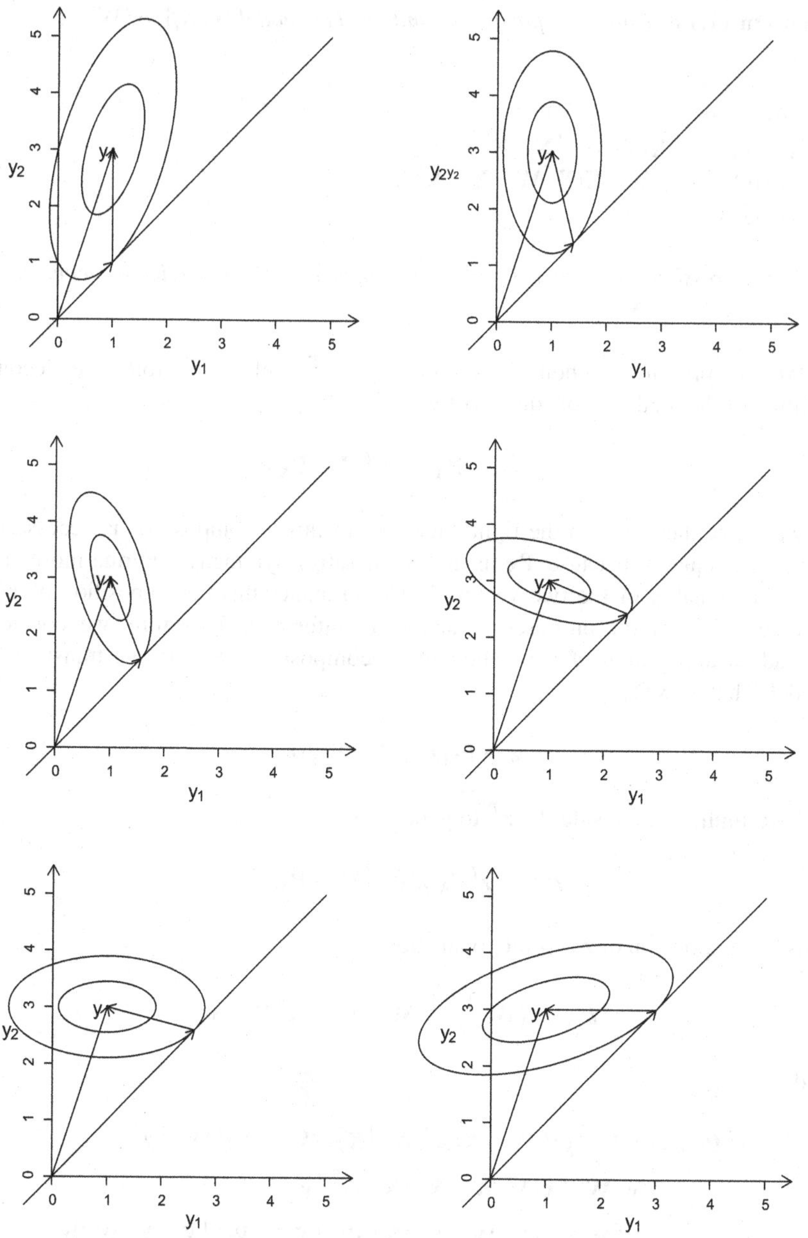

Fig. 11.1 Skew projections of $\mathbf{y} = (1, 3)^T$ onto $\mathcal{C}(\mathbf{1}_2)$. Upper left panel, $\mathbf{W} = \begin{pmatrix} 1 & 1 \\ 1 & 4 \end{pmatrix}$; upper right panel, $\mathbf{W} = \begin{pmatrix} 1 & 0 \\ 0 & 4 \end{pmatrix}$; middle left panel, $\mathbf{W} = \begin{pmatrix} 1 & -1 \\ -1 & 4 \end{pmatrix}$; middle right panel, $\mathbf{W} = \begin{pmatrix} 4 & -1 \\ -1 & 1 \end{pmatrix}$; lower left panel, $\mathbf{W} = \begin{pmatrix} 4 & 0 \\ 0 & 1 \end{pmatrix}$; lower right panel, $\mathbf{W} = \begin{pmatrix} 4 & 1 \\ 1 & 1 \end{pmatrix}$. The ellipses in each panel depict sets $\{\mathbf{v} : (\mathbf{y} - \mathbf{v})^T \mathbf{W}^{-1} (\mathbf{y} - \mathbf{v}) = c^2\}$ for two values of c

Theorem 11.1.8 *Under the positive definite Aitken model* $\{\mathbf{y}, \mathbf{X}\boldsymbol{\beta}, \sigma^2\mathbf{W}\}$:

(a) $E(\tilde{\mathbf{y}}) = \mathbf{X}\boldsymbol{\beta}$;
(b) $E(\tilde{\mathbf{e}}) = \mathbf{0}$;
(c) $var(\tilde{\mathbf{y}}) = \sigma^2\mathbf{X}(\mathbf{X}^T\mathbf{W}^{-1}\mathbf{X})^-\mathbf{X}^T$;
(d) $var(\tilde{\mathbf{e}}) = \sigma^2[\mathbf{W} - \mathbf{X}(\mathbf{X}^T\mathbf{W}^{-1}\mathbf{X})^-\mathbf{X}^T]$;
(e) $cov(\tilde{\mathbf{y}}, \tilde{\mathbf{e}}) = \mathbf{0}$.

The expressions in parts (c) and (d) are invariant to the choice of generalized inverse of $\mathbf{X}^T\mathbf{W}^{-1}\mathbf{X}$.

We may pre-multiply both sides of (11.1) by $\mathbf{y}^T$, yielding the following decomposition of the total sum of squares of $\mathbf{y}$:

$$\mathbf{y}^T\mathbf{y} = \mathbf{y}^T\tilde{\mathbf{P}}_{\mathbf{X}}\mathbf{y} + \mathbf{y}^T(\mathbf{I} - \tilde{\mathbf{P}}_{\mathbf{X}})\mathbf{y}.$$

However, neither term on the right-hand side of this decomposition is necessarily a sum of squares because $\tilde{\mathbf{P}}_{\mathbf{X}}$ is not (generally) symmetric. Furthermore, the non-orthogonality of the two vectors in (11.1) makes this decomposition of $\mathbf{y}^T\mathbf{y}$ relatively useless for purposes of statistical inference. Therefore, we consider instead an application of the orthogonal decomposition (8.1) to the transformed model. That is, write

$$\mathbf{z} = \mathbf{P}_{\mathbf{M}}\mathbf{z} + (\mathbf{I} - \mathbf{P}_{\mathbf{M}})\mathbf{z}$$

and pre-multiply both sides by $\mathbf{z}^T$ to yield

$$\mathbf{z}^T\mathbf{z} = \mathbf{z}^T\mathbf{P}_{\mathbf{M}}\mathbf{z} + \mathbf{z}^T(\mathbf{I} - \mathbf{P}_{\mathbf{M}})\mathbf{z}.$$

This decomposition of $\mathbf{z}$ is orthogonal. Now

$$\mathbf{z}^T\mathbf{z} = (\mathbf{W}^{-\frac{1}{2}}\mathbf{y})^T(\mathbf{W}^{-\frac{1}{2}}\mathbf{y}) = \mathbf{y}^T\mathbf{W}^{-1}\mathbf{y}$$

and

$$\begin{aligned}
\mathbf{z}^T\mathbf{P}_{\mathbf{M}}\mathbf{z} &= (\mathbf{W}^{-\frac{1}{2}}\mathbf{y})^T\mathbf{W}^{-\frac{1}{2}}\mathbf{X}(\mathbf{X}^T\mathbf{W}^{-1}\mathbf{X})^-(\mathbf{W}^{-\frac{1}{2}}\mathbf{X})^T\mathbf{W}^{-\frac{1}{2}}\mathbf{y} \\
&= \mathbf{y}^T\mathbf{W}^{-1}\mathbf{X}(\mathbf{X}^T\mathbf{W}^{-1}\mathbf{X})^-\mathbf{X}^T\mathbf{W}^{-1}\mathbf{y} \\
&= \mathbf{y}^T\mathbf{W}^{-1}\mathbf{X}(\mathbf{X}^T\mathbf{W}^{-1}\mathbf{X})^-\mathbf{X}^T\mathbf{W}^{-1}\mathbf{X}(\mathbf{X}^T\mathbf{W}^{-1}\mathbf{X})^-\mathbf{X}^T\mathbf{W}^{-1}\mathbf{y} \\
&= \mathbf{y}^T\tilde{\mathbf{P}}_{\mathbf{X}}^T\mathbf{W}^{-1}\tilde{\mathbf{P}}_{\mathbf{X}}\mathbf{y},
\end{aligned}$$

where we used Theorem 11.1.6a for the third equality. By subtraction,

$$\mathbf{z}^T(\mathbf{I} - \mathbf{P}_{\mathbf{M}})\mathbf{z} = \mathbf{y}^T\mathbf{W}^{-1}\mathbf{y} - \mathbf{y}^T\tilde{\mathbf{P}}_{\mathbf{X}}^T\mathbf{W}^{-1}\tilde{\mathbf{P}}_{\mathbf{X}}\mathbf{y}.$$

Thus, the decomposition of the sum of squares $\mathbf{z}^T\mathbf{z}$ may be written in terms of $\mathbf{y}$, $\mathbf{W}$, and the skew projection matrix $\tilde{\mathbf{P}}_{\mathbf{X}}$ in tabular form as follows:

Source	Rank	Generalized sum of squares
Model	rank($\mathbf{X}$)	$\mathbf{y}^T\tilde{\mathbf{P}}_{\mathbf{X}}^T\mathbf{W}^{-1}\tilde{\mathbf{P}}_{\mathbf{X}}\mathbf{y}$
Residual	n − rank($\mathbf{X}$)	$\mathbf{y}^T(\mathbf{W}^{-1}-\tilde{\mathbf{P}}_{\mathbf{X}}^T\mathbf{W}^{-1}\tilde{\mathbf{P}}_{\mathbf{X}})\mathbf{y}$
Total	n	$\mathbf{y}^T\mathbf{W}^{-1}\mathbf{y}$

We have inserted the word "Generalized" into the heading of the third column because the entries in this column are not sums of squares of vectors that comprise the decomposition of $\mathbf{y}$ either into $\mathbf{P}_{\mathbf{X}}\mathbf{y}$ and $(\mathbf{I}-\mathbf{P}_{\mathbf{X}})\mathbf{y}$ or into $\tilde{\mathbf{P}}_{\mathbf{X}}\mathbf{y}$ and $(\mathbf{I}-\tilde{\mathbf{P}}_{\mathbf{X}})\mathbf{y}$. Rather, they are sums of squares of the decomposition of the transformed vector $\mathbf{W}^{-\frac{1}{2}}\mathbf{y}$ into $\mathbf{P}_{\mathbf{W}^{-\frac{1}{2}}\mathbf{X}}\mathbf{W}^{-\frac{1}{2}}\mathbf{y}$ and $(\mathbf{I}-\mathbf{P}_{\mathbf{W}^{-\frac{1}{2}}\mathbf{X}})\mathbf{W}^{-\frac{1}{2}}\mathbf{y}$. For the same reason, we refer to this overall ANOVA as the *generalized overall ANOVA*.

The generalized sums of squares in the table above may be re-expressed in various ways. For example, the Model generalized sum of squares may be written alternatively as either $\tilde{\boldsymbol{\beta}}^T\mathbf{X}^T\mathbf{W}^{-1}\mathbf{X}\tilde{\boldsymbol{\beta}}$, $\tilde{\boldsymbol{\beta}}^T\mathbf{X}^T\mathbf{W}^{-1}\mathbf{y}$, $\mathbf{y}^T\mathbf{W}^{-1}\mathbf{X}(\mathbf{X}^T\mathbf{W}^{-1}\mathbf{X})^-\mathbf{X}^T\mathbf{W}^{-1}\mathbf{y}$, or $\tilde{\mathbf{y}}^T\mathbf{W}^{-1}\tilde{\mathbf{y}}$. (The first three of these expressions may be obtained by directly applying formulas from Chap. 8 for the Gauss–Markov Model sum of squares to the transformed model, and the fourth follows from the first by the definition of $\tilde{\mathbf{y}}$.) The Residual generalized sum of squares may be expressed as the generalized residual sum of squares function evaluated at $\boldsymbol{\beta}=\tilde{\boldsymbol{\beta}}$, i.e., as $(\mathbf{y}-\mathbf{X}\tilde{\boldsymbol{\beta}})^T\mathbf{W}^{-1}(\mathbf{y}-\mathbf{X}\tilde{\boldsymbol{\beta}})$, or equivalently as either $\mathbf{y}^T\mathbf{E}\mathbf{y}$ or $\tilde{\mathbf{e}}^T\mathbf{W}^{-1}\tilde{\mathbf{e}}$.

The remaining theorems in this section extend Theorems 8.2.1, 8.2.2, and 8.2.3 (which pertained to the overall ANOVA) to the generalized overall ANOVA.

Theorem 11.1.9 *Under the positive definite Aitken model* $\{\mathbf{y}, \mathbf{X}\boldsymbol{\beta}, \sigma^2\mathbf{W}\}$*:*

(a) $E(\mathbf{y}^T\tilde{\mathbf{P}}_{\mathbf{X}}^T\mathbf{W}^{-1}\tilde{\mathbf{P}}_{\mathbf{X}}\mathbf{y}) = \boldsymbol{\beta}^T\mathbf{X}^T\mathbf{W}^{-1}\mathbf{X}\boldsymbol{\beta}+p^*\sigma^2$;
(b) $E[\mathbf{y}^T(\mathbf{W}^{-1}-\tilde{\mathbf{P}}_{\mathbf{X}}^T\mathbf{W}^{-1}\tilde{\mathbf{P}}_{\mathbf{X}})\mathbf{y}] = (n-p^*)\sigma^2$.

Proof By Theorems 4.2.4, 2.10.2, 11.1.7b, and 3.3.5,

$$
\begin{aligned}
\mathrm{E}(\mathbf{y}^T\tilde{\mathbf{P}}_{\mathbf{X}}^T\mathbf{W}^{-1}\tilde{\mathbf{P}}_{\mathbf{X}}\mathbf{y}) &= \boldsymbol{\beta}^T\mathbf{X}^T\tilde{\mathbf{P}}_{\mathbf{X}}^T\mathbf{W}^{-1}\tilde{\mathbf{P}}_{\mathbf{X}}\mathbf{X}\boldsymbol{\beta}+\sigma^2\mathrm{tr}(\tilde{\mathbf{P}}_{\mathbf{X}}^T\mathbf{W}^{-1}\tilde{\mathbf{P}}_{\mathbf{X}}\mathbf{W})\\
&= \boldsymbol{\beta}^T\mathbf{X}^T\mathbf{W}^{-1}\mathbf{X}\boldsymbol{\beta}+\sigma^2\mathrm{tr}[\tilde{\mathbf{P}}_{\mathbf{X}}^T\mathbf{W}^{-1}\mathbf{X}(\mathbf{X}^T\mathbf{W}^{-1}\mathbf{X})^-\mathbf{X}^T\mathbf{W}^{-1}\mathbf{W}]\\
&= \boldsymbol{\beta}^T\mathbf{X}^T\mathbf{W}^{-1}\mathbf{X}\boldsymbol{\beta}+\sigma^2\mathrm{tr}[\mathbf{X}^T\mathbf{W}^{-1}\mathbf{X}(\mathbf{X}^T\mathbf{W}^{-1}\mathbf{X})^-]\\
&= \boldsymbol{\beta}^T\mathbf{X}^T\mathbf{W}^{-1}\mathbf{X}\boldsymbol{\beta}+p^*\sigma^2.
\end{aligned}
$$

This completes the proof of part (a). Part (b) may be proved similarly (see the exercises). □

Corollary 11.1.9.1 *Under the positive definite Aitken model* $\{\mathbf{y}, \mathbf{X}\boldsymbol{\beta}, \sigma^2\mathbf{W}\}$, *if* $n > p^*$ *then the generalized residual mean square given by any of the following equivalent expressions,*

$$\tilde{\sigma}^2 = \frac{\mathbf{y}^T(\mathbf{W}^{-1} - \tilde{\mathbf{P}}_{\mathbf{X}}^T\mathbf{W}^{-1}\tilde{\mathbf{P}}_{\mathbf{X}})\mathbf{y}}{n - p^*} = \frac{\mathbf{y}^T\mathbf{E}\mathbf{y}}{n - p^*} = \frac{\tilde{\mathbf{e}}^T\mathbf{W}^{-1}\tilde{\mathbf{e}}}{n - p^*},$$

is an unbiased estimator of σ^2.

It follows from Corollary 11.1.9.1 and Theorem 11.1.3 that $\tilde{\sigma}^2\mathbf{C}^T(\mathbf{X}^T\mathbf{W}^{-1}\mathbf{X})^-\mathbf{C}$ is an unbiased estimator [under the positive definite Aitken model $\{\mathbf{y}, \mathbf{X}\boldsymbol{\beta}, \sigma^2\mathbf{W}\}$] of the variance–covariance matrix of the generalized least squares estimator of an estimable vector $\mathbf{C}^T\boldsymbol{\beta}$ associated with that model.

Theorem 11.1.10 *Let* $\mathbf{C}^T\boldsymbol{\beta}$ *be an estimable vector and let* $\mathbf{C}^T\tilde{\boldsymbol{\beta}}$ *be its generalized least squares estimator associated with the positive definite Aitken model* $\{\mathbf{y}, \mathbf{X}\boldsymbol{\beta}, \sigma^2\mathbf{W}\}$. *Suppose that* $n > p^*$, *and let* $\tilde{\sigma}^2$ *be the generalized residual mean square for the same model.*

(a) *If the skewness matrix of* $\mathbf{y}$ *is null, then* $\text{cov}(\mathbf{C}^T\tilde{\boldsymbol{\beta}}, \tilde{\sigma}^2) = \mathbf{0}$ *under the specified model;*
(b) *If the skewness and excess kurtosis matrices of* $\mathbf{y}$ *are null, then* $\text{var}(\tilde{\sigma}^2) = 2\sigma^4/(n - p^*)$ *under the specified model.*

Proof Left as an exception. □

Example 11.1-2. Generalized Least Squares Estimation Based on a Vector of Means

Consider once more the positive definite Aitken model for a vector of means, introduced in Example 5.2.2-1:

$$\bar{\mathbf{y}} = \check{\mathbf{X}}\boldsymbol{\beta} + \bar{\mathbf{e}}$$

where

$$\check{\mathbf{X}} = \begin{pmatrix} \mathbf{x}_1^T \\ \vdots \\ \mathbf{x}_m^T \end{pmatrix}, \qquad \bar{\mathbf{e}} = \begin{pmatrix} \bar{e}_{1\cdot} \\ \vdots \\ \bar{e}_{m\cdot} \end{pmatrix},$$

$\text{E}(\bar{\mathbf{e}}) = \mathbf{0}$, and $\text{var}(\bar{\mathbf{e}}) = \sigma^2[\oplus_{i=1}^m(1/n_i)]$. Thus $\mathbf{W} = \oplus_{i=1}^m(1/n_i)$ here, which is clearly positive definite. The generalized least squares estimator of any estimable function $\mathbf{c}^T\boldsymbol{\beta}$ may then be expressed as

$$\mathbf{c}^T\tilde{\boldsymbol{\beta}} = \mathbf{c}^T[\breve{\mathbf{X}}^T(\oplus n_i)\breve{\mathbf{X}}]^-\breve{\mathbf{X}}^T(\oplus n_i)\bar{\mathbf{y}}$$
$$= \mathbf{c}^T\left(\sum_{i=1}^m n_i\mathbf{x}_i\mathbf{x}_i^T\right)^-\left(\sum_{i=1}^m n_i\bar{y}_{i\cdot}\mathbf{x}_i\right).$$

Furthermore, if $m > p^*$, the generalized residual mean square (which estimates σ^2 unbiasedly) may be expressed as

$$\tilde{\sigma}^2 = \bar{\mathbf{y}}^T\left[(\oplus n_i) - (\oplus n_i)\breve{\mathbf{X}}\left(\sum_{i=1}^m n_i\mathbf{x}_i\mathbf{x}_i^T\right)^-\breve{\mathbf{X}}^T(\oplus n_i)\right]\bar{\mathbf{y}}\Big/(m-p^*)$$
$$= \left[\sum_{i=1}^m n_i\bar{y}_{i\cdot}^2 - \left(\sum_{i=1}^m n_i\bar{y}_{i\cdot}\mathbf{x}_i^T\right)\left(\sum_{i=1}^m n_i\mathbf{x}_i\mathbf{x}_i^T\right)^-\left(\sum_{i=1}^m n_i\bar{y}_{i\cdot}\mathbf{x}_i\right)\right]\Big/(m-p^*).$$

In fact, the generalized overall ANOVA is as follows:

Source	Rank	Generalized sum of squares
Model	$\text{rank}(\breve{\mathbf{X}})$	$\left(\sum_{i=1}^m n_i\bar{y}_{i\cdot}\mathbf{x}_i^T\right)\left(\sum_{i=1}^m n_i\mathbf{x}_i\mathbf{x}_i^T\right)^-\left(\sum_{i=1}^m n_i\bar{y}_{i\cdot}\mathbf{x}_i\right)$
Residual	$n - \text{rank}(\breve{\mathbf{X}})$	by subtraction
Total	n	$\sum_{i=1}^m n_i\bar{y}_{i\cdot}^2$

■

Example 11.1-3. Generalized Least Squares Estimation of the Mean of a Stationary First-Order Autoregressive Model

Consider once more the autoregressive model of order one introduced in Example 5.2.4-1, which, if the parameter ρ is known and satisfies $-1 < \rho < 1$, can be written as a positive definite Aitken model with variance–covariance matrix as follows:

$$\mathbf{y} = \mu\mathbf{1}_n + \mathbf{e}, \qquad \text{var}(\mathbf{e}) = \sigma^2\mathbf{W},$$

where

$$\mathbf{W} = \frac{1}{1-\rho^2}\begin{pmatrix} 1 & \rho & \rho^2 & \rho^3 & \cdots & \rho^{n-1} \\ & 1 & \rho & \rho^2 & \cdots & \rho^{n-2} \\ & & 1 & \rho & \cdots & \rho^{n-3} \\ & & & 1 & \cdots & \rho^{n-4} \\ & & & & \ddots & \vdots \\ & & & & & 1 \end{pmatrix}.$$

W is a patterned matrix whose inverse is well known (and easily verified) to equal

$$\mathbf{W}^{-1} = \begin{pmatrix} 1 & -\rho & 0 & 0 & \cdots & 0 & 0 \\ & 1+\rho^2 & -\rho & 0 & \cdots & 0 & 0 \\ & & 1+\rho^2 & -\rho & \cdots & 0 & 0 \\ & & & 1+\rho^2 & \cdots & 0 & 0 \\ & & & & \ddots & \vdots & \vdots \\ & & & & & 1+\rho^2 & -\rho \\ & & & & & & 1 \end{pmatrix}.$$

Suppose that $n \geq 3$. Then, by Theorem 11.1.1, the generalized least squares estimator of μ is

$$\tilde{\mu} = \left[\mathbf{1}^T \begin{pmatrix} 1 & -\rho & 0 & \cdots & 0 & 0 \\ & 1+\rho^2 & -\rho & \cdots & 0 & 0 \\ & & 1+\rho^2 & \cdots & 0 & 0 \\ & & & & \vdots & \vdots \\ & & & & 1+\rho^2 & -\rho \\ & & & & & 1 \end{pmatrix} \mathbf{1} \right]^{-1} \mathbf{1}^T \begin{pmatrix} 1 & -\rho & 0 & \cdots & 0 & 0 \\ & 1+\rho^2 & -\rho & \cdots & 0 & 0 \\ & & 1+\rho^2 & \cdots & 0 & 0 \\ & & & \ddots & \vdots & \vdots \\ & & & & 1+\rho^2 & -\rho \\ & & & & & 1 \end{pmatrix} \mathbf{y}$$

$$= \frac{(1-\rho)(y_1 + y_n) + (1 - 2\rho + \rho^2)\sum_{i=2}^{n-1} y_i}{n - 2(n-1)\rho + (n-2)\rho^2}$$

$$= \frac{(1-\rho)(y_1 + y_n) + (1-\rho)^2\sum_{i=2}^{n-1} y_i}{n(1-\rho)^2 + 2\rho(1-\rho)}$$

$$= \frac{y_1 + y_n + (1-\rho)\sum_{i=2}^{n-1} y_i}{2 + (1-\rho)(n-2)}.$$

Thus $\tilde{\mu}$ is a weighted mean of the y_i's: the first and last observations are weighted equally, but the "interior" observations $y_2, \ldots, y_{n-1}$ are all weighted by $1 - \rho$, thereby getting less weight than the first and last if ρ is positive, or more weight than the first and last if ρ is negative. Furthermore, by Theorem 11.1.3,

$$\text{var}(\tilde{\mu}) = \sigma^2(\mathbf{1}^T\mathbf{W}^{-1}\mathbf{1})^{-1} = \frac{\sigma^2}{(1-\rho)[2 + (1-\rho)(n-2)]}.$$

This variance is greater than σ^2/n if ρ is positive, and smaller than σ^2/n if ρ is negative. ■

Best linear unbiased estimation of estimable linear functions of $\boldsymbol{\beta}$ under the constrained positive definite Aitken model $\{\mathbf{y}, \mathbf{X}\boldsymbol{\beta}, \sigma^2\mathbf{W} : \mathbf{A}\boldsymbol{\beta} = \mathbf{h}\}$ can be accomplished with little additional difficulty beyond that required for best linear unbiased estimation under the constrained Gauss–Markov model $\{\mathbf{y}, \mathbf{X}\boldsymbol{\beta}, \sigma^2\mathbf{I} : \mathbf{A}\boldsymbol{\beta} = \mathbf{h}\}$. This topic is considered in an exercise.

11.2 The General Case

We now consider the best linear unbiased estimation of an estimable function $\mathbf{c}^T\boldsymbol{\beta}$ for the most general case of the Aitken model $\{\mathbf{y}, \mathbf{X}\boldsymbol{\beta}, \sigma^2\mathbf{W}\}$, for which $\mathbf{W}$ is merely nonnegative definite, hence possibly positive semidefinite. The "transformed model" approach that was taken when $\mathbf{W}$ is positive definite is no longer applicable because $\mathbf{W}^{-\frac{1}{2}}$ does not exist when $\mathbf{W}$ is positive semidefinite. Instead, we take an approach that minimizes the variance of a linear unbiased estimator of an arbitrary estimable function rather than the model's (weighted) residual sum of squares. This alternative approach allows $\mathbf{W}$ to be positive semidefinite, but also yields results that specialize to those presented in the previous section when $\mathbf{W}$ is positive definite.

As in the previous section, we let $\mathbf{W}^{\frac{1}{2}}$ represent the (unique) nonnegative definite square root of $\mathbf{W}$ and let $\mathbf{c}^T\boldsymbol{\beta}$ be an estimable function. For any linear estimator $t_0 + \mathbf{t}^T\mathbf{y}$ of $\mathbf{c}^T\boldsymbol{\beta}$,

$$\text{var}(t_0 + \mathbf{t}^T\mathbf{y}) = \sigma^2\mathbf{t}^T\mathbf{W}\mathbf{t}.$$

Because t_0 must equal 0 and $\mathbf{t}^T\mathbf{X}$ must equal $\mathbf{c}^T$ for $t_0 + \mathbf{t}^T\mathbf{y}$ to be unbiased (Theorem 6.1.1), the BLUE of $\mathbf{c}^T\boldsymbol{\beta}$ is $\tilde{\mathbf{t}}^T\mathbf{y}$, where $\tilde{\mathbf{t}}$ is any vector that minimizes $\mathbf{t}^T\mathbf{W}\mathbf{t}$ subject to the condition $\mathbf{t}^T\mathbf{X} = \mathbf{c}^T$. The Lagrangian function for this minimization problem is

$$L(\mathbf{t}, \boldsymbol{\lambda}) = \mathbf{t}^T\mathbf{W}\mathbf{t} + 2\boldsymbol{\lambda}^T(\mathbf{X}^T\mathbf{t} - \mathbf{c}).$$

Partial derivatives of the Lagrangian are

$$\frac{\partial L}{\partial \mathbf{t}} = 2\mathbf{W}\mathbf{t} + 2\mathbf{X}\boldsymbol{\lambda} \quad \text{and} \quad \frac{\partial L}{\partial \boldsymbol{\lambda}} = 2(\mathbf{X}^T\mathbf{t} - \mathbf{c}),$$

and equating these to zero yields

$$\mathbf{W}\mathbf{t} + \mathbf{X}\boldsymbol{\lambda} = \mathbf{0}, \qquad \mathbf{X}^T\mathbf{t} = \mathbf{c}.$$

Therefore, a necessary condition for $\mathbf{t}^T\mathbf{W}\mathbf{t}$ to attain a minimum at $\mathbf{t} = \tilde{\mathbf{t}}$, for $\mathbf{t}$ satisfying $\mathbf{X}^T\mathbf{t} = \mathbf{c}$, is that $\tilde{\mathbf{t}}$ be the first n-component subvector of a solution to the system of equations

$$\begin{pmatrix} \mathbf{W} & \mathbf{X} \\ \mathbf{X}^T & \mathbf{0} \end{pmatrix} \begin{pmatrix} \mathbf{t} \\ \boldsymbol{\lambda} \end{pmatrix} = \begin{pmatrix} \mathbf{0} \\ \mathbf{c} \end{pmatrix}.$$

We call this system the *best linear unbiased estimating equations*, or *BLUE equations*, for $\mathbf{c}^T\boldsymbol{\beta}$. The first theorem of this section establishes that this system has a solution. This is accomplished by showing that the system is compatible.

Theorem 11.2.1 *The BLUE equations for any function $\mathbf{c}^T\boldsymbol{\beta}$ that is estimable under the Aitken model $\{\mathbf{y}, \mathbf{X}\boldsymbol{\beta}, \sigma^2\mathbf{W}\}$ are consistent.*

Proof Because $\mathbf{c}^T\boldsymbol{\beta}$ is estimable under the specified model, a vector $\mathbf{a}$ exists such that $\mathbf{c}^T = \mathbf{a}^T\mathbf{X}$. Let $\boldsymbol{\ell}_1$ and $\boldsymbol{\ell}_2$ represent any n-vector and any p-vector, respectively, such that

$$(\boldsymbol{\ell}_1^T, \boldsymbol{\ell}_2^T)\begin{pmatrix} \mathbf{W} & \mathbf{X} \\ \mathbf{X}^T & \mathbf{0} \end{pmatrix} = \mathbf{0}^T,$$

or equivalently, such that

$$\boldsymbol{\ell}_1^T\mathbf{W} + \boldsymbol{\ell}_2^T\mathbf{X}^T = \mathbf{0}^T \quad \text{and} \quad \boldsymbol{\ell}_1^T\mathbf{X} = \mathbf{0}^T. \tag{11.2}$$

Post-multiplying both sides of the first system of equations in (11.2) by $\boldsymbol{\ell}_1$ yields

$$\boldsymbol{\ell}_1^T\mathbf{W}\boldsymbol{\ell}_1 + \boldsymbol{\ell}_2^T\mathbf{X}^T\boldsymbol{\ell}_1 = 0,$$

which, using the second system of equations in (11.2), can be reduced to the equation $\boldsymbol{\ell}_1^T\mathbf{W}\boldsymbol{\ell}_1 = 0$. It follows from Corollary 2.10.4.1 that $\mathbf{W}^{\frac{1}{2}}\boldsymbol{\ell}_1 = \mathbf{0}$, implying that

$$\mathbf{W}\boldsymbol{\ell}_1 = \mathbf{0}. \tag{11.3}$$

Putting (11.3) back into the first system of equations in (11.2) yields the system $\boldsymbol{\ell}_2^T\mathbf{X}^T = \mathbf{0}^T$. Then

$$(\boldsymbol{\ell}_1^T, \boldsymbol{\ell}_2^T)\begin{pmatrix} \mathbf{0} \\ \mathbf{c} \end{pmatrix} = \boldsymbol{\ell}_2^T\mathbf{c} = \boldsymbol{\ell}_2^T\mathbf{X}^T\mathbf{a} = \mathbf{0}^T\mathbf{a} = 0.$$

Because compatibility implies consistency, the theorem is proved. □

Just prior to Theorem 11.2.1, we established that necessary conditions for a linear estimator $t_0 + \mathbf{t}^T\mathbf{y}$ to be a BLUE of an estimable function $\mathbf{c}^T\boldsymbol{\beta}$ under the nonnegative definite Aitken model $\{\mathbf{y}, \mathbf{X}\boldsymbol{\beta}, \sigma^2\mathbf{W}\}$ are that $t_0 = 0$ and $\mathbf{t} = \tilde{\mathbf{t}}$ where $\tilde{\mathbf{t}}$ is the first n-component subvector of a solution to the BLUE equations for that function. The second theorem of this section establishes that these conditions are not only necessary but sufficient. The theorem could be regarded as an extension of the Gauss–Markov theorem to the general Aitken model.

Theorem 11.2.2 *Under the Aitken model $\{\mathbf{y}, \mathbf{X}\boldsymbol{\beta}, \sigma^2\mathbf{W}\}$, $t_0 + \mathbf{t}^T\mathbf{y}$ is a BLUE of an estimable function $\mathbf{c}^T\boldsymbol{\beta}$ if and only if $t_0 = 0$ and $\mathbf{t} = \tilde{\mathbf{t}}$, where $\tilde{\mathbf{t}}$ is the first n-component subvector of a solution to the BLUE equations for $\mathbf{c}^T\boldsymbol{\beta}$.*

Proof Let $(\tilde{\mathbf{t}}^T, \tilde{\boldsymbol{\lambda}}^T)^T$ be any solution to the BLUE equations for $\mathbf{c}^T\boldsymbol{\beta}$ (such a solution exists by Theorem 11.2.1). Then $\mathbf{X}^T\tilde{\mathbf{t}} = \mathbf{c}$, implying (by Theorem 6.1.1) that if $t_0 = 0$, then $t_0 + \tilde{\mathbf{t}}^T\mathbf{y}$ is a linear unbiased estimator of $\mathbf{c}^T\boldsymbol{\beta}$. Next, let $\ell_0 + \boldsymbol{\ell}^T\mathbf{y}$ be any linear unbiased estimator of $\mathbf{c}^T\boldsymbol{\beta}$. Then by Theorem 6.1.1, $\ell_0 = 0$ and $\boldsymbol{\ell}^T\mathbf{X} = \mathbf{c}^T$. Furthermore,

$$\begin{aligned}
\text{cov}[\tilde{\mathbf{t}}^T\mathbf{y}, (\boldsymbol{\ell} - \tilde{\mathbf{t}})^T\mathbf{y}] &= \sigma^2\tilde{\mathbf{t}}^T\mathbf{W}(\boldsymbol{\ell} - \tilde{\mathbf{t}}) \\
&= -\sigma^2\tilde{\boldsymbol{\lambda}}^T\mathbf{X}^T(\boldsymbol{\ell} - \tilde{\mathbf{t}}) \\
&= -\sigma^2\tilde{\boldsymbol{\lambda}}^T(\mathbf{X}^T\boldsymbol{\ell} - \mathbf{X}^T\tilde{\mathbf{t}}) \\
&= -\sigma^2\tilde{\boldsymbol{\lambda}}^T(\mathbf{c} - \mathbf{c}) \\
&= 0,
\end{aligned}$$

so that

$$\begin{aligned}
\text{var}(\ell_0 + \boldsymbol{\ell}^T\mathbf{y}) &= \text{var}[\tilde{\mathbf{t}}^T\mathbf{y} + (\boldsymbol{\ell} - \tilde{\mathbf{t}})^T\mathbf{y}] \\
&= \text{var}(\tilde{\mathbf{t}}^T\mathbf{y}) + \text{var}[(\boldsymbol{\ell} - \tilde{\mathbf{t}})^T\mathbf{y}] \\
&= \text{var}(\tilde{\mathbf{t}}^T\mathbf{y}) + \sigma^2(\boldsymbol{\ell} - \tilde{\mathbf{t}})^T\mathbf{W}(\boldsymbol{\ell} - \tilde{\mathbf{t}}) \\
&= \text{var}(\tilde{\mathbf{t}}^T\mathbf{y}) + \sigma^2(\boldsymbol{\ell} - \tilde{\mathbf{t}})^T\mathbf{W}^{\frac{1}{2}}\mathbf{W}^{\frac{1}{2}}(\boldsymbol{\ell} - \tilde{\mathbf{t}}).
\end{aligned}$$

Thus, $\text{var}(\ell_0 + \boldsymbol{\ell}^T\mathbf{y}) \geq \text{var}(\tilde{\mathbf{t}}^T\mathbf{y})$, with equality holding if and only if $\mathbf{W}^{\frac{1}{2}}(\boldsymbol{\ell} - \tilde{\mathbf{t}}) = \mathbf{0}$, i.e., if and only if $\mathbf{W}(\boldsymbol{\ell} - \tilde{\mathbf{t}}) = \mathbf{0}$. i.e., if and only if $\mathbf{W}\boldsymbol{\ell} = \mathbf{W}\tilde{\mathbf{t}}$, i.e., if and only if $(\boldsymbol{\ell}^T, \tilde{\boldsymbol{\lambda}}^T)^T$ is also a solution to the BLUE equations for $\mathbf{c}^T\boldsymbol{\beta}$. □

Corollary 11.2.2.1 *Under the Aitken model* $\{\mathbf{y}, \mathbf{X}\boldsymbol{\beta}, \sigma^2\mathbf{W}\}$, $t_0 + \mathbf{t}^T\mathbf{y}$ *is a BLUE of an estimable function* $\mathbf{c}^T\boldsymbol{\beta}$ *if and only if: (1)* $t_0 = 0$, *(2)* $\mathbf{t}^T\mathbf{X} = \mathbf{c}^T$, *and (3)* $\mathbf{W}\mathbf{t} \in \mathcal{C}(\mathbf{X})$.

Proof If $t_0 + \mathbf{t}^T\mathbf{y}$ is a BLUE of an estimable function $\mathbf{c}^T\boldsymbol{\beta}$, then the three conditions follow immediately from the necessity part of Theorem 11.2.2. Conversely, suppose that $\mathbf{c}^T\boldsymbol{\beta}$ is estimable and that the three conditions hold. Then by the third condition, $\mathbf{W}\mathbf{t} = \mathbf{X}\mathbf{b}$ for some p-vector $\mathbf{b}$. This and the second condition imply that

$$\begin{pmatrix} \mathbf{W} & \mathbf{X} \\ \mathbf{X}^T & \mathbf{0} \end{pmatrix} \begin{pmatrix} \mathbf{t} \\ -\mathbf{b} \end{pmatrix} = \begin{pmatrix} \mathbf{0} \\ \mathbf{c} \end{pmatrix},$$

establishing that $\mathbf{t}$ is the first n-component subvector of a solution to the BLUE equations for $\mathbf{c}^T\boldsymbol{\beta}$. This and the first condition imply, by the sufficiency part of Theorem 11.2.2, that $t_0 + \mathbf{t}^T\mathbf{y}$ is a BLUE of $\mathbf{c}^T\boldsymbol{\beta}$. □

Corollary 11.2.2.2 *Under the Aitken model* $\{\mathbf{y}, \mathbf{X}\boldsymbol{\beta}, \sigma^2\mathbf{W}\}$, $\mathbf{t}^T\mathbf{y}$ *is a BLUE of its expectation if and only if* $\mathbf{W}\mathbf{t} \in \mathcal{C}(\mathbf{X})$.

The proof of Theorem 11.2.2 reveals that when $\mathbf{W}$ is positive semidefinite, BLUEs generally are not strictly unique; that is, $\mathbf{t}^T\mathbf{y}$ and $\boldsymbol{\ell}^T\mathbf{y}$—with $\mathbf{t} \neq \boldsymbol{\ell}$—can both be BLUEs of the same estimable function. (For a numerical example, see the exercises.) However, by the following additional corollary to Theorem 11.2.2, the BLUEs of that function could be regarded as "essentially unique," in the sense that they equal one another with probability one.

Corollary 11.2.2.3 *If $\mathbf{t}^T\mathbf{y}$ and $\boldsymbol{\ell}^T\mathbf{y}$ are both BLUEs of the same estimable function $\mathbf{c}^T\boldsymbol{\beta}$ under the Aitken model $\{\mathbf{y}, \mathbf{X}\boldsymbol{\beta}, \sigma^2\mathbf{W}\}$, then $\mathbf{t}^T\mathbf{y} = \boldsymbol{\ell}^T\mathbf{y}$ (with probability one).*

Proof Because $\mathbf{t}^T\mathbf{y}$ and $\boldsymbol{\ell}^T\mathbf{y}$ are both unbiased for $\mathbf{c}^T\boldsymbol{\beta}$, we have

$$\mathrm{E}(\mathbf{t}^T\mathbf{y} - \boldsymbol{\ell}^T\mathbf{y}) = \mathrm{E}[(\mathbf{t} - \boldsymbol{\ell})^T\mathbf{y}] = (\mathbf{t}^T - \boldsymbol{\ell}^T)\mathbf{X}\boldsymbol{\beta} = (\mathbf{c}^T - \mathbf{c}^T)\boldsymbol{\beta} = 0. \tag{11.4}$$

Furthermore, because $\mathbf{t}^T\mathbf{y}$ and $\boldsymbol{\ell}^T\mathbf{y}$ are both BLUEs of $\mathbf{c}^T\boldsymbol{\beta}$, from the proof of Theorem 11.2.2 we have $\mathbf{W}(\mathbf{t} - \boldsymbol{\ell}) = \mathbf{0}$, which implies that $(\mathbf{t} - \boldsymbol{\ell})^T\mathbf{W}(\mathbf{t} - \boldsymbol{\ell}) = 0$, which implies further that

$$\mathrm{var}(\mathbf{t}^T\mathbf{y} - \boldsymbol{\ell}^T\mathbf{y}) = \mathrm{var}[(\mathbf{t} - \boldsymbol{\ell})^T\mathbf{y}] = 0. \tag{11.5}$$

Together, (11.4) and (11.5) imply the desired result. □

What does a solution to the BLUE equations for an estimable function look like? And what is the covariance between BLUEs of two estimable functions? And finally how, if at all, may we estimate σ^2 unbiasedly (as we were able to do under the positive definite Aitken model)? The next theorem addresses these three questions.

Theorem 11.2.3 *Consider the Aitken model $\{\mathbf{y}, \mathbf{X}\boldsymbol{\beta}, \sigma^2\mathbf{W}\}$. Let*

$$\begin{pmatrix} \mathbf{G}_{11} & \mathbf{G}_{12} \\ \mathbf{G}_{21} & \mathbf{G}_{22} \end{pmatrix}$$

represent a generalized inverse of the matrix

$$\begin{pmatrix} \mathbf{W} & \mathbf{X} \\ \mathbf{X}^T & \mathbf{0} \end{pmatrix}.$$

(a) If $\mathbf{c}^T\boldsymbol{\beta}$ is an estimable function, then the collection of BLUEs of $\mathbf{c}^T\boldsymbol{\beta}$ consists of all quantities of the form $\tilde{\mathbf{t}}^T\mathbf{y}$, where

$$\tilde{\mathbf{t}} = \mathbf{G}_{12}\mathbf{c} + (\mathbf{I}_n - \mathbf{G}_{11}\mathbf{W} - \mathbf{G}_{12}\mathbf{X}^T, \ -\mathbf{G}_{11}\mathbf{X})\mathbf{z}, \quad \mathbf{z} \in \mathbb{R}^{n+p}.$$

In particular, $\mathbf{c}^T\mathbf{G}_{12}^T\mathbf{y}$ is a BLUE of $\mathbf{c}^T\boldsymbol{\beta}$, and so is $\mathbf{c}^T\mathbf{G}_{21}\mathbf{y}$.

(b) If $\mathbf{C}_1^T\boldsymbol{\beta}$ and $\mathbf{C}_2^T\boldsymbol{\beta}$ are vectors of estimable functions, then the matrix of covariances between a vector of BLUEs of $\mathbf{C}_1^T\boldsymbol{\beta}$ and a vector of BLUEs of $\mathbf{C}_2^T\boldsymbol{\beta}$

is $-\sigma^2 \mathbf{C}_1^T \mathbf{G}_{22} \mathbf{C}_2$*, and this expression is invariant to the choice of generalized inverse of* $\begin{pmatrix} \mathbf{W} & \mathbf{X} \\ \mathbf{X}^T & \mathbf{0} \end{pmatrix}$*. In particular, the variance–covariance matrix of any vector of BLUEs of an estimable vector* $\mathbf{C}^T \boldsymbol{\beta}$ *is* $-\sigma^2 \mathbf{C}^T \mathbf{G}_{22} \mathbf{C}$*.*

(c) *Provided that* $rank(\mathbf{W}, \mathbf{X}) > rank(\mathbf{X})$*,*

$$\tilde{\sigma}^2 = (\mathbf{y}^T \mathbf{G}_{11} \mathbf{y})/[rank(\mathbf{W}, \mathbf{X}) - rank(\mathbf{X})]$$

is an unbiased estimator of σ^2*, regardless of the choice of generalized inverse of* $\begin{pmatrix} \mathbf{W} & \mathbf{X} \\ \mathbf{X}^T & \mathbf{0} \end{pmatrix}$*.*

Proof If $\mathbf{c}^T \boldsymbol{\beta}$ is estimable, then the BLUE equations for it are consistent (by Theorem 11.2.1), so by Theorem 3.2.3 all solutions to those equations are given by

$$\begin{pmatrix} \mathbf{G}_{11} & \mathbf{G}_{12} \\ \mathbf{G}_{21} & \mathbf{G}_{22} \end{pmatrix} \begin{pmatrix} \mathbf{0} \\ \mathbf{c} \end{pmatrix} + \left[\mathbf{I}_{n+p} - \begin{pmatrix} \mathbf{G}_{11} & \mathbf{G}_{12} \\ \mathbf{G}_{21} & \mathbf{G}_{22} \end{pmatrix} \begin{pmatrix} \mathbf{W} & \mathbf{X} \\ \mathbf{X}^T & \mathbf{0} \end{pmatrix} \right] \mathbf{z},$$

where $\mathbf{z}$ ranges over $\mathbb{R}^{n+p}$. The first n-component subvector of any such solution is

$$\mathbf{G}_{12}\mathbf{c} + (\mathbf{I}_n - \mathbf{G}_{11}\mathbf{W} - \mathbf{G}_{12}\mathbf{X}^T, \ -\mathbf{G}_{11}\mathbf{X})\mathbf{z},$$

which reduces to $\mathbf{G}_{12}\mathbf{c}$ when $\mathbf{z} = \mathbf{0}$. That $\mathbf{c}^T \mathbf{G}_{12}^T \mathbf{y}$ is a BLUE of $\mathbf{c}^T \boldsymbol{\beta}$ then follows by Theorem 11.2.2. Because $\begin{pmatrix} \mathbf{W} & \mathbf{X} \\ \mathbf{X}^T & \mathbf{0} \end{pmatrix}$ is symmetric, another generalized inverse of it (by Corollary 3.3.1.1) is $\begin{pmatrix} \mathbf{G}_{11}^T & \mathbf{G}_{21}^T \\ \mathbf{G}_{12}^T & \mathbf{G}_{22}^T \end{pmatrix}$, implying that $\mathbf{G}_{21}^T \mathbf{c}$ is the first n-component subvector of another solution to the BLUE equations for $\mathbf{c}^T \boldsymbol{\beta}$. Hence $\mathbf{c}^T \mathbf{G}_{21} \mathbf{y}$ is also a BLUE of $\mathbf{c}^T \boldsymbol{\beta}$.

For part (b), let $\mathbf{L}_1$ and $\mathbf{L}_2$ represent matrices such that $\mathbf{C}_1^T = \mathbf{L}_1 \mathbf{X}$ and $\mathbf{C}_2^T = \mathbf{L}_2 \mathbf{X}$. By part (a), arbitrary vectors of BLUEs of $\mathbf{C}_1^T \boldsymbol{\beta}$ and $\mathbf{C}_2^T \boldsymbol{\beta}$ may be written as

$$[\mathbf{C}_1^T \mathbf{G}_{12}^T + \mathbf{Z}_{11}^T (\mathbf{I}_n - \mathbf{W}\mathbf{G}_{11}^T - \mathbf{X}\mathbf{G}_{12}^T) - \mathbf{Z}_{12}^T \mathbf{X}^T \mathbf{G}_{11}^T] \mathbf{y}$$

and

$$[\mathbf{C}_2^T \mathbf{G}_{12}^T + \mathbf{Z}_{21}^T (\mathbf{I}_n - \mathbf{W}\mathbf{G}_{11}^T - \mathbf{X}\mathbf{G}_{12}^T) - \mathbf{Z}_{22}^T \mathbf{X}^T \mathbf{G}_{11}^T] \mathbf{y},$$

respectively, for suitably chosen matrices $\mathbf{Z}_{11}$, $\mathbf{Z}_{12}$, $\mathbf{Z}_{21}$, $\mathbf{Z}_{22}$. The matrix of covariances between these two vectors of BLUEs is

$$\begin{aligned}\sigma^2[\mathbf{C}_1^T\mathbf{G}_{12}^T + \mathbf{Z}_{11}^T(\mathbf{I}_n - \mathbf{W}\mathbf{G}_{11}^T - \mathbf{X}\mathbf{G}_{12}^T) - \mathbf{Z}_{12}^T\mathbf{X}^T\mathbf{G}_{11}^T] \\ \times \mathbf{W}[\mathbf{G}_{12}\mathbf{C}_2 + (\mathbf{I}_n - \mathbf{G}_{11}\mathbf{W} - \mathbf{G}_{12}\mathbf{X}^T)\mathbf{Z}_{21} - \mathbf{G}_{11}\mathbf{X}\mathbf{Z}_{22}].\end{aligned}$$

Now

$$\begin{aligned}\mathbf{C}_1^T\mathbf{G}_{12}^T\mathbf{W}\mathbf{G}_{12}\mathbf{C}_2 &= \mathbf{L}_1\mathbf{X}\mathbf{G}_{12}^T\mathbf{W}\mathbf{G}_{12}\mathbf{X}^T\mathbf{L}_2^T \\ &= -\mathbf{L}_1\mathbf{X}\mathbf{G}_{12}^T\mathbf{X}\mathbf{G}_{22}\mathbf{X}^T\mathbf{L}_2^T \\ &= -\mathbf{L}_1\mathbf{X}\mathbf{G}_{22}\mathbf{X}^T\mathbf{L}_2^T \\ &= -\mathbf{C}_1^T\mathbf{G}_{22}\mathbf{C}_2,\end{aligned}$$

where parts (a) and (b) of Theorem 3.3.8 (with $\mathbf{W}$ and $\mathbf{X}$ playing the roles of $\mathbf{A}$ and $\mathbf{B}$, respectively, in that theorem) were used to obtain the second and third equalities. Furthermore, Theorem 3.3.8f tells us that $\mathbf{X}\mathbf{G}_{22}\mathbf{X}^T$ is invariant to the choice of generalized inverse of $\begin{pmatrix} \mathbf{W} & \mathbf{X} \\ \mathbf{X}^T & \mathbf{0} \end{pmatrix}$, hence so is $-\mathbf{C}_1^T\mathbf{G}_{22}\mathbf{C}_2$. Next,

$$\begin{aligned}\mathbf{Z}_{11}^T(\mathbf{I}_n - \mathbf{W}\mathbf{G}_{11}^T - \mathbf{X}\mathbf{G}_{12}^T)\mathbf{W} &= \mathbf{Z}_{11}^T(\mathbf{W} - \mathbf{W}\mathbf{G}_{11}^T\mathbf{W} - \mathbf{X}\mathbf{G}_{12}^T\mathbf{W}) \\ &= \mathbf{Z}_{11}^T(-\mathbf{X}\mathbf{G}_{22}^T\mathbf{X}^T + \mathbf{X}\mathbf{G}_{22}\mathbf{X}^T) \\ &= \mathbf{0},\end{aligned}$$

where we used Theorem 3.3.8b, d for the second equality and Theorem 3.3.8f for the third equality. Similarly, $\mathbf{W}(\mathbf{I}_n - \mathbf{G}_{11}\mathbf{W} - \mathbf{G}_{12}\mathbf{X}^T)\mathbf{Z}_{21} = \mathbf{0}$. Finally, by Theorem 3.3.8c, $\mathbf{Z}_{12}^T\mathbf{X}^T\mathbf{G}_{11}^T\mathbf{W} = \mathbf{0}$, with a similar result for $\mathbf{W}\mathbf{G}_{11}\mathbf{X}\mathbf{Z}_{22}$. Putting all of these results together yields the desired expression for the matrix of covariances.

For part (c) we obtain, using Theorems 4.2.4 and 3.3.8c, e,

$$\begin{aligned}\mathrm{E}(\mathbf{y}^T\mathbf{G}_{11}\mathbf{y}) &= \boldsymbol{\beta}^T\mathbf{X}^T\mathbf{G}_{11}\mathbf{X}\boldsymbol{\beta} + \sigma^2\mathrm{tr}(\mathbf{G}_{11}\mathbf{W}) \\ &= \sigma^2\mathrm{tr}(\mathbf{W}\mathbf{G}_{11}) \\ &= \sigma^2[\mathrm{rank}(\mathbf{W}, \mathbf{X}) - \mathrm{rank}(\mathbf{X})],\end{aligned}$$

regardless of the choice of generalized inverse of $\begin{pmatrix} \mathbf{W} & \mathbf{X} \\ \mathbf{X}^T & \mathbf{0} \end{pmatrix}$. Part (c) follows immediately. □

It follows from Theorem 11.2.3b, c that $-\tilde{\sigma}^2\mathbf{C}^T\mathbf{G}_{22}\mathbf{C}$ is an unbiased estimator (under the Aitken model $\{\mathbf{y}, \mathbf{X}\boldsymbol{\beta}, \sigma^2\mathbf{W}\}$) of the variance–covariance matrix of any vector of BLUEs of an estimable vector $\mathbf{C}^T\boldsymbol{\beta}$.

The next theorem in this section is the general Aitken analogue of Theorem 11.1.10.

Theorem 11.2.4 *Let* $\mathbf{C}^T\boldsymbol{\beta}$ *be an estimable vector and let* $\tilde{\mathbf{t}}^T\mathbf{y}$ *be a BLUE of it under the Aitken model* $\{\mathbf{y}, \mathbf{X}\boldsymbol{\beta}, \sigma^2\mathbf{W}\}$. *Suppose that rank*$(\mathbf{W}, \mathbf{X}) >$ *rank*$(\mathbf{X})$.

(a) If the skewness matrix of $\mathbf{y}$ *is null, then* $cov(\tilde{\mathbf{t}}^T\mathbf{y}, \tilde{\sigma}^2) = \mathbf{0}$;
(b) If the skewness and excess kurtosis matrices of $\mathbf{y}$ *are null, then*

$$var(\tilde{\sigma}^2) = \frac{2\sigma^4}{rank(\mathbf{W}, \mathbf{X}) - rank(\mathbf{X})}.$$

Proof Left as an exercise. □

Next, we show that the expressions given in Theorem 11.2.3 coincide with those given in theorems in the previous section when $\mathbf{W}$ is positive definite. By Theorem 3.3.7a, when $\mathbf{W}$ is positive definite,

$$\begin{pmatrix} \mathbf{W}^{-1} - \mathbf{W}^{-1}\mathbf{X}(\mathbf{X}^T\mathbf{W}^{-1}\mathbf{X})^-\mathbf{X}^T\mathbf{W}^{-1} & \mathbf{W}^{-1}\mathbf{X}(\mathbf{X}^T\mathbf{W}^{-1}\mathbf{X})^- \\ (\mathbf{X}^T\mathbf{W}^{-1}\mathbf{X})^-\mathbf{X}^T\mathbf{W}^{-1} & -(\mathbf{X}^T\mathbf{W}^{-1}\mathbf{X})^- \end{pmatrix}$$

is a generalized inverse of

$$\begin{pmatrix} \mathbf{W} & \mathbf{X} \\ \mathbf{X}^T & \mathbf{0} \end{pmatrix}.$$

Thus, by Theorem 11.2.3a, the collection of BLUEs of an estimable function $\mathbf{c}^T\boldsymbol{\beta}$ is given by

$$\begin{aligned} &\Bigg\{\mathbf{c}^T(\mathbf{X}^T\mathbf{W}^{-1}\mathbf{X})^-\mathbf{X}^T\mathbf{W}^{-1}\mathbf{y} \\ &\quad + \mathbf{z}^T \begin{pmatrix} \mathbf{I}_n - \mathbf{W}(\mathbf{W}^{-1} - \mathbf{W}^{-1}\widetilde{\mathbf{P}}_{\mathbf{X}})^T - \widetilde{\mathbf{P}}_{\mathbf{X}} \\ -\mathbf{X}^T(\mathbf{W}^{-1} - \mathbf{W}^{-1}\widetilde{\mathbf{P}}_{\mathbf{X}})^T \end{pmatrix} \mathbf{y} : \mathbf{z} \in \mathbb{R}^{n+p}\Bigg\} \\ &= \mathbf{c}^T(\mathbf{X}^T\mathbf{W}^{-1}\mathbf{X})^-\mathbf{X}^T\mathbf{W}^{-1}\mathbf{y}, \end{aligned}$$

where we have used Theorem 11.1.6a, d and the fact that $\mathbf{c}^T(\mathbf{X}^T\mathbf{W}^{-1}\mathbf{X})^-\mathbf{X}^T\mathbf{W}^{-1}\mathbf{y}$ is invariant to the choice of generalized inverse of $\mathbf{X}^T\mathbf{W}^{-1}\mathbf{X}$. Thus, when $\mathbf{W}$ is positive definite every estimable function $\mathbf{c}^T\boldsymbol{\beta}$ has a unique BLUE, and this BLUE coincides with the generalized least squares estimator of $\mathbf{c}^T\boldsymbol{\beta}$ given in Theorem 11.1.1. Similarly, by Theorem 11.2.3b, the matrix of covariances between the vector of BLUEs of the estimable vector $\mathbf{C}_1^T\boldsymbol{\beta}$ and the vector of BLUEs of the estimable vector $\mathbf{C}_2^T\boldsymbol{\beta}$ is

$$-\sigma^2\mathbf{C}_1^T[-(\mathbf{X}^T\mathbf{W}^{-1}\mathbf{X})^-]\mathbf{C}_2 = \sigma^2\mathbf{C}_1^T(\mathbf{X}^T\mathbf{W}^{-1}\mathbf{X})^-\mathbf{C}_2,$$

which matches the expression given in Theorem 11.1.3. Finally, because $\text{rank}(\mathbf{W}, \mathbf{X}) = \text{rank}(\mathbf{W}) = n$ when $\mathbf{W}$ is positive definite, the condition stated in Theorem 11.2.3c specializes to $n > \text{rank}(\mathbf{X})$, and the expression for the unbiased estimator $\tilde{\sigma}^2$ of σ^2 in that theorem specializes to

$$\frac{\mathbf{y}^T[\mathbf{W}^{-1} - \mathbf{W}^{-1}\mathbf{X}(\mathbf{X}^T\mathbf{W}^{-1}\mathbf{X})^-\mathbf{X}^T\mathbf{W}^{-1}]\mathbf{y}}{n - \text{rank}(\mathbf{X})},$$

which coincides with one of the expressions for $\tilde{\sigma}^2$ given in Corollary 11.1.9.1.

Because (as was just shown) solving the BLUE equations for an estimable function $\mathbf{c}^T\boldsymbol{\beta}$ under the Aitken model $\{\mathbf{y}, \mathbf{X}\boldsymbol{\beta}, \sigma^2\mathbf{W}\}$ reduces to solving the Aitken equations for the special case of the positive definite Aitken model $\{\mathbf{y}, \mathbf{X}\boldsymbol{\beta}, \sigma^2\mathbf{W}\}$ [and, as a further special case, to solving the normal equations for the Gauss–Markov model $\{\mathbf{y}, \mathbf{X}\boldsymbol{\beta}, \sigma^2\mathbf{I}\}$], it is natural to ask, "For every Aitken model, can we obtain BLUEs of all estimable functions by solving the *generalized Aitken equations*

$$\mathbf{X}^T\mathbf{W}^-\mathbf{X}\boldsymbol{\beta} = \mathbf{X}^T\mathbf{W}^-\mathbf{y},$$

where $\mathbf{W}^-$ is any generalized inverse of $\mathbf{W}$?" The following toy example reveals that the answer to this question is no, but that if the question is modified by replacing the phrase "any generalized inverse of $\mathbf{W}$" with "some generalized inverse of $\mathbf{W}$," the answer might be yes.

Example 11.2-1. Solutions to Generalized Aitken Equations for a Toy Example

Consider the following Aitken model for two observations:

$$\begin{pmatrix} y_1 \\ y_2 \end{pmatrix} = \begin{pmatrix} 1 \\ 0 \end{pmatrix}\beta + \mathbf{e}, \quad \text{E}(\mathbf{e}) = \mathbf{0}, \quad \text{var}(\mathbf{e}) = \sigma^2\mathbf{W} = \sigma^2\mathbf{J}_2.$$

Here, β is estimable and $\mathbf{W}$ is positive semidefinite. It is easily verified that any 2×2 matrix whose elements sum to one, and only such a matrix, is a generalized inverse of $\mathbf{W}$ (this was the topic of Exercise 3.1h). Thus, the matrix $\mathbf{W}^@ = \begin{pmatrix} 0 & \frac{1}{2} \\ \frac{1}{2} & 0 \end{pmatrix}$ is a generalized inverse of $\mathbf{W}$. Furthermore,

$$\mathbf{X}^T\mathbf{W}^@\mathbf{X} = 0 \quad \text{and} \quad \mathbf{X}^T\mathbf{W}^@\mathbf{y} = y_2/2,$$

so the generalized Aitken "equations" (there is just one equation here) corresponding to $\mathbf{W}^{@}$ are

$$0 \cdot \beta = y_2/2,$$

which has no solution (unless $y_2 = 0$). It is apparent, therefore, that for this model, solving the generalized Aitken equations that correspond to an arbitrary generalized inverse of $\mathbf{W}$ will not necessarily yield BLUEs of all estimable functions.

However, consider the generalized Aitken equations corresponding to $\mathbf{W}^{@@} = \begin{pmatrix} \frac{1}{2} & 0 \\ 0 & \frac{1}{2} \end{pmatrix}$ and $\mathbf{W}^{@@@} = \begin{pmatrix} 1 & -1 \\ -1 & 2 \end{pmatrix}$, which are two other generalized inverses of $\mathbf{W}$. The generalized Aitken "equations" corresponding to $\mathbf{W}^{@@}$ are

$$\frac{1}{2} \cdot \beta = y_1/2,$$

which has the unique solution $\tilde{\beta}_{@@} = y_1$, and the generalized Aitken "equations" corresponding to $\mathbf{W}^{@@@}$ are

$$1 \cdot \beta = y_1 - y_2,$$

which has the unique solution $\tilde{\beta}_{@@@} = y_1 - y_2$. Might either or both of these solutions be BLUEs?

An answer may be found using Corollary 11.2.2.1. Writing $\tilde{\beta}_{@@} = \mathbf{t}_{@@}^T\mathbf{y}$ and $\tilde{\beta}_{@@@} = \mathbf{t}_{@@@}^T\mathbf{y}$ where $\mathbf{t}_{@@} = (1, 0)^T$ and $\mathbf{t}_{@@@} = (1, -1)^T$, we have

$$\mathbf{t}_{@@}^T\mathbf{X} = \mathbf{t}_{@@@}^T\mathbf{X} = 1, \quad \mathbf{W}\mathbf{t}_{@@} = \mathbf{1}_2 \notin \mathcal{C}(\mathbf{X}), \quad \mathbf{W}\mathbf{t}_{@@@} = \mathbf{0}_2 \in \mathcal{C}(\mathbf{X}).$$

Thus, by Corollary 11.2.2.1 $\tilde{\beta}_{@@}$ is not a BLUE, but $\tilde{\beta}_{@@@}$ is a BLUE. Thus we see that, at least for this model, the generalized Aitken equations corresponding to some generalized inverses of $\mathbf{W}$ do not yield BLUEs, but the generalized Aitken equations corresponding to at least one other generalized inverse of $\mathbf{W}$ does yield a BLUE.

Next consider another two-observation model with the same variance–covariance matrix but with model matrix $\mathbf{X} = \mathbf{1}_2$. Because the elements of every generalized inverse of $\mathbf{W}$ must sum to one but are otherwise arbitrary, $\mathbf{X}^T\mathbf{W}^-\mathbf{X} = 1$ and $\mathbf{X}^T\mathbf{W}^-\mathbf{y} = ay_1 + (1-a)y_2$ for some a, regardless of the choice of generalized inverse (although a can depend on that choice). Thus the generalized Aitken equations for any generalized inverse of $\mathbf{W}$ are consistent and have solutions $\{\tilde{\beta} = ay_1 + (1-a)y_2 : a \in \mathbb{R}\}$. Are these solutions BLUEs? Using Corollary 11.2.2.1 again, it is easily verified that indeed these solutions, and only these solutions, are BLUEs. Thus, for this model, the generalized Aitken equations corresponding to all generalized inverses of $\mathbf{W}$ yield BLUEs. ■

Example 11.2-1 prompts many additional questions, of which we address the following three:

1. Why did the generalized Aitken equations corresponding to all generalized inverses of $\mathbf{W}$ yield BLUEs for the second model considered in the example, while the generalized Aitken equations corresponding to only a proper subset of generalized inverses of $\mathbf{W}$ yielded BLUEs for the first model?
2. Does a collection of generalized inverses of $\mathbf{W}$ for which the corresponding generalized Aitken equations yield BLUEs of all estimable functions exist for every Aitken model?
3. If the answer to Question 2 is yes, can the collection be described explicitly?

The answer to the first question lies in how $\mathcal{C}(\mathbf{X})$ is related to $\mathcal{C}(\mathbf{W})$ and is given by the next theorem.

Theorem 11.2.5 *Consider the Aitken model* $\{\mathbf{y}, \mathbf{X}\boldsymbol{\beta}, \sigma^2\mathbf{W}\}$, *and let* $\mathbf{c}^T\boldsymbol{\beta}$ *be an estimable function. If* $\mathcal{C}(\mathbf{X}) \subseteq \mathcal{C}(\mathbf{W})$, *then the generalized Aitken equations corresponding to any generalized inverse of* $\mathbf{W}$ *are consistent, and if* $\tilde{\boldsymbol{\beta}}$ *is any solution to those equations, then* $\mathbf{c}^T\tilde{\boldsymbol{\beta}}$ *is a BLUE of* $\mathbf{c}^T\boldsymbol{\beta}$.

Proof Because $\mathcal{C}(\mathbf{X}) \subseteq \mathcal{C}(\mathbf{W})$, by Theorem 3.3.7

$$\begin{pmatrix} \mathbf{W}^- - \mathbf{W}^-\mathbf{X}(\mathbf{X}^T\mathbf{W}^-\mathbf{X})^-\mathbf{X}^T\mathbf{W}^- & \mathbf{W}^-\mathbf{X}(\mathbf{X}^T\mathbf{W}^-\mathbf{X})^- \\ (\mathbf{X}^T\mathbf{W}^-\mathbf{X})^-\mathbf{X}^T\mathbf{W}^- & -(\mathbf{X}^T\mathbf{W}^-\mathbf{X})^- \end{pmatrix}$$

is a generalized inverse of $\begin{pmatrix} \mathbf{W} & \mathbf{X} \\ \mathbf{X}^T & \mathbf{0} \end{pmatrix}$. Furthermore, by Theorem 3.3.8a,

$$\mathbf{X}^T\mathbf{W}^-\mathbf{X}(\mathbf{X}^T\mathbf{W}^-\mathbf{X})^-\mathbf{X}^T = \mathbf{X}^T \tag{11.6}$$

and

$$\mathbf{X}(\mathbf{X}^T\mathbf{W}^-\mathbf{X})^-\mathbf{X}^T\mathbf{W}^-\mathbf{X} = \mathbf{X}. \tag{11.7}$$

Let $\overline{\mathbf{P}}_{\mathbf{X}} = \mathbf{X}(\mathbf{X}^T\mathbf{W}^-\mathbf{X})^-\mathbf{X}^T\mathbf{W}^-$. It follows by Theorem 11.2.3a that the collection of all BLUEs of $\mathbf{c}^T\boldsymbol{\beta}$ is given by

$$\begin{aligned} &\Big\{\mathbf{c}^T(\mathbf{X}^T\mathbf{W}^-\mathbf{X})^-\mathbf{X}^T\mathbf{W}^-\mathbf{y} \\ &+\mathbf{z}^T\begin{pmatrix} \mathbf{I}_n - \mathbf{W}(\mathbf{W}^- - \mathbf{W}^-\overline{\mathbf{P}}_{\mathbf{X}})^T - \mathbf{X}[(\mathbf{X}^T\mathbf{W}^-\mathbf{X})^-]^T\mathbf{X}^T\mathbf{W}^{-1} \\ -\mathbf{X}^T(\mathbf{W}^- - \mathbf{W}^-\overline{\mathbf{P}}_{\mathbf{X}})^T \end{pmatrix}\mathbf{y} : \mathbf{z} \in \mathbb{R}^{n+p}\Big\} \\ &= \{\mathbf{c}^T(\mathbf{X}^T\mathbf{W}^-\mathbf{X})^-\mathbf{X}^T\mathbf{W}^-\mathbf{y} + \mathbf{z}_1^T(\mathbf{I}_n - \mathbf{W}\mathbf{W}^-)\mathbf{y} :\ \mathbf{z}_1 \in \mathbb{R}^n\} \end{aligned} \tag{11.8}$$

where $\mathbf{z}_1$ is the first n-component subvector of $\mathbf{z}$ and we have used Theorem 2.6.1 and (11.6) to simplify the representation of the collection.

Now consider the generalized Aitken equations corresponding to $\mathbf{W}^-$. Post-multiplying (11.6) by $\mathbf{W}^-\mathbf{y}$, we see that $(\mathbf{X}^T\mathbf{W}^-\mathbf{X})^-\mathbf{X}^T\mathbf{W}^-\mathbf{y}$ is a solution to these equations. Thus, the generalized Aitken equations corresponding to $\mathbf{W}^-$ are consistent, implying by Theorem 3.2.3 that the collection of all solutions to them is given by

$$\{(\mathbf{X}^T\mathbf{W}^-\mathbf{X})^-\mathbf{X}^T\mathbf{W}^-\mathbf{y} + [\mathbf{I} - (\mathbf{X}^T\mathbf{W}^-\mathbf{X})^-\mathbf{X}^T\mathbf{W}^-\mathbf{X}]\mathbf{z} : \ \mathbf{z} \in \mathbb{R}^p\}.$$

So, if $\tilde{\boldsymbol{\beta}}$ is a solution to the generalized Aitken equations corresponding to $\mathbf{W}^-$ and $\tilde{\mathbf{z}}$ is its corresponding $\mathbf{z}$, then

$$\begin{aligned}\mathbf{c}^T\tilde{\boldsymbol{\beta}} &= \{\mathbf{c}^T(\mathbf{X}^T\mathbf{W}^-\mathbf{X})^-\mathbf{X}^T\mathbf{W}^-\mathbf{y} + \mathbf{a}^T\mathbf{X}[\mathbf{I} - (\mathbf{X}^T\mathbf{W}^-\mathbf{X})^-\mathbf{X}^T\mathbf{W}^-\mathbf{X}]\tilde{\mathbf{z}} \\ &= \mathbf{c}^T(\mathbf{X}^T\mathbf{W}^-\mathbf{X})^-\mathbf{X}^T\mathbf{W}^-\mathbf{y}\end{aligned}$$

where $\mathbf{a}$ is any vector such that $\mathbf{a}^T\mathbf{X} = \mathbf{c}^T$; the last equality holds by (11.7). Thus, $\mathbf{c}^T\tilde{\boldsymbol{\beta}}$ belongs to the collection of BLUEs specified by (11.8). □

Note that $\mathcal{C}(\mathbf{X}) \subseteq \mathcal{C}(\mathbf{W})$ for the second model considered in Example 11.2-1, but not for the first model. In fact, the first model suggests that if $\mathcal{C}(\mathbf{X}) \nsubseteq \mathcal{C}(\mathbf{W})$, then the generalized Aitken equations corresponding to some generalized inverses of $\mathbf{W}$ may not be consistent, and even if they are consistent, not all of their solutions may yield BLUEs of estimable functions. This leads us to consider the existence question (Question 2) posed previously. The following two theorems pertain to this question, although their relevance may not be clear immediately.

Theorem 11.2.6 *For any Aitken model* $\{\mathbf{y}, \mathbf{X}\boldsymbol{\beta}, \sigma^2\mathbf{W}\}$, *a matrix* $\mathbf{U}$ *exists such that* $\mathcal{R}(\mathbf{X}^T) \subseteq \mathcal{R}(\mathbf{W} + \mathbf{X}\mathbf{U}\mathbf{X}^T)$ *and* $\mathcal{C}(\mathbf{X}) \subseteq \mathcal{C}(\mathbf{W} + \mathbf{X}\mathbf{U}\mathbf{X}^T)$, *and for any such* $\mathbf{U}$ *a generalized inverse* $\begin{pmatrix} \mathbf{G}_{11} & \mathbf{G}_{12} \\ \mathbf{G}_{21} & \mathbf{G}_{22} \end{pmatrix}$ *of* $\begin{pmatrix} \mathbf{W} & \mathbf{X} \\ \mathbf{X}^T & \mathbf{0} \end{pmatrix}$ *may be obtained by taking*

$$\begin{aligned}\mathbf{G}_{11} &= (\mathbf{W}+\mathbf{X}\mathbf{U}\mathbf{X}^T)^- - (\mathbf{W}+\mathbf{X}\mathbf{U}\mathbf{X}^T)^-\mathbf{X}[\mathbf{X}^T(\mathbf{W}+\mathbf{X}\mathbf{U}\mathbf{X}^T)^-\mathbf{X}]^-\mathbf{X}^T(\mathbf{W}+\mathbf{X}\mathbf{U}\mathbf{X}^T)^-, \\ \mathbf{G}_{12} &= (\mathbf{W}+\mathbf{X}\mathbf{U}\mathbf{X}^T)^-\mathbf{X}[\mathbf{X}^T(\mathbf{W}+\mathbf{X}\mathbf{U}\mathbf{X}^T)^-\mathbf{X}]^- \\ \mathbf{G}_{21} &= [\mathbf{X}^T(\mathbf{W}+\mathbf{X}\mathbf{U}\mathbf{X}^T)^-\mathbf{X}]^-\mathbf{X}^T(\mathbf{W}+\mathbf{X}\mathbf{U}\mathbf{X}^T)^- \\ \mathbf{G}_{22} &= -[\mathbf{X}^T(\mathbf{W}+\mathbf{X}\mathbf{U}\mathbf{X}^T)^-\mathbf{X}]^- + \mathbf{U}.\end{aligned}$$

Proof Let k be any nonzero scalar. Then

$$\mathbf{W} + \mathbf{X}(k^2\mathbf{I})\mathbf{X}^T = \mathbf{W} + k^2\mathbf{X}\mathbf{X}^T = (\mathbf{W}^{\frac{1}{2}}, \ k\mathbf{X})\begin{pmatrix} \mathbf{W}^{\frac{1}{2}} \\ k\mathbf{X}^T \end{pmatrix},$$

implying (by Theorem 6.1.3) that

$$\mathcal{R}(\mathbf{X}^T) \subseteq \mathcal{R}\begin{pmatrix} \mathbf{W}^{\frac{1}{2}} \\ k\mathbf{X}^T \end{pmatrix} = \mathcal{R}(\mathbf{W} + k^2\mathbf{X}\mathbf{X}^T).$$

Because $\mathbf{W} + k^2\mathbf{X}\mathbf{X}^T$ is symmetric, it is also the case that $\mathcal{C}(\mathbf{X}) \subseteq \mathcal{C}(\mathbf{W} + k^2\mathbf{X}\mathbf{X}^T)$. Thus, the row and column space conditions of the theorem are satisfied by taking $\mathbf{U} = k^2\mathbf{I}_p$.

Next, observe that those same row and column space conditions are sufficient, by Theorem 3.3.7a, for $\begin{pmatrix} \mathbf{H}_{11} & \mathbf{H}_{12} \\ \mathbf{H}_{21} & \mathbf{H}_{22} \end{pmatrix}$ to be a generalized inverse of $\begin{pmatrix} \mathbf{W} + \mathbf{X}\mathbf{U}\mathbf{X}^T & \mathbf{X} \\ \mathbf{X}^T & \mathbf{0} \end{pmatrix}$, where

$$\mathbf{H}_{11} = (\mathbf{W} + \mathbf{X}\mathbf{U}\mathbf{X}^T)^- - (\mathbf{W} + \mathbf{X}\mathbf{U}\mathbf{X}^T)^-\mathbf{X}[\mathbf{X}^T(\mathbf{W} + \mathbf{X}\mathbf{U}\mathbf{X}^T)^-\mathbf{X}]^-\mathbf{X}^T(\mathbf{W} + \mathbf{X}\mathbf{U}\mathbf{X}^T)^-,$$
$$\mathbf{H}_{12} = (\mathbf{W} + \mathbf{X}\mathbf{U}\mathbf{X}^T)^-\mathbf{X}[\mathbf{X}^T(\mathbf{W} + \mathbf{X}\mathbf{U}\mathbf{X}^T)^-\mathbf{X}]^-$$
$$\mathbf{H}_{21} = [\mathbf{X}^T(\mathbf{W} + \mathbf{X}\mathbf{U}\mathbf{X}^T)^-\mathbf{X}]^-\mathbf{X}^T(\mathbf{W} + \mathbf{X}\mathbf{U}\mathbf{X}^T)^-$$
$$\mathbf{H}_{22} = -[\mathbf{X}^T(\mathbf{W} + \mathbf{X}\mathbf{U}\mathbf{X}^T)^-\mathbf{X}]^-.$$

The theorem then follows immediately upon taking **A** and **B** in Theorem 3.3.9b to be **W** and **X**, respectively. □

Theorem 11.2.7 *Consider the Aitken model* $\{\mathbf{y}, \mathbf{X}\boldsymbol{\beta}, \sigma^2\mathbf{W}\}$ *and let* **U** *be any matrix such that* $\mathcal{R}(\mathbf{X}^T) \subseteq \mathcal{R}(\mathbf{W} + \mathbf{X}\mathbf{U}\mathbf{X}^T)$ *and* $\mathcal{C}(\mathbf{X}) \subseteq \mathcal{C}(\mathbf{W} + \mathbf{X}\mathbf{U}\mathbf{X}^T)$. *Then:*

(a) The system of equations

$$\mathbf{X}^T(\mathbf{W} + \mathbf{X}\mathbf{U}\mathbf{X}^T)^-\mathbf{X}\boldsymbol{\beta} = \mathbf{X}^T(\mathbf{W} + \mathbf{X}\mathbf{U}\mathbf{X}^T)^-\mathbf{y} \tag{11.9}$$

is consistent.

(b) If $\mathbf{c}^T\boldsymbol{\beta}$ *is an estimable function, then for any solution* $\tilde{\boldsymbol{\beta}}$ *to the equations in part (a),* $\mathbf{c}^T\tilde{\boldsymbol{\beta}}$ *is a BLUE of* $\mathbf{c}^T\boldsymbol{\beta}$.

(c) If $\mathbf{C}_1^T\boldsymbol{\beta}$ *and* $\mathbf{C}_2^T\boldsymbol{\beta}$ *are vectors of estimable functions, then*

$$cov(\mathbf{C}_1^T\tilde{\boldsymbol{\beta}}, \mathbf{C}_2^T\tilde{\boldsymbol{\beta}}) = \sigma^2\mathbf{C}_1^T\{[\mathbf{X}^T(\mathbf{W} + \mathbf{X}\mathbf{U}\mathbf{X}^T)^-\mathbf{X}]^- - \mathbf{U}\}\mathbf{C}_2$$

and

$$var(\mathbf{C}_1^T\tilde{\boldsymbol{\beta}}) = \sigma^2\mathbf{C}_1^T\{[\mathbf{X}^T(\mathbf{W} + \mathbf{X}\mathbf{U}\mathbf{X}^T)^-\mathbf{X}]^- - \mathbf{U}\}\mathbf{C}_1,$$

and these expressions are invariant to the choices of generalized inverses of $\mathbf{W} + \mathbf{X}\mathbf{U}\mathbf{X}^T$ *and* $\mathbf{X}^T(\mathbf{W} + \mathbf{X}\mathbf{U}\mathbf{X}^T)^-\mathbf{X}$.

(d) Provided that $rank(\mathbf{W}, \mathbf{X}) > rank(\mathbf{X})$,

$$\tilde{\sigma}^2 = \frac{\mathbf{y}^T\{(\mathbf{W}+\mathbf{XUX}^T)^- - (\mathbf{W}+\mathbf{XUX}^T)^-\mathbf{X}[\mathbf{X}^T(\mathbf{W}+\mathbf{XUX}^T)^-\mathbf{X}]^-\mathbf{X}^T(\mathbf{W}+\mathbf{XUX}^T)^-\}\mathbf{y}}{rank(\mathbf{W}, \mathbf{X}) - rank(\mathbf{X})}$$

is an unbiased estimator of σ^2*, regardless of the choices of generalized inverses of* $\mathbf{W} + \mathbf{XUX}^T$ *and* $\mathbf{X}^T(\mathbf{W} + \mathbf{XUX}^T)^-\mathbf{X}$.

Proof Let $\begin{pmatrix} \mathbf{G}_{11} & \mathbf{G}_{12} \\ \mathbf{G}_{21} & \mathbf{G}_{22} \end{pmatrix}$ be defined as in Theorem 11.2.6. Theorems 11.2.6 and 3.3.8a (with $\mathbf{W}$ and $\mathbf{X}$ here playing the roles of $\mathbf{A}$ and $\mathbf{B}$ in the latter theorem) yield the identities

$$\mathbf{X}^T(\mathbf{W} + \mathbf{XUX}^T)^-\mathbf{X}[\mathbf{X}^T(\mathbf{W} + \mathbf{XUX}^T)^-\mathbf{X}]^-\mathbf{X}^T = \mathbf{X}^T \tag{11.10}$$

and

$$\mathbf{X}[\mathbf{X}^T(\mathbf{W} + \mathbf{XUX}^T)^-\mathbf{X}]^-\mathbf{X}^T(\mathbf{W} + \mathbf{XUX}^T)^-\mathbf{X} = \mathbf{X}. \tag{11.11}$$

Post-multiplying both sides of (11.10) by $(\mathbf{W} + \mathbf{XUX}^T)^-\mathbf{y}$ yields

$$\mathbf{X}^T(\mathbf{W}+\mathbf{XUX}^T)^-\mathbf{X}[\mathbf{X}^T(\mathbf{W}+\mathbf{XUX}^T)^-\mathbf{X}]^-\mathbf{X}^T(\mathbf{W}+\mathbf{XUX}^T)^-\mathbf{y} = \mathbf{X}^T(\mathbf{W}+\mathbf{XUX}^T)^-\mathbf{y}.$$

This shows that one solution to the system of equations given by (11.9) is

$$[\mathbf{X}^T(\mathbf{W} + \mathbf{XUX}^T)^-\mathbf{X}]^-\mathbf{X}^T(\mathbf{W} + \mathbf{XUX}^T)^-\mathbf{y};$$

hence the system is consistent. By Theorem 3.2.3 and (11.11), for any solution $\tilde{\boldsymbol{\beta}}$ to the system it is the case that for some $\mathbf{z} \in \mathbb{R}^p$,

$$\begin{aligned}
\mathbf{c}^T\tilde{\boldsymbol{\beta}} &= \mathbf{c}^T[\mathbf{X}^T(\mathbf{W} + \mathbf{XUX}^T)^-\mathbf{X}]^-\mathbf{X}^T(\mathbf{W} + \mathbf{XUX}^T)^-\mathbf{y} \\
&\quad + \mathbf{c}^T\{\mathbf{I} - [\mathbf{X}^T(\mathbf{W} + \mathbf{XUX}^T)^-\mathbf{X}]^-[\mathbf{X}^T(\mathbf{W} + \mathbf{XUX}^T)^-\mathbf{X}]\}\mathbf{z} \\
&= \mathbf{c}^T[\mathbf{X}^T(\mathbf{W} + \mathbf{XUX}^T)^-\mathbf{X}]^-\mathbf{X}^T(\mathbf{W} + \mathbf{XUX}^T)^-\mathbf{y} \\
&= \mathbf{c}^T\mathbf{G}_{21}\mathbf{y}.
\end{aligned}$$

But this last expression is a BLUE of $\mathbf{c}^T\boldsymbol{\beta}$ by Theorem 11.2.3a. Furthermore, by Theorem 11.2.3b,

$$\begin{aligned}
\text{cov}(\mathbf{C}_1^T\tilde{\boldsymbol{\beta}}, \mathbf{C}_2^T\tilde{\boldsymbol{\beta}}) &= -\sigma^2\mathbf{C}_1^T\{-[\mathbf{X}^T(\mathbf{W} + \mathbf{XUX}^T)^-\mathbf{X}]^- + \mathbf{U}\}\mathbf{C}_2 \\
&= \sigma^2\mathbf{C}_1^T\{[\mathbf{X}^T(\mathbf{W} + \mathbf{XUX}^T)^-\mathbf{X}]^- - \mathbf{U}\}\mathbf{C}_2.
\end{aligned}$$

This proves parts (a), (b), and (c); part (d) follows immediately from Theorem 11.2.3c. □

How do Theorems 11.2.6 and 11.2.7 relate to the original question of existence of a collection of generalized inverses of $\mathbf{W}$ for which a solution to the generalized Aitken equations yields BLUEs of all estimable functions? Observe that if $(\mathbf{W} + \mathbf{XUX}^T)^-$ was a generalized inverse of $\mathbf{W}$, then the system of equations given by (11.9)—which by Theorem 11.2.7 are consistent and for which any solution yields a BLUE of any estimable function—would coincide with the generalized Aitken equations corresponding to that generalized inverse. As a case in point, for the first model described in Example 11.2-1, taking $\mathbf{U} = \mathbf{I}_2$ yields $\mathbf{W} + \mathbf{XUX}^T = \begin{pmatrix} 2 & 1 \\ 1 & 1 \end{pmatrix}$, which is nonsingular, so its only generalized inverse is its ordinary inverse, $\begin{pmatrix} 1 & -1 \\ -1 & 2 \end{pmatrix}$. This matrix is a generalized inverse of $\mathbf{W}$; in fact, it coincides with the last of the three specific generalized inverses of $\mathbf{W}$ that were considered in the example. Furthermore, for $\mathbf{U} = \mathbf{I}_2$ the row and column space conditions of Theorem 11.2.7 are satisfied. And so it was seen in the example that the generalized Aitken equations corresponding to this generalized inverse are consistent and their solution yielded a BLUE.

In general, however, $(\mathbf{W} + \mathbf{XUX}^T)^-$ (where $\mathbf{U}$ is such that the conditions of Theorem 11.2.7 are satisfied) is not necessarily a generalized inverse of $\mathbf{W}$. To illustrate, suppose that the model matrix in Example 11.2-1 was changed to $\mathbf{X} = \mathbf{I}_2$, and we keep $\mathbf{U} = \mathbf{I}_2$ as well. Then $\mathbf{W} + \mathbf{XUX}^T = \begin{pmatrix} 2 & 1 \\ 1 & 2 \end{pmatrix}$ (which is nonsingular), $\mathcal{R}(\mathbf{X}^T) \subseteq \mathcal{R}(\mathbf{W} + \mathbf{XUX}^T)$, and $\mathcal{C}(\mathbf{X}) \subseteq \mathcal{C}(\mathbf{W} + \mathbf{XUX}^T)$, but $(\mathbf{W} + \mathbf{XUX}^T)^{-1} = \frac{1}{3}\begin{pmatrix} 2 & -1 \\ -1 & 2 \end{pmatrix}$ so

$$\mathbf{W}(\mathbf{W} + \mathbf{XUX}^T)^{-1}\mathbf{W} = \begin{pmatrix} 1 & 1 \\ 1 & 1 \end{pmatrix} \frac{1}{3} \begin{pmatrix} 2 & -1 \\ -1 & 2 \end{pmatrix} \begin{pmatrix} 1 & 1 \\ 1 & 1 \end{pmatrix} = \frac{2}{3} \begin{pmatrix} 1 & 1 \\ 1 & 1 \end{pmatrix} \neq \mathbf{W}.$$

However, as indicated by the following theorem, if $\mathbf{U}$ is such that some additional conditions beyond those required in Theorem 11.2.7 are satisfied, then any generalized inverse of $\mathbf{W} + \mathbf{XUX}^T$ is indeed a generalized inverse of $\mathbf{W}$, so that we can obtain BLUEs of estimable functions by solving the generalized Aitken equations corresponding to such a generalized inverse.

Theorem 11.2.8 *For any Aitken model* $\{\mathbf{y}, \mathbf{X}\boldsymbol{\beta}, \sigma^2\mathbf{W}\}$, *a matrix* $\mathbf{U}$ *exists such that* $\mathcal{R}(\mathbf{W}) \cap \mathcal{R}(\mathbf{XUX}^T) = \mathbf{0}$ *and* $\mathcal{C}(\mathbf{W}) \cap \mathcal{C}(\mathbf{XUX}^T) = \mathbf{0}$, *and such that the row and column space conditions of Theorem 11.2.7 are also satisfied. Furthermore, if* $\mathbf{W}^*$ *is any generalized inverse of* $\mathbf{W} + \mathbf{XUX}^T$ *for such a* $\mathbf{U}$, *then:*

(a) $\mathbf{W}^*$ *is also a generalized inverse of* $\mathbf{W}$;
(b) The system of equations $\mathbf{X}^T\mathbf{W}^*\mathbf{X} = \mathbf{X}^T\mathbf{W}^*\mathbf{y}$ *is consistent, and if* $\tilde{\boldsymbol{\beta}}$ *is any solution to the system and* $\mathbf{c}^T\boldsymbol{\beta}$ *is estimable, then* $\mathbf{c}^T\tilde{\boldsymbol{\beta}}$ *is a BLUE of* $\mathbf{c}^T\boldsymbol{\beta}$;

(c) *If* $\mathbf{C}_1^T\boldsymbol{\beta}$ *and* $\mathbf{C}_2^T\boldsymbol{\beta}$ *are vectors of estimable functions, then the matrix of covariances between* $\mathbf{C}_1^T\tilde{\boldsymbol{\beta}}$ *and* $\mathbf{C}_2^T\tilde{\boldsymbol{\beta}}$ *is*

$$\sigma^2\mathbf{C}_1^T[(\mathbf{X}^T\mathbf{W}^*\mathbf{X})^- - \mathbf{U}]^-\mathbf{C}_2$$

and the variance–covariance matrix of $\mathbf{C}_1^T\tilde{\boldsymbol{\beta}}$ *is*

$$\sigma^2\mathbf{C}_1^T[(\mathbf{X}^T\mathbf{W}^*\mathbf{X})^- - \mathbf{U}]^-\mathbf{C}_1,$$

and these expressions are invariant to the choice of generalized inverses of $\mathbf{X}^T\mathbf{W}^*\mathbf{X}$ *and* $(\mathbf{X}^T\mathbf{W}^*\mathbf{X})^- - \mathbf{U}$;

(d) *Provided that rank*$(\mathbf{W},\mathbf{X}) >$ *rank*$(\mathbf{X})$,

$$\tilde{\sigma}^2 = \frac{\mathbf{y}^T[\mathbf{W}^* - \mathbf{W}^*\mathbf{X}(\mathbf{X}^T\mathbf{W}^*\mathbf{X})^-\mathbf{X}^T\mathbf{W}^*]\mathbf{y}}{rank(\mathbf{W},\mathbf{X}) - rank(\mathbf{X})}$$

is an unbiased estimator of σ^2*, regardless of the choice of generalized inverse of* $\mathbf{X}^T\mathbf{W}^*\mathbf{X}$.

Proof Let $\mathbf{X}_1$ represent any matrix whose columns are any rank$(\mathbf{W},\mathbf{X})$ − rank$(\mathbf{W})$ columns of $\mathbf{X}$ for which rank$(\mathbf{W},\mathbf{X}_1)$ = rank$(\mathbf{W},\mathbf{X})$; let k be any nonzero scalar; and let $\mathbf{D}$ represent a diagonal matrix with main diagonal elements equal to either 0 or k^2 such that $\mathbf{W}+\mathbf{X}\mathbf{D}\mathbf{X}^T = \mathbf{W}+k^2\mathbf{X}_1\mathbf{X}_1^T$. By construction, the columns of $\mathbf{X}_1$ are linearly independent, implying that

$$\text{rank}(\mathbf{X}_1) = \text{rank}(\mathbf{W},\mathbf{X}) - \text{rank}(\mathbf{W}) = \text{rank}(\mathbf{W},\mathbf{X}_1) - \text{rank}(\mathbf{W}),$$

or equivalently, that

$$\text{rank}(\mathbf{W},\mathbf{X}_1) = \text{rank}(\mathbf{W}) + \text{rank}(\mathbf{X}_1).$$

Thus, by Theorem 2.8.6,

$$\mathcal{R}(\mathbf{W}) \cap \mathcal{R}(\mathbf{X}_1) = \{\mathbf{0}\} \quad \text{and} \quad \mathcal{C}(\mathbf{W}) \cap \mathcal{C}(\mathbf{X}_1) = \{\mathbf{0}\},$$

implying further (by Theorem 6.1.3) that

$$\mathcal{R}(\mathbf{W}) \cap \mathcal{R}(k^2\mathbf{X}_1\mathbf{X}_1^T) = \{\mathbf{0}\} \quad \text{and} \quad \mathcal{C}(\mathbf{W}) \cap \mathcal{C}(k^2\mathbf{X}_1\mathbf{X}_1^T) = \{\mathbf{0}\}$$

or equivalently, if $\mathbf{U}$ is taken to be $\mathbf{D}$, that

$$\mathcal{R}(\mathbf{W}) \cap \mathcal{R}(\mathbf{X}\mathbf{U}\mathbf{X}^T) = \{\mathbf{0}\} \quad \text{and} \quad \mathcal{C}(\mathbf{W}) \cap \mathcal{C}(\mathbf{X}\mathbf{U}\mathbf{X}^T) = \{\mathbf{0}\}.$$

Furthermore, for this choice of $\mathbf{U}$,

$$\mathcal{R}(\mathbf{X}^T) \subseteq \mathcal{R}\begin{pmatrix} \mathbf{W} \\ \mathbf{X}^T \end{pmatrix} = \mathcal{R}\begin{pmatrix} \mathbf{W} \\ \mathbf{X}_1^T \end{pmatrix} = \mathcal{R}\begin{pmatrix} (\mathbf{W}^{\frac{1}{2}})^T(\mathbf{W}^{\frac{1}{2}}) \\ \mathbf{X}_1^T \end{pmatrix} = \mathcal{R}\begin{pmatrix} \mathbf{W}^{\frac{1}{2}} \\ k\mathbf{X}_1^T \end{pmatrix}$$

$$= \mathcal{R}\left((\mathbf{W}^{\frac{1}{2}}, k\mathbf{X}_1)\begin{pmatrix} \mathbf{W}^{\frac{1}{2}} \\ k\mathbf{X}_1^T \end{pmatrix}\right) = \mathcal{R}(\mathbf{W} + k^2\mathbf{X}_1\mathbf{X}_1^T) = \mathcal{R}(\mathbf{W} + \mathbf{XUX}^T),$$

and a similar argument establishes that $\mathcal{C}(\mathbf{X}) \subseteq \mathcal{C}(\mathbf{W} + \mathbf{XUX}^T)$. Thus, a $\mathbf{U}$ exists such that all the specified conditions are satisfied.

It then follows from Theorem 3.3.12 that, taking $\mathbf{U} = \mathbf{D}$, $\mathbf{W}^*$ is a generalized inverse not only of $\mathbf{W} + \mathbf{XUX}^T$, but also of $\mathbf{W}$. This proves part (a). Parts (b), (c), and (d) follow immediately upon substituting $\mathbf{W}^*$ for the arbitrary generalized inverse of $\mathbf{W}+\mathbf{XUX}^T$ in parts (a), (b), and (c), respectively, of Theorem 11.2.7. □

The results on BLUE presented in this section may be summarized by observing, following Christensen (2011), that Aitken models can be divided into four categories, according to the degree to which $\mathbf{W}$ is specialized:

1. $\mathbf{W} = \mathbf{I}$ (the Gauss–Markov model);
2. $\mathbf{W}$ is positive definite (the positive definite Aitken model);
3. $\mathcal{C}(\mathbf{X}) \subseteq \mathcal{C}(\mathbf{W})$;
4. $\mathbf{W}$ is nonnegative definite.

Each category is more general than the one that precedes it. To obtain BLUEs of estimable functions under the Aitken model corresponding to each category, an ever-more-general system of equations must be solved: for the first category it is the normal equations; for the second, the Aitken equations; for the third, the generalized Aitken equations; and for the fourth, either the system of equations given by (11.9) (for a suitably chosen $\mathbf{U}$) or the generalized Aitken equations corresponding to the particular subclass of generalized inverses of $\mathbf{W}$ specified in Theorem 11.2.8.

11.3 Conditions Under Which the Ordinary Least Squares Estimator is a BLUE

Suppose that $\mathbf{y}$ follows the Aitken model $\{\mathbf{y}, \mathbf{X}\boldsymbol{\beta}, \sigma^2\mathbf{W}\}$, and that $\mathbf{c}^T\boldsymbol{\beta}$ is an estimable function under this model. Recall from Chap. 7 that the least squares estimator of $\mathbf{c}^T\boldsymbol{\beta}$, which is $\mathbf{c}^T\hat{\boldsymbol{\beta}}$ where $\hat{\boldsymbol{\beta}}$ is any solution to

$$\mathbf{X}^T\mathbf{X}\boldsymbol{\beta} = \mathbf{X}^T\mathbf{y},$$

is a linear unbiased estimator of $\mathbf{c}^T\boldsymbol{\beta}$. However, in general, the least squares estimator will not be a BLUE under the model $\{\mathbf{y}, \mathbf{X}\boldsymbol{\beta}, \sigma^2\mathbf{W}\}$, i.e., it will not be a linear unbiased estimator of $\mathbf{c}^T\boldsymbol{\beta}$ with smallest variance. That distinction is reserved for estimators given by Theorem 11.2.2 (which, as we showed in the previous section, reduce to the generalized least squares estimator of $\mathbf{c}^T\boldsymbol{\beta}$ when $\mathbf{W}$ is positive definite).

In some circumstances, however, the least squares estimator of $\mathbf{c}^T\boldsymbol{\beta}$ is indeed a BLUE. Interestingly, the circumstances are considerably more general than merely $\mathbf{W} = \mathbf{I}$. In this section we give a theorem and two corollaries that specify the circumstances. These results are due primarily to Zyskind (1967) and Rao (1967). Henceforth, for additional clarity, we refer to the least squares estimator as the *ordinary least squares estimator*.

Theorem 11.3.1 *Let* $\mathbf{c}^T(\mathbf{X}^T\mathbf{X})^-\mathbf{X}^T\mathbf{y}$ *represent the ordinary least squares estimator of an estimable function* $\mathbf{c}^T\boldsymbol{\beta}$. *Under the Aitken model* $\{\mathbf{y}, \mathbf{X}\boldsymbol{\beta}, \sigma^2\mathbf{W}\}$, $\mathbf{c}^T(\mathbf{X}^T\mathbf{X})^-\mathbf{X}^T\mathbf{y}$ *is a BLUE of* $\mathbf{c}^T\boldsymbol{\beta}$ *if and only if* $\mathbf{W}\mathbf{X}(\mathbf{X}^T\mathbf{X})^-\mathbf{c} \in \mathcal{C}(\mathbf{X})$.

Proof The ordinary least squares estimator is linear and unbiased under the specified Aitken model (Theorem 7.2.1). The theorem follows immediately by substituting $\mathbf{X}[(\mathbf{X}^T\mathbf{X})^-]^T\mathbf{c}$ for $\mathbf{t}$ in Corollary 11.2.2.2 and observing that $\mathbf{X}[(\mathbf{X}^T\mathbf{X})^-]^T\mathbf{c} = \mathbf{X}(\mathbf{X}^T\mathbf{X})^-\mathbf{c}$ (because $\mathbf{c} = \mathbf{X}^T\mathbf{a}$ for some $\mathbf{a}$). □

Under what circumstances, if any, is the ordinary least squares estimator of *every* estimable function $\mathbf{c}^T\boldsymbol{\beta}$ a BLUE of the corresponding function? One answer is given by the following corollary to Theorem 11.3.1.

Corollary 11.3.1.1 *Under the Aitken model* $\{\mathbf{y}, \mathbf{X}\boldsymbol{\beta}, \sigma^2\mathbf{W}\}$, *the ordinary least squares estimator of an estimable function* $\mathbf{c}^T\boldsymbol{\beta}$ *is a BLUE of that function, for every estimable* $\mathbf{c}^T\boldsymbol{\beta}$, *if and only if* $\mathcal{C}(\mathbf{W}\mathbf{X}) \subseteq \mathcal{C}(\mathbf{X})$.

Proof Denote the columns of $\mathbf{X}$ by $\mathbf{x}_1, \ldots, \mathbf{x}_p$. If the ordinary least squares estimator of $\mathbf{c}^T\boldsymbol{\beta}$ is a BLUE for every estimable function $\mathbf{c}^T\boldsymbol{\beta}$, then by Theorem 11.3.1, for every $\mathbf{c}^T \in \mathcal{R}(\mathbf{X})$ a vector $\mathbf{q}$ (possibly depending on $\mathbf{c}$) exists such that $\mathbf{W}\mathbf{X}(\mathbf{X}^T\mathbf{X})^-\mathbf{c} = \mathbf{X}\mathbf{q}$. Equivalently, for every $\mathbf{a} \in \mathbb{R}^n$ a vector $\mathbf{q}$ (possibly depending on $\mathbf{a}$) exists such that $\mathbf{W}\mathbf{X}(\mathbf{X}^T\mathbf{X})^-\mathbf{X}^T\mathbf{a} = \mathbf{X}\mathbf{q}$. Substituting each of $\mathbf{x}_1, \ldots, \mathbf{x}_p$ in turn for $\mathbf{a}$ in this last system of equations yields

$$\mathbf{W}\mathbf{X}(\mathbf{X}^T\mathbf{X})^-\mathbf{X}^T(\mathbf{x}_1, \ldots, \mathbf{x}_p) = \mathbf{X}(\mathbf{q}_1, \ldots, \mathbf{q}_p)$$

for some vectors $\mathbf{q}_1, \ldots, \mathbf{q}_p$. By Theorem 3.3.3a, this last matrix equation reduces to $\mathbf{W}\mathbf{X} = \mathbf{X}\mathbf{Q}$ for $\mathbf{Q} = (\mathbf{q}_1, \ldots, \mathbf{q}_p)$, implying that $\mathcal{C}(\mathbf{W}\mathbf{X}) \subseteq \mathcal{C}(\mathbf{X})$.

Conversely, if $\mathcal{C}(\mathbf{W}\mathbf{X}) \subseteq \mathcal{C}(\mathbf{X})$, then for every estimable function $\mathbf{c}^T\boldsymbol{\beta}$, $\mathbf{W}\mathbf{X}(\mathbf{X}^T\mathbf{X})^-\mathbf{c} \in \mathcal{C}(\mathbf{X})$. □

In order to use Corollary 11.3.1.1 to establish the equivalence or nonequivalence of ordinary least squares estimators and BLUEs of all estimable functions, one has to either find a matrix $\mathbf{Q}$ such that $\mathbf{WX} = \mathbf{XQ}$, or show that such a $\mathbf{Q}$ does not exist. For cases in which this is not straightforward, it may be desirable to have a more direct way to establish the aforementioned equivalence or nonequivalence. This motivates the next corollary.

Corollary 11.3.1.2 *Under the Aitken model* $\{\mathbf{y}, \mathbf{X}\boldsymbol{\beta}, \sigma^2\mathbf{W}\}$, *the ordinary least squares estimator of an estimable function* $\mathbf{c}^T\boldsymbol{\beta}$ *is a BLUE of that function for every estimable* $\mathbf{c}^T\boldsymbol{\beta}$ *if and only if* $\mathbf{W}\mathbf{P_X} = \mathbf{P_X}\mathbf{W}$, *i.e., if and only if* $\mathbf{W}\mathbf{P_X}$ *is a symmetric matrix.*

Proof Suppose that $\mathbf{W}\mathbf{P_X} = \mathbf{P_X}\mathbf{W}$. Then $\mathbf{W}\mathbf{P_X}\mathbf{X} = \mathbf{P_X}\mathbf{W}\mathbf{X}$, i.e., $\mathbf{WX} = \mathbf{X}(\mathbf{X}^T\mathbf{X})^-\mathbf{X}^T\mathbf{WX}$, i.e., $\mathcal{C}(\mathbf{WX}) \subseteq \mathcal{C}(\mathbf{X})$. Conversely, if $\mathcal{C}(\mathbf{WX}) \subseteq \mathcal{C}(\mathbf{X})$, then $\mathbf{WX} = \mathbf{XQ}$ for some $\mathbf{Q}$, implying further that $\mathbf{WX}(\mathbf{X}^T\mathbf{X})^-\mathbf{X}^T = \mathbf{XQ}(\mathbf{X}^T\mathbf{X})^-\mathbf{X}^T$, i.e., $\mathbf{W}\mathbf{P_X} = \mathbf{XQ}(\mathbf{X}^T\mathbf{X})^-\mathbf{X}^T$. Pre-multiplication of both sides of this last equality by $\mathbf{P_X}$ yields

$$\mathbf{P_X}\mathbf{W}\mathbf{P_X} = \mathbf{P_X}[\mathbf{XQ}(\mathbf{X}^T\mathbf{X})^-\mathbf{X}^T] = \mathbf{XQ}(\mathbf{X}^T\mathbf{X})^-\mathbf{X}^T = \mathbf{W}\mathbf{P_X}.$$

Because $\mathbf{P_X}\mathbf{W}\mathbf{P_X}$ is symmetric, the corollary is established. □

Theorem 11.3.1 and its two corollaries are also valid under the general mixed linear model if the variance–covariance matrix $\mathbf{V}(\boldsymbol{\theta})$ of that model is substituted for $\sigma^2\mathbf{W}$, provided that the condition(s) of each theorem/corollary are then required to hold for all $\boldsymbol{\theta} \in \Theta$; for example, in the case of the first corollary, the required condition is $\mathcal{C}[\mathbf{V}(\boldsymbol{\theta})\mathbf{X}] \subseteq \mathcal{C}(\mathbf{X})$ for all $\boldsymbol{\theta} \in \Theta$.

Now consider the geometry of the projection of $\mathbf{y}$ onto $\mathcal{C}(\mathbf{X})$ when the ordinary least squares estimator of every estimable function is a BLUE of that function. Because $\mathbf{X}\boldsymbol{\beta}$ is estimable and its ordinary least squares estimator is the *orthogonal* projection of $\mathbf{y}$ onto $\mathcal{C}(\mathbf{X})$, in this case the BLUE of $\mathbf{X}\boldsymbol{\beta}$ must likewise be that projection, even if $\mathbf{W} \neq \mathbf{I}$. Figure 11.2 illustrates this geometry for several choices of $\mathbf{W} \neq \mathbf{I}$ for a toy example with the same $\mathbf{y}$ and $\mathbf{X}$ as in Example 11.1-1.

Example 11.3-1. Equivalence of Ordinary Least Squares Estimators and BLUEs for a Simple Linear Regression Toy Example

Consider a simple linear regression setting with three observations, taken at $x_1 = -1$, $x_2 = 0$, and $x_3 = 1$. For which variance–covariance matrices $\sigma^2\mathbf{W}$ will the ordinary least squares estimator of the slope be the best linear unbiased estimator? By Theorem 11.3.1, the answer to this question is: those for which

$$\mathbf{W}\begin{pmatrix} 1 & -1 \\ 1 & 0 \\ 1 & 1 \end{pmatrix}\begin{pmatrix} 3 & 0 \\ 0 & 2 \end{pmatrix}^{-1}\begin{pmatrix} 0 \\ 1 \end{pmatrix} \in \mathcal{C}(\mathbf{X}),$$

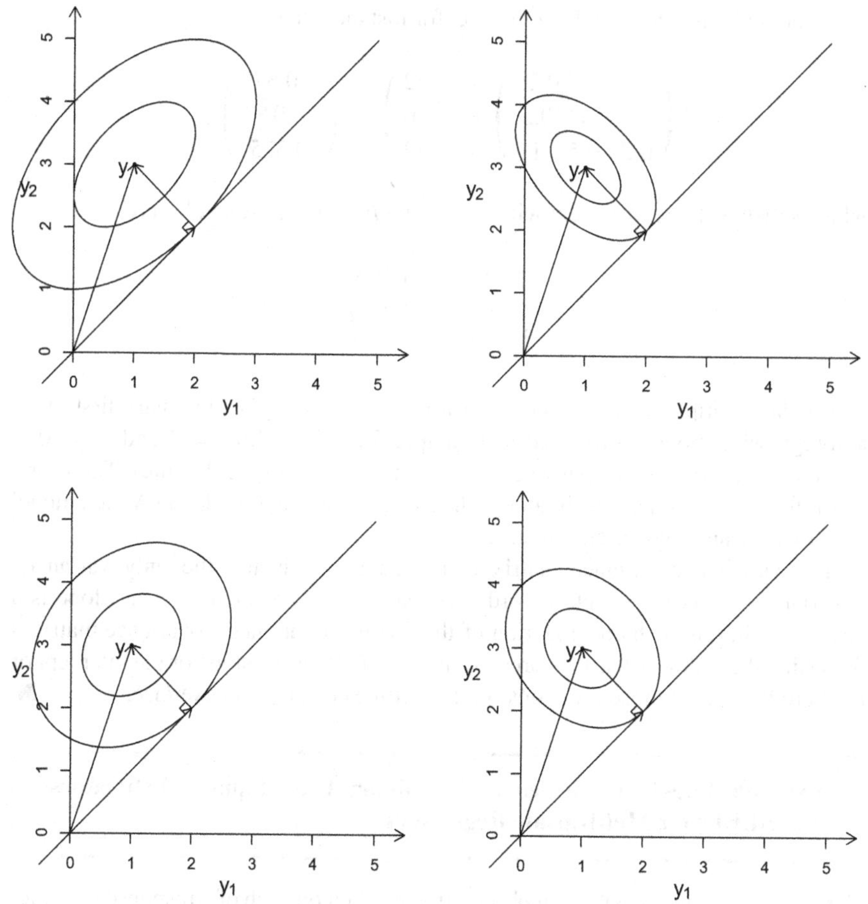

Fig. 11.2 Orthogonal projections of $\mathbf{y} = (1, 3)^T$ onto $\mathcal{C}(\mathbf{1}_2)$ for matrices $\mathbf{W} \neq \mathbf{I}$. Upper left panel, $\mathbf{W} = \begin{pmatrix} 1.0 & 0.5 \\ 0.5 & 1.0 \end{pmatrix}$; upper right panel, $\mathbf{W} = \begin{pmatrix} 1.0 & -0.5 \\ -0.5 & 1.0 \end{pmatrix}$; lower left panel, $\mathbf{W} = \begin{pmatrix} 1.0 & 0.25 \\ 0.25 & 1.0 \end{pmatrix}$; lower right panel, $\mathbf{W} = \begin{pmatrix} 1.0 & -0.25 \\ -0.25 & 1.0 \end{pmatrix}$. The ellipses in each panel depict sets $\{\mathbf{v} : (\mathbf{y} - \mathbf{v})^T \mathbf{W}^{-1} (\mathbf{y} - \mathbf{v}) = c^2\}$ for two values of c

i.e., those nonnegative definite matrices $\mathbf{W} = (w_{ij})$ for which

$$\begin{pmatrix} w_{11} & w_{12} & w_{13} \\ w_{12} & w_{22} & w_{23} \\ w_{13} & w_{23} & w_{33} \end{pmatrix} \begin{pmatrix} -1/2 \\ 0 \\ 1/2 \end{pmatrix} = \begin{pmatrix} a - b \\ a \\ a + b \end{pmatrix}$$

for some real numbers a and b. Observe, for instance, that

$$\begin{pmatrix} 1 & 0.5 & 0.25 \\ 0.5 & 1 & 0.5 \\ 0.25 & 0.5 & 1 \end{pmatrix} \begin{pmatrix} -1/2 \\ 0 \\ 1/2 \end{pmatrix} = \begin{pmatrix} -0.375 \\ 0 \\ 0.375 \end{pmatrix},$$

which satisfies the requirement with $a = 0$ and $b = 0.375$. Also observe that

$$\begin{pmatrix} 1 & 0.5 & 0.25 \\ 0.5 & 1 & 0.5 \\ 0.25 & 0.5 & 1 \end{pmatrix}$$

is a scalar multiple of the variance–covariance matrix of the stationary first-order autoregressive model introduced in Example 5.2.4-1, with $n = 3$ and $\rho = 0.5$. Thus, this matrix is nonnegative definite (in fact it is positive definite). Therefore, the ordinary least squares estimator of the slope is the BLUE under an Aitken model with this variance–covariance matrix.

The variance–covariance matrix displayed above is not the only variance–covariance matrix for which the ordinary least squares estimator of the slope is a BLUE. A thorough characterization of the set of all variance–covariance matrices for which the ordinary least squares estimators of either the slope or the intercept or both are BLUEs of those functions (in this setting) is left as an exercise. ■

Example 11.3-2. Equivalence of Ordinary Least Squares Estimators and BLUEs for Multivariate Regression

Consider a situation in which an observation is taken on each of s response variables for each of n_0 units. (For example, the units could be people and the response variables could be measures of aerobic fitness, e.g., resting pulse rate, lung capacity, pulse recovery time, etc.). Let $\mathbf{y}_i = (y_{i1}, \ldots, y_{in_0})^T$ represent the n_0-vector whose elements are the observations of the ith response variable over all units. Suppose that each $\mathbf{y}_i$ follows the linear model

$$\mathbf{y}_i = \mathbf{X}_0\boldsymbol{\beta}_i + \mathbf{e}_i \qquad (i = 1, \ldots, s),$$

where $\mathbf{X}_0$ is a specified $n_0 \times p_0$ matrix and $\boldsymbol{\beta}_i$ is a p_0-vector of unknown parameters. (Observe that the model matrix is the same for each response variable.) Suppose further that

$$\text{var}\begin{pmatrix} y_{1j} \\ \vdots \\ y_{sj} \end{pmatrix} = \begin{pmatrix} \sigma_{11} & \cdots & \sigma_{1s} \\ \vdots & \ddots & \vdots \\ \sigma_{s1} & \cdots & \sigma_{ss} \end{pmatrix} \equiv \mathbf{V}_0 \quad (j = 1, \ldots, n_0),$$

where the σ_{ij}'s are unknown parameters with parameter space such that $\mathbf{V}_0$ is nonnegative definite, and that $\text{cov}(y_{ij}, y_{i'j'}) = 0$ if $j \neq j'$, i.e., observations on different units are uncorrelated, but observations of different response variables on same unit may be correlated. Then, this model can be viewed as a special case of the general mixed linear model $\mathbf{y} = \mathbf{X}\boldsymbol{\beta} + \mathbf{e}$, where

$$\mathbf{y} = \begin{pmatrix} \mathbf{y}_1 \\ \vdots \\ \mathbf{y}_s \end{pmatrix}, \quad \mathbf{e} = \begin{pmatrix} \mathbf{e}_1 \\ \vdots \\ \mathbf{e}_s \end{pmatrix}, \quad \boldsymbol{\beta} = \begin{pmatrix} \boldsymbol{\beta}_1 \\ \vdots \\ \boldsymbol{\beta}_s \end{pmatrix}, \quad \mathbf{X} = \begin{pmatrix} \mathbf{X}_0 & \mathbf{0} & \cdots & \mathbf{0} \\ \mathbf{0} & \mathbf{X}_0 & \cdots & \mathbf{0} \\ \vdots & \vdots & \ddots & \vdots \\ \mathbf{0} & \mathbf{0} & \cdots & \mathbf{X}_0 \end{pmatrix} = \mathbf{I}_s \otimes \mathbf{X}_0$$

and

$$\mathbf{V}(\boldsymbol{\theta}) = \begin{pmatrix} \sigma_{11}\mathbf{I}_{n_0} & \sigma_{12}\mathbf{I}_{n_0} & \cdots & \sigma_{1s}\mathbf{I}_{n_0} \\ \vdots & & & \vdots \\ \sigma_{s1}\mathbf{I}_{n_0} & \sigma_{s2}\mathbf{I}_{n_0} & \cdots & \sigma_{ss}\mathbf{I}_{n_0} \end{pmatrix} = \mathbf{V}_0 \otimes \mathbf{I}_{n_0},$$

and where the elements of $\boldsymbol{\theta}$ consist of the unknown σ_{ij}'s.

Consider now the estimation of an estimable function $\mathbf{c}^T\boldsymbol{\beta}$, which in this context may be written as $\sum_{i=1}^{s} \mathbf{c}_i^T \boldsymbol{\beta}_i$ where $\mathbf{c}^T = (\mathbf{c}_1^T, \ldots, \mathbf{c}_s^T)$. Specifically consider the following two estimators:

1. The ordinary least squares estimator

$$\mathbf{c}^T\hat{\boldsymbol{\beta}} = \mathbf{c}^T(\mathbf{X}^T\mathbf{X})^-\mathbf{X}^T\mathbf{y}$$
$$= \mathbf{c}^T[(\mathbf{I}_s \otimes \mathbf{X}_0)^T(\mathbf{I}_s \otimes \mathbf{X}_0)]^-(\mathbf{I}_s \otimes \mathbf{X}_0)^T\mathbf{y}$$
$$= \mathbf{c}^T[\mathbf{I}_s \otimes (\mathbf{X}_0^T\mathbf{X}_0)^-\mathbf{X}_0^T]\mathbf{y},$$

where the third equality follows from the second because $\mathbf{I}_s \otimes (\mathbf{X}_0^T\mathbf{X}_0)^-$ is a generalized inverse of $\mathbf{I}_s \otimes (\mathbf{X}_0^T\mathbf{X}_0)$ and $\mathbf{c}^T\hat{\boldsymbol{\beta}}$ is invariant to the choice of generalized inverse of the latter matrix;

2. A BLUE $\tilde{\mathbf{t}}^T\mathbf{y}$, where $\tilde{\mathbf{t}}$ is the first component of a solution to the BLUE equations for $\mathbf{c}^T\boldsymbol{\beta}$.

Note, from the Kronecker-product form of the ordinary least squares estimator given above, that it may be obtained by applying ordinary least squares separately to the data on each response variable, i.e.,

$$\mathbf{c}^T\hat{\boldsymbol{\beta}} = \sum_{i=1}^{s} \mathbf{c}_i^T \hat{\boldsymbol{\beta}}_i,$$

where $\hat{\boldsymbol{\beta}}_i$ is any solution to the system of equations $\mathbf{X}_0^T\mathbf{X}_0\boldsymbol{\beta}_i = \mathbf{X}_0^T\mathbf{y}_i$.

It might seem that the ordinary least squares estimator and the BLUE of $\mathbf{c}^T\boldsymbol{\beta}$ would be different because the former completely ignores the different variabilities of, and correlations between, the s response variables while the latter takes these into account. However, observe that for any (sp_0)-vector $\mathbf{a}$,

$$\mathbf{V}(\boldsymbol{\theta})\mathbf{X}\mathbf{a} = (\mathbf{V}_0 \otimes \mathbf{I}_{n_0})(\mathbf{I}_s \otimes \mathbf{X}_0)\mathbf{a} = (\mathbf{V}_0 \otimes \mathbf{X}_0)\mathbf{a} = (\mathbf{I}_s \otimes \mathbf{X}_0)(\mathbf{V}_0 \otimes \mathbf{I}_{p_0})\mathbf{a} = \mathbf{X}(\mathbf{V}_0 \otimes \mathbf{I}_{p_0})\mathbf{a},$$

demonstrating that $\mathcal{C}[\mathbf{V}(\boldsymbol{\theta})\mathbf{X}] \subseteq \mathcal{C}(\mathbf{X})$ for all $\{\sigma_{ij}\}$, which implies (by the extension of Corollary 11.3.1.1 to general mixed linear models) that the ordinary least squares estimator of $\mathbf{c}^T\boldsymbol{\beta}$ is a BLUE. This result would not necessarily be true if the model matrix was different for some response variates than for others, as would be the case if at least one (but at most $p_0 - 1$) of the explanatory variables were known to have no effect on at least one (but at most $s - 1$) of the response variables, or if there were missing values on at least one (but not all) of the explanatory variables for at least one (but not all) of the units. ■

Example 11.3-3. Equivalence of Ordinary Least Squares Estimators and BLUEs for Models with an Overall Intercept and Compound Symmetric Variance–Covariance Structure

Consider a linear model that has an overall intercept, so that we can write the model matrix $\mathbf{X}$ as $(\mathbf{1}_n, \mathbf{X}_{-1})$, and errors that have the following variance–covariance structure:

$$\text{cov}(e_i, e_j) = \begin{cases} \sigma^2 & \text{for } i = j, \\ \sigma^2\rho & \text{otherwise,} \end{cases}$$

where $\sigma^2 > 0$ and $-\frac{1}{n-1} < \rho < 1$. The corresponding variance–covariance matrix of $\mathbf{e}$ may be written as $\mathbf{V}(\sigma^2, \rho) = \sigma^2[(1-\rho)\mathbf{I}_n + \rho\mathbf{J}_n]$, which we recognize as being compound symmetric. As noted in Example 4.1-1, $\mathbf{V}(\sigma^2, \rho)$ is positive definite over the entire specified parameter space for σ^2 and ρ.

Now observe that

$$\begin{aligned} \mathbf{V}(\sigma^2, \rho)\mathbf{X} &= \sigma^2[(1-\rho)\mathbf{I}_n + \rho\mathbf{J}_n](\mathbf{1}_n, \mathbf{X}_{-1}) \\ &= \sigma^2[(1-\rho+n\rho)\mathbf{1}_n, \ (1-\rho)\mathbf{X}_{-1} + \rho\mathbf{1}_n\mathbf{1}_n^T\mathbf{X}_{-1}] \\ &= (\mathbf{1}_n, \mathbf{X}_{-1})\sigma^2 \begin{pmatrix} 1-\rho+n\rho & \rho\mathbf{1}_n^T\mathbf{X}_{-1} \\ \mathbf{0}_{p-1} & (1-\rho)\mathbf{I}_{p-1} \end{pmatrix}, \end{aligned}$$

implying that $\mathcal{C}[\mathbf{V}(\sigma^2, \rho)\mathbf{X}] \subseteq \mathcal{C}(\mathbf{X})$. Thus, by Corollary 11.3.1.1 the ordinary least squares estimator of every estimable function of $\boldsymbol{\beta}$ will be the BLUE of that function regardless of the value of ρ.

In fact, this same result holds under the slightly weaker condition that $\mathbf{1}_n \in \mathcal{C}(\mathbf{X})$. For under this condition,

$$\begin{aligned}
\mathbf{V}(\sigma^2, \rho)\mathbf{P_X} &= \sigma^2[(1-\rho)\mathbf{I}_n + \rho\mathbf{1}_n\mathbf{1}_n^T]\mathbf{P_X} \\
&= \sigma^2(1-\rho)\mathbf{P_X} + \rho\mathbf{1}_n\mathbf{1}_n^T \\
&= \sigma^2\mathbf{P_X}[(1-\rho)\mathbf{I}_n + \rho\mathbf{1}_n\mathbf{1}_n^T] \\
&= \mathbf{P_X}\mathbf{V}(\sigma^2, \rho),
\end{aligned}$$

and the result follows from Corollary 11.3.1.2. ■

The final theorem of this section gives an explicit form for all Aitken model variance–covariance matrices $\sigma^2\mathbf{W}$ for which the corresponding BLUEs of all estimable functions coincide with ordinary least squares estimators of those functions. This form is known as *Rao's structure*; it is named after C. R. Rao, who first established (Rao, 1967) the result stated in the theorem.

Theorem 11.3.2 *Under the Aitken model* $\{\mathbf{y}, \mathbf{X}\boldsymbol{\beta}, \sigma^2\mathbf{W}\}$, *the ordinary least squares estimator of every estimable function is a BLUE of that function if and only if* $\mathbf{W} = \mathbf{I} + \mathbf{P_X}\mathbf{A}\mathbf{P_X} + (\mathbf{I} - \mathbf{P_X})\mathbf{B}(\mathbf{I} - \mathbf{P_X})$ *for some* $n \times n$ *symmetric matrices* $\mathbf{A}$ *and* $\mathbf{B}$ *such that* $\mathbf{W}$ *is nonnegative definite.*

Proof By Corollary 11.3.1.2, it suffices to show that $\mathbf{W}\mathbf{P_X} = \mathbf{P_X}\mathbf{W}$ if and only if $\mathbf{W}$ has the specified form. Suppose that $\mathbf{W} = \mathbf{I} + \mathbf{P_X}\mathbf{A}\mathbf{P_X} + (\mathbf{I} - \mathbf{P_X})\mathbf{B}(\mathbf{I} - \mathbf{P_X})$ for some $n \times n$ symmetric matrices $\mathbf{A}$ and $\mathbf{B}$ such that $\mathbf{W}$ is nonnegative definite. Then

$$\begin{aligned}
\mathbf{W}\mathbf{P_X} &= [\mathbf{I} + \mathbf{P_X}\mathbf{A}\mathbf{P_X} + (\mathbf{I} - \mathbf{P_X})\mathbf{B}(\mathbf{I} - \mathbf{P_X})]\mathbf{P_X} \\
&= \mathbf{P_X} + \mathbf{P_X}\mathbf{A}\mathbf{P_X} \\
&= \mathbf{P_X}[\mathbf{I} + \mathbf{P_X}\mathbf{A}\mathbf{P_X} + (\mathbf{I} - \mathbf{P_X})\mathbf{B}(\mathbf{I} - \mathbf{P_X})] \\
&= \mathbf{P_X}\mathbf{W}.
\end{aligned}$$

Conversely, suppose that $\mathbf{W}$ is a nonnegative definite matrix for which $\mathbf{W}\mathbf{P_X} = \mathbf{P_X}\mathbf{W}$. Post-multiplication of both sides of this matrix equation by $\mathbf{P_X}$ yields $\mathbf{W}\mathbf{P_X} = \mathbf{P_X}\mathbf{W}\mathbf{P_X}$, or equivalently $(\mathbf{I} - \mathbf{P_X})\mathbf{W}\mathbf{P_X} = \mathbf{0}$. Then we have the identity

$$\mathbf{W} = [\mathbf{P_X} + (\mathbf{I} - \mathbf{P_X})]\mathbf{W}[\mathbf{P_X} + (\mathbf{I} - \mathbf{P_X})],$$

which may be expressed alternatively as

$$\begin{aligned}
\mathbf{W} &= \mathbf{P_X}\mathbf{W}\mathbf{P_X} + (\mathbf{I} - \mathbf{P_X})\mathbf{W}(\mathbf{I} - \mathbf{P_X}) \\
&= \mathbf{I} + \mathbf{P_X}(\mathbf{W} - \mathbf{I})\mathbf{P_X} + (\mathbf{I} - \mathbf{P_X})(\mathbf{W} - \mathbf{I})(\mathbf{I} - \mathbf{P_X}).
\end{aligned}$$

□

11.4 Empirical BLUE (E-BLUE)

Throughout this chapter, it was assumed that $\mathbf{y}$ follows an Aitken model, for which var($\mathbf{y}$) is known up to a scalar multiple (σ^2). If $\mathbf{y}$ follows a more general model, such as the random effects, mixed effects, or general mixed linear model, then var($\mathbf{y}$) has additional unknown parameters, with the consequence that the procedures presented in this chapter for estimating a vector of estimable functions $\mathbf{C}^T\boldsymbol{\beta}$ and the variance–covariance matrix of that estimator under the Aitken model are not directly applicable. However, those same procedures suggest what would seem to be a sensible strategy for accomplishing the same goals under more general models. For example, under the positive definite general mixed linear model $\{\mathbf{y}, \mathbf{X}\boldsymbol{\beta}, \mathbf{V}(\boldsymbol{\theta})\}$, it would seem reasonable to first obtain an estimate $\hat{\boldsymbol{\theta}}$ of $\boldsymbol{\theta}$ and then estimate an estimable vector of functions $\mathbf{C}^T\boldsymbol{\beta}$ by substituting $\mathbf{V}(\hat{\boldsymbol{\theta}})$ for $\mathbf{W}$ in the expression for the generalized least squares estimator (which is the unique BLUE in this case). Furthermore, it would be convenient, though perhaps not as reasonable, to estimate the variance–covariance matrix of that estimator by substituting $\mathbf{V}(\hat{\boldsymbol{\theta}})$ for $\tilde{\sigma}^2\mathbf{W}$ in the expression for the estimated variance–covariance matrix of the vector of generalized least squares estimators. That is, we might estimate $\mathbf{C}^T\boldsymbol{\beta}$ by $\mathbf{C}^T[\mathbf{X}^T\mathbf{V}^{-1}(\hat{\boldsymbol{\theta}})\mathbf{X}]^-\mathbf{X}^T\mathbf{V}^{-1}(\hat{\boldsymbol{\theta}})\mathbf{y}$, and we might estimate the variance–covariance matrix of this estimator by $\mathbf{C}^T[\mathbf{X}^T\mathbf{V}^{-1}(\hat{\boldsymbol{\theta}})\mathbf{X}]^{-1}\mathbf{C}$. This approach to estimation is known as "estimated generalized least squares" or, more generally, as estimated BLUE or *empirical BLUE* (E-BLUE) ; we use the latter term. The estimators are called E-BLUEs, though the term is somewhat of a misnomer because E-BLUEs in general are neither linear nor unbiased, nor are they known to be "best" in any sense. We postpone further discussion of E-BLUE and its properties until Chap. 17, after we have considered (in Chap. 16) methods for estimating $\boldsymbol{\theta}$.

11.5 Exercises

1. Prove Theorem 11.1.8.
2. Prove Theorem 11.1.9b.
3. Prove Theorem 11.1.10.
4. Obtain $\tilde{\beta}$ (the generalized least squares estimator of β) and its variance under an Aitken model for which $\mathbf{X} = \mathbf{1}_n$ and $\mathbf{W} = (1-\rho)\mathbf{I}_n + \rho\mathbf{J}_n$, where ρ is a known scalar satisfying $-\frac{1}{n-1} < \rho < 1$. How does var($\tilde{\beta}$) compare to σ^2/n?
5. Obtain $\tilde{\beta}$ (the generalized least squares estimator of β) and its variance under an Aitken model for which $\mathbf{X} = \mathbf{1}_n$ and $\mathbf{W} = \mathbf{I} + \mathbf{a}\mathbf{a}^T$, where $\mathbf{a}$ is a known n-vector whose elements sum to 0. How does var($\tilde{\beta}$) compare to σ^2/n?
6. Consider the model

$$y_i = \beta x_i + e_i \quad (i = 1, \ldots, n)$$

where $x_i = i$ for all i and $(e_1, \ldots, e_n)^T$ has mean $\mathbf{0}$ and positive definite variance–covariance matrix

$$\sigma^2\mathbf{W} = \sigma^2 \begin{pmatrix} 1 & 1 & 1 & \cdots & 1 \\ 1 & 2 & 2 & \cdots & 2 \\ 1 & 2 & 3 & \cdots & 3 \\ \vdots & \vdots & \vdots & \ddots & \vdots \\ 1 & 2 & 3 & \cdots & n \end{pmatrix}.$$

Obtain specialized expressions for the BLUE of β and its variance. (Hint: Observe that the model matrix is one of the columns of $\mathbf{W}$, which makes it possible to compute the BLUE without obtaining an expression for $\mathbf{W}^{-1}$.)

7. Consider the problem of best linear unbiased estimation of estimable linear functions of $\boldsymbol{\beta}$ under the constrained positive definite Aitken model $\{\mathbf{y}, \mathbf{X}\boldsymbol{\beta}, \sigma^2\mathbf{W} : \mathbf{A}\boldsymbol{\beta} = \mathbf{h}\}$.
   (a) Derive the system of equations that must be solved to obtain the BLUE (called the *constrained generalized least squares estimator*) under this model.
   (b) Generalize Theorems 10.1.4, 10.1.5, and 10.1.12 to give explicit expressions for the constrained generalized least squares estimator of a vector $\mathbf{C}^T\boldsymbol{\beta}$ of estimable functions and its variance–covariance matrix, and for an unbiased estimator of σ^2, in terms of a generalized inverse of the coefficient matrix for the system of equations derived in part (a).
   (c) If the constraints are jointly nonestimable, will the constrained generalized least squares estimators of functions that are estimable under the constrained model coincide with the (unconstrained) generalized least squares estimators of those functions? Explain.
8. Consider an Aitken model in which

$$\mathbf{X} = \begin{pmatrix} 1 \\ 1 \\ -2 \end{pmatrix} \quad \text{and} \quad \mathbf{W} = \begin{pmatrix} 1 & -0.5 & -0.5 \\ -0.5 & 1 & -0.5 \\ -0.5 & -0.5 & 1 \end{pmatrix}.$$

   Observe that the columns of $\mathbf{W}$ sum to $\mathbf{0}$; thus $\mathbf{W}$ is singular.
   (a) Show that $\mathbf{t}^T\mathbf{y}$ and $\boldsymbol{\ell}^T\mathbf{y}$, where $\mathbf{t} = (\frac{1}{6}, \frac{1}{6}, -\frac{1}{3})^T$ and $\boldsymbol{\ell} = (\frac{1}{2}, \frac{1}{2}, 0)^T$, are BLUEs of β, thus establishing that there is no unique BLUE of β under this model.
   (b) Using Theorem 3.3.7, find a generalized inverse of $\begin{pmatrix} \mathbf{W} & \mathbf{X} \\ \mathbf{X}^T & \mathbf{0} \end{pmatrix}$ and use Theorem 11.2.3a to characterize the collection of all BLUEs of β.
9. Prove Theorem 11.2.4.
10. Results on constrained least squares estimation under the constrained Gauss–Markov model $\{\mathbf{y}, \mathbf{X}\boldsymbol{\beta}, \sigma^2\mathbf{I} : \mathbf{A}\boldsymbol{\beta} = \mathbf{h}\}$ that were established in Chap. 10 may be used to obtain results for best linear unbiased estimation under the uncon-

strained Aitken model $\{\mathbf{y}, \mathbf{X}\boldsymbol{\beta}, \sigma^2\mathbf{W}\}$, as follows. Consider the constrained Gauss–Markov model $\{\mathbf{y}, \mathbf{X}\boldsymbol{\beta}, \sigma^2\mathbf{I} : \mathbf{A}\boldsymbol{\beta} = \mathbf{h}\}$. Regard the vector $\mathbf{h}$ on the right-hand side of the system of constraint equations as a vector of q "pseudo-observations" and append it to the vector of actual observations $\mathbf{y}$, and then consider the unconstrained Aitken model

$$\begin{pmatrix} \mathbf{y} \\ \mathbf{h} \end{pmatrix} = \begin{pmatrix} \mathbf{X} \\ \mathbf{A} \end{pmatrix} \boldsymbol{\beta} + \begin{pmatrix} \mathbf{e} \\ \mathbf{0} \end{pmatrix}, \qquad \text{var} \begin{pmatrix} \mathbf{e} \\ \mathbf{0} \end{pmatrix} = \sigma^2 \begin{pmatrix} \mathbf{I} & \mathbf{0} \\ \mathbf{0} & \mathbf{0} \end{pmatrix}.$$

Prove that the constrained least squares estimator of a function $\mathbf{c}^T\boldsymbol{\beta}$ that is estimable under the constrained Gauss–Markov model is a BLUE of $\mathbf{c}^T\boldsymbol{\beta}$ under the corresponding unconstrained Aitken model.

11. Consider the same simple linear regression setting described in Example 11.3-1.
    (a) Assuming that the elements of $\mathbf{W}$ are such that $\mathbf{W}$ is nonnegative definite, determine additional conditions on those elements that are necessary and sufficient for the ordinary least squares estimator of the slope to be a BLUE of that parameter. (Express your conditions as a set of equations that the w_{ij}'s must satisfy, for example, $w_{11} + w_{12} = 3w_{22}$, etc.)
    (b) Assuming that the elements of $\mathbf{W}$ are such that $\mathbf{W}$ is nonnegative definite, determine additional conditions on those elements that are necessary and sufficient for the ordinary least squares estimator of the intercept to be a BLUE of that parameter.
    (c) Assuming that the elements of $\mathbf{W}$ are such that $\mathbf{W}$ is nonnegative definite, determine additional conditions on those elements that are necessary and sufficient for the ordinary least squares estimator of every estimable function to be a BLUE of that function. [Hint: Combine your results from parts (a) and (b).]
    (d) Give numerical entries of a positive definite matrix $\mathbf{W} \neq \mathbf{I}$ for which:
        (i) the ordinary least squares estimator of the slope, but not of the intercept, is a BLUE;
        (ii) the ordinary least squares estimator of the intercept, but not of the slope, is a BLUE;
        (iii) the ordinary least squares estimators of the slope and intercept are BLUEs.
12. Determine the most general form that the variance–covariance matrix of an Aitken one-factor model can have in order for the ordinary least squares estimator of every estimable function under the model to be a BLUE of that function.
13. Consider a k-part general mixed linear model

$$\mathbf{y} = \mathbf{X}_1\boldsymbol{\beta}_1 + \mathbf{X}_2\boldsymbol{\beta}_2 + \cdots \mathbf{X}_k\boldsymbol{\beta}_k + \mathbf{e}$$

for which $\mathbf{V}(\boldsymbol{\theta}) = \theta_0\mathbf{I} + \theta_1\mathbf{X}_1\mathbf{X}_1^T + \theta_2\mathbf{X}_2\mathbf{X}_2^T + \cdots \theta_k\mathbf{X}_k\mathbf{X}_k^T$, where $\boldsymbol{\theta} = (\theta_0, \theta_1, \theta_2, \ldots, \theta_k)^T$ is an unknown parameter belonging to the subset of $\mathbb{R}^{k+1}$ within which $\mathbf{V}(\boldsymbol{\theta})$ is positive definite for every $\boldsymbol{\theta}$.

(a) Show that the ordinary least squares estimator of every estimable function under this model is a BLUE of that function.

(b) For the special case of a two-way main effects model with two levels of Factor A, three levels of Factor B, and one observation per cell, determine numerical entries of a $\mathbf{V}(\boldsymbol{\theta}) \neq \mathbf{I}$ for which the ordinary least squares estimator of every estimable function is a BLUE. [Hint: Use part (a).]

(c) For the special case of a two-factor nested model with two levels of Factor A, two levels of Factor B within each level of Factor A, and two observations per cell, determine numerical entries of a $\mathbf{V}(\boldsymbol{\theta}) \neq \mathbf{I}$ for which the ordinary least squares estimator of every estimable function is a BLUE.

14. Consider the positive definite Aitken model $\{\mathbf{y}, \mathbf{X}\boldsymbol{\beta}, \sigma^2\mathbf{W}\}$, and suppose that $n > p^*$.

(a) Prove that the ordinary least squares estimator and the generalized least squares estimator of every estimable function corresponding to this Aitken model are equal if and only if $\mathbf{P_X} = \tilde{\mathbf{P}}_\mathbf{X}$.

(b) Define $\hat{\sigma}^2 = \mathbf{y}^T(\mathbf{I} - \mathbf{P_X})\mathbf{y}/(n - p^*)$; note that this would be the residual mean square under a Gauss–Markov model. Also define

$$\tilde{\sigma}^2 = \mathbf{y}^T[\mathbf{W}^{-1} - \mathbf{W}^{-1}\mathbf{X}(\mathbf{X}^T\mathbf{W}^{-1}\mathbf{X})^-\mathbf{X}^T\mathbf{W}^{-1}]\mathbf{y}/(n - p^*),$$

which is the generalized residual mean square. Suppose that the ordinary least squares estimator and generalized least squares estimator of every estimable function are equal. Prove that $\hat{\sigma}^2 = \tilde{\sigma}^2$ (for all $\mathbf{y}$) if and only if $\mathbf{W}(\mathbf{I} - \mathbf{P_X}) = \mathbf{I} - \mathbf{P_X}$.

(c) Suppose that $\mathbf{W} = \mathbf{I} + \mathbf{P_X}\mathbf{A}\mathbf{P_X}$ for some $n \times n$ matrix $\mathbf{A}$ such that $\mathbf{W}$ is positive definite. Prove that in this case the ordinary least squares estimator and generalized least squares estimator (corresponding to this Aitken model) of every estimable function are equal and $\hat{\sigma}^2 = \tilde{\sigma}^2$ (for all $\mathbf{y}$).

References

Christensen, R. (2011). *Plane answers to complex questions* (4th ed.). New York: Springer.

Rao, C. R. (1967). Least squares theory using an estimated dispersion matrix and its application to measurement of signals. In *Proceedings of the Fifth Berkeley Symposium on Mathematical Statistics and Probability* (Vol. 1, pp. 355–372).

Zyskind, G. (1967). On canonical forms, nonnegative covariance matrices, and best and simple least squares estimators in linear models. *Annals of Statistics, 38*, 1092–1110.

Model Misspecification

12

Beginning in Chap. 7, we derived least squares estimators of estimable functions and estimators of residual variance corresponding to several linear models, and we derived various sampling properties of those estimators. With few exceptions, the sampling properties were derived under the same models that the estimators were derived. For example, it was established (in Theorems 7.2.1 and 7.2.2) that the ordinary least squares estimator of an estimable function $\mathbf{c}^T\boldsymbol{\beta}$ in a Gauss–Markov model with model matrix $\mathbf{X}$ is unbiased and has variance $\sigma^2\mathbf{c}^T(\mathbf{X}^T\mathbf{X})^-\mathbf{c}$ under that model. What happens to these properties of the ordinary least squares estimator if the model is incorrectly specified? Answering this question and others like it is the primary objective of this chapter.

Either the mean structure ($\mathbf{X}\boldsymbol{\beta}$) or the variance–covariance structure of the model (or both) may be misspecified. We consider each possibility in turn.

12.1 Misspecification of the Mean Structure

In this section we consider situations in which the model's variance–covariance structure is correctly specified, but its mean structure is misspecified. For simplicity we assume initially that the variance–covariance structure satisfies Gauss–Markov assumptions. Thus, we suppose that the data analyst fits the Gauss–Markov model $\{\mathbf{y}, \mathbf{X}\boldsymbol{\beta}, \sigma^2\mathbf{I}\}$, when in reality the correct model is another Gauss–Markov model $\{\mathbf{y}, \mathbf{X}^*\boldsymbol{\beta}^*, \sigma^2\mathbf{I}\}$ where $\mathcal{C}(\mathbf{X}) \neq \mathcal{C}(\mathbf{X}^*)$. If nothing further is known about how $\mathcal{C}(\mathbf{X})$ compares to $\mathcal{C}(\mathbf{X}^*)$, little can be said about how estimability, least squares estimation of estimable functions, and estimation of residual variance are affected by the misspecification. If, however, the column space of either model matrix is a subspace of the column space of the other, then some useful results can be obtained.

In the *underspecified mean structure scenario*, the data analyst fits the model $\{\mathbf{y}, \mathbf{X}_1\boldsymbol{\beta}_1, \sigma^2\mathbf{I}\}$ when in reality the correct model is $\{\mathbf{y}, \mathbf{X}_1\boldsymbol{\beta}_1 + \mathbf{X}_2\boldsymbol{\beta}_2, \sigma^2\mathbf{I}\}$. Here, $\mathbf{X}_1$ and $\mathbf{X}_2$ have $p_1 \geq 1$ and $p_2 \geq 1$ columns, respectively, and it is assumed that

D. L. Zimmerman, *Linear Model Theory*,
https://doi.org/10.1007/978-3-030-52063-2_12

$\mathcal{C}(\mathbf{X}_1)$ is a proper subspace of $\mathcal{C}(\mathbf{X}_1, \mathbf{X}_2)$. In this scenario, $\hat{\boldsymbol{\beta}}_1^{(U)}$ represents a solution to the normal equations $\mathbf{X}_1^T\mathbf{X}_1\boldsymbol{\beta}_1 = \mathbf{X}_1^T\mathbf{y}$ for the underspecified model (where "U" stands for "underspecified"), and $\hat{\boldsymbol{\beta}}_1$ represents the first p_1-component subvector of a solution to the two-part normal equations

$$\begin{pmatrix} \mathbf{X}_1^T\mathbf{X}_1 & \mathbf{X}_1^T\mathbf{X}_2 \\ \mathbf{X}_2^T\mathbf{X}_1 & \mathbf{X}_2^T\mathbf{X}_2 \end{pmatrix} \begin{pmatrix} \boldsymbol{\beta}_1 \\ \boldsymbol{\beta}_2 \end{pmatrix} = \begin{pmatrix} \mathbf{X}_1^T\mathbf{y} \\ \mathbf{X}_2^T\mathbf{y} \end{pmatrix}$$

for the correct model.

In contrast, in the *overspecified mean structure scenario*, the data analyst fits the model $\{\mathbf{y}, \mathbf{X}_1\boldsymbol{\beta}_1 + \mathbf{X}_2\boldsymbol{\beta}_2, \sigma^2\mathbf{I}\}$ when in reality the correct model is $\{\mathbf{y}, \mathbf{X}_1\boldsymbol{\beta}_1, \sigma^2\mathbf{I}\}$, where $\mathbf{X}_1$ and $\mathbf{X}_2$ satisfy the same assumptions as in the underspecified mean structure scenario. In this scenario, $\hat{\boldsymbol{\beta}}_1^{(O)}$ (where "O" stands for "overspecified") represents the first p_1-component subvector of a solution to the two-part normal equations given in the previous paragraph, and $\hat{\boldsymbol{\beta}}_1$ represents a solution to the normal equations $\mathbf{X}_1^T\mathbf{X}_1\boldsymbol{\beta}_1 = \mathbf{X}_1^T\mathbf{y}$ for the correct model. Thus $\hat{\boldsymbol{\beta}}_1$ represents a solution to a different set of equations in this scenario than in the previous scenario.

Throughout this chapter, as in Chap. 9, let $\mathbf{P}_1$ and $\mathbf{P}_2$ denote the orthogonal projection matrices onto the column spaces of $\mathbf{X}_1$ and $\mathbf{X}_2$, respectively, and let $\mathbf{P}_{12}$ denote the orthogonal projection matrix on the column space of $(\mathbf{X}_1, \mathbf{X}_2)$.

12.1.1 Underspecification

In the underspecified mean structure scenario, the data analyst, because of his/her (mistaken) belief about the mean structure, would be concerned with estimating only those functions $\mathbf{c}_1^T\boldsymbol{\beta}_1$ that are estimable under the underspecified model. But such functions may not be estimable under the correct model. In fact, $\mathbf{c}_1^T\boldsymbol{\beta}_1$ is estimable under the correct model only for those $\mathbf{c}_1^T$ that belong to a particular subspace of $\mathcal{R}(\mathbf{X}_1)$. The following theorem specifies this subspace.

Theorem 12.1.1 *In the underspecified mean structure scenario,* $\mathbf{c}_1^T\boldsymbol{\beta}_1$ *is estimable under the correct model if and only if* $\mathbf{c}_1^T \in \mathcal{R}[(\mathbf{I} - \mathbf{P}_2)\mathbf{X}_1]$.

Proof By Theorem 6.1.2, a function $\mathbf{c}^T\boldsymbol{\beta} \equiv \mathbf{c}_1^T\boldsymbol{\beta}_1 + \mathbf{c}_2^T\boldsymbol{\beta}_2$ is estimable under the correct model if and only if $(\mathbf{c}_1^T, \mathbf{c}_2^T) = \mathbf{a}^T(\mathbf{X}_1, \mathbf{X}_2)$ for some $\mathbf{a}$, i.e., if and only if $\mathbf{c}_1^T = \mathbf{a}^T\mathbf{X}_1$ and $\mathbf{c}_2^T = \mathbf{a}^T\mathbf{X}_2$ for some $\mathbf{a}$. Thus $\mathbf{c}_1^T\boldsymbol{\beta}_1$ is estimable under the correct model if and only if a vector $\mathbf{a}$ exists such that $\mathbf{a}^T\mathbf{X}_1 = \mathbf{c}_1^T$ and $\mathbf{a}^T\mathbf{X}_2 = \mathbf{0}^T$. The last equality is equivalent to $\mathbf{a} \in \mathcal{N}(\mathbf{X}_2^T)$, which in turn is equivalent to $\mathbf{a} = (\mathbf{I} - \mathbf{P}_2)\mathbf{z}$ for some vector $\mathbf{z}$. Thus, $\mathbf{c}_1^T\boldsymbol{\beta}_1$ is estimable under the correct model if and only if $\mathbf{c}_1^T = \mathbf{z}^T(\mathbf{I} - \mathbf{P}_2)\mathbf{X}_1$ for some $\mathbf{z} \in \mathbb{R}^n$, or equivalently, if and only if $\mathbf{c}_1^T \in \mathcal{R}[(\mathbf{I} - \mathbf{P}_2)\mathbf{X}_1]$. □

Although $\mathbf{c}_1^T\boldsymbol{\beta}_1$ for a $\mathbf{c}_1^T \in \mathcal{R}(\mathbf{X}_1)$ is not necessarily estimable under the correct model in the underspecified mean structure scenario, the quantity $\mathbf{c}_1^T\hat{\boldsymbol{\beta}}_1^{(U)}$ that a data analyst would obtain by applying least squares to the underspecified model is uniquely determined for such a function. The next theorem considers some sampling properties of $\mathbf{c}_1^T\hat{\boldsymbol{\beta}}_1^{(U)}$. The theorem and the two that follow it extend some results originally given by Rao (1971) for underspecified full-rank models to underspecified less-than-full-rank models; see also Hocking (1976).

Theorem 12.1.2 *In the underspecified mean structure scenario, if* $\mathbf{c}_1^T \in \mathcal{R}(\mathbf{X}_1)$, *then:*

(a) $E(\mathbf{c}_1^T\hat{\boldsymbol{\beta}}_1^{(U)}) = \mathbf{c}_1^T\boldsymbol{\beta}_1 + \mathbf{c}_1^T(\mathbf{X}_1^T\mathbf{X}_1)^-\mathbf{X}_1^T\mathbf{X}_2\boldsymbol{\beta}_2$, *implying that* $\mathbf{c}_1^T\hat{\boldsymbol{\beta}}_1^{(U)}$ *is unbiased for* $\mathbf{c}_1^T\boldsymbol{\beta}_1$ *if and only if* $\mathbf{c}_1^T(\mathbf{X}_1^T\mathbf{X}_1)^-\mathbf{X}_1^T\mathbf{X}_2 = \mathbf{0}^T$;

(b) $var(\mathbf{c}_1^T\hat{\boldsymbol{\beta}}_1^{(U)}) = \sigma^2\mathbf{c}_1^T(\mathbf{X}_1^T\mathbf{X}_1)^-\mathbf{c}_1$; *and*

(c) $MSE(\mathbf{c}_1^T\hat{\boldsymbol{\beta}}_1^{(U)}) = \sigma^2\mathbf{c}_1^T(\mathbf{X}_1^T\mathbf{X}_1)^-\mathbf{c}_1 + [\mathbf{c}_1^T(\mathbf{X}_1^T\mathbf{X}_1)^-\mathbf{X}_1^T\mathbf{X}_2\boldsymbol{\beta}_2]^2$.

All of these expressions and conditions are invariant to the choice of generalized inverse of $\mathbf{X}_1^T\mathbf{X}_1$.

Proof

$$\begin{aligned} E(\mathbf{c}_1^T\hat{\boldsymbol{\beta}}_1^{(U)}) &= E[\mathbf{c}_1^T(\mathbf{X}_1^T\mathbf{X}_1)^-\mathbf{X}_1^T\mathbf{y}] = \mathbf{c}_1^T(\mathbf{X}_1^T\mathbf{X}_1)^-\mathbf{X}_1^T(\mathbf{X}_1\boldsymbol{\beta}_1 + \mathbf{X}_2\boldsymbol{\beta}_2) \\ &= \mathbf{c}_1^T\boldsymbol{\beta}_1 + \mathbf{c}_1^T(\mathbf{X}_1^T\mathbf{X}_1)^-\mathbf{X}_1^T\mathbf{X}_2\boldsymbol{\beta}_2, \end{aligned}$$

where we used the given row space condition and Theorem 3.3.3a for the last equality. The bias is zero if and only if $\mathbf{c}_1^T(\mathbf{X}_1^T\mathbf{X}_1)^-\mathbf{X}_1^T\mathbf{X}_2\boldsymbol{\beta}_2 = 0$ for all $\boldsymbol{\beta}_2$, i.e., if and only if $\mathbf{c}_1^T(\mathbf{X}_1^T\mathbf{X}_1)^-\mathbf{X}_1^T\mathbf{X}_2 = \mathbf{0}^T$. This establishes part (a). The proof of part (b) is exactly the same as the proof of Theorem 7.2.2 (with $\mathbf{X}_1$ here playing the role of $\mathbf{X}$ in that theorem), and part (c) follows immediately from the fact that mean squared error is equal to variance plus squared bias. Invariance of these expressions and conditions with respect to the choice of generalized inverse of $\mathbf{X}_1^T\mathbf{X}_1$ follows from the invariance of $\mathbf{P}_1$ with respect to said generalized inverse and the fact that $\mathbf{c}_1^T(\mathbf{X}_1^T\mathbf{X}_1)^-\mathbf{X}_1^T = \mathbf{a}^T\mathbf{P}_1$, where $\mathbf{a}$ is any vector such that $\mathbf{a}^T\mathbf{X}_1 = \mathbf{c}_1^T$. □

By Theorem 12.1.2a, for $\mathbf{c}_1^T \in \mathcal{R}(\mathbf{X}_1)$ the estimator of $\mathbf{c}_1^T\boldsymbol{\beta}_1$ obtained by applying least squares to the underspecified model generally is biased. Of course, for $\mathbf{c}_1^T \in \mathcal{R}[(\mathbf{I} - \mathbf{P}_2)\mathbf{X}_1]$, the estimator of $\mathbf{c}_1^T\boldsymbol{\beta}_1$ obtained by applying least squares to the correct model, i.e., $\mathbf{c}_1^T\hat{\boldsymbol{\beta}}_1$, is the *bona fide* least squares estimator of $\mathbf{c}_1^T\boldsymbol{\beta}_1$ and is therefore unbiased. Theorem 12.1.2a tells us further that $\mathbf{c}_1^T\hat{\boldsymbol{\beta}}_1^{(U)}$ and $\mathbf{c}_1^T\hat{\boldsymbol{\beta}}_1$ have the same bias (which is zero) if and only if $\mathbf{c}_1^T(\mathbf{X}_1^T\mathbf{X}_1)^-\mathbf{X}_1^T\mathbf{X}_2 = \mathbf{0}^T$. In the special case in which $\text{rank}[(\mathbf{I} - \mathbf{P}_1)\mathbf{X}_2] = p_2$, it can be shown (see the exercises) that this condition for zero bias is also necessary and sufficient for $\mathbf{c}_1^T\hat{\boldsymbol{\beta}}_1^{(U)}$ and $\mathbf{c}_1^T\hat{\boldsymbol{\beta}}_1$ to coincide. Thus, in that case $\mathbf{c}_1^T\hat{\boldsymbol{\beta}}_1^{(U)}$ is unbiased for such a function if and only if it coincides with $\mathbf{c}_1^T\hat{\boldsymbol{\beta}}_1$.

The following theorem gives a sufficient condition for $\mathbf{c}_1^T\hat{\boldsymbol{\beta}}_1^{(U)}$ and $\mathbf{c}_1^T\hat{\boldsymbol{\beta}}_1$ to coincide for *all* functions $\mathbf{c}_1^T\boldsymbol{\beta}_1$ that are estimable under *either* model.

Theorem 12.1.3 *In the underspecified mean structure scenario, suppose that* $\mathbf{X}_1^T\mathbf{X}_2 = \mathbf{0}$ *and that* $\mathbf{c}_1^T\boldsymbol{\beta}_1$ *is estimable under the underspecified model. Then* $\mathbf{c}_1^T\boldsymbol{\beta}_1$ *is also estimable under the correct model; furthermore,* $\mathbf{c}_1^T\hat{\boldsymbol{\beta}}_1^{(U)}$ *and* $\mathbf{c}_1^T\hat{\boldsymbol{\beta}}_1$ *coincide.*

Proof Because $\mathbf{X}_1^T\mathbf{X}_2 = \mathbf{0}$, $\mathbf{P}_2\mathbf{X}_1 = \mathbf{X}_2(\mathbf{X}_2^T\mathbf{X}_2)^-\mathbf{X}_2^T\mathbf{X}_1 = \mathbf{0}$, implying that $\mathcal{R}[(\mathbf{I}-\mathbf{P}_2)\mathbf{X}_1] = \mathcal{R}(\mathbf{X}_1)$. The theorem's estimability claim follows immediately by Theorem 12.1.1. Furthermore, $\mathbf{c}_1^T(\mathbf{X}_1^T\mathbf{X}_1)^-\mathbf{X}_1^T\mathbf{X}_2 = \mathbf{0}^T$ for all $\mathbf{c}_1^T \in \mathcal{R}(\mathbf{X}_1)$, so $\mathbf{c}_1^T\hat{\boldsymbol{\beta}}_1^{(U)}$ and $\mathbf{c}_1^T\hat{\boldsymbol{\beta}}_1$ coincide. □

Because $\mathbf{c}_1^T\hat{\boldsymbol{\beta}}_1$ is the BLUE of $\mathbf{c}_1^T\boldsymbol{\beta}_1$ for $\mathbf{c}_1^T \in \mathcal{R}[(\mathbf{I}-\mathbf{P}_2)\mathbf{X}_1]$, $\mathbf{c}_1^T\hat{\boldsymbol{\beta}}_1$ is preferable to $\mathbf{c}_1^T\hat{\boldsymbol{\beta}}_1^{(U)}$ if the data analyst insists on using an unbiased estimator. But is it possible that $\mathbf{c}_1^T\hat{\boldsymbol{\beta}}_1^{(U)}$, despite being biased, could have smaller mean squared error than $\mathbf{c}_1^T\hat{\boldsymbol{\beta}}_1$? The next theorem addresses this question.

Theorem 12.1.4 *In the underspecified mean structure scenario, if* $\mathbf{c}_1^T \in \mathcal{R}[(\mathbf{I}-\mathbf{P}_2)\mathbf{X}_1]$, *then:*

(a) *var*$(\mathbf{c}_1^T\hat{\boldsymbol{\beta}}_1^{(U)}) \le$ *var*$(\mathbf{c}_1^T\hat{\boldsymbol{\beta}}_1)$ *and*
(b) *MSE*$(\mathbf{c}_1^T\hat{\boldsymbol{\beta}}_1^{(U)}) \le$ *MSE*$(\mathbf{c}_1^T\hat{\boldsymbol{\beta}}_1)$ *if and only if* $[\mathbf{c}_1^T(\mathbf{X}_1^T\mathbf{X}_1)^-\mathbf{X}_1^T\mathbf{X}_2\boldsymbol{\beta}_2]^2 \le \sigma^2\mathbf{c}_1^T(\mathbf{X}_1^T\mathbf{X}_1)^-\mathbf{X}_1^T\mathbf{X}_2[\mathbf{X}_2^T(\mathbf{I}-\mathbf{P}_1)\mathbf{X}_2]^-\mathbf{X}_2^T\mathbf{X}_1(\mathbf{X}_1^T\mathbf{X}_1)^-\mathbf{c}_1$.
(c) *Furthermore, in the special case in which rank*$[(\mathbf{I}-\mathbf{P}_1)\mathbf{X}_2] = p_2$, *the inequality in part (a) is an equality if and only if* $\mathbf{c}_1^T(\mathbf{X}_1^T\mathbf{X}_1)^-\mathbf{X}_1^T\mathbf{X}_2 = \mathbf{0}^T$.

All of these conditions are invariant to the choices of the generalized inverses.

Proof By Theorem 3.3.7a, the upper left $p_1 \times p_1$ block of one generalized inverse of the coefficient matrix of the two-part normal equations corresponding to the correct model is

$$(\mathbf{X}_1^T\mathbf{X}_1)^- + (\mathbf{X}_1^T\mathbf{X}_1)^-\mathbf{X}_1^T\mathbf{X}_2[\mathbf{X}_2^T(\mathbf{I}-\mathbf{P}_1)\mathbf{X}_2]^-\mathbf{X}_2^T\mathbf{X}_1(\mathbf{X}_1^T\mathbf{X}_1)^-.$$

Then, letting $\mathbf{a}$ represent any vector such that $\mathbf{c}_1^T = \mathbf{a}^T(\mathbf{I}-\mathbf{P}_2)\mathbf{X}_1$, by Theorems 7.2.2 and 12.1.2b we obtain

$$\begin{aligned}
&\mathrm{var}(\mathbf{c}_1^T\hat{\boldsymbol{\beta}}_1)\\
&\quad = \sigma^2\mathbf{c}_1^T(\mathbf{X}_1^T\mathbf{X}_1)^-\mathbf{c}_1 + \sigma^2\mathbf{c}_1^T(\mathbf{X}_1^T\mathbf{X}_1)^-\mathbf{X}_1^T\mathbf{X}_2[\mathbf{X}_2^T(\mathbf{I}-\mathbf{P}_1)\mathbf{X}_2]^-\mathbf{X}_2^T\mathbf{X}_1(\mathbf{X}_1^T\mathbf{X}_1)^-\mathbf{c}_1\\
&\quad = \mathrm{var}(\mathbf{c}_1^T\hat{\boldsymbol{\beta}}_1^{(U)}) + \sigma^2\mathbf{a}^T(\mathbf{I}-\mathbf{P}_2)\mathbf{P}_1\mathbf{X}_2[\mathbf{X}_2^T(\mathbf{I}-\mathbf{P}_1)\mathbf{X}_2]^-\mathbf{X}_2^T\mathbf{P}_1(\mathbf{I}-\mathbf{P}_2)\mathbf{a}\\
&\quad = \mathrm{var}(\mathbf{c}_1^T\hat{\boldsymbol{\beta}}_1^{(U)}) + \sigma^2\mathbf{a}^T(\mathbf{I}-\mathbf{P}_2)(\mathbf{I}-\mathbf{P}_1)\mathbf{X}_2[\mathbf{X}_2^T(\mathbf{I}-\mathbf{P}_1)\mathbf{X}_2]^-\mathbf{X}_2^T(\mathbf{I}-\mathbf{P}_1)(\mathbf{I}-\mathbf{P}_2)\mathbf{a}\\
&\quad = \mathrm{var}(\mathbf{c}_1^T\hat{\boldsymbol{\beta}}_1^{(U)}) + \sigma^2\mathbf{a}^T(\mathbf{I}-\mathbf{P}_2)(\mathbf{P}_{12}-\mathbf{P}_1)(\mathbf{I}-\mathbf{P}_2)\mathbf{a}, \qquad (12.1)
\end{aligned}$$

where Theorem 9.1.2 was used for the last equality. Now, $\mathbf{P}_{12} - \mathbf{P}_1$ is nonnegative definite by Theorem 9.1.1e, f and Corollary 2.15.9.1; therefore, the second summand on the right-hand side of (12.1) is nonnegative. This establishes part (a). For part (b), note that $\mathrm{MSE}(\mathbf{c}_1^T\hat{\boldsymbol{\beta}}_1) = \mathrm{var}(\mathbf{c}_1^T\hat{\boldsymbol{\beta}}_1)$ because $\mathbf{c}_1^T\hat{\boldsymbol{\beta}}_1$ is unbiased, and by Theorem 12.1.2b, c

$$\mathrm{MSE}(\mathbf{c}_1^T\hat{\boldsymbol{\beta}}_1^{(U)}) = \mathrm{var}(\mathbf{c}_1^T\hat{\boldsymbol{\beta}}_1^{(U)}) + [\mathbf{c}_1^T(\mathbf{X}_1^T\mathbf{X}_1)^-\mathbf{X}_1^T\mathbf{X}_2\boldsymbol{\beta}_2]^2. \qquad (12.2)$$

Part (b) follows upon comparing (12.2) to (12.1). Part (c) follows by Theorems 2.15.6 and 2.15.9, since in this case $\mathbf{X}_2^T(\mathbf{I}-\mathbf{P}_1)\mathbf{X}_2$ is positive definite so its ordinary inverse exists and is positive definite. Invariance of these expressions and conditions with respect to the choice of generalized inverse of $\mathbf{X}_1^T\mathbf{X}_1$ holds for the same reason as that given in the proof of Theorem 12.1.2, and the invariance with respect to the choice of generalized inverse of $\mathbf{X}_2^T(\mathbf{I}-\mathbf{P}_1)\mathbf{X}_2$ holds because $\mathbf{P}_{12}-\mathbf{P}_1$ is invariant to that choice. □

Thus, the answer to the question posed immediately ahead of Theorem 12.1.4 is that even if the least squares estimator obtained by fitting the underspecified model is biased, it can indeed have smaller mean squared error than the least squares estimator obtained by fitting the correct model.

The following two examples illustrate some aspects of Theorems 12.1.2–12.1.4.

Example 12.1.1-1. Least Squares Estimation in an Underspecified Simple Linear Regression Model

Consider an underspecified mean structure scenario in which the correct model is the simple linear regression model

$$y_i = \beta_1 + \beta_2 x_i + e_i \quad (i = 1, \ldots, n)$$

and the underspecified model is the intercept-only model

$$y_i = \beta_1 + e_i \quad (i = 1, \ldots, n)$$

(both with Gauss–Markov assumptions). Here, β_1 is estimable under both models, and its least squares estimator obtained by fitting the underspecified model is $\hat{\beta}_1^{(U)} = \bar{y}$. By Theorems 12.1.1 and 12.1.2, $\mathrm{E}(\hat{\beta}_1^{(U)}) = \beta_1 + \beta_2\bar{x}$, $\mathrm{var}(\hat{\beta}_1^{(U)}) = \sigma^2/n$, and $\mathrm{MSE}(\hat{\beta}_1^{(U)}) = (\sigma^2/n) + \beta_2^2\bar{x}^2$. Thus, $\hat{\beta}_1^{(U)}$ is unbiased if and only if $\bar{x} = 0$; moreover, by Theorem 12.1.4b $\mathrm{MSE}(\hat{\beta}_1^{(U)}) \leq \mathrm{MSE}(\hat{\beta}_1)$ if and only if $\beta_2^2\bar{x}^2 \leq \sigma^2\bar{x}^2(1/SXX)$. If $\bar{x} \neq 0$, this last inequality is equivalent to

$$\frac{|\beta_2|}{\sqrt{\mathrm{var}(\hat{\beta}_2)}} \leq 1. \tag{12.3}$$

Thus, provided that $\bar{x} \neq 0$, including the explanatory variable and its corresponding slope parameter in the model is beneficial (in the sense that the mean squared error of the estimated intercept is reduced) if and only if the absolute value of the slope exceeds the standard deviation of its least squares estimator under the correct model. Of course, in practice neither of these two quantities is known, but it is interesting to observe that if β_2 and σ^2 in (12.3) are replaced by the least squares estimator of β_2 and the residual mean square from the fit of the correct model, respectively, the expression on the left-hand side coincides with the expression for the t test statistic, to be described in Chap. 15, for testing the null hypothesis $H_0 : \beta_2 = 0$ versus the alternative hypothesis $H_a : \beta_2 \neq 0$. Thus (12.3) suggests that we should fit the larger model if and only if the t test statistic exceeds 1.0.

An example with the same correct model, but for which the underspecified model is a no-intercept simple linear regression model, is considered in an exercise. ■

Example 12.1.1-2. Least Squares Estimation of Level Means and Differences in a One-Factor Model Lacking an Important Covariate

Consider an underspecified mean structure scenario in which the correct model is the one-factor analysis of covariance model

$$y_{ij} = \mu + \alpha_i + \gamma z_{ij} + e_{ij} \quad (i = 1, \ldots, q;\ j = 1, \ldots, n_i)$$

and the underspecified model is the one-factor model

$$y_{ij} = \mu + \alpha_i + e_{ij} \quad (i = 1, \ldots, q;\ j = 1, \ldots, n_i)$$

(both with Gauss–Markov assumptions). In this scenario, $\mu + \alpha_i$ and $\alpha_i - \alpha_{i'}$ are estimable under both models, and their least squares estimators obtained by fitting the underspecified model are $\widehat{\mu + \alpha_i}^{(U)} = \bar{y}_{i\cdot}$ and $\widehat{\alpha_i - \alpha_{i'}}^{(U)} = \bar{y}_{i\cdot} - \bar{y}_{i'\cdot}$, respectively, for $i' > i = 1, \ldots, n$. By Theorem 12.1.2, the biases of these estimators are $\gamma \bar{z}_{i\cdot}$ and $\gamma(\bar{z}_{i\cdot} - \bar{z}_{i'\cdot})$, respectively. Thus, $\widehat{\alpha_i - \alpha_{i'}}^{(U)}$ is unbiased for all i and i' if and only if $\bar{z}_{i\cdot} = \bar{z}_{i'\cdot}$ for all i and i', or in words, if and only if the level means of the covariate are equal across levels. The bias could be large if these means are quite different and γ is large. If the levels represent treatment groups in a designed experiment and the number of units per group is relatively large, then *randomly* assigning the units to treatments will, with high probability, result in covariate level means that are nearly equal, implying that the bias of $\widehat{\alpha_i - \alpha_{i'}}^{(U)}$ will be relatively small even if the covariate that is lacking is unknown or unobservable. This is one rationale for experimental randomization. ∎

The final results of this subsection concern the estimation of σ^2 in the underspecified mean structure scenario. The residual mean square from the fit of the underspecified model is

$$\hat{\sigma}^2_U \equiv \frac{\mathbf{y}^T(\mathbf{I} - \mathbf{P}_1)\mathbf{y}}{n - \text{rank}(\mathbf{X}_1)}.$$

Theorem 12.1.5 *In the underspecified mean structure scenario, if $n > \text{rank}(\mathbf{X}_1)$, then:*

(a) $E(\hat{\sigma}^2_U) = \sigma^2 + \{1/[n - \text{rank}(\mathbf{X}_1)]\}\boldsymbol{\beta}_2^T\mathbf{X}_2^T(\mathbf{I} - \mathbf{P}_1)\mathbf{X}_2\boldsymbol{\beta}_2$;
(b) if the skewness matrix of $\mathbf{y}$ is null, then $cov(\mathbf{c}_1^T\hat{\boldsymbol{\beta}}_1^{(U)}, \hat{\sigma}^2_U) = 0$; *and*
(c) if the skewness and excess kurtosis matrices of $\mathbf{y}$ are null, then

$$var(\hat{\sigma}^2_U) = \frac{2\sigma^4}{n - rank(\mathbf{X}_1)} + \frac{4\sigma^2\boldsymbol{\beta}_2^T\mathbf{X}_2^T(\mathbf{I} - \mathbf{P}_1)\mathbf{X}_2\boldsymbol{\beta}_2}{[n - rank(\mathbf{X}_1)]^2}.$$

Proof By Theorem 4.2.4,

$$\begin{aligned} \text{E}(\hat{\sigma}^2_U) &= \{1/[n - \text{rank}(\mathbf{X}_1)]\}\{(\mathbf{X}_1\boldsymbol{\beta}_1 + \mathbf{X}_2\boldsymbol{\beta}_2)^T(\mathbf{I} - \mathbf{P}_1)(\mathbf{X}_1\boldsymbol{\beta}_1 + \mathbf{X}_2\boldsymbol{\beta}_2) + \text{tr}[(\mathbf{I} - \mathbf{P}_1)\sigma^2\mathbf{I}]\} \\ &= \{1/[n - \text{rank}(\mathbf{X}_1)]\}[\boldsymbol{\beta}_2^T\mathbf{X}_2^T(\mathbf{I} - \mathbf{P}_1)\mathbf{X}_2\boldsymbol{\beta}_2] + \sigma^2. \end{aligned}$$

This establishes (a). Parts (b) and (c) are left as exercises. □

Because $\mathbf{I} - \mathbf{P}_1$ is nonnegative definite, it follows from Theorem 12.1.5a that $\text{E}(\hat{\sigma}^2_U) \geq \sigma^2$; thus $\hat{\sigma}^2_U$ is nonnegatively biased in the underspecified mean structure scenario by an amount $\boldsymbol{\beta}_2^T\mathbf{X}_2^T(\mathbf{I} - \mathbf{P}_1)\mathbf{X}_2\boldsymbol{\beta}_2/[n - \text{rank}(\mathbf{X}_1)]$. Observe that $\hat{\sigma}^2_U$ is biased even if $\mathbf{X}_1^T\mathbf{X}_2 = \mathbf{0}$, although the expression for the bias in this case does simplify to $\boldsymbol{\beta}_2^T\mathbf{X}_2^T\mathbf{X}_2\boldsymbol{\beta}_2/[n - \text{rank}(\mathbf{X}_1)]$. Also observe from Theorems 12.1.5c and 8.2.3 that var($\hat{\sigma}_U^2$) can be either larger or smaller than the variance of the residual mean square from the correct model, depending on rank($\mathbf{X}_1$) and $\mathbf{X}_2\boldsymbol{\beta}_2$, among other quantities.

12.1.2 Overspecification

In the overspecified mean structure scenario, a data analyst typically will be interested in estimating a linear function $\mathbf{c}^T\boldsymbol{\beta} \equiv \mathbf{c}_1^T\boldsymbol{\beta}_1 + \mathbf{c}_2^T\boldsymbol{\beta}_2$, which is estimable under the correct model if and only a linear estimator $\mathbf{a}^T\mathbf{y}$ exists such that $\mathrm{E}(\mathbf{a}^T\mathbf{y}) = \mathbf{c}_1^T\boldsymbol{\beta}_1 + \mathbf{c}_2^T\boldsymbol{\beta}_2 = \mathbf{c}_1^T\boldsymbol{\beta}_1$ for all $\boldsymbol{\beta}_1$. (The last equality holds because $\boldsymbol{\beta}_2 = \mathbf{0}$ under the correct model.) Thus, $\mathbf{c}_1^T\boldsymbol{\beta}_1 + \mathbf{c}_2^T\boldsymbol{\beta}_2$ is estimable under the correct model if and only if $\mathbf{c}_1^T \in \mathcal{R}(\mathbf{X}_1)$; the value of $\mathbf{c}_2$ makes no difference. Moreover, if $\mathbf{c}_1^T\boldsymbol{\beta}_1$ is estimable under the correct model, then it is estimable under the overspecified model, and vice versa. Consequently, in this scenario we refer to a function $\mathbf{c}_1^T\boldsymbol{\beta}_1$ for which $\mathbf{c}_1^T \in \mathcal{R}(\mathbf{X}_1)$ simply as estimable (without referring to either of the models).

The *bona fide* least squares estimator of an estimable function $\mathbf{c}_1^T\boldsymbol{\beta}_1$ in this scenario is $\mathbf{c}_1^T\hat{\boldsymbol{\beta}}_1$ (where it may be recalled that $\hat{\boldsymbol{\beta}}_1$ is any solution to the normal equations $\mathbf{X}_1^T\mathbf{X}_1\boldsymbol{\beta}_1 = \mathbf{X}_1^T\mathbf{y}$). The estimator of the same function obtained by applying least squares to the overspecified model is $\mathbf{c}_1^T\hat{\boldsymbol{\beta}}_1^{(O)}$.

The following two theorems extend results originally given by Rao (1971) for overspecified full-rank models to overspecified less-than-full-rank models.

Theorem 12.1.6 *In the overspecified mean structure scenario, if $\mathbf{c}_1^T\boldsymbol{\beta}_1$ is estimable, then:*

(a) $E(\mathbf{c}_1^T\hat{\boldsymbol{\beta}}_1^{(O)}) = \mathbf{c}_1^T\boldsymbol{\beta}_1$, *i.e.,* $\mathbf{c}_1^T\hat{\boldsymbol{\beta}}_1^{(O)}$ *is unbiased for* $\mathbf{c}_1^T\boldsymbol{\beta}_1$;
(b) $var(\mathbf{c}_1^T\hat{\boldsymbol{\beta}}_1^{(O)}) = \sigma^2\mathbf{c}_1^T(\mathbf{X}_1^T\mathbf{X}_1)^-\mathbf{c}_1 + \sigma^2\mathbf{c}_1^T(\mathbf{X}_1^T\mathbf{X}_1)^-\mathbf{X}_1^T\mathbf{X}_2[\mathbf{X}_2^T(\mathbf{I}-\mathbf{P}_1)\mathbf{X}_2]^- \times \mathbf{X}_2^T\mathbf{X}_1(\mathbf{X}_1^T\mathbf{X}_1)^-\mathbf{c}_1$, *invariant to the choices of generalized inverses; and*
(c) $MSE(\mathbf{c}_1^T\hat{\boldsymbol{\beta}}_1^{(O)}) = var(\mathbf{c}_1^T\hat{\boldsymbol{\beta}}_1^{(O)})$.

Proof Let $\mathbf{a}$ represent any vector such that $\mathbf{a}^T\mathbf{X}_1 = \mathbf{c}_1^T$. By Theorem 3.3.7a, the upper left $p_1 \times p_1$ block and upper right $p_1 \times p_2$ block of one generalized inverse of the coefficient matrix of the two-part normal equations corresponding to the correct model are

$$(\mathbf{X}_1^T\mathbf{X}_1)^- + (\mathbf{X}_1^T\mathbf{X}_1)^-\mathbf{X}_1^T\mathbf{X}_2[\mathbf{X}_2^T(\mathbf{I}-\mathbf{P}_1)\mathbf{X}_2]^-\mathbf{X}_2^T\mathbf{X}_1(\mathbf{X}_1^T\mathbf{X}_1)^-$$

and

$$-(\mathbf{X}_1^T\mathbf{X}_1)^-\mathbf{X}_1^T\mathbf{X}_2[\mathbf{X}_2^T(\mathbf{I}-\mathbf{P}_1)\mathbf{X}_2]^-,$$

respectively. Then

$$\begin{aligned}
\mathrm{E}(\mathbf{c}_1^T\hat{\boldsymbol{\beta}}_1^{(O)}) &= \mathrm{E}\left[(\mathbf{c}_1^T, \mathbf{0}^T)\begin{pmatrix}\hat{\boldsymbol{\beta}}_1^{(O)} \\ \hat{\boldsymbol{\beta}}_2^{(O)}\end{pmatrix}\right] \\
&= \mathrm{E}[(\mathbf{c}_1^T, \mathbf{0}^T)(\mathbf{X}^T\mathbf{X})^-\mathbf{X}^T\mathbf{y}] \\
&= (\mathbf{c}_1^T, \mathbf{0}^T)(\mathbf{X}^T\mathbf{X})^-\mathbf{X}^T\mathbf{X}\begin{pmatrix}\boldsymbol{\beta}_1 \\ \mathbf{0}\end{pmatrix} \\
&= \mathbf{c}_1^T\{(\mathbf{X}_1^T\mathbf{X}_1)^-\mathbf{X}_1^T\mathbf{X}_1 + (\mathbf{X}_1^T\mathbf{X}_1)^-\mathbf{X}_1^T\mathbf{X}_2[\mathbf{X}_2^T(\mathbf{I}-\mathbf{P}_1)\mathbf{X}_2]^-\mathbf{X}_2^T\mathbf{X}_1(\mathbf{X}_1^T\mathbf{X}_1)^-\mathbf{X}_1^T\mathbf{X}_1 \\
&\quad -(\mathbf{X}_1^T\mathbf{X}_1)^-\mathbf{X}_1^T\mathbf{X}_2[\mathbf{X}_2^T(\mathbf{I}-\mathbf{P}_1)\mathbf{X}_2]^-\mathbf{X}_2^T\mathbf{X}_1\}\boldsymbol{\beta}_1 \\
&= \mathbf{a}^T\mathbf{X}_1(\mathbf{X}_1^T\mathbf{X}_1)^-\mathbf{X}_1^T\mathbf{X}_1\boldsymbol{\beta}_1 \\
&= \mathbf{a}^T\mathbf{X}_1\boldsymbol{\beta}_1 \\
&= \mathbf{c}_1^T\boldsymbol{\beta}_1
\end{aligned}$$

where Theorem 3.3.3a was used for the fifth and sixth equalities. This establishes part (a). For part (b), using the upper left $p_1 \times p_1$ block of the same generalized inverse of $\mathbf{X}^T\mathbf{X}$ described earlier in this proof, we obtain

$$\begin{aligned}
&\mathrm{var}(\mathbf{c}_1^T\hat{\boldsymbol{\beta}}_1^{(O)}) \\
&\quad = \mathrm{var}[(\mathbf{c}_1^T, \mathbf{0}^T)\hat{\boldsymbol{\beta}}^{(O)}] \\
&\quad = \sigma^2\mathbf{c}_1^T(\mathbf{X}_1^T\mathbf{X}_1)^-\mathbf{c}_1 + \sigma^2\mathbf{c}_1^T(\mathbf{X}_1^T\mathbf{X}_1)^-\mathbf{X}_1^T\mathbf{X}_2[\mathbf{X}_2^T(\mathbf{I}-\mathbf{P}_1)\mathbf{X}_2]^-\mathbf{X}_2^T\mathbf{X}_1(\mathbf{X}_1^T\mathbf{X}_1)^-\mathbf{c}_1,
\end{aligned}$$

which is invariant to the choices of generalized inverses of $\mathbf{X}_1^T\mathbf{X}_1$ and $\mathbf{X}_2^T(\mathbf{I}-\mathbf{P}_1)\mathbf{X}_2$ for the same reasons given in the proofs of Theorems 12.1.2 and 12.1.4. Part (c) follows immediately from part (a). □

Theorem 12.1.7 *In the overspecified mean structure scenario, if $\mathbf{c}_1^T\boldsymbol{\beta}_1$ is estimable, then:*

(a) $var(\mathbf{c}_1^T\hat{\boldsymbol{\beta}}_1^{(O)}) \geq var(\mathbf{c}_1^T\hat{\boldsymbol{\beta}}_1)$ and
(b) $MSE(\mathbf{c}_1^T\hat{\boldsymbol{\beta}}_1^{(O)}) \geq MSE(\mathbf{c}_1^T\hat{\boldsymbol{\beta}}_1)$.
(c) Furthermore, in the special case in which $rank[(\mathbf{I}-\mathbf{P}_1)\mathbf{X}_2] = p_2$, the inequalities in parts (a) and (b) are equalities if and only if $\mathbf{c}_1^T(\mathbf{X}_1^T\mathbf{X}_1)^-\mathbf{X}_1^T\mathbf{X}_2 = \mathbf{0}^T$.

All of these conditions are invariant to the choice of generalized inverse of $\mathbf{X}_1^T\mathbf{X}_1$.

Proof By Theorems 12.1.6b and 7.2.2,

$$\text{var}(\mathbf{c}_1^T\hat{\boldsymbol{\beta}}_1^{(O)}) = \text{var}(\mathbf{c}_1^T\hat{\boldsymbol{\beta}}_1)+\sigma^2\mathbf{c}_1^T(\mathbf{X}_1^T\mathbf{X}_1)^-\mathbf{X}_1^T\mathbf{X}_2[\mathbf{X}_2^T(\mathbf{I}-\mathbf{P}_1)\mathbf{X}_2]^-\mathbf{X}_2^T\mathbf{X}_1(\mathbf{X}_1^T\mathbf{X}_1)^-\mathbf{c}_1,$$

and this expression is invariant to the choices of generalized inverses of $\mathbf{X}_1^T\mathbf{X}_1$ and $\mathbf{X}_2^T(\mathbf{I} - \mathbf{P}_1)\mathbf{X}_2$ for the same reasons given in the proofs of Theorems 12.1.2 and 12.1.4. Then, by the same argument used in the proof of Theorem 12.1.3, the second summand on the right-hand side is nonnegative. This establishes part (a). Part (b) follows immediately from part (a) and Theorem 12.1.6a, while part (c) follows by the same argument used in the proof of part (c) of Theorem 12.1.4. □

Theorem 12.1.7b reveals that it is never beneficial (in the sense of reducing the mean squared error of the least squares estimator of $\mathbf{c}_1^T\boldsymbol{\beta}_1$) to overfit the model. This is in contrast to the underfitting scenario, where Theorem 12.1.4b indicated that there are conditions under which it is beneficial to underfit.

Although it is never beneficial to overfit the model, there are conditions under which overfitting incurs no penalty because $\mathbf{c}_1^T\hat{\boldsymbol{\beta}}_1^{(O)}$ and $\mathbf{c}_1^T\hat{\boldsymbol{\beta}}_1$ coincide. The next theorem, whose proof is left as an exercise, gives a sufficient condition.

Theorem 12.1.8 *In the overspecified mean structure scenario, suppose that* $\mathbf{X}_1^T\mathbf{X}_2 = \mathbf{0}$ *and that* $\mathbf{c}_1^T\boldsymbol{\beta}_1$ *is estimable. Then* $\mathbf{c}_1^T\hat{\boldsymbol{\beta}}_1^{(O)}$ *and* $\mathbf{c}_1^T\hat{\boldsymbol{\beta}}_1$ *coincide.*

The practical value of Theorem 12.1.8 is that the inclusion of unnecessary explanatory variables in the fitted model will not affect least squares estimators of linear functions of the parameters associated with the necessary explanatory variables if the unnecessary variables are orthogonal to the necessary ones. This is one nice property of orthogonal explanatory variables; another, namely order-invariance of the sequential ANOVA of the overspecified model, was established previously (see Theorem 9.3.1).

Example 12.1.2-1. Least Squares Estimation in an Overspecified Simple Linear Regression Model

Consider an overspecified mean structure scenario in which the correct model is the intercept-only model

$$y_i = \beta_1 + e_i \quad (i = 1, \ldots, n)$$

and the overspecified model is the simple linear regression model

$$y_i = \beta_1 + \beta_2 x_i + e_i \quad (i = 1, \ldots, n)$$

(both with Gauss–Markov assumptions). Here, β_1 is estimable, $\hat{\beta}_1 = \bar{y}$, and $\hat{\beta}_1^{(O)} = \bar{y} - \hat{\beta}_2\bar{x}$. Furthermore, as a special case of Theorem 12.1.6, $\hat{\beta}_1^{(O)}$ is unbiased under the correct model and its variance (and mean squared error) is $\sigma^2[(1/n) + (\bar{x}^2/SXX)]$. By comparison, the variance (and mean squared error) of $\hat{\beta}_1$ is σ^2/n, which is smaller than the variance of $\hat{\beta}_1^{(O)}$ unless $\bar{x} = 0$, in which case $\hat{\beta}_1$ and $\hat{\beta}_1^{(O)}$ coincide (exemplifying Theorems 12.1.7 and 12.1.8).

The complementary case with the same overspecified model but for which the correct model is the no-intercept model is considered in an exercise. ■

The residual mean square from the fit of the overspecified model is

$$\hat{\sigma}_O^2 = \frac{\mathbf{y}^T(\mathbf{I} - \mathbf{P}_{12})\mathbf{y}}{n - \text{rank}(\mathbf{X}_1, \mathbf{X}_2)}.$$

The last theorem in this subsection is the counterpart of Theorem 12.1.5 for the overspecified mean structure scenario. Its proof is left as an exercise.

Theorem 12.1.9 *In the overspecified mean structure scenario, if $n > \text{rank}(\mathbf{X}_1, \mathbf{X}_2)$, then:*

(a) $E(\hat{\sigma}_O^2) = \sigma^2$;
(b) if the skewness matrix of $\mathbf{y}$ is null, then $cov(\mathbf{c}_1^T\hat{\boldsymbol{\beta}}_1^{(O)}, \hat{\sigma}_O^2) = 0$; and
(c) if the skewness and excess kurtosis matrices of $\mathbf{y}$ are null, then

$$var(\hat{\sigma}_O^2) = 2\sigma^4/[n - rank(\mathbf{X}_1, \mathbf{X}_2)].$$

Observe from Theorems 12.1.9a, c and 8.2.3 that although $\hat{\sigma}_O^2$ is unbiased, its variance is larger than that of the residual mean square from the correct model because $\text{rank}(\mathbf{X}_1, \mathbf{X}_2) > \text{rank}(\mathbf{X}_1)$.

Methodological Interlude #6: Model Selection Criteria for "Best Subsets Regression"

An important problem in applied regression analysis arises when it is not known which, if any, of the available explanatory variables actually help to explain the response variable. The problem, known as *model selection*, is to select the helpful explanatory variables from among the available ones. A number of criteria are commonly used to inform the selection. The simplest of these include the coefficient of determination,

$$R^2 = \frac{\sum_{i=1}^n (\hat{y}_i - \bar{y})^2}{\sum_{i=1}^n (y_i - \bar{y})^2} = 1 - \frac{\sum_{i=1}^n (y_i - \hat{y}_i)^2}{\sum_{i=1}^n (y_i - \bar{y})^2},$$

which was introduced in Sect. 9.3.3; *adjusted* R^2, defined by

$$\bar{R}^2 = 1 - \frac{\sum_{i=1}^n (y_i - \hat{y}_i)^2/(n-p^*)}{\sum_{i=1}^n (y_i - \bar{y})^2/(n-1)} = 1 - \frac{\hat{\sigma}^2(n-1)}{\sum_{i=1}^n (y_i - \bar{y})^2}; \qquad (12.4)$$

and the residual mean square, $\hat{\sigma}^2$. For a collection of available explanatory variables, the "best subsets regression" approach computes these (and possibly other) criteria for a Gauss–Markov linear model with mean structure consisting of each possible subset of the available explanatory variables, then compares them across subsets and selects the model that is best with respect to the criterion (largest R^2 or $\bar{R}^2$, smallest $\hat{\sigma}^2$). If one subset is contained within another, R^2 for the larger subset will be at least as large as that for the smaller subset because the residual sum of squares $\sum_{i=1}^n (y_i - \hat{y}_i)^2$ cannot increase as variables are added to the mean structure. Adjusted R^2 improves upon R^2 in this respect by accounting for not only the reduction in residual sum of squares but also the reduction in the rank of $\mathbf{I} - \mathbf{P_X}$ (i.e., $n - p^*$) as variables are added to the mean structure. Note from (12.4) that $\bar{R}^2$ and $\hat{\sigma}^2$ are perfectly inversely related, so they order the models identically. By Theorem 12.1.5a, the residual mean square from the fit of an underspecified model will tend to be larger than that from the fit of the correct model, but this is not guaranteed because both of these quantities are random variables. Furthermore, by Theorem 12.1.9a, the residual mean square from the fit of an overspecified model does *not* tend to be larger than that from the fit of the correct model, which leads to frequent selection of an overspecified model. For this reason, the use of only the residual mean square (and $\bar{R}^2$) to select models should be avoided.

A fourth popular model selection criterion for best subsets regression is the sum of squared PRESS residuals,

$$PRESS = \sum_{i=1}^n \hat{e}_{i,-i}^2 = \sum_{i=1}^n \left(\frac{e_{ii}}{1-h_{ii}}\right)^2,$$

where the PRESS residuals $\hat{e}_{i,-i}$ $(i = 1, \ldots, n)$ were defined in Methodological Interlude #3. *PRESS* measures, in aggregate, how well each response is predicted by its fitted value (which is also its BLUP) based on all the remaining data. The moniker "*PRESS*" is an abbreviation of "Prediction Sum of Squares," which was introduced by D. Allen (1974).

The final prevalent model selection criterion we mention here is known as *Mallow's* C_p, named after its originator, C. Mallows (Mallows, 1973), though a more transparent name for it would be the *standardized total mean squared error (STMSE)* of fitted values. By definition,

$$STMSE = \sum_{i=1}^n \frac{MSE(\hat{y}_i)}{\sigma^2}$$

but using the theorems in this section, many alternative expressions can be given. Let us write

$$STMSE = \sum_{i=1}^{n} \frac{\mathrm{var}(\hat{y}_i)}{\sigma^2} + \sum_{i=1}^{n} \frac{[\mathrm{bias}(\hat{y}_i)]^2}{\sigma^2},$$

and consider the variance and squared bias terms separately in an underspecified mean structure scenario. In such a scenario, $\hat{y}_i = \mathbf{x}_{1i}^T \hat{\boldsymbol{\beta}}_1^{(U)}$ where $\mathbf{x}_{1i}^T$ is the ith row of $\mathbf{X}_1$, and the corresponding estimand is $\mathrm{E}(y_i) = \mathbf{x}_{1i}^T \boldsymbol{\beta}_1 + \mathbf{x}_{2i}^T \boldsymbol{\beta}_2$ where $\mathbf{x}_{2i}^T$ is the ith row of $\mathbf{X}_2$. Denoting the ith diagonal element of $\mathbf{P}_1$ by $p_{1,ii}$, we have

$$\sum_{i=1}^{n} \frac{\mathrm{var}(\hat{y}_i)}{\sigma^2} = \sum_{i=1}^{n} \mathbf{x}_{1i}^T (\mathbf{X}_1^T \mathbf{X}_1)^- \mathbf{x}_{1i} = \sum_{i=1}^{n} p_{1,ii} = \mathrm{rank}(\mathbf{X}_1),$$

where we used Theorem 12.1.2b for the first equality and Theorem 8.1.6a for the last. Furthermore, by Theorem 12.1.2b the bias of $\mathbf{x}_{1i}^T \hat{\boldsymbol{\beta}}_1^{(U)}$ is

$$\mathrm{E}(\mathbf{x}_{1i}^T \hat{\boldsymbol{\beta}}_1^{(U)}) - (\mathbf{x}_{1i}^T \boldsymbol{\beta}_1 + \mathbf{x}_{2i}^T \boldsymbol{\beta}_2) = \mathbf{x}_{1i}^T (\mathbf{X}_1^T \mathbf{X}_1)^- \mathbf{X}_1^T \mathbf{X}_2 \boldsymbol{\beta}_2 - \mathbf{x}_{2i}^T \boldsymbol{\beta}_2.$$

Thus,

$$\begin{aligned}
&\sum_{i=1}^{n} [\mathrm{bias}(\mathbf{x}_{1i}^T \hat{\boldsymbol{\beta}}_1^{(U)})]^2 \\
&= \sum_{i=1}^{n} \boldsymbol{\beta}_2^T [\mathbf{X}_2^T \mathbf{X}_1 (\mathbf{X}_1^T \mathbf{X}_1)^- \mathbf{x}_{1i} - \mathbf{x}_{2i}][\mathbf{x}_{1i}^T (\mathbf{X}_1^T \mathbf{X}_1)^- \mathbf{X}_1^T \mathbf{X}_2 - \mathbf{x}_{2i}^T] \boldsymbol{\beta}_2 \\
&= \boldsymbol{\beta}_2^T \mathbf{X}_2^T \mathbf{X}_1 (\mathbf{X}_1^T \mathbf{X}_1)^- \left(\sum_{i=1}^{n} \mathbf{x}_{1i} \mathbf{x}_{1i}^T \right) (\mathbf{X}_1^T \mathbf{X}_1)^- \mathbf{X}_1^T \mathbf{X}_2 \boldsymbol{\beta}_2 + \boldsymbol{\beta}_2^T \left(\sum_{i=1}^{n} \mathbf{x}_{2i} \mathbf{x}_{2i}^T \right) \boldsymbol{\beta}_2 \\
&\quad - \boldsymbol{\beta}_2^T \mathbf{X}_2^T \mathbf{X}_1 (\mathbf{X}_1^T \mathbf{X}_1)^- \left(\sum_{i=1}^{n} \mathbf{x}_{1i} \mathbf{x}_{2i}^T \right) \boldsymbol{\beta}_2 - \boldsymbol{\beta}_2^T \left(\sum_{i=1}^{n} \mathbf{x}_{2i} \mathbf{x}_{1i}^T \right) (\mathbf{X}_1^T \mathbf{X}_1)^- \mathbf{X}_1^T \mathbf{X}_2 \boldsymbol{\beta}_2 \\
&= \boldsymbol{\beta}_2^T \mathbf{X}_2^T \mathbf{X}_1 (\mathbf{X}_1^T \mathbf{X}_1)^- \mathbf{X}_1^T \mathbf{X}_2 \boldsymbol{\beta}_2 + \boldsymbol{\beta}_2^T \mathbf{X}_2^T \mathbf{X}_2 \boldsymbol{\beta}_2 - 2\boldsymbol{\beta}_2^T \mathbf{X}_2^T \mathbf{X}_1 (\mathbf{X}_1^T \mathbf{X}_1)^- \mathbf{X}_1^T \mathbf{X}_2 \boldsymbol{\beta}_2 \\
&= \boldsymbol{\beta}_2^T \mathbf{X}_2^T (\mathbf{I} - \mathbf{P}_1) \mathbf{X}_2 \boldsymbol{\beta}_2 \\
&= [n - \mathrm{rank}(\mathbf{X}_1)][\mathrm{E}(\hat{\sigma}_U^2) - \sigma^2],
\end{aligned}$$

where we used Theorem 12.1.5a for the last equality. Therefore,

$$STMSE = \mathrm{rank}(\mathbf{X}_1) + \frac{[n - \mathrm{rank}(\mathbf{X}_1)](\hat{\sigma}_U^2 - \sigma^2)}{\sigma^2},$$

which is the classical form of Mallow's C_p (in slightly different notation). This model selection criterion cannot be used as it stands, however, because it depends on the unknown parameter σ^2. In practice, the residual mean square from the largest model under consideration is usually substituted for σ^2; hence, all models are evaluated for underspecification relative to this model. Small values of *STMSE* are desirable, with values close to rank($\mathbf{X}_1$) indicating that there is little bias in the fitted values.

None of the model selection criteria discussed in this interlude requires the observations to have any particular probability distribution. In Chap. 15 we consider additional criteria for model selection that are appropriate when $\mathbf{y}$ has a multivariate normal distribution. ■

The results obtained for misspecified mean structure scenarios in this section assumed that all models under consideration are Gauss–Markov models, but the results can be generalized with no difficulty whatsoever to scenarios in which all models are positive definite Aitken models with the same variance–covariance matrix $\sigma^2\mathbf{W}$. These models can be transformed to Gauss–Markov models by pre-multiplying $\mathbf{y}$, $\mathbf{X}_1$, and $\mathbf{X}_2$ by $\mathbf{W}^{-\frac{1}{2}}$ as in Sect. 11.1, after which Theorems 12.1.1–12.1.9 can be applied to the transformed models. For example, in an underspecified mean structure scenario, if $\mathbf{c}_1^T\boldsymbol{\beta}_1$ is estimable under the underspecified model, then the bias of its estimator obtained by applying generalized least squares to the underspecified model is $\mathbf{c}_1^T(\mathbf{X}_1^T\mathbf{W}^{-1}\mathbf{X}_1)^-\mathbf{X}_1^T\mathbf{W}^{-1}\mathbf{X}_2\boldsymbol{\beta}_2$.

12.2 Misspecification of the Variance–Covariance Structure

In this section, we assume that the model's mean structure is correctly specified and examine the implications of misspecifying the variance–covariance structure. Specifically, we suppose that the data analyst fits the positive definite Aitken model $\{\mathbf{y}, \mathbf{X}\boldsymbol{\beta}, \sigma^2\mathbf{W}_1\}$ when in fact the correct model is the Aitken model (not necessarily positive definite) $\{\mathbf{y}, \mathbf{X}\boldsymbol{\beta}, \sigma^2\mathbf{W}_2\}$, where $\mathbf{W}_2$ is not a scalar multiple of $\mathbf{W}_1$. We call this scenario the *misspecified variance–covariance structure scenario*. Two important special cases are $\{\mathbf{W}_1 = \mathbf{I}$, $\mathbf{W}_2$ is an arbitrary nonnegative definite matrix$\}$ (i.e., the analyst incorrectly fits a Gauss–Markov model), and $\{\mathbf{W}_1$ is an arbitrary positive definite matrix, $\mathbf{W}_2 = \mathbf{I}\}$ (i.e., the analyst fits a more complicated variance–covariance structure when a Gauss–Markov structure would suffice). Because the two models have the same model matrix, all functions of $\boldsymbol{\beta}$ that are estimable under one of the models are also estimable under the other.

In this scenario, let $\mathbf{c}^T\boldsymbol{\beta}$ be an estimable function, let $\tilde{\boldsymbol{\beta}}_I$ represent a solution to the Aitken equations $\mathbf{X}^T\mathbf{W}_1^{-1}\mathbf{X}\boldsymbol{\beta} = \mathbf{X}^T\mathbf{W}_1^{-1}\mathbf{y}$ for the incorrect model, and let $\tilde{\mathbf{t}}_C$ represent the first n-component subvector of a solution to the BLUE equations for $\mathbf{c}^T\boldsymbol{\beta}$ associated with the correct model. The BLUE of $\mathbf{c}^T\boldsymbol{\beta}$ is $\tilde{\mathbf{t}}_C^T\mathbf{y}$, and the estimator of $\mathbf{c}^T\boldsymbol{\beta}$ obtained by applying generalized least squares to the incorrect model is $\mathbf{c}^T\tilde{\boldsymbol{\beta}}_I$. Let $\tilde{\sigma}_I^2$ and $\tilde{\sigma}_C^2$ represent the generalized residual mean square obtained by fitting the incorrect model and the unbiased estimator of σ^2 defined in Theorem 11.2.3c, respectively.

The next theorem summarizes several relevant results. All expectations, variances, and covariances in the theorem are obtained under the correct model.

Theorem 12.2.1 *In the misspecified variance–covariance structure scenario, let* $\mathbf{c}^T\boldsymbol{\beta}$ *be an estimable function. Then:*

(a) $\mathbf{c}^T\tilde{\boldsymbol{\beta}}_I$ *and* $\tilde{\mathbf{t}}_C^T\mathbf{y}$ *coincide if and only if* $\mathbf{W}_2\mathbf{W}_1^{-1}\mathbf{X}(\mathbf{X}^T\mathbf{W}_1^{-1}\mathbf{X})^-\mathbf{c} \in \mathcal{C}(\mathbf{X})$*;*
(b) $E(\mathbf{c}^T\tilde{\boldsymbol{\beta}}_I) = \mathbf{c}^T\boldsymbol{\beta}$*, i.e.,* $\mathbf{c}^T\tilde{\boldsymbol{\beta}}_I$ *is unbiased;*
(c) $var(\mathbf{c}^T\tilde{\boldsymbol{\beta}}_I) = \sigma^2\mathbf{c}^T(\mathbf{X}^T\mathbf{W}_1^{-1}\mathbf{X})^-\mathbf{X}^T\mathbf{W}_1^{-1}\mathbf{W}_2\mathbf{W}_1^{-1}\mathbf{X}(\mathbf{X}^T\mathbf{W}_1^{-1}\mathbf{X})^-\mathbf{c} \geq var(\tilde{\mathbf{t}}_C^T\mathbf{y})$*, and the inequality is strict unless* $\mathbf{c}^T\tilde{\boldsymbol{\beta}}_I$ *and* $\tilde{\mathbf{t}}_C^T\mathbf{y}$ *coincide;*
(d) If $n > p^*$*, then*

$$E(\tilde{\sigma}_I^2) = \sigma^2\frac{tr\{[\mathbf{W}_1^{-1} - \mathbf{W}_1^{-1}\mathbf{X}(\mathbf{X}^T\mathbf{W}_1^{-1}\mathbf{X})^-\mathbf{X}^T\mathbf{W}_1^{-1}]\mathbf{W}_2\}}{n-p^*};$$

(e) If $n > p^*$ *and the skewness matrix of* $\mathbf{y}$ *is null, then* $cov(\mathbf{c}^T\tilde{\boldsymbol{\beta}}_I, \tilde{\sigma}_I^2) = 0$*; and*
(f) If $n > p^*$ *and the skewness and excess kurtosis matrices of* $\mathbf{y}$ *are null, then*

$$\begin{aligned} var(\tilde{\sigma}_I^2) = \frac{2\sigma^4}{(n-p^*)^2}tr\{[\mathbf{W}_1^{-1} - \mathbf{W}_1^{-1}\mathbf{X}(\mathbf{X}^T\mathbf{W}_1^{-1}\mathbf{X})^-\mathbf{X}^T\mathbf{W}_1^{-1}] \\ \times \mathbf{W}_2[\mathbf{W}_1^{-1} - \mathbf{W}_1^{-1}\mathbf{X}(\mathbf{X}^T\mathbf{W}_1^{-1}\mathbf{X})^-\mathbf{X}^T\mathbf{W}_1^{-1}]\mathbf{W}_2\}. \end{aligned}$$

All of the expressions above involving a generalized inverse of $\mathbf{X}^T\mathbf{W}_1^{-1}\mathbf{X}$ *are invariant to the choice of that generalized inverse.*

Proof The proof of part (a) is very similar to that of Theorem 11.3.1 and is left as an exercise. For part (b), let $\mathbf{a}$ represent any vector such that $\mathbf{a}^T\mathbf{X} = \mathbf{c}^T$. Then by Theorem 11.1.6a,

$$\begin{aligned} \mathrm{E}(\mathbf{c}^T\tilde{\boldsymbol{\beta}}_I) &= \mathbf{a}^T\mathbf{X}(\mathbf{X}^T\mathbf{W}_1^{-1}\mathbf{X})^-\mathbf{X}^T\mathbf{W}_1^{-1}\mathbf{X}\boldsymbol{\beta} \\ &= \mathbf{a}^T\mathbf{X}\boldsymbol{\beta} \\ &= \mathbf{c}^T\boldsymbol{\beta}, \end{aligned}$$

which establishes part (b). For part (c), we have

$$\begin{aligned} \mathrm{var}(\mathbf{c}^T\tilde{\boldsymbol{\beta}}_I) &= \mathrm{var}[\mathbf{a}^T\mathbf{X}(\mathbf{X}^T\mathbf{W}_1^{-1}\mathbf{X})^-\mathbf{X}^T\mathbf{W}_1^{-1}\mathbf{y}] \\ &= \mathbf{a}^T\mathbf{X}(\mathbf{X}^T\mathbf{W}_1^{-1}\mathbf{X})^-\mathbf{X}^T\mathbf{W}_1^{-1}(\sigma^2\mathbf{W}_2)\mathbf{W}_1^{-1}\mathbf{X}[(\mathbf{X}^T\mathbf{W}_1^{-1}\mathbf{X})^-]^T\mathbf{X}^T\mathbf{a} \end{aligned}$$

$$= \sigma^2 \mathbf{c}^T(\mathbf{X}^T\mathbf{W}_1^{-1}\mathbf{X})^-\mathbf{X}^T\mathbf{W}_1^{-1}\mathbf{W}_2\mathbf{W}_1^{-1}\mathbf{X}(\mathbf{X}^T\mathbf{W}_1^{-1}\mathbf{X})^-\mathbf{c}$$
$$\geq \text{var}(\tilde{\mathbf{t}}_C^T\mathbf{y}),$$

where we used Theorem 11.1.6b for the third equality and Theorem 11.2.2 for the inequality. That the inequality is strict unless $\mathbf{c}^T\tilde{\boldsymbol{\beta}}_I$ and $\tilde{\mathbf{t}}_C^T\mathbf{y}$ coincide also follows immediately from Theorem 11.2.2. Next, we have

$$\begin{aligned}(n-p^*)\text{E}(\tilde{\sigma}_I^2) &= \text{E}\{\mathbf{y}^T[\mathbf{W}_1^{-1} - \mathbf{W}_1^{-1}\mathbf{X}(\mathbf{X}^T\mathbf{W}_1^{-1}\mathbf{X})^-\mathbf{X}^T\mathbf{W}_1^{-1}]\mathbf{y}\}\\ &= \boldsymbol{\beta}^T\mathbf{X}^T[\mathbf{W}_1^{-1} - \mathbf{W}_1^{-1}\mathbf{X}(\mathbf{X}^T\mathbf{W}_1^{-1}\mathbf{X})^-\mathbf{X}^T\mathbf{W}_1^{-1}]\mathbf{X}\boldsymbol{\beta}\\ &\quad + \text{tr}\{[\mathbf{W}_1^{-1} - \mathbf{W}_1^{-1}\mathbf{X}(\mathbf{X}^T\mathbf{W}_1^{-1}\mathbf{X})^-\mathbf{X}^T\mathbf{W}_1^{-1}](\sigma^2\mathbf{W}_2)\}\\ &= 0 + \sigma^2\text{tr}\{[\mathbf{W}_1^{-1} - \mathbf{W}_1^{-1}\mathbf{X}(\mathbf{X}^T\mathbf{W}_1^{-1}\mathbf{X})^-\mathbf{X}^T\mathbf{W}_1^{-1}]\mathbf{W}_2\},\end{aligned}$$

which immediately yields part (d). Next, by Corollary 4.2.5.1 and Theorem 11.1.6a,

$$\begin{aligned}\text{cov}(\mathbf{c}^T\tilde{\boldsymbol{\beta}}_I, \tilde{\sigma}_I^2) &= 2\mathbf{c}^T(\mathbf{X}^T\mathbf{W}_1^{-1}\mathbf{X})^-\mathbf{X}^T\mathbf{W}_1^{-1}(\sigma^2\mathbf{W}_2)\\ &\quad \times[\mathbf{W}_1^{-1} - \mathbf{W}_1^{-1}\mathbf{X}(\mathbf{X}^T\mathbf{W}_1^{-1}\mathbf{X})^-\mathbf{X}^T\mathbf{W}_1^{-1}]\mathbf{X}\boldsymbol{\beta}/(n-p^*)\\ &= 2\mathbf{c}^T(\mathbf{X}^T\mathbf{W}_1^{-1}\mathbf{X})^-\mathbf{X}^T\mathbf{W}_1^{-1}(\sigma^2\mathbf{W}_2)(\mathbf{W}_1^{-1}\mathbf{X} - \mathbf{W}_1^{-1}\mathbf{X})\boldsymbol{\beta}/(n-p^*)\\ &= 0,\end{aligned}$$

establishing part (e). Finally, by Corollary 4.2.6.3,

$$\begin{aligned}\text{var}(\tilde{\sigma}_I^2) &= 2\text{tr}\{[\mathbf{W}_1^{-1} - \mathbf{W}_1^{-1}\mathbf{X}(\mathbf{X}^T\mathbf{W}_1^{-1}\mathbf{X})^-\mathbf{X}^T\mathbf{W}_1^{-1}](\sigma^2\mathbf{W}_2)\\ &\quad [\mathbf{W}_1^{-1} - \mathbf{W}_1^{-1}\mathbf{X}(\mathbf{X}^T\mathbf{W}_1^{-1}\mathbf{X})^-\mathbf{X}^T\mathbf{W}_1^{-1}]\cdot(\sigma^2\mathbf{W}_2)\}/(n-p^*)^2\\ &= \frac{2\sigma^4}{(n-p^*)^2}\text{tr}\{[\mathbf{W}_1^{-1} - \mathbf{W}_1^{-1}\mathbf{X}(\mathbf{X}^T\mathbf{W}_1^{-1}\mathbf{X})^-\mathbf{X}^T\mathbf{W}_1^{-1}]\\ &\quad \times\mathbf{W}_2[\mathbf{W}_1^{-1} - \mathbf{W}_1^{-1}\mathbf{X}(\mathbf{X}^T\mathbf{W}_1^{-1}\mathbf{X})^-\mathbf{X}^T\mathbf{W}_1^{-1}]\mathbf{W}_2\}.\end{aligned}$$

The invariance of the expressions involving a generalized inverse of $\mathbf{X}^T\mathbf{W}_1^{-1}\mathbf{X}$ to the choice of that generalized inverse follows immediately from Theorem 11.1.6b. □

The upshot of part (a) of Theorem 12.2.1 is that it is possible for the BLUEs corresponding to different variance–covariance matrices to coincide; indeed, the special case in which one of the variance–covariance matrices is $\sigma^2\mathbf{I}$ was the topic of Sect. 11.3. Part (b) of the theorem indicates that even in cases in which the two estimators do not coincide, the one obtained using the incorrect variance–covariance structure is unbiased, but by part (c) its variance is strictly larger than that of the estimator obtained using the correct variance–covariance structure. Unfortunately, however, parts (d) and (f) of the theorem reveal that similar definitive statements cannot be made about the bias and variance of the generalized residual mean square obtained using an incorrect variance–covariance structure. It is not difficult to construct examples for which this bias is negative and others for which it is positive, or for which the variance of $\tilde{\sigma}_I^2$ is less than or greater than the variance of $\tilde{\sigma}_C^2$. Some examples of this type are considered in the exercises.

The expression for the variance of $\mathbf{c}^T\tilde{\boldsymbol{\beta}}_I$ given in Theorem 12.2.1c may be used to evaluate the loss of efficiency due to applying generalized least squares to a model with incorrect variance–covariance structure. The following is an example of such an evaluation.

Example 12.2-1. Efficiency of the Ordinary Least Squares Estimator of Slope Under a No-Intercept Regression Model with Compound Symmetric Variance–Covariance Structure

Consider an example of the misspecified variance–covariance structure scenario in which the incorrect model is the Gauss–Markov no-intercept simple linear regression model $\{\mathbf{y}, \mathbf{x}\beta, \sigma^2\mathbf{I}\}$, and the correct model is $\{\mathbf{y}, \mathbf{x}\beta, \sigma^2[(1-\rho)\mathbf{I} + \rho\mathbf{J}]\}$, where $-\frac{1}{n-1} < \rho < 1$. Thus, β is estimated by ordinary least squares when it should be estimated by generalized least squares with a positive definite compound symmetric variance–covariance structure. Specializing the expressions in Theorems 12.2.1c and 11.1.3, we obtain

$$\begin{aligned}\text{var}(\tilde{\beta}_I) &= \sigma^2\left(\sum_{i=1}^n x_i^2\right)^{-1}\mathbf{x}^T[(1-\rho)\mathbf{I}+\rho\mathbf{J}]\mathbf{x}\left(\sum_{i=1}^n x_i^2\right)^{-1}\\ &= \sigma^2\left(\sum_{i=1}^n x_i^2\right)^{-2}\left[(1-\rho)\left(\sum_{i=1}^n x_i^2\right)+\rho\left(\sum_{i=1}^n x_i\right)^2\right]\end{aligned}$$

and

$$\begin{aligned}\text{var}(\tilde{\beta}_C) &= \sigma^2\{\mathbf{x}^T[(1-\rho)\mathbf{I}+\rho\mathbf{J}]^{-1}\mathbf{x}\}^{-1}\\ &= \sigma^2\left\{\mathbf{x}^T\left[\frac{1}{1-\rho}\mathbf{I}-\frac{\rho}{(1-\rho)(1-\rho+n\rho)}\mathbf{J}\right]\mathbf{x}\right\}^{-1}\\ &= \sigma^2(1-\rho)\left[\sum_{i=1}^n x_i^2-\left(\frac{\rho}{1-\rho+n\rho}\right)\left(\sum_{i=1}^n x_i\right)^2\right]^{-1},\end{aligned}$$

where we used expression (2.1) for the inverse of a nonsingular compound symmetric matrix. Thus the efficiency of $\tilde{\beta}_I$ relative to $\tilde{\beta}_C$ is

$$
\begin{aligned}
Eff &= \frac{\text{var}(\tilde{\beta}_C)}{\text{var}(\tilde{\beta}_I)} \\
&= \frac{(1-\rho)(\sum_{i=1}^n x_i^2)^2}{[(1-\rho)(\sum_{i=1}^n x_i^2) + \rho(\sum_{i=1}^n x_i)^2][\sum_{i=1}^n x_i^2 - \frac{\rho}{1-\rho+n\rho}(\sum_{i=1}^n x_i)^2]}.
\end{aligned}
$$

In order to obtain some numerical results, now suppose that $n = 5$ and $x_i = i$ $(i = 1, 2, 3, 4, 5)$. Then it is easily verified that

$$
Eff = \frac{3025(1-\rho)}{(55+170\rho)(55-\frac{225\rho}{1+4\rho})}.
$$

Figure 12.1 plots *Eff* as a function of the correlation parameter ρ. As expected, the ordinary least squares estimator is only fully efficient when $\rho = 0$, and its efficiency decreases monotonically to zero as ρ either increases or decreases away from zero. ■

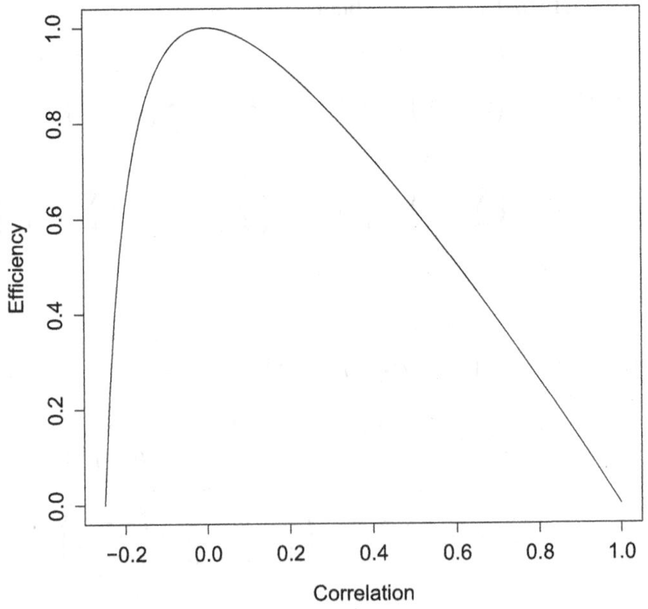

Fig. 12.1 Efficiency of the ordinary least squares estimator of slope in Example 12.2-1

12.3 Exercises

1. Consider the underspecified mean structure scenario, and suppose that $\mathbf{c}_1^T\boldsymbol{\beta}_1$ is estimable under the correct model and $\text{rank}[(\mathbf{I}-\mathbf{P}_1)\mathbf{X}_2] = p_2$. Prove that $\mathbf{c}_1^T\hat{\boldsymbol{\beta}}_1^{(U)}$ and $\mathbf{c}_1^T\hat{\boldsymbol{\beta}}_1$ coincide if and only if $\mathbf{c}_1^T(\mathbf{X}_1^T\mathbf{X}_1)^-\mathbf{X}_1^T\mathbf{X}_2 = \mathbf{0}^T$.
2. Consider an underspecified mean structure scenario, where both models satisfy Gauss–Markov assumptions. Suppose that $\mathbf{c}_1^T\boldsymbol{\beta}_1$ is estimable under the underspecified model and that $\mathbf{c}_2^T\boldsymbol{\beta}_2$ is estimable under the correct model. Prove that $\mathbf{c}_1^T\hat{\boldsymbol{\beta}}_1^{(U)}$ and $\mathbf{c}_2^T\hat{\boldsymbol{\beta}}_2$ are uncorrelated under the correct model.
3. Suppose that the Gauss–Markov no-intercept simple linear regression model is fit to some data, but the correct model is the Gauss–Markov simple linear regression model.
   (a) Determine the bias of $\hat{\beta}_2^{(U)}$, and determine necessary and sufficient conditions for the bias to equal 0.
   (b) Determine necessary and sufficient conditions for the mean squared error of $\hat{\beta}_2^{(U)}$ to be less than or equal to the mean squared error of the least squares estimator of β_2 under the correct model.
4. Consider a situation in which the true model for three observations is the two-way main effects Gauss–Markov model

$$y_{ij} = \mu + \alpha_i + \gamma_j + e_{ij} \quad (i, j) \in \{(1, 1)\,(1, 2),\ (2, 1)\}$$

   but the model fitted to the observations is the (underspecified) one-factor Gauss–Markov model

$$y_{ij} = \mu + \alpha_i + e_{ij} \quad (i, j) \in \{(1, 1)\,(1, 2),\ (2, 1)\}.$$

   Recall that the true model for such observations was considered in Examples 7.1-3 and 7.2-3.
   (a) Is $\mu + \alpha_1$ estimable under the true model? Regardless of the answer to that question, obtain its least squares estimator under the fitted model. Also obtain the expectation, variance, and mean squared error of that estimator under the true model.
   (b) Is $\alpha_1 - \alpha_2$ estimable under the true model? Regardless of the answer to that question, obtain its least squares estimator under the fitted model. Also obtain the expectation, variance, and mean squared error of that estimator under the true model.
   (c) Obtain the least squares estimator of $\alpha_1 - \alpha_2$ under the true model, and obtain its expectation, variance, and mean squared error under the true model.
   (d) Compare the variances of the estimators of $\alpha_1 - \alpha_2$ obtained in parts (b) and (c). Which theorem does this exemplify?

(e) Obtain an explicit condition on γ_1, γ_2, and σ^2 for the mean squared error of the estimator of $\alpha_1 - \alpha_2$ defined in part (b) to be smaller than the mean squared error of the estimator of $\alpha_1 - \alpha_2$ defined in part (c).

5. Consider a situation in which the true model for three observations is the one-factor Gauss–Markov model

$$y_{ij} = \mu + \alpha_i + e_{ij} \quad (i, j) \in \{(1, 1)\,(1, 2),\ (2, 1)\}$$

but the model fitted to the observations is the (overspecified) two-way main effects Gauss–Markov model

$$y_{ij} = \mu + \alpha_i + \gamma_j + e_{ij} \quad (i, j) \in \{(1, 1)\,(1, 2),\ (2, 1)\}.$$

Recall that the fitted model for such observations was considered in Examples 7.1-3 and 7.2-3.

(a) Obtain the least squares estimator of $\alpha_1 - \alpha_2$ under the fitted model, and obtain the expectation, variance, and mean squared error of that estimator under the true model.

(b) Obtain the least squares estimator of $\alpha_1 - \alpha_2$ under the true model, and obtain the expectation, variance, and mean squared error of that estimator under the true model.

(c) Which estimator has larger mean squared error? Which theorem does this exemplify?

6. Prove Theorem 12.1.5b, c.

7. In the underspecified simple linear regression settings considered in Example 12.1.1-1 and Exercise 12.3, obtain the bias and variance of $\hat{\sigma}_U^2$, and compare the latter to the variance of $\hat{\sigma}^2$ (assuming that $n > 2$ and that the skewness and excess kurtosis matrices of $\mathbf{y}$ are null).

8. Prove Theorem 12.1.8.

9. Suppose that the Gauss–Markov simple linear regression model is fit to some data, but the correct model is the no-intercept version of the same model. Determine the mean squared error of $\hat{\beta}_2^{(O)}$ and compare it to that of $\hat{\beta}_2$.

10. Prove Theorem 12.1.9.

11. In the overspecified simple linear regression settings considered in Example 12.1.2-1 and Exercise 12.9, obtain the variance of $\hat{\sigma}_O^2$ (assuming that $n > 2$ and that the skewness and excess kurtosis matrices of $\mathbf{y}$ are null), and compare it to the variance of $\hat{\sigma}^2$.

12. Prove Theorem 12.2.1a.

13. Consider a misspecified variance–covariance structure scenario with $\mathbf{W}_1 = \mathbf{I}$, in which case $\tilde{\sigma}_I^2$ is the ordinary residual mean square.

(a) Show that $0 \leq \mathrm{E}(\tilde{\sigma}_I^2) \leq \frac{n}{n-p^*}\bar{\sigma}^2$, where $\bar{\sigma}^2 = (1/n)\sum_{i=1}^{n}\sigma^2 w_{2,ii}$ and $w_{2,ii}$ is the ith main diagonal element of $\mathbf{W}_2$. (Hint: Use Theorem 2.15.13.)

(b) Prove the following lemma, and then use it to show that the lower bound in part (a) is attained if and only if $\mathbf{W}_2 = \mathbf{P_X}\mathbf{B}\mathbf{P_X}$ for some nonnegative definite matrix $\mathbf{B}$, and that the upper bound is attained if and only if $\mathbf{W}_2 = (\mathbf{I}-\mathbf{P_X})\mathbf{C}(\mathbf{I}-\mathbf{P_X})$ for some nonnegative definite matrix $\mathbf{C}$. (Hint: To prove the lemma, use Theorems 3.2.3 and 2.15.9.)

Lemma Let $\mathbf{A}$ and $\mathbf{Q}$ represent an $m \times n$ matrix and an $n \times n$ nonnegative definite matrix, respectively, and let $\mathbf{A}^-$ represent any generalized inverse of $\mathbf{A}$. Then $\mathbf{AQ} = \mathbf{0}$ if and only if $\mathbf{Q} = (\mathbf{I} - \mathbf{A}^-\mathbf{A})\mathbf{B}(\mathbf{I} - \mathbf{A}^-\mathbf{A})^T$ for some $n \times n$ nonnegative definite matrix $\mathbf{B}$.

[Note: This exercise was inspired by results from Dufour (1986).]

14. Consider the Aitken model $\{\mathbf{y}, \mathbf{X}\boldsymbol{\beta}, \sigma^2[\mathbf{I}+\mathbf{P_X}\mathbf{A}\mathbf{P_X}+(\mathbf{I}-\mathbf{P_X})\mathbf{B}(\mathbf{I}-\mathbf{P_X})]\}$, where $\mathbf{B}$ is a specified $n \times n$ nonnegative definite matrix and $\mathbf{A}$ is a specified $n \times n$ matrix such that var($\mathbf{e}$) is nonnegative definite. Suppose that $n > p^*$.
   (a) Let $\tilde{\sigma}_I^2 = \mathbf{y}^T(\mathbf{I}-\mathbf{P_X})\mathbf{y}/(n-p^*)$. Obtain a simplified expression for the bias of $\tilde{\sigma}_I^2$ under this model, and determine necessary and sufficient conditions on $\mathbf{A}$ and $\mathbf{B}$ for this bias to equal 0.
   (b) Suppose that the skewness and excess kurtosis matrices of $\mathbf{y}$ are null. Obtain a simplified expression for var($\tilde{\sigma}_I^2$). How does var($\tilde{\sigma}_I^2$) compare to var($\tilde{\sigma}_C^2$)?
15. Consider a special case of the misspecified variance–covariance structure scenario in which $\mathbf{W}_1 = \mathbf{I}$ and $\mathbf{W}_2$ is a nonnegative definite matrix such that $\text{tr}(\mathbf{W}_2) = n$ but is otherwise arbitrary.
   (a) Show that $\tilde{\sigma}_I^2 \equiv \mathbf{y}^T(\mathbf{I}-\mathbf{P_X})\mathbf{y}/(n-p^*)$ is an unbiased estimator of σ^2 under the correct model if and only if $\text{tr}(\mathbf{P_X}\mathbf{W}_2) = p^*$.
   (b) As an even more special case, suppose that

$$\mathbf{X} = \begin{pmatrix} 1 & 1 \\ 1 & -1 \\ 1 & 1 \\ 1 & -1 \end{pmatrix} \quad \text{and} \quad \mathbf{W}_2 = \begin{pmatrix} 1 & \rho & 0 & 0 \\ \rho & 1 & \rho & 0 \\ 0 & \rho & 1 & \rho \\ 0 & 0 & \rho & 1 \end{pmatrix}.$$

   Here ρ is a specified real number for which $\mathbf{W}_2$ is nonnegative definite. Using part (a), determine whether $\tilde{\sigma}_I^2$ is an unbiased estimator of σ^2 in this case.
16. Consider a misspecified variance–covariance structure scenario in which the correct model is a heteroscedastic no-intercept linear regression model $\{\mathbf{y}, \mathbf{x}\beta, \sigma^2\text{diag}(x_1^2, x_2^2, \ldots, x_n^2)\}$ and the incorrect model is its Gauss–Markov counterpart $\{\mathbf{y}, \mathbf{x}\beta, \sigma^2\mathbf{I}\}$.
   (a) Obtain specialized expressions for var($\tilde{\beta}_I$) and var($\tilde{\beta}_C$).
   (b) Suppose that $x_i = 100/(101-i)$ $(i = 1, \ldots, 100)$. Evaluate the expression for var($\tilde{\beta}_I$) obtained in part (a) for each $n = 1, \ldots, 100$. (You should write a short computer program to do this.) Are you surprised by the behavior of var($\tilde{\beta}_I$) as the sample size increases? Explain.

   [Note: This exercise was inspired by results from Meng and Xie (2014).]
17. Consider a misspecified variance–covariance structure scenario in which the incorrect model is $\{\mathbf{y}, \mathbf{1}\mu, \sigma^2\mathbf{I}\}$ and the correct model has the same mean

structure but a variance–covariance matrix given by (5.13). Obtain a general expression for the efficiency (the ratio of variances) of the (incorrect) least squares estimator of μ relative to the (correct) generalized least squares estimator, and evaluate it for $n = 10$ and $\rho = -0.9, -0.7, -0.5, \ldots, 0.9$.

References

Allen, D. M. (1974). The relationship between variable selection and data augmentation and a method for prediction. *Technometrics, 16*, 125–127.

Dufour, J. (1986). Bias of S^2 in linear regressions with dependent errors. *The American Statistician, 40*, 284–285.

Hocking, R. R. (1976). The analysis and selection of variables in linear regression. *Biometrics, 32*, 1–49.

Mallows, C. L. (1973). Some comments on C_p. *Technometrics, 15*, 661–675.

Meng, X., & Xie, X. (2014). I got more data, my model is more refined, but my estimator is getting worse! Am I just dumb? *Econometric Reviews, 33*, 218–250.

Rao, P. (1971). Some notes on misspecification in multiple regressions. *The American Statistician, 25*, 37–39.

Best Linear Unbiased Prediction 13

Suppose, as in Chap. 11, that the model for $\mathbf{y}$ is an Aitken model. In this chapter, however, rather than considering the problem of estimating $\mathbf{c}^T\boldsymbol{\beta}$ under that model (which we have already dealt with), we consider the problem of estimating or, to state it more accurately, *predicting* $\tau \equiv \mathbf{c}^T\boldsymbol{\beta} + u$, where u is a random variable satisfying

$$\mathrm{E}(u) = 0, \quad \mathrm{var}(u) = \sigma^2 h, \quad \text{and } \mathrm{cov}(\mathbf{y}, u) = \sigma^2\mathbf{k}.$$

Here h is a specified nonnegative scalar and $\mathbf{k}$ is a specified n-vector such that the matrix $\begin{pmatrix} \mathbf{W} & \mathbf{k} \\ \mathbf{k}^T & h \end{pmatrix}$, which is equal to $(1/\sigma^2)$ times the variance–covariance matrix of $\begin{pmatrix} \mathbf{y} \\ u \end{pmatrix}$, is nonnegative definite. We speak of "predicting τ" rather than "estimating τ" because τ is now a random variable rather than a parametric function (although one of the summands in its definition, namely $\mathbf{c}^T\boldsymbol{\beta}$, *is* a parametric function). We refer to the joint model for $\mathbf{y}$ and u just described as the *prediction-extended Aitken model*, and to the inference problem as the *general prediction problem* (under that model). In the degenerate case in which $u = 0$ with probability one, the general prediction problem reduces to the estimation problem considered in Chap. 11.

One relatively simple and well-known example of the general prediction problem is the prediction of a "new" observation in classical full-rank multiple linear regression. In that setting, $\tau \equiv y_{n+1} = \mathbf{x}_{n+1}^T\boldsymbol{\beta} + e_{n+1} \equiv \mathbf{c}^T\boldsymbol{\beta} + u$ where $\mathbf{c} \equiv \mathbf{x}_{n+1}$ is a specified p-vector of explanatory variables, $\tau \equiv y_{n+1}$ is the response variable to be observed at $\mathbf{x} = \mathbf{x}_{n+1}$, and $u \equiv e_{n+1}$ is the corresponding error (residual). Furthermore, in that setting $\mathbf{W} = \mathbf{I}$ and it is often also assumed that $h = 1$ and $\mathbf{k} = \mathbf{0}$, in which case $\mathbf{y}$ and y_{n+1} jointly satisfy Gauss–Markov model assumptions. Other special cases of the general prediction problem (under prediction-extended Aitken and even more general linear models) arise in forecasting time series, in kriging spatial data, in the inter-block analysis of an incomplete block design, in

D. L. Zimmerman, *Linear Model Theory*,
https://doi.org/10.1007/978-3-030-52063-2_13

the imputation of missing values in an experimental layout, and in the calculation of breeding values in genetic evaluation. In fact, it was primarily for applications to animal breeding that the theory of best linear unbiased prediction was first developed by C. R. Henderson in the 1950s [though he did not publish the results until somewhat later; see Henderson (1963)]. Additional original contributors to the material presented here were Goldberger (1962) and Harville (1976).

The first section of this chapter defines a best linear unbiased predictor and derives a system of equations whose solution (under the most general conditions possible) yields such a predictor. The second section considers the problem of best linear unbiased prediction for the special case of the prediction-extended Aitken model in which $\mathbf{W}$ is positive definite; results are simplest for this case. The third section extends results to the general prediction-extended Aitken model. The fourth section considers best linear unbiased prediction for mixed (and random) effects models with positive definite matrices $\mathbf{R}(\boldsymbol{\psi})$ and $\mathbf{G}(\boldsymbol{\psi})$. The final section briefly considers an attempt to extend best linear unbiased prediction to the general mixed linear model.

13.1 The BLUP Equations

Definition 13.1.1 A predictor $p(\mathbf{y})$ of $\mathbf{c}^T\boldsymbol{\beta} + u$ associated with the model $\left\{\begin{pmatrix}\mathbf{y}\\u\end{pmatrix}, \begin{pmatrix}\mathbf{X}\boldsymbol{\beta}\\0\end{pmatrix}\right\}$ is said to be linear if $p(\mathbf{y}) \equiv t_0 + \mathbf{t}^T\mathbf{y}$ for some constant t_0 and some n-vector $\mathbf{t}$ of constants.

Definition 13.1.2 A predictor $p(\mathbf{y})$ of $\mathbf{c}^T\boldsymbol{\beta} + u$ associated with the model $\left\{\begin{pmatrix}\mathbf{y}\\u\end{pmatrix}, \begin{pmatrix}\mathbf{X}\boldsymbol{\beta}\\0\end{pmatrix}\right\}$ is said to be *unbiased* under that model if $\mathrm{E}[p(\mathbf{y})] = \mathbf{c}^T\boldsymbol{\beta}$ for all $\boldsymbol{\beta}$.

Theorem 13.1.1 *A linear predictor $t_0 + \mathbf{t}^T\mathbf{y}$ of $\mathbf{c}^T\boldsymbol{\beta} + u$ associated with the model $\left\{\begin{pmatrix}\mathbf{y}\\u\end{pmatrix}, \begin{pmatrix}\mathbf{X}\boldsymbol{\beta}\\0\end{pmatrix}\right\}$ is unbiased under any unconstrained model with model matrix $\mathbf{X}$ if and only if $t_0 = 0$ and $\mathbf{t}^T\mathbf{X} = \mathbf{c}^T$.*

Proof Because the condition defining an unbiased predictor of $\mathbf{c}^T\boldsymbol{\beta} + u$ specified in Definition 13.1.2 is identical to the condition defining an unbiased estimator of $\mathbf{c}^T\boldsymbol{\beta}$ specified in Definition 6.1.2, the theorem can be proved in exactly the same way that Theorem 6.1.1 was proved. □

Definition 13.1.3 A function $\mathbf{c}^T\boldsymbol{\beta} + u$ is said to be *predictable* under the model $\left\{\begin{pmatrix}\mathbf{y}\\u\end{pmatrix}, \begin{pmatrix}\mathbf{X}\boldsymbol{\beta}\\0\end{pmatrix}\right\}$ if a linear predictor $t_0 + \mathbf{t}^T\mathbf{y}$ exists that predicts it unbiasedly under that model. Otherwise, the function is said to be *nonpredictable* (under that model).

Theorem 13.1.2 *A function* $\tau = \mathbf{c}^T\boldsymbol{\beta} + u$ *is predictable under the model* $\left\{\begin{pmatrix}\mathbf{y}\\ u\end{pmatrix}, \begin{pmatrix}\mathbf{X}\boldsymbol{\beta}\\ 0\end{pmatrix}\right\}$ *if and only if* $\mathbf{c}^T\boldsymbol{\beta}$ *is estimable under the model* $\{\mathbf{y}, \mathbf{X}\boldsymbol{\beta}\}$, *i.e., if and only if* $\mathbf{c}^T \in \mathcal{R}(\mathbf{X})$.

Proof Left as an exercise. □

Definition 13.1.4 The *prediction error* associated with a predictor $p(\mathbf{y})$ of a predictable function τ is given by $p(\mathbf{y}) - \tau$.

Both terms in the prediction error, $p(\mathbf{y})$ and τ, are random variables. Thus,

$$\begin{aligned}\text{var}[p(\mathbf{y}) - \tau] &= \text{var}[p(\mathbf{y}) - (\mathbf{c}^T\boldsymbol{\beta} + u)]\\ &= \text{var}[p(\mathbf{y})] + \text{var}(u) - 2\cdot\text{cov}[p(\mathbf{y}), u]\\ &\neq \text{var}[p(\mathbf{y})],\end{aligned}$$

in general. Thus, the variance of the prediction error does not generally coincide with the variance of the predictor. This is a considerably different situation from that for estimation, in which the estimand $\mathbf{c}^T\boldsymbol{\beta}$ is a nonrandom quantity and the variance of the estimation error associated with an estimator is identical to the variance of that estimator.

Definition 13.1.5 A *best linear unbiased predictor*, or *BLUP*, of a predictable function τ is a linear unbiased predictor that minimizes the variance of prediction error among all linear unbiased predictors. A BLUP of a vector $\boldsymbol{\tau}$ of predictable functions is a vector of BLUPs of those functions.

It cannot be overemphasized that what a BLUP of a predictable function minimizes is not the variance of the predictor, but the variance of the prediction error, among all linear unbiased predictors of that function.

Now we describe an approach for obtaining a BLUP of a predictable function. This approach is similar to that used in Sect. 11.2 to obtain a BLUE of an estimable function under the (general) Aitken model. Take the model to be the prediction-extended Aitken model $\left\{\begin{pmatrix}\mathbf{y}\\ u\end{pmatrix}, \begin{pmatrix}\mathbf{X}\boldsymbol{\beta}\\ 0\end{pmatrix}, \sigma^2\begin{pmatrix}\mathbf{W} & \mathbf{k}\\ \mathbf{k}^T & h\end{pmatrix}\right\}$, and let $\mathbf{c}^T\boldsymbol{\beta} + u$ be a predictable function of interest. For any linear predictor $t_0 + \mathbf{t}^T\mathbf{y}$ of $\mathbf{c}^T\boldsymbol{\beta} + u$, we have

$$\text{var}[(t_0 + \mathbf{t}^T\mathbf{y}) - (\mathbf{c}^T\boldsymbol{\beta} + u)] = \text{var}(\mathbf{t}^T\mathbf{y} - u) = \sigma^2(\mathbf{t}^T\mathbf{W}\mathbf{t} - 2\mathbf{t}^T\mathbf{k} + h).$$

Because t_0 must equal 0 and $\mathbf{t}^T\mathbf{X}$ must equal $\mathbf{c}^T$ for $t_0 + \mathbf{t}^T\mathbf{y}$ to be to be an unbiased predictor (Theorem 13.1.1), a BLUP of τ is given by $\mathbf{t}^T\mathbf{y}$, where $\mathbf{t}$ is any vector that minimizes $\mathbf{t}^T\mathbf{W}\mathbf{t} - 2\mathbf{t}^T\mathbf{k} + h$ subject to the condition $\mathbf{t}^T\mathbf{X} = \mathbf{c}^T$. The Lagrangian for this minimization problem is

$$L(\mathbf{t}, \boldsymbol{\lambda}) = \mathbf{t}^T\mathbf{W}\mathbf{t} - 2\mathbf{t}^T\mathbf{k} + h + 2\boldsymbol{\lambda}^T(\mathbf{X}^T\mathbf{t} - \mathbf{c}).$$

We have

$$\frac{\partial L}{\partial \mathbf{t}} = 2\mathbf{W}\mathbf{t} - 2\mathbf{k} + 2\mathbf{X}\boldsymbol{\lambda}, \qquad \frac{\partial L}{\partial \boldsymbol{\lambda}} = 2(\mathbf{X}^T\mathbf{t} - \mathbf{c}).$$

Equating these partial derivatives to zero yields

$$\mathbf{W}\mathbf{t} + \mathbf{X}\boldsymbol{\lambda} = \mathbf{k}, \qquad \mathbf{X}^T\mathbf{t} = \mathbf{c}.$$

Therefore, a necessary condition for the prediction error variance $\sigma^2(\mathbf{t}^T\mathbf{W}\mathbf{t} - 2\mathbf{t}^T\mathbf{k} + h)$ to attain a minimum at $\mathbf{t} = \tilde{\mathbf{t}}$, for $\mathbf{t}$ satisfying $\mathbf{X}^T\mathbf{t} = \mathbf{c}$, is that $\tilde{\mathbf{t}}$ be the first n-component subvector of a solution to the system of equations

$$\begin{pmatrix} \mathbf{W} & \mathbf{X} \\ \mathbf{X}^T & \mathbf{0} \end{pmatrix} \begin{pmatrix} \mathbf{t} \\ \boldsymbol{\lambda} \end{pmatrix} = \begin{pmatrix} \mathbf{k} \\ \mathbf{c} \end{pmatrix}.$$

We call this system the *best linear unbiased predicting equations*, or *BLUP equations*, for $\mathbf{c}^T\boldsymbol{\beta} + u$ associated with the prediction-extended Aitken model $\left\{\begin{pmatrix} \mathbf{y} \\ u \end{pmatrix}, \begin{pmatrix} \mathbf{X}\boldsymbol{\beta} \\ 0 \end{pmatrix}, \sigma^2 \begin{pmatrix} \mathbf{W} & \mathbf{k} \\ \mathbf{k}^T & h \end{pmatrix}\right\}$. Observe that the coefficient matrix of this system is identical to the coefficient matrix of the BLUE equations for $\mathbf{c}^T\boldsymbol{\beta}$; in fact, the only difference between the BLUP equations for $\mathbf{c}^T\boldsymbol{\beta} + u$ and the BLUE equations for $\mathbf{c}^T\boldsymbol{\beta}$ is that $\mathbf{k}$ replaces $\mathbf{0}$ as the first n-component subvector of the right-hand side vector. The next theorem establishes sufficient conditions for the BLUP equations to have a solution.

Theorem 13.1.3 *Under the prediction-extended Aitken model* $\left\{\begin{pmatrix} \mathbf{y} \\ u \end{pmatrix}, \begin{pmatrix} \mathbf{X}\boldsymbol{\beta} \\ 0 \end{pmatrix}, \sigma^2 \begin{pmatrix} \mathbf{W} & \mathbf{k} \\ \mathbf{k}^T & h \end{pmatrix}\right\}$, *if* $\mathbf{c}^T\boldsymbol{\beta} + u$ *is predictable and* $\mathbf{k} \in \mathcal{C}(\mathbf{W}, \mathbf{X})$, *then the BLUP equations for* $\mathbf{c}^T\boldsymbol{\beta} + u$ *are consistent.*

Proof We show that the BLUP equations for $\mathbf{c}^T\boldsymbol{\beta} + u$ are compatible, from which the consistency will follow by Theorem 3.2.1. As in the proof of Theorem 11.2.1, let $\boldsymbol{\ell}_1$ and $\boldsymbol{\ell}_2$ represent any n-vector and any p-vector, respectively, such that

$$(\boldsymbol{\ell}_1^T, \boldsymbol{\ell}_2^T) \begin{pmatrix} \mathbf{W} & \mathbf{X} \\ \mathbf{X}^T & \mathbf{0} \end{pmatrix} = \mathbf{0}^T.$$

Then, as shown in the same proof,

$$\boldsymbol{\ell}_1^T \mathbf{X} = \mathbf{0}^T, \qquad \boldsymbol{\ell}_1^T \mathbf{W} = \mathbf{0}^T, \quad \text{and} \quad \boldsymbol{\ell}_2^T \mathbf{X}^T = \mathbf{0}^T.$$

Thus, an $(n + p)$-vector $\mathbf{a}_1$ and an n-vector $\mathbf{a}_2$ exist such that

$$(\boldsymbol{\ell}_1^T, \boldsymbol{\ell}_2^T)\begin{pmatrix}\mathbf{k}\\ \mathbf{c}\end{pmatrix} = \boldsymbol{\ell}_1^T \mathbf{k} + \boldsymbol{\ell}_2^T \mathbf{c} = \boldsymbol{\ell}_1^T (\mathbf{W}, \mathbf{X})\mathbf{a}_1 + \boldsymbol{\ell}_2^T \mathbf{X}^T \mathbf{a}_2 = \mathbf{0}^T \mathbf{a}_1 + \mathbf{0}^T \mathbf{a}_2 = 0.$$

This establishes the compatibility (and hence consistency) of the BLUP equations. □

The next theorem is an extension of the Gauss–Markov theorem to the prediction setting. It establishes that the aforementioned necessary condition for the prediction error variance to attain a minimum is also sufficient.

Theorem 13.1.4 *Under a prediction-extended Aitken model* $\left\{\begin{pmatrix}\mathbf{y}\\ u\end{pmatrix}, \begin{pmatrix}\mathbf{X}\boldsymbol{\beta}\\ 0\end{pmatrix}, \sigma^2\begin{pmatrix}\mathbf{W} & \mathbf{k}\\ \mathbf{k}^T & h\end{pmatrix}\right\}$ *for which* $\mathbf{k} \in \mathcal{C}(\mathbf{W}, \mathbf{X})$, $t_0 + \mathbf{t}^T\mathbf{y}$ *is a BLUP of a predictable function* $\mathbf{c}^T\boldsymbol{\beta} + u$ *if and only if* $t_0 = 0$ *and* $\mathbf{t} = \tilde{\mathbf{t}}$, *where* $\tilde{\mathbf{t}}$ *is the first* n*-component subvector of a solution to the BLUP equations for* $\mathbf{c}^T\boldsymbol{\beta} + u$.

Proof Let $(\tilde{\mathbf{t}}^T, \tilde{\boldsymbol{\lambda}}^T)^T$ be any solution to the BLUP equations for $\mathbf{c}^T\boldsymbol{\beta} + u$ (such a solution exists by Theorem 13.1.3). Then $\mathbf{X}^T\tilde{\mathbf{t}} = \mathbf{c}$, implying (by Theorem 13.1.1) that if $t_0 = 0$ then $t_0 + \tilde{\mathbf{t}}^T\mathbf{y}$ is an unbiased predictor of $\mathbf{c}^T\boldsymbol{\beta} + u$. Next, let $\ell_0 + \boldsymbol{\ell}^T\mathbf{y}$ be any linear unbiased predictor of $\mathbf{c}^T\boldsymbol{\beta} + u$. Then by Theorem 13.1.1, $\ell_0 = 0$ and $\boldsymbol{\ell}^T\mathbf{X} = \mathbf{c}^T$. Furthermore,

$$\begin{aligned}
\mathrm{cov}[\tilde{\mathbf{t}}^T\mathbf{y} - u, (\boldsymbol{\ell} - \tilde{\mathbf{t}})^T\mathbf{y}] &= \mathrm{cov}[\tilde{\mathbf{t}}^T\mathbf{y}, (\boldsymbol{\ell} - \tilde{\mathbf{t}})^T\mathbf{y})] - \mathrm{cov}[u, (\boldsymbol{\ell} - \tilde{\mathbf{t}})^T\mathbf{y}]\\
&= \sigma^2[\tilde{\mathbf{t}}^T\mathbf{W}(\boldsymbol{\ell} - \tilde{\mathbf{t}}) - \mathbf{k}^T(\boldsymbol{\ell} - \tilde{\mathbf{t}})]\\
&= \sigma^2(\tilde{\mathbf{t}}^T\mathbf{W} - \mathbf{k}^T)(\boldsymbol{\ell} - \tilde{\mathbf{t}})\\
&= \sigma^2(\mathbf{k}^T - \tilde{\boldsymbol{\lambda}}^T\mathbf{X}^T - \mathbf{k}^T)(\boldsymbol{\ell} - \tilde{\mathbf{t}})\\
&= -\sigma^2\tilde{\boldsymbol{\lambda}}^T(\mathbf{X}^T\boldsymbol{\ell} - \mathbf{X}^T\tilde{\mathbf{t}})\\
&= -\sigma^2\tilde{\boldsymbol{\lambda}}^T(\mathbf{c} - \mathbf{c})\\
&= 0,
\end{aligned}$$

so that

$$\begin{aligned}
\mathrm{var}(\boldsymbol{\ell}^T\mathbf{y} - u) &= \mathrm{var}[\tilde{\mathbf{t}}^T\mathbf{y} + (\boldsymbol{\ell} - \tilde{\mathbf{t}})^T\mathbf{y} - u]\\
&= \mathrm{var}(\tilde{\mathbf{t}}^T\mathbf{y} - u) + \mathrm{var}[(\boldsymbol{\ell} - \tilde{\mathbf{t}})^T\mathbf{y}]
\end{aligned}$$

$$= \text{var}(\tilde{\mathbf{t}}^T\mathbf{y} - u) + \sigma^2(\boldsymbol{\ell} - \tilde{\mathbf{t}})^T\mathbf{W}(\boldsymbol{\ell} - \tilde{\mathbf{t}})$$
$$= \text{var}(\tilde{\mathbf{t}}^T\mathbf{y} - u) + \sigma^2(\boldsymbol{\ell} - \tilde{\mathbf{t}})^T\mathbf{W}^{\frac{1}{2}}\mathbf{W}^{\frac{1}{2}}(\boldsymbol{\ell} - \tilde{\mathbf{t}}).$$

Thus, $\text{var}(\boldsymbol{\ell}^T\mathbf{y}-u) \geq \text{var}(\tilde{\mathbf{t}}^T\mathbf{y}-u)$, with equality holding if and only if $\mathbf{W}^{\frac{1}{2}}(\boldsymbol{\ell}-\tilde{\mathbf{t}}) = \mathbf{0}$, i.e., if and only if $\mathbf{W}(\boldsymbol{\ell} - \tilde{\mathbf{t}}) = \mathbf{0}$. i.e., if and only if $\mathbf{W}\boldsymbol{\ell} = \mathbf{W}\tilde{\mathbf{t}}$, i.e., if and only if $(\boldsymbol{\ell}^T, \tilde{\boldsymbol{\lambda}}^T)^T$ is also a solution to the BLUP equations for $\mathbf{c}^T\boldsymbol{\beta} + u$. □

Corollary 13.1.4.1 *Under a prediction-extended Aitken model* $\left\{\begin{pmatrix}\mathbf{y}\\ u\end{pmatrix}, \begin{pmatrix}\mathbf{X}\boldsymbol{\beta}\\ 0\end{pmatrix}, \sigma^2\begin{pmatrix}\mathbf{W} & \mathbf{k}\\ \mathbf{k}^T & h\end{pmatrix}\right\}$ *for which* $\mathbf{k} \in \mathcal{C}(\mathbf{W}, \mathbf{X})$, $t_0 + \mathbf{t}^T\mathbf{y}$ *is a BLUP of a predictable function* $\mathbf{c}^T\boldsymbol{\beta} + u$ *if and only if* $t_0 = 0$, $\mathbf{t}^T\mathbf{X} = \mathbf{c}^T$, *and* $\mathbf{W}\mathbf{t} - \mathbf{k} \in \mathcal{C}(\mathbf{X})$.

The proof of Theorem 13.1.4 reveals that BLUPs of a predictable function, like BLUEs of an estimable function, generally are not unique. However, also like BLUEs, they are "essentially unique," in the sense that any two BLUPs of a predictable function differ at most on a set of probability zero.

Often, we wish to predict a *vector* of predictable functions rather than just one. In that case, the prediction problem (in the context of an Aitken model) is that of predicting $\boldsymbol{\tau} = \mathbf{C}^T\boldsymbol{\beta} + \mathbf{u}$, where $\mathbf{u}$ is an s-vector of random variables satisfying

$$\text{E}(\mathbf{u}) = \mathbf{0}, \quad \text{var}(\mathbf{u}) = \sigma^2\mathbf{H}, \quad \text{cov}(\mathbf{y}, \mathbf{u}) = \sigma^2\mathbf{K}.$$

Here, $\mathbf{H}$ and $\mathbf{K}$ are $s \times s$ and $n \times s$ specified matrices, respectively, such that the matrix $\begin{pmatrix}\mathbf{W} & \mathbf{K}\\ \mathbf{K}^T & \mathbf{H}\end{pmatrix}$ is nonnegative definite. Using the shorthand notation introduced in Sect. 5.3, the model for this "multiple-predictand" scenario may be written as

$$\left\{\begin{pmatrix}\mathbf{y}\\ \mathbf{u}\end{pmatrix}, \begin{pmatrix}\mathbf{X}\boldsymbol{\beta}\\ \mathbf{0}\end{pmatrix}, \sigma^2\begin{pmatrix}\mathbf{W} & \mathbf{K}\\ \mathbf{K}^T & \mathbf{H}\end{pmatrix}\right\}. \tag{13.1}$$

By Definition 13.1.5, the BLUP of a vector of predictable functions $\mathbf{C}^T\boldsymbol{\beta} + \mathbf{u}$ is the vector of BLUPs of each scalar component of $\mathbf{C}^T\boldsymbol{\beta} + \mathbf{u}$. Thus, an expression for the BLUP of $\mathbf{C}^T\boldsymbol{\beta} + \mathbf{u}$ associated with the model (13.1) is given in the following additional corollary to Theorem 13.1.4.

Corollary 13.1.4.2 *Under a multiple-predictand prediction-extended Aitken model (13.1) for which* $\mathcal{C}(\mathbf{K}) \subseteq \mathcal{C}(\mathbf{W}, \mathbf{X})$, $\mathbf{t}_0 + \mathbf{T}^T\mathbf{y}$ *is a BLUP of the vector of predictable functions* $\mathbf{C}^T\boldsymbol{\beta} + \mathbf{u}$ *if and only if* $\mathbf{t}_0 = \mathbf{0}$ *and* $\mathbf{T} = \tilde{\mathbf{T}}$, *where the kth column of* $\tilde{\mathbf{T}}$ *is the first n-component subvector of a solution to the BLUP equations for the kth element of* $\mathbf{C}^T\boldsymbol{\beta} + \mathbf{u}$.

13.2 The Case of the Prediction-Extended Positive Definite Aitken Model

Now let us consider the case of the prediction-extended positive definite Aitken model for a single predictand given by

$$\left\{\begin{pmatrix}\mathbf{y}\\ u\end{pmatrix}, \begin{pmatrix}\mathbf{X}\boldsymbol{\beta}\\ 0\end{pmatrix}, \sigma^2\begin{pmatrix}\mathbf{W} & \mathbf{k}\\ \mathbf{k}^T & h\end{pmatrix}\right\},$$

which is used very often in practice. Because $\mathbf{W}$ is positive definite, $\mathbf{k} \in \mathcal{C}(\mathbf{W}, \mathbf{X})$ for every $\mathbf{k} \in \mathbb{R}^n$. Let $\mathbf{c}^T\boldsymbol{\beta}+u$ be a predictable function under this model. Furthermore, recall from Sect. 11.2 that

$$\begin{pmatrix}\mathbf{W}^{-1} - \mathbf{W}^{-1}\mathbf{X}(\mathbf{X}^T\mathbf{W}^{-1}\mathbf{X})^-\mathbf{X}^T\mathbf{W}^{-1} & \mathbf{W}^{-1}\mathbf{X}(\mathbf{X}^T\mathbf{W}^{-1}\mathbf{X})^-\\ (\mathbf{X}^T\mathbf{W}^{-1}\mathbf{X})^-\mathbf{X}^T\mathbf{W}^{-1} & -(\mathbf{X}^T\mathbf{W}^{-1}\mathbf{X})^-\end{pmatrix} \tag{13.2}$$

is a generalized inverse of

$$\begin{pmatrix}\mathbf{W} & \mathbf{X}\\ \mathbf{X}^T & \mathbf{0}\end{pmatrix}.$$

Thus, by Theorems 13.1.3 and 3.2.2 there is a solution to the BLUP equations for $\mathbf{c}^T\boldsymbol{\beta}+u$ that has

$$[\mathbf{W}^{-1} - \mathbf{W}^{-1}\mathbf{X}(\mathbf{X}^T\mathbf{W}^{-1}\mathbf{X})^-\mathbf{X}^T\mathbf{W}^{-1}]\mathbf{k} + \mathbf{W}^{-1}\mathbf{X}(\mathbf{X}^T\mathbf{W}^{-1}\mathbf{X})^-\mathbf{c}$$

as its first n-component subvector. The BLUP corresponding to that solution is

$$\{\mathbf{c}^T(\mathbf{X}^T\mathbf{W}^{-1}\mathbf{X})^-\mathbf{X}^T\mathbf{W}^{-1} + \mathbf{k}^T[\mathbf{W}^{-1} - \mathbf{W}^{-1}\mathbf{X}(\mathbf{X}^T\mathbf{W}^{-1}\mathbf{X})^-\mathbf{X}^T\mathbf{W}^{-1}]^T\}\mathbf{y},$$

or equivalently,

$$\mathbf{c}^T\tilde{\boldsymbol{\beta}} + \tilde{u} \quad \text{where } \tilde{u} = \mathbf{k}^T\mathbf{E}\mathbf{y} = \mathbf{k}^T\mathbf{W}^{-1}\tilde{\mathbf{e}}, \tag{13.3}$$

$\mathbf{E} = \mathbf{W}^{-1} - \mathbf{W}^{-1}\mathbf{X}(\mathbf{X}^T\mathbf{W}^{-1}\mathbf{X})^-\mathbf{X}^T\mathbf{W}^{-1}$, and $\tilde{\mathbf{e}} = \mathbf{y} - \mathbf{X}\tilde{\boldsymbol{\beta}}$. (Here we have used the facts that $\mathbf{c}^T[(\mathbf{X}^T\mathbf{W}^{-1}\mathbf{X})^-]^T\mathbf{X}^T = \mathbf{c}^T(\mathbf{X}^T\mathbf{W}^{-1}\mathbf{X})^-\mathbf{X}^T$ and $\mathbf{E}$ is symmetric, according to Theorem 11.1.6b, d.) In fact,

$$\tilde{\tau} = \mathbf{c}^T\tilde{\boldsymbol{\beta}} + \tilde{u}$$

is *the* BLUP of $\mathbf{c}^T\boldsymbol{\beta} + u$ in this case; there is no other. That this is so derives from the proof of Theorem 13.1.4, which revealed that for any two solutions $(\boldsymbol{\ell}^T, \tilde{\boldsymbol{\lambda}}^T)^T$ and $(\tilde{\mathbf{t}}^T, \tilde{\boldsymbol{\lambda}}^T)^T$ to the BLUP equations for $\mathbf{c}^T\boldsymbol{\beta} + u$, $\mathbf{W}\boldsymbol{\ell} = \mathbf{W}\tilde{\mathbf{t}}$ and vice versa. Premultiplying both sides by $\mathbf{W}^{-1}$ yields $\boldsymbol{\ell} = \tilde{\mathbf{t}}$.

By this development and Corollary 13.1.4.2, we have the following theorem.

Theorem 13.2.1 *Under the prediction-extended positive definite Aitken model*

$$\left\{\begin{pmatrix}\mathbf{y}\\ \mathbf{u}\end{pmatrix}, \begin{pmatrix}\mathbf{X}\boldsymbol{\beta}\\ \mathbf{0}\end{pmatrix}, \sigma^2\begin{pmatrix}\mathbf{W} & \mathbf{K}\\ \mathbf{K}^T & \mathbf{H}\end{pmatrix}\right\},$$

the BLUP of a vector $\boldsymbol{\tau} = \mathbf{C}^T\boldsymbol{\beta} + \mathbf{u}$ *of predictable functions is* $\tilde{\boldsymbol{\tau}} = \mathbf{C}^T\tilde{\boldsymbol{\beta}} + \tilde{\mathbf{u}}$, *where* $\tilde{\boldsymbol{\beta}}$ *is any solution to the Aitken equations and* $\tilde{\mathbf{u}} = \mathbf{K}^T\mathbf{E}\mathbf{y} = \mathbf{K}^T\mathbf{W}^{-1}\tilde{\mathbf{e}}$.

The following corollary to Theorem 13.2.1 is analogous to Corollary 7.1.4.1. Its proof is left as an exercise.

Corollary 13.2.1.1 *Under the prediction-extended positive definite Aitken model described in Theorem 13.2.1, if* $\mathbf{C}^T\boldsymbol{\beta} + \mathbf{u}$ *is a vector of predictable functions and* $\mathbf{L}^T$ *is a matrix of constants having the same number of columns as* $\mathbf{C}^T$ *has rows, then* $\mathbf{L}^T(\mathbf{C}^T\boldsymbol{\beta} + \mathbf{u})$ *is predictable and its BLUP is* $\mathbf{L}^T(\mathbf{C}^T\tilde{\boldsymbol{\beta}} + \tilde{\mathbf{u}})$.

Next, we obtain an expression for the variance–covariance matrix of prediction errors associated with the BLUP of a vector of predictable functions.

Theorem 13.2.2 *Let* $\tilde{\boldsymbol{\tau}} = \mathbf{C}^T\tilde{\boldsymbol{\beta}} + \tilde{\mathbf{u}}$ *be the BLUP of an s-vector of predictable functions* $\mathbf{C}^T\boldsymbol{\beta} + \mathbf{u}$ *under the prediction-extended positive definite Aitken model described in Theorem 13.2.1. Then:*

(a) The variance–covariance matrix of prediction errors associated with $\tilde{\boldsymbol{\tau}}$ *is*

$$\begin{aligned} var(\tilde{\boldsymbol{\tau}} - \boldsymbol{\tau}) &= \sigma^2[\mathbf{H} - \mathbf{K}^T\mathbf{W}^{-1}\mathbf{K} + (\mathbf{C}^T - \mathbf{K}^T\mathbf{W}^{-1}\mathbf{X})(\mathbf{X}^T\mathbf{W}^{-1}\mathbf{X})^-(\mathbf{C}^T - \mathbf{K}^T\mathbf{W}^{-1}\mathbf{X})^T] \\ &\equiv \sigma^2\mathbf{Q}, \end{aligned}$$

say, where $\mathbf{Q}$ *is invariant to the choice of generalized inverse of* $\mathbf{X}^T\mathbf{W}^{-1}\mathbf{X}$;

(b) $\mathbf{Q}$ *is positive definite (hence nonsingular) if either* $\begin{pmatrix}\mathbf{W} & \mathbf{K}\\ \mathbf{K}^T & \mathbf{H}\end{pmatrix}$ *is positive definite or* $rank(\mathbf{C}^T - \mathbf{K}^T\mathbf{W}^{-1}\mathbf{X}) = s$.

Proof By Theorem 4.2.2b,

$$\begin{aligned} &\mathrm{var}(\tilde{\boldsymbol{\tau}} - \boldsymbol{\tau}) \\ &= \mathrm{var}[(\mathbf{C}^T\tilde{\boldsymbol{\beta}} + \tilde{\mathbf{u}}) - (\mathbf{C}^T\boldsymbol{\beta} + \mathbf{u})] \\ &= \mathrm{var}(\mathbf{C}^T\tilde{\boldsymbol{\beta}}) + \mathrm{var}(\tilde{\mathbf{u}}) + \mathrm{var}(\mathbf{u}) + \mathrm{cov}(\mathbf{C}^T\tilde{\boldsymbol{\beta}}, \tilde{\mathbf{u}}) + [\mathrm{cov}(\mathbf{C}^T\tilde{\boldsymbol{\beta}}, \tilde{\mathbf{u}})]^T - \mathrm{cov}(\mathbf{C}^T\tilde{\boldsymbol{\beta}}, \mathbf{u}) \\ &\quad - [\mathrm{cov}(\mathbf{C}^T\tilde{\boldsymbol{\beta}}, \mathbf{u})]^T - \mathrm{cov}(\tilde{\mathbf{u}}, \mathbf{u}) - [\mathrm{cov}(\tilde{\mathbf{u}}, \mathbf{u})]^T. \end{aligned}$$

Now $\mathrm{var}(\mathbf{C}^T\tilde{\boldsymbol{\beta}}) = \sigma^2\mathbf{C}^T(\mathbf{X}^T\mathbf{W}^{-1}\mathbf{X})^-\mathbf{C}$ by Theorem 11.1.3 and $\mathrm{var}(\mathbf{u}) = \sigma^2\mathbf{H}$ by definition. Letting $\mathbf{A}$ represent any matrix such that $\mathbf{C}^T = \mathbf{A}^T\mathbf{X}$, the remaining terms in this expression may be evaluated as follows:

$$\begin{aligned}\mathrm{var}(\tilde{\mathbf{u}}) &= \mathrm{var}(\mathbf{K}^T\mathbf{W}^{-1}\tilde{\mathbf{e}}) \\ &= \mathbf{K}^T\mathbf{W}^{-1}\{\sigma^2[\mathbf{W} - \mathbf{X}(\mathbf{X}^T\mathbf{W}^{-1}\mathbf{X})^-\mathbf{X}^T]\}\mathbf{W}^{-1}\mathbf{K} \\ &= \sigma^2[\mathbf{K}^T\mathbf{W}^{-1}\mathbf{K} - \mathbf{K}^T\mathbf{W}^{-1}\mathbf{X}(\mathbf{X}^T\mathbf{W}^{-1}\mathbf{X})^-\mathbf{X}^T\mathbf{W}^{-1}\mathbf{K}]\end{aligned}$$

where we used Theorem 11.1.8d for the second equality;

$$\begin{aligned}\mathrm{cov}(\mathbf{C}^T\tilde{\boldsymbol{\beta}}, \tilde{\mathbf{u}}) &= \mathrm{cov}(\mathbf{A}^T\mathbf{X}\tilde{\boldsymbol{\beta}}, \mathbf{K}^T\mathbf{W}^{-1}\tilde{\mathbf{e}}) \\ &= \mathrm{cov}(\mathbf{A}^T\tilde{\mathbf{y}}, \mathbf{K}^T\mathbf{W}^{-1}\tilde{\mathbf{e}}) \\ &= \mathbf{A}^T\mathrm{cov}(\tilde{\mathbf{y}}, \tilde{\mathbf{e}})\mathbf{W}^{-1}\mathbf{K} \\ &= \mathbf{0}\end{aligned}$$

where we used Theorem 11.1.8e for the last equality;

$$\begin{aligned}\mathrm{cov}(\mathbf{C}^T\tilde{\boldsymbol{\beta}}, \mathbf{u}) &= \mathrm{cov}[\mathbf{C}^T(\mathbf{X}^T\mathbf{W}^{-1}\mathbf{X})^-\mathbf{X}^T\mathbf{W}^{-1}\mathbf{y}, \mathbf{u}] \\ &= \sigma^2\mathbf{C}^T(\mathbf{X}^T\mathbf{W}^{-1}\mathbf{X})^-\mathbf{X}^T\mathbf{W}^{-1}\mathbf{K};\end{aligned}$$

and

$$\begin{aligned}\mathrm{cov}(\tilde{\mathbf{u}}, \mathbf{u}) &= \mathrm{cov}[\mathbf{K}^T\mathbf{W}^{-1}(\mathbf{y} - \mathbf{X}\tilde{\boldsymbol{\beta}}), \mathbf{u}] \\ &= \mathrm{cov}\{[\mathbf{K}^T\mathbf{W}^{-1} - \mathbf{K}^T\mathbf{W}^{-1}\mathbf{X}(\mathbf{X}^T\mathbf{W}^{-1}\mathbf{X})^-\mathbf{X}^T\mathbf{W}^{-1}]\mathbf{y}, \mathbf{u}\} \\ &= \sigma^2[\mathbf{K}^T\mathbf{W}^{-1}\mathbf{K} - \mathbf{K}^T\mathbf{W}^{-1}\mathbf{X}(\mathbf{X}^T\mathbf{W}^{-1}\mathbf{X})^-\mathbf{X}^T\mathbf{W}^{-1}\mathbf{K}].\end{aligned}$$

Substituting all of these results into the expression for $\mathrm{var}(\tilde{\boldsymbol{\tau}} - \boldsymbol{\tau})$ yields

$$\begin{aligned}\mathrm{var}(\tilde{\boldsymbol{\tau}} - \boldsymbol{\tau}) &= \sigma^2\{\mathbf{C}^T(\mathbf{X}^T\mathbf{W}^{-1}\mathbf{X})^-\mathbf{C} + \mathbf{K}^T\mathbf{W}^{-1}\mathbf{K} - \mathbf{K}^T\mathbf{W}^{-1}\mathbf{X}(\mathbf{X}^T\mathbf{W}^{-1}\mathbf{X})^-\mathbf{X}^T\mathbf{W}^{-1}\mathbf{K} + \mathbf{H} \\ &\quad - \mathbf{C}^T(\mathbf{X}^T\mathbf{W}^{-1}\mathbf{X})^-\mathbf{X}^T\mathbf{W}^{-1}\mathbf{K} - \mathbf{K}^T\mathbf{W}^{-1}\mathbf{X}[(\mathbf{X}^T\mathbf{W}^{-1}\mathbf{X})^-]^T\mathbf{C} \\ &\quad - [\mathbf{K}^T\mathbf{W}^{-1}\mathbf{K} - \mathbf{K}^T\mathbf{W}^{-1}\mathbf{X}(\mathbf{X}^T\mathbf{W}^{-1}\mathbf{X})^-\mathbf{X}^T\mathbf{W}^{-1}\mathbf{K}] \\ &\quad - \{\mathbf{K}^T\mathbf{W}^{-1}\mathbf{K} - \mathbf{K}^T\mathbf{W}^{-1}\mathbf{X}[(\mathbf{X}^T\mathbf{W}^{-1}\mathbf{X})^-]^T\mathbf{X}^T\mathbf{W}^{-1}\mathbf{K}\} \\ &= \sigma^2[\mathbf{H} - \mathbf{K}^T\mathbf{W}^{-1}\mathbf{K} + (\mathbf{C}^T - \mathbf{K}^T\mathbf{W}^{-1}\mathbf{X})(\mathbf{X}^T\mathbf{W}^{-1}\mathbf{X})^-(\mathbf{C}^T - \mathbf{K}^T\mathbf{W}^{-1}\mathbf{X})^T].\end{aligned}$$

The invariance of this expression with respect to the choice of generalized inverse of $\mathbf{X}^T\mathbf{W}^{-1}\mathbf{X}$ follows from Theorem 11.1.6b and the predictability of $\mathbf{C}^T\boldsymbol{\beta} + \mathbf{u}$. This completes the proof of part (a).

For part (b), suppose first that $\begin{pmatrix} \mathbf{W} & \mathbf{K} \\ \mathbf{K}^T & \mathbf{H} \end{pmatrix}$ is positive definite. Then, it follows from Theorem 2.15.7a that $\mathbf{H} - \mathbf{K}^T\mathbf{W}^{-1}\mathbf{K}$ is positive definite. Now again letting $\mathbf{A}$ represent any matrix such that $\mathbf{C}^T = \mathbf{A}^T\mathbf{X}$, we have

$$\begin{aligned}
&(\mathbf{C}^T - \mathbf{K}^T\mathbf{W}^{-1}\mathbf{X})(\mathbf{X}^T\mathbf{W}^{-1}\mathbf{X})^-(\mathbf{C}^T - \mathbf{K}^T\mathbf{W}^{-1}\mathbf{X})^T \\
&= (\mathbf{A}^T - \mathbf{K}^T\mathbf{W}^{-1})\mathbf{X}(\mathbf{X}^T\mathbf{W}^{-1}\mathbf{X})^-\mathbf{X}^T(\mathbf{A}^T - \mathbf{K}^T\mathbf{W}^{-1})^T \\
&= (\mathbf{A}^T - \mathbf{K}^T\mathbf{W}^{-1})\mathbf{W}^{\frac{1}{2}}\mathbf{W}^{-\frac{1}{2}}\mathbf{X}(\mathbf{X}^T\mathbf{W}^{-1}\mathbf{X})^-\mathbf{X}^T\mathbf{W}^{-\frac{1}{2}}\mathbf{W}^{\frac{1}{2}}(\mathbf{A}^T - \mathbf{K}^T\mathbf{W}^{-1})^T \\
&= (\mathbf{A}^T - \mathbf{K}^T\mathbf{W}^{-1})\mathbf{W}^{\frac{1}{2}}\mathbf{P}_{\mathbf{W}^{-\frac{1}{2}}\mathbf{X}}\mathbf{W}^{\frac{1}{2}}(\mathbf{A}^T - \mathbf{K}^T\mathbf{W}^{-1})^T \\
&= (\mathbf{A}^T - \mathbf{K}^T\mathbf{W}^{-1})\mathbf{W}^{\frac{1}{2}}\mathbf{P}^T_{\mathbf{W}^{-\frac{1}{2}}\mathbf{X}}\mathbf{P}_{\mathbf{W}^{-\frac{1}{2}}\mathbf{X}}\mathbf{W}^{\frac{1}{2}}(\mathbf{A}^T - \mathbf{K}^T\mathbf{W}^{-1})^T, \qquad (13.4)
\end{aligned}$$

which is of the form $\mathbf{M}^T\mathbf{M}$ for a suitably chosen matrix $\mathbf{M}$ and therefore is nonnegative definite by Theorem 2.15.9. The positive definiteness of $\mathbf{Q}$ then follows by Theorem 2.15.3.

Alternatively, instead of supposing that $\begin{pmatrix} \mathbf{W} & \mathbf{K} \\ \mathbf{K}^T & \mathbf{H} \end{pmatrix}$ is positive definite, suppose that $\text{rank}(\mathbf{C}^T - \mathbf{K}^T\mathbf{W}^{-1}\mathbf{X}) = s$. Then

$$\begin{aligned}
s &= \text{rank}[(\mathbf{A}^T - \mathbf{K}^T\mathbf{W}^{-1})\mathbf{X}] \\
&= \text{rank}[(\mathbf{A}^T - \mathbf{K}^T\mathbf{W}^{-1})\mathbf{X}(\mathbf{X}^T\mathbf{W}^{-1}\mathbf{X})^-\mathbf{X}^T\mathbf{W}^{-1}\mathbf{X}] \\
&\leq \text{rank}[(\mathbf{A}^T - \mathbf{K}^T\mathbf{W}^{-1})\mathbf{X}(\mathbf{X}^T\mathbf{W}^{-1}\mathbf{X})^-\mathbf{X}^T\mathbf{W}^{-\frac{1}{2}}] \\
&= \text{rank}[(\mathbf{A}^T - \mathbf{K}^T\mathbf{W}^{-1})\mathbf{W}^{\frac{1}{2}}\mathbf{P}_{\mathbf{W}^{-\frac{1}{2}}\mathbf{X}}] \\
&\leq s
\end{aligned}$$

where the first inequality holds by Theorem 2.8.4 and the last inequality holds because $(\mathbf{A}^T - \mathbf{K}^T\mathbf{W}^{-1})\mathbf{W}^{\frac{1}{2}}\mathbf{P}_{\mathbf{W}^{-\frac{1}{2}}\mathbf{X}}$ has s rows. Thus, $\text{rank}[(\mathbf{A}^T - \mathbf{K}^T\mathbf{W}^{-1})\mathbf{W}^{\frac{1}{2}}\mathbf{P}_{\mathbf{W}^{-\frac{1}{2}}\mathbf{X}}] = s$. This, together with (13.4) and Theorem 2.15.9, implies that $(\mathbf{C}^T - \mathbf{K}^T\mathbf{W}^{-1}\mathbf{X})(\mathbf{X}^T\mathbf{W}^{-1}\mathbf{X})^-(\mathbf{C}^T - \mathbf{K}^T\mathbf{W}^{-1}\mathbf{X})^T$ is positive definite. Furthermore, $\mathbf{H} - \mathbf{K}^T\mathbf{W}^{-1}\mathbf{K}$ is nonnegative definite by Theorem 2.15.7b. The positive definiteness of $\mathbf{Q}$ then follows by Theorem 2.15.3. □

Assume now that $n > p^*$. Then recall from Corollary 11.1.9.1 that

$$\tilde{\sigma}^2 = \frac{(\mathbf{y} - \mathbf{X}\tilde{\boldsymbol{\beta}})^T\mathbf{W}^{-1}(\mathbf{y} - \mathbf{X}\tilde{\boldsymbol{\beta}})}{n - p^*} \qquad (13.5)$$

is an unbiased estimator of σ^2 under the positive definite Aitken model $\{\mathbf{y}, \mathbf{X}\boldsymbol{\beta}, \sigma^2\mathbf{W}\}$. By substituting $\tilde{\sigma}^2$ for σ^2 in the expression for $\text{var}(\tilde{\boldsymbol{\tau}} - \boldsymbol{\tau})$ in Theorem 13.2.2, we obtain an unbiased estimator of $\text{var}(\tilde{\boldsymbol{\tau}} - \boldsymbol{\tau})$, which can be written as $\tilde{\sigma}^2\mathbf{Q}$, under the prediction-extended positive definite Aitken model $\left\{\begin{pmatrix}\mathbf{y}\\\mathbf{u}\end{pmatrix}, \begin{pmatrix}\mathbf{X}\boldsymbol{\beta}\\\mathbf{0}\end{pmatrix}, \sigma^2\begin{pmatrix}\mathbf{W} & \mathbf{K}\\\mathbf{K}^T & \mathbf{H}\end{pmatrix}\right\}$.

Theorem 13.2.3 *Under a prediction-extended positive definite Aitken model for which $n > p^*$ and the skewness matrix of the joint distribution of* $\mathbf{y}$ *and* $\mathbf{u}$ *is null,* $\text{cov}(\tilde{\boldsymbol{\tau}} - \boldsymbol{\tau}, \tilde{\sigma}^2) = \mathbf{0}$.

Proof Left as an exercise. □

Example 13.2-1. Prediction of a "New" Observation in Full-Rank Multiple and Simple Linear Regression

Consider the problem of predicting a "new" observation in classical full-rank multiple linear regression, which was described briefly in the preamble of this chapter. That is, suppose that we wish to predict $\tau = y_{n+1} = \mathbf{x}_{n+1}^T\boldsymbol{\beta} + e_{n+1}$ where $\mathbf{x}_{n+1}$ is a specified value of the vector of explanatory variables, y_{n+1} is the new response to be observed at $\mathbf{x} = \mathbf{x}_{n+1}$, and e_{n+1} is the corresponding residual; in addition, assume that $\text{var}(e_{n+1}) = \sigma^2$ and $\text{cov}(\mathbf{y}, e_{n+1}) = \mathbf{0}$, so that $\mathbf{y}$ and y_{n+1} jointly satisfy Gauss–Markov assumptions. Plainly, using (13.3) the BLUP of y_{n+1} is $\mathbf{x}_{n+1}^T\hat{\boldsymbol{\beta}}$, where $\hat{\boldsymbol{\beta}}$ is any solution to the normal equations. Furthermore, by Theorem 13.2.2 the variance of the BLUP's prediction error is

$$\text{var}(\mathbf{x}_{n+1}^T\hat{\boldsymbol{\beta}} - y_{n+1}) = \sigma^2[1 + \mathbf{x}_{n+1}^T(\mathbf{X}^T\mathbf{X})^{-1}\mathbf{x}_{n+1}].$$

Thus, the BLUP of a new observation is identical to the ordinary least squares fitted value at the corresponding combination of the explanatory variables, but its prediction error variance is larger than the variance of the corresponding ordinary least squares fitted value by an amount equal to σ^2. In the special case of simple linear regression, the BLUP of y_{n+1} is $\hat{\beta}_1 + \hat{\beta}_2 x_{n+1}$ and, using results from Example 8.1-2, the prediction error variance is $\sigma^2[1 + (1/n) + (x_{n+1} - \bar{x})^2/SXX]$. ■

Example 13.2-2. Forecasting Under a First-Order Autoregressive Model

Consider once again the first-order stationary autoregressive model introduced in Example 5.2.4-1 and featured also in Example 11.1-2. Now consider the problem of predicting, or *forecasting*, the response variable t units of time into the future. That is, suppose that we wish to predict y_{n+t}, where

$$\begin{pmatrix} \mathbf{y} \\ y_{n+t} \end{pmatrix} = \mu \mathbf{1}_{n+1} + \begin{pmatrix} \mathbf{e} \\ e_{n+t} \end{pmatrix},$$

$$\text{var}\begin{pmatrix} \mathbf{e} \\ e_{n+t} \end{pmatrix} = \sigma^2 \begin{pmatrix} \mathbf{W} & \mathbf{k} \\ \mathbf{k}^T & h \end{pmatrix},$$

$\mathbf{k}^T = [1/(1-\rho)](\rho^{n+t-1}, \rho^{n+t-2}, \ldots, \rho^t)$, $h = 1/(1-\rho)$, and $\mathbf{W}$ is the positive definite matrix specified in Example 11.1-2. By Theorem 13.2.1, the BLUP of y_{n+t} is

$$\tilde{\mu} + \mathbf{k}^T \mathbf{W}^{-1}(\mathbf{y} - \tilde{\mu}\mathbf{1}_n),$$

where an expression for $\tilde{\mu}$ was given in Example 11.1-2. Observe that $\mathbf{k}^T$ is equal to ρ^t times the last row of $\mathbf{W}$, implying that $\mathbf{k}^T \mathbf{W}^{-1} = \rho^t \mathbf{u}_n^T$. Hence, the BLUP of y_{n+t} may be expressed more simply as

$$\tilde{\mu} + \rho^t(y_n - \tilde{\mu}) = (1-\rho^t)\tilde{\mu} + \rho^t y_n,$$

which reveals that the BLUP is a weighted linear combination of the BLUE of the overall mean and the last observation, with more weight given to the latter as ρ increases from 0 to 1. By Theorem 13.2.2, the prediction error variance of the BLUP is

$$\begin{aligned} &\sigma^2[h - \rho^t \mathbf{u}_n^T \mathbf{k} + (1 - \rho^t \mathbf{u}_n^T \mathbf{1})(\mathbf{1}^T \mathbf{W}^{-1}\mathbf{1})^{-1}(1 - \rho^t \mathbf{u}_n^T \mathbf{1})^T] \\ &= \left(\frac{\sigma^2}{1-\rho}\right)\left(1 - \rho^{2t} + \frac{(1-\rho^t)^2(1+\rho)}{n(1-\rho)+2\rho}\right). \end{aligned}$$

This last expression increases with t, which is to be expected: the further into the future that we forecast, the less reliable the prediction should be. ■

Example 13.2-3. Spatial Prediction (Kriging)

Suppose that $\mathbf{y}$ is a vector of observations of a response variable made at locations in space, specifically at the six points of a square grid in the plane that are circled in Fig. 13.1. Suppose further that there is also a value of the response variable, call it y_7, at the grid point marked by an asterisk, but that we have not observed this value and want to predict it. (In the spatial statistics literature, this prediction problem is known as *kriging*.) Assume that $(\mathbf{y}, y_7)^T$ has mean vector $\mu\mathbf{1}$ and variance–covariance matrix $\sigma^2 \begin{pmatrix} \mathbf{W} & \mathbf{k} \\ \mathbf{k}^T & h \end{pmatrix}$, with ijth element equal to $\exp(-d_{ij}/2)$ where d_{ij}

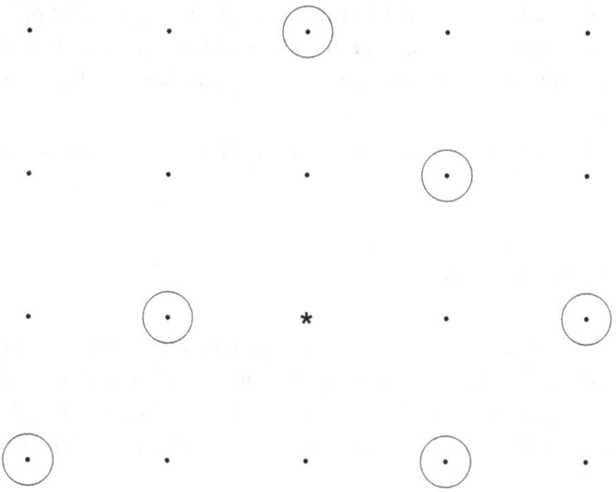

Fig. 13.1 Spatial configuration of data sites (circled dots) and prediction site (an asterisk) for Example 13.2-3

is the (Euclidean) distance between locations corresponding to variables i and j $(i, j = 1, \dots, 7)$ in units of the grid spacing. Thus the variance of the variable at each location is equal to σ^2 and the correlation between variables at any two locations is a decreasing function of the distance between the locations. Without loss of generality, label the six observations as $y_1, \dots, y_6$ from left to right, with the fourth and fifth observations defined so that the fourth's location is below that of the fifth. Then, it can be verified that

$$\mathbf{W} = \begin{pmatrix} 1.000 & 0.493 & 0.165 & 0.223 & 0.165 & 0.127 \\ 0.493 & 1.000 & 0.327 & 0.327 & 0.327 & 0.223 \\ 0.165 & 0.327 & 1.000 & 0.206 & 0.493 & 0.243 \\ 0.223 & 0.327 & 0.206 & 1.000 & 0.368 & 0.493 \\ 0.165 & 0.327 & 0.493 & 0.368 & 1.000 & 0.493 \\ 0.127 & 0.223 & 0.243 & 0.493 & 0.493 & 1.000 \end{pmatrix},$$

$$\mathbf{k}^T = (0.327\ 0.607\ 0.368\ 0.493\ 0.493\ 0.368),$$

and $h = 1.000$. Using a computer to invert $\mathbf{W}$ and perform other necessary calculations, it can also be verified that the BLUP of y_7 is

$$\tilde{y}_7 = 0.017y_1 + 0.422y_2 + 0.065y_3 + 0.247y_4 + 0.218y_5 + 0.030y_6,$$

with corresponding prediction error variance $\text{var}(\tilde{y}_7 - y_7) = 0.478\sigma^2$. Often, the closer in space an element of $\mathbf{y}$ is to y_7, the greater the corresponding coefficient, or "weight," for that element in the expression for the BLUP; for example, y_2 (which lies closest to y_7) has the largest coefficient. Intuitively, this result seems reasonable because those observations closest to the location of y_7 are more highly correlated with it. However, the correspondence between distance and weight is not perfect; for example, y_4 and y_5 are equidistant from y_7 but the weights corresponding to them are not equal.

Some further analysis of and variations on this example are considered in the exercises. ■

13.3 The General Case

For the sake of greater generality, we now characterize the collection of all BLUPs of an arbitrary predictable function under a general prediction-extended Aitken model. We also give an expression for the variance–covariance matrix of prediction errors associated with BLUPs of a vector of predictable functions under this model.

Theorem 13.3.1 *For the prediction-extended Aitken model* $\left\{\begin{pmatrix}\mathbf{y}\\ \mathbf{u}\end{pmatrix}, \begin{pmatrix}\mathbf{X}\boldsymbol{\beta}\\ \mathbf{0}\end{pmatrix}, \sigma^2\begin{pmatrix}\mathbf{W} & \mathbf{K}\\ \mathbf{K}^T & \mathbf{H}\end{pmatrix}\right\}$, *let* $\begin{pmatrix}\mathbf{G}_{11} & \mathbf{G}_{12}\\ \mathbf{G}_{21} & \mathbf{G}_{22}\end{pmatrix}$ *represent a generalized inverse of* $\begin{pmatrix}\mathbf{W} & \mathbf{X}\\ \mathbf{X}^T & \mathbf{0}\end{pmatrix}$, *let* $\boldsymbol{\tau} = \mathbf{C}^T\boldsymbol{\beta} + \mathbf{u}$ *be an s-vector of predictable functions, and suppose that* $\mathcal{C}(\mathbf{K}) \subseteq \mathcal{C}(\mathbf{W}, \mathbf{X})$. *Then:*

(a) the collection of BLUPs of $\boldsymbol{\tau}$ *consists of all quantities of the form* $\tilde{\mathbf{T}}^T\mathbf{y}$, *where*

$$\tilde{\mathbf{T}} = \mathbf{G}_{12}\mathbf{C} + \mathbf{G}_{11}\mathbf{K} + (\mathbf{I}_n - \mathbf{G}_{11}\mathbf{W} - \mathbf{G}_{12}\mathbf{X}^T,\ -\mathbf{G}_{11}\mathbf{X})\mathbf{Z},$$

where $\mathbf{Z}$ *is an arbitrary* $(n + p) \times s$ *matrix. In particular,* $(\mathbf{C}^T\mathbf{G}_{12}^T + \mathbf{K}^T\mathbf{G}_{11}^T)\mathbf{y}$ *is a BLUP of* $\boldsymbol{\tau}$*; and*

(b) the variance–covariance matrix of prediction errors associated with a vector of BLUPs $\tilde{\boldsymbol{\tau}} = (\mathbf{C}^T\mathbf{G}_{12}^T + \mathbf{K}^T\mathbf{G}_{11}^T)\mathbf{y}$ *of a vector of predictable functions* $\boldsymbol{\tau} = \mathbf{C}^T\boldsymbol{\beta} + \mathbf{u}$ *is*

$$var(\tilde{\boldsymbol{\tau}} - \boldsymbol{\tau}) = \sigma^2\left[\mathbf{H} - (\mathbf{K}^T, \mathbf{C}^T)\begin{pmatrix}\mathbf{G}_{11} & \mathbf{G}_{12}\\ \mathbf{G}_{21} & \mathbf{G}_{22}\end{pmatrix}\begin{pmatrix}\mathbf{K}\\ \mathbf{C}\end{pmatrix}\right],$$

invariant to the choice of generalized inverse.

Proof A proof of part (a) may be constructed in a manner similar to the proof of Theorem 11.2.3a and is left as an exercise. For part (b), we have

$$\begin{aligned}\text{var}(\tilde{\boldsymbol{\tau}} - \boldsymbol{\tau}) &= \text{var}[(\mathbf{C}^T\mathbf{G}_{12}^T + \mathbf{K}^T\mathbf{G}_{11}^T)\mathbf{y} - \mathbf{u}]\\ &= (\mathbf{C}^T\mathbf{G}_{12}^T + \mathbf{K}^T\mathbf{G}_{11}^T,\ -\mathbf{I})\sigma^2\begin{pmatrix}\mathbf{W} & \mathbf{K}\\ \mathbf{K}^T & \mathbf{H}\end{pmatrix}\begin{pmatrix}\mathbf{G}_{12}\mathbf{C} + \mathbf{G}_{11}\mathbf{K}\\ -\mathbf{I}\end{pmatrix}\end{aligned}$$

$$= \sigma^2(\mathbf{C}^T\mathbf{G}_{12}^T\mathbf{W}\mathbf{G}_{12}\mathbf{C} + \mathbf{K}^T\mathbf{G}_{11}^T\mathbf{W}\mathbf{G}_{12}\mathbf{C} + \mathbf{C}^T\mathbf{G}_{12}^T\mathbf{W}\mathbf{G}_{11}\mathbf{K} + \mathbf{K}^T\mathbf{G}_{11}^T\mathbf{W}\mathbf{G}_{11}\mathbf{K}$$
$$-\mathbf{K}^T\mathbf{G}_{12}\mathbf{C} - \mathbf{K}^T\mathbf{G}_{11}\mathbf{K} - \mathbf{C}^T\mathbf{G}_{12}^T\mathbf{K} - \mathbf{K}^T\mathbf{G}_{11}^T\mathbf{K} + \mathbf{H}).$$

Now letting $\mathbf{A}$ represent any matrix such that $\mathbf{C}^T = \mathbf{A}^T\mathbf{X}$, we have

$$\begin{aligned}\mathbf{C}^T\mathbf{G}_{12}^T\mathbf{W}\mathbf{G}_{12}\mathbf{C} &= \mathbf{A}^T\mathbf{X}\mathbf{G}_{12}^T\mathbf{W}\mathbf{G}_{12}\mathbf{X}^T\mathbf{A} = \mathbf{A}^T\mathbf{X}\mathbf{G}_{12}^T(-\mathbf{X}\mathbf{G}_{22}\mathbf{X}^T)\mathbf{A}\\ &= -\mathbf{A}^T\mathbf{X}\mathbf{G}_{22}\mathbf{X}^T\mathbf{A} = -\mathbf{C}^T\mathbf{G}_{22}\mathbf{C},\end{aligned}$$

where we have used Theorem 3.3.8a, b (with $\mathbf{W}$ and $\mathbf{X}$ here playing the role of $\mathbf{A}$ and $\mathbf{B}$ in the theorem). Letting $\mathbf{F}$ and $\mathbf{L}$ represent matrices such that $\mathbf{K} = \mathbf{W}\mathbf{F} + \mathbf{X}\mathbf{L}$, by Theorem 3.3.8b, c, f we also have

$$\begin{aligned}\mathbf{K}^T\mathbf{G}_{11}^T\mathbf{W}\mathbf{G}_{12}\mathbf{C} &= (\mathbf{F}^T\mathbf{W} + \mathbf{L}^T\mathbf{X}^T)\mathbf{G}_{11}^T\mathbf{W}\mathbf{G}_{12}\mathbf{X}^T\mathbf{A}\\ &= \mathbf{F}^T\mathbf{W}\mathbf{G}_{11}^T\mathbf{W}\mathbf{G}_{12}\mathbf{X}^T\mathbf{A}\\ &= \mathbf{F}^T\mathbf{W}\mathbf{G}_{11}\mathbf{W}\mathbf{G}_{12}\mathbf{X}^T\mathbf{A}\\ &= \mathbf{F}^T\mathbf{W}\mathbf{G}_{11}(-\mathbf{X}\mathbf{G}_{22}\mathbf{X}^T)\mathbf{A}\\ &= \mathbf{0}.\end{aligned}$$

It follows easily that $\mathbf{C}^T\mathbf{G}_{12}^T\mathbf{W}\mathbf{G}_{11}\mathbf{K} = \mathbf{0}$. Furthermore,

$$\begin{aligned}\mathbf{C}^T\mathbf{G}_{12}^T\mathbf{K} &= \mathbf{A}^T\mathbf{X}\mathbf{G}_{12}^T(\mathbf{W}\mathbf{F} + \mathbf{X}\mathbf{L})\\ &= \mathbf{A}^T\mathbf{X}\mathbf{G}_{12}^T\mathbf{W}\mathbf{F} + \mathbf{A}^T\mathbf{X}\mathbf{G}_{12}^T\mathbf{X}\mathbf{L}\\ &= \mathbf{A}^T(\mathbf{W}\mathbf{G}_{12}\mathbf{X}^T)^T\mathbf{F} + \mathbf{A}^T(\mathbf{X}^T\mathbf{G}_{12}\mathbf{X}^T)^T\mathbf{L}\\ &= \mathbf{A}^T\mathbf{W}\mathbf{G}_{12}\mathbf{X}^T\mathbf{F} + \mathbf{A}^T\mathbf{X}\mathbf{L}\\ &= \mathbf{A}^T\mathbf{X}\mathbf{G}_{21}\mathbf{W}\mathbf{F} + \mathbf{A}^T\mathbf{X}\mathbf{G}_{21}\mathbf{X}\mathbf{L}\\ &= \mathbf{C}^T\mathbf{G}_{21}(\mathbf{W}\mathbf{F} + \mathbf{X}\mathbf{L})\\ &= \mathbf{C}^T\mathbf{G}_{21}\mathbf{K},\end{aligned}$$

where we used Theorem 3.3.8a, f for the fourth equality and Theorem 3.3.8a, b for the fifth equality. Finally, by Theorem 3.3.8c, d,

$$\begin{aligned}\mathbf{K}^T\mathbf{G}_{11}^T\mathbf{W}\mathbf{G}_{11}\mathbf{K} - \mathbf{K}^T\mathbf{G}_{11}^T\mathbf{K} &= (\mathbf{F}^T\mathbf{W} + \mathbf{L}^T\mathbf{X}^T)\mathbf{G}_{11}^T\mathbf{W}\mathbf{G}_{11}(\mathbf{W}\mathbf{F} + \mathbf{X}\mathbf{L})\\ &\quad -(\mathbf{F}^T\mathbf{W} + \mathbf{L}^T\mathbf{X}^T)\mathbf{G}_{11}^T(\mathbf{W}\mathbf{F} + \mathbf{X}\mathbf{L})\\ &= \mathbf{F}^T\mathbf{W}\mathbf{G}_{11}^T\mathbf{W}\mathbf{G}_{11}\mathbf{W}\mathbf{F} - \mathbf{F}^T\mathbf{W}\mathbf{G}_{11}^T\mathbf{W}\mathbf{F}\end{aligned}$$

$$= \mathbf{F}^T\mathbf{W}\mathbf{G}_{11}^T(\mathbf{W}\mathbf{G}_{11}\mathbf{W} - \mathbf{W})\mathbf{F}$$
$$= \mathbf{F}^T\mathbf{W}\mathbf{G}_{11}^T(\mathbf{X}\mathbf{G}_{22}\mathbf{X}^T)\mathbf{F}$$
$$= \mathbf{0}.$$

Upon collecting terms, the given expression for $\text{var}(\tilde{\boldsymbol{\tau}} - \boldsymbol{\tau})$ is obtained. Establishing the invariance with respect to the choice of generalized inverse is left as an exercise. □

13.4 The Case of Positive Definite Mixed (and Random) Effects Models

In this section we consider BLUP for predictands in random effects models and mixed effects models for which the variance–covariance matrices of the random effects are positive definite. We give all results in terms of mixed effects models; however, because random effects models are a subclass of mixed effects models (with either $\mathbf{X}\boldsymbol{\beta} = \mathbf{0}$ or $\mathbf{X}\boldsymbol{\beta} = \mathbf{1}\mu$), all of the results obtained herein for mixed effects models specialize to random effects models as well.

13.4.1 BLUP (and BLUE)

Recall from Sect. 5.2.3 that the mixed effects model may be written as

$$\mathbf{y} = \mathbf{X}\boldsymbol{\beta} + \mathbf{Z}\mathbf{b} + \mathbf{d},$$

where $\mathbf{X}$ and $\boldsymbol{\beta}$ are defined as in a fixed effects model, $\mathbf{Z}$ is a specified $n \times q$ matrix, $\mathbf{b}$ is a q-vector of zero-mean random variables called *random effects* in this context, and $\mathbf{d}$ is an n-vector of zero-mean random variables. Furthermore, it is assumed that

$$\text{cov}(\mathbf{b}, \mathbf{d}) = \mathbf{0}, \qquad \text{var}(\mathbf{b}) = \sigma^2\mathbf{G}(\boldsymbol{\psi}), \qquad \text{var}(\mathbf{d}) = \sigma^2\mathbf{R}(\boldsymbol{\psi}),$$

where $\sigma^2 > 0$ and $\mathbf{R}(\boldsymbol{\psi})$ and $\mathbf{G}(\boldsymbol{\psi})$ are nonnegative definite or positive definite matrices of dimensions $n \times n$ and $q \times q$, respectively; throughout this section we assume that $\mathbf{R}(\boldsymbol{\psi})$ and $\mathbf{G}(\boldsymbol{\psi})$ are positive definite. In general, the elements of $\mathbf{R}(\boldsymbol{\psi})$ and $\mathbf{G}(\boldsymbol{\psi})$ are not fully specified, but instead are specified functions of an $(m-1)$-vector of unknown parameters $\boldsymbol{\psi}$. The parameter space for σ^2 and $\boldsymbol{\psi}$ is $\Theta \equiv \{\boldsymbol{\theta} = (\sigma^2, \boldsymbol{\psi}^T)^T : \sigma^2 > 0, \boldsymbol{\psi} \in \Psi\}$, where Ψ is a specified subset of $\mathbb{R}^{m-1}$ such that

$\mathbf{R}(\boldsymbol{\psi})$ and $\mathbf{G}(\boldsymbol{\psi})$ are positive definite for all $\boldsymbol{\psi} \in \Psi$. Under this model,

$$\text{var}(\mathbf{y}) = \text{var}(\mathbf{Zb} + \mathbf{d}) = \sigma^2[\mathbf{R}(\boldsymbol{\psi}) + \mathbf{ZG}(\boldsymbol{\psi})\mathbf{Z}^T],$$

which is positive definite but not a positive scalar multiple of a fully specified matrix unless $\boldsymbol{\psi}$ is known.

In conjunction with the mixed (or random) effects model, consider the problem of predicting an s-vector of predictable linear functions

$$\boldsymbol{\tau} = \mathbf{C}^T\boldsymbol{\beta} + \mathbf{F}^T\mathbf{b}.$$

Suppose, until noted otherwise, that $\boldsymbol{\psi}$ is known; the inference procedures that we obtain by proceeding as though $\boldsymbol{\psi}$ is known will suggest procedures that could be used when it is unknown. In this case we write $\mathbf{G} = \mathbf{G}(\boldsymbol{\psi})$ and $\mathbf{R} = \mathbf{R}(\boldsymbol{\psi})$. Then the problem of predicting $\boldsymbol{\tau}$ can be regarded as a special case of the general prediction problem considered in Sect. 13.2, i.e., that of predicting a (predictable) vector $\boldsymbol{\tau} = \mathbf{C}^T\boldsymbol{\beta} + \mathbf{u}$ under the prediction-extended positive definite Aitken model. Specifically, it is the special case where

$$\mathbf{e} = \mathbf{Zb} + \mathbf{d},$$

$$\mathbf{u} = \mathbf{F}^T\mathbf{b},$$

$$\mathbf{H} = (1/\sigma^2)\mathbf{F}^T[\text{var}(\mathbf{b})]\mathbf{F} = \mathbf{F}^T\mathbf{GF}, \tag{13.6}$$

$$\mathbf{K} = (1/\sigma^2)\text{cov}(\mathbf{Zb} + \mathbf{d},\ \mathbf{F}^T\mathbf{b}) = \mathbf{ZGF}, \tag{13.7}$$

$$\mathbf{W} = (1/\sigma^2)\text{var}(\mathbf{Zb} + \mathbf{d}) = \mathbf{R} + \mathbf{ZGZ}^T. \tag{13.8}$$

Theorem 13.4.1 *Let* $\boldsymbol{\tau} = \mathbf{C}^T\boldsymbol{\beta} + \mathbf{F}^T\mathbf{b}$ *be an s-vector of predictable linear functions. Then, under the positive definite mixed effects model with known* $\boldsymbol{\psi}$*:*

(a) the BLUP of $\boldsymbol{\tau}$ *is*

$$\tilde{\boldsymbol{\tau}} = \mathbf{C}^T\tilde{\boldsymbol{\beta}} + \mathbf{F}^T\tilde{\mathbf{b}},$$

where $\tilde{\boldsymbol{\beta}}$ *is any solution to the Aitken equations*

$$\mathbf{X}^T(\mathbf{R} + \mathbf{ZGZ}^T)^{-1}\mathbf{X}\boldsymbol{\beta} = \mathbf{X}^T(\mathbf{R} + \mathbf{ZGZ}^T)^{-1}\mathbf{y}$$

and

$$\tilde{\mathbf{b}} = \mathbf{GZ}^T(\mathbf{R} + \mathbf{ZGZ}^T)^{-1}\tilde{\mathbf{e}};$$

(b) *the variance–covariance matrix of the BLUP's prediction error vector, i.e., var*($\tilde{\boldsymbol{\tau}} - \boldsymbol{\tau}$), *is* $\sigma^2\mathbf{Q}$, *where*

$$\begin{aligned}\mathbf{Q} &= \mathbf{F}^T\mathbf{GF} - \mathbf{F}^T\mathbf{GZ}^T(\mathbf{R}+\mathbf{ZGZ}^T)^{-1}\mathbf{ZGF} \\ &\quad + [\mathbf{C}^T - \mathbf{F}^T\mathbf{GZ}^T(\mathbf{R}+\mathbf{ZGZ}^T)^{-1}\mathbf{X}][\mathbf{X}^T(\mathbf{R}+\mathbf{ZGZ}^T)^{-1}\mathbf{X}]^- \\ &\quad \times [\mathbf{C}^T - \mathbf{F}^T\mathbf{GZ}^T(\mathbf{R}+\mathbf{ZGZ}^T)^{-1}\mathbf{X}]^T,\end{aligned}$$

invariant to the choice of generalized inverse of $\mathbf{X}^T(\mathbf{R}+\mathbf{ZGZ}^T)^{-1}\mathbf{X}$; *and*

(c) $\mathbf{Q}$ *is positive definite if rank*($\mathbf{F}$) $= s$ *or rank*$[\mathbf{C}^T - \mathbf{F}^T\mathbf{GZ}^T(\mathbf{R}+\mathbf{ZGZ}^T)^{-1}\mathbf{X}] = s$; *and*

(d) *if* $n > p^*$, *then* $\tilde{\sigma}^2\mathbf{Q}$ *is an unbiased estimator of var*($\tilde{\boldsymbol{\tau}} - \boldsymbol{\tau}$), *where*

$$\tilde{\sigma}^2 = \frac{(\mathbf{y}-\mathbf{X}\tilde{\boldsymbol{\beta}})^T(\mathbf{R}+\mathbf{ZGZ}^T)^{-1}(\mathbf{y}-\mathbf{X}\tilde{\boldsymbol{\beta}})}{n-p^*}.$$

Proof Parts (a), (b), and (d) follow immediately upon substituting expressions (13.6)–(13.8) into Theorems 13.2.1 and 13.2.2a and expression (13.5); likewise the sufficiency of the second condition in part (c) follows immediately from that of the second condition in Theorem 13.2.2b. For the sufficiency of the first condition in part (c), observe that

$$\begin{aligned}\begin{pmatrix}\mathbf{W} & \mathbf{K}\\ \mathbf{K}^T & \mathbf{H}\end{pmatrix} &= \begin{pmatrix}\mathbf{R}+\mathbf{ZGZ}^T & \mathbf{ZGF}\\ \mathbf{F}^T\mathbf{GZ}^T & \mathbf{F}^T\mathbf{GF}\end{pmatrix}\\ &= \begin{pmatrix}\mathbf{ZG}^{\frac{1}{2}} & \mathbf{R}^{\frac{1}{2}}\\ \mathbf{F}^T\mathbf{G}^{\frac{1}{2}} & \mathbf{0}\end{pmatrix}\begin{pmatrix}\mathbf{ZG}^{\frac{1}{2}} & \mathbf{R}^{\frac{1}{2}}\\ \mathbf{F}^T\mathbf{G}^{\frac{1}{2}} & \mathbf{0}\end{pmatrix}^T.\end{aligned}$$

If rank($\mathbf{F}$) $= s$, then (by Theorem 2.8.9) rank($\mathbf{F}^T\mathbf{G}^{\frac{1}{2}}$) $= s$, implying that the rank of $\begin{pmatrix}\mathbf{ZG}^{\frac{1}{2}} & \mathbf{R}^{\frac{1}{2}}\\ \mathbf{F}^T\mathbf{G}^{\frac{1}{2}} & \mathbf{0}\end{pmatrix}$ is $n + s$, implying further (by Theorem 2.8.8) that the rank of $\begin{pmatrix}\mathbf{W} & \mathbf{K}\\ \mathbf{K}^T & \mathbf{H}\end{pmatrix}$ is $n + s$. Thus $\begin{pmatrix}\mathbf{W} & \mathbf{K}\\ \mathbf{K}^T & \mathbf{H}\end{pmatrix}$ is positive definite, by which the sufficiency of the first condition in part (c) follows from that of the first condition in Theorem 13.2.2b. □

Although Theorem 13.4.1 is couched in the terminology of prediction, it also yields results obtained previously for best linear unbiased estimation of a vector $\mathbf{C}^T\boldsymbol{\beta}$ of estimable functions merely by setting $\mathbf{F} = \mathbf{0}$ in the expressions.

13.4.2 More Computationally Efficient Representations

The expressions in Theorem 13.4.1 for the BLUP, its prediction error variance–covariance matrix, and an unbiased estimator of that matrix require the inversion of the positive definite matrix $\mathbf{R}+\mathbf{Z}\mathbf{G}\mathbf{Z}^T$, which has dimensions $n \times n$. Inversion of an arbitrary $n \times n$ positive definite matrix requires $O(n^3)$ computations, so if n is large, the computational requirements of this inversion could be prohibitive. However, the positive definite mixed effects model has enough additional structure that the inverse may be obtained more efficiently, provided that $\mathbf{R}$ is easy to invert and that q is much smaller than n (which is usually the case in practice). The computational efficiency is realized by applying Theorems 2.9.7 (for the inverse of the sum of two matrices) and 2.15.16. Specifically, using Theorem 2.9.7 with $\mathbf{R}$, $\mathbf{Z}\mathbf{G}$, $\mathbf{G}^{-1}$, and $\mathbf{G}\mathbf{Z}^T$ here playing the roles of $\mathbf{A}$, $\mathbf{B}$, $\mathbf{C}$, and $\mathbf{D}$ in the theorem, we find that

$$(\mathbf{R}+\mathbf{Z}\mathbf{G}\mathbf{Z}^T)^{-1} = \mathbf{R}^{-1} - \mathbf{R}^{-1}\mathbf{Z}(\mathbf{G}^{-1}+\mathbf{Z}^T\mathbf{R}^{-1}\mathbf{Z})^{-1}\mathbf{Z}^T\mathbf{R}^{-1}. \tag{13.9}$$

And, using Theorems 2.9.4 and 2.15.16 (with $\mathbf{G}$ and $\mathbf{R}^{-\frac{1}{2}}\mathbf{Z}$ here playing the roles of $\mathbf{A}$ and $\mathbf{B}$ in the latter theorem), we find that

$$\begin{aligned}
\mathbf{G}\mathbf{Z}^T(\mathbf{R}+\mathbf{Z}\mathbf{G}\mathbf{Z}^T)^{-1} &= \mathbf{G}\mathbf{Z}^T[\mathbf{R}^{\frac{1}{2}}(\mathbf{I}+\mathbf{R}^{-\frac{1}{2}}\mathbf{Z}\mathbf{G}\mathbf{Z}^T\mathbf{R}^{-\frac{1}{2}})\mathbf{R}^{\frac{1}{2}}]^{-1} \\
&= \mathbf{G}\mathbf{Z}^T\mathbf{R}^{-\frac{1}{2}}(\mathbf{I}+\mathbf{R}^{-\frac{1}{2}}\mathbf{Z}\mathbf{G}\mathbf{Z}^T\mathbf{R}^{-\frac{1}{2}})^{-1}\mathbf{R}^{-\frac{1}{2}} \\
&= (\mathbf{G}^{-1}+\mathbf{Z}^T\mathbf{R}^{-1}\mathbf{Z})^{-1}\mathbf{Z}^T\mathbf{R}^{-1}.
\end{aligned} \tag{13.10}$$

By (13.10) and the identity

$$(\mathbf{G}^{-1}+\mathbf{Z}^T\mathbf{R}^{-1}\mathbf{Z})^{-1}\mathbf{G}^{-1}+(\mathbf{G}^{-1}+\mathbf{Z}^T\mathbf{R}^{-1}\mathbf{Z})^{-1}\mathbf{Z}^T\mathbf{R}^{-1}\mathbf{Z}=\mathbf{I},$$

it also follows that

$$\begin{aligned}
\mathbf{G}-\mathbf{G}\mathbf{Z}^T(\mathbf{R}+\mathbf{Z}\mathbf{G}\mathbf{Z}^T)^{-1}\mathbf{Z}\mathbf{G} &= \mathbf{G}-(\mathbf{G}^{-1}+\mathbf{Z}^T\mathbf{R}^{-1}\mathbf{Z})^{-1}\mathbf{Z}^T\mathbf{R}^{-1}\mathbf{Z}\mathbf{G} \\
&= \mathbf{G}-[\mathbf{I}-(\mathbf{G}^{-1}+\mathbf{Z}^T\mathbf{R}^{-1}\mathbf{Z})^{-1}\mathbf{G}^{-1}]\mathbf{G} \\
&= (\mathbf{G}^{-1}+\mathbf{Z}^T\mathbf{R}^{-1}\mathbf{Z})^{-1}.
\end{aligned} \tag{13.11}$$

Results (13.9), (13.10), and (13.11) may be used to obtain expressions for $\tilde{\boldsymbol{\tau}}$, $\text{var}(\tilde{\boldsymbol{\tau}}-\boldsymbol{\tau})$, and $\tilde{\sigma}^2$ that are in terms of the matrix inverses $\mathbf{R}^{-1}$, $\mathbf{G}^{-1}$, and $(\mathbf{G}^{-1}+\mathbf{Z}^T\mathbf{R}^{-1}\mathbf{Z})^{-1}$ rather than the matrix inverse $\mathbf{W}^{-1}=(\mathbf{R}+\mathbf{Z}\mathbf{G}\mathbf{Z}^T)^{-1}$. The following theorem gives the expressions so obtained. Note that $\mathbf{G}$ and $\mathbf{G}^{-1}+\mathbf{Z}^T\mathbf{R}^{-1}\mathbf{Z}$ are $q \times q$ matrices, whereas $\mathbf{W}$ is an $n \times n$ matrix. In many applications, $\mathbf{R}$ is easy to invert (such as when $\mathbf{R}=\mathbf{I}$) and the number q of random effects is considerably smaller than the number n of observations, so it is more efficient computationally to invert $\mathbf{R}$ and the two $q \times q$ matrices ($\mathbf{G}$ and $\mathbf{G}^{-1}+\mathbf{Z}^T\mathbf{R}^{-1}\mathbf{Z}$) and then combine them according to (13.9), than it is to invert the one $n \times n$ matrix $\mathbf{W}$.

Theorem 13.4.2 *Let* $\boldsymbol{\tau} = \mathbf{C}^T\boldsymbol{\beta} + \mathbf{F}^T\mathbf{b}$ *be an s-vector of predictable linear functions. Then, under the positive definite mixed effects model with known* $\boldsymbol{\psi}$:

(a) the BLUP of $\boldsymbol{\tau}$ *is*

$$\tilde{\boldsymbol{\tau}} = \mathbf{C}^T\tilde{\boldsymbol{\beta}} + \mathbf{F}^T\tilde{\mathbf{b}},$$

where $\tilde{\boldsymbol{\beta}}$ *is any solution to the Aitken equations*

$$\mathbf{X}^T[\mathbf{R}^{-1} - \mathbf{R}^{-1}\mathbf{Z}(\mathbf{G}^{-1} + \mathbf{Z}^T\mathbf{R}^{-1}\mathbf{Z})^{-1}\mathbf{Z}^T\mathbf{R}^{-1}]\mathbf{X}\boldsymbol{\beta}$$
$$= \mathbf{X}^T[\mathbf{R}^{-1} - \mathbf{R}^{-1}\mathbf{Z}(\mathbf{G}^{-1} + \mathbf{Z}^T\mathbf{R}^{-1}\mathbf{Z})^{-1}\mathbf{Z}^T\mathbf{R}^{-1}]\mathbf{y}$$

and

$$\tilde{\mathbf{b}} = (\mathbf{G}^{-1} + \mathbf{Z}^T\mathbf{R}^{-1}\mathbf{Z})^{-1}\mathbf{Z}^T\mathbf{R}^{-1}(\mathbf{y} - \mathbf{X}\tilde{\boldsymbol{\beta}});$$

(b) the variance–covariance matrix of the BLUP's prediction error vector, i.e., var$(\tilde{\boldsymbol{\tau}} - \boldsymbol{\tau})$, *is* $\sigma^2\mathbf{Q}$, *where*

$$\mathbf{Q} = \mathbf{F}^T(\mathbf{G}^{-1} + \mathbf{Z}^T\mathbf{R}^{-1}\mathbf{Z})^{-1}\mathbf{F} + [\mathbf{C}^T - \mathbf{F}^T(\mathbf{G}^{-1} + \mathbf{Z}^T\mathbf{R}^{-1}\mathbf{Z})^{-1}\mathbf{Z}^T\mathbf{R}^{-1}\mathbf{X}]$$
$$\times\{\mathbf{X}^T[\mathbf{R}^{-1} - \mathbf{R}^{-1}\mathbf{Z}(\mathbf{G}^{-1} + \mathbf{Z}^T\mathbf{R}^{-1}\mathbf{Z})^{-1}\mathbf{Z}^T\mathbf{R}^{-1}]\mathbf{X}\}^-$$
$$\times[\mathbf{C}^T - \mathbf{F}^T(\mathbf{G}^{-1} + \mathbf{Z}^T\mathbf{R}^{-1}\mathbf{Z})^{-1}\mathbf{Z}^T\mathbf{R}^{-1}\mathbf{X}]^T,$$

invariant to the choice of generalized inverse of $\mathbf{X}^T[\mathbf{R}^{-1} - \mathbf{R}^{-1}\mathbf{Z}(\mathbf{G}^{-1} + \mathbf{Z}^T\mathbf{R}^{-1}\mathbf{Z})^{-1}\mathbf{Z}^T\mathbf{R}^{-1}]\mathbf{X}$; *and*

(c) if $n > p^*$, *then* $\tilde{\sigma}^2\mathbf{Q}$ *is an unbiased estimator of var*$(\tilde{\boldsymbol{\tau}} - \boldsymbol{\tau})$, *where*

$$\tilde{\sigma}^2 = \frac{(\mathbf{y} - \mathbf{X}\tilde{\boldsymbol{\beta}})^T[\mathbf{R}^{-1} - \mathbf{R}^{-1}\mathbf{Z}(\mathbf{G}^{-1} + \mathbf{Z}^T\mathbf{R}^{-1}\mathbf{Z})^{-1}\mathbf{Z}^T\mathbf{R}^{-1}](\mathbf{y} - \mathbf{X}\tilde{\boldsymbol{\beta}})}{n - p^*}.$$

Example 13.4.2-1. Prediction of the Slope in a No-Intercept Random-Slope Simple Linear Regression Model

Consider the model

$$y_i = bz_i + d_i \quad (i = 1, \ldots, n),$$

where the z_i's are the values of a quantitative explanatory variable, $\mathrm{E}(b) = 0$, $\mathrm{var}(b) = \sigma_b^2$, and the d_i's are zero-mean random variables with common variance σ^2 that are uncorrelated with each other and with b. This model may be written in matrix notation as a random effects model

$$\mathbf{y} = \mathbf{z}b + \mathbf{d},$$

where $\text{var}(b) = \sigma^2\mathbf{G}(\boldsymbol{\psi})$ with $\mathbf{G}(\boldsymbol{\psi}) = \psi \equiv \sigma_b^2/\sigma^2$ (a scalar) and $\text{var}(\mathbf{d}) = \sigma^2\mathbf{I}$. The parameter space for the model is $\Theta = \{(\sigma^2, \psi) : \sigma^2 > 0, \ \psi > 0\}$.

Suppose that we wish to predict b. If we act as though ψ is known, then this model is a special case of the Aitken model with $\mathbf{X} = \mathbf{0}$ and positive definite variance–covariance matrix $\sigma^2\mathbf{W} = \sigma^2(\mathbf{I} + \psi\mathbf{z}\mathbf{z}^T)$. Specializing Theorem 13.4.2a, we obtain the BLUP of b under this model as

$$\tilde{b} = \frac{\sum_{i=1}^n z_i y_i}{\psi^{-1} + \sum_{i=1}^n z_i^2}.$$

Observe that the value of $\tilde{b}$ lies between 0 and the ordinary least squares estimator of slope (for a no-intercept fixed-slope model). The smaller that ψ is, the closer that $\tilde{b}$ is to 0, and the larger that ψ is, the closer that $\tilde{b}$ is to the ordinary least squares estimator of slope in a no-intercept fixed-slope model.

It is left as an exercise to show that

$$\text{var}(\tilde{b} - b) = \frac{\sigma^2}{\psi^{-1} + \sum_{i=1}^n z_i^2}$$

and that an unbiased estimator of this prediction error variance is obtained by replacing its numerator with the estimator

$$\tilde{\sigma}^2 = \left(\sum_{i=1}^n y_i^2 - \frac{\left(\sum_{i=1}^n z_i y_i\right)^2}{\psi^{-1} + \sum_{i=1}^n z_i^2}\right) \bigg/ n.$$

■

Example 13.4.2-2. Prediction Under a Balanced One-Factor Random Effects Model

Consider the balanced one-factor random effects model

$$y_{ij} = \mu + b_i + d_{ij} \quad (i = 1, \ldots, q; \ j = 1, \ldots, r),$$

where $\text{E}(b_i) = \text{E}(d_{ij}) = 0$ for all i and j, the b_i's are uncorrelated with common variance $\sigma_b^2 > 0$, and the d_{ij}'s are uncorrelated (with each other and with the b_i's) with common variance $\sigma^2 > 0$. This model may be written in matrix notation as

$$\mathbf{y} = \mathbf{X}\boldsymbol{\beta} + \mathbf{Z}\mathbf{b} + \mathbf{d},$$

where

$$\mathbf{X} = \mathbf{1}_{qr}, \quad \boldsymbol{\beta} = \mu, \quad \mathbf{Z} = \oplus_{i=1}^{q}\mathbf{1}_r, \quad \mathbf{b} = (b_1, \ldots, b_q)^T, \quad \mathbf{G}(\boldsymbol{\psi}) = \psi\mathbf{I}_q,$$
$$\psi = \sigma_b^2/\sigma^2, \quad \mathbf{R} = \mathbf{R} = \mathbf{I}_{qr}.$$

The parameter space for the model is $\{(\mu, \sigma^2, \psi) : \mu \in \mathbb{R}, \sigma^2 > 0, \psi > 0\}$.

Suppose that we wish to predict the quantities $\mu + b_i$ $(i = 1, \ldots, q)$. If we act as though ψ is known, then this model is a special case of the Aitken model with positive definite variance–covariance matrix

$$\begin{aligned}\sigma^2\mathbf{W} &= \sigma^2(\mathbf{R} + \mathbf{Z}\mathbf{G}\mathbf{Z}^T) = \sigma^2[\mathbf{I}_{qr} + (\oplus_{i=1}^{q}\mathbf{1}_r)(\psi\mathbf{I}_q)(\oplus_{i=1}^{q}\mathbf{1}_r)^T] \\ &= \sigma^2(\mathbf{I}_{qr} + \psi \oplus_{i=1}^{q} \mathbf{J}_r).\end{aligned}$$

Then

$$\mathbf{W}\mathbf{X} = (\mathbf{I}_{qr} + \psi \oplus_{i=1}^{q} \mathbf{J}_r)\mathbf{1}_{qr} = (1 + r\psi)\mathbf{1}_{qr} \in \mathcal{C}(\mathbf{X}),$$

so by Corollary 11.3.1.1, the BLUE of μ is $\bar{y}_{\cdot\cdot}$ (the ordinary least squares estimator). By Theorem 13.4.2a, the BLUP of $\mu + b_i$ under this model is $\mathbf{C}^T\tilde{\boldsymbol{\beta}} + \mathbf{F}^T(\mathbf{G}^{-1} + \mathbf{Z}^T\mathbf{R}^{-1}\mathbf{Z})^{-1}\mathbf{Z}^T\mathbf{R}^{-1}(\mathbf{y} - \mathbf{X}\tilde{\boldsymbol{\beta}})$, where $\mathbf{C} = 1$ and $\mathbf{F}^T$ is the ith unit q-vector. Thus, the BLUP of $\mu + b_i$ is

$$\begin{aligned}\widetilde{\mu + b_i} &= \bar{y}_{\cdot\cdot} + \mathbf{u}_i^T(\psi^{-1}\mathbf{I} + r\mathbf{I})^{-1}\begin{pmatrix} r\bar{y}_{1\cdot} - r\bar{y}_{\cdot\cdot} \\ \vdots \\ r\bar{y}_{q\cdot} - r\bar{y}_{\cdot\cdot}\end{pmatrix} \\ &= \bar{y}_{\cdot\cdot} + (\psi^{-1} + r)^{-1}(r\bar{y}_{i\cdot} - r\bar{y}_{\cdot\cdot}) \\ &= \bar{y}_{\cdot\cdot} + [r\psi/(r\psi + 1)](\bar{y}_{i\cdot} - \bar{y}_{\cdot\cdot}) \\ &= \left(\frac{r\psi}{r\psi + 1}\right)\bar{y}_{i\cdot} + \left(\frac{1}{r\psi + 1}\right)\bar{y}_{\cdot\cdot}.\end{aligned}$$

Observe that $\widetilde{\mu + b_i}$ is a weighted linear combination of the ith level mean and the overall mean, with weights that sum to one. The larger that ψ is, the closer that $\widetilde{\mu + b_i}$ is to the ith level mean, and the smaller that ψ is, the closer that $\widetilde{\mu + b_i}$ is to the overall mean.

From the development above and Corollary 13.2.1.1, we find that the BLUP of $b_i - b_{i'}$ $(i \neq i')$ is

$$\widetilde{b_i - b_{i'}} = \widetilde{\mu + b_i} - \widetilde{\mu + b_{i'}} = \frac{r\psi}{r\psi + 1}(\bar{y}_{i\cdot} - \bar{y}_{i'\cdot}),$$

which is plainly a multiple of the BLUE of $\alpha_i - \alpha_{i'}$ under the one-factor fixed effects model. The multiplier is a number between 0 and 1, revealing that the BLUP of $b_i - b_{i'}$ under the random effects model is a "shrunken" (towards 0) version of the fixed effects model's BLUE of $\alpha_i - \alpha_{i'}$. Deriving the variances and covariances of the prediction errors, $\{\widetilde{b_i - b_{i'}} - (b_i - b_{i'}) : i < i'\}$, is left as an exercise. ■

13.4.3 The Mixed-Model Equations

In this subsection we show that alternative representations (to those given in Theorem 13.4.1) of the BLUP and its estimated variance of prediction errors can be given in terms of solutions to the system of equations

$$\begin{pmatrix} \mathbf{X}^T\mathbf{R}^{-1}\mathbf{X} & \mathbf{X}^T\mathbf{R}^{-1}\mathbf{Z} \\ \mathbf{Z}^T\mathbf{R}^{-1}\mathbf{X} & \mathbf{G}^{-1} + \mathbf{Z}^T\mathbf{R}^{-1}\mathbf{Z} \end{pmatrix} \begin{pmatrix} \boldsymbol{\beta} \\ \mathbf{b} \end{pmatrix} = \begin{pmatrix} \mathbf{X}^T\mathbf{R}^{-1}\mathbf{y} \\ \mathbf{Z}^T\mathbf{R}^{-1}\mathbf{y} \end{pmatrix}.$$

This system, known as the *mixed-model equations*, closely resembles what would be the ordinary two-part Aitken equations if $\mathbf{b}$ was regarded as a vector of unknown parameters rather than random variables. The coefficient matrix of the mixed-model equations differs from that of the just-mentioned ordinary two-part Aitken equations only by the addition of $\mathbf{G}^{-1}$ to the lower right block of the matrix.

The relationship between the BLUP of $\boldsymbol{\tau}$ and the mixed-model equations is given by the following theorem.

Theorem 13.4.3 *Consider the positive definite mixed effects model with known* $\boldsymbol{\psi}$. *If* $\tilde{\boldsymbol{\beta}}$ *is a solution to the Aitken equations* $\mathbf{X}^T(\mathbf{R} + \mathbf{Z}\mathbf{G}\mathbf{Z}^T)^{-1}\mathbf{X}\boldsymbol{\beta} = \mathbf{X}^T(\mathbf{R} + \mathbf{Z}\mathbf{G}\mathbf{Z}^T)^{-1}\mathbf{y}$, *then the mixed-model equations have a solution whose first component is* $\tilde{\boldsymbol{\beta}}$ *and whose second component is* $\tilde{\mathbf{b}} = \mathbf{G}\mathbf{Z}^T(\mathbf{R}+\mathbf{Z}\mathbf{G}\mathbf{Z}^T)^{-1}(\mathbf{y}-\mathbf{X}\tilde{\boldsymbol{\beta}})$. *Conversely, if* $\dot{\boldsymbol{\beta}}$ *and* $\dot{\mathbf{b}}$ *are the components of any solution to the mixed-model equations, then* $\dot{\boldsymbol{\beta}}$ *is a solution to the Aitken equations and* $\dot{\mathbf{b}} = \mathbf{G}\mathbf{Z}^T(\mathbf{R} + \mathbf{Z}\mathbf{G}\mathbf{Z}^T)^{-1}(\mathbf{y} - \mathbf{X}\dot{\boldsymbol{\beta}})$.

Proof Let $\tilde{\boldsymbol{\beta}}$ represent any solution to the Aitken equations, and define $\tilde{\mathbf{b}} = \mathbf{G}\mathbf{Z}^T(\mathbf{R} + \mathbf{Z}\mathbf{G}\mathbf{Z}^T)^{-1}(\mathbf{y} - \mathbf{X}\tilde{\boldsymbol{\beta}})$. Then by (13.10),

$$\tilde{\mathbf{b}} = (\mathbf{G}^{-1} + \mathbf{Z}^T\mathbf{R}^{-1}\mathbf{Z})^{-1}\mathbf{Z}^T\mathbf{R}^{-1}(\mathbf{y} - \mathbf{X}\tilde{\boldsymbol{\beta}}),$$

and by (13.9),

$$\begin{aligned} &\mathbf{X}^T[\mathbf{R}^{-1} - \mathbf{R}^{-1}\mathbf{Z}(\mathbf{G}^{-1} + \mathbf{Z}^T\mathbf{R}^{-1}\mathbf{Z})^{-1}\mathbf{Z}^T\mathbf{R}^{-1}]\mathbf{X}\tilde{\boldsymbol{\beta}} \\ &\quad = \mathbf{X}^T[\mathbf{R}^{-1} - \mathbf{R}^{-1}\mathbf{Z}(\mathbf{G}^{-1} + \mathbf{Z}^T\mathbf{R}^{-1}\mathbf{Z})^{-1}\mathbf{Z}^T\mathbf{R}^{-1}]\mathbf{y}, \end{aligned}$$

i.e.,

$$\mathbf{X}^T\mathbf{R}^{-1}\mathbf{X}\tilde{\boldsymbol{\beta}} + \mathbf{X}^T\mathbf{R}^{-1}\mathbf{Z}(\mathbf{G}^{-1} + \mathbf{Z}^T\mathbf{R}^{-1}\mathbf{Z})^{-1}\mathbf{Z}^T\mathbf{R}^{-1}(\mathbf{y} - \mathbf{X}\tilde{\boldsymbol{\beta}}) = \mathbf{X}^T\mathbf{R}^{-1}\mathbf{y},$$

i.e.,

$$\mathbf{X}^T\mathbf{R}^{-1}\mathbf{X}\tilde{\boldsymbol{\beta}} + \mathbf{X}^T\mathbf{R}^{-1}\mathbf{Z}\tilde{\mathbf{b}} = \mathbf{X}^T\mathbf{R}^{-1}\mathbf{y}.$$

Furthermore,

$$\begin{aligned}\mathbf{Z}^T\mathbf{R}^{-1}\mathbf{X}\tilde{\boldsymbol{\beta}} + (\mathbf{G}^{-1} + \mathbf{Z}^T\mathbf{R}^{-1}\mathbf{Z})\tilde{\mathbf{b}} &= \mathbf{Z}^T\mathbf{R}^{-1}\mathbf{X}\tilde{\boldsymbol{\beta}} + (\mathbf{G}^{-1} + \mathbf{Z}^T\mathbf{R}^{-1}\mathbf{Z}) \\ &\quad \times (\mathbf{G}^{-1} + \mathbf{Z}^T\mathbf{R}^{-1}\mathbf{Z})^{-1}\mathbf{Z}^T\mathbf{R}^{-1}(\mathbf{y} - \mathbf{X}\tilde{\boldsymbol{\beta}}) \\ &= \mathbf{Z}^T\mathbf{R}^{-1}\mathbf{y}.\end{aligned}$$

Thus, $\begin{pmatrix}\tilde{\boldsymbol{\beta}} \\ \tilde{\mathbf{b}}\end{pmatrix}$ satisfies the mixed-model equations.

Conversely, if $\begin{pmatrix}\dot{\boldsymbol{\beta}} \\ \dot{\mathbf{b}}\end{pmatrix}$ satisfies the mixed-model equations, then according to the "bottom" subset of mixed-model equations,

$$\mathbf{Z}^T\mathbf{R}^{-1}\mathbf{X}\dot{\boldsymbol{\beta}} + (\mathbf{G}^{-1} + \mathbf{Z}^T\mathbf{R}^{-1}\mathbf{Z})\dot{\mathbf{b}} = \mathbf{Z}^T\mathbf{R}^{-1}\mathbf{y},$$

implying that

$$\begin{aligned}\dot{\mathbf{b}} &= (\mathbf{G}^{-1} + \mathbf{Z}^T\mathbf{R}^{-1}\mathbf{Z})^{-1}\mathbf{Z}^T\mathbf{R}^{-1}(\mathbf{y} - \mathbf{X}\dot{\boldsymbol{\beta}}) \\ &= \mathbf{G}\mathbf{Z}^T(\mathbf{R} + \mathbf{Z}\mathbf{G}\mathbf{Z}^T)^{-1}(\mathbf{y} - \mathbf{X}\dot{\boldsymbol{\beta}}),\end{aligned}$$

where we used (13.10) for the last equality. Also, substitution of the first of these expressions for $\dot{\mathbf{b}}$ in the "top" subset of mixed-model equations gives

$$\mathbf{X}^T\mathbf{R}^{-1}\mathbf{X}\dot{\boldsymbol{\beta}} + \mathbf{X}^T\mathbf{R}^{-1}\mathbf{Z}(\mathbf{G}^{-1} + \mathbf{Z}^T\mathbf{R}^{-1}\mathbf{Z})^{-1}\mathbf{Z}^T\mathbf{R}^{-1}(\mathbf{y} - \mathbf{X}\dot{\boldsymbol{\beta}}) = \mathbf{X}^T\mathbf{R}^{-1}\mathbf{y}$$

or equivalently,

$$\begin{aligned}&\mathbf{X}^T[\mathbf{R}^{-1} - \mathbf{R}^{-1}\mathbf{Z}(\mathbf{G}^{-1} + \mathbf{Z}^T\mathbf{R}^{-1}\mathbf{Z})^{-1}\mathbf{Z}^T\mathbf{R}^{-1}]\mathbf{X}\dot{\boldsymbol{\beta}} \\ &\quad = \mathbf{X}^T[\mathbf{R}^{-1} - \mathbf{R}^{-1}\mathbf{Z}(\mathbf{G}^{-1} + \mathbf{Z}^T\mathbf{R}^{-1}\mathbf{Z})^{-1}\mathbf{Z}^T\mathbf{R}^{-1}]\mathbf{y},\end{aligned}$$

or equivalently [by (13.9)],

$$\mathbf{X}^T(\mathbf{R} + \mathbf{Z}\mathbf{G}\mathbf{Z}^T)^{-1}\mathbf{X}\dot{\boldsymbol{\beta}} = \mathbf{X}^T(\mathbf{R} + \mathbf{Z}\mathbf{G}\mathbf{Z}^T)^{-1}\mathbf{y}.$$

Thus $\dot{\boldsymbol{\beta}}$ satisfies the Aitken equations. □

The following corollary is an immediate consequence of Theorem 13.4.1 and the converse part of Theorem 13.4.3.

Corollary 13.4.3.1 *Under the positive definite mixed effects model with known* $\boldsymbol{\psi}$*, the BLUP of a vector of predictable linear functions* $\boldsymbol{\tau} = \mathbf{C}^T\boldsymbol{\beta} + \mathbf{F}^T\mathbf{b}$ *is*

$$\tilde{\boldsymbol{\tau}} = \mathbf{C}^T\dot{\boldsymbol{\beta}} + \mathbf{F}^T\dot{\mathbf{b}},$$

where $\dot{\boldsymbol{\beta}}$ *and* $\dot{\mathbf{b}}$ *are the components of any solution to the mixed-model equations.*

The ordinary unbiased estimator of σ^2 (i.e., the generalized residual mean square) can also be computed from a solution to the mixed-model equations, as indicated by the following additional corollary to Theorem 13.4.3.

Corollary 13.4.3.2 *Under the positive definite mixed effects model with known* $\boldsymbol{\psi}$*, assume that* $n > p^*$ *and let* $\tilde{\sigma}^2$ *denote the generalized residual mean square, i.e.,* $\tilde{\sigma}^2 = (\mathbf{y} - \mathbf{X}\tilde{\boldsymbol{\beta}})^T(\mathbf{R} + \mathbf{Z}\mathbf{G}\mathbf{Z}^T)^{-1}(\mathbf{y} - \mathbf{X}\tilde{\boldsymbol{\beta}})/(n - p^*)$*, where* $\tilde{\boldsymbol{\beta}}$ *is any solution to the Aitken equations. Then an alternative expression for* $\tilde{\sigma}^2$ *is*

$$\tilde{\sigma}^2 = \frac{\mathbf{y}^T\mathbf{R}^{-1}\mathbf{y} - \dot{\boldsymbol{\beta}}^T\mathbf{X}^T\mathbf{R}^{-1}\mathbf{y} - \dot{\mathbf{b}}^T\mathbf{Z}^T\mathbf{R}^{-1}\mathbf{y}}{n - p^*},$$

where $\dot{\boldsymbol{\beta}}$ *and* $\dot{\mathbf{b}}$ *are the components of any solution to the mixed-model equations.*

Proof Using Corollary 11.1.9.1, (13.9), the definition of $\tilde{\mathbf{b}}$, and the converse part of Theorem 13.4.3, we obtain

$$\begin{aligned}
\tilde{\sigma}^2 &= \frac{\mathbf{y}^T(\mathbf{R} + \mathbf{Z}\mathbf{G}\mathbf{Z}^T)^{-1}(\mathbf{y} - \mathbf{X}\tilde{\boldsymbol{\beta}})}{n - p^*} \\
&= \frac{\mathbf{y}^T[\mathbf{R}^{-1} - \mathbf{R}^{-1}\mathbf{Z}(\mathbf{G}^{-1} + \mathbf{Z}^T\mathbf{R}^{-1}\mathbf{Z})^{-1}\mathbf{Z}^T\mathbf{R}^{-1}](\mathbf{y} - \mathbf{X}\tilde{\boldsymbol{\beta}})}{n - p^*} \\
&= \frac{\mathbf{y}^T\mathbf{R}^{-1}\mathbf{y} - \mathbf{y}^T\mathbf{R}^{-1}\mathbf{X}\tilde{\boldsymbol{\beta}} - \mathbf{y}^T\mathbf{R}^{-1}\mathbf{Z}(\mathbf{G}^{-1} + \mathbf{Z}^T\mathbf{R}^{-1}\mathbf{Z})^{-1}\mathbf{Z}^T\mathbf{R}^{-1}(\mathbf{y} - \mathbf{X}\tilde{\boldsymbol{\beta}})}{n - p^*} \\
&= \frac{\mathbf{y}^T\mathbf{R}^{-1}\mathbf{y} - \tilde{\boldsymbol{\beta}}^T\mathbf{X}^T\mathbf{R}^{-1}\mathbf{y} - \tilde{\mathbf{b}}^T\mathbf{Z}^T\mathbf{R}^{-1}\mathbf{y}}{n - p^*} \\
&= \frac{\mathbf{y}^T\mathbf{R}^{-1}\mathbf{y} - \dot{\boldsymbol{\beta}}^T\mathbf{X}^T\mathbf{R}^{-1}\mathbf{y} - \dot{\mathbf{b}}^T\mathbf{Z}^T\mathbf{R}^{-1}\mathbf{y}}{n - p^*}.
\end{aligned}$$

□

The variance–covariance matrix of BLUP prediction errors, i.e., var($\tilde{\boldsymbol{\tau}} - \boldsymbol{\tau}$) or equivalently $\sigma^2\mathbf{Q}$, and its unbiased estimator $\tilde{\sigma}^2\mathbf{Q}$ may also be expressed in terms of the mixed-model equations, as indicated by the following theorem. The proof of the theorem is left as an exercise.

Theorem 13.4.4 *Under the positive definite mixed effects model with known* $\boldsymbol{\psi}$, *let* $\dot{\boldsymbol{\beta}}$ *and* $\dot{\mathbf{b}}$ *be the components of any solution to the mixed-model equations; let* $\boldsymbol{\tau} = \mathbf{C}^T\boldsymbol{\beta} + \mathbf{F}^T\mathbf{b}$ *represent a vector of predictable linear functions; and let* $\tilde{\boldsymbol{\tau}} = \mathbf{C}^T\dot{\boldsymbol{\beta}} + \mathbf{F}^T\dot{\mathbf{b}}$ *be the BLUP of* $\boldsymbol{\tau}$. *Then*

$$var(\tilde{\boldsymbol{\tau}} - \boldsymbol{\tau}) = \sigma^2 \begin{pmatrix} \mathbf{C} \\ \mathbf{F} \end{pmatrix}^T \begin{pmatrix} \mathbf{X}^T\mathbf{R}^{-1}\mathbf{X} & \mathbf{X}^T\mathbf{R}^{-1}\mathbf{Z} \\ \mathbf{Z}^T\mathbf{R}^{-1}\mathbf{X} & \mathbf{G}^{-1} + \mathbf{Z}^T\mathbf{R}^{-1}\mathbf{Z} \end{pmatrix}^{-} \begin{pmatrix} \mathbf{C} \\ \mathbf{F} \end{pmatrix},$$

invariant to the choice of generalized inverse of the coefficient matrix of the mixed model equations.

13.4.4 The Mixed-Model ANOVA

Associated with any special case of a mixed effects model in which $\mathbf{R} = \mathbf{I}$ is an analysis of variance, here called the *mixed-model ANOVA*. In this ANOVA, the total sum of squares of $\mathbf{y}$ is partitioned exactly as it would be for the sequential ANOVA of the ordered two-part Gauss–Markov fixed effects model $\mathbf{y} = \mathbf{X}\boldsymbol{\beta}_1 + \mathbf{Z}\boldsymbol{\beta}_2 + \mathbf{e}$, or equivalently by acting as though $\mathbf{b}$ in the mixed effects model was a vector of *fixed* effects. Thus, the ANOVA table (up to the mean square column) is as follows:

Source	Rank	Sum of squares	Mean square
$\mathbf{X}$	rank($\mathbf{X}$)	$\mathbf{y}^T\mathbf{P_X}\mathbf{y}$	$\mathbf{y}^T\mathbf{P_X}\mathbf{y}/\text{rank}(\mathbf{X})$
$\mathbf{Z}\vert\mathbf{X}$	rank($\mathbf{X}, \mathbf{Z}$) − rank($\mathbf{X}$)	$\mathbf{y}^T(\mathbf{P_{X,Z}} - \mathbf{P_X})\mathbf{y}$	$\mathbf{y}^T(\mathbf{P_{X,Z}} - \mathbf{P_X})\mathbf{y}/[\text{rank}(\mathbf{X}, \mathbf{Z}) - \text{rank}(\mathbf{X})]$
Residual	n − rank($\mathbf{X}, \mathbf{Z}$)	$\mathbf{y}^T(\mathbf{I} - \mathbf{P_{X,Z}})\mathbf{y}$	$\mathbf{y}^T(\mathbf{I} - \mathbf{P_{X,Z}})\mathbf{y}/[n - \text{rank}(\mathbf{X}, \mathbf{Z})]$
Total	n	$\mathbf{y}^T\mathbf{y}$	

However, expected sums of squares (and expected mean squares) for this ANOVA are obtained under the mixed effects model and, as the next theorem indicates, are generally different than they would be under a Gauss–Markov fixed effects model.

Theorem 13.4.5 *Under the special case of a mixed effects model in which* $\mathbf{R} = \mathbf{I}$:

(a) $E(\mathbf{y}^T\mathbf{P_X}\mathbf{y}) = \boldsymbol{\beta}^T\mathbf{X}^T\mathbf{X}\boldsymbol{\beta} + \sigma^2 rank(\mathbf{X}) + \sigma^2 tr[\mathbf{G}(\boldsymbol{\psi})\mathbf{Z}^T\mathbf{P_X}\mathbf{Z}]$;
(b) $E[\mathbf{y}^T(\mathbf{P_{X,Z}} - \mathbf{P_X})\mathbf{y}] = \sigma^2[rank(\mathbf{X}, \mathbf{Z}) - rank(\mathbf{X})] + \sigma^2 tr[\mathbf{G}(\boldsymbol{\psi})\mathbf{Z}^T(\mathbf{I} - \mathbf{P_X})\mathbf{Z}]$; *and*
(c) $E[\mathbf{y}^T(\mathbf{I} - \mathbf{P_{X,Z}})\mathbf{y}] = \sigma^2[n - rank(\mathbf{X}, \mathbf{Z})]$.

Proof Using Theorem 4.2.4 and other well-known (by now) results repeatedly, we obtain

$$\begin{aligned} E(\mathbf{y}^T\mathbf{P_X}\mathbf{y}) &= (\mathbf{X}\boldsymbol{\beta})^T\mathbf{P_X}(\mathbf{X}\boldsymbol{\beta}) + \text{tr}[\mathbf{P_X}\sigma^2(\mathbf{I} + \mathbf{Z}\mathbf{G}(\boldsymbol{\psi})\mathbf{Z}^T)] \\ &= \boldsymbol{\beta}^T\mathbf{X}^T\mathbf{X}\boldsymbol{\beta} + \sigma^2\text{rank}(\mathbf{P_X}) + \sigma^2\text{tr}(\mathbf{P_X}\mathbf{Z}\mathbf{G}(\boldsymbol{\psi})\mathbf{Z}^T) \end{aligned}$$

$$
\begin{aligned}
&= \boldsymbol{\beta}^T\mathbf{X}^T\mathbf{X}\boldsymbol{\beta} + \sigma^2\text{rank}(\mathbf{X}) + \sigma^2\text{tr}[\mathbf{G}(\boldsymbol{\psi})\mathbf{Z}^T\mathbf{P}_\mathbf{X}\mathbf{Z}], \\
\text{E}[\mathbf{y}^T(\mathbf{P}_{\mathbf{X},\mathbf{Z}} - \mathbf{P}_\mathbf{X})\mathbf{y}] &= (\mathbf{X}\boldsymbol{\beta})^T(\mathbf{P}_{\mathbf{X},\mathbf{Z}} - \mathbf{P}_\mathbf{X})(\mathbf{X}\boldsymbol{\beta}) + \text{tr}[(\mathbf{P}_{\mathbf{X},\mathbf{Z}} - \mathbf{P}_\mathbf{X})\sigma^2(\mathbf{I} + \mathbf{Z}\mathbf{G}(\boldsymbol{\psi})\mathbf{Z}^T)] \\
&= \sigma^2\text{rank}(\mathbf{P}_{\mathbf{X},\mathbf{Z}} - \mathbf{P}_\mathbf{X}) + \sigma^2\text{tr}[(\mathbf{I} - \mathbf{P}_\mathbf{X})\mathbf{Z}\mathbf{G}(\boldsymbol{\psi})\mathbf{Z}^T] \\
&= \sigma^2[\text{rank}(\mathbf{X}, \mathbf{Z}) - \text{rank}(\mathbf{X})] + \sigma^2\text{tr}[\mathbf{G}(\boldsymbol{\psi})\mathbf{Z}^T(\mathbf{I} - \mathbf{P}_\mathbf{X})\mathbf{Z}], \\
\text{E}[\mathbf{y}^T(\mathbf{I} - \mathbf{P}_{\mathbf{X},\mathbf{Z}})\mathbf{y}] &= (\mathbf{X}\boldsymbol{\beta})^T(\mathbf{I} - \mathbf{P}_{\mathbf{X},\mathbf{Z}})(\mathbf{X}\boldsymbol{\beta}) + \text{tr}[(\mathbf{I} - \mathbf{P}_{\mathbf{X},\mathbf{Z}})\sigma^2(\mathbf{I} + \mathbf{Z}\mathbf{G}(\boldsymbol{\psi})\mathbf{Z}^T)] \\
&= \sigma^2[n - \text{rank}(\mathbf{X}, \mathbf{Z})].
\end{aligned}
$$

□

One important aspect of Theorem 13.4.5 is that while the expectation of $\mathbf{y}^T\mathbf{P}_\mathbf{X}\mathbf{y}$ is a function of all three of $\boldsymbol{\beta}$, σ^2, and $\boldsymbol{\psi}$, the expectations of the other two sums of squares depend only on σ^2 and $\boldsymbol{\psi}$. In Chap. 16, sums of squares free of $\boldsymbol{\beta}$ from the mixed-model ANOVA, and from extensions of it in which $\mathbf{Z}$ itself is partitioned, will serve as a basis for obtaining estimators of the unknown parameters in $\mathbf{G}(\boldsymbol{\psi})$. For now, we merely state a corollary to Theorem 13.4.5 that gives an unbiased estimator of σ^2 obtainable from the mixed-model ANOVA.

Corollary 13.4.5.1 *Under the special case of a mixed effects model in which* $\mathbf{R} = \mathbf{I}$, *if* $n > \text{rank}(\mathbf{X}, \mathbf{Z})$, *then*

$$
\bar{\sigma}^2 = \frac{\mathbf{y}^T(\mathbf{I} - \mathbf{P}_{\mathbf{X},\mathbf{Z}})\mathbf{y}}{n - \text{rank}(\mathbf{X}, \mathbf{Z})}
$$

is an unbiased estimator of σ^2.

Observe that $\bar{\sigma}^2$ generally is a different estimator of σ^2 than the generalized residual mean square, $\tilde{\sigma}^2$. Indeed, upon comparing the expression for $\bar{\sigma}^2$ in the previous corollary to that for $\tilde{\sigma}^2$ in Corollary 13.4.3.2, we see that both the numerators and denominators of these two estimators are generally different, even when $\mathbf{R} = \mathbf{I}$.

13.4.5 Split-Plot ANOVAs

Recall from Chap. 5 that a split-plot design consists of two types of "plots" (observational units): whole plots and split plots. The split plots are nested, and equal in number, within whole plots. The model matrix may be partitioned as $\mathbf{X} = (\mathbf{X}_w, \mathbf{X}_s)$ where the columns of $\mathbf{X}_w$ consist of only those variables that describe the model for means of the response over split plots within whole plots, and $\mathbf{X}_s$ consists of columns for the remaining variables. Two distinct errors, whole-plot errors and split-plot errors, are assigned to each type of plot. As a consequence, the model is a mixed model, with whole-plot errors comprising $\mathbf{Zb}$ and split-plot errors comprising $\mathbf{d}$. Thus, a split-plot model has an associated mixed-model ANOVA;

however, owing to the relationship between the partitioning of $\mathbf{X}$ and the two types of errors, this ANOVA is sufficiently specialized that we give it its own name—a *split-plot ANOVA*. The following is a concrete example.

Example 13.4.5-1. A One Whole-Plot-Factor, One Split-Plot-Factor ANOVA

Consider a split-plot design in which whole plots are assigned to levels of a single factor A and, likewise, split plots are assigned to the levels of a single factor B. Suppose that there are q levels of Factor A, each replicated on r whole plots, and m levels of Factor B, each replicated only once within each whole plot. Then the model may be written as

$$y_{ijk} = \mu+\alpha_i+b_{ij}+\gamma_k+\xi_{ik}+d_{ijk} \qquad (i = 1,\ldots,q;\ j = 1,\ldots,r;\ k = 1,\ldots,m).$$

Here the b_{ij}'s and d_{ijk}s are uncorrelated random variables such that $\mathrm{E}(b_{ij}) = \mathrm{E}(d_{ijk}) = 0$ for all (i, j, k), $\mathrm{var}(b_{ij}) = \sigma_b^2$ for all i and j, and $\mathrm{var}(d_{ijk}) = \sigma^2$ for all (i, j, k). All other model effects are fixed. In matrix notation, the model may be written as

$$\mathbf{y} = \mathbf{1}_{qrm}\mu + \mathbf{X}_\alpha\boldsymbol{\alpha} + \mathbf{Z}\mathbf{b} + \mathbf{X}_\gamma\boldsymbol{\gamma} + \mathbf{X}_\xi\boldsymbol{\xi} + \mathbf{d},$$

where

$$\mathbf{X}_\alpha = \mathbf{I}_q \otimes \mathbf{1}_{rm}, \quad \mathbf{Z} = \mathbf{I}_{qr} \otimes \mathbf{1}_m, \quad \mathbf{X}_\gamma = \mathbf{1}_{qr} \otimes \mathbf{I}_m, \quad \mathbf{X}_\xi = \mathbf{I}_q \otimes \mathbf{1}_r \otimes \mathbf{I}_m.$$

Furthermore,

$$\mathrm{var}(\mathbf{y}) = \sigma^2\mathbf{I}_{qrm} + \mathbf{Z}(\sigma_b^2\mathbf{I}_{qr})\mathbf{Z}^T = \sigma^2[\mathbf{I}_{qrm} + (\sigma_b^2/\sigma^2)(\mathbf{I}_{qr} \otimes \mathbf{J}_m)].$$

Now let $\mathbf{P_1}$ and $\mathbf{P_Z}$ denote the orthogonal projection matrices onto the column spaces of $\mathbf{1}$ and $\mathbf{Z}$, respectively, and in an abuse of notation, let $\mathbf{P}_\alpha$, $\mathbf{P}_\gamma$, and $\mathbf{P}_\xi$ denote the orthogonal projection matrices onto the column spaces of $\mathbf{X}_\alpha$, $\mathbf{X}_\gamma$, and $\mathbf{X}_\xi$. Observe that, corresponding to the model, $\mathbf{y}$ admits the decomposition

$$\begin{aligned}\mathbf{y} &= \mathbf{P_1}\mathbf{y} + (\mathbf{P}_\alpha - \mathbf{P_1})\mathbf{y} + (\mathbf{P_Z} - \mathbf{P}_\alpha)\mathbf{y} + (\mathbf{P}_\gamma - \mathbf{P_1})\mathbf{y} + (\mathbf{P}_\xi - \mathbf{P}_\alpha - \mathbf{P}_\gamma + \mathbf{P_1})\mathbf{y}\\ &\quad + (\mathbf{I} - \mathbf{P_Z} + \mathbf{P}_\alpha - \mathbf{P}_\xi)\mathbf{y},\end{aligned}$$

so that (upon pre-multiplying both sides by $\mathbf{y}^T$) we have the following decomposition of the total sum of squares:

$$\begin{aligned}\mathbf{y}^T\mathbf{y} &= \mathbf{y}^T\mathbf{P_1}\mathbf{y} + \mathbf{y}^T(\mathbf{P}_\alpha - \mathbf{P_1})\mathbf{y} + \mathbf{y}^T(\mathbf{P_Z} - \mathbf{P}_\alpha)\mathbf{y} + \mathbf{y}^T(\mathbf{P}_\gamma - \mathbf{P_1})\mathbf{y}\\ &\quad + \mathbf{y}^T(\mathbf{P}_\xi - \mathbf{P}_\alpha - \mathbf{P}_\gamma + \mathbf{P_1})\mathbf{y} + \mathbf{y}^T(\mathbf{I} - \mathbf{P_Z} + \mathbf{P}_\alpha - \mathbf{P}_\xi)\mathbf{y}.\end{aligned} \qquad (13.12)$$

Furthermore, it is easily verified (see the exercises) that

$$\mathbf{P}_1 = (1/qrm)\mathbf{J}_{qrm}, \quad \mathbf{P}_\alpha = (1/rm)(\mathbf{I}_q \otimes \mathbf{J}_{rm}), \quad \mathbf{P}_\mathbf{Z} = (1/m)(\mathbf{I}_{qr} \otimes \mathbf{J}_m),$$
$$\mathbf{P}_\gamma = (1/qr)(\mathbf{J}_{qr} \otimes \mathbf{I}_m), \quad \mathbf{P}_\xi = (1/r)(\mathbf{I}_q \otimes \mathbf{J}_r \otimes \mathbf{I}_m);$$

that

$$\mathbf{P}_\alpha\mathbf{P}_1 = \mathbf{P}_1, \quad \mathbf{P}_\mathbf{Z}\mathbf{P}_1 = \mathbf{P}_1, \quad \mathbf{P}_\gamma\mathbf{P}_1 = \mathbf{P}_1, \quad \mathbf{P}_\xi\mathbf{P}_1 = \mathbf{P}_1, \quad \mathbf{P}_\alpha\mathbf{P}_\mathbf{Z} = \mathbf{P}_\alpha,$$
$$\mathbf{P}_\alpha\mathbf{P}_\gamma = \mathbf{P}_1, \quad \mathbf{P}_\alpha\mathbf{P}_\xi = \mathbf{P}_\alpha, \quad \mathbf{P}_\mathbf{Z}\mathbf{P}_\gamma = \mathbf{P}_1, \quad \mathbf{P}_\mathbf{Z}\mathbf{P}_\xi = \mathbf{P}_\alpha, \quad \mathbf{P}_\gamma\mathbf{P}_\xi = \mathbf{P}_\gamma;$$

and that for each of the six quadratic forms in the decomposition of $\mathbf{y}^T\mathbf{y}$, the matrix of the quadratic form is idempotent. Thus,

$$\text{rank}(\mathbf{P}_1) = 1, \quad \text{rank}(\mathbf{P}_\alpha - \mathbf{P}_1) = q - 1, \quad \text{rank}(\mathbf{P}_\mathbf{Z} - \mathbf{P}_\alpha) = q(r - 1),$$
$$\text{rank}(\mathbf{P}_\gamma - \mathbf{P}_1) = m - 1, \quad \text{rank}(\mathbf{P}_\xi - \mathbf{P}_\alpha - \mathbf{P}_\gamma + \mathbf{P}_1) = (q - 1)(m - 1),$$
$$\text{rank}(\mathbf{I} - \mathbf{P}_\mathbf{Z} + \mathbf{P}_\alpha - \mathbf{P}_\xi) = q(r - 1)(m - 1).$$

We may therefore write the corrected-for-the-mean ANOVA without using matrix notation as follows:

Source	Rank	Sum of squares
Factor A	$q - 1$	$rm\sum_{i=1}^{q}(\bar{y}_{i\cdot\cdot} - \bar{y}_{\cdot\cdot\cdot})^2$
Whole-plot error	$q(r - 1)$	$m\sum_{i=1}^{q}\sum_{j=1}^{r}(\bar{y}_{ij\cdot} - \bar{y}_{i\cdot\cdot})^2$
Factor B	$m - 1$	$qr\sum_{k=1}^{m}(\bar{y}_{\cdot\cdot k} - \bar{y}_{\cdot\cdot\cdot})^2$
AB interaction	$(q - 1)(m - 1)$	$r\sum_{i=1}^{q}\sum_{k=1}^{m}(\bar{y}_{i\cdot k} - \bar{y}_{i\cdot\cdot} - \bar{y}_{\cdot\cdot k} + \bar{y}_{\cdot\cdot\cdot})^2$
Split-plot error	$q(r - 1)(m - 1)$	$\sum_{i=1}^{q}\sum_{j=1}^{r}\sum_{k=1}^{m}(y_{ijk} - \bar{y}_{ij\cdot} - \bar{y}_{i\cdot k} + \bar{y}_{i\cdot\cdot})^2$
Total	$qrm - 1$	$\sum_{i=1}^{q}\sum_{j=1}^{r}\sum_{k=1}^{m}(y_{ijk} - \bar{y}_{\cdot\cdot\cdot})^2$

Verifying the following expressions for the expected mean squares is left as an exercise:

$$EMS(\text{Factor A}) = \sigma^2 + m\sigma_b^2 + \frac{rm}{q-1}\sum_{i=1}^{q}[(\alpha_i - \bar{\alpha}_\cdot) + (\bar{\xi}_{i\cdot} - \bar{\xi}_{\cdot\cdot})]^2,$$

$$EMS(\text{Whole-plot error}) = \sigma^2 + m\sigma_b^2,$$

$$EMS(\text{Factor B}) = \sigma^2 + \frac{qr}{m-1}\sum_{k=1}^{m}[(\gamma_k - \bar{\gamma}_\cdot) + (\bar{\xi}_{\cdot k} - \bar{\xi}_{\cdot\cdot})]^2,$$

$$EMS(\text{AB interaction}) = \sigma^2 + \frac{r}{(q-1)(m-1)} \sum_{i=1}^{q} \sum_{k=1}^{m} (\xi_{ik} - \bar{\xi}_{i\cdot} - \bar{\xi}_{\cdot k} + \bar{\xi}_{\cdot\cdot})^2,$$

$$EMS(\text{Split-plot error}) = \sigma^2.$$

■

Many types of split-plot models somewhat different than the one featured in the previous example are possible. For example, the split-plot factor may be replicated with split plots, or there may be two or more factors involved in either the whole-plot portion or split-plot portion of the model, or quantitative covariates may be observed on either type of plot. The distinguishing features of a split-plot design are that there are two types of observational units and that a different error term is associated with each type.

13.5 Empirical BLUP (E-BLUP)

Throughout this chapter, as an extension of the previous one, it was assumed that **y** and **u** follow a prediction-extended Aitken model. Under this model, the variance–covariance matrix of **y** and **u** is assumed to be known up to the scalar multiple, σ^2. In practice it will often be more realistic to allow the variance–covariance matrix to be unknown, but to have a structure specified by a more general model such as a mixed effects model or general mixed linear model. For the prediction-extended version of the latter model,

$$\text{var}\begin{pmatrix} \mathbf{y} \\ \mathbf{u} \end{pmatrix} = \begin{pmatrix} \mathbf{V}_{yy}(\boldsymbol{\theta}) & \mathbf{V}_{yu}(\boldsymbol{\theta}) \\ \mathbf{V}_{yu}(\boldsymbol{\theta})^T & \mathbf{V}_{uu}(\boldsymbol{\theta}) \end{pmatrix}.$$

How might prediction be accomplished under this model? First observe that if $\boldsymbol{\theta}$ was known and $\mathbf{V}_{yy}(\boldsymbol{\theta})$ was positive definite, then the model would essentially be a special case of the prediction-extended positive definite Aitken model, so the BLUP of a vector of predictable linear functions $\boldsymbol{\tau} = \mathbf{C}^T\boldsymbol{\beta} + \mathbf{u}$ would, as a special case of Theorem 13.2.1, be given by

$$\mathbf{C}^T\tilde{\boldsymbol{\beta}} + \mathbf{V}_{yu}(\boldsymbol{\theta})^T\mathbf{E}(\boldsymbol{\theta})\mathbf{y},$$

where $\tilde{\boldsymbol{\beta}}$ is any solution to the Aitken equations $\mathbf{X}^T[\mathbf{V}_{yy}(\boldsymbol{\theta})]^{-1}\mathbf{X}\boldsymbol{\beta} = \mathbf{X}^T[\mathbf{V}_{yy}(\boldsymbol{\theta})]^{-1}\mathbf{y}$ and $\mathbf{E}(\boldsymbol{\theta}) = [\mathbf{V}_{yy}(\boldsymbol{\theta})]^{-1} - [\mathbf{V}_{yy}(\boldsymbol{\theta})]^{-1}\mathbf{X}\{\mathbf{X}^T[\mathbf{V}_{yy}(\boldsymbol{\theta})]^{-1}\mathbf{X}\}^-\mathbf{X}^T [\mathbf{V}_{yy}(\boldsymbol{\theta})]^{-1}$; furthermore, the variance–covariance matrix of its vector of prediction errors would, as a special case of Theorem 13.2.2, be given by

$$\begin{aligned}\mathbf{M}(\boldsymbol{\theta}) \equiv \mathbf{V}_{uu}(\boldsymbol{\theta}) &- [\mathbf{V}_{yu}(\boldsymbol{\theta})]^T[\mathbf{V}_{yy}(\boldsymbol{\theta})]^{-1}\mathbf{V}_{yu}(\boldsymbol{\theta}) \\ &+ \{\mathbf{C}^T - [\mathbf{V}_{yu}(\boldsymbol{\theta})]^T[\mathbf{V}_{yy}(\boldsymbol{\theta})]^{-1}\mathbf{X}\}\{\mathbf{X}^T[\mathbf{V}_{yy}(\boldsymbol{\theta})]^{-1}\mathbf{X}\}^- \\ &\times\{\mathbf{C}^T - [\mathbf{V}_{yu}(\boldsymbol{\theta})]^T[\mathbf{V}_{yy}(\boldsymbol{\theta})]^{-1}\mathbf{X}\}^T.\end{aligned}$$

This suggests that when $\boldsymbol{\theta}$ is unknown, we might predict $\boldsymbol{\tau}$ by first obtaining an estimate $\hat{\boldsymbol{\theta}}$ of $\boldsymbol{\theta}$ by some procedure and then proceeding as though this estimate was the true $\boldsymbol{\theta}$, i.e., as though the model was the prediction-extended positive definite Aitken model $\left\{\begin{pmatrix}\mathbf{y}\\ \mathbf{u}\end{pmatrix}, \begin{pmatrix}\mathbf{X}\boldsymbol{\beta}\\ \mathbf{0}\end{pmatrix}, \begin{pmatrix}\hat{\mathbf{V}}_{yy} & \hat{\mathbf{V}}_{yu}\\ \hat{\mathbf{V}}_{yu}^T & \hat{\mathbf{V}}_{uu}\end{pmatrix}\right\}$, where $\hat{\mathbf{V}}_{yy} = \mathbf{V}_{yy}(\hat{\boldsymbol{\theta}})$, $\hat{\mathbf{V}}_{yu} = \mathbf{V}_{yu}(\hat{\boldsymbol{\theta}})$, and $\hat{\mathbf{V}}_{uu} = \mathbf{V}_{uu}(\hat{\boldsymbol{\theta}})$. That is, we might predict $\boldsymbol{\tau}$ by $\mathbf{C}^T\tilde{\hat{\boldsymbol{\beta}}} + \hat{\mathbf{V}}_{yu}^T\hat{\mathbf{E}}\mathbf{y}$, where $\tilde{\hat{\boldsymbol{\beta}}}$ is any solution to the "empirical" Aitken equations

$$\mathbf{X}^T\hat{\mathbf{V}}_{yy}^{-1}\mathbf{X}\boldsymbol{\beta} = \mathbf{X}^T\hat{\mathbf{V}}_{yy}^{-1}\mathbf{y}$$

and $\hat{\mathbf{E}} = \hat{\mathbf{V}}_{yy}^{-1} - \hat{\mathbf{V}}_{yy}^{-1}\mathbf{X}(\mathbf{X}^T\hat{\mathbf{V}}_{yy}^{-1}\mathbf{X})^-\mathbf{X}^T\hat{\mathbf{V}}_{yy}^{-1}$. Along the same lines, we might also estimate the variance–covariance matrix of prediction errors of this predictor by

$$\mathbf{M}(\hat{\boldsymbol{\theta}}) = \hat{\mathbf{V}}_{uu} - \hat{\mathbf{V}}_{yu}^T\hat{\mathbf{V}}_{yy}^{-1}\hat{\mathbf{V}}_{yu} + (\mathbf{C}^T - \hat{\mathbf{V}}_{yu}^T\hat{\mathbf{V}}_{yy}^{-1}\mathbf{X})(\mathbf{X}^T\hat{\mathbf{V}}_{yy}^{-1}\mathbf{X})^-(\mathbf{C}^T - \hat{\mathbf{V}}_{yu}^T\hat{\mathbf{V}}_{yy}^{-1}\mathbf{X})^T.$$

This approach is, in fact, the conventional approach to prediction under a prediction-extended positive definite general mixed linear model. It is often called plug-in prediction, two-stage prediction, or *empirical best linear unbiased prediction* (E-BLUP); we use the last term (and acronym) exclusively, and we use the same acronym for the corresponding predictor. Similar to the E-BLUE, the E-BLUP generally is neither linear nor unbiased, nor is it known to be best in any sense. Properties of the E-BLUP will be considered in Chap. 17, after methods for estimating $\boldsymbol{\theta}$ are presented in Chap. 16.

13.6 Exercises

1. Prove Theorem 13.1.2.
2. Suppose that $n > p^*$. Find a nonnegative definite matrix $\begin{pmatrix}\mathbf{W} & \mathbf{k}\\ \mathbf{k}^T & h\end{pmatrix}$ for which the BLUP equations for a predictable function $\mathbf{c}^T\boldsymbol{\beta}+u$ are not consistent. (Hint: Take $\mathbf{W} = \mathbf{X}\mathbf{X}^T$ and determine a suitable $\mathbf{k}$.)
3. Prove Corollary 13.2.1.1.
4. Prove Theorem 13.2.3.
5. Prove Theorem 13.3.1a.
6. Prove the invariance result in Theorem 13.3.1b.
7. In this exercise, you are to consider using the BLUE $\mathbf{c}^T\tilde{\boldsymbol{\beta}}$ of $\mathbf{c}^T\boldsymbol{\beta}$ as a predictor of a predictable $\tau = \mathbf{c}^T\boldsymbol{\beta} + u$ associated with the prediction-extended positive definite Aitken model $\left\{\begin{pmatrix}\mathbf{y}\\ u\end{pmatrix}, \begin{pmatrix}\mathbf{X}\boldsymbol{\beta}\\ 0\end{pmatrix}, \sigma^2\begin{pmatrix}\mathbf{W} & \mathbf{k}\\ \mathbf{k}^T & h\end{pmatrix}\right\}$. Let us call this predictor the "BLUE-predictor."
   (a) Show that the BLUE-predictor is a linear unbiased predictor of τ.

(b) Obtain an expression for the variance of the prediction error $\mathbf{c}^T\tilde{\boldsymbol{\beta}} - \tau$ corresponding to the BLUE-predictor.
(c) Because [from part (a)] the BLUE-predictor is a linear unbiased predictor, its prediction error variance [obtained in part (b)] must be at least as large as the prediction error variance of the BLUP of τ. Give an expression for how much larger it is, i.e., give an expression for $\text{var}(\mathbf{c}^T\tilde{\boldsymbol{\beta}} - \tau) - \text{var}(\tilde{\tau} - \tau)$.
(d) Determine a necessary and sufficient condition for your answer to part (c) to equal 0. Your condition should take the form $\mathbf{k} \in S$, where S is a certain set of vectors. Thus you will have established a necessary and sufficient condition for the BLUE of $\mathbf{c}^T\boldsymbol{\beta}$ to also be the BLUP of $\mathbf{c}^T\boldsymbol{\beta} + u$.

8. Verify all results for the spatial prediction problem considered in Example 13.2-3. Then, obtain the BLUP of y_7 and the corresponding prediction error variance after making each one of the following modifications to the spatial configuration or model. For each modification, compare the weights corresponding to the elements of $\mathbf{y}$ with the weights in Example 13.2-3, and use your intuition to explain the notable differences. (Note: It is expected that you will use a computer to do this exercise.)
(a) Suppose that y_7 is to be observed at the grid point in the second row (from the bottom) of the first column (from the left).
(b) Suppose that the (i, j)th element of $\mathbf{W}$ is equal to $\exp(-d_{ij}/4)$.
(c) Suppose that the (i, j)th element of $\mathbf{W}$ is equal to 1 if $i = j$, or $0.5\exp(-d_{ij}/2)$ otherwise.

9. Observe that in Example 13.2-3 and Exercise 13.8, the "weights" (the coefficients on the elements of $\mathbf{y}$ in the expression for the BLUP) sum to one (apart from roundoff error). Explain why this is so.

10. Suppose that observations (x_i, y_i) follow the simple linear regression model

$$y_i = \beta_1 + \beta_2 x_i + e_i \quad (i = 1, \ldots, n)$$

where $n \geq 3$. Consider the problem of predicting an unobserved y-value, y_{n+1}, corresponding to a specified x-value, x_{n+1}. Assume that y_{n+1} follows the same basic model as the observed responses; that is, the joint model for y_{n+1} and the observed responses can be written as

$$y_i = \beta_1 + \beta_2 x_i + e_i \qquad (i = 1, \ldots, n, n+1),$$

where the e_i's satisfy Gauss–Markov assumptions with common variance $\sigma^2 > 0$. Example 13.2-1 established that the best linear unbiased predictor of y_{n+1} under this model is

$$\tilde{y}_{n+1} = \hat{\beta}_1 + \hat{\beta}_2 x_{n+1}$$

and its mean squared prediction error (MSPE) is $\sigma^2[1 + (1/n) + (x_{n+1} - \bar{x})^2/SXX]$. Although $\tilde{y}_{n+1}$ has smallest MSPE among all *unbiased* linear predictors of y_{n+1}, *biased* linear predictors of y_{n+1} exist that may have smaller MSPE than $\tilde{y}_{n+1}$. One such predictor is $\bar{y}$, the sample mean of the y_i's. For $x_{n+1} \neq \bar{x}$, show that $\bar{y}$ has smaller MSPE than $\tilde{y}_{n+1}$ if and only if $\beta_2^2 SXX/\sigma^2 < 1$.

11. Obtain the expressions for $\text{var}(\tilde{b} - b)$ and its unbiased estimator given in Example 13.4.2-1.
12. Consider the random no-intercept simple linear regression model

$$y_i = bz_i + d_i \quad (i = 1, \ldots, n)$$

where $b, d_1, d_2, \ldots, d_n$ are uncorrelated zero-mean random variables such that $\text{var}(b) = \sigma_b^2 > 0$ and $\text{var}(d_i) = \sigma^2 > 0$ for all i. Let $\psi = \sigma_b^2/\sigma^2$ and suppose that ψ is known. Let z represent a specified real number.

(a) Write out the mixed-model equations in as simple a form as possible, and solve them.

(b) Give a nonmatrix expression, simplified as much as possible, for the BLUP of bz.

(c) Give a nonmatrix expression, simplified as much as possible, for the prediction error variance of the BLUP of bz.

13. Consider the following random-slope simple linear regression model:

$$y_i = \beta + bz_i + d_i \quad (i = 1, \cdots, n),$$

where β is an unknown parameter, b is a zero-mean random variable with variance $\sigma_b^2 > 0$, and the d_i's are uncorrelated (with each other and with b) zero-mean random variables with common variance $\sigma^2 > 0$. Let $\psi = \sigma_b^2/\sigma^2$ and suppose that ψ is known. The model equation may be written in matrix form as

$$\mathbf{y} = \beta\mathbf{1} + b\mathbf{z} + \mathbf{d}.$$

(a) Determine $\sigma^2\mathbf{W} \equiv \text{var}(\mathbf{y})$.

(b) Write out the mixed-model equations in as simple a form as possible and solve them.

(c) Consider predicting the predictable function $\tau \equiv \beta + bz$, where z is a specified real number. Give an expression for the BLUP $\tilde{\tau}$ of τ in terms of a solution $(\dot{\beta}, \dot{b})^T$ to the mixed-model equations.

(d) Give a nonmatrix expression for $\tilde{\sigma}^2$, the generalized residual mean square, in terms of a solution to the mixed-model equations mentioned in part (c).

(e) Obtain a nonmatrix expression for the prediction error variance associated with $\tilde{\tau}$.

14. Consider the following random-intercept simple linear regression model:

$$y_i = b + \beta x_i + d_i \quad (i = 1, \cdots, n),$$

where β is an unknown parameter, b is a zero-mean random variable with variance $\sigma_b^2 > 0$, and the d_i's are uncorrelated (with each other and with b) zero-mean random variables with common variance $\sigma^2 > 0$. Let $\psi = \sigma_b^2/\sigma^2$ and suppose that ψ is known. The model equation may be written in matrix form as

$$\mathbf{y} = b\mathbf{1} + \beta\mathbf{x} + \mathbf{d}.$$

(a) Determine $\sigma^2\mathbf{W} \equiv \text{var}(\mathbf{y})$.
(b) Write out the mixed-model equations in as simple a form as possible and solve them.
(c) Consider predicting the predictable function $\tau \equiv b + \beta x$, where x is a specified real number. Give an expression for the BLUP $\tilde{\tau}$ of τ in terms of a solution $(\dot{\beta}, \dot{b})^T$ to the mixed-model equations.
(d) Give a nonmatrix expression for $\tilde{\sigma}^2$, the generalized residual mean square, in terms of a solution to the mixed-model equations mentioned in part (c).
(e) Obtain a nonmatrix expression for the prediction error variance associated with $\tilde{\tau}$.

15. Consider the following random simple linear regression model:

$$y_i = b_1 + b_2 z_i + d_i \quad (i = 1, \cdots, n),$$

where b_1 and b_2 are uncorrelated zero-mean random variables with variances $\sigma_{b_1}^2 > 0$ and $\sigma_{b_2}^2 > 0$, respectively, and the d_i's are uncorrelated (with each other and with b_1 and b_2) zero-mean random variables with common variance $\sigma^2 > 0$. Let $\psi_1 = \sigma_{b_1}^2/\sigma^2$ and $\psi_2 = \sigma_{b_2}^2/\sigma^2$, and suppose that ψ_1 and ψ_2 are known. The model equation may be written in matrix form as

$$\mathbf{y} = b_1\mathbf{1} + b_2\mathbf{z} + \mathbf{d}.$$

(a) Determine $\sigma^2\mathbf{W} \equiv \text{var}(\mathbf{y})$.
(b) Write out the mixed-model equations in as simple a form as possible and solve them.
(c) Consider predicting the predictable function $\tau \equiv b_1 + b_2 z$, where z is a specified real number. Give an expression for the BLUP $\tilde{\tau}$ of τ in terms of a solution $(\dot{b}_1, \dot{b}_2)^T$ to the mixed-model equations.
(d) Give a nonmatrix expression for $\tilde{\sigma}^2$, the generalized residual mean square, in terms of a solution to the mixed-model equations mentioned in part (c).
(e) Obtain a nonmatrix expression for the prediction error variance associated with $\tilde{\tau}$.

16. For the balanced one-factor random effects model considered in Example 13.4.2-2:
    (a) Obtain specialized expressions for the variances and covariances of the prediction errors, $\{\widetilde{b_i - b_{i'}} - (b_i - b_{i'}) : i < i' = 1, \ldots, q\}$.
    (b) Obtain a specialized expression for the generalized residual mean square, $\tilde{\sigma}^2$.
17. Consider a two-way layout with two rows and two columns, and one observation in three of the four cells, labeled as y_{11}, y_{12}, and y_{21} as depicted in the sketch below:

y_{11}	y_{12}
y_{21}	

No response is observed in the other cell, but we would like to use the existing data to predict such a response, which we label as y_{22} (the response, not its predictor). Obtain simplified expressions for the BLUP of y_{22} and its mean squared prediction error, under each of the following three models. Note: The following suggestions may make your work easier in this exercise:

- For part (a), reparameterize the model in such a way that the parameter vector is given by

$$\begin{pmatrix} \mu + \alpha_2 + \gamma_2 \\ \gamma_1 - \gamma_2 \\ \alpha_1 - \alpha_2 \end{pmatrix}.$$

  For part (b), reparameterize the model in such a way that the parameter vector is given by

$$\begin{pmatrix} \mu + \alpha_2 \\ \alpha_1 - \alpha_2 \end{pmatrix}.$$

- The following matrix inverses may be useful:

$$\begin{pmatrix} 3 & 2 & 2 \\ 2 & 2 & 1 \\ 2 & 1 & 2 \end{pmatrix}^{-1} = \begin{pmatrix} 3 & -2 & -2 \\ -2 & 2 & 1 \\ -2 & 1 & 2 \end{pmatrix},$$

$$\begin{pmatrix} 2 & 0 & 1 \\ 0 & 2 & 0 \\ 1 & 0 & 2 \end{pmatrix}^{-1} = \begin{pmatrix} \frac{2}{3} & 0 & -\frac{1}{3} \\ 0 & \frac{1}{2} & 0 \\ -\frac{1}{3} & 0 & \frac{2}{3} \end{pmatrix},$$

$$\begin{pmatrix} 3 & 1 & 1 \\ 1 & 3 & 0 \\ 1 & 0 & 3 \end{pmatrix}^{-1} = \begin{pmatrix} \frac{3}{7} & -\frac{1}{7} & -\frac{1}{7} \\ -\frac{1}{7} & \frac{8}{21} & \frac{1}{21} \\ -\frac{1}{7} & \frac{1}{21} & \frac{8}{21} \end{pmatrix}.$$

(a) The Gauss–Markov two-way main effects model $y_{ij} = \mu + \alpha_i + \gamma_j + e_{ij}$,

$$\mathrm{E}\begin{pmatrix} e_{11} \\ e_{12} \\ e_{21} \\ e_{22} \end{pmatrix} = \mathbf{0}, \quad \mathrm{var}\begin{pmatrix} e_{11} \\ e_{12} \\ e_{21} \\ e_{22} \end{pmatrix} = \sigma^2 \mathbf{I}.$$

(b) A mixed two-way main effects model $y_{ij} = \mu + \alpha_i + b_j + d_{ij}$,

$$\mathrm{E}\begin{pmatrix} b_1 \\ b_2 \\ d_{11} \\ d_{12} \\ d_{21} \\ d_{22} \end{pmatrix} = \mathbf{0}, \quad \mathrm{var}\begin{pmatrix} b_1 \\ b_2 \\ d_{11} \\ d_{12} \\ d_{21} \\ d_{22} \end{pmatrix} = \begin{pmatrix} \sigma_b^2 \mathbf{I}_2 & \mathbf{0} \\ \mathbf{0} & \sigma^2 \mathbf{I}_4 \end{pmatrix},$$

where $\sigma_b^2/\sigma^2 = 1$.

(c) A random two-way main effects model $y_{ij} = \mu + a_i + b_j + d_{ij}$,

$$\mathrm{E}\begin{pmatrix} a_1 \\ a_2 \\ b_1 \\ b_2 \\ d_{11} \\ d_{12} \\ d_{21} \\ d_{22} \end{pmatrix} = \mathbf{0}, \quad \mathrm{var}\begin{pmatrix} a_1 \\ a_2 \\ b_1 \\ b_2 \\ d_{11} \\ d_{12} \\ d_{21} \\ d_{22} \end{pmatrix} = \begin{pmatrix} \sigma_a^2 \mathbf{I}_2 & \mathbf{0} & \mathbf{0} \\ \mathbf{0} & \sigma_b^2 \mathbf{I}_2 & \\ \mathbf{0} & \mathbf{0} & \sigma^2 \mathbf{I}_4 \end{pmatrix},$$

where $\sigma_b^2/\sigma^2 = \sigma_a^2/\sigma^2 = 1$.

18. Consider the following mixed linear model for two observations, y_1 and y_2:

$$y_1 = 2\beta + b + d_1$$
$$y_2 = \beta + 3b + d_2,$$

where β is a fixed unknown (and unrestricted) parameter, and b, d_1, d_2 are independent random variables with zero means and variances $\sigma_b^2 = \mathrm{var}(b)$ and $\sigma^2 = \mathrm{var}(d_1) = \mathrm{var}(d_2)$. Suppose that $\sigma_b^2/\sigma^2 = 1/2$.

(a) Write this model in the matrix form $\mathbf{y} = \mathbf{X}\boldsymbol{\beta} + \mathbf{Z}\mathbf{b} + \mathbf{d}$ and give an expression for var($\mathbf{y}$) in which σ^2 is the only unknown parameter.

(b) Compute the BLUE of β and the BLUP of b.

19. Consider the balanced mixed two-way main effects model

$$y_{ijk} = \mu + \alpha_i + b_j + d_{ijk} \quad (i = 1, \ldots, q;\ j = 1, \ldots, m;\ k = 1, \ldots, r)$$

where $\mu, \alpha_1, \ldots, \alpha_q$ are unknown parameters, the b_j's are uncorrelated zero-mean random variables with common variance $\sigma_b^2 > 0$, and the d_{ijk}'s are uncorrelated (with each other and with the b_j's) zero-mean random variables with common variance $\sigma^2 > 0$. Let $\psi = \sigma_b^2/\sigma^2$ and suppose that ψ is known.
   (a) One solution to the Aitken equations for this model is $\tilde{\boldsymbol{\beta}}=(0, \bar{y}_{1\cdot\cdot}, \ldots, \bar{y}_{q\cdot\cdot})^T$. Using this solution, obtain specialized expressions for the BLUEs of $\mu+\alpha_i$ and $\alpha_i - \alpha_{i'}$ $(i' > i = 1, \ldots, q)$.
   (b) Obtain a specialized expression for the variance–covariance matrix of the BLUEs of $\mu + \alpha_i$, and do likewise for the BLUEs of $\alpha_i - \alpha_{i'}$ $(i' > i = 1, \ldots, q)$.
   a. Obtain specialized expressions for the BLUPs of $\mu + \alpha_i + b_j$ $(i = 1, \ldots, q;\ j = 1, \ldots, m)$ and $b_j - b_{j'}$ $(j' > j = 1, \ldots, m)$.
   (c) Obtain a specialized expression for the variance–covariance matrix of the BLUPs of $\mu + \alpha_i + b_j$, and do likewise for the BLUPs of $b_j - b_{j'}$ $(i = 1, \ldots, q;\ j' > j = 1, \ldots, m)$.
20. Consider a random two-way partially crossed model analogous to the fixed-effects model introduced in Example 5.1.4-1, with one observation per cell, i.e.,

$$y_{ij} = \mu + b_i - b_j + d_{ij} \qquad (i \neq j = 1, \ldots, q)$$

where μ is an unknown parameter, the b_i's are uncorrelated zero-mean random variables with common variance $\sigma_b^2 > 0$, and the d_{ij}'s are uncorrelated (with each other and with the b_i's) zero-mean random variables with common variance $\sigma^2 > 0$. Let $\psi = \sigma_b^2/\sigma^2$ and suppose that ψ is known.
   (a) This model is a special case of the general mixed linear model

$$\mathbf{y} = \mathbf{X}\boldsymbol{\beta} + \mathbf{Z}\mathbf{b} + \mathbf{d}$$

   where $\text{var}(\mathbf{d}) = \sigma^2\mathbf{I}$ and $\text{var}(\mathbf{b}) = \sigma^2\mathbf{G}$ for some positive definite matrix $\mathbf{G}$. Write out the elements of the matrices $\mathbf{X}$, $\mathbf{Z}$, and $\mathbf{G}$ for this case.
   (b) Write out the mixed-model equations for this model and solve them.
   (c) Determine $\mathbf{W} \equiv (1/\sigma^2)\text{var}(\mathbf{y})$ and determine $\mathbf{W}^{-1}$.
   (d) Give expressions, in as simple a form as possible, for the BLUPs of $b_i - b_j$ $(j > i = 1, \ldots, q)$ and for the variance–covariance matrix of the prediction errors associated with these BLUPs. Specialize that variance–covariance matrix for the case $q = 4$.
21. Consider the balanced mixed two-factor nested model

$$y_{ijk} = \mu + \alpha_i + b_{ij} + d_{ijk} \qquad (i = 1, \ldots, q;\ j = 1, \ldots, m;\ k = 1, \ldots, r),$$

where $\mu, \alpha_1, \ldots, \alpha_q$ are unknown parameters, the b_{ij}'s are uncorrelated zero-mean random variables with common variance $\sigma_b^2 > 0$, and the d_{ijk}'s are uncorrelated (with each other and with the b_{ij}'s) zero-mean random variables with common variance $\sigma^2 > 0$. Let $\psi = \sigma_b^2/\sigma^2$ and suppose that ψ is known.

(a) This model is a special case of the general mixed linear model

$$\mathbf{y} = \mathbf{X}\boldsymbol{\beta} + \mathbf{Zb} + \mathbf{d},$$

where $\text{var}(\mathbf{d}) = \sigma^2\mathbf{I}$ and $\text{var}(\mathbf{b}) = \sigma^2\mathbf{G}$ for some positive definite matrix $\mathbf{G}$. Specialize the matrices $\mathbf{X}$, $\mathbf{Z}$, and $\mathbf{G}$ for this model.

(b) Is $\alpha_1 - \alpha_2 + b_{31}$ (assuming that $q \geq 3$) a predictable function? Explain why or why not.

(c) Specialize the mixed-model equations for this model.

(d) Obtain specialized expressions for $\mathbf{W} = (1/\sigma^2)\text{var}(\mathbf{y})$ and $\mathbf{W}^{-1}$.

(e) It can be shown that one solution to the Aitken equations corresponding to this model is $\tilde{\boldsymbol{\beta}} = (0, \bar{y}_{1\cdot\cdot}, \ldots, \bar{y}_{q\cdot\cdot})$. Using this fact, give a specialized expression for the BLUP of $\mu + \alpha_i + b_{ij}$.

(f) Obtain a specialized expression for the prediction error variance of the BLUP of $\mu + \alpha_i + b_{ij}$.

22. Consider a mixed effects model with known $\boldsymbol{\psi}$, where $\mathbf{G} = \mathbf{G}(\boldsymbol{\psi})$ and $\mathbf{R} = \mathbf{R}(\boldsymbol{\psi})$ are positive definite. Let $\mathbf{B}$ represent the unique $q \times q$ positive definite matrix for which $\mathbf{B}^T\mathbf{B} = \mathbf{G}^{-1}$ (such a matrix exists by Theorem 2.15.12). Suppose that we transform this model by pre-multiplying all of its terms by $\mathbf{R}^{-\frac{1}{2}}$, then augment the transformed model with the model $\mathbf{y}^* = \mathbf{Bb} + \mathbf{h}$, where $\mathbf{y}^* = \mathbf{0}_q$ and $\mathbf{h}$ is a random q-vector satisfying $\text{E}(\mathbf{h}) = \mathbf{0}$ and $\text{var}(\mathbf{h}) = \sigma^2\mathbf{I}$. Show that any solution to the normal equations for this transformed/augmented model is also a solution to the mixed-model equations associated with the original mixed effects model, thus establishing that any computer software that can perform ordinary least squares regression can be coerced into obtaining BLUPs under a mixed effects model.

23. Prove Theorem 13.4.4. (Hint: For the invariance, use Theorem 3.2.1.)

24. For Example 13.4.5-1:

(a) Verify the given expressions for $\mathbf{P_1}$, $\mathbf{P}_\alpha$, $\mathbf{P_Z}$, $\mathbf{P}_\gamma$, and $\mathbf{P}_\xi$.

(b) Verify all the expressions for the pairwise products of $\mathbf{P_1}$, $\mathbf{P}_\alpha$, $\mathbf{P_Z}$, $\mathbf{P}_\gamma$, and $\mathbf{P}_\xi$.

(c) Verify that the matrices of the six quadratic forms in decomposition (13.12) are idempotent.

(d) Verify the given expressions for the expected mean squares.

25. For the split-plot model considered in Example 13.4.5-1, obtain simplified expressions for:

(a) $\text{var}(\bar{y}_{i\cdot\cdot} - \bar{y}_{i'\cdot\cdot})$ $(i \neq i')$;

(b) $\text{var}(\bar{y}_{\cdot\cdot k} - \bar{y}_{\cdot\cdot k'})$ $(k \neq k')$;

(c) $\text{var}(\bar{y}_{i\cdot k} - \bar{y}_{i\cdot k'})$ $(k \neq k')$;

(d) $\text{var}(\bar{y}_{i\cdot k} - \bar{y}_{i'\cdot k'})$ $(i \neq i')$.

References

Goldberger (1962). Best linear unbiased prediction in the generalized linear regression model. *Journal of the American Statistical Association, 57*, 369–375.

Harville, D. A. (1976). Extension of the Gauss–Markov theorem to include the estimation of random effects. *Annals of Statistics, 4*, 384–395.

Henderson, C. R. (1963). Selection index and expected genetic advance. In W. D. Hanson & H. F. Robinson (Eds.), *Statistical genetics and plant breeding*. Publication 982 (pp. 141–163). Washington: National Academy of Sciences — National Research Council.

14 Distribution Theory

Much of classical statistical inference for linear models is based on special cases of those models for which the response vector **y** has a multivariate normal distribution. Before we can present those inferential methods, therefore, we must precisely define the multivariate normal distribution and the related noncentral chi-square, t, and F distributions, and describe some of their important properties. These are the topics of this chapter. It is possible to extend some of the results presented in this chapter and in the remainder of the book to linear models in which **y** has a distribution from the more general family of "elliptical" distributions, but we do not consider these extensions here. The reader who is interested in such extensions is referred to Ravishanker and Dey (2002, Sections 5.5.3 and 7.5.2) and Harville (2018, Sections 5.9cd and various parts of Chapter 6).

14.1 The Multivariate Normal Distribution

Unless noted otherwise, in this chapter **x** represents a generic random n-vector, rather than the vector of explanatory variables in a linear model. In examples, random vectors encountered in earlier chapters such as the response vector **y**, the fitted residuals vector $\hat{\mathbf{e}}$, and the predictable predictand $\mathbf{C}^T\boldsymbol{\beta} + \mathbf{u}$ will exemplify some of the results obtained here for the generic random vector **x**.

Definition 14.1.1 The *moment generating function* (mgf) of the random n-vector **x** is the function

$$m_{\mathbf{x}}(\mathbf{t}) = \mathrm{E}(e^{\mathbf{t}^T\mathbf{x}}), \quad \mathbf{t} \in \mathbb{R}^n,$$

provided that the expectation exists for **t** in an open rectangle within $\mathbb{R}^n$ that contains the origin. If this expectation does not exist in such a rectangle, then **x** has no mgf.

D. L. Zimmerman, *Linear Model Theory*,
https://doi.org/10.1007/978-3-030-52063-2_14

It is shown in elementary mathematical statistics courses that if the mgf of a random vector $\mathbf{x}$ exists, then the gth-order moments ($g = 1, 2, \ldots$) of $\mathbf{x}$ may be obtained by gth-order partial differentiation of the mgf of $\mathbf{x}$ with respect to $\mathbf{t}$, followed by evaluation of the partial derivatives at $\mathbf{t} = \mathbf{0}$.

Proofs of the following two theorems can be found in textbooks on probability and mathematical statistics; for example, a proof of Theorem 14.1.1 can be found in Chung (1974), and a proof of the bivariate case of Theorem 14.1.2, which can be extended easily to more than two variables, is given in Hogg et al. (2013, pp. 114–115).

Theorem 14.1.1 (Uniqueness Theorem) *Consider two random n-vectors $\mathbf{x}_1$ and $\mathbf{x}_2$, and suppose that the mgf of each exists. The cumulative distribution functions (cdfs) of $\mathbf{x}_1$ and $\mathbf{x}_2$ are identical if and only if the mgfs are equal for all values of $\mathbf{t}$ in an open rectangle that contains the origin.*

Theorem 14.1.2 *Partition a random n-vector $\mathbf{x}$ as $\mathbf{x} = (\mathbf{x}_1^T, \ldots, \mathbf{x}_m^T)^T$, and denote the mgfs of $\mathbf{x}$ and its subvectors by $m_{\mathbf{x}}(\cdot)$, $m_{\mathbf{x}_1}(\cdot)$, ..., $m_{\mathbf{x}_m}(\cdot)$. Then $\mathbf{x}_1, \ldots, \mathbf{x}_m$ are mutually independent if and only if*

$$m_{\mathbf{x}}(\mathbf{t}) = m_{\mathbf{x}_1}(\mathbf{t}_1)m_{\mathbf{x}_2}(\mathbf{t}_2)\cdots m_{\mathbf{x}_m}(\mathbf{t}_m)$$

for all $\mathbf{t} = (\mathbf{t}_1^T, \ldots, \mathbf{t}_m^T)^T$ in an open rectangle that contains the origin.

In a first or second calculus course, students typically learn the following integral calculus results:

$$\int_{-\infty}^{\infty} e^{-x^2/2}dx = (2\pi)^{\frac{1}{2}}, \tag{14.1}$$

$$\int_{-\infty}^{\infty} xe^{-x^2/2}dx = 0, \tag{14.2}$$

$$\int_{-\infty}^{\infty} x^2e^{-x^2/2}dx = (2\pi)^{\frac{1}{2}}. \tag{14.3}$$

We shall need multivariate generalizations of these results. The following theorem is proved by Harville (1997, Section 15.12).

Theorem 14.1.3 *Let $\mathbf{A}$ represent a symmetric $n \times n$ matrix, let $\mathbf{B}$ represent a positive definite $n \times n$ matrix, let $\mathbf{a}$ and $\mathbf{b}$ represent two arbitrary n-vectors, and let a_0 and b_0 represent two arbitrary scalars. Then*

$$\int_{-\infty}^{\infty}\cdots\int_{-\infty}^{\infty}(a_0 + \mathbf{a}^T\mathbf{x} + \mathbf{x}^T\mathbf{A}\mathbf{x})\exp[-(b_0 + \mathbf{b}^T\mathbf{x} + \mathbf{x}^T\mathbf{B}\mathbf{x})]dx_1\cdots dx_n$$

$$= (1/2)\pi^{\frac{n}{2}}|\mathbf{B}|^{-\frac{1}{2}}[tr(\mathbf{A}\mathbf{B}^{-1}) - \mathbf{a}^T\mathbf{B}^{-1}\mathbf{b} + (1/2)\mathbf{b}^T\mathbf{B}^{-1}\mathbf{A}\mathbf{B}^{-1}\mathbf{b} + 2a_0]\{\exp[(1/4)\mathbf{b}^T\mathbf{B}^{-1}\mathbf{b} - b_0]\}.$$

Definition 14.1.2 A random variable z is said to have a *univariate standard normal distribution* if the probability density function (pdf) of z exists and is given by

$$f(z) = (2\pi)^{-\frac{1}{2}} \exp(-z^2/2), \quad -\infty < z < \infty.$$

It follows easily from this definition that if random variables $z_1, \ldots, z_n$ are mutually independent and each has a univariate standard normal distribution, then the joint pdf of the random n-vector $\mathbf{z} = (z_1, \ldots, z_n)^T$ is

$$\phi(\mathbf{z}) = (2\pi)^{-\frac{n}{2}} \exp(-\mathbf{z}^T\mathbf{z}/2), \quad \mathbf{z} \in \mathbb{R}^n. \tag{14.4}$$

Definition 14.1.3 A random n-vector $\mathbf{z}$ is said to have an *n-variate standard normal distribution* if its pdf exists and is given by (14.4).

By (14.2), the mean of a univariate standard normal distribution is 0. This, together with (14.3), implies that the variance of a univariate standard normal distribution is 1. It follows easily that the mean vector and variance–covariance matrix of an n-variate standard normal vector are $\mathbf{0}_n$ and $\mathbf{I}_n$, respectively.

Now we define a more general n-variate normal distribution.

Definition 14.1.4 Let $\mathbf{x}$ represent a random n-vector with mean vector $\boldsymbol{\mu}$ and variance–covariance matrix $\boldsymbol{\Sigma}$, where $0 \le r \equiv \text{rank}(\boldsymbol{\Sigma}) \le n$. Let $\boldsymbol{\Gamma}$ represent an $r \times n$ matrix such that $\boldsymbol{\Sigma} = \boldsymbol{\Gamma}^T\boldsymbol{\Gamma}$. (Such a matrix exists by Theorem 2.15.14.) Then, $\mathbf{x}$ is said to have an *n-variate normal distribution of rank r* if the cdf of $\mathbf{x}$ is the same as that of the random vector $\boldsymbol{\mu} + \boldsymbol{\Gamma}^T\mathbf{z}$, where $\mathbf{z}$ has an r-variate standard normal distribution.

In Definition 14.1.4, $\mathbf{x}$ and $\boldsymbol{\mu} + \boldsymbol{\Gamma}^T\mathbf{z}$ have the same mean vector and variance–covariance matrix, whether or not their cdfs are the same. Furthermore, the factorization of $\boldsymbol{\Sigma}$ as $\boldsymbol{\Gamma}^T\boldsymbol{\Gamma}$ may not be unique, so the n-variate normal distribution of rank r could potentially be ill-defined; however, we shall see shortly that there is no such problem with the definition.

We use the notation $\text{N}_n(\boldsymbol{\mu}, \boldsymbol{\Sigma})$ to represent the cdf of an n-variate normal distribution with mean vector $\boldsymbol{\mu}$ and variance–covariance matrix $\boldsymbol{\Sigma}$; furthermore, if $\mathbf{x}$ is a random n-vector having this distribution, we write $\mathbf{x} \sim \text{N}_n(\boldsymbol{\mu}, \boldsymbol{\Sigma})$). The subscript n may be omitted if the dimensionality is not important in a particular context. The rank of $\boldsymbol{\Sigma}$ is not generally specified by the notation, so if necessary it will be specified in a separate statement. The following theorem gives the mgf of a random vector having this cdf.

Theorem 14.1.4 *Suppose that* $\mathbf{x} \sim N_n(\boldsymbol{\mu}, \boldsymbol{\Sigma})$. *Then*

$$m_{\mathbf{x}}(\mathbf{t}) = \exp(\mathbf{t}^T\boldsymbol{\mu} + \mathbf{t}^T\boldsymbol{\Sigma}\mathbf{t}/2), \quad \mathbf{t} \in \mathbb{R}^n.$$

Proof Let $r = \text{rank}(\boldsymbol{\Sigma})$, let $\boldsymbol{\Gamma}$ represent an $r \times n$ matrix such that $\boldsymbol{\Sigma} = \boldsymbol{\Gamma}^T\boldsymbol{\Gamma}$, and let $\mathbf{z} \sim \text{N}_r(\mathbf{0}, \mathbf{I})$. Then,

$$\begin{aligned} m_{\mathbf{x}}(\mathbf{t}) &= m_{\boldsymbol{\mu}+\boldsymbol{\Gamma}^T\mathbf{z}}(\mathbf{t}) \\ &= \text{E}\{\exp[\mathbf{t}^T(\boldsymbol{\mu} + \boldsymbol{\Gamma}^T\mathbf{z})]\} \\ &= (2\pi)^{-\frac{r}{2}} \int_{-\infty}^{\infty} \cdots \int_{-\infty}^{\infty} \exp(\mathbf{t}^T\boldsymbol{\mu} + \mathbf{t}^T\boldsymbol{\Gamma}^T\mathbf{z} - \mathbf{z}^T\mathbf{z}/2)d\mathbf{z} \\ &= \exp(\mathbf{t}^T\boldsymbol{\mu} + \mathbf{t}^T\boldsymbol{\Sigma}\mathbf{t}/2), \quad \mathbf{t} \in \mathbb{R}^n, \end{aligned}$$

where the last equality results from putting $a_0 = 1$, $\mathbf{a} = \mathbf{0}$, $\mathbf{A} = \mathbf{0}$, $b_0 = -\mathbf{t}^T\boldsymbol{\mu}$, $\mathbf{b} = -\boldsymbol{\Gamma}\mathbf{t}$, and $\mathbf{B} = \frac{1}{2}\mathbf{I}$ into Theorem 14.1.3. □

By Theorem 14.1.4, if $\mathbf{x} \sim \text{N}_n(\boldsymbol{\mu}, \boldsymbol{\Sigma})$, then its mgf is the same no matter what factorization of $\boldsymbol{\Sigma}$ is used; thus, by Theorem 14.1.1, the n-variate normal distribution of rank r is well defined. However, its pdf may or may not exist. The following theorem gives a condition under which the pdf exists.

Theorem 14.1.5 *Suppose that* $\mathbf{x} \sim N_n(\boldsymbol{\mu}, \boldsymbol{\Sigma})$ *where rank*$(\boldsymbol{\Sigma}) = n$. *Then the pdf of* $\mathbf{x}$ *exists and is given by*

$$f(\mathbf{x}) = (2\pi)^{-\frac{n}{2}}|\boldsymbol{\Sigma}|^{-\frac{1}{2}}\exp[-(\mathbf{x}-\boldsymbol{\mu})^T\boldsymbol{\Sigma}^{-1}(\mathbf{x}-\boldsymbol{\mu})/2], \quad \mathbf{x} \in \mathbb{R}^n.$$

Proof Because $\text{rank}(\boldsymbol{\Sigma}) = n$, $\boldsymbol{\Sigma}$ is positive definite and $|\boldsymbol{\Sigma}|^{-\frac{1}{2}}$ is positive (by Theorem 2.15.8). It follows that $f(\mathbf{x})$ is positive for all $\mathbf{x}$. Furthermore, using Theorem 14.1.3 with $a_0 = 1$, $\mathbf{a} = \mathbf{0}$, $\mathbf{A} = \mathbf{0}$, $b_0 = \frac{1}{2}\boldsymbol{\mu}^T\boldsymbol{\Sigma}^{-1}\boldsymbol{\mu}$, $\mathbf{b} = -\boldsymbol{\Sigma}^{-1}\boldsymbol{\mu}$, and $\mathbf{B} = \frac{1}{2}\boldsymbol{\Sigma}^{-1}$, we find that $f(\mathbf{x})$ integrates to 1. Thus, $f(\mathbf{x})$ is the pdf of some random n-vector. The mgf of such a random vector, if it exists, is given by

$$\begin{aligned} m_{\mathbf{x}}(\mathbf{t}) &= \int_{-\infty}^{\infty} \cdots \int_{-\infty}^{\infty} (2\pi)^{-\frac{n}{2}}|\boldsymbol{\Sigma}|^{-\frac{1}{2}}\exp[\mathbf{t}^T\mathbf{x} - (\mathbf{x}-\boldsymbol{\mu})^T\boldsymbol{\Sigma}^{-1}(\mathbf{x}-\boldsymbol{\mu})/2]dx_1\cdots dx_n \\ &= (2\pi)^{-\frac{n}{2}}|\boldsymbol{\Sigma}|^{-\frac{1}{2}}(1/2)\pi^{\frac{n}{2}}|(1/2)\boldsymbol{\Sigma}^{-1}|^{-\frac{1}{2}} \\ &\quad \times \exp[(1/4)(\mathbf{t}+\boldsymbol{\Sigma}^{-1}\boldsymbol{\mu})^T(2\boldsymbol{\Sigma})(\mathbf{t}+\boldsymbol{\Sigma}^{-1}\boldsymbol{\mu}) - (1/2)\boldsymbol{\mu}^T\boldsymbol{\Sigma}^{-1}\boldsymbol{\mu}](2) \\ &= (2\pi)^{-\frac{n}{2}}|\boldsymbol{\Sigma}|^{-\frac{1}{2}}\pi^{\frac{n}{2}}(1/2)^{-\frac{n}{2}}|\boldsymbol{\Sigma}|^{\frac{1}{2}} \\ &\quad \times \exp[(1/2)(\mathbf{t}^T\boldsymbol{\Sigma}\mathbf{t} + \boldsymbol{\mu}^T\mathbf{t} + \mathbf{t}^T\boldsymbol{\mu} + \boldsymbol{\mu}^T\boldsymbol{\Sigma}^{-1}\boldsymbol{\mu}) - (1/2)\boldsymbol{\mu}^T\boldsymbol{\Sigma}^{-1}\boldsymbol{\mu}] \\ &= \exp(\mathbf{t}^T\boldsymbol{\mu} + \mathbf{t}^T\boldsymbol{\Sigma}\mathbf{t}/2) \end{aligned}$$

for all $\mathbf{t} \in \mathbb{R}^n$, where we used Theorem 14.1.3 with $a_0 = 1$, $\mathbf{a} = \mathbf{0}$, $\mathbf{A} = \mathbf{0}$, $b_0 = \frac{1}{2}\boldsymbol{\mu}^T\boldsymbol{\Sigma}^{-1}\boldsymbol{\mu}$, $\mathbf{b} = -\mathbf{t} - \boldsymbol{\Sigma}^{-1}\boldsymbol{\mu}$, and $\mathbf{B} = \frac{1}{2}\boldsymbol{\Sigma}^{-1}$ to obtain the second equality, and Theorems 2.11.4 and 2.11.6 to obtain the third equality. The result follows by Theorem 14.1.1 upon recognizing that this mgf matches the mgf specified in Theorem 14.1.4. □

Henceforth, we refer to a linear model in which the response vector $\mathbf{y}$ (or equivalently, the residual vector $\mathbf{e}$) has a multivariate normal distribution as a *normal* linear model. Thus we may refer to a normal Gauss–Markov model, a normal Aitken model, a normal constrained Gauss–Markov model, etc.

Example 14.1-1. Maximum Likelihood Estimators of Estimable Functions and Residual Variance Under the Normal Positive Definite Aitken Model

Consider the problem of obtaining maximum likelihood estimators of an estimable function $\mathbf{c}^T\boldsymbol{\beta}$ and the residual variance σ^2 under the normal positive definite Aitken model $\{\mathbf{y}, \mathbf{X}\boldsymbol{\beta}, \sigma^2\mathbf{W}\}$. By definition, maximum likelihood estimators of $\boldsymbol{\beta}$ and σ^2 are those values that maximize the likelihood function (the joint pdf, viewed as a function of the unknown parameters) $L(\boldsymbol{\beta}, \sigma^2)$, or equivalently those values that maximize its natural logarithm, with respect to $\boldsymbol{\beta}$ and σ^2 over the parameter space $\{(\boldsymbol{\beta}.\sigma^2) : \boldsymbol{\beta} \in \mathbb{R}^p,\ \sigma^2 > 0\}$. By Theorem 14.1.5, the log-likelihood function is

$$\begin{aligned}\log L(\boldsymbol{\beta}, \sigma^2) &= -\frac{n}{2}\log(2\pi) - \frac{1}{2}\log|\sigma^2\mathbf{W}| - \frac{1}{2}(\mathbf{y} - \mathbf{X}\boldsymbol{\beta})^T(\sigma^2\mathbf{W})^{-1}(\mathbf{y} - \mathbf{X}\boldsymbol{\beta}) \\ &= -\frac{n}{2}\log(2\pi) - \frac{n}{2}\log\sigma^2 - \frac{1}{2}\log|\mathbf{W}| - \frac{Q_{\mathbf{W}}(\boldsymbol{\beta})}{2\sigma^2},\end{aligned}$$

where $Q_{\mathbf{W}}(\boldsymbol{\beta}) = (\mathbf{y}-\mathbf{X}\boldsymbol{\beta})^T\mathbf{W}^{-1}(\mathbf{y}-\mathbf{X}\boldsymbol{\beta})$ is the generalized residual sum of squares function, and where we have used Theorem 2.11.4. Now recall from the definition of the generalized least squares estimator in Sect. 11.1 that the generalized residual sum of squares function $Q_{\mathbf{W}}(\boldsymbol{\beta})$ is minimized at, and only at, those values of $\boldsymbol{\beta}$ that are solutions to the Aitken equations, and that the Aitken equations are free of σ^2. Thus, for each fixed value of σ^2, $\log L(\boldsymbol{\beta}, \sigma^2)$ is maximized with respect to $\boldsymbol{\beta}$ at $\tilde{\boldsymbol{\beta}}$, where $\tilde{\boldsymbol{\beta}}$ is any solution to the Aitken equations. Thus the maximum likelihood estimator of $\boldsymbol{\beta}$ is generally not unique, but by the invariance property of maximum likelihood estimation, the maximum likelihood estimator of an estimable function $\mathbf{c}^T\boldsymbol{\beta}$ is $\mathbf{c}^T\tilde{\boldsymbol{\beta}}$, which is uniquely determined and coincides with the generalized least squares estimator of that function. It remains to maximize

$$\log L(\tilde{\boldsymbol{\beta}}, \sigma^2) = -\frac{n}{2}\log(2\pi) - \frac{n}{2}\log\sigma^2 - \frac{1}{2}\log|\mathbf{W}| - \frac{Q_{\mathbf{W}}(\tilde{\boldsymbol{\beta}})}{2\sigma^2}$$

with respect to σ^2. Taking the first derivative with respect to σ^2 yields

$$\frac{\partial \log L(\tilde{\boldsymbol{\beta}}, \sigma^2)}{\partial \sigma^2} = -\frac{n}{2\sigma^2} + \frac{Q_{\mathbf{W}}(\tilde{\boldsymbol{\beta}})}{2\sigma^4} = \left(\frac{n}{2\sigma^4}\right)(\bar{\sigma}^2 - \sigma^2),$$

where $\bar{\sigma}^2 = \frac{Q_{\mathbf{W}}(\tilde{\boldsymbol{\beta}})}{n}$. Furthermore,

$$\frac{\partial \log L(\tilde{\boldsymbol{\beta}}, \sigma^2)}{\partial \sigma^2} \begin{cases} > 0 \text{ for } \sigma^2 < \bar{\sigma}^2, \\ = 0 \text{ for } \sigma^2 = \bar{\sigma}^2, \\ < 0 \text{ for } \sigma^2 > \bar{\sigma}^2, \end{cases}$$

so that $L(\boldsymbol{\beta}, \sigma^2)$ attains a global maximum at $(\boldsymbol{\beta}, \sigma^2) = (\tilde{\boldsymbol{\beta}}, \bar{\sigma}^2)$ [unless $Q_{\mathbf{W}}(\tilde{\boldsymbol{\beta}}) = 0$, in which case $L(\boldsymbol{\beta}, \sigma^2)$ does not have a maximum.[1]] Thus, the maximum likelihood estimator of σ^2 is $(\mathbf{y} - \mathbf{X}\tilde{\boldsymbol{\beta}})^T \mathbf{W}^{-1}(\mathbf{y} - \mathbf{X}\tilde{\boldsymbol{\beta}})/n = \mathbf{y}^T[\mathbf{W}^{-1} - \mathbf{W}^{-1}\mathbf{X}(\mathbf{X}^T\mathbf{W}^{-1}\mathbf{X})^-\mathbf{X}^T\mathbf{W}^{-1}]\mathbf{y}/n$, which does not coincide with the unbiased estimator $\tilde{\sigma}^2$ defined in Corollary 11.1.9.1. However, the two are related by multiplication:

$$\bar{\sigma}^2 = \left(\frac{n - p^*}{n}\right)\tilde{\sigma}^2. \qquad \blacksquare$$

The next theorem indicates that the vector resulting from an arbitrary linear transformation of a multivariate normal random vector is another multivariate normal random vector.

Theorem 14.1.6 *Suppose that* $\mathbf{x} \sim N_n(\boldsymbol{\mu}, \boldsymbol{\Sigma})$. *Let* $\mathbf{w} = \mathbf{A}\mathbf{x} + \mathbf{a}$, *where* $\mathbf{A}$ *is a* $q \times n$ *matrix of constants and* $\mathbf{a}$ *is a* q*-vector of constants. Then,* $\mathbf{w} \sim N_q(\mathbf{A}\boldsymbol{\mu} + \mathbf{a}, \mathbf{A}\boldsymbol{\Sigma}\mathbf{A}^T)$.

Proof The mgf of $\mathbf{w}$, if it exists, is

$$\begin{aligned} m_{\mathbf{w}}(\mathbf{t}) &= \mathrm{E}[\exp(\mathbf{t}^T\mathbf{w})] \\ &= \mathrm{E}[\exp(\mathbf{t}^T\mathbf{A}\mathbf{x} + \mathbf{t}^T\mathbf{a})] \\ &= \exp(\mathbf{t}^T\mathbf{a}) m_{\mathbf{x}}(\mathbf{A}^T\mathbf{t}) \\ &= \exp(\mathbf{t}^T\mathbf{a}) \exp(\mathbf{t}^T\mathbf{A}\boldsymbol{\mu} + \mathbf{t}^T\mathbf{A}\boldsymbol{\Sigma}\mathbf{A}^T\mathbf{t}/2) \\ &= \exp[\mathbf{t}^T(\mathbf{A}\boldsymbol{\mu} + \mathbf{a}) + \mathbf{t}^T(\mathbf{A}\boldsymbol{\Sigma}\mathbf{A}^T)\mathbf{t}/2], \end{aligned}$$

which (by Theorem 14.1.4) we recognize as the mgf of a $\mathrm{N}_q(\mathbf{A}\boldsymbol{\mu} + \mathbf{a}, \mathbf{A}\boldsymbol{\Sigma}\mathbf{A}^T)$ distribution. The result follows by Theorem 14.1.1. $\square$

[1] This is an event of probability 0, however.

Suppose that $\mathbf{x} \sim N_n(\boldsymbol{\mu}, \boldsymbol{\Sigma})$, where $\text{rank}(\boldsymbol{\Sigma}) = r \leq n$. Let $\boldsymbol{\Gamma}$ represent an $r \times n$ matrix of rank r such that $\boldsymbol{\Sigma} = \boldsymbol{\Gamma}^T\boldsymbol{\Gamma}$. Then, it follows from Theorem 14.1.6 that $(\boldsymbol{\Gamma}\boldsymbol{\Gamma}^T)^{-1}\boldsymbol{\Gamma}(\mathbf{x} - \boldsymbol{\mu})$ has an r-variate standard normal distribution. In this way, it is always possible to standardize a multivariate normal random vector. As a special case, if $r = n$, then $\boldsymbol{\Sigma}^{-\frac{1}{2}}(\mathbf{x} - \boldsymbol{\mu})$ has an n-variate standard normal distribution.

Theorem 14.1.7 *Suppose that* $\mathbf{x} \sim N(\boldsymbol{\mu}, \boldsymbol{\Sigma})$. *Partition* $\mathbf{x}$, $\boldsymbol{\mu}$, *and* $\boldsymbol{\Sigma}$ *conformably as*

$$\mathbf{x} = \begin{pmatrix} \mathbf{x}_1 \\ \mathbf{x}_2 \end{pmatrix}, \quad \boldsymbol{\mu} = \begin{pmatrix} \boldsymbol{\mu}_1 \\ \boldsymbol{\mu}_2 \end{pmatrix}, \quad \boldsymbol{\Sigma} = \begin{pmatrix} \boldsymbol{\Sigma}_{11} & \boldsymbol{\Sigma}_{12} \\ \boldsymbol{\Sigma}_{21} & \boldsymbol{\Sigma}_{22} \end{pmatrix}.$$

Then, $\mathbf{x}_1 \sim N(\boldsymbol{\mu}_1, \boldsymbol{\Sigma}_{11})$.

Proof The result follows immediately by putting $\mathbf{a} = \mathbf{0}$ and $\mathbf{A} = (\mathbf{I}, \mathbf{0})$ into Theorem 14.1.6. □

Theorem 14.1.8 *Suppose that* $\mathbf{x} \sim N_n(\boldsymbol{\mu}, \boldsymbol{\Sigma})$, *and partition* $\mathbf{x}$, $\boldsymbol{\mu}$, *and* $\boldsymbol{\Sigma}$ *conformably as*

$$\mathbf{x} = \begin{pmatrix} \mathbf{x}_1 \\ \vdots \\ \mathbf{x}_m \end{pmatrix}, \quad \boldsymbol{\mu} = \begin{pmatrix} \boldsymbol{\mu}_1 \\ \vdots \\ \boldsymbol{\mu}_m \end{pmatrix}, \quad \boldsymbol{\Sigma} = \begin{pmatrix} \boldsymbol{\Sigma}_{11} & \cdots & \boldsymbol{\Sigma}_{1m} \\ \vdots & & \vdots \\ \boldsymbol{\Sigma}_{m1} & \cdots & \boldsymbol{\Sigma}_{mm} \end{pmatrix}.$$

Then, $\mathbf{x}_1, \ldots, \mathbf{x}_m$ *are mutually independent if and only if* $\boldsymbol{\Sigma}_{ij} = \mathbf{0}$ *for* $j \neq i = 1, \ldots, m$.

Proof Left as an exercise. □

According to Theorem 14.1.8, two or more variables whose joint distribution is multivariate normal are independent if they are uncorrelated. It is not true, however, that two or more uncorrelated univariate normal variables are necessarily independent. For some counterexamples, see Melnick and Tenenbein (1982).

Corollary 14.1.8.1 *Suppose that* $\mathbf{x} \sim N_n(\boldsymbol{\mu}, \boldsymbol{\Sigma})$, $\mathbf{A}_1$ *and* $\mathbf{A}_2$ *are* $q_1 \times n$ *and* $q_2 \times n$ *matrices of constants, and* $\mathbf{a}_1$ *and* $\mathbf{a}_2$ *are a* q_1*-vector and a* q_2*-vector of constants. Then the random vectors* $\mathbf{w}_1 = \mathbf{A}_1\mathbf{x} + \mathbf{a}_1$ *and* $\mathbf{w}_2 = \mathbf{A}_2\mathbf{x} + \mathbf{a}_2$ *are independent if and only if* $\mathbf{A}_1\boldsymbol{\Sigma}\mathbf{A}_2^T = \mathbf{0}$.

Example 14.1-2. Normality and Independence of the Least Squares Fitted Values Vector and Least Squares Fitted Residuals Vector Under The Normal Gauss–Markov Model

Recall from Chap. 8 that under the Gauss–Markov model $\{\mathbf{y}, \mathbf{X}\boldsymbol{\beta}, \sigma^2\mathbf{I}\}$, the least squares fitted values vector and least squares fitted residuals vector are defined as

$$\hat{\mathbf{y}} = \mathbf{X}\hat{\boldsymbol{\beta}} = \mathbf{P}_{\mathbf{X}}\mathbf{y}, \qquad \hat{\mathbf{e}} = \mathbf{y} - \mathbf{X}\hat{\boldsymbol{\beta}} = (\mathbf{I} - \mathbf{P}_{\mathbf{X}})\mathbf{y},$$

where $\hat{\boldsymbol{\beta}}$ is any solution to the normal equations. If $\mathbf{y}$ is assumed to have a multivariate normal distribution, then the joint distribution of $\hat{\mathbf{y}}$ and $\hat{\mathbf{e}}$ may be determined, using Theorem 14.1.6, as follows:

$$\begin{pmatrix} \hat{\mathbf{y}} \\ \hat{\mathbf{e}} \end{pmatrix} = \begin{pmatrix} \mathbf{P}_{\mathbf{X}} \\ \mathbf{I} - \mathbf{P}_{\mathbf{X}} \end{pmatrix} \mathbf{y} \sim \mathrm{N}\left(\begin{pmatrix} \mathbf{P}_{\mathbf{X}}(\mathbf{X}\boldsymbol{\beta}) \\ (\mathbf{I} - \mathbf{P}_{\mathbf{X}})(\mathbf{X}\boldsymbol{\beta}) \end{pmatrix}, \begin{pmatrix} \mathbf{P}_{\mathbf{X}} \\ \mathbf{I} - \mathbf{P}_{\mathbf{X}} \end{pmatrix} (\sigma^2\mathbf{I}) \begin{pmatrix} \mathbf{P}_{\mathbf{X}} \\ \mathbf{I} - \mathbf{P}_{\mathbf{X}} \end{pmatrix}^T \right)$$

i.e.,

$$\begin{pmatrix} \hat{\mathbf{y}} \\ \hat{\mathbf{e}} \end{pmatrix} \sim \mathrm{N}\left(\begin{pmatrix} \mathbf{X}\boldsymbol{\beta} \\ \mathbf{0} \end{pmatrix}, \sigma^2 \begin{pmatrix} \mathbf{P}_{\mathbf{X}} & \mathbf{0} \\ \mathbf{0} & \mathbf{I} - \mathbf{P}_{\mathbf{X}} \end{pmatrix} \right),$$

where we have simplified using various parts of Theorems 8.1.2 and 8.1.3. By Theorem 14.1.8, $\hat{\mathbf{y}}$ and $\hat{\mathbf{e}}$ are independent. ■

Methodological Interlude #7: Normal Probability Plots of Fitted Residuals

Example 14.1-2 established that the fitted residuals are normally distributed under the assumed Gauss–Markov model. In applications of such models, the validity of the normality assumption for $\mathbf{y}$ is often assessed by attempting to establish its validity for the fitted residuals, or for the standardized residuals defined (using the variances of fitted residuals specified in Example 14.1-2) as $\dot{e}_i = \hat{e}_i/\sqrt{1 - p_{ii}}$. This is simply a plot of the points $(\dot{e}_{(i)}, \pi_i)$ where $\dot{e}_{(i)}$ is the ith order statistic among the standardized residuals and π_i is the expected value of the ith order statistic of a random sample from a standard normal distribution. If all of the assumptions of the model, including normality, are valid, then the points in the plot should fall approximately on a straight line. A substantial departure from a straight line is taken as evidence of a model misspecification; the misspecification may be ascribed to nonnormality if all other model assumptions hold.

Further guidance on the use and interpretation of normal probability plots can be found in many applied regression analysis textbooks. ■

Example 14.1-3. Normality of the BLUE, BLUP, BLUP Prediction Error, and Generalized Least Squares Residual Vector Under the Normal Prediction-Extended Positive Definite Aitken Model

Recall from Theorem 11.1.1 that the BLUE of a vector of estimable functions $\mathbf{C}^T\boldsymbol{\beta}$ under the positive definite Aitken model $\{\mathbf{y}, \mathbf{X}\boldsymbol{\beta}, \sigma^2\mathbf{W}\}$ may be expressed as $\mathbf{C}^T(\mathbf{X}^T\mathbf{W}^{-1}\mathbf{X})^-\mathbf{X}^T\mathbf{W}^{-1}\mathbf{y}$, invariant to the choice of generalized inverse of $\mathbf{X}^T\mathbf{W}^{-1}\mathbf{X}$. If normality is added to the model's assumptions, then, by Theorem 14.1.6, this BLUE is normally distributed with mean vector $\mathbf{C}^T\boldsymbol{\beta}$ (by Theorem 11.1.2) and variance–covariance matrix $\sigma^2\mathbf{C}^T(\mathbf{X}^T\mathbf{W}^{-1}\mathbf{X})^-\mathbf{C}$ (by Theorem 11.1.3).

Furthermore, suppose that $\mathbf{u}$ is a random vector, and that the joint distribution of $\mathbf{e}$ and $\mathbf{u}$ is

$$\begin{pmatrix} \mathbf{e} \\ \mathbf{u} \end{pmatrix} \sim \mathrm{N}\left(\begin{pmatrix} \mathbf{0} \\ \mathbf{0} \end{pmatrix}, \sigma^2\begin{pmatrix} \mathbf{W} & \mathbf{K} \\ \mathbf{K}^T & \mathbf{H} \end{pmatrix}\right).$$

Then, using (13.3) and Theorem 13.2.2 and various results obtained while proving them, it can be shown (see the exercises) that the distribution of the BLUP $\tilde{\boldsymbol{\tau}} = \mathbf{C}^T\tilde{\boldsymbol{\beta}} + \tilde{\mathbf{u}}$ of $\boldsymbol{\tau} = \mathbf{C}^T\boldsymbol{\beta} + \mathbf{u}$ is

$$\mathrm{N}\left(\mathbf{C}^T\boldsymbol{\beta}, \sigma^2\{\mathbf{C}^T(\mathbf{X}^T\mathbf{W}^{-1}\mathbf{X})^-\mathbf{C} + \mathbf{K}^T[\mathbf{W}^{-1} - \mathbf{W}^{-1}\mathbf{X}(\mathbf{X}^T\mathbf{W}^{-1}\mathbf{X})^-\mathbf{X}^T\mathbf{W}^{-1}]\mathbf{K}\}\right) \tag{14.5}$$

and that

$$\begin{pmatrix} \tilde{\boldsymbol{\tau}} - \boldsymbol{\tau} \\ \tilde{\mathbf{e}} \end{pmatrix} = \begin{pmatrix} \mathbf{C}^T\tilde{\boldsymbol{\beta}} + \tilde{\mathbf{u}} - (\mathbf{C}^T\boldsymbol{\beta} + \mathbf{u}) \\ \tilde{\mathbf{e}} \end{pmatrix} \sim \mathrm{N}\left(\begin{pmatrix} \mathbf{0} \\ \mathbf{0} \end{pmatrix}, \sigma^2\begin{pmatrix} \mathbf{Q} & \mathbf{0} \\ \mathbf{0} & \mathbf{W} - \mathbf{X}(\mathbf{X}^T\mathbf{W}^{-1}\mathbf{X})^-\mathbf{X}^T \end{pmatrix}\right). \tag{14.6}$$

This last result indicates (by Theorem 14.1.8) that the BLUP's prediction error vector and the generalized least squares residual vector are independent. ■

Theorem 14.1.9 *Suppose that* $\mathbf{x} \sim N_n(\boldsymbol{\mu}, \boldsymbol{\Sigma})$, *where rank*$(\boldsymbol{\Sigma}) = n$. *Partition* $\mathbf{x}$, $\boldsymbol{\mu}$, *and* $\boldsymbol{\Sigma}$ *conformably as*

$$\mathbf{x} = \begin{pmatrix} \mathbf{x}_1 \\ \mathbf{x}_2 \end{pmatrix}, \quad \boldsymbol{\mu} = \begin{pmatrix} \boldsymbol{\mu}_1 \\ \boldsymbol{\mu}_2 \end{pmatrix}, \quad \boldsymbol{\Sigma} = \begin{pmatrix} \boldsymbol{\Sigma}_{11} & \boldsymbol{\Sigma}_{12} \\ \boldsymbol{\Sigma}_{21} & \boldsymbol{\Sigma}_{22} \end{pmatrix},$$

where $\mathbf{x}_1$ *is an* n_1*-vector and* $\mathbf{x}_2$ *is an* n_2*-vector. Then, the conditional distribution of* $\mathbf{x}_1$, *given that* $\mathbf{x}_2 = \mathbf{x}_2^*$, *is* $N_{n_1}(\boldsymbol{\mu}_1 + \boldsymbol{\Sigma}_{12}\boldsymbol{\Sigma}_{22}^{-1}(\mathbf{x}_2^* - \boldsymbol{\mu}_2), \boldsymbol{\Sigma}_{11} - \boldsymbol{\Sigma}_{12}\boldsymbol{\Sigma}_{22}^{-1}\boldsymbol{\Sigma}_{21})$.

Proof Using Theorems 14.1.5 and 14.1.7, the pdf of the conditional distribution of $\mathbf{x}_1$ given $\mathbf{x}_2$ is

$$\frac{(2\pi)^{-n/2}|\boldsymbol{\Sigma}|^{-\frac{1}{2}}\exp[-(\mathbf{x}-\boldsymbol{\mu})^T\boldsymbol{\Sigma}^{-1}(\mathbf{x}-\boldsymbol{\mu})/2]}{(2\pi)^{-n_2/2}|\boldsymbol{\Sigma}_{22}|^{-\frac{1}{2}}\exp[-(\mathbf{x}_2-\boldsymbol{\mu}_2)^T\boldsymbol{\Sigma}_{22}^{-1}(\mathbf{x}_2-\boldsymbol{\mu}_2)/2]} = (2\pi)^{-\frac{n_1}{2}}c\exp(-Q/2),$$

where $c = |\boldsymbol{\Sigma}|^{-\frac{1}{2}}|\boldsymbol{\Sigma}_{22}|^{\frac{1}{2}}$ and

$$Q = (\mathbf{x}-\boldsymbol{\mu})^T\boldsymbol{\Sigma}^{-1}(\mathbf{x}-\boldsymbol{\mu}) - (\mathbf{x}_2-\boldsymbol{\mu}_2)^T\boldsymbol{\Sigma}_{22}^{-1}(\mathbf{x}_2-\boldsymbol{\mu}_2).$$

By Theorem 2.11.7,

$$\begin{aligned} c &= |\boldsymbol{\Sigma}_{22}|^{-\frac{1}{2}}|\boldsymbol{\Sigma}_{11}-\boldsymbol{\Sigma}_{12}\boldsymbol{\Sigma}_{22}^{-1}\boldsymbol{\Sigma}_{21}|^{-\frac{1}{2}}|\boldsymbol{\Sigma}_{22}|^{\frac{1}{2}} \\ &= |\boldsymbol{\Sigma}_{11}-\boldsymbol{\Sigma}_{12}\boldsymbol{\Sigma}_{22}^{-1}\boldsymbol{\Sigma}_{21}|^{-\frac{1}{2}}. \end{aligned}$$

Let $\mathbf{R} = \boldsymbol{\Sigma}_{11}-\boldsymbol{\Sigma}_{12}\boldsymbol{\Sigma}_{22}^{-1}\boldsymbol{\Sigma}_{21}$. By Theorem 2.9.5b,

$$\begin{aligned} Q &= [(\mathbf{x}_1-\boldsymbol{\mu}_1)^T\mathbf{R}^{-1}(\mathbf{x}_1-\boldsymbol{\mu}_1) - (\mathbf{x}_1-\boldsymbol{\mu}_1)^T\mathbf{R}^{-1}\boldsymbol{\Sigma}_{12}\boldsymbol{\Sigma}_{22}^{-1}(\mathbf{x}_2-\boldsymbol{\mu}_2) \\ &\quad -(\mathbf{x}_2-\boldsymbol{\mu}_2)^T\boldsymbol{\Sigma}_{22}^{-1}\boldsymbol{\Sigma}_{21}\mathbf{R}^{-1}(\mathbf{x}_1-\boldsymbol{\mu}_1) + (\mathbf{x}_2-\boldsymbol{\mu}_2)^T(\boldsymbol{\Sigma}_{22}^{-1}+\boldsymbol{\Sigma}_{22}^{-1}\boldsymbol{\Sigma}_{21}\mathbf{R}^{-1}\boldsymbol{\Sigma}_{12}\boldsymbol{\Sigma}_{22}^{-1}) \\ &\quad \times(\mathbf{x}_2-\boldsymbol{\mu}_2)] - (\mathbf{x}_2-\boldsymbol{\mu}_2)^T\boldsymbol{\Sigma}_{22}^{-1}(\mathbf{x}_2-\boldsymbol{\mu}_2) \\ &= [\mathbf{x}_1-\boldsymbol{\mu}_1-\boldsymbol{\Sigma}_{12}\boldsymbol{\Sigma}_{22}^{-1}(\mathbf{x}_2-\boldsymbol{\mu}_2)]^T\mathbf{R}^{-1}[\mathbf{x}_1-\boldsymbol{\mu}_1-\boldsymbol{\Sigma}_{12}\boldsymbol{\Sigma}_{22}^{-1}(\mathbf{x}_2-\boldsymbol{\mu}_2)]. \end{aligned}$$

Thus, the conditional distribution of $\mathbf{x}_1$, given that $\mathbf{x}_2 = \mathbf{x}_2^*$, is $\mathrm{N}_{n_1}(\boldsymbol{\mu}_1 + \boldsymbol{\Sigma}_{12}\boldsymbol{\Sigma}_{22}^{-1}(\mathbf{x}_2^*-\boldsymbol{\mu}_2), \boldsymbol{\Sigma}_{11}-\boldsymbol{\Sigma}_{12}\boldsymbol{\Sigma}_{22}^{-1}\boldsymbol{\Sigma}_{21})$. □

From the multivariate normal distribution's mgf (Theorem 14.1.4), it can be seen that a multivariate normal distribution is completely specified by its mean vector $\boldsymbol{\mu}$ and variance–covariance matrix $\boldsymbol{\Sigma}$. Thus, all higher-order moments of the distribution must be functions of $\boldsymbol{\mu}$ and $\boldsymbol{\Sigma}$. The next theorem gives specifics for the skewnesses and kurtoses of the distribution.

Theorem 14.1.10 *Suppose that* $\mathbf{x} \sim N_n(\boldsymbol{\mu}, \boldsymbol{\Sigma})$. *Then:*
(a) the skewness matrix, $\boldsymbol{\Lambda}$, *is null; and*
(b) the kurtosis matrix, $\boldsymbol{\Gamma}$, *equals* $\boldsymbol{\Gamma} = (\gamma_{ijkl}) = (\sigma_{ij}\sigma_{kl}+\sigma_{ik}\sigma_{jl}+\sigma_{il}\sigma_{jk})$, *implying further that the excess kurtosis matrix,* $\boldsymbol{\Omega} = (\gamma_{ijkl} - \sigma_{ij}\sigma_{kl} - \sigma_{ik}\sigma_{jl} - \sigma_{il}\sigma_{jk})$, *is null.*

Proof By Theorem 14.1.6 (with $\mathbf{A} = \mathbf{I}$ and $\mathbf{a} = -\boldsymbol{\mu}$), $(\mathbf{x}-\boldsymbol{\mu}) \sim \mathrm{N}(\mathbf{0}, \boldsymbol{\Sigma})$. Another application of Theorem 14.1.6, this time with $\mathbf{A} = -\mathbf{I}$ and $\mathbf{a} = \boldsymbol{\mu}$, reveals that $-(\mathbf{x}-\boldsymbol{\mu}) \sim \mathrm{N}(\mathbf{0}, \boldsymbol{\Sigma})$ also. Part (a) follows immediately by Theorem 4.1.3b. For part (b), persistence yields the following expression for the fourth-order partial derivatives of the mgf of $\mathbf{x} - \boldsymbol{\mu}$, verification of which is left as an exercise:

$$\begin{aligned}\frac{\partial^4 m_{\mathbf{x}-\boldsymbol{\mu}}(\mathbf{t})}{\partial t_i \partial t_j \partial t_k \partial t_l} = \exp(\mathbf{t}^T \boldsymbol{\Sigma}\mathbf{t}/2)\Bigg[&\left(\sum_{a=1}^n \sigma_{ia} t_a\right)\left(\sum_{b=1}^n \sigma_{jb} t_b\right)\left(\sum_{c=1}^n \sigma_{kc} t_c\right)\left(\sum_{d=1}^n \sigma_{ld} t_d\right) \\ &+\sigma_{il}\left(\sum_{b=1}^n \sigma_{jb} t_b\right)\left(\sum_{c=1}^n \sigma_{kc} t_c\right) + \sigma_{jl}\left(\sum_{a=1}^n \sigma_{ia} t_a\right)\left(\sum_{c=1}^n \sigma_{kc} t_c\right) \\ &+\sigma_{kl}\left(\sum_{a=1}^n \sigma_{ia} t_a\right)\left(\sum_{b=1}^n \sigma_{jb} t_b\right) + \sigma_{ik}\left(\sum_{b=1}^n \sigma_{jb} t_b\right)\left(\sum_{d=1}^n \sigma_{ld} t_d\right) \\ &+\sigma_{jk}\left(\sum_{a=1}^n \sigma_{ia} t_a\right)\left(\sum_{d=1}^n \sigma_{ld} t_d\right) + \sigma_{ij}\left(\sum_{c=1}^n \sigma_{kc} t_c\right)\left(\sum_{d=1}^n \sigma_{ld} t_d\right) \\ &+\sigma_{ij}\sigma_{kl} + \sigma_{ik}\sigma_{jl} + \sigma_{il}\sigma_{jk}\Bigg].\end{aligned}$$

Evaluating this partial derivative at $\mathbf{t} = \mathbf{0}$ yields the specified expression for γ_{ijkl}. □

Corollary 14.1.10.1 *Suppose that* $\mathbf{x} \sim N_n(\boldsymbol{\mu}, \boldsymbol{\Sigma})$. *Let* $\mathbf{b}$ *represent an n-vector of constants, and let* $\mathbf{A}_1$ *and* $\mathbf{A}_2$ *represent symmetric* $n \times n$ *matrices of constants. Then:*

(a) $cov(\mathbf{b}^T\mathbf{x}, \mathbf{x}^T\mathbf{A}_1\mathbf{x}) = 2\mathbf{b}^T\boldsymbol{\Sigma}\mathbf{A}_1\boldsymbol{\mu}$ *and*
(b) $cov(\mathbf{x}^T\mathbf{A}_1\mathbf{x}, \mathbf{x}^T\mathbf{A}_2\mathbf{x}) = 2tr(\mathbf{A}_1\boldsymbol{\Sigma}\mathbf{A}_2\boldsymbol{\Sigma}) + 4\boldsymbol{\mu}^T\mathbf{A}_1\boldsymbol{\Sigma}\mathbf{A}_2\boldsymbol{\mu}$.

Proof These results follow immediately from combining the results of Theorem 14.1.10 with those of Corollaries 4.2.5.1 and 4.2.6.3. □

Example 14.1-4. Expectations, Variances, and Covariance of the Mean and Variance of a Random Sample from a Normal Population

In Example 4.2-1 the following results were obtained for the mean and variance of a random sample $x_1, x_2, \ldots, x_n$ taken from a population with mean μ, variance σ^2, skewness λ, and kurtosis γ:

$$\mathrm{E}(\overline{x}) = \mu, \quad \mathrm{E}(s^2) = \sigma^2, \quad \mathrm{var}(\overline{x}) = \sigma^2/n, \quad \mathrm{var}(s^2) = [(\gamma - 3\sigma^4)/n] + 2\sigma^4/(n-1),$$

$$\mathrm{cov}(\overline{x}, s^2) = \lambda/n.$$

Now suppose that the population being sampled is normal. Then by Theorem 14.1.10, the latter two moments specialize to

$$\text{var}(s^2) = 2\sigma^4/(n-1), \quad \text{cov}(\overline{x}, s^2) = 0. \qquad \blacksquare$$

14.2 The Noncentral Chi-Square Distribution

Among many results presented in the previous section was a result on the distribution of a linear function of a multivariate normal random vector, which said that such a function also had a multivariate normal distribution. There are many important applications of that result in linear models, some of which we noted in examples. There are many other important applications in linear models that require knowledge of the distribution of quadratic forms (e.g., sums of squares), rather than linear forms, in a multivariate normal random vector. In this section, we introduce a distribution that such quadratic forms may have, under certain circumstances, and we characterize the circumstances. The distribution is known as the *noncentral chi-square distribution*.

Definition 14.2.1 A random variable u is said to have a *noncentral chi-square distribution with degrees of freedom* ν *and noncentrality parameter* λ if it has pdf

$$f(u; \nu, \lambda) = \begin{cases} \sum_{j=0}^{\infty} \left(\frac{\exp(-\lambda)\lambda^j}{j!} \right) \left(\frac{u^{\frac{\nu+2j-2}{2}} \exp(-\frac{u}{2})}{\Gamma(\frac{\nu+2j}{2}) 2^{j+\frac{\nu}{2}}} \right) & \text{for } u \geq 0, \\ 0 & \text{otherwise,} \end{cases}$$

where $\lambda \geq 0$ and $\nu > 0$.

It can be seen from this definition that, in general, the pdf of a noncentral chi-square distribution is an infinite weighted linear combination of central chi-square pdfs with ν, $\nu+2$, $\nu+4, \ldots$ degrees of freedom, with weights equal to probabilities from a Poisson distribution with mean equal to the noncentrality parameter. We do not require that ν be an integer; we only require that it be positive. In the special case $\lambda = 0$, all weights in the infinite sum equal 0 except the first (for which we define $0^0 = 1$), so the distribution in that case reduces to the *central chi-square distribution*, or more simply the *chi-square distribution, with degrees of freedom* ν. In this case the pdf is

$$f(u; \nu) = \begin{cases} \frac{u^{\frac{\nu-2}{2}} \exp(-\frac{u}{2})}{\Gamma(\frac{\nu}{2}) 2^{\frac{\nu}{2}}}, & \text{for } u \geq 0, \\ 0, & \text{otherwise.} \end{cases}$$

Subsequently, we denote the noncentral and central chi-square distributions, with parameters as noted above, by $\chi^2(\nu, \lambda)$ and $\chi^2(\nu)$, respectively. The first theorem of this section obtains the mgf of the noncentral chi-square distribution.

Theorem 14.2.1 *Suppose that $u \sim \chi^2(\nu, \lambda)$. Then, the mgf of u is*

$$m_u(t) = (1-2t)^{-\frac{\nu}{2}} \exp\left(\frac{2t\lambda}{1-2t}\right), \quad \textit{for } t < \frac{1}{2}.$$

Proof By definition, the mgf of u is

$$m_u(t) = \int_0^\infty e^{tu} \sum_{j=0}^{\infty} \left(\frac{e^{-\lambda}\lambda^j}{j!}\right) \left(\frac{u^{\frac{\nu+2j-2}{2}} e^{-u/2}}{\Gamma(\frac{\nu+2j}{2}) 2^{j+\frac{\nu}{2}}}\right) du$$

provided that this integral exists for all t in an interval containing the origin. Interchanging integration and summation[2] yields

$$m_u(t) = \sum_{j=0}^{\infty} \left(\frac{e^{-\lambda}\lambda^j}{j!\Gamma\left(\frac{\nu+2j}{2}\right) 2^{j+\frac{\nu}{2}}}\right) \int_0^\infty u^{(\nu+2j-2)/2} e^{-(1-2t)u/2} du.$$

We recognize that the integral resembles a gamma function; in particular, recall from, say, Casella and Berger (2002, p. 100), that for any $\alpha, \beta > 0$,

$$\int_0^\infty u^{\alpha-1} e^{-\beta u} du = \Gamma(\alpha) \left(\frac{1}{\beta}\right)^\alpha. \tag{14.7}$$

Thus we have, for $t < 1/2$,

$$\begin{aligned}
m_u(t) &= \sum_{j=0}^{\infty} \left(\frac{e^{-\lambda}\lambda^j}{j!\Gamma\left(\frac{\nu+2j}{2}\right) 2^{j+\frac{\nu}{2}}}\right) \Gamma\left(\frac{\nu+2j}{2}\right) \left(\frac{2}{1-2t}\right)^{\frac{\nu+2j}{2}} \\
&= (1-2t)^{-\nu/2} \sum_{j=0}^{\infty} \frac{e^{-\lambda}[\lambda/(1-2t)]^j}{j!} \\
&= (1-2t)^{-\nu/2} \sum_{j=0}^{\infty} \frac{e^{-\lambda/(1-2t)}[\lambda/(1-2t)]^j}{j!} e^{2t\lambda/(1-2t)} \\
&= (1-2t)^{-\frac{\nu}{2}} \exp\left(\frac{2t\lambda}{1-2t}\right).
\end{aligned}$$

□

[2]Interchanging integration and summation can be justified by the dominated convergence theorem.

The next two theorems, taken together, establish that the sum of squares of a multivariate normal vector with arbitrary mean vector and an identity variance–covariance matrix has a certain noncentral chi-square distribution.

Theorem 14.2.2 *Suppose that* $\mathbf{x} \sim N_n(\boldsymbol{\mu}, \mathbf{I})$. *Then, for* $t < 1/2$, *the mgf of* $\mathbf{x}^T\mathbf{x}$ *is*

$$m_{\mathbf{x}^T\mathbf{x}}(t) = (1-2t)^{-n/2} \exp\left[\left(\frac{t}{1-2t}\right)\boldsymbol{\mu}^T\boldsymbol{\mu}\right].$$

Proof Let $\mathbf{z}$ represent an n-variate standard normal vector. Then $\boldsymbol{\mu}+\mathbf{z}$ has the same distribution as $\mathbf{x}$, so

$$\begin{aligned} m_{\mathbf{x}^T\mathbf{x}}(t) &= m_{(\boldsymbol{\mu}+\mathbf{z})^T(\boldsymbol{\mu}+\mathbf{z})}(t) \\ &= (2\pi)^{-\frac{n}{2}} \int_{-\infty}^{\infty} \cdots \int_{-\infty}^{\infty} \exp[t(\boldsymbol{\mu}^T\boldsymbol{\mu} + 2\mathbf{z}^T\boldsymbol{\mu} + \mathbf{z}^T\mathbf{z}) - \frac{1}{2}\mathbf{z}^T\mathbf{z}]d\mathbf{z} \\ &= (2\pi)^{-\frac{n}{2}} \int_{-\infty}^{\infty} \cdots \int_{-\infty}^{\infty} \exp\left[-\left(\frac{1-2t}{2}\right)\mathbf{z}^T\mathbf{z} + 2t\mathbf{z}^T\boldsymbol{\mu} + t\boldsymbol{\mu}^T\boldsymbol{\mu}\right]d\mathbf{z}. \end{aligned}$$

Observe that for $t < 1/2$, the matrix $\frac{1}{2}(1-2t)\mathbf{I}_n$ is positive definite. Therefore, by Theorem 14.1.3 with $\mathbf{A} = \mathbf{0}$, $\mathbf{a} = \mathbf{0}$, $a_0 = 1$, $\mathbf{B} = \frac{1}{2}(1-2t)\mathbf{I}_n$, $\mathbf{b} = -2t\boldsymbol{\mu}$, and $b_0 = -t\boldsymbol{\mu}^T\boldsymbol{\mu}$, we have, for $t < 1/2$,

$$\begin{aligned} m_{\mathbf{x}^T\mathbf{x}}(t) &= (2\pi)^{-\frac{n}{2}}(1/2)\pi^{\frac{n}{2}}|(1/2)(1-2t)\mathbf{I}_n|^{-\frac{1}{2}} \exp\{(1/4)(-2t\boldsymbol{\mu})^T \\ &\quad \times[(1/2)(1-2t)]^{-1}(-2t\boldsymbol{\mu}) + t\boldsymbol{\mu}^T\boldsymbol{\mu}\} \cdot 2 \\ &= (1-2t)^{-n/2}\exp[2t^2(1-2t)^{-1}\boldsymbol{\mu}^T\boldsymbol{\mu} + t\boldsymbol{\mu}^T\boldsymbol{\mu}] \\ &= (1-2t)^{-n/2}\exp\left[\left(\frac{t}{1-2t}\right)\boldsymbol{\mu}^T\boldsymbol{\mu}\right]. \end{aligned}$$

□

Theorem 14.2.3 *Suppose that* $\mathbf{x} \sim N_n(\boldsymbol{\mu}, \mathbf{I})$, *and define* $w = \mathbf{x}^T\mathbf{x}$. *Then* $w \sim \chi^2(n, \frac{1}{2}\boldsymbol{\mu}^T\boldsymbol{\mu})$.

Proof By Theorem 14.2.2, for $t < 1/2$,

$$m_w(t) = m_{\mathbf{x}^T\mathbf{x}}(t) = (1-2t)^{-\frac{n}{2}}\exp\left(\frac{t\boldsymbol{\mu}^T\boldsymbol{\mu}}{1-2t}\right) = (1-2t)^{-\frac{n}{2}}\exp\left(\frac{2t(\frac{\boldsymbol{\mu}^T\boldsymbol{\mu}}{2})}{1-2t}\right).$$

The result follows by Theorems 14.2.1 and 14.1.1. □

The following corollary to Theorem 14.2.3 gives a result typically taught in elementary mathematical statistics courses.

Corollary 14.2.3.1 *If* $\mathbf{x} \sim N_n(\mathbf{0}, \mathbf{I})$*, then* $\mathbf{x}^T\mathbf{x} \sim \chi^2(n)$.

In addition to being of interest on its own, Theorem 14.2.3 provides a simple way to obtain the mean and variance of the $\chi^2(\nu, \lambda)$ distribution when ν is an integer. By Theorem 14.2.3, the distribution of w is the same as that of $\mathbf{x}^T\mathbf{x}$ where $\mathbf{x}$ is a random vector whose distribution is $\mathrm{N}_\nu((\frac{2\lambda}{\nu})^{\frac{1}{2}}\mathbf{1}_\nu, \mathbf{I}_\nu)$. Thus, by Theorem 4.2.4 and Corollary 14.1.10.1b,

$$\mathrm{E}(w) = \mathrm{E}(\mathbf{x}^T\mathbf{x}) = \left[\left(\frac{2\lambda}{\nu}\right)^{\frac{1}{2}}\mathbf{1}_\nu\right]^T \mathbf{I}_\nu \left[\left(\frac{2\lambda}{\nu}\right)^{\frac{1}{2}}\mathbf{1}_\nu\right] + \mathrm{tr}(\mathbf{I}_\nu\mathbf{I}_\nu) = \nu + 2\lambda$$

and

$$\mathrm{var}(w) = \mathrm{var}(\mathbf{x}^T\mathbf{x}) = 2\mathrm{tr}(\mathbf{I}_\nu\mathbf{I}_\nu\mathbf{I}_\nu\mathbf{I}_\nu) + 4\left[\left(\frac{2\lambda}{\nu}\right)^{\frac{1}{2}}\mathbf{1}_\nu\right]^T \mathbf{I}_\nu\mathbf{I}_\nu\mathbf{I}_\nu \left[\left(\frac{2\lambda}{\nu}\right)^{\frac{1}{2}}\mathbf{1}_\nu\right] = 2\nu + 8\lambda.$$

By evaluating the first- and second-order derivatives of the mgf of a noncentral chi-square distribution with respect to t at the origin, it can be shown (see the exercises) that these same expressions apply when ν is any positive number.

The following theorem is a generalization of Theorem 14.2.3 that allows the elements of the random vector $\mathbf{x}$ to have unequal variances and nonzero correlations, but the particular quadratic form is functionally dependent on those variances and correlations.

Theorem 14.2.4 *Suppose that* $\mathbf{x} \sim N_n(\boldsymbol{\mu}, \boldsymbol{\Sigma})$*, where rank*$(\boldsymbol{\Sigma}) = n$*. Define* $w = \mathbf{x}^T\boldsymbol{\Sigma}^{-1}\mathbf{x}$*. Then,*

$$w \sim \chi^2(n, \frac{1}{2}\boldsymbol{\mu}^T\boldsymbol{\Sigma}^{-1}\boldsymbol{\mu}).$$

Proof A unique positive definite matrix $\boldsymbol{\Sigma}^{\frac{1}{2}}$ exists such that $\boldsymbol{\Sigma} = \boldsymbol{\Sigma}^{\frac{1}{2}}\boldsymbol{\Sigma}^{\frac{1}{2}}$ (Theorem 2.15.12). Let $\mathbf{z} = \boldsymbol{\Sigma}^{-\frac{1}{2}}\mathbf{x}$. By Theorem 14.1.6, $\mathbf{z} \sim \mathrm{N}_n(\boldsymbol{\Sigma}^{-\frac{1}{2}}\boldsymbol{\mu}, \mathbf{I})$. Also,

$$w = \mathbf{x}^T\boldsymbol{\Sigma}^{-1}\mathbf{x} = \mathbf{x}^T(\boldsymbol{\Sigma}^{-\frac{1}{2}})^T\boldsymbol{\Sigma}^{-\frac{1}{2}}\mathbf{x} = \mathbf{z}^T\mathbf{z}.$$

Thus, by Theorem 14.2.3,

$$w \sim \chi^2[n, \frac{1}{2}(\boldsymbol{\Sigma}^{-\frac{1}{2}}\boldsymbol{\mu})^T\boldsymbol{\Sigma}^{-\frac{1}{2}}\boldsymbol{\mu}],$$

i.e.,

$$w \sim \chi^2(n, \frac{1}{2}\boldsymbol{\mu}^T\boldsymbol{\Sigma}^{-1}\boldsymbol{\mu}).$$

□

Example 14.2-1. Distribution of the (Suitably Scaled) Unconstrained|Constrained Second-Order Polynomial Under the Normal Gauss–Markov Model with Linearly Independent Estimable Constraints

Consider a constrained Gauss–Markov model $\{\mathbf{y}, \mathbf{X}\boldsymbol{\beta}, \sigma^2\mathbf{I} : \mathbf{A}\boldsymbol{\beta} = \mathbf{h}\}$, where the q constraints are linearly independent and estimable. It was shown in Sect. 10.3 that the total sum of squares, $\mathbf{y}^T\mathbf{y}$, may be written as the sum of three second-order polynomial terms, of which the term that was labeled Unconstrained|Constrained can be written as

$$(\mathbf{A}\hat{\boldsymbol{\beta}} - \mathbf{h})^T[\mathbf{A}(\mathbf{X}^T\mathbf{X})^-\mathbf{A}^T]^{-1}(\mathbf{A}\hat{\boldsymbol{\beta}} - \mathbf{h})$$

[where $\hat{\boldsymbol{\beta}}$ is a solution to the normal equations corresponding to the unconstrained model and where we have replaced the generalized inverse of $\mathbf{A}(\mathbf{X}^T\mathbf{X})^-\mathbf{A}^T$ with an ordinary inverse because the constraints are assumed here to be linearly independent]. Although this is a second-order polynomial function in $\mathbf{y}$, it is a quadratic form in $\mathbf{A}\hat{\boldsymbol{\beta}} - \mathbf{h}$. Now using Theorem 14.1.6 and results from Chap. 7, $\mathbf{A}\hat{\boldsymbol{\beta}} - \mathbf{h} \sim \mathrm{N}(\mathbf{A}\boldsymbol{\beta} - \mathbf{h}, \sigma^2\mathbf{A}(\mathbf{X}^T\mathbf{X})^-\mathbf{A}^T)$. Thus, by Theorem 14.2.4 the Unconstrained|Constrained second-order polynomial, divided by σ^2, has a chi-square distribution with q degrees of freedom and noncentrality parameter $(1/2\sigma^2)(\mathbf{A}\boldsymbol{\beta} - \mathbf{h})^T[\mathbf{A}(\mathbf{X}^T\mathbf{X})^-\mathbf{A}^T]^{-1}(\mathbf{A}\boldsymbol{\beta} - \mathbf{h})$. ■

In applications of the theory of linear models, we often encounter quadratic forms that do not have the particular form featured in Theorem 14.2.4 (i.e., $\mathbf{x}^T\boldsymbol{\Sigma}^{-1}\mathbf{x}$). Therefore, we wish to extend Theorem 14.2.4 to more general quadratic forms. The second of the next two theorems provides such an extension; the first theorem is of some importance on its own but is primarily a "stepping stone" to the second.

Theorem 14.2.5 *Suppose that* $\mathbf{x} \sim N_n(\mathbf{0}, \mathbf{I})$. *Define* $w = \mathbf{x}^T\mathbf{A}\mathbf{x} + 2\boldsymbol{\ell}^T\mathbf{x} + c$, *where* $\mathbf{A}$ *is a symmetric nonnull matrix of constants,* $\boldsymbol{\ell}$ *is a vector of constants, and* c *is a constant.*

(a) If
  (1) $\mathbf{A}^2 = \mathbf{A}$,
  (2) $\mathbf{A}\boldsymbol{\ell} = \boldsymbol{\ell}$, *and*
  (3) $\boldsymbol{\ell}^T\boldsymbol{\ell} = c$,
  then $w \sim \chi^2[tr(\mathbf{A}), \frac{1}{2}\boldsymbol{\ell}^T\boldsymbol{\ell}]$.

(b) Conversely, if $w \sim \chi^2(\nu, \lambda)$ *for some* ν *and* λ, *then the three conditions in part (a) are satisfied; in addition,* $\nu = tr(\mathbf{A})$ *and* $\lambda = \frac{1}{2}\boldsymbol{\ell}^T\boldsymbol{\ell}$.

Proof We prove part (a) first. By condition (1), an $n \times r$ matrix $\mathbf{P}$ exists such that $\mathbf{A} = \mathbf{P}\mathbf{P}^T$ and $\mathbf{P}^T\mathbf{P} = \mathbf{I}_r$, where $r = \text{rank}(\mathbf{A})$ (Theorem 2.15.15). Then, using conditions (2) and (3), we have

$$\begin{aligned} w &= \mathbf{x}^T\mathbf{A}\mathbf{x} + 2\boldsymbol{\ell}^T\mathbf{x} + \boldsymbol{\ell}^T\boldsymbol{\ell} \\ &= \mathbf{x}^T\mathbf{A}\mathbf{x} + 2\boldsymbol{\ell}^T\mathbf{A}\mathbf{x} + \boldsymbol{\ell}^T\mathbf{A}\boldsymbol{\ell} \\ &= (\mathbf{x}+\boldsymbol{\ell})^T\mathbf{A}(\mathbf{x}+\boldsymbol{\ell}) \\ &= \mathbf{v}^T\mathbf{v}, \quad \text{say,} \quad \text{where } \mathbf{v} = \mathbf{P}^T(\mathbf{x}+\boldsymbol{\ell}). \end{aligned}$$

By Theorem 14.1.6, $\mathbf{v} \sim \text{N}_r(\mathbf{P}^T\boldsymbol{\ell}, \mathbf{P}^T\mathbf{P})$, i.e., $\mathbf{v} \sim \text{N}_r(\mathbf{P}^T\boldsymbol{\ell}, \mathbf{I}_r)$. Thus, by Theorem 14.2.3, $\mathbf{v}^T\mathbf{v} \sim \chi^2(r, \lambda)$, where

$$\lambda = \frac{1}{2}(\mathbf{P}^T\boldsymbol{\ell})^T(\mathbf{P}^T\boldsymbol{\ell}) = \frac{1}{2}\boldsymbol{\ell}^T\mathbf{P}\mathbf{P}^T\boldsymbol{\ell} = \frac{1}{2}\boldsymbol{\ell}^T\mathbf{A}\boldsymbol{\ell} = \frac{1}{2}\boldsymbol{\ell}^T\boldsymbol{\ell},$$

and because $\mathbf{A}$ is idempotent, $r = \text{tr}(\mathbf{A})$.

The proof of part (b) requires knowledge of the mathematics of polynomial functions beyond the scope of this presentation. See Harville (2018, Section 6.7d) for more details. □

While Theorem 14.2.5 extends the types of quadratic forms (or even second-order polynomials) that can have noncentral chi-square distributions, the mean and variance–covariance matrix of the normal random vector in the theorem are highly specialized. The next theorem allows the mean and variance–covariance matrix of the normal random vector to be completely general.

Theorem 14.2.6 *Suppose that $\mathbf{x} \sim N_n(\boldsymbol{\mu}, \boldsymbol{\Sigma})$. Define $w = \mathbf{x}^T\mathbf{A}\mathbf{x}$, where $\mathbf{A}$ is a symmetric matrix of constants such that $\boldsymbol{\Sigma}\mathbf{A}\boldsymbol{\Sigma}$ is nonnull.*

(a) If
(1) $\boldsymbol{\Sigma}\mathbf{A}\boldsymbol{\Sigma}\mathbf{A}\boldsymbol{\Sigma} = \boldsymbol{\Sigma}\mathbf{A}\boldsymbol{\Sigma}$,
(2) $\boldsymbol{\Sigma}\mathbf{A}\boldsymbol{\Sigma}\mathbf{A}\boldsymbol{\mu} = \boldsymbol{\Sigma}\mathbf{A}\boldsymbol{\mu}$, and
(3) $\boldsymbol{\mu}^T\mathbf{A}\boldsymbol{\Sigma}\mathbf{A}\boldsymbol{\mu} = \boldsymbol{\mu}^T\mathbf{A}\boldsymbol{\mu}$,
then

$$w \sim \chi^2[tr(\mathbf{A}\boldsymbol{\Sigma}), \frac{1}{2}\boldsymbol{\mu}^T\mathbf{A}\boldsymbol{\mu}].$$

(b) Conversely, if $w \sim \chi^2(\nu, \lambda)$ for some ν and λ, then the three conditions in part (a) are satisfied, $\nu = tr(\mathbf{A}\boldsymbol{\Sigma})$, and $\lambda = \frac{1}{2}\boldsymbol{\mu}^T\mathbf{A}\boldsymbol{\mu}$.

Proof First we prove part (a). Let $r = \text{rank}(\boldsymbol{\Sigma})$ and let $\boldsymbol{\Gamma}$ represent an $n \times r$ matrix such that $\boldsymbol{\Sigma} = \boldsymbol{\Gamma}\boldsymbol{\Gamma}^T$. (The existence of such a $\boldsymbol{\Gamma}$ is guaranteed by Theorem 2.15.14.) Then it is easy to show (using Theorem 14.1.6) that $\mathbf{x}$ has the same distribution as

$\boldsymbol{\mu} + \boldsymbol{\Gamma}\mathbf{z}$, where $\mathbf{z} \sim \mathrm{N}_r(\mathbf{0}, \mathbf{I})$. Thus w has the same distribution as $(\boldsymbol{\mu} + \boldsymbol{\Gamma}\mathbf{z})^T\mathbf{A}(\boldsymbol{\mu} + \boldsymbol{\Gamma}\mathbf{z})$ or equivalently as $\boldsymbol{\mu}^T\mathbf{A}\boldsymbol{\mu} + 2\boldsymbol{\mu}^T\mathbf{A}\boldsymbol{\Gamma}\mathbf{z} + \mathbf{z}^T\boldsymbol{\Gamma}^T\mathbf{A}\boldsymbol{\Gamma}\mathbf{z}$. Applying Theorem 14.2.5 (with the current $\boldsymbol{\mu}^T\mathbf{A}\boldsymbol{\mu}$, $\boldsymbol{\Gamma}^T\mathbf{A}\boldsymbol{\mu}$, and $\boldsymbol{\Gamma}^T\mathbf{A}\boldsymbol{\Gamma}$ playing the roles of c, $\boldsymbol{\ell}$, and $\mathbf{A}$ in that theorem), we find that if $\boldsymbol{\Gamma}^T\mathbf{A}\boldsymbol{\Sigma}\mathbf{A}\boldsymbol{\Gamma} = \boldsymbol{\Gamma}^T\mathbf{A}\boldsymbol{\Gamma}$, $\boldsymbol{\Gamma}^T\mathbf{A}\boldsymbol{\Sigma}\mathbf{A}\boldsymbol{\mu} = \boldsymbol{\Gamma}^T\mathbf{A}\boldsymbol{\mu}$, and $\boldsymbol{\mu}^T\mathbf{A}\boldsymbol{\Sigma}\mathbf{A}\boldsymbol{\mu} = \boldsymbol{\mu}^T\mathbf{A}\boldsymbol{\mu}$, then $w \sim \chi^2[\mathrm{tr}(\boldsymbol{\Gamma}^T\mathbf{A}\boldsymbol{\Gamma}), \frac{1}{2}\boldsymbol{\mu}^T\mathbf{A}\boldsymbol{\mu}]$. But

$$\boldsymbol{\Gamma}^T\mathbf{A}\boldsymbol{\Sigma}\mathbf{A}\boldsymbol{\Gamma} = \boldsymbol{\Gamma}^T\mathbf{A}\boldsymbol{\Gamma} \Leftrightarrow \boldsymbol{\Gamma}\boldsymbol{\Gamma}^T\mathbf{A}\boldsymbol{\Sigma}\mathbf{A}\boldsymbol{\Gamma} = \boldsymbol{\Gamma}\boldsymbol{\Gamma}^T\mathbf{A}\boldsymbol{\Gamma}$$

$$\Leftrightarrow \boldsymbol{\Sigma}\mathbf{A}\boldsymbol{\Sigma}\mathbf{A}\boldsymbol{\Gamma}\boldsymbol{\Gamma}^T = \boldsymbol{\Sigma}\mathbf{A}\boldsymbol{\Gamma}\boldsymbol{\Gamma}^T \Leftrightarrow \boldsymbol{\Sigma}\mathbf{A}\boldsymbol{\Sigma}\mathbf{A}\boldsymbol{\Sigma} = \boldsymbol{\Sigma}\mathbf{A}\boldsymbol{\Sigma}$$

(where the forward implications are trivial and Theorem 3.3.2 may be used for the reverse implications). By similar arguments,

$$\boldsymbol{\Gamma}^T\mathbf{A}\boldsymbol{\Sigma}\mathbf{A}\boldsymbol{\mu} = \boldsymbol{\Gamma}^T\mathbf{A}\boldsymbol{\mu} \Leftrightarrow \boldsymbol{\Sigma}\mathbf{A}\boldsymbol{\Sigma}\mathbf{A}\boldsymbol{\mu} = \boldsymbol{\Sigma}\mathbf{A}\boldsymbol{\mu}.$$

Furthermore, by Theorem 2.10.3,

$$\mathrm{tr}(\boldsymbol{\Gamma}^T\mathbf{A}\boldsymbol{\Gamma}) = \mathrm{tr}(\mathbf{A}\boldsymbol{\Gamma}\boldsymbol{\Gamma}^T) = \mathrm{tr}(\mathbf{A}\boldsymbol{\Sigma}).$$

Next, we prove part (b). If $w \sim \chi^2(\nu, \lambda)$, where as indicated above, $w \sim \boldsymbol{\mu}^T\mathbf{A}\boldsymbol{\mu} + 2\boldsymbol{\mu}^T\mathbf{A}\boldsymbol{\Gamma}\mathbf{z} + \mathbf{z}^T\boldsymbol{\Gamma}^T\mathbf{A}\boldsymbol{\Gamma}\mathbf{z}$, then by the converse to Theorem 14.2.5 the conditions of the present theorem are satisfied, $\nu = \mathrm{tr}(\boldsymbol{\Gamma}^T\mathbf{A}\boldsymbol{\Gamma}) = \mathrm{tr}(\mathbf{A}\boldsymbol{\Sigma})$, and $\lambda = \frac{1}{2}\boldsymbol{\mu}^T\mathbf{A}\boldsymbol{\mu}$. □

Example 14.2-2. Distribution of the (Suitably Scaled) Residual Sum of Squares in the Normal Constrained Gauss–Markov Model

Consider the constrained Gauss–Markov model $\{\mathbf{y}, \mathbf{X}\boldsymbol{\beta}, \sigma^2\mathbf{I} : \mathbf{A}\boldsymbol{\beta} = \mathbf{h}\}$, for which point estimation of an estimable function $\mathbf{c}^T\boldsymbol{\beta}$ and the residual variance σ^2 were presented in Sect. 10.1. As in that section, let $(\breve{\boldsymbol{\beta}}^T, \breve{\boldsymbol{\lambda}}^T)^T$ be any solution to the constrained normal equations for that model, let $p_{\mathbf{A}}^* = \mathrm{rank}\begin{pmatrix}\mathbf{X}\\ \mathbf{A}\end{pmatrix} - \mathrm{rank}(\mathbf{A})$ and, assuming that $n > p_{\mathbf{A}}^*$, let $\breve{\sigma}^2 = Q(\breve{\boldsymbol{\beta}})/(n - p_{\mathbf{A}}^*)$.

Suppose now that the errors of the model are normally distributed. Then by Theorem 10.1.10, $Q(\breve{\boldsymbol{\beta}}) = \breve{\mathbf{e}}^T\breve{\mathbf{e}}$ where $\breve{\mathbf{e}} \sim \mathrm{N}(\mathbf{0}, \sigma^2(\mathbf{I} - \mathbf{P}_{\mathbf{X}(\mathbf{I}-\mathbf{A}^-\mathbf{A})}))$. Consider the distribution of $\breve{\mathbf{e}}^T\breve{\mathbf{e}}$. Applying Theorem 14.2.6 with $\mathbf{0}_n$, $\sigma^2(\mathbf{I} - \mathbf{P}_{\mathbf{X}(\mathbf{I}-\mathbf{A}^-\mathbf{A})})$ and $(1/\sigma^2)\mathbf{I}_n$ here playing the roles of $\boldsymbol{\mu}$, $\boldsymbol{\Sigma}$, and $\mathbf{A}$ in the theorem, and noting, as a consequence of Theorem 2.12.1 that $\mathbf{I} - \mathbf{P}_{\mathbf{X}(\mathbf{I}-\mathbf{A}^-\mathbf{A})}$ is idempotent, we see that $Q(\breve{\boldsymbol{\beta}})/\sigma^2$ has a central chi-square distribution with degrees of freedom equal to

$$\mathrm{tr}(\mathbf{I} - \mathbf{P}_{\mathbf{X}(\mathbf{I}-\mathbf{A}^-\mathbf{A})}) = n - \mathrm{rank}(\mathbf{P}_{\mathbf{X}(\mathbf{I}-\mathbf{A}^-\mathbf{A})}) = n - p_{\mathbf{A}}^*,$$

where we used Corollary 10.1.9.1 to obtain the final expression. ■

The previous example notwithstanding, by far the most frequent applications of Theorem 14.2.6 in linear models are to situations in which the variance–covariance matrix $\boldsymbol{\Sigma}$ is positive definite. Therefore, we give the following corollary to Theorem 14.2.6, which streamlines the conditions as well as the results of the theorem for this case.

Corollary 14.2.6.1 *Suppose that* $\mathbf{x} \sim N_n(\boldsymbol{\mu}, \boldsymbol{\Sigma})$, *where* $rank(\boldsymbol{\Sigma}) = n$, *and that* $\mathbf{A}$ *is a nonnull symmetric* $n \times n$ *matrix of constants.*

(a) If $\mathbf{A}\boldsymbol{\Sigma}$ *is idempotent, then* $\mathbf{x}^T\mathbf{A}\mathbf{x} \sim \chi^2[rank(\mathbf{A}), \frac{1}{2}\boldsymbol{\mu}^T\mathbf{A}\boldsymbol{\mu}]$.
(b) If $\mathbf{x}^T\mathbf{A}\mathbf{x} \sim \chi^2(\nu, \lambda)$, *then* $\mathbf{A}\boldsymbol{\Sigma}$ *is idempotent,* $\nu = rank(\mathbf{A})$, *and* $\lambda = \frac{1}{2}\boldsymbol{\mu}^T\mathbf{A}\boldsymbol{\mu}$.

Proof If $\mathbf{A}\boldsymbol{\Sigma}$ is idempotent, then clearly the first condition of Theorem 14.2.6a is satisfied. Furthermore, by pre-multiplying and post-multiplying that condition by $\boldsymbol{\Sigma}^{-1}$, we obtain $\mathbf{A}\boldsymbol{\Sigma}\mathbf{A} = \mathbf{A}$, from which the other two conditions of Theorem 14.2.6a follow easily. Also, by the idempotency of $\mathbf{A}\boldsymbol{\Sigma}$ and the nonsingularity of $\boldsymbol{\Sigma}$, Theorems 2.12.2 and 2.8.9 yield $\text{tr}(\mathbf{A}\boldsymbol{\Sigma}) = \text{rank}(\mathbf{A}\boldsymbol{\Sigma}) = \text{rank}(\mathbf{A})$. This proves part (a).

Conversely, if $\mathbf{x}^T\mathbf{A}\mathbf{x} \sim \chi^2(\nu, \lambda)$ for some s and λ, then by Theorem 14.2.6b the first condition of Theorem 14.2.6a holds: $\boldsymbol{\Sigma}\mathbf{A}\boldsymbol{\Sigma}\mathbf{A}\boldsymbol{\Sigma} = \boldsymbol{\Sigma}\mathbf{A}\boldsymbol{\Sigma}$. Pre-multiplication of both sides by $\boldsymbol{\Sigma}^{-1}$ yields $\mathbf{A}\boldsymbol{\Sigma}\mathbf{A}\boldsymbol{\Sigma} = \mathbf{A}\boldsymbol{\Sigma}$, which establishes that $\mathbf{A}\boldsymbol{\Sigma}$ is idempotent. The rest then follows by part (a). This proves part (b). □

Example 14.2-3. Distributions of (Suitably Scaled) Sums of Squares in the Overall and Sequential ANOVAs Under the Normal Gauss–Markov and Positive Definite Aitken Models

Recall from Chap. 8 that the overall ANOVA of the model $\{\mathbf{y}, \mathbf{X}\boldsymbol{\beta}\}$ may be laid out in a table as follows:

Source	Rank	Sum of squares
Model	p^*	$\mathbf{y}^T\mathbf{P_X}\mathbf{y}$
Residual	$n - p^*$	$\mathbf{y}^T(\mathbf{I} - \mathbf{P_X})\mathbf{y}$
Total	n	$\mathbf{y}^T\mathbf{y}$

Now assume that $\mathbf{y}$ follows the normal Gauss–Markov model, so that $\mathbf{y} \sim \mathrm{N}_n(\mathbf{X}\boldsymbol{\beta}, \sigma^2\mathbf{I})$. Then $(1/\sigma)\mathbf{y} \sim \mathrm{N}_n((1/\sigma)\mathbf{X}\boldsymbol{\beta}, \mathbf{I})$, and upon applying Corollary 14.2.6.1a with $\mathbf{A} = \mathbf{P_X}$, we obtain (because $\mathbf{P_X}\mathbf{I}$ is idempotent),

$$[(1/\sigma)\mathbf{y}]^T\mathbf{P_X}[(1/\sigma)\mathbf{y}] \sim \chi^2(\text{rank}(\mathbf{P_X}), [(1/\sigma)\mathbf{X}\boldsymbol{\beta}]^T\mathbf{P_X}[(1/\sigma)\mathbf{X}\boldsymbol{\beta}]/2),$$

i.e.,

$$\mathbf{y}^T \mathbf{P_X} \mathbf{y}/\sigma^2 \sim \chi^2(p^*, \boldsymbol{\beta}^T \mathbf{X}^T \mathbf{X}\boldsymbol{\beta}/(2\sigma^2)).$$

Furthermore, because $(\mathbf{I} - \mathbf{P_X})\mathbf{I}$ is idempotent, the same corollary also yields

$$[(1/\sigma)\mathbf{y}]^T(\mathbf{I}-\mathbf{P_X})[(1/\sigma)\mathbf{y}] \sim \chi^2(\text{rank}(\mathbf{I}-\mathbf{P_X}), [(1/\sigma)\mathbf{X}\boldsymbol{\beta}]^T(\mathbf{I}-\mathbf{P_X})[(1/\sigma)\mathbf{X}\boldsymbol{\beta}]/2),$$

i.e.,

$$\mathbf{y}^T(\mathbf{I} - \mathbf{P_X})\mathbf{y}/\sigma^2 \sim \chi^2(n - p^*). \tag{14.8}$$

Similar results may be obtained for the sums of squares in the sequential ANOVA. Recall from Sect. 9.3.1 that the sequential ANOVA of an ordered k-part model may be laid out as follows:

Source	Rank	Sum of squares
$\mathbf{X}_1$	p_1^*	$\mathbf{y}^T\mathbf{P}_1\mathbf{y}$
$\mathbf{X}_2\|\mathbf{X}_1$	$p_{12}^* - p_1^*$	$\mathbf{y}^T(\mathbf{P}_{12} - \mathbf{P}_1)\mathbf{y}$
$\mathbf{X}_3\|(\mathbf{X}_1, \mathbf{X}_2)$	$p_{123}^* - p_{12}^*$	$\mathbf{y}^T(\mathbf{P}_{123} - \mathbf{P}_{12})\mathbf{y}$
$\vdots$	$\vdots$	$\vdots$
$\mathbf{X}_k\|(\mathbf{X}_1, \ldots, \mathbf{X}_{k-1})$	$p_{12\cdots k}^* - p_{12\cdots k-1}^*$	$\mathbf{y}^T(\mathbf{P}_{12\cdots k} - \mathbf{P}_{12\cdots k-1})\mathbf{y}$
Residual	$n - p^*$	$\mathbf{y}^T(\mathbf{I} - \mathbf{P}_{12\cdots k})\mathbf{y}$
Total	n	$\mathbf{y}^T\mathbf{y}$

where $p_{12\cdots j}^* = \text{rank}(\mathbf{X}_1, \mathbf{X}_2, \ldots, \mathbf{X}_j)$ for $j = 1, \ldots, k$. Each matrix of the quadratic form in the sums of squares column is idempotent, so the same reasoning used for the overall ANOVA yields the following results:

$$\begin{aligned}
\mathbf{y}^T\mathbf{P}_1\mathbf{y}/\sigma^2 &\sim \chi^2[p_1^*, \boldsymbol{\beta}^T\mathbf{X}^T\mathbf{P}_1\mathbf{X}\boldsymbol{\beta}/(2\sigma^2)],\\
\mathbf{y}^T(\mathbf{P}_{12} - \mathbf{P}_1)\mathbf{y}/\sigma^2 &\sim \chi^2[p_{12}^* - p_1^*, \boldsymbol{\beta}^T\mathbf{X}^T(\mathbf{P}_{12} - \mathbf{P}_1)\mathbf{X}\boldsymbol{\beta}/(2\sigma^2)],\\
&\vdots\\
\mathbf{y}^T(\mathbf{P}_{12\cdots k} - \mathbf{P}_{12\cdots k-1})\mathbf{y}/\sigma^2 &\sim \chi^2[p_{12\cdots k}^* - p_{12\cdots k-1}^*, \boldsymbol{\beta}^T\mathbf{X}^T(\mathbf{P}_{12\cdots k} - \mathbf{P}_{12\cdots k-1})\mathbf{X}\boldsymbol{\beta}/(2\sigma^2)],\\
\mathbf{y}^T(\mathbf{I} - \mathbf{P}_{12\cdots k})\mathbf{y}/\sigma^2 &\sim \chi^2(n - p^*),
\end{aligned}$$

where the last result is just a restatement of the result for the residual sum of squares in the overall ANOVA given previously.

Because each entry of the column in the overall and sequential ANOVA tables labeled "Rank" coincides with the degrees of freedom of the chi-square distribution of the scaled sum of squares in that row, this column will henceforth be renamed "degrees of freedom," or "df" for short.

Of course, these results can be specialized to specific cases of normal Gauss–Markov linear models. For example, from Example 9.3.3-1 the corrected ANOVA for full-rank simple linear regression is

Source	df	Sum of squares	Expected mean square
Model	1	$\hat{\beta}_2^2 SXX$	$\sigma^2 + \beta_2^2 SXX$
Residual	$n-2$	$\sum_{i=1}^n (y_i - \bar{y})^2 - \hat{\beta}_2^2 SXX$	σ^2
Corrected Total	$n-1$	$\sum_{i=1}^n (y_i - \bar{y})^2$	

and using the results just described, we obtain

$$\hat{\beta}_2^2 SXX/\sigma^2 \sim \chi^2\left(1, \beta_2^2 SXX/(2\sigma^2)\right),$$

$$\left(\sum_{i=1}^n (y_i - \bar{y})^2 - \hat{\beta}_2^2 SXX\right) \Big/ \sigma^2 \sim \chi^2(n-2).$$

Along the same lines, the corrected ANOVA for the one-factor model is, from Example 9.3.3-2,

Source	df	Sum of squares	Expected mean square
Model	$q-1$	$\sum_{i=1}^q n_i(\bar{y}_{i\cdot} - \bar{y}_{\cdot\cdot})^2$	$\sigma^2 + \frac{1}{q-1}\sum_{i=1}^q n_i\left(\alpha_i - \frac{\sum_{k=1}^q n_k\alpha_k}{n}\right)^2$
Residual	$n-q$	$\sum_{i=1}^q \sum_{j=1}^{n_i} (y_{ij} - \bar{y}_{i\cdot})^2$	σ^2
Corrected Total	$n-1$	$\sum_{i=1}^q \sum_{j=1}^{n_i} (y_{ij} - \bar{y}_{\cdot\cdot})^2$	

to which we may now add the distributional results

$$\sum_{i=1}^q n_i(\bar{y}_{i\cdot} - \bar{y}_{\cdot\cdot})^2/\sigma^2 \sim \chi^2\left(q-1, \sum_{i=1}^q n_i\left(\alpha_i - \frac{\sum_{k=1}^q n_k\alpha_k}{n}\right)^2 \Big/ (2\sigma^2)\right),$$

$$\sum_{i=1}^q \sum_{j=1}^{n_i} (y_{ij} - \bar{y}_{i\cdot})^2/\sigma^2 \sim \chi^2(n-q).$$

Finally, similar results can be obtained for sums of squares under the positive definite Aitken model. We merely show that $(n-p^*)\tilde{\sigma}^2 \sim \chi^2(n-p^*)$ under such a model with $n > p^*$. Recall that

$$(n-p^*)\tilde{\sigma}^2 = \mathbf{y}^T\mathbf{E}\mathbf{y}$$

where $\mathbf{E} = \mathbf{W}^{-1} - \mathbf{W}^{-1}\mathbf{X}(\mathbf{X}^T\mathbf{W}^{-1}\mathbf{X})^-\mathbf{X}^T\mathbf{W}^{-1}$, and

$$(1/\sigma)\mathbf{y} \sim \mathrm{N}((1/\sigma)\mathbf{X}\boldsymbol{\beta}, \mathbf{W}).$$

Thus, by Corollary 14.2.6.1a the desired distributional result will follow if **EW** is idempotent. But according to Theorem 11.1.6d, $\mathbf{EWE} = \mathbf{E}$, implying (by post-multiplying both sides by **W**) that **EW** is indeed idempotent. ■

Example 14.2-4. Distributions of (Suitably Scaled) Lack-of-Fit and Pure Error Sums of Squares Under the Normal Gauss–Markov Model

Recall from Sect. 9.5 that when some of the y-observations correspond to identical combinations of the explanatory variables, it is possible to decompose the residual sum of squares into two parts according to the table below:

Source	df	Sum of squares
Model	p^*	$\mathbf{y}^T\mathbf{P_X}\mathbf{y}$
Pure error	$n - m$	$\mathbf{y}^T(\mathbf{I} - \bar{\mathbf{J}})\mathbf{y}$
Lack of fit	$m - p^*$	$\mathbf{y}^T(\bar{\mathbf{J}} - \mathbf{P_X})\mathbf{y}$
Total	n	$\mathbf{y}^T\mathbf{y}$

(Definitions of $\bar{\mathbf{J}}$ and m may be found in Sect. 9.5.) Also recall, from Theorem 9.5.1b, f, that $\mathbf{I} - \bar{\mathbf{J}}$ and $\bar{\mathbf{J}} - \mathbf{P_X}$ are idempotent. Therefore, under the normal Gauss–Markov model and by the same rationale used in Example 14.2-3,

$$\mathbf{y}^T(\mathbf{I} - \bar{\mathbf{J}})\mathbf{y}/\sigma^2 \sim \chi^2(n - m)$$

and

$$\mathbf{y}^T(\bar{\mathbf{J}} - \mathbf{P_X})\mathbf{y}/\sigma^2 \sim \chi^2(m - p^*).$$

Here we have used Theorem 9.5.1c to show that both noncentrality parameters are zero, i.e.,

$$[(1/\sigma)\mathbf{X}\boldsymbol{\beta}]^T(\mathbf{I} - \bar{\mathbf{J}})[(1/\sigma)\mathbf{X}\boldsymbol{\beta}]/2 = [(1/\sigma)\mathbf{X}\boldsymbol{\beta}]^T(\mathbf{X} - \bar{\mathbf{J}}\mathbf{X})[(1/\sigma)\mathbf{X}\boldsymbol{\beta}]/2 = 0$$

and

$$[(1/\sigma)\mathbf{X}\boldsymbol{\beta}]^T(\bar{\mathbf{J}} - \mathbf{P_X})[(1/\sigma)\mathbf{X}\boldsymbol{\beta}]/2 = [(1/\sigma)\mathbf{X}\boldsymbol{\beta}]^T(\bar{\mathbf{J}}\mathbf{X} - \mathbf{P_X}\mathbf{X})[(1/\sigma)\mathbf{X}\boldsymbol{\beta}]/2 = 0.$$

Here again, because each entry in the "Rank" column coincides with the degrees of freedom of the chi-square distribution of the corresponding scaled sum of squares, we renamed that column "df." ■

We end this section with two more theorems pertaining to noncentral chi-square variables that are useful in some situations.

Theorem 14.2.7 *If $u \sim \chi^2(\nu)$ where $\nu > 2$, then $E(1/u) = 1/(\nu - 2)$.*

Proof

$$
\begin{aligned}
\mathrm{E}\left(\frac{1}{u}\right) &= \int_0^\infty \frac{1}{u}\,\frac{1}{\Gamma(\nu/2)2^{\nu/2}} u^{(\nu-2)/2} e^{-u/2}\, du \\
&= \frac{1}{\Gamma(\nu/2)2^{\nu/2}} \int_0^\infty u^{[(\nu-2)/2]-1} e^{-u/2}\, du \\
&= \frac{\Gamma[(\nu-2)/2]2^{(\nu-2)/2}}{\Gamma(\nu/2)2^{\nu/2}} \\
&= \frac{\Gamma[(\nu-2)/2]}{2[(\nu-2)/2]\Gamma[(\nu-2)/2]} \\
&= \frac{1}{\nu-2},
\end{aligned}
$$

where we used (14.7) for the third equality. □

Theorem 14.2.8 *Suppose that $u_1, \ldots, u_k$ are independent random variables such that $u_i \sim \chi^2(\nu_i, \lambda_i)$ $(i = 1, \ldots, k)$. Then the random variable u defined by $u = \sum_{i=1}^k u_i$ has a noncentral chi-square distribution with $\nu = \sum_{i=1}^k \nu_i$ degrees of freedom and noncentrality parameter $\lambda = \sum_{i=1}^k \lambda_i$.*

Proof By Theorems 14.1.2 and 14.2.1,

$$
\begin{aligned}
m_u(t) &= \mathrm{E}\{\exp[t(u_1 + \cdots + u_k)]\} \\
&= \prod_{i=1}^k \mathrm{E}[\exp(tu_i)] \\
&= \prod_{i=1}^k (1-2t)^{-\nu_i/2} \exp\left(\frac{2t\lambda_i}{1-2t}\right) \\
&= (1-2t)^{-\nu/2} \exp\left(\frac{2t\lambda}{1-2t}\right).
\end{aligned}
$$

□

14.3 Independence of Linear and Quadratic Forms in a Multivariate Normal Vector

In Chap. 4 we obtained expressions for the covariance between linear or quadratic forms in a random vector $\mathbf{x}$ whose fourth-order moments exist but whose distribution is otherwise arbitrary. Earlier in this chapter, we specialized those expressions to the case where the distribution of $\mathbf{x}$ is multivariate normal, obtaining Corollary 14.1.10.1. Such expressions may be used to determine conditions under which the linear/quadratic forms are uncorrelated. For such functions to be independent it is necessary that they be uncorrelated, but if at least one of the functions is a quadratic form, uncorrelatedness is not sufficient for independence even if $\mathbf{x}$ has a multivariate normal distribution. In this section, we obtain necessary and sufficient conditions for two or more linear or quadratic forms in a multivariate normal vector to be distributed independently.

Theorem 14.3.1 *Suppose that $\mathbf{x} \sim N_n(\boldsymbol{\mu}, \boldsymbol{\Sigma})$, $\mathbf{A}_i$ is a symmetric $n \times n$ matrix of constants, and $\boldsymbol{\ell}_i$ is an n-vector of constants ($i = 1, \ldots, k$). Define $w_i = 2\boldsymbol{\ell}_i^T \mathbf{x} + \mathbf{x}^T \mathbf{A}_i \mathbf{x}$ (note that w_i could include constant terms as well because such terms would have no effect on independence). Then, $w_1, w_2, \ldots, w_k$ are distributed independently if and only if*

(1) $\boldsymbol{\Sigma}\mathbf{A}_i\boldsymbol{\Sigma}\mathbf{A}_j\boldsymbol{\Sigma} = \mathbf{0}$,
(2) $\boldsymbol{\Sigma}\mathbf{A}_i\boldsymbol{\Sigma}(\boldsymbol{\ell}_j + \mathbf{A}_j\boldsymbol{\mu}) = \mathbf{0}$, and
(3) $(\boldsymbol{\ell}_i + \mathbf{A}_i\boldsymbol{\mu})^T\boldsymbol{\Sigma}(\boldsymbol{\ell}_j + \mathbf{A}_j\boldsymbol{\mu}) = 0$

for all $j \neq i = 1, \ldots, k$.

Proof We begin by proving the sufficiency of the three conditions. Let $r = \text{rank}(\boldsymbol{\Sigma})$, let $\mathbf{d}_i = \boldsymbol{\ell}_i + \mathbf{A}_i\boldsymbol{\mu}$ ($i = 1, \ldots, k$), and let $\boldsymbol{\Gamma}$ represent any $n \times r$ matrix such that $\boldsymbol{\Sigma} = \boldsymbol{\Gamma}\boldsymbol{\Gamma}^T$. Then $\mathbf{x}$ has the same distribution as $\boldsymbol{\mu} + \boldsymbol{\Gamma}\mathbf{z}$, where $\mathbf{z} \sim \text{N}_r(\mathbf{0}, \mathbf{I})$. Thus, w_i has the same distribution as

$$\begin{aligned}
2\boldsymbol{\ell}_i^T(\boldsymbol{\mu} + \boldsymbol{\Gamma}\mathbf{z}) + (\boldsymbol{\mu} + \boldsymbol{\Gamma}\mathbf{z})^T\mathbf{A}_i(\boldsymbol{\mu} + \boldsymbol{\Gamma}\mathbf{z}) &= 2\boldsymbol{\ell}_i^T\boldsymbol{\mu} + \boldsymbol{\mu}^T\mathbf{A}_i\boldsymbol{\mu} + 2\mathbf{d}_i^T\boldsymbol{\Gamma}\mathbf{z} + \mathbf{z}^T\boldsymbol{\Gamma}^T\mathbf{A}_i\boldsymbol{\Gamma}\mathbf{z} \\
&= 2\boldsymbol{\ell}_i^T\boldsymbol{\mu} + \boldsymbol{\mu}^T\mathbf{A}_i\boldsymbol{\mu} + 2\mathbf{d}_i^T\boldsymbol{\Gamma}\mathbf{z} \\
&\quad + \mathbf{z}^T\boldsymbol{\Gamma}^T\mathbf{A}_i\boldsymbol{\Gamma}(\boldsymbol{\Gamma}^T\mathbf{A}_i\boldsymbol{\Gamma})^-\boldsymbol{\Gamma}^T\mathbf{A}_i\boldsymbol{\Gamma}\mathbf{z},
\end{aligned}$$

which depends on $\mathbf{z}$ only through the vector $(\boldsymbol{\Gamma}^T\mathbf{d}_i, \boldsymbol{\Gamma}^T\mathbf{A}_i\boldsymbol{\Gamma})^T\mathbf{z}$. Clearly, the joint distribution of the vectors $(\boldsymbol{\Gamma}^T\mathbf{d}_1, \boldsymbol{\Gamma}^T\mathbf{A}_1\boldsymbol{\Gamma})^T\mathbf{z}, \ldots, (\boldsymbol{\Gamma}^T\mathbf{d}_k, \boldsymbol{\Gamma}^T\mathbf{A}_k\boldsymbol{\Gamma})^T\mathbf{z}$ is multivariate normal. Thus, to show the independence of $w_1, \ldots, w_k$, it suffices to show that these vectors are independent, for which, in turn, it suffices (by Theorem 14.1.8) to show that for all $i \neq j$,

$$\text{cov}[(\boldsymbol{\Gamma}^T\mathbf{d}_i, \boldsymbol{\Gamma}^T\mathbf{A}_i\boldsymbol{\Gamma})^T\mathbf{z}, \ldots, (\boldsymbol{\Gamma}^T\mathbf{d}_j, \boldsymbol{\Gamma}^T\mathbf{A}_j\boldsymbol{\Gamma})^T\mathbf{z}] = \mathbf{0},$$

or equivalently (by Corollary 14.1.8.1) that for all $i \neq j$,

(1′) $\boldsymbol{\Gamma}^T\mathbf{A}_i\boldsymbol{\Sigma}\mathbf{A}_j\boldsymbol{\Gamma} = \mathbf{0}$,
(2′) $\boldsymbol{\Gamma}^T\mathbf{A}_i\boldsymbol{\Sigma}\mathbf{d}_j = \mathbf{0}$, and
(3′) $\mathbf{d}_i^T\boldsymbol{\Sigma}\mathbf{d}_j = 0$.

But (3′) is identical to (3), and by Theorem 3.3.2, (1′) and (2′) are equivalent to (1) and (2), respectively.

Proof of the necessity of the three conditions requires knowledge of the mathematics of polynomial functions beyond the scope of this presentation. See Harville (2018, Section 6.8e) for more details. □

Corollary 14.3.1.1 *Suppose, in addition to the conditions of Theorem 14.3.1, that rank($\boldsymbol{\Sigma}$) $= n$. Then, $w_1, \ldots, w_k$ are distributed independently if and only if, for all $j \neq i = 1, \ldots, k$,*

(1) $\mathbf{A}_i\boldsymbol{\Sigma}\mathbf{A}_j = \mathbf{0}$,
(2) $\mathbf{A}_i\boldsymbol{\Sigma}\boldsymbol{\ell}_j = \mathbf{0}$, *and*
(3) $\boldsymbol{\ell}_i^T\boldsymbol{\Sigma}\boldsymbol{\ell}_j = 0$.

Proof Pre- and/or post-multiplying conditions (1), (2), and (3) of Theorem 14.3.1 by $\boldsymbol{\Sigma}^{-1}$ and then combining the results yield the three conditions of the corollary. Conversely, the three conditions of Theorem 14.3.1 follow easily from the three conditions of the corollary. □

Corollary 14.3.1.2 *Suppose that* $\mathbf{x} \sim N_n(\boldsymbol{\mu}, \boldsymbol{\Sigma})$, *where rank*($\boldsymbol{\Sigma}$) $= n$, $\mathbf{L}_1$ *and* $\mathbf{L}_2$ *are* $n \times m$ *matrices of constants, and* $\mathbf{A}_1$ *and* $\mathbf{A}_2$ *are symmetric* $n \times n$ *matrices of constants. Then:*

(a) $\mathbf{x}^T\mathbf{A}_1\mathbf{x}$ *and* $\mathbf{x}^T\mathbf{A}_2\mathbf{x}$ *are independent if and only if* $\mathbf{A}_1\boldsymbol{\Sigma}\mathbf{A}_2 = \mathbf{0}$;
(b) $\mathbf{x}^T\mathbf{A}_1\mathbf{x}$ *and* $\mathbf{L}_1^T\mathbf{x}$ *are independent if and only if* $\mathbf{A}_1\boldsymbol{\Sigma}\mathbf{L}_1 = \mathbf{0}$;
(c) $\mathbf{L}_1^T\mathbf{x}$ *and* $\mathbf{L}_2^T\mathbf{x}$ *are independent if and only if* $\mathbf{L}_1^T\boldsymbol{\Sigma}\mathbf{L}_2 = \mathbf{0}$.

Proof Part (a) is obtained by putting $k = 2$ and $\boldsymbol{\ell}_1 = \boldsymbol{\ell}_2 = \mathbf{0}$ into Corollary 1, while part (b) is obtained by putting $k = m + 1$, $\mathbf{A}_2 = \mathbf{A}_3 = \cdots = \mathbf{A}_{m+1} = \mathbf{0}$, $\boldsymbol{\ell}_1 = \mathbf{0}$,

and

$$\mathbf{L}_1^T = \begin{pmatrix} \boldsymbol{\ell}_2^T \\ \boldsymbol{\ell}_3^T \\ \vdots \\ \boldsymbol{\ell}_{m+1}^T \end{pmatrix}$$

into the same corollary. Part (c) is a special case of Corollary 14.1.8.1. □

Recall, from Corollary 14.1.10.1, that the conditions for $\mathbf{x}^T\mathbf{A}_1\mathbf{x}$ and $\mathbf{x}^T\mathbf{A}_2\mathbf{x}$ to be uncorrelated, and for $\mathbf{x}^T\mathbf{A}_1\mathbf{x}$ and $\mathbf{L}_1^T\mathbf{x}$ to be uncorrelated are, respectively,

$$\text{tr}(\mathbf{A}_1\boldsymbol{\Sigma}\mathbf{A}_2\boldsymbol{\Sigma}) = -2\boldsymbol{\mu}^T\mathbf{A}_1\boldsymbol{\Sigma}\mathbf{A}_2\boldsymbol{\mu}$$

and

$$\mathbf{L}_1^T\boldsymbol{\Sigma}\mathbf{A}_1\boldsymbol{\mu} = \mathbf{0}.$$

Clearly, these are not as restrictive as the conditions for independence given by parts (a) and (b) of Corollary 14.3.1.2.

Example 14.3-1. Independence of the Generalized Least Squares Estimator and the Generalized Residual Mean Square Under the Normal Positive Definite Aitken Model

We have noted previously (Theorem 11.1.10a) that under the positive definite Aitken model $\{\mathbf{y}, \mathbf{X}\boldsymbol{\beta}, \sigma^2\mathbf{W}\}$ with skewness matrix $\mathbf{0}$ (as is the case under normality), the generalized least squares estimator $\mathbf{C}^T\tilde{\boldsymbol{\beta}}$ of an estimable vector of functions $\mathbf{C}^T\boldsymbol{\beta}$ and the generalized residual mean square $\tilde{\sigma}^2 = \mathbf{y}^T\mathbf{E}\mathbf{y}/(n-p^*)$ from the generalized overall ANOVA are uncorrelated. In fact, from Corollary 14.3.1.2b and Theorem 11.1.6d, we may see now that they are not merely uncorrelated but independent because

$$\begin{aligned} &[\mathbf{E}/(n-p^*)](\sigma^2\mathbf{W})\mathbf{W}^{-1}\mathbf{X}[(\mathbf{X}^T\mathbf{W}^{-1}\mathbf{X})^-]^T\mathbf{C} \\ &= [\sigma^2/(n-p^*)]\mathbf{E}\mathbf{X}[(\mathbf{X}^T\mathbf{W}^{-1}\mathbf{X})^-]^T\mathbf{C} \\ &= \mathbf{0}. \end{aligned}$$

■

Example 14.3-2. Independence of the BLUP's Prediction Error and the Generalized Residual Mean Square Under the Normal Prediction-Extended Positive Definite Aitken Model

Recall from Example 14.1-3 that under the normal prediction-extended positive definite Aitken model $\left\{\begin{pmatrix}\mathbf{y}\\ \mathbf{u}\end{pmatrix}, \begin{pmatrix}\mathbf{X}\boldsymbol{\beta}\\ \mathbf{0}\end{pmatrix}, \sigma^2\begin{pmatrix}\mathbf{W} & \mathbf{K}\\ \mathbf{K}^T & \mathbf{H}\end{pmatrix}\right\}$, the BLUP's prediction error vector, $\tilde{\boldsymbol{\tau}} - \boldsymbol{\tau}$, and the generalized least squares residual vector, $\tilde{\mathbf{e}} = \mathbf{y} - \mathbf{X}\tilde{\boldsymbol{\beta}}$, are independent. Because the generalized residual mean square depends on $\mathbf{y}$ and $\mathbf{u}$ only as a function of the generalized least squares residual vector, i.e., $\tilde{\sigma}^2 = \tilde{\mathbf{e}}^T\mathbf{W}^{-1}\tilde{\mathbf{e}}/(n - p^*)$, the BLUP's prediction error vector and $\tilde{\sigma}^2$ likewise are independent.

In this case, it was demonstrated that a certain linear form and a certain quadratic form are independent by showing that two linear forms are independent and that the quadratic form is a function of only one of the linear forms, rather than by verifying the conditions of Corollary 14.3.1.2b. This approach often is useful in other settings as well. ■

Example 14.3-3. Independence of Sums of Squares in the Overall and Sequential ANOVAs Under the Normal Gauss–Markov Model

The two sums of squares in the overall ANOVA, $\mathbf{y}^T\mathbf{P_X}\mathbf{y}$ and $\mathbf{y}^T(\mathbf{I} - \mathbf{P_X})\mathbf{y}$, are independent under the normal Gauss–Markov model $\{\mathbf{y}, \mathbf{X}\boldsymbol{\beta}, \sigma^2\mathbf{I}\}$ as a direct application of Corollary 14.3.1.2a and Theorem 8.1.4:

$$\mathbf{P_X}(\sigma^2\mathbf{I})(\mathbf{I} - \mathbf{P_X}) = \mathbf{0}.$$

By the same corollary and Theorem 9.1.3d, e, independence holds between any pair of sums of squares in the sequential ANOVA of an ordered k-part normal Gauss–Markov model. These results also extend to overall and sequential generalized ANOVAs corresponding to a normal positive definite Aitken model. ■

Example 14.3-4. Independence of Lack-of-Fit and Pure Error Sums of Squares Under the Normal Gauss–Markov Model

Applying Corollary 14.3.1.2a to the pure error sum of squares $\mathbf{y}^T(\mathbf{I} - \bar{\mathbf{J}})\mathbf{y}$ and lack-of-fit sum of squares $\mathbf{y}^T(\bar{\mathbf{J}} - \mathbf{P_X})\mathbf{y}$, we see that they are independent under the normal Gauss–Markov model $\{\mathbf{y}, \mathbf{X}\boldsymbol{\beta}, \sigma^2\mathbf{I}\}$ because

$$(\mathbf{I} - \bar{\mathbf{J}})(\sigma^2\mathbf{I})(\bar{\mathbf{J}} - \mathbf{P_X}) = \mathbf{0}$$

by Theorem 9.5.1e. ■

14.4 Cochran's Theorem

In the previous two sections, we established, among other things, necessary and sufficient conditions for a quadratic form in a multivariate normal vector to have a chi-square distribution (Theorem 14.2.6), and necessary and sufficient conditions for two or more quadratic forms in a multivariate normal vector to be independent (Theorem 14.3.1). Furthermore, we specialized those results to the case where the variance–covariance matrix of the multivariate normal vector is positive definite (Corollaries 14.2.6.1 and 14.3.1.1) because this case is so prevalent in applications of linear models. In this section we prove and apply a theorem that gives necessary and sufficient conditions for two or more quadratic forms in a multivariate normal vector with positive definite variance–covariance matrix to have chi-square distributions *and* be independent. Of course, we could simply combine the conditions of the two aforementioned corollaries, but the combined conditions would contain some redundancies. We will eliminate the redundancies, arriving at a generalized version of a result that is known as Cochran's theorem.

Theorem 14.4.1 *Suppose that* $\mathbf{x} \sim N_n(\boldsymbol{\mu}, \boldsymbol{\Sigma})$, *where rank*$(\boldsymbol{\Sigma}) = n$. *Let* $\mathbf{A}_1, \ldots, \mathbf{A}_k$ *represent symmetric matrices of constants, and define* $\mathbf{A} = \sum_{i=1}^{k} \mathbf{A}_i$. *If* $\mathbf{A}\boldsymbol{\Sigma}$ *is idempotent and rank*$(\mathbf{A}) = \sum_{i=1}^{k}$ *rank*$(\mathbf{A}_i)$, *then* $\mathbf{x}^T\mathbf{A}_1\mathbf{x}, \ldots, \mathbf{x}^T\mathbf{A}_k\mathbf{x}$ *are distributed independently as* $\chi^2[\text{rank}(\mathbf{A}_i), \boldsymbol{\mu}^T\mathbf{A}_i\boldsymbol{\mu}/2]$.

Proof The given condition on the ranks together with the nonsingularity of $\boldsymbol{\Sigma}$ yield (by Theorem 2.8.9)

$$\text{rank}(\mathbf{A}\boldsymbol{\Sigma}) = \text{rank}(\mathbf{A}) = \sum_{i=1}^{k} \text{rank}(\mathbf{A}_i) = \sum_{i=1}^{k} \text{rank}(\mathbf{A}_i\boldsymbol{\Sigma}). \tag{14.9}$$

Now we apply Theorem 2.12.3 to the matrices $\mathbf{A}_1\boldsymbol{\Sigma}, \ldots, \mathbf{A}_k\boldsymbol{\Sigma}$. Upon observing that the sum of these matrices is $\mathbf{A}\boldsymbol{\Sigma}$, which is idempotent by hypothesis, (14.9) implies that: (a) $\mathbf{A}_i\boldsymbol{\Sigma}$ is idempotent for all $i = 1, \ldots, k$, and (b) $\mathbf{A}_i\boldsymbol{\Sigma}\mathbf{A}_j\boldsymbol{\Sigma} = \mathbf{0}$ for all $j \neq i = 1, \ldots, k$. Note that the latter simplifies to $\mathbf{A}_i\boldsymbol{\Sigma}\mathbf{A}_j = \mathbf{0}$ for all $j \neq i = 1, \ldots, k$ because $\boldsymbol{\Sigma}$ is nonsingular. The claimed distributions and independence then follow from Corollaries 14.2.6.1a and 14.3.1.1, respectively. □

Example 14.4-1. Independent Chi-Square Scaled Sums of Squares in the Sequential ANOVA for the Ordered *k*-Part Normal Gauss–Markov Model

Under the normal Gauss–Markov model $\{\mathbf{y}, \mathbf{X}\boldsymbol{\beta}, \sigma^2\mathbf{I}\}$, $(1/\sigma)\mathbf{y} \sim \mathrm{N}_n[(1/\sigma)\mathbf{X}\boldsymbol{\beta}, \mathbf{I}]$. Now the matrices $\mathbf{P}_1, \mathbf{P}_{12} - \mathbf{P}_1, \ldots, \mathbf{P}_{12\cdots k} - \mathbf{P}_{12\cdots k-1}, \mathbf{I} - \mathbf{P}_{12\cdots k}$ are symmetric matrices that sum to $\mathbf{I}_n$, which is idempotent. Furthermore,

$$\text{rank}(\mathbf{I}_n) = n = p_1^* + \sum_{j=2}^{k}(p_{1\cdots j}^* - p_{1\cdots j-1}^*) + (n - p_{1\cdots k}^*),$$

showing that the conditions of Theorem 14.4.1 hold in this situation. Therefore, the sums of squares in the sequential ANOVA for the ordered k-part model, scaled by σ^2, are distributed independently as follows:

$$\begin{aligned}
\mathbf{y}^T\mathbf{P}_1\mathbf{y}/\sigma^2 &\sim \chi^2[p_1^*, \boldsymbol{\beta}^T\mathbf{X}^T\mathbf{P}_1\mathbf{X}\boldsymbol{\beta}/(2\sigma^2)],\\
\mathbf{y}^T(\mathbf{P}_{12} - \mathbf{P}_1)\mathbf{y}/\sigma^2 &\sim \chi^2[p_{12}^* - p_1^*, \boldsymbol{\beta}^T\mathbf{X}^T(\mathbf{P}_{12} - \mathbf{P}_1)\mathbf{X}\boldsymbol{\beta}/(2\sigma^2)],\\
&\vdots\\
\mathbf{y}^T(\mathbf{P}_{12\cdots k} - \mathbf{P}_{12\cdots k-1})\mathbf{y}/\sigma^2 &\sim \chi^2[p_{12\cdots k}^* - p_{12\cdots k-1}^*, \boldsymbol{\beta}^T\mathbf{X}^T(\mathbf{P}_{12\cdots k} - \mathbf{P}_{12\cdots k-1})\mathbf{X}\boldsymbol{\beta}/(2\sigma^2)],\\
\mathbf{y}^T(\mathbf{I} - \mathbf{P}_{12\cdots k})\mathbf{y}/\sigma^2 &\sim \chi^2(n - p^*).
\end{aligned}$$

Of course, this merely repeats results that we had obtained previously in Examples 14.2-3 and 14.3-3. However, obtaining the same results via Theorem 14.4.1 is somewhat more efficient, in the sense that overall there are fewer conditions to check. ■

Example 14.4-2. Independent Chi-Square Scaled Sums of Squares in a Split-Plot ANOVA Under Normality

Consider the split-plot ANOVA of the one whole-plot-factor, one split-plot-factor model introduced in Example 13.4.5-1, and now assume that the distribution of $\mathbf{y}$ is multivariate normal. Recall from that example that $\text{var}(\mathbf{y}) = \sigma^2\mathbf{I}_{qrm} + \sigma_b^2(\mathbf{I}_{qr} \otimes \mathbf{J}_m)$, but now also observe that $\text{var}(\mathbf{y})$ may be expressed alternatively as $\sigma^2\mathbf{I}_{qrm} + m\sigma_b^2\mathbf{P_Z}$ where, as defined in that same example, $\mathbf{P_Z} = \frac{1}{m}(\mathbf{I}_{qr} \otimes \mathbf{J}_m)$. Observe also that the matrices of the first three quadratic forms in decomposition (13.12), when divided by $\sigma^2 + m\sigma_b^2$, sum to $[1/(\sigma^2 + m\sigma_b^2)]\mathbf{P_Z}$, and that

$$\begin{aligned}
[1/(\sigma^2 + m\sigma_b^2)]\mathbf{P_Z}(\sigma^2\mathbf{I}_{qrm} + m\sigma_b^2\mathbf{P_Z}) &= [1/(\sigma^2 + m\sigma_b^2)](\sigma^2\mathbf{P_Z} + m\sigma_b^2\mathbf{P_Z})\\
&= \mathbf{P_Z},
\end{aligned}$$

which is idempotent. Observe still further that the ranks of the matrices of the first three quadratic forms in decomposition (13.12) sum to qr, which coincides with rank($\mathbf{P_Z}$) by Theorem 2.17.8. Thus, by Theorem 14.4.1,

$$\frac{rm}{\sigma^2+m\sigma_b^2}\sum_{i=1}^{q}(\bar{y}_{i\cdot\cdot}-\bar{y}_{\cdot\cdot\cdot})^2 \sim \chi^2\left(q-1,\ \frac{rm}{2(q-1)(\sigma^2+m\sigma_b^2)}\sum_{i=1}^{q}[(\alpha_i-\bar{\alpha}_{\cdot})+(\bar{\xi}_{i\cdot}-\bar{\xi}_{\cdot\cdot})]^2\right)$$

and

$$\frac{m}{\sigma^2+m\sigma_b^2}\sum_{i=1}^{q}\sum_{j=1}^{r}(\bar{y}_{ij\cdot}-\bar{y}_{i\cdot\cdot})^2 \sim \chi^2[q(r-1)],$$

and these two quantities are independent. Furthermore, the last three quadratic forms in decomposition (13.12), when divided by σ^2, sum to $(1/\sigma^2)(\mathbf{I}-\mathbf{P_Z})$, and

$$(1/\sigma^2)(\mathbf{I}-\mathbf{P_Z})(\sigma^2\mathbf{I}_{qrm}+m\sigma_b^2\mathbf{P_Z})=\mathbf{I}-\mathbf{P_Z},$$

which is idempotent. The rank of $\mathbf{I}-\mathbf{P_Z}$ is $qrm - qr = qr(m-1)$, which equals the sum of the ranks of the last three quadratic forms. Therefore, again by Theorem 14.4.1,

$$\frac{qr}{\sigma^2}\sum_{k=1}^{m}(\bar{y}_{\cdot\cdot k}-\bar{y}_{\cdot\cdot\cdot})^2,\quad \frac{r}{\sigma^2}\sum_{i=1}^{q}\sum_{k=1}^{m}(\bar{y}_{i\cdot k}-\bar{y}_{i\cdot\cdot}-\bar{y}_{\cdot\cdot k}+\bar{y}_{\cdot\cdot\cdot})^2,$$

and

$$(1/\sigma^2)\sum_{i=1}^{q}\sum_{j=1}^{r}\sum_{k=1}^{m}(y_{ijk}-\bar{y}_{ij\cdot}-\bar{y}_{i\cdot k}+\bar{y}_{i\cdot\cdot})^2$$

are distributed independently as

$$\chi^2\left(m-1,\ \frac{qr}{2(m-1)\sigma^2}\sum_{k=1}^{m}[(\gamma_k-\bar{\gamma}_{\cdot})+(\bar{\xi}_{\cdot k}-\bar{\xi}_{\cdot\cdot})]^2\right),$$

$$\chi^2\left((q-1)(m-1),\ \frac{r}{2(q-1)(m-1)\sigma^2}\sum_{i=1}^{q}\sum_{k=1}^{m}(\xi_{ik}-\bar{\xi}_{i\cdot}-\bar{\xi}_{\cdot k}+\bar{\xi}_{\cdot\cdot})^2\right),$$

$$\chi^2[q(r-1)(m-1)].$$

■

14.5 The Noncentral F and Noncentral t Distributions

To this point we have introduced two important distributions in the theory of linear models: the multivariate normal and noncentral chi-square distributions. In this section, we introduce two more extremely useful distributions that are derived from the normal and/or noncentral chi-square distributions: the noncentral F and noncentral t distributions.

Definition 14.5.1 Suppose that $u_1 \sim \chi^2(\nu_1, \lambda)$ and $u_2 \sim \chi^2(\nu_2)$, and that u_1 and u_2 are independent. Then the distribution of the random variable w defined by

$$w = \frac{u_1/\nu_1}{u_2/\nu_2}$$

is called the *noncentral F distribution with degrees of freedom* ν_1 *and* ν_2 *and noncentrality parameter* λ, and is denoted by $\mathrm{F}(\nu_1, \nu_2, \lambda)$.

Definition 14.5.2 Suppose that $z \sim \mathrm{N}(\mu, 1)$ and $u \sim \chi^2(\nu)$, and that z and u are independent. Then the distribution of the random variable t defined by

$$t = \frac{z}{\sqrt{u/\nu}}$$

is called the *noncentral t distribution with* ν *degrees of freedom and noncentrality parameter* μ, and is denoted by $\mathrm{t}(\nu, \mu)$. If $\mu = 0$, t is said to have the *central t distribution with* ν *degrees of freedom*, which is denoted by $\mathrm{t}(\nu)$.

There is a close relationship between the noncentral F and noncentral t distributions, as described by the following theorem. Its proof is left as an exercise.

Theorem 14.5.1 *If* $t \sim t(\nu, \mu)$, *then* $t^2 \sim F(1, \nu, \mu^2/2)$.

We now consider several quantities arising in linear models that have noncentral F and t distributions. The quantities to be considered involve the residual mean square, so for the remainder of this section it is assumed that $n > p^*$. One quantity that has a noncentral F distribution can be constructed from the residual mean square and a vector $\mathbf{C}^T\hat{\boldsymbol{\beta}}$ of least squares estimators of s linearly independent estimable functions $\mathbf{C}^T\boldsymbol{\beta}$. Under the normal Gauss–Markov model,

$$\mathbf{C}^T\hat{\boldsymbol{\beta}} \sim \mathrm{N}_s[\mathbf{C}^T\boldsymbol{\beta}, \sigma^2\mathbf{C}^T(\mathbf{X}^T\mathbf{X})^-\mathbf{C}]$$

by Theorems 7.2.1, 7.2.2, and 14.1.6. Furthermore, because the functions in $\mathbf{C}^T\boldsymbol{\beta}$ are linearly independent, by Theorem 7.2.2 $\sigma^2\mathbf{C}^T(\mathbf{X}^T\mathbf{X})^-\mathbf{C}$ is positive definite (hence invertible) and $\text{rank}[\sigma^2\mathbf{C}^T(\mathbf{X}^T\mathbf{X})^-\mathbf{C}] = s$. Thus, by Theorem 14.2.4,

$$(\mathbf{C}^T\hat{\boldsymbol{\beta}})^T[\sigma^2\mathbf{C}^T(\mathbf{X}^T\mathbf{X})^-\mathbf{C}]^{-1}(\mathbf{C}^T\hat{\boldsymbol{\beta}}) \sim \chi^2[s, (\mathbf{C}^T\boldsymbol{\beta})^T[\mathbf{C}^T(\mathbf{X}^T\mathbf{X})^-\mathbf{C}]^{-1}(\mathbf{C}^T\boldsymbol{\beta})/(2\sigma^2)]$$

and

$$(\mathbf{C}^T\hat{\boldsymbol{\beta}} - \mathbf{C}^T\boldsymbol{\beta})^T[\sigma^2\mathbf{C}^T(\mathbf{X}^T\mathbf{X})^-\mathbf{C}]^{-1}(\mathbf{C}^T\hat{\boldsymbol{\beta}} - \mathbf{C}^T\boldsymbol{\beta}) \sim \chi^2(s).$$

Furthermore, as a special case of the result derived in Example 14.3-1, $(\mathbf{C}^T\hat{\boldsymbol{\beta}})^T[\sigma^2\mathbf{C}^T(\mathbf{X}^T\mathbf{X})^-\mathbf{C}]^{-1}(\mathbf{C}^T\hat{\boldsymbol{\beta}})$ and $\hat{\sigma}^2$ are independently distributed. By Definition 14.5.1, therefore, we have the following theorem.

Theorem 14.5.2 *If $\mathbf{C}^T\boldsymbol{\beta}$ is an s-vector of linearly independent estimable functions, then under the normal Gauss–Markov model $\{\mathbf{y}, \mathbf{X}\boldsymbol{\beta}, \sigma^2\mathbf{I}\}$,*

$$\begin{aligned}\frac{(\mathbf{C}^T\hat{\boldsymbol{\beta}})^T[\mathbf{C}^T(\mathbf{X}^T\mathbf{X})^-\mathbf{C}]^{-1}(\mathbf{C}^T\hat{\boldsymbol{\beta}})}{s\hat{\sigma}^2} &= \frac{(\mathbf{C}^T\hat{\boldsymbol{\beta}})^T[\sigma^2\mathbf{C}^T(\mathbf{X}^T\mathbf{X})^-\mathbf{C}]^{-1}(\mathbf{C}^T\hat{\boldsymbol{\beta}})/s}{\frac{(n-p^*)\hat{\sigma}^2/\sigma^2}{n-p^*}} \\ &\sim F\left(s, n-p^*, \frac{(\mathbf{C}^T\boldsymbol{\beta})^T[\mathbf{C}^T(\mathbf{X}^T\mathbf{X})^-\mathbf{C}]^{-1}(\mathbf{C}^T\boldsymbol{\beta})}{2\sigma^2}\right)\end{aligned}$$

and

$$\frac{(\mathbf{C}^T\hat{\boldsymbol{\beta}} - \mathbf{C}^T\boldsymbol{\beta})^T[\mathbf{C}^T(\mathbf{X}^T\mathbf{X})^-\mathbf{C}]^{-1}(\mathbf{C}^T\hat{\boldsymbol{\beta}} - \mathbf{C}^T\boldsymbol{\beta})}{s\hat{\sigma}^2} \sim F(s, n-p^*).$$

An important set of quantities that have (noncentral) F distributions is the ratios of mean squares from the sequential ANOVA of an ordered k-part normal Gauss–Markov model, as we now show. Recall from Example 14.4-1 that under that model,

$$\begin{aligned}\mathbf{y}^T\mathbf{P}_1\mathbf{y}/\sigma^2 &\sim \chi^2[p_1^*, \boldsymbol{\beta}^T\mathbf{X}^T\mathbf{P}_1\mathbf{X}\boldsymbol{\beta}/(2\sigma^2)], \\ \mathbf{y}^T(\mathbf{P}_{12} - \mathbf{P}_1)\mathbf{y}/\sigma^2 &\sim \chi^2[p_{12}^* - p_1^*, \boldsymbol{\beta}^T\mathbf{X}^T(\mathbf{P}_{12} - \mathbf{P}_1)\mathbf{X}\boldsymbol{\beta}/(2\sigma^2)], \\ &\vdots \\ \mathbf{y}^T(\mathbf{P}_{12\cdots k} - \mathbf{P}_{12\cdots k-1})\mathbf{y}/\sigma^2 &\sim \chi^2[p_{12\cdots k}^* - p_{12\cdots k-1}^*, \boldsymbol{\beta}^T\mathbf{X}^T(\mathbf{P}_{12\cdots k} - \mathbf{P}_{12\cdots k-1})\mathbf{X}\boldsymbol{\beta}/(2\sigma^2)], \\ \mathbf{y}^T(\mathbf{I} - \mathbf{P}_{12\cdots k})\mathbf{y}/\sigma^2 &\sim \chi^2(n-p^*),\end{aligned}$$

and that all these scaled quadratic forms are independent. Thus, we have the following theorem, again as a consequence of Definition 14.5.1.

Theorem 14.5.3 *Under the ordered k-part normal Gauss–Markov model* $\{\mathbf{y}, \sum_{j=1}^{k} \mathbf{X}_j\boldsymbol{\beta}_j, \sigma^2\mathbf{I}\}$,

$$\begin{aligned}
&\frac{\mathbf{y}^T(\mathbf{P}_{12\cdots j} - \mathbf{P}_{12\cdots j-1})\mathbf{y}/(p^*_{12\cdots j} - p^*_{12\cdots j-1})}{\mathbf{y}^T(\mathbf{I} - \mathbf{P}_{12\cdots k})\mathbf{y}/(n - p^*)} \\
&\quad = \frac{\mathbf{y}^T(\mathbf{P}_{12\cdots j} - \mathbf{P}_{12\cdots j-1})\mathbf{y}/[(p^*_{12\cdots j} - p^*_{12\cdots j-1})\sigma^2]}{\mathbf{y}^T(\mathbf{I} - \mathbf{P}_{12\cdots k})\mathbf{y}/[(n - p^*)\sigma^2]} \\
&\quad \sim F(p^*_{12\cdots j} - p^*_{12\cdots j-1}, n - p^*, \boldsymbol{\beta}^T\mathbf{X}^T(\mathbf{P}_{12\cdots j} - \mathbf{P}_{12\cdots j-1})\mathbf{X}\boldsymbol{\beta}/2\sigma^2)
\end{aligned}$$

for $j = 1, \ldots, k$, *where* $\mathbf{P}_{12\cdots j-1} = \mathbf{0}$ *when* $j = 1$.

As a special case of Theorem 14.5.3, the ratio of the model mean square to the residual mean square in the overall ANOVA has a noncentral F distribution under the normal Gauss–Markov model, i.e.,

$$\frac{\mathbf{y}^T\mathbf{P_X}\mathbf{y}/p^*}{\mathbf{y}^T(\mathbf{I} - \mathbf{P_X})\mathbf{y}/(n - p^*)} \sim \mathrm{F}[p^*, n - p^*, \boldsymbol{\beta}^T\mathbf{X}^T\mathbf{X}\boldsymbol{\beta}/(2\sigma^2)].$$

Quantities that have noncentral t distributions under the normal Gauss–Markov model can be constructed from the least squares estimator of an estimable function and the residual mean square. Recall from earlier in this section that if $\mathbf{c}^T\hat{\boldsymbol{\beta}}$ is the least squares estimator of an estimable function $\mathbf{c}^T\boldsymbol{\beta}$, then under the normal Gauss–Markov model,

$$\mathbf{c}^T\hat{\boldsymbol{\beta}} \sim \mathrm{N}[\mathbf{c}^T\boldsymbol{\beta}, \sigma^2\mathbf{c}^T(\mathbf{X}^T\mathbf{X})^-\mathbf{c}].$$

Thus, upon partial standardization and assuming that $\mathbf{c} \neq \mathbf{0}$,

$$\frac{\mathbf{c}^T\hat{\boldsymbol{\beta}}}{\sigma[\mathbf{c}^T(\mathbf{X}^T\mathbf{X})^-\mathbf{c}]^{\frac{1}{2}}} \sim \mathrm{N}\left(\frac{\mathbf{c}^T\boldsymbol{\beta}}{\sigma[\mathbf{c}^T(\mathbf{X}^T\mathbf{X})^-\mathbf{c}]^{\frac{1}{2}}}, 1\right),$$

and upon complete standardization,

$$\frac{\mathbf{c}^T\hat{\boldsymbol{\beta}} - \mathbf{c}^T\boldsymbol{\beta}}{\sigma[\mathbf{c}^T(\mathbf{X}^T\mathbf{X})^-\mathbf{c}]^{\frac{1}{2}}} \sim \mathrm{N}(0, 1).$$

As a special case of the result given in Example 14.3-1, $\mathbf{c}^T\hat{\boldsymbol{\beta}}$ and $\hat{\sigma}^2$ are independent. We therefore have the following theorem.

Theorem 14.5.4 *Let* $\mathbf{c}^T\boldsymbol{\beta}$ *(where* $\mathbf{c} \neq \mathbf{0}$*) be an estimable function under the normal Gauss–Markov model* $\{\mathbf{y}, \mathbf{X}\boldsymbol{\beta}, \sigma^2\mathbf{I}\}$*. Then,*

$$\frac{\mathbf{c}^T\hat{\boldsymbol{\beta}}}{\hat{\sigma}[\mathbf{c}^T(\mathbf{X}^T\mathbf{X})^-\mathbf{c}]^{\frac{1}{2}}} = \frac{\frac{\mathbf{c}^T\hat{\boldsymbol{\beta}}}{\sigma[\mathbf{c}^T(\mathbf{X}^T\mathbf{X})^-\mathbf{c}]^{\frac{1}{2}}}}{\sqrt{\frac{(n-p^*)\hat{\sigma}^2}{(n-p^*)\sigma^2}}} \sim t\left(n-p^*, \frac{\mathbf{c}^T\boldsymbol{\beta}}{\sigma[\mathbf{c}^T(\mathbf{X}^T\mathbf{X})^-\mathbf{c}]^{\frac{1}{2}}}\right)$$

and

$$\frac{\mathbf{c}^T\hat{\boldsymbol{\beta}} - \mathbf{c}^T\boldsymbol{\beta}}{\hat{\sigma}[\mathbf{c}^T(\mathbf{X}^T\mathbf{X})^-\mathbf{c}]^{\frac{1}{2}}} \sim t(n-p^*).$$

Still more quantities that have noncentral F or noncentral t distributions arise in prediction, as established by the next theorem. Its proof is left as an exercise.

Theorem 14.5.5 *Consider the prediction of a vector of s predictable functions* $\boldsymbol{\tau} = \mathbf{C}^T\boldsymbol{\beta} + \mathbf{u}$ *under the normal prediction-extended positive definite Aitken model* $\left\{\begin{pmatrix}\mathbf{y}\\ \mathbf{u}\end{pmatrix}, \begin{pmatrix}\mathbf{X}\boldsymbol{\beta}\\ \mathbf{0}\end{pmatrix}, \sigma^2\begin{pmatrix}\mathbf{W} & \mathbf{K}\\ \mathbf{K}^T & \mathbf{H}\end{pmatrix}\right\}$*. Recall from Theorem 13.2.1 that* $\tilde{\boldsymbol{\tau}} = \mathbf{C}^T\tilde{\boldsymbol{\beta}} + \mathbf{K}^T\mathbf{E}\mathbf{y}$ *is the vector of BLUPs of the elements of* $\boldsymbol{\tau}$*, where* $\tilde{\boldsymbol{\beta}}$ *is any solution to the Aitken equations and* $\mathbf{E} = \mathbf{W}^{-1} - \mathbf{W}^{-1}\mathbf{X}(\mathbf{X}^T\mathbf{W}^{-1}\mathbf{X})^-\mathbf{X}^T\mathbf{W}^{-1}$*, and that* $var(\tilde{\boldsymbol{\tau}} - \boldsymbol{\tau}) = \sigma^2\mathbf{Q}$*, where an expression for* $\mathbf{Q}$ *was given in Theorem 13.2.2a. If* $\mathbf{Q}$ *is positive definite (as is the case under either of the conditions of Theorem 13.2.2b), then*

$$\frac{(\tilde{\boldsymbol{\tau}} - \boldsymbol{\tau})^T\mathbf{Q}^{-1}(\tilde{\boldsymbol{\tau}} - \boldsymbol{\tau})}{s\tilde{\sigma}^2} \sim F(s, n-p^*).$$

If $s = 1$*, then*

$$\frac{\tilde{\tau} - \tau}{\tilde{\sigma}\sqrt{Q}} \sim t(n-p^*), \tag{14.10}$$

where Q *is the scalar version of* $\mathbf{Q}$*.*

Example 14.5-1. Quantities that Have t or F Distributions in Normal Full-Rank Multiple and Simple Linear Regression

Consider the normal Gauss–Markov full-rank multiple regression model. For this case, Theorem 14.5.2 specializes to yield

$$\frac{\hat{\boldsymbol{\beta}}^T\mathbf{X}^T\mathbf{X}\hat{\boldsymbol{\beta}}}{p\hat{\sigma}^2} \sim \mathrm{F}(p, n-p, \boldsymbol{\beta}^T\mathbf{X}^T\mathbf{X}\boldsymbol{\beta}/(2\sigma^2))$$

and

$$\frac{(\hat{\boldsymbol{\beta}} - \boldsymbol{\beta})^T \mathbf{X}^T \mathbf{X} (\hat{\boldsymbol{\beta}} - \boldsymbol{\beta})}{p\hat{\sigma}^2} \sim \mathrm{F}(p, n - p).$$

For this same case, Theorem 14.5.4 yields, for $j = 1, \ldots, p$,

$$\frac{\hat{\beta}_j}{\hat{\sigma}\sqrt{v_{jj}}} \sim \mathrm{t}(n - p, \frac{\beta_j}{\sigma\sqrt{v_{jj}}})$$

and

$$\frac{\hat{\beta}_j - \beta_j}{\hat{\sigma}\sqrt{v_{jj}}} \sim \mathrm{t}(n - p),$$

where v_{jj} is the jth main diagonal element of $(\mathbf{X}^T\mathbf{X})^{-1}$.

As a further special case, consider the normal Gauss–Markov simple linear regression model. For this case, the first part of Theorem 14.5.2 specializes to

$$\frac{\sum_{i=1}^n (\hat{\beta}_1 + \hat{\beta}_2 x_i)^2}{2\hat{\sigma}^2} \sim \mathrm{F}\left(2, n - 2, \frac{\sum_{i=1}^n (\beta_1 + \beta_2 x_i)^2}{2\sigma^2}\right).$$

More importantly, Theorem 14.5.3 indicates that the ratio of the corrected-for-the-mean model mean square to the residual mean square is distributed as follows:

$$\frac{\hat{\beta}_2^2 SXX}{\hat{\sigma}^2} \sim \mathrm{F}\left(1, n - 2, \frac{\beta_2^2 SXX}{2\sigma^2}\right),$$

where it may be recalled from Example 7.1-1 that $SXX = \sum_{i=1}^n (x_i - \bar{x})^2$. Also, by Theorem 14.5.4, we obtain the following results:

$$\frac{\hat{\beta}_1}{\hat{\sigma}\sqrt{(1/n) + (\bar{x}^2/SXX)}} \sim \mathrm{t}\left(n - 2, \frac{\beta_1}{\sigma\sqrt{(1/n) + (\bar{x}^2/SXX)}}\right),$$

$$\frac{\hat{\beta}_2}{\hat{\sigma}/\sqrt{SXX}} \sim \mathrm{t}\left(n - 2, \frac{\beta_2}{\sigma/\sqrt{SXX}}\right),$$

$$\frac{\hat{\beta}_1 + \hat{\beta}_2 x}{\hat{\sigma}\sqrt{(1/n) + (x - \bar{x})^2/SXX}} \sim \mathrm{t}\left(n - 2, \frac{\beta_1 + \beta_2 x}{\sigma\sqrt{(1/n) + (x - \bar{x})^2/SXX}}\right),$$

where, in the last result, x is any real number. Finally, recall the problem of predicting, under the prediction-extended version of this model, a new observation y_{n+1} corresponding to a value x_{n+1} of the explanatory variable. Recall from

Example 13.2-1 that the BLUP of y_{n+1} is $\hat{\beta}_1 + \hat{\beta}_2 x_{n+1}$ and that the variance of its prediction error is $\sigma^2[1 + (1/n) + (x_{n+1} - \bar{x})^2/SXX]$. Specializing (14.10), we obtain

$$\frac{\hat{\beta}_1 + \hat{\beta}_2 x_{n+1} - y_{n+1}}{\hat{\sigma}\sqrt{1 + (1/n) + (x_{n+1} - \bar{x})^2/SXX}} \sim \mathrm{t}(n-2).$$ ■

Example 14.5-2. Quantities that Have t or F Distributions Under the Normal One-Factor Model

Using results from Example 9.3.3-2 and Theorem 14.5.3, we find that under the normal Gauss–Markov one-factor model, the ratio of the corrected-for-the-mean model mean square to the residual mean square is distributed as follows:

$$\frac{\sum_{i=1}^{q} n_i(\bar{y}_{i\cdot} - \bar{y}_{\cdot\cdot})^2}{(q-1)\hat{\sigma}^2} \sim \mathrm{F}\left(q-1, n-q, \sum_{i=1}^{q} n_i \left(\alpha_i - \frac{\sum_{k=1}^{q} n_k \alpha_k}{n}\right)^2 \Big/ (2\sigma^2)\right).$$

Furthermore, specializing (14.10) we obtain

$$\frac{\bar{y}_{i\cdot}}{\hat{\sigma}\sqrt{1/n_i}} \sim \mathrm{t}\left(n-q, \frac{\mu + \alpha_i}{\sigma\sqrt{1/n_i}}\right)$$

for $i = 1, \ldots, q$, and

$$\frac{\bar{y}_{i\cdot} - \bar{y}_{i'\cdot}}{\hat{\sigma}\sqrt{(1/n_i) + (1/n_{i'})}} \sim \mathrm{t}\left(n-q, \frac{\alpha_i - \alpha_{i'}}{\sigma\sqrt{(1/n_i) + (1/n_{i'})}}\right)$$

for $i \neq i'$. ■

Example 14.5-3. A Quantity that Has a t Distribution Under the Normal Constrained Gauss–Markov Model

Consider once again the normal constrained Gauss–Markov model considered in Example 14.2-2, and let $\begin{pmatrix} \mathbf{G}_{11} & \mathbf{G}_{12} \\ \mathbf{G}_{21} & \mathbf{G}_{22} \end{pmatrix}$, where $\mathbf{G}_{11}$ has dimensions $p \times p$, be any generalized inverse of the coefficient matrix of the constrained normal equations. Let $\mathbf{c}^T\boldsymbol{\beta}$ be an estimable function, implying (by Theorem 6.2.2) that $\mathbf{c}^T = \mathbf{a}_1^T\mathbf{X} + \mathbf{a}_2^T\mathbf{A}$ for some n-vector $\mathbf{a}_1$ and some q-vector $\mathbf{a}_2$. By Theorem 10.1.4, the constrained least squares estimator of $\mathbf{c}^T\boldsymbol{\beta}$ is unbiased and is given by $\mathbf{c}^T\breve{\boldsymbol{\beta}}$. Furthermore, its variance, according to Theorem 10.1.5, is $\sigma^2\mathbf{c}^T\mathbf{G}_{11}\mathbf{c}$. Therefore,

$\mathbf{c}^T\breve{\boldsymbol{\beta}} \sim \mathrm{N}(\mathbf{c}^T\boldsymbol{\beta}, \sigma^2\mathbf{c}^T\mathbf{G}_{11}\mathbf{c})$. Partial standardization yields

$$\frac{\mathbf{c}^T\breve{\boldsymbol{\beta}}}{\sigma\sqrt{\mathbf{c}^T\mathbf{G}_{11}\mathbf{c}}} \sim \mathrm{N}\left(\frac{\mathbf{c}^T\boldsymbol{\beta}}{\sigma\sqrt{\mathbf{c}^T\mathbf{G}_{11}\mathbf{c}}}, 1\right).$$

It was shown in Example 14.2-2 that $\breve{\mathbf{e}}^T\breve{\mathbf{e}}/\sigma^2$ is distributed as $\chi^2(n - p^*_{\mathbf{A}})$. Furthermore, according to Theorem 10.1.10d, $\breve{\mathbf{e}}$ and $\mathbf{X}\breve{\boldsymbol{\beta}}$ are uncorrelated, hence (since their joint distribution is multivariate normal) independent. Thus $\mathbf{c}^T\breve{\boldsymbol{\beta}}$ ($= \mathbf{a}_1^T\mathbf{X}\breve{\boldsymbol{\beta}} + \mathbf{a}_2^T\mathbf{h}$) and $\breve{\sigma}^2$ are independent, implying further that the quantity

$$\frac{\frac{\mathbf{c}^T\breve{\boldsymbol{\beta}}}{\sigma\sqrt{\mathbf{c}^T\mathbf{G}_{11}\mathbf{c}}}}{\sqrt{\{(n - p^*_{\mathbf{A}})\breve{\sigma}^2/\sigma^2\}\Big/(n - p^*_{\mathbf{A}})}} = \frac{\mathbf{c}^T\breve{\boldsymbol{\beta}}}{\breve{\sigma}\sqrt{\mathbf{c}^T\mathbf{G}_{11}\mathbf{c}}}$$

has a noncentral t distribution with degrees of freedom equal to $n - p^*_{\mathbf{A}}$ and noncentrality parameter $\frac{\mathbf{c}^T\boldsymbol{\beta}}{\sigma\sqrt{\mathbf{c}^T\mathbf{G}_{11}\mathbf{c}}}$. ∎

14.6 Two More Distributions

We close this chapter by introducing two more distributions derived from multivariate normal vectors and a central chi-square variable: the multivariate t distribution and the studentized range distribution. These distributions will be used in the next chapter to construct simultaneous confidence intervals for estimable functions of $\boldsymbol{\beta}$.

Definition 14.6.1 Suppose that w and $\mathbf{x}$ are independently distributed as $\chi^2(\nu)$ and $\mathrm{N}_r(\mathbf{0}, \mathbf{R})$, respectively, where the diagonal elements of $\mathbf{R}$ all equal one (i.e., where $\mathbf{R}$ is a correlation matrix). Then the distribution of the random vector $\mathbf{t}$ defined by

$$\mathbf{t} = \left(\frac{w}{\nu}\right)^{-\frac{1}{2}}\mathbf{x}$$

is called the *r-variate t distribution with ν degrees of freedom and correlation matrix parameter* $\mathbf{R}$*, and is denoted by* $t(r, \nu, \mathbf{R})$.

The multivariate t distribution is symmetric about $\mathbf{0}$. Furthermore, when $r = 1$, the multivariate t distribution reduces to the (central) t distribution; more precisely, $\mathrm{t}(1, \nu, 1)$ and $\mathrm{t}(\nu)$ are the same distribution.

Definition 14.6.2 Suppose that w and $\mathbf{z} = (z_1, \ldots, z_k)^T$ are independently distributed as $\chi^2(\nu)$ and $\mathrm{N}(\mathbf{0}, \mathbf{I})$, respectively. Then the distribution of the random variable

$$\frac{\max_{i=1,\ldots,k} z_i - \min_{i=1,\ldots,k} z_i}{\sqrt{w/v}}$$

is called the *studentized range distribution.*

14.7 Exercises

1. Let $\mathbf{x}$ represent an n-dimensional random vector with mean vector $\boldsymbol{\mu}$ and variance–covariance matrix $\boldsymbol{\Sigma}$. Use Theorem 14.1.1 to show that $\mathbf{x}$ has a n-variate normal distribution if and only if, for every n-vector of constants $\mathbf{a}$, $\mathbf{a}^T\mathbf{x}$ has a (univariate) normal distribution.
2. Use Theorem 14.1.3 to construct an alternate proof of Theorem 4.2.4 for the special case in which $\mathbf{x} \sim \mathrm{N}(\boldsymbol{\mu}, \boldsymbol{\Sigma})$ where $\boldsymbol{\Sigma}$ is positive definite.
3. Suppose that $\mathbf{y}$ follows the normal positive definite Aitken model $\{\mathbf{y}, \mathbf{X}\boldsymbol{\beta}, \sigma^2\mathbf{W}\}$, and that $\mathbf{X}$ has full column rank. Use the "Factorization Theorem" [e.g., Theorem 6.2.6 of Casella and Berger (2002)] to show that the maximum likelihood estimators of $\boldsymbol{\beta}$ and σ^2 derived in Example 14.1-1 are sufficient statistics for those parameters.
4. Suppose that $\mathbf{y}$ follows the normal constrained Gauss–Markov model $\{\mathbf{y}, \mathbf{X}\boldsymbol{\beta}, \sigma^2\mathbf{I} : \mathbf{A}\boldsymbol{\beta} = \mathbf{h}\}$. Let $\mathbf{c}^T\boldsymbol{\beta}$ be an estimable function under this model. Obtain maximum likelihood estimators of $\mathbf{c}^T\boldsymbol{\beta}$ and σ^2.
5. Prove Theorem 14.1.8.
6. Derive the distributions specified by (14.5) and (14.6) in Example 14.1-3.
7. Suppose that $\mathbf{x} \sim \mathrm{N}_4(\boldsymbol{\mu}, \sigma^2\mathbf{W})$ where $\boldsymbol{\mu} = (1, 2, 3, 4)^T$ and $\mathbf{W} = \frac{1}{2}\mathbf{I}_4 + \frac{1}{2}\mathbf{1}_4\mathbf{1}_4^T$. Find the conditional distribution of $\mathbf{Ax}$ given that $\mathbf{Bx} = \mathbf{c}$, where

$$\mathbf{A} = \begin{pmatrix} 2 & 0 & -1 & -1 \end{pmatrix}, \quad \mathbf{B} = \begin{pmatrix} 0 & 0 & 2 & -2 \\ 1 & 1 & 1 & 1 \end{pmatrix}, \quad \mathbf{c} = \begin{pmatrix} 2 \\ 2 \end{pmatrix}.$$

8. Suppose that $(y_1, y_2, y_3)^T$ satisfies the stationary autoregressive model of order one described in Example 5.2.4-1 with $n = 3$, and assume that the joint distribution of $(y_1, y_2, y_3)^T$ is trivariate normal. Determine the conditional distribution of $(y_1, y_3)^T$ given that $y_2 = 1$. What can you conclude about the dependence of y_1 and y_3, conditional on y_2?
9. Let $\mathbf{x} = (x_i)$ be a random n-vector whose distribution is multivariate normal with mean vector $\boldsymbol{\mu} = (\mu_i)$ and variance–covariance matrix $\boldsymbol{\Sigma} = (\sigma_{ij})$. Show, by differentiating the moment generating function, that

$$E[(x_i - \mu_i)(x_j - \mu_j)(x_k - \mu_k)(x_l - \mu_l)] = \sigma_{ij}\sigma_{kl} + \sigma_{ik}\sigma_{jl} + \sigma_{il}\sigma_{jk}$$

for $i, j, k, l = 1, \ldots, n$.

10. Let u and v be random variables whose joint distribution is $\mathrm{N}_2(\mathbf{0}, \boldsymbol{\Sigma})$, where $\mathrm{rank}(\boldsymbol{\Sigma}) = 2$. Show that $\mathrm{corr}(u^2, v^2) = [\mathrm{corr}(u, v)]^2$.
11. Suppose that $\mathbf{x} \sim \mathrm{N}_n(\mathbf{a}, \sigma^2\mathbf{I})$ where $\mathbf{a} = (a_i)$ is an n-vector of constants such that $\mathbf{a}^T\mathbf{a} = 1$. Let $\mathbf{b}$ represent another n-vector such that $\mathbf{b}^T\mathbf{b} = 1$ and $\mathbf{a}^T\mathbf{b} = 0$. Define $\mathbf{P_a} = \mathbf{aa}^T/(\mathbf{a}^T\mathbf{a})$ and $\mathbf{P_b} = \mathbf{bb}^T/(\mathbf{b}^T\mathbf{b})$.
    (a) Find the distribution of $\mathbf{b}^T\mathbf{x}$.
    (b) Find the distribution of $\mathbf{P_a}\mathbf{x}$.
    (c) Find the distribution of $\mathbf{P_b}\mathbf{x}$.
    (d) Find the conditional distribution of x_1 (the first element of $\mathbf{x}$) given that $\mathbf{a}^T\mathbf{x} = 0$. Assume that $n \geq 2$.
12. Use the mgf of a $\chi^2(\nu, \lambda)$ distribution to obtain its mean and variance.
13. Suppose that $\mathbf{x} \sim \mathrm{N}_n(\boldsymbol{\mu}, \boldsymbol{\Sigma})$, where $\mathrm{rank}(\boldsymbol{\Sigma}) = n$. Let $\mathbf{A}$ represent a nonnull nonnegative definite matrix of constants such that $\mathbf{A}\boldsymbol{\Sigma}$ is not idempotent. It has been proposed that the distribution of $\mathbf{x}^T\mathbf{A}\mathbf{x}$ may be approximated by a multiple of a central chi-square distribution, specifically by $c\chi^2(f)$ where c and f are chosen so that $\mathbf{x}^T\mathbf{A}\mathbf{x}$ and $c\chi^2(f)$ have the same mean and variance.
    (a) Determine expressions for the appropriate c and f.
    (b) Under the normal Gauss–Markov model $\{\mathbf{y}, \mathbf{X}\boldsymbol{\beta}, \sigma^2\mathbf{I}\}$, suppose that $\mathbf{L} \neq \mathbf{P_X}$ is a matrix such that $\mathrm{E}(\mathbf{Ly}) = \mathbf{X}\boldsymbol{\beta}$. Define $\breve{\mathbf{e}} = (\mathbf{I} - \mathbf{L})\mathbf{y}$. Using part (a), determine an approximation to the distribution of $\breve{\mathbf{e}}^T\breve{\mathbf{e}}$.
14. Suppose that the conditions of part (a) of Corollary 14.2.6.1 hold, i.e., suppose that $\mathbf{x} \sim \mathrm{N}_n(\boldsymbol{\mu}, \boldsymbol{\Sigma})$ where $\mathrm{rank}(\boldsymbol{\Sigma}) = n$ and that $\mathbf{A}$ is a nonnull symmetric $n \times n$ matrix of constants such that $\mathbf{A}\boldsymbol{\Sigma}$ is idempotent. Prove that $\mathbf{A}$ is nonnegative definite.
15. Consider the special case of Theorem 14.3.1 in which there are only two functions of interest and they are quadratic forms, i.e., $w_1 = \mathbf{x}^T\mathbf{A}_1\mathbf{x}$ and $w_2 = \mathbf{x}^T\mathbf{A}_2\mathbf{x}$. Show that, in the following cases, necessary and sufficient conditions for these two quadratic forms to be distributed independently are:
    (a) $\boldsymbol{\Sigma}\mathbf{A}_1\boldsymbol{\Sigma}\mathbf{A}_2\boldsymbol{\Sigma} = \mathbf{0}$, $\boldsymbol{\Sigma}\mathbf{A}_1\boldsymbol{\Sigma}\mathbf{A}_2\boldsymbol{\mu} = \mathbf{0}$, $\boldsymbol{\Sigma}\mathbf{A}_2\boldsymbol{\Sigma}\mathbf{A}_1\boldsymbol{\mu} = \mathbf{0}$, and $\boldsymbol{\mu}^T\mathbf{A}_1\boldsymbol{\Sigma}\mathbf{A}_2\boldsymbol{\mu} = 0$, when there are no restrictions on $\mathbf{A}_1$ and $\mathbf{A}_2$ beyond those stated in the theorem.
    (b) $\mathbf{A}_1\boldsymbol{\Sigma}\mathbf{A}_2\boldsymbol{\Sigma} = \mathbf{0}$ and $\mathbf{A}_1\boldsymbol{\Sigma}\mathbf{A}_2\boldsymbol{\mu} = \mathbf{0}$ when $\mathbf{A}_1$ is nonnegative definite.
    (c) $\mathbf{A}_1\boldsymbol{\Sigma}\mathbf{A}_2 = \mathbf{0}$ when $\mathbf{A}_1$ and $\mathbf{A}_2$ are nonnegative definite.
16. Suppose that $\mathbf{x}_1, \mathbf{x}_2$, and $\mathbf{x}_3$ are independent and identically distributed as $\mathrm{N}_n(\boldsymbol{\mu}, \boldsymbol{\Sigma})$, where $\mathrm{rank}(\boldsymbol{\Sigma}) = n$. Define

$$\mathbf{Q} = (\mathbf{x}_1, \mathbf{x}_2, \mathbf{x}_3)\mathbf{A}(\mathbf{x}_1, \mathbf{x}_2, \mathbf{x}_3)^T,$$

where $\mathbf{A}$ is a symmetric idempotent 3×3 matrix of constants. Let $r = \mathrm{rank}(\mathbf{A})$ and let $\mathbf{c}$ represent a nonzero n-vector of constants. Determine the distribution of $\mathbf{c}^T\mathbf{Q}\mathbf{c}/(\mathbf{c}^T\boldsymbol{\Sigma}\mathbf{c})$. [Hint: First obtain the distribution of $(\mathbf{x}_1, \mathbf{x}_2, \mathbf{x}_3)^T\mathbf{c}$.]

17. Suppose that $\mathbf{x} \sim N_n(\boldsymbol{\mu}, \boldsymbol{\Sigma})$, where $\text{rank}(\boldsymbol{\Sigma}) = n$, and partition $\mathbf{x}$, $\boldsymbol{\mu}$, and $\boldsymbol{\Sigma}$ conformably as follows:

$$\mathbf{x} = \begin{pmatrix} \mathbf{x}_1 \\ \mathbf{x}_2 \end{pmatrix}, \quad \boldsymbol{\mu} = \begin{pmatrix} \boldsymbol{\mu}_1 \\ \boldsymbol{\mu}_2 \end{pmatrix}, \quad \boldsymbol{\Sigma} = \begin{pmatrix} \boldsymbol{\Sigma}_{11} & \boldsymbol{\Sigma}_{12} \\ \boldsymbol{\Sigma}_{21} & \boldsymbol{\Sigma}_{22} \end{pmatrix},$$

where $\mathbf{x}_1$ is $n_1 \times 1$ and $\mathbf{x}_2$ is $n_2 \times 1$.
   (a) Obtain the distribution of $Q_1 = (\mathbf{x}_1 - \boldsymbol{\Sigma}_{12}\boldsymbol{\Sigma}_{22}^{-1}\mathbf{x}_2)^T(\boldsymbol{\Sigma}_{11} - \boldsymbol{\Sigma}_{12}\boldsymbol{\Sigma}_{22}^{-1}\boldsymbol{\Sigma}_{21})^{-1}$ $(\mathbf{x}_1 - \boldsymbol{\Sigma}_{12}\boldsymbol{\Sigma}_{22}^{-1}\mathbf{x}_2)$.
   (b) Obtain the distribution of $Q_2 = \mathbf{x}_2^T\boldsymbol{\Sigma}_{22}^{-1}\mathbf{x}_2$.
   (c) Suppose that $\boldsymbol{\mu}_2 = \mathbf{0}$. Obtain the distribution of Q_1/Q_2 (suitably scaled).
18. Suppose that $\mathbf{x} \sim N_n(\mathbf{0}, \mathbf{I})$ and that $\mathbf{A}$ and $\mathbf{B}$ are $n \times n$ symmetric idempotent matrices of ranks p and q, respectively, where $p > q$. Show, by completing the following three steps, that $\mathbf{x}^T(\mathbf{A} - \mathbf{B})\mathbf{x} \sim \chi^2(p - q)$ if $\text{corr}(\mathbf{x}^T\mathbf{A}\mathbf{x}, \mathbf{x}^T\mathbf{B}\mathbf{x}) = \sqrt{q/p}$.
   (a) Show that $\text{tr}[(\mathbf{I} - \mathbf{A})\mathbf{B}] = 0$ if and only if $\text{corr}(\mathbf{x}^T\mathbf{A}\mathbf{x}, \mathbf{x}^T\mathbf{B}\mathbf{x}) = \sqrt{q/p}$.
   (b) Show that $(\mathbf{I} - \mathbf{A})\mathbf{B} = \mathbf{0}$ if and only if $\text{tr}[(\mathbf{I} - \mathbf{A})\mathbf{B}] = 0$.
   (c) Show that $\mathbf{x}^T(\mathbf{A} - \mathbf{B})\mathbf{x} \sim \chi^2(p - q)$ if $(\mathbf{I} - \mathbf{A})\mathbf{B} = \mathbf{0}$.
19. Suppose that $\mathbf{x} \sim N_n(\boldsymbol{\mu}, \boldsymbol{\Sigma})$, where $\text{rank}(\boldsymbol{\Sigma}) = n$. Partition $\mathbf{x}$, $\boldsymbol{\mu}$, and $\boldsymbol{\Sigma}$ conformably as

$$\begin{pmatrix} \mathbf{x}_1 \\ \mathbf{x}_2 \end{pmatrix}, \quad \begin{pmatrix} \boldsymbol{\mu}_1 \\ \boldsymbol{\mu}_2 \end{pmatrix}, \quad \text{and} \quad \begin{pmatrix} \boldsymbol{\Sigma}_{11} & \boldsymbol{\Sigma}_{12} \\ \boldsymbol{\Sigma}_{21} & \boldsymbol{\Sigma}_{22} \end{pmatrix},$$

where $\mathbf{x}_1$ is $n_1 \times 1$ and $\mathbf{x}_2$ is $n_2 \times 1$. Let $\mathbf{A}$ and $\mathbf{B}$ represent symmetric $n_1 \times n_1$ and $n_2 \times n_2$ matrices of constants, respectively. Determine a necessary and sufficient condition for $\mathbf{x}_1^T\mathbf{A}\mathbf{x}_1$ and $\mathbf{x}_2^T\mathbf{B}\mathbf{x}_2$ to be independent.
20. Suppose that $\mathbf{x} \sim N_n(\mathbf{a}, \sigma^2\mathbf{I})$ where $\mathbf{a}$ is a nonzero n-vector of constants. Let $\mathbf{b}$ represent another nonzero n-vector. Define $\mathbf{P_a} = \mathbf{a}\mathbf{a}^T/(\mathbf{a}^T\mathbf{a})$ and $\mathbf{P_b} = \mathbf{b}\mathbf{b}^T/(\mathbf{b}^T\mathbf{b})$.
   (a) Determine the distribution of $\mathbf{x}^T\mathbf{P_a}\mathbf{x}/\sigma^2$. (Simplify as much as possible here and in all parts of this exercise.)
   (b) Determine the distribution of $\mathbf{x}^T\mathbf{P_b}\mathbf{x}/\sigma^2$.
   (c) Determine, in as simple a form as possible, a necessary and sufficient condition for the two quadratic forms in parts (a) and (b) to be independent.
   (d) Under the condition in part (c), can the expressions for the parameters in either or both of the distributions in parts (a) and (b) be simplified further? If so, how?
21. Suppose that $\mathbf{y}$ follows the normal Aitken model $\{\mathbf{y}, \mathbf{X}\boldsymbol{\beta}, \sigma^2\mathbf{W}\}$. Let $\mathbf{A}$ represent any $n \times n$ nonnegative definite matrix such that $\mathbf{y}^T\mathbf{A}\mathbf{y}$ is an unbiased estimator of σ^2.
   (a) Show that $\mathbf{y}^T\mathbf{A}\mathbf{y}$ is uncorrelated with every linear function $\mathbf{a}^T\mathbf{y}$.
   (b) Suppose further that $\mathbf{a}^T\mathbf{y}$ is a BLUE of its expectation. Show that $\mathbf{y}^T\mathbf{A}\mathbf{y}$ and $\mathbf{a}^T\mathbf{y}$ are distributed independently.

(c) The unbiasedness of $\mathbf{y}^T\mathbf{A}\mathbf{y}$ for σ^2 implies that the main diagonal elements of $\mathbf{AW}$ satisfy a certain property. Give this property. If $\mathbf{W} = k\mathbf{I} + (1-k)\mathbf{J}$ for $k \in [0, 1]$, what must k equal (in terms of $\mathbf{A}$) in order for this property to be satisfied?

22. Suppose that $\mathbf{x} \sim \mathrm{N}_n(\mathbf{1}\mu, \sigma^2[(1-\rho)\mathbf{I} + \rho\mathbf{J}])$ where $n \geq 2$, $-\infty < \mu < \infty$, $\sigma^2 > 0$, and $0 < \rho < 1$. Define $\mathbf{A} = \frac{1}{n}\mathbf{J}$ and $\mathbf{B} = \frac{1}{n-1}(\mathbf{I} - \frac{1}{n}\mathbf{J})$.
    (a) Determine $\mathrm{var}(\mathbf{x}^T\mathbf{A}\mathbf{x})$.
    (b) Determine $\mathrm{var}(\mathbf{x}^T\mathbf{B}\mathbf{x})$.
    (c) Are $\mathbf{x}^T\mathbf{A}\mathbf{x}$ and $\mathbf{x}^T\mathbf{B}\mathbf{x}$ independent? Explain.
    (d) Suppose that n is even. Let $\mathbf{x}_1$ represent the vector consisting of the first $n/2$ elements of $\mathbf{x}$ and let $\mathbf{x}_2$ represent the vector consisting of the remaining $n/2$ elements of $\mathbf{x}$. Determine the distribution of $\mathbf{x}_1 - \mathbf{x}_2$.
    (e) Suppose that $n = 3$, in which case $\mathbf{x} = (x_1, x_2, x_3)^T$. Determine the conditional distribution of $(x_1, x_2)^T$ given that $x_3 = 1$.
23. Suppose that $\mathbf{x} \sim \mathrm{N}_n(\boldsymbol{\mu}, c\mathbf{I} + \mathbf{b}\mathbf{1}_n^T + \mathbf{1}_n\mathbf{b}^T)$, where $c\mathbf{I} + \mathbf{b}\mathbf{1}_n^T + \mathbf{1}_n\mathbf{b}^T$ is positive definite.
    (a) Determine the distribution of $\sum_{i=1}^n (x_i - \bar{x})^2$ (suitably scaled), and fully specify the parameters of that distribution.
    (b) Determine, in as simple a form as possible, a necessary and sufficient condition on $\mathbf{b}$ for $\sum_{i=1}^n (x_i - \bar{x})^2$ and $\bar{x}^2$ to be independent.
24. Suppose that

$$\begin{pmatrix} \mathbf{x}_1 \\ \mathbf{x}_2 \end{pmatrix} \sim \mathrm{N}\left(\begin{pmatrix} \boldsymbol{\mu}_1 \\ \boldsymbol{\mu}_2 \end{pmatrix}, \begin{pmatrix} \boldsymbol{\Sigma}_{11} & \boldsymbol{\Sigma}_{12} \\ \boldsymbol{\Sigma}_{21} & \boldsymbol{\Sigma}_{22} \end{pmatrix} \right),$$

where $\mathbf{x}_1$ and $\boldsymbol{\mu}_1$ are n_1-vectors, $\mathbf{x}_2$ and $\boldsymbol{\mu}_2$ are n_2-vectors, $\boldsymbol{\Sigma}_{11}$ is $n_1 \times n_1$, and $\boldsymbol{\Sigma}$ is nonsingular. Let $\mathbf{A}$ represent an $n_1 \times n_1$ symmetric matrix of constants and let $\mathbf{B}$ be a matrix of constants having n_2 columns.
    (a) Determine, in as simple a form as possible, a necessary and sufficient condition for $\mathbf{x}_1^T\mathbf{A}\mathbf{x}_1$ and $\mathbf{B}\mathbf{x}_2$ to be uncorrelated.
    (b) Determine, in as simple a form as possible, a necessary and sufficient condition for $\mathbf{x}_1^T\mathbf{A}\mathbf{x}_1$ and $\mathbf{B}\mathbf{x}_2$ to be independent.
    (c) Using your answer to part (a), give a necessary and sufficient condition for $\mathbf{x}_1^T\mathbf{x}_1$ and $\mathbf{x}_2$ to be uncorrelated.
    (d) Using your answer to part (b), give a necessary and sufficient condition for $\mathbf{x}_1^T\mathbf{x}_1$ and $\mathbf{x}_2$ to be independent.
    (e) Consider the special case in which $\boldsymbol{\Sigma} = (1-\rho)\mathbf{I} + \rho\mathbf{J}$, where $0 < \rho < 1$. Can $\mathbf{x}_1^T\mathbf{x}_1$ and $\mathbf{x}_2$ be uncorrelated? Can $\mathbf{x}_1^T\mathbf{x}_1$ and $\mathbf{x}_2$ be independent? If the answer to either or both of these questions is yes, give a necessary and sufficient condition for the result to hold.
25. Suppose that $\mathbf{x} \sim \mathrm{N}_n(\boldsymbol{\mu}, \boldsymbol{\Sigma})$, where $\mathrm{rank}(\boldsymbol{\Sigma}) = n$. Let $\mathbf{A}$ represent a nonnull symmetric idempotent matrix such that $\mathbf{A}\boldsymbol{\Sigma} = k\mathbf{A}$ for some $k \neq 0$. Show that:
    (a) $\mathbf{x}^T\mathbf{A}\mathbf{x}/k \sim \chi^2(\nu, \lambda)$ for some ν and λ, and determine ν and λ.
    (b) $\mathbf{x}^T\mathbf{A}\mathbf{x}$ and $\mathbf{x}^T(\mathbf{I} - \mathbf{A})\mathbf{x}$ are independent.

(c) $\mathbf{Ax}$ and $(\mathbf{I}-\mathbf{A})\mathbf{x}$ are independent.
(d) $k = \text{tr}(\mathbf{A\Sigma A})/\text{rank}(\mathbf{A})$.

26. Suppose that $\mathbf{x} \sim \text{N}_n(\mu\mathbf{1}, \mathbf{\Sigma})$, where μ is an unknown parameter, $\mathbf{\Sigma} = \mathbf{I}+\mathbf{CC}^T$ where $\mathbf{C}$ is known, $\mathbf{C} \neq \mathbf{0}$, and $\mathbf{C}^T\mathbf{1} = \mathbf{0}$. Let $u_1 = \mathbf{x}^T\mathbf{\Sigma}^{-1}\mathbf{x}$ and let $u_i = \mathbf{x}^T\mathbf{A}_i\mathbf{x}$ $(i = 2, 3)$ where $\mathbf{A}_2$ and $\mathbf{A}_3$ are nonnull symmetric matrices.
   (a) Prove or disprove: There is no nonnull symmetric matrix $\mathbf{A}_2$ for which u_1 and u_2 are independent.
   (b) Find two nonnull symmetric matrices $\mathbf{A}_2$ and $\mathbf{A}_3$ such that u_2 and u_3 are independent.
   (c) Determine which, if either, of u_2 and u_3 [as you defined them in part (b)] have noncentral chi-square distributions. If either of them does have a such a distribution, give the parameters of the distribution in as simple a form as possible.
27. Suppose that the model $\{\mathbf{y}, \mathbf{X}\boldsymbol{\beta}\}$ is fit to $n > \text{rank}(\mathbf{X})$ observations by ordinary least squares. Let $\mathbf{c}^T\boldsymbol{\beta}$ be an estimable function under the model, and let $\hat{\boldsymbol{\beta}}$ represent a solution to the normal equations. Furthermore, let $\hat{\sigma}^2$ denote the residual mean square from the fit, i.e., $\hat{\sigma}^2 = \mathbf{y}^T(\mathbf{I}-\mathbf{P_X})\mathbf{y}/[n-\text{rank}(\mathbf{X})]$. Now, suppose that the distribution of $\mathbf{y}$ is $\text{N}(\mathbf{X}\boldsymbol{\beta}, \mathbf{\Sigma})$ where $\boldsymbol{\beta}$ is an unknown vector of parameters and $\mathbf{\Sigma}$ is a specified positive definite matrix having one of the following four forms:

   - Form 1: $\mathbf{\Sigma} = \sigma^2\mathbf{I}$.
   - Form 2: $\mathbf{\Sigma} = \sigma^2(\mathbf{I}+\mathbf{P_X A P_X})$ for an arbitrary $n \times n$ matrix $\mathbf{A}$.
   - Form 3: $\mathbf{\Sigma} = \sigma^2[\mathbf{I}+(\mathbf{I}-\mathbf{P_X})\mathbf{B}(\mathbf{I}-\mathbf{P_X})]$ for an arbitrary $n \times n$ matrix $\mathbf{B}$.
   - Form 4: $\mathbf{\Sigma} = \sigma^2[\mathbf{I}+\mathbf{P_X A P_X}+(\mathbf{I}-\mathbf{P_X})\mathbf{B}(\mathbf{I}-\mathbf{P_X})]$ for arbitrary $n \times n$ matrices $\mathbf{A}$ and $\mathbf{B}$.

   (a) For which of the four forms of $\mathbf{\Sigma}$, if any, will $\mathbf{c}^T\hat{\boldsymbol{\beta}}$ and the fitted residuals vector $\hat{\mathbf{e}}$ be independent? Justify your answer.
   (b) For which of the four forms of $\mathbf{\Sigma}$, if any, will $[n-\text{rank}(\mathbf{X})]\hat{\sigma}^2/\sigma^2$ have a noncentral chi-square distribution? Justify your answer, and for those forms for which the quantity does have a noncentral chi-square distribution, give the parameters of the distribution.
   (c) Suppose that the ordinary least squares estimator of every estimable function $\mathbf{c}^T\boldsymbol{\beta}$ is a BLUE of that function. Will $\mathbf{c}^T\hat{\boldsymbol{\beta}}$ and $\hat{\mathbf{e}}$ be independent for every estimable function $\mathbf{c}^T\boldsymbol{\beta}$? Will $[n-\text{rank}(\mathbf{X})]\hat{\sigma}^2/\sigma^2$ have a noncentral chi-square distribution? Justify your answers.
28. Suppose that $w \sim \text{F}(\nu_1, \nu_2, \lambda)$, where $\nu_2 > 2$. Find $\text{E}(w)$.
29. Prove Theorem 14.5.1.
30. Prove Theorem 14.5.5.
31. Let $\boldsymbol{\mu}_1$ and $\boldsymbol{\mu}_2$ represent orthonormal nonrandom nonnull n-vectors. Suppose that $\mathbf{x} \sim \text{N}_n(\boldsymbol{\mu}_1, \sigma^2\mathbf{I})$. Determine the distribution of $(\mathbf{x}^T\boldsymbol{\mu}_1\boldsymbol{\mu}_1^T\mathbf{x})/(\mathbf{x}^T\boldsymbol{\mu}_2\boldsymbol{\mu}_2^T\mathbf{x})$.
32. Suppose that $\mathbf{y}$ follows the normal Gauss–Markov model $\{\mathbf{y}, \mathbf{X}\boldsymbol{\beta}, \sigma^2\mathbf{I}\}$. Let $\mathbf{c}^T\boldsymbol{\beta}$ be an estimable function, and let $\mathbf{a}^T\mathbf{y}$ be a linear unbiased estimator (not

necessarily the least squares estimator) of $\mathbf{c}^T\boldsymbol{\beta}$. Let $\hat{\sigma}^2 = \mathbf{y}^T(\mathbf{I}-\mathbf{P_X})\mathbf{y}/(n-p^*)$ be the residual mean square and let w represent a positive constant. Consider the quantity

$$t^* = \frac{\mathbf{a}^T\mathbf{y} - \mathbf{c}^T\boldsymbol{\beta}}{\hat{\sigma}w}.$$

(a) What general conditions on **a** and w are sufficient for t^* to be distributed as a central t random variable with $n - p^*$ degrees of freedom? Express the conditions in as simple a form as possible.
(b) If $\mathbf{a}^T\mathbf{y}$ was not unbiased but everything else was as specified above, and if the conditions found in part (a) held, what would be the distribution of t^*? Be as specific as possible.
(c) Repeat your answers to parts (a) and (b), but this time supposing that **y** follows a positive definite Aitken model (with variance–covariance matrix $\sigma^2\mathbf{W}$) and with $\tilde{\sigma}^2 = \mathbf{y}^T\mathbf{W}^{-1}(\mathbf{I}-\tilde{\mathbf{P}}_\mathbf{X})\mathbf{y}/(n-p^*)$ in place of $\hat{\sigma}^2$.

33. Consider the normal prediction-extended positive definite Aitken model $\left\{\begin{pmatrix}\mathbf{y}\\ \mathbf{u}\end{pmatrix}, \begin{pmatrix}\mathbf{X}\boldsymbol{\beta}\\ \mathbf{0}\end{pmatrix}, \sigma^2\begin{pmatrix}\mathbf{W} & \mathbf{K}\\ \mathbf{K}^T & \mathbf{H}\end{pmatrix}\right\}$. Let $\tilde{\boldsymbol{\beta}}$ be a solution to the Aitken equations; let $\boldsymbol{\tau} = \mathbf{C}^T\boldsymbol{\beta} + \mathbf{u}$ be a vector of s predictable functions; let $\tilde{\boldsymbol{\tau}}$ be the BLUP of $\boldsymbol{\tau}$; let $\tilde{\sigma}^2$ be the generalized residual mean square, i.e., $\tilde{\sigma}^2 = (\mathbf{y}-\mathbf{X}\tilde{\boldsymbol{\beta}})^T\mathbf{W}^{-1}(\mathbf{y}-\mathbf{X}\tilde{\boldsymbol{\beta}})/(n-p^*)$; and let $\mathbf{Q} = \text{var}(\tilde{\boldsymbol{\tau}} - \boldsymbol{\tau})$. Assume that **Q** and

$$\begin{pmatrix}\mathbf{W} & \mathbf{K}\\ \mathbf{K}^T & \mathbf{H}\end{pmatrix}$$

are positive definite. By Theorem 14.5.5,

$$F \equiv \frac{(\tilde{\boldsymbol{\tau}}-\boldsymbol{\tau})^T\mathbf{Q}^{-1}(\tilde{\boldsymbol{\tau}}-\boldsymbol{\tau})}{s\tilde{\sigma}^2} \sim \mathrm{F}(s, n-p^*).$$

Suppose that $\mathbf{L}^T\mathbf{y}$ is some other linear unbiased predictor of $\boldsymbol{\tau}$; that is, $\mathbf{L}^T\mathbf{y}$ is a LUP but not the BLUP of $\boldsymbol{\tau}$. A quantity F^* analogous to F may be obtained by replacing $\tilde{\boldsymbol{\tau}}$ with $\mathbf{L}^T\mathbf{y}$ and replacing **Q** with $\mathbf{L}^T\mathbf{W}\mathbf{L}+\mathbf{H}-\mathbf{L}^T\mathbf{K}-\mathbf{K}^T\mathbf{L}$ in the definition of F, i.e.,

$$F^* \equiv \frac{(\mathbf{L}^T\mathbf{y}-\boldsymbol{\tau})^T(\mathbf{L}^T\mathbf{W}\mathbf{L}+\mathbf{H}-\mathbf{L}^T\mathbf{K}-\mathbf{K}^T\mathbf{L})^{-1}(\mathbf{L}^T\mathbf{y}-\boldsymbol{\tau})}{s\tilde{\sigma}^2}.$$

Does F^* have a central F-distribution in general? If so, verify it; if not, explain why not and give a necessary and sufficient condition on $\mathbf{L}^T$ for F^* to have a central F-distribution.

34. Consider the normal Aitken model $\{\mathbf{y}, \mathbf{X}\boldsymbol{\beta}, \sigma^2\mathbf{W}\}$, for which point estimation of an estimable function $\mathbf{c}^T\boldsymbol{\beta}$ and the residual variance σ^2 were presented in

Sect. 11.2. Suppose that $\text{rank}(\mathbf{W}, \mathbf{X}) > \text{rank}(\mathbf{X})$. Let $\begin{pmatrix} \tilde{\mathbf{t}} \\ \tilde{\boldsymbol{\lambda}} \end{pmatrix}$ be any solution to the BLUE equations

$$\begin{pmatrix} \mathbf{W} & \mathbf{X} \\ \mathbf{X}^T & \mathbf{0} \end{pmatrix} \begin{pmatrix} \mathbf{t} \\ \boldsymbol{\lambda} \end{pmatrix} = \begin{pmatrix} \mathbf{0} \\ \mathbf{c} \end{pmatrix}$$

for $\mathbf{c}^T\boldsymbol{\beta}$, and let $\tilde{\sigma}^2 = \mathbf{y}^T\mathbf{G}_{11}\mathbf{y}/[\text{rank}(\mathbf{W}, \mathbf{X}) - \text{rank}(\mathbf{X})]$ where $\mathbf{G}_{11}$ is the upper left $n \times n$ block of any symmetric generalized inverse $\begin{pmatrix} \mathbf{G}_{11} & \mathbf{G}_{12} \\ \mathbf{G}_{21} & \mathbf{G}_{22} \end{pmatrix}$ of the coefficient matrix of the BLUE equations. Show that

$$\frac{\tilde{\mathbf{t}}^T\mathbf{y}}{\tilde{\sigma}\sqrt{\tilde{\mathbf{t}}^T\mathbf{W}\tilde{\mathbf{t}}}}$$

has a noncentral t distribution with degrees of freedom equal to $\text{rank}(\mathbf{W}, \mathbf{X}) - \text{rank}(\mathbf{X})$ and noncentrality parameter $\frac{\mathbf{c}^T\boldsymbol{\beta}}{\sigma\sqrt{\tilde{\mathbf{t}}^T\mathbf{W}\tilde{\mathbf{t}}}}$.

35. Let

$$\mathbf{M} = \begin{pmatrix} m_{11} & m_{12} \\ m_{21} & m_{22} \end{pmatrix}$$

represent a random matrix whose elements $m_{11}, m_{12}, m_{21}, m_{22}$ are independent N(0, 1) random variables.
    (a) Show that $m_{11} - m_{22}$, $m_{12} + m_{21}$, and $m_{12} - m_{21}$ are mutually independent, and determine their distributions.
    (b) Show that $\Pr[\{\text{tr}(\mathbf{M})\}^2 - 4|\mathbf{M}| > 0] = \Pr(W > \frac{1}{2})$, where $W \sim \text{F}(2, 1)$. [Hint: You may find it helpful to recall that $(u + v)^2 - (u - v)^2 = 4uv$ for all real numbers u and v.]

36. Using the results of Example 14.4-2, show that the ratio of the Factor A mean square to the whole-plot error mean square, the ratio of the Factor B mean square to the split-plot error mean square, and the ratio of the AB interaction mean square to the split-plot error mean square have noncentral F distributions, and give the parameters of those distributions.

References

Casella, G. & Berger, R. L. (2002). *Statistical inference* (2nd ed.). Pacific Grove, CA: Duxbury.
Chung, K. L. (1974). *A course in probability theory*. New York: Academic Press.
Harville, D. A. (1997). *Matrix algebra from a statistician's perspective*. New York: Springer.
Harville, D. A. (2018). *Linear models and the relevant distributions and matrix algebra*. Boca Raton, FL: CRC Press.

Hogg, R. V., McKean, J., & Craig, A. T. (2013). *Introduction to mathematical statistics* (7th ed.). Boston: Pearson.
Melnick, E. L. & Tenenbein, A. (1982). Misspecifications of the normal distribution. *The American Statistician, 36*, 372–373.
Ravishanker, N. & Dey, D. K. (2002). *A first course in linear model theory*. Boca Raton, FL: Chapman & Hall/CRC Press.

Inference for Estimable and Predictable Functions

15

As regards the estimation of estimable functions of $\boldsymbol{\beta}$ and the prediction of predictable functions of $\boldsymbol{\beta}$ and $\mathbf{u}$, we have, to this point in our development, considered only point estimation and prediction. Now that we have added distribution theory related to the multivariate normal distribution to our arsenal, we may consider interval and "regional" estimation/prediction and hypothesis testing for such functions under normal linear models. Furthermore, we may use hypothesis testing or other procedures that require a normality assumption to choose the linear model upon which further inferences may be based. This chapter takes up all of these topics.

15.1 Confidence and Prediction Intervals and Ellipsoids

We begin this section by defining notation for quantiles, or "cutoff points," of central t and central F distributions. Define $t_{\xi,\nu}$ to be the upper ξ cutoff point of a central t distribution with ν degrees of freedom, i.e., if $t \sim \mathrm{t}(\nu)$, then $\Pr(t > t_{\xi,\nu}) = \xi$. Similarly, let F_{ξ,ν_1,ν_2} denote the upper ξ cutoff point of a central F distribution with ν_1 and ν_2 degrees of freedom, i.e., if $w \sim \mathrm{F}(\nu_1, \nu_2)$, then $\Pr(w > F_{\xi,\nu_1,\nu_2}) = \xi$. Figure 15.1 illustrates these cutoff points, which may be obtained easily, for given ξ and degrees of freedom, using the `qt` and `qf` functions in `R`.

Unless noted otherwise, in this chapter we take the model for $\mathbf{y}$ to be the normal Gauss–Markov model $\{\mathbf{y}, \mathbf{X}\boldsymbol{\beta}, \sigma^2\mathbf{I}\}$. Let $\hat{\boldsymbol{\beta}}$ represent any solution to the normal equations for that model, suppose that $n > p^*$, and let $\hat{\sigma}^2$ be the residual mean square.

Suppose that $\mathbf{c}^T\boldsymbol{\beta}$, where $\mathbf{c} \neq \mathbf{0}$, is an estimable function. Recall from Theorem 14.5.4 that

$$\frac{\mathbf{c}^T\hat{\boldsymbol{\beta}} - \mathbf{c}^T\boldsymbol{\beta}}{\hat{\sigma}[\mathbf{c}^T(\mathbf{X}^T\mathbf{X})^-\mathbf{c}]^{\frac{1}{2}}} \sim \mathrm{t}(n - p^*).$$

D. L. Zimmerman, *Linear Model Theory*,
https://doi.org/10.1007/978-3-030-52063-2_15

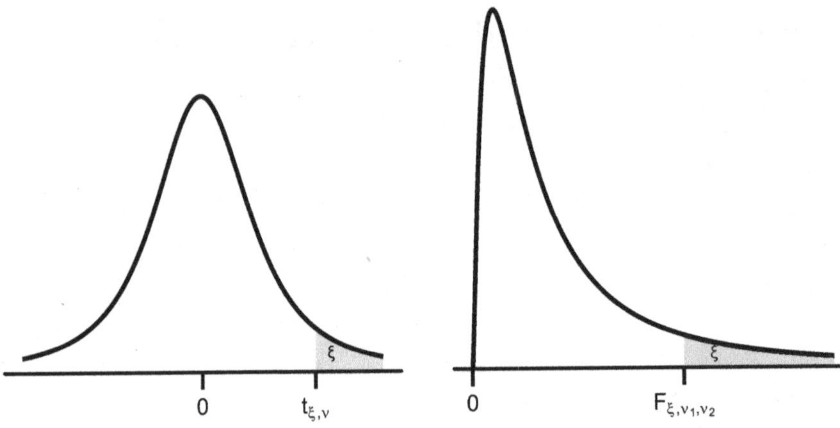

Fig. 15.1 Cutoff points for the t(ν) (left panel) and F(ν_1, ν_2) (right panel) distributions

Because the pdf of the t distribution is symmetric about zero, we have

$$\Pr\left[-t_{\xi_1,n-p^*} \le \frac{\mathbf{c}^T\hat{\boldsymbol{\beta}} - \mathbf{c}^T\boldsymbol{\beta}}{\hat{\sigma}\sqrt{\mathbf{c}^T(\mathbf{X}^T\mathbf{X})^-\mathbf{c}}} \le t_{\xi_2,n-p^*}\right] = 1 - \xi,$$

where $\xi_1 + \xi_2 = \xi$, or equivalently (after some algebraic manipulation)

$$\Pr\left[\mathbf{c}^T\hat{\boldsymbol{\beta}} - t_{\xi_2,n-p^*}\hat{\sigma}\sqrt{\mathbf{c}^T(\mathbf{X}^T\mathbf{X})^-\mathbf{c}} \le \mathbf{c}^T\boldsymbol{\beta} \le \mathbf{c}^T\hat{\boldsymbol{\beta}} + t_{\xi_1,n-p^*}\hat{\sigma}\sqrt{\mathbf{c}^T(\mathbf{X}^T\mathbf{X})^-\mathbf{c}}\right] = 1 - \xi.$$

Thus, the interval

$$\left[\mathbf{c}^T\hat{\boldsymbol{\beta}} - t_{\xi_2,n-p^*}\hat{\sigma}\sqrt{\mathbf{c}^T(\mathbf{X}^T\mathbf{X})^-\mathbf{c}},\ \mathbf{c}^T\hat{\boldsymbol{\beta}} + t_{\xi_1,n-p^*}\hat{\sigma}\sqrt{\mathbf{c}^T(\mathbf{X}^T\mathbf{X})^-\mathbf{c}}\right]$$

is a $100(1-\xi)\%$ confidence interval for $\mathbf{c}^T\boldsymbol{\beta}$. The length of this interval is $(t_{\xi_1,n-p^*} + t_{\xi_2,n-p^*})\hat{\sigma}\sqrt{\mathbf{c}^T(\mathbf{X}^T\mathbf{X})^-\mathbf{c}}$. The most common choice for (ξ_1, ξ_2) is $\xi_1 = \xi_2 = \frac{\xi}{2}$, which is easily seen to be the choice that minimizes the length of the confidence interval. For this choice, the confidence interval is *symmetric*, i.e., centered on the point estimator. Henceforth, unless noted otherwise we use the symmetric $100(1 - \xi)\%$ confidence interval

$$\mathbf{c}^T\hat{\boldsymbol{\beta}} \pm t_{\xi/2,n-p^*}\hat{\sigma}\sqrt{\mathbf{c}^T(\mathbf{X}^T\mathbf{X})^-\mathbf{c}} \tag{15.1}$$

for $\mathbf{c}^T\boldsymbol{\beta}$ without explicit use of the word "symmetric."

Now take $\mathbf{C}^T\boldsymbol{\beta}$ to be a vector of s linearly independent estimable functions. Suppose that we want to obtain a set of values of $\mathbf{C}^T\boldsymbol{\beta}$, say $S(\mathbf{y})$ (depending on $\mathbf{y}$), such that

$$\Pr[\mathbf{C}^T\boldsymbol{\beta} \in S(\mathbf{y})] = 1 - \xi.$$

Such a set is called a $100(1-\xi)\%$ *confidence set for* $\mathbf{C}^T\boldsymbol{\beta}$. One such set can be based on Theorem 14.5.2, which said that

$$\frac{(\mathbf{C}^T\hat{\boldsymbol{\beta}} - \mathbf{C}^T\boldsymbol{\beta})^T[\mathbf{C}^T(\mathbf{X}^T\mathbf{X})^-\mathbf{C}]^{-1}(\mathbf{C}^T\hat{\boldsymbol{\beta}} - \mathbf{C}^T\boldsymbol{\beta})}{s\hat{\sigma}^2} \sim \mathrm{F}(s, n-p^*).$$

By this distributional result, we have

$$\Pr[(\mathbf{C}^T\hat{\boldsymbol{\beta}} - \mathbf{C}^T\boldsymbol{\beta})^T[\mathbf{C}^T(\mathbf{X}^T\mathbf{X})^-\mathbf{C}]^{-1}(\mathbf{C}^T\hat{\boldsymbol{\beta}} - \mathbf{C}^T\boldsymbol{\beta}) \leq s\hat{\sigma}^2 F_{\xi,s,n-p^*}] = 1 - \xi.$$

Consequently, the set $S(\mathbf{y})$ of values of $\mathbf{C}^T\boldsymbol{\beta}$ that satisfy the inequality

$$(\mathbf{C}^T\hat{\boldsymbol{\beta}} - \mathbf{C}^T\boldsymbol{\beta})^T[\mathbf{C}^T(\mathbf{X}^T\mathbf{X})^-\mathbf{C}]^{-1}(\mathbf{C}^T\hat{\boldsymbol{\beta}} - \mathbf{C}^T\boldsymbol{\beta}) \leq s\hat{\sigma}^2 F_{\xi,s,n-p^*} \tag{15.2}$$

comprise a $100(1-\xi)\%$ confidence set for $\mathbf{C}^T\boldsymbol{\beta}$. In $\mathbb{R}^s$, this set takes the form of an s-dimensional ellipsoid centered at $\mathbf{C}^T\hat{\boldsymbol{\beta}}$, so it is called a $100(1-\xi)\%$ *confidence ellipsoid for* $\mathbf{C}^T\boldsymbol{\beta}$.

In the special case $s = 1$, (15.2) simplifies to the symmetric confidence interval (15.1) because $\mathrm{t}^2(n-p^*)$ is the same distribution as $\mathrm{F}(1, n-p^*)$ (Theorem 14.5.1). In the special case in which $\mathbf{X}$ has full rank and $\mathbf{C} = \mathbf{I}$, (15.2) reduces to

$$(\hat{\boldsymbol{\beta}} - \boldsymbol{\beta})^T\mathbf{X}^T\mathbf{X}(\hat{\boldsymbol{\beta}} - \boldsymbol{\beta}) \leq p\hat{\sigma}^2 F_{\xi,p,n-p}. \tag{15.3}$$

The development so far has applied to the normal Gauss–Markov model only. However, (15.1) and (15.2) may be extended easily to the normal positive definite Aitken model. We give this extension as a theorem; its proof is an immediate consequence of the special case of Theorem 14.5.5 in which $\mathbf{u} \equiv \mathbf{0}$.

Theorem 15.1.1 *If* $\mathbf{C}^T\boldsymbol{\beta}$ *is a vector of* s *linearly independent estimable functions, then under the normal positive definite Aitken model* $\{\mathbf{y}, \mathbf{X}\boldsymbol{\beta}, \sigma^2\mathbf{W}\}$,

$$(\mathbf{C}^T\tilde{\boldsymbol{\beta}} - \mathbf{C}^T\boldsymbol{\beta})^T[\mathbf{C}^T(\mathbf{X}^T\mathbf{W}^{-1}\mathbf{X})^-\mathbf{C}]^{-1}(\mathbf{C}^T\tilde{\boldsymbol{\beta}} - \mathbf{C}^T\boldsymbol{\beta}) \leq s\tilde{\sigma}^2 F_{\xi,s,n-p^*} \tag{15.4}$$

is a $100(1-\xi)\%$ *confidence ellipsoid for* $\mathbf{C}^T\boldsymbol{\beta}$. *When* $s = 1$, *this ellipsoid reduces to the symmetric* $100(1-\xi)\%$ *confidence interval*

$$\mathbf{c}^T\tilde{\boldsymbol{\beta}} \pm t_{\xi/2,n-p^*}\tilde{\sigma}\sqrt{\mathbf{c}^T(\mathbf{X}^T\mathbf{W}^{-1}\mathbf{X})^-\mathbf{c}}.$$

Example 15.1-1. Confidence Intervals for the Intercept, Slope, and Mean of y at a Given x in Normal Simple Linear Regression

Consider the normal Gauss–Markov simple linear regression model. Using results from Example 14.5-1, $100(1-\xi)\%$ confidence intervals of form (15.1) for the slope β_2, the intercept β_1, and the mean of y at a given x (i.e., $\beta_1 + \beta_2 x$) are, respectively,

$$\hat{\beta}_2 \pm t_{\xi/2,n-2}\hat{\sigma}/\sqrt{SXX},$$

$$\hat{\beta}_1 \pm t_{\xi/2,n-2}\hat{\sigma}\sqrt{(1/n) + \bar{x}^2/SXX},$$

$$(\hat{\beta}_1 + \hat{\beta}_2 x) \pm t_{\xi/2,n-2}\hat{\sigma}\sqrt{(1/n) + (x-\bar{x})^2/SXX}.$$

The F-based $100(1-\xi)\%$ confidence ellipsoid (15.3) for the intercept and slope is the set of (β_1, β_2)-values for which

$$\begin{pmatrix} \hat{\beta}_1 - \beta_1 \\ \hat{\beta}_2 - \beta_2 \end{pmatrix}^T \begin{pmatrix} n & \sum_{i=1}^n x_i \\ \sum_{i=1}^n x_i & \sum_{i=1}^n x_i^2 \end{pmatrix} \begin{pmatrix} \hat{\beta}_1 - \beta_1 \\ \hat{\beta}_2 - \beta_2 \end{pmatrix} \leq 2\hat{\sigma}^2 F_{\xi,2,n-2}. \quad \blacksquare$$

Example 15.1-2. Confidence Intervals for Level Means and Contrasts in the Normal Gauss–Markov One-Factor Model

Consider the normal Gauss–Markov one-factor model, with q levels. Using results from Example 14.5-2, the symmetric $100(1-\xi)\%$ confidence interval (15.1) for the ith level mean is

$$\bar{y}_{i\cdot} \pm t_{\xi/2,n-q}\hat{\sigma}\sqrt{1/n_i},$$

while that for the difference in the ith and i'th level means is

$$(\bar{y}_{i\cdot} - \bar{y}_{i'\cdot}) \pm t_{\xi/2,n-q}\hat{\sigma}\sqrt{(1/n_i) + (1/n_{i'})}.$$

More generally, the symmetric $100(1-\xi)\%$ confidence interval (15.1) for a contrast in level means, $\sum_{i=1}^q d_i\alpha_i$ (where $\sum_{i=1}^q d_i = 0$), is

$$\sum_{i=1}^q d_i\bar{y}_{i\cdot} \pm t_{\xi/2,n-q}\hat{\sigma}\left(\sum_{i=1}^q \frac{d_i^2}{n_i}\right)^{\frac{1}{2}}. \quad \blacksquare$$

The derivation of confidence intervals for estimable functions of $\boldsymbol{\beta}$ can be extended to the normal constrained Gauss–Markov model and the normal Aitken model with little difficulty. Recall from Example 14.5-3 that if $\mathbf{c}^T\boldsymbol{\beta}$ is estimable under the constrained Gauss–Markov model $\{\mathbf{y}, \mathbf{X}\boldsymbol{\beta}, \sigma^2\mathbf{I} : \mathbf{A}\boldsymbol{\beta} = \mathbf{h}\}$, $\breve{\boldsymbol{\beta}}$ is the first p-component subvector of a solution to the constrained normal equations, $\mathbf{G}_{11}$ is the upper left $p \times p$ block of any generalized inverse of the coefficient matrix of those equations, $n > p_{\mathbf{A}}^*$ where $p_{\mathbf{A}}^* = \text{rank}\begin{pmatrix}\mathbf{X}\\ \mathbf{A}\end{pmatrix} - \text{rank}(\mathbf{A})$, and $\breve{\sigma}^2$ is the constrained residual mean square, then $\frac{\mathbf{c}^T\breve{\boldsymbol{\beta}}-\mathbf{c}^T\boldsymbol{\beta}}{\breve{\sigma}\sqrt{\mathbf{c}^T\mathbf{G}_{11}\mathbf{c}}}$ has a central t distribution with $n - p_{\mathbf{A}}^*$ degrees of freedom. It follows that

$$\mathbf{c}^T\breve{\boldsymbol{\beta}} \pm t_{\xi/2,n-p_{\mathbf{A}}^*}\breve{\sigma}\sqrt{\mathbf{c}^T\mathbf{G}_{11}\mathbf{c}}$$

is a $100(1-\xi)\%$ confidence interval for $\mathbf{c}^T\boldsymbol{\beta}$. Similarly, recall from Exercise 14.34 that if $\mathbf{c}^T\boldsymbol{\beta}$ is estimable under the normal Aitken model $\{\mathbf{y}, \mathbf{X}\boldsymbol{\beta}, \sigma^2\mathbf{W}\}$, $\tilde{\mathbf{t}}$ is the first n-component subvector of a solution to the BLUE equations for $\mathbf{c}^T\boldsymbol{\beta}$, $\text{rank}(\mathbf{W}, \mathbf{X}) > \text{rank}(\mathbf{X})$, and $\tilde{\sigma}^2$ is the unbiased estimator of σ^2 defined in Theorem 11.2.3, then $\frac{\tilde{\mathbf{t}}^T\mathbf{y}-\mathbf{c}^T\boldsymbol{\beta}}{\breve{\sigma}\sqrt{\tilde{\mathbf{t}}^T\mathbf{W}\tilde{\mathbf{t}}}}$ has a central t distribution with $\text{rank}(\mathbf{W}, \mathbf{X}) - \text{rank}(\mathbf{X})$ degrees of freedom. It follows that

$$\tilde{\mathbf{t}}^T\mathbf{y} \pm t_{\xi/2,\text{rank}(\mathbf{W},\mathbf{X})-\text{rank}(\mathbf{X})}\tilde{\sigma}\sqrt{\tilde{\mathbf{t}}^T\mathbf{W}\tilde{\mathbf{t}}}$$

is a $100(1-\xi)\%$ confidence interval for $\mathbf{c}^T\boldsymbol{\beta}$.

Along the same lines, we may extend the foregoing development to the prediction setting. Suppose that $\boldsymbol{\tau} = \mathbf{C}^T\boldsymbol{\beta} + \mathbf{u}$ is a vector of s predictable functions, and that we want to obtain a set of values of $\boldsymbol{\tau}$, say $S(\mathbf{y})$, such that

$$\Pr[\boldsymbol{\tau} \in S(\mathbf{y})] = 1 - \xi.$$

Such a set is called a $100(1-\xi)\%$ *prediction set for* $\boldsymbol{\tau}$. If the set is ellipsoidal, it is called a *prediction ellipsoid*, and if $s = 1$ and the set is an interval it is called a *prediction interval*. The following theorem is an immediate consequence of Theorem 14.5.5.

Theorem 15.1.2 *Consider the prediction of a vector of s predictable functions $\boldsymbol{\tau} = \mathbf{C}^T\boldsymbol{\beta} + \mathbf{u}$ under the normal prediction-extended positive definite Aitken model with variance–covariance matrix $\sigma^2\begin{pmatrix}\mathbf{W} & \mathbf{K}\\ \mathbf{K}^T & \mathbf{H}\end{pmatrix}$. Recall that $\tilde{\boldsymbol{\tau}}$ is the vector of BLUPs of the elements of $\boldsymbol{\tau}$, and that var$(\tilde{\boldsymbol{\tau}} - \boldsymbol{\tau}) = \sigma^2\mathbf{Q}$, where an expression for $\mathbf{Q}$ was* *given in Theorem 13.2.2a. If* $\mathbf{Q}$ *is positive definite (as is the case under either of the conditions of Theorem 13.2.2b), then*

$$(\tilde{\boldsymbol{\tau}} - \boldsymbol{\tau})^T \mathbf{Q}^{-1} (\tilde{\boldsymbol{\tau}} - \boldsymbol{\tau}) \leq s\tilde{\sigma}^2 F_{\xi,s,n-p^*} \tag{15.5}$$

is a $100(1-\xi)\%$ *prediction ellipsoid for* $\boldsymbol{\tau}$. *When* $s = 1$, *this ellipsoid reduces to the symmetric* $100(1-\xi)\%$ *prediction interval*

$$\tilde{\tau} \pm t_{\xi/2,n-p^*}\tilde{\sigma}\sqrt{Q} \tag{15.6}$$

where Q *is the scalar version of* $\mathbf{Q}$.

Example 15.1-3. Prediction Intervals and Ellipsoids for New Observations in Normal Simple Linear Regression

Suppose that we wish to obtain a $100(1-\xi)\%$ prediction interval for a new observation $y_{n+1} = \beta_1 + \beta_2 x_{n+1} + e_{n+1}$ under a normal prediction-extended Gauss–Markov simple linear regression model for $\mathbf{y}$ and y_{n+1}. The symmetric $100(1-\xi)\%$ prediction interval given by (15.6) specializes in this case to

$$(\hat{\beta}_1 + \hat{\beta}_2 x_{n+1}) \pm t_{\xi/2,n-2}\hat{\sigma}\sqrt{1 + (1/n) + (x_{n+1} - \bar{x})^2/SXX}.$$

Suppose instead that under the same model we wish to obtain a $100(1-\xi)\%$ prediction ellipsoid for s new observations $y_{n+1}, \ldots, y_{n+s}$, where $y_{n+j} = \beta_1 + \beta_2 x_{n+j} + e_{n+j}$, $(j = 1, \ldots, s)$. Because $\mathbf{W} = \mathbf{I}_n$, $\mathbf{K} = \mathbf{0}_{n\times s}$, and $\mathbf{H} = \mathbf{I}_s$ in this case, the variance–covariance matrix $\sigma^2\mathbf{Q}$ of prediction errors is, using Theorem 13.2.2a,

$$\sigma^2\mathbf{Q} = \sigma^2[\mathbf{I}_s - \mathbf{0} + (\breve{\mathbf{X}}^T - \mathbf{0})(\mathbf{X}^T\mathbf{I}_n^{-1}\mathbf{X})^-(\breve{\mathbf{X}}^T - \mathbf{0})^T] = \sigma^2[\mathbf{I}_s + \breve{\mathbf{X}}(\mathbf{X}^T\mathbf{X})^{-1}\breve{\mathbf{X}}^T]$$

where

$$\breve{\mathbf{X}} = \begin{pmatrix} 1 & x_{n+1} \\ \vdots & \vdots \\ 1 & x_{n+s} \end{pmatrix}.$$

It can be verified rather easily that the ijth element of $\mathbf{I}_s + \breve{\mathbf{X}}(\mathbf{X}^T\mathbf{X})^{-1}\breve{\mathbf{X}}^T$ is $1 + (1/n) + (x_{n+i} - \bar{x})^2/SXX$ if $i = j$, or $(1/n) + (x_{n+i} - \bar{x})(x_{n+j} - \bar{x})/SXX$ if $i \neq j$.

Thus, the prediction ellipsoid given by (15.5) specializes here to the set of values of $(y_{n+1}, \ldots, y_{n+s})^T$ such that

$$\begin{pmatrix} \hat{\beta}_1 + \hat{\beta}_2 x_{n+1} - y_{n+1} \\ \vdots \\ \hat{\beta}_1 + \hat{\beta}_2 x_{n+s} - y_{n+s} \end{pmatrix}^T [\mathbf{I}_s + \breve{\mathbf{X}}(\mathbf{X}^T\mathbf{X})^{-1}\breve{\mathbf{X}}^T]^{-1} \begin{pmatrix} \hat{\beta}_1 + \hat{\beta}_2 x_{n+1} - y_{n+1} \\ \vdots \\ \hat{\beta}_1 + \hat{\beta}_2 x_{n+s} - y_{n+s} \end{pmatrix}$$
$$\leq s\hat{\sigma}^2 F_{\xi,s,n-2}.$$

■

15.2 Hypothesis Testing

The previous section presented confidence intervals (and regions) for estimable linear functions of $\boldsymbol{\beta}$ in normal linear models. Another important type of inference for such functions is hypothesis testing, which is the topic of this section.

Definition 15.2.1 Let $\mathbf{C}$ represent a specified $s \times p$ matrix, and let $\mathbf{c}_0 \in \mathcal{R}(\mathbf{C})$ be a specified s-vector such that the system of equations $\mathbf{C}\boldsymbol{\beta} = \mathbf{c}_0$ is consistent. Under any unconstrained linear model, the null hypothesis $H_0 : \mathbf{C}^T\boldsymbol{\beta} = \mathbf{c}_0$ is said to be *testable* if $\mathcal{R}(\mathbf{C}^T) \subseteq \mathcal{R}(\mathbf{X})$.

Upon comparing Definitions 15.2.1 and 6.1.3, it is clear that the condition for a null hypothesis to be testable is identical to the condition for the vector of linear functions in the hypothesis to be estimable.

Suppose that $\mathbf{y}$ follows the normal Gauss–Markov model $\{\mathbf{y}, \mathbf{X}\boldsymbol{\beta}, \sigma^2\mathbf{I}\}$ and that $n > p^*$. Let $\mathbf{C}^T\boldsymbol{\beta}$ be a vector of s linearly independent estimable functions. Consider the problem of testing the null hypothesis $H_0 : \mathbf{C}^T\boldsymbol{\beta} = \mathbf{c}_0$ versus the alternative hypothesis $H_a : \mathbf{C}^T\boldsymbol{\beta} \neq \mathbf{c}_0$, where $\mathbf{c}_0$ is a specified s-vector for which the system $\mathbf{C}\boldsymbol{\beta} = \mathbf{c}_0$ is consistent. Clearly, this null hypothesis is testable. If H_0 is true, then by Theorem 14.5.2,

$$\frac{(\mathbf{C}^T\hat{\boldsymbol{\beta}} - \mathbf{c}_0)^T[\mathbf{C}^T(\mathbf{X}^T\mathbf{X})^-\mathbf{C}]^{-1}(\mathbf{C}^T\hat{\boldsymbol{\beta}} - \mathbf{c}_0)}{s\hat{\sigma}^2} \sim \mathrm{F}(s, n - p^*).$$

Thus,

$$\xi = \Pr_{H_0}\left[\frac{(\mathbf{C}^T\hat{\boldsymbol{\beta}} - \mathbf{c}_0)^T[\mathbf{C}^T(\mathbf{X}^T\mathbf{X})^-\mathbf{C}]^{-1}(\mathbf{C}^T\hat{\boldsymbol{\beta}} - \mathbf{c}_0)}{s\hat{\sigma}^2} > F_{\xi,s,n-p^*}\right]$$

where $\Pr_{H_0}$ may be read as "the probability under H_0." This establishes the following theorem.

Theorem 15.2.1 *A size-ξ test of $H_0 : \mathbf{C}^T\boldsymbol{\beta} = \mathbf{c}_0$ versus $H_a : \mathbf{C}^T\boldsymbol{\beta} \neq \mathbf{c}_0$ under the normal Gauss–Markov model $\{\mathbf{y}, \mathbf{X}\boldsymbol{\beta}, \sigma^2\mathbf{I}\}$ is obtained by rejecting H_0 if and only if*

$$\frac{(\mathbf{C}^T\hat{\boldsymbol{\beta}} - \mathbf{c}_0)^T[\mathbf{C}^T(\mathbf{X}^T\mathbf{X})^-\mathbf{C}]^{-1}(\mathbf{C}^T\hat{\boldsymbol{\beta}} - \mathbf{c}_0)}{s\hat{\sigma}^2} > F_{\xi,s,n-p^*}. \tag{15.7}$$

This test is called the size-ξ F test *of H_0 versus H_a.*

Comparing (15.7) to (15.2), we see that the decision rule for the F test is equivalent to rejecting H_0 if $\mathbf{c}_0$ lies outside the $100(1-\xi)\%$ F-based confidence ellipsoid for $\mathbf{C}^T\boldsymbol{\beta}$, and not rejecting H_0 otherwise.

Corollary 15.2.1.1 *When $s = 1$, (15.7) may be written as*

$$\frac{(\mathbf{c}^T\hat{\boldsymbol{\beta}} - c_0)^2}{\hat{\sigma}^2[\mathbf{c}^T(\mathbf{X}^T\mathbf{X})^-\mathbf{c}]} > F_{\xi,1,n-p^*},$$

or equivalently as

$$\frac{|\mathbf{c}^T\hat{\boldsymbol{\beta}} - c_0|}{\hat{\sigma}\sqrt{\mathbf{c}^T(\mathbf{X}^T\mathbf{X})^-\mathbf{c}}} > t_{\xi/2,n-p^*}. \tag{15.8}$$

The test of $H_0 : \mathbf{c}^T\boldsymbol{\beta} = c_0$ versus $H_a : \mathbf{c}^T\boldsymbol{\beta} \neq c_0$ in which H_0 is rejected if and only if the inequality in (15.8) is satisfied is called the size-ξ *t test* of H_0 versus H_a.

The next theorem establishes that the test described by Theorem 15.2.1 is the likelihood ratio test.

Theorem 15.2.2 *The size-ξ F test described by Theorem 15.2.1 is equivalent to the size-ξ likelihood ratio test of $H_0 : \mathbf{C}^T\boldsymbol{\beta} = \mathbf{c}_0$ versus $H_a : \mathbf{C}^T\boldsymbol{\beta} \neq \mathbf{c}_0$.*

Proof Let $\Omega = \{(\boldsymbol{\beta}, \sigma^2) : \boldsymbol{\beta} \in \mathbb{R}^p, \sigma^2 > 0\}$ and $\Omega_0 = \{(\boldsymbol{\beta}, \sigma^2) : \mathbf{C}^T\boldsymbol{\beta} = \mathbf{c}_0, \sigma^2 > 0\}$. Ω is the unconstrained parameter space, while Ω_0 is the parameter space under H_0. From Example 14.1-1, the likelihood function is

$$L(\boldsymbol{\beta}, \sigma^2; \mathbf{y}) = (2\pi\sigma^2)^{-n/2}\exp[-(\mathbf{y} - \mathbf{X}\boldsymbol{\beta})^T(\mathbf{y} - \mathbf{X}\boldsymbol{\beta})/(2\sigma^2)], \quad (\boldsymbol{\beta}, \sigma^2) \in \Omega.$$

By definition, the likelihood ratio test statistic for testing $H_0 : \mathbf{C}^T\boldsymbol{\beta} = \mathbf{c}_0$ versus $H_a : \mathbf{C}^T\boldsymbol{\beta} \neq \mathbf{c}_0$ is

$$\frac{\sup_{\Omega_0} L(\boldsymbol{\beta}, \sigma^2)}{\sup_{\Omega} L(\boldsymbol{\beta}, \sigma^2)}.$$

The decision rule for the likelihood ratio test is to reject H_0 if and only if this test statistic is less than a constant k, where k is chosen so that the test has the desired size.

In Example 14.1-1, we obtained maximum likelihood estimators of $\boldsymbol{\beta}$ and σ^2 for the unconstrained normal positive definite Aitken model. Specialized to the normal Gauss–Markov model, those estimators are $\hat{\boldsymbol{\beta}}$ and $\dot{\sigma}^2$, where $\hat{\boldsymbol{\beta}}$ is any solution to the normal equations and $\dot{\sigma}^2 = Q(\hat{\boldsymbol{\beta}})/n$. Thus,

$$\begin{aligned}\sup_{\Omega} L(\boldsymbol{\beta}, \sigma^2) &= L(\hat{\boldsymbol{\beta}}, \dot{\sigma}^2) = (2\pi\dot{\sigma}^2)^{-n/2} \exp[-(\mathbf{y} - \mathbf{X}\hat{\boldsymbol{\beta}})^T(\mathbf{y} - \mathbf{X}\hat{\boldsymbol{\beta}})/(2\dot{\sigma}^2)] \\ &= (2\pi\dot{\sigma}^2)^{-n/2} \exp(-n/2).\end{aligned}$$

A very similar development to that presented in Example 14.1-1 but for the normal constrained Gauss–Markov model establishes, with the aid of Theorem 10.1.2, that the maximum likelihood estimators of $\boldsymbol{\beta}$ and σ^2 under the model with constraints $\mathbf{C}^T\boldsymbol{\beta} = \mathbf{c}_0$ are $\breve{\boldsymbol{\beta}}$ and $\ddot{\sigma}^2$, where $\breve{\boldsymbol{\beta}}$ is the first p-component subvector of any solution to the constrained normal equations

$$\begin{pmatrix} \mathbf{X}^T\mathbf{X} & \mathbf{C} \\ \mathbf{C}^T & \mathbf{0} \end{pmatrix} \begin{pmatrix} \boldsymbol{\beta} \\ \boldsymbol{\lambda} \end{pmatrix} = \begin{pmatrix} \mathbf{X}^T\mathbf{y} \\ \mathbf{c}_0 \end{pmatrix}$$

and $\ddot{\sigma}^2 = Q(\breve{\boldsymbol{\beta}})/n$; in fact, this was the topic of Exercise 14.4. (Here, $\mathbf{C}^T$ and $\mathbf{c}_0$ are playing the roles of $\mathbf{A}$ and $\mathbf{h}$ in Theorem 10.1.2.) Thus,

$$\begin{aligned}\sup_{\Omega_0} L(\boldsymbol{\beta}, \sigma^2) &= L(\breve{\boldsymbol{\beta}}, \ddot{\sigma}^2) = (2\pi\ddot{\sigma}^2)^{-n/2} \exp[-(\mathbf{y} - \mathbf{X}\breve{\boldsymbol{\beta}})^T(\mathbf{y} - \mathbf{X}\breve{\boldsymbol{\beta}})/(2\ddot{\sigma}^2)] \\ &= (2\pi\ddot{\sigma}^2)^{-n/2} \exp(-n/2).\end{aligned}$$

Therefore, the likelihood ratio test statistic is

$$\Lambda = \frac{L(\breve{\boldsymbol{\beta}}, \ddot{\sigma}^2)}{L(\hat{\boldsymbol{\beta}}, \dot{\sigma}^2)} = \frac{(2\pi\ddot{\sigma}^2)^{-n/2}}{(2\pi\dot{\sigma}^2)^{-n/2}} = \left(\frac{\ddot{\sigma}^2}{\dot{\sigma}^2}\right)^{-n/2} = \left(\frac{Q(\hat{\boldsymbol{\beta}})}{Q(\breve{\boldsymbol{\beta}})}\right)^{n/2}.$$

Now, provided that $n > p^*$, the following statements are all equivalent:

$$\begin{aligned}\left(\frac{Q(\hat{\boldsymbol{\beta}})}{Q(\breve{\boldsymbol{\beta}})}\right)^{n/2} < k \Leftrightarrow \frac{Q(\hat{\boldsymbol{\beta}})}{Q(\breve{\boldsymbol{\beta}})} < k^{2/n} \Leftrightarrow \frac{Q(\breve{\boldsymbol{\beta}})}{Q(\hat{\boldsymbol{\beta}})} > k^{-2/n} \Leftrightarrow \frac{[Q(\breve{\boldsymbol{\beta}}) - Q(\hat{\boldsymbol{\beta}})]}{Q(\hat{\boldsymbol{\beta}})} > k^{-2/n} - 1 \\ \Leftrightarrow \frac{[Q(\breve{\boldsymbol{\beta}}) - Q(\hat{\boldsymbol{\beta}})]/s}{Q(\hat{\boldsymbol{\beta}})/(n - p^*)} > \left(\frac{n - p^*}{s}\right)(k^{-2/n} - 1).\end{aligned}$$

So the likelihood ratio test rejects H_0 if and only if

$$\frac{[Q(\breve{\boldsymbol{\beta}}) - Q(\hat{\boldsymbol{\beta}})]/s}{Q(\hat{\boldsymbol{\beta}})/(n-p^*)} > k^*,$$

where k^* is a constant chosen so that the test has the desired size. Because the null hypothesis is testable, the constraints are estimable, so by (10.17) the likelihood ratio test rejects H_0 if and only if

$$\frac{(\mathbf{C}^T\hat{\boldsymbol{\beta}} - \mathbf{c}_0)^T[\mathbf{C}^T(\mathbf{X}^T\mathbf{X})^-\mathbf{C}]^{-1}(\mathbf{C}^T\hat{\boldsymbol{\beta}} - \mathbf{c}_0)/s}{\hat{\sigma}^2} > k^*.$$

By Theorem 15.2.1, $k^* = F_{\xi,s,n-p^*}$. □

What is the distribution of the F test statistic under the alternative hypothesis? Under H_a, $\mathrm{E}(\mathbf{C}^T\hat{\boldsymbol{\beta}} - \mathbf{c}_0) = \mathbf{C}^T\boldsymbol{\beta} - \mathbf{c}_0 \neq \mathbf{0}$; therefore, the numerator of the test statistic, divided by σ^2, has a noncentral chi-square distribution with s degrees of freedom and noncentrality parameter $(\mathbf{C}^T\boldsymbol{\beta} - \mathbf{c}_0)^T[\mathbf{C}^T(\mathbf{X}^T\mathbf{X})^-\mathbf{C}]^{-1}(\mathbf{C}^T\boldsymbol{\beta} - \mathbf{c}_0)/(2\sigma^2)$ (Theorem 14.2.4). Thus, under H_a, the F test statistic has a noncentral F distribution, namely $\mathrm{F}[s, n-p^*, (\mathbf{C}^T\boldsymbol{\beta}-\mathbf{c}_0)^T[\mathbf{C}^T(\mathbf{X}^T\mathbf{X})^-\mathbf{C}]^{-1}(\mathbf{C}^T\boldsymbol{\beta}-\mathbf{c}_0)/(2\sigma^2)]$. This distributional result is useful for evaluating the power of the size-ξ F test, i.e., the probability of rejecting H_0 as a function of $\boldsymbol{\beta}$ and σ^2, i.e.,

$$\pi(\boldsymbol{\beta}, \sigma^2) = \Pr(w > F_{\xi,s,n-p^*})$$

where $w \sim \mathrm{F}[s, n-p^*, (\mathbf{C}^T\boldsymbol{\beta} - \mathbf{c}_0)^T[\mathbf{C}^T(\mathbf{X}^T\mathbf{X})^-\mathbf{C}]^{-1}(\mathbf{C}^T\boldsymbol{\beta} - \mathbf{c}_0)/(2\sigma^2)]$. When $s = 1$, this may be written alternatively as $\Pr(|u| > t_{\xi/2,n-p^*})$ where $u \sim \mathrm{t}\left(n-p^*, \frac{\mathbf{c}^T\boldsymbol{\beta}-c_0}{\sigma\sqrt{\mathbf{c}^T(\mathbf{X}^T\mathbf{X})^-\mathbf{c}}}\right)$.

It can be shown that the F test is the uniformly most powerful size-ξ test among all "invariant" size-ξ tests of $H_0 : \mathbf{C}^T\boldsymbol{\beta} = \mathbf{c}_0$ versus $H_a : \mathbf{C}^T\boldsymbol{\beta} \neq \mathbf{c}_0$ [see Lehmann and Romano (2005, Sections 7.1–7.3)], and that it has several other optimality properties as well (for example, it is also the likelihood ratio test, as Theorem 15.2.2 noted). It can also be shown that there is no uniformly most powerful test of these hypotheses.

When $\mathbf{c}_0 = \mathbf{0}$, the F test statistic in (15.7) has a certain invariance property, as described by the following theorem.

Theorem 15.2.3 *Let $\mathbf{C}_1^T\boldsymbol{\beta}$ be a vector of s linearly independent estimable functions, and let $\mathbf{C}_2^T\boldsymbol{\beta}$ be likewise. If $\mathcal{R}(\mathbf{C}_1^T) = \mathcal{R}(\mathbf{C}_2^T)$, then the F test statistic for testing $H_0 : \mathbf{C}_1^T\boldsymbol{\beta} = \mathbf{0}$ versus $H_a : \mathbf{C}_1^T\boldsymbol{\beta} \neq \mathbf{0}$ is equal to the F test statistic for testing $H_0 : \mathbf{C}_2^T\boldsymbol{\beta} = \mathbf{0}$ versus $H_a : \mathbf{C}_2^T\boldsymbol{\beta} \neq \mathbf{0}$.*

Proof Because $\mathcal{R}(\mathbf{C}_1^T) = \mathcal{R}(\mathbf{C}_2^T)$, $\mathbf{C}_1^T = \mathbf{B}\mathbf{C}_2^T$ for some nonsingular $s \times s$ matrix $\mathbf{B}$. Then

$$\begin{aligned}\mathbf{C}_1[\mathbf{C}_1^T(\mathbf{X}^T\mathbf{X})^-\mathbf{C}_1]^{-1}\mathbf{C}_1^T &= \mathbf{C}_2\mathbf{B}^T[\mathbf{B}\mathbf{C}_2^T(\mathbf{X}^T\mathbf{X})^-\mathbf{C}_2\mathbf{B}^T]^{-1}\mathbf{B}\mathbf{C}_2^T \\ &= \mathbf{C}_2\mathbf{B}^T(\mathbf{B}^T)^{-1}[\mathbf{C}_2^T(\mathbf{X}^T\mathbf{X})^-\mathbf{C}_2]^{-1}\mathbf{B}^{-1}\mathbf{B}\mathbf{C}_2^T \\ &= \mathbf{C}_2[\mathbf{C}_2^T(\mathbf{X}^T\mathbf{X})^-\mathbf{C}_2]^{-1}\mathbf{C}_2^T.\end{aligned}$$

Because $\mathbf{c}_0 = \mathbf{0}$, it follows that the numerator in the F test statistic in (15.7) is invariant to whether the null hypothesis is specified using $\mathbf{C}_1$ or $\mathbf{C}_2$. □

An important special case of the F test described in Theorem 15.2.1 is related to the sequential ANOVA corresponding to an ordered k-part model. Consider a normal Gauss–Markov ordered two-part model $\{\mathbf{y}, \mathbf{X}_1\boldsymbol{\beta}_1+\mathbf{X}_2\boldsymbol{\beta}_2, \sigma^2\mathbf{I}\}$, and suppose that we wish to evaluate the importance of $\mathbf{X}_2\boldsymbol{\beta}_2$ in that model. By Theorem 14.5.3, the distribution of the ratio of the mean square for $\mathbf{X}_2|\mathbf{X}_1$ to the residual mean square is

$$\begin{aligned}\frac{\text{SS}(\mathbf{X}_2|\mathbf{X}_1)/(p_{12}^* - p_1^*)}{\text{Residual mean square}} &= \frac{\mathbf{y}^T(\mathbf{P}_{12} - \mathbf{P}_1)\mathbf{y}/(p_{12}^* - p_1^*)}{\mathbf{y}^T(\mathbf{I} - \mathbf{P}_{12})\mathbf{y}/(n - p_{12}^*)} \\ &\sim \text{F}[p_{12}^* - p_1^*, n - p_{12}^*, \boldsymbol{\beta}^T\mathbf{X}^T(\mathbf{P}_{12} - \mathbf{P}_1)\mathbf{X}\boldsymbol{\beta}/(2\sigma^2)].\end{aligned}$$

If $(\mathbf{P}_{21} - \mathbf{P}_1)\mathbf{X}\boldsymbol{\beta} = \mathbf{0}$, then the noncentrality parameter of this F-distribution equals 0. Thus, if the equality $(\mathbf{P}_{21} - \mathbf{P}_1)\mathbf{X}\boldsymbol{\beta} = \mathbf{0}$ corresponds to a complete lack of explanatory value ascribable to $\mathbf{X}_2\boldsymbol{\beta}_2$ in the ordered two-part model $\mathbf{y} = \mathbf{X}_1\boldsymbol{\beta}_1+\mathbf{X}_2\boldsymbol{\beta}_2+\mathbf{e}$, then a size-$\xi$ test of the null hypothesis of that lack of explanatory value, versus the alternative hypothesis that $\mathbf{X}_2\boldsymbol{\beta}_2$ has some explanatory value in the same ordered two-part model, is obtained by rejecting the null hypothesis if and only if the ratio of the mean square for $\mathbf{X}_2|\mathbf{X}_1$ to the residual mean square exceeds $F_{\xi, p_{12}^*-p_1^*, n-p_{12}^*}$. The following theorem shows that such a test is completely equivalent to a particular case of the F test for $H_0 : \mathbf{C}^T\boldsymbol{\beta} = \mathbf{c}_0$ versus $H_a : \mathbf{C}^T\boldsymbol{\beta} \neq \mathbf{c}_0$ given by (15.7). This particular case is $\mathbf{C}^T = (\mathbf{0}^T, \mathbf{C}_2^T)$ and $\mathbf{c}_0 = \mathbf{0}$, where the rows of $\mathbf{C}_2^T$ constitute any basis for $\mathcal{R}[(\mathbf{I} - \mathbf{P}_1)\mathbf{X}_2]$.

Theorem 15.2.4 *Let T_1 denote the ratio of the mean square for $\mathbf{X}_2|\mathbf{X}_1$ to the residual mean square in the normal Gauss–Markov ordered two-part model $\{\mathbf{y}, \mathbf{X}_1\boldsymbol{\beta}_1 + \mathbf{X}_2\boldsymbol{\beta}_2, \sigma^2\mathbf{I}\}$, i.e.,*

$$T_1 = \frac{\mathbf{y}^T(\mathbf{P}_{12} - \mathbf{P}_1)\mathbf{y}/(p_{12}^* - p_1^*)}{\mathbf{y}^T(\mathbf{I} - \mathbf{P}_{12})\mathbf{y}/(n - p_{12}^*)}.$$

Let T_2 denote the F test statistic for testing $H_0 : \mathbf{C}^T\boldsymbol{\beta} = \mathbf{0}$ versus $H_a : \mathbf{C}^T\boldsymbol{\beta} \neq \mathbf{0}$ where $\mathbf{C}^T = (\mathbf{0}^T, \mathbf{C}_2^T)$, $s = \text{rank}(\mathbf{C})$, and the rows of $\mathbf{C}_2^T$ constitute any basis for $\mathcal{R}[(\mathbf{I} - \mathbf{P}_1)\mathbf{X}_2]$, i.e.,

$$T_2 = \frac{(\mathbf{C}^T\hat{\boldsymbol{\beta}})^T[\mathbf{C}^T(\mathbf{X}^T\mathbf{X})^-\mathbf{C}]^{-1}(\mathbf{C}^T\hat{\boldsymbol{\beta}})}{s\hat{\sigma}^2}.$$

Then, $T_1 = T_2$.

Proof First observe that H_0 is testable because $(\mathbf{I} - \mathbf{P}_1)\mathbf{X}_2\boldsymbol{\beta}_2$ is estimable by Theorem 12.1.1. Next, observe that

$$s = \text{rank}(\mathbf{C}) = \text{rank}(\mathbf{C}_2) = \text{rank}[(\mathbf{I} - \mathbf{P}_1)\mathbf{X}_2] = p_{12}^* - p_1^*$$

and $\hat{\sigma}^2 = \mathbf{y}^T(\mathbf{I} - \mathbf{P}_{12})\mathbf{y}/(n - p_{12}^*)$. Therefore, to establish that $T_1 = T_2$ it suffices to show that

$$\mathbf{y}^T(\mathbf{P}_{12} - \mathbf{P}_1)\mathbf{y} = (\mathbf{C}^T\hat{\boldsymbol{\beta}})^T[\mathbf{C}^T(\mathbf{X}^T\mathbf{X})^-\mathbf{C}]^{-1}(\mathbf{C}^T\hat{\boldsymbol{\beta}}).$$

Equivalently, because $\mathbf{C}^T\hat{\boldsymbol{\beta}} = \mathbf{C}^T(\mathbf{X}^T\mathbf{X})^-\mathbf{X}^T\mathbf{y}$, invariant to the choice of generalized inverse of $\mathbf{X}^T\mathbf{X}$ (Theorem 7.1.4), it suffices to show that

$$\mathbf{y}^T(\mathbf{P}_{12} - \mathbf{P}_1)\mathbf{y} = \mathbf{y}^T\mathbf{X}(\mathbf{X}^T\mathbf{X})^-\mathbf{C}[\mathbf{C}^T(\mathbf{X}^T\mathbf{X})^-\mathbf{C}]^{-1}\mathbf{C}^T(\mathbf{X}^T\mathbf{X})^-\mathbf{X}^T\mathbf{y}. \tag{15.9}$$

Now because $\mathbf{C}_2^T = \mathbf{A}_2^T(\mathbf{I} - \mathbf{P}_1)\mathbf{X}_2$ for some $s \times n$ matrix $\mathbf{A}_2^T$, we may write $\mathbf{C}^T$ as $\mathbf{A}^T\mathbf{X}$ for $\mathbf{A}^T = \mathbf{A}_2^T(\mathbf{I} - \mathbf{P}_1)$. Therefore, the matrix of the quadratic form in (15.9) may be re-expressed as follows:

$$\begin{aligned}\mathbf{X}(\mathbf{X}^T\mathbf{X})^-\mathbf{C}[\mathbf{C}^T(\mathbf{X}^T\mathbf{X})^-\mathbf{C}]^{-1}\mathbf{C}^T(\mathbf{X}^T\mathbf{X})^-\mathbf{X}^T &= \mathbf{P_X}\mathbf{A}[\mathbf{A}^T\mathbf{P_X}\mathbf{A}]^{-1}\mathbf{A}^T\mathbf{P_X} \\ &= (\mathbf{P_X}\mathbf{A})[(\mathbf{P_X}\mathbf{A})^T(\mathbf{P_X}\mathbf{A})]^{-1}(\mathbf{P_X}\mathbf{A})^T.\end{aligned}$$

This establishes that the matrix of the quadratic form in (15.9) is the orthogonal projection matrix onto $\mathcal{C}(\mathbf{P_X}\mathbf{A})$ or equivalently the orthogonal projection matrix onto $\mathcal{C}[\mathbf{X}(\mathbf{X}^T\mathbf{X})^-\mathbf{C}]$. Because $\mathbf{P}_{12} - \mathbf{P}_1$ is the orthogonal projection matrix onto $\mathcal{C}[(\mathbf{I} - \mathbf{P}_1)\mathbf{X}_2]$ (Theorem 9.1.2), it suffices to show that $\mathcal{C}[\mathbf{X}(\mathbf{X}^T\mathbf{X})^-\mathbf{C}] = \mathcal{C}[(\mathbf{I} - \mathbf{P}_1)\mathbf{X}_2]$. To that end, let $\mathbf{X}^* = [\mathbf{X}_1, (\mathbf{I} - \mathbf{P}_1)\mathbf{X}_2]$, in which case $\mathcal{C}(\mathbf{X}^*) = \mathcal{C}(\mathbf{X})$ and hence $\mathbf{P}_{\mathbf{X}^*} = \mathbf{P_X}$. Then

$$\begin{aligned}\mathbf{X}(\mathbf{X}^T\mathbf{X})^-\mathbf{C} &= \mathbf{P_X}\mathbf{A} \\ &= \mathbf{P}_{\mathbf{X}^*}(\mathbf{I} - \mathbf{P}_1)\mathbf{A}_2 \\ &= \left(\mathbf{X}_1 \ (\mathbf{I} - \mathbf{P}_1)\mathbf{X}_2\right)\begin{pmatrix}(\mathbf{X}_1^T\mathbf{X}_1)^- & \mathbf{0} \\ \mathbf{0} & [\mathbf{X}_2^T(\mathbf{I} - \mathbf{P}_1)\mathbf{X}_2]^-\end{pmatrix}\end{aligned}$$

$$\times \begin{pmatrix} \mathbf{X}_1^T \\ \mathbf{X}_2^T(\mathbf{I}-\mathbf{P}_1) \end{pmatrix} (\mathbf{I}-\mathbf{P}_1)\mathbf{A}_2$$
$$= \{\mathbf{P}_1 + (\mathbf{I}-\mathbf{P}_1)\mathbf{X}_2[\mathbf{X}_2^T(\mathbf{I}-\mathbf{P}_1)\mathbf{X}_2]^-\mathbf{X}_2^T(\mathbf{I}-\mathbf{P}_1)\}(\mathbf{I}-\mathbf{P}_1)\mathbf{A}_2$$
$$= (\mathbf{I}-\mathbf{P}_1)\mathbf{X}_2[\mathbf{X}_2^T(\mathbf{I}-\mathbf{P}_1)\mathbf{X}_2]^-\mathbf{X}_2^T(\mathbf{I}-\mathbf{P}_1)\mathbf{A}_2,$$

showing that $\mathcal{C}[\mathbf{X}(\mathbf{X}^T\mathbf{X})^-\mathbf{C}] \subseteq \mathcal{C}[(\mathbf{I}-\mathbf{P}_1)\mathbf{X}_2]$. Furthermore,

$$s = \text{rank}[\mathbf{C}^T(\mathbf{X}^T\mathbf{X})^-\mathbf{C}] = \text{rank}[\mathbf{A}^T\mathbf{X}(\mathbf{X}^T\mathbf{X})^-\mathbf{C}] \leq \text{rank}[\mathbf{X}(\mathbf{X}^T\mathbf{X})^-\mathbf{C}] \leq \text{rank}(\mathbf{C}) = s,$$

implying that $\text{rank}[\mathbf{X}(\mathbf{X}^T\mathbf{X})^-\mathbf{C}] = s$. It follows from Theorem 2.8.5 that $\mathcal{C}[\mathbf{X}(\mathbf{X}^T\mathbf{X})^-\mathbf{C}] = \mathcal{C}[(\mathbf{I}-\mathbf{P}_1)\mathbf{X}_2]$, which establishes the desired result. □

Although Theorem 15.2.4 pertained to an ordered two-part model only, it may be generalized to the ordered k-part model for general k with little difficulty. Furthermore, all of the theorems presented in this section may be extended to the normal positive definite Aitken model $\{\mathbf{y}, \mathbf{X}\boldsymbol{\beta}, \sigma^2\mathbf{W}\}$. For example, using Theorem 15.1.1, Theorem 15.2.1 may be extended to state that a size-ξ test of $H_0 : \mathbf{C}^T\boldsymbol{\beta} = \mathbf{c}_0$ versus $H_a : \mathbf{C}^T\boldsymbol{\beta} \neq \mathbf{c}_0$ is obtained by rejecting H_0 if and only if

$$\frac{(\mathbf{C}^T\tilde{\boldsymbol{\beta}} - \mathbf{c}_0)^T[\mathbf{C}^T(\mathbf{X}^T\mathbf{W}^{-1}\mathbf{X})^-\mathbf{C}]^{-1}(\mathbf{C}^T\tilde{\boldsymbol{\beta}} - \mathbf{c}_0)}{s\tilde{\sigma}^2} > F_{\xi,s,n-p^*}. \tag{15.10}$$

Example 15.2-1. Tests for Nonzero Slope and Intercept in Normal Simple Linear Regression

Consider the normal Gauss–Markov simple linear regression model, and suppose that we wish to test $H_0 : \beta_2 = 0$ versus $H_a : \beta_2 \neq 0$. This can be accomplished with either an F test from the corrected-for-the-mean overall ANOVA or a t test. Using a result from Example 14.5-1, the size-ξ F test rejects H_0 if and only if

$$\frac{\hat{\beta}_2^2 SXX}{\hat{\sigma}^2} > F_{\xi,1,n-2}.$$

As a special case of (15.8), the size-ξ t test, which is equivalent to the F test just given, rejects H_0 if and only if

$$\frac{|\hat{\beta}_2|}{\hat{\sigma}/\sqrt{SXX}} > t_{\xi/2,n-2}.$$

As another special case of (15.8), the size-ξ t test of $H_0 : \beta_1 = 0$ versus $H_a : \beta_1 \neq 0$ rejects H_0 if and only if

$$\frac{|\hat{\beta}_1|}{\hat{\sigma}\sqrt{(1/n) + \bar{x}^2/SXX}} > t_{\xi/2,n-2}.$$

■

Example 15.2-2. Test for Nonzero Factor Effects in a Normal Gauss–Markov One-Factor Model

Consider testing

$$H_0 : \alpha_1 = \alpha_2 = \cdots = \alpha_q \quad \text{versus} \quad H_a : \text{not all } \alpha_i \text{ are equal}$$

in the normal Gauss–Markov one-factor model

$$y_{ij} = \mu + \alpha_i + e_{ij} \quad (i = 1, \ldots, q;\ j = 1, \ldots, n_i).$$

The null hypothesis may be written alternatively as $H_0 : \mathbf{C}^T\boldsymbol{\beta} = \mathbf{c}_0$ with $\mathbf{c}_0 = \mathbf{0}_{q-1}$, $\boldsymbol{\beta} = (\mu, \alpha_1, \ldots, \alpha_q)^T$, and

$$\mathbf{C}^T = \begin{pmatrix} 0 & 1 & -1 & 0 & \cdots & & 0 \\ 0 & 0 & 1 & -1 & 0 & \cdots & 0 \\ \vdots & & & & & & \\ 0 & \cdots & & & & 1 & -1 \end{pmatrix}_{(q-1)\times(q+1)}.$$

This matrix $\mathbf{C}^T$ is not the only suitable matrix for specifying H_0; another would be

$$\breve{\mathbf{C}}^T = \begin{pmatrix} 0 & 1 & -1 & 0 & \cdots & & 0 \\ 0 & 1 & 0 & -1 & 0 & \cdots & 0 \\ \vdots & & & & & & \\ 0 & 1 & \cdots & & & 0 & -1 \end{pmatrix}_{(q-1)\times(q+1)}.$$

Note that $\breve{\mathbf{C}}^T$ has the same row space as $\mathbf{C}^T$, and that the rows of each of $\mathbf{C}_2^T$ and $\breve{\mathbf{C}}_2^T$ [where $(\mathbf{0}_{q-1}, \mathbf{C}_2^T) = \mathbf{C}^T$ and $(\mathbf{0}_{q-1}, \breve{\mathbf{C}}_2^T) = \breve{\mathbf{C}}^T$] constitute a basis for $\mathcal{R}[(\mathbf{I} - \mathbf{P}_1)\mathbf{X}_2]$ in an ordered two-part representation of the model with the overall mean fitted first [in which case $\mathbf{P}_1 = (1/n)\mathbf{J}_n$]. Thus the F test using $\breve{\mathbf{C}}^T$ in place of $\mathbf{C}^T$ (and indeed, the F test using any $(q-1) \times (q+1)$ matrix with the same row space as $\mathbf{C}^T$ in place of it) in (15.7) is identical to the F test using $\mathbf{C}^T$, by Theorem 15.2.3. Furthermore, it follows from Theorem 15.2.4 that $(\mathbf{C}^T\hat{\boldsymbol{\beta}} - \mathbf{0})^T[\mathbf{C}^T(\mathbf{X}^T\mathbf{X})^-\mathbf{C}]^{-1}(\mathbf{C}^T\hat{\boldsymbol{\beta}} - \mathbf{0})$ specializes to $\sum_{i=1}^q n_i(\bar{y}_{i\cdot} - \bar{y}_{\cdot\cdot})^2$, the model sum of squares (corrected for the mean) under this model. Thus, the size-ξ F test rejects H_0 if and only if

$$\frac{\sum_{i=1}^{q} n_i(\bar{y}_{i\cdot} - \bar{y}_{\cdot\cdot})^2}{(q-1)\hat{\sigma}^2} > F_{\xi,q-1,n-q}. \qquad \blacksquare$$

Methodological Interlude #8: Stepwise Regression Variable Selection Procedures

Consider the case of a normal Gauss–Markov full-rank multiple regression model, with an overall intercept and $k = p - 1$ n-vectors of explanatory variables $\mathbf{x}_1, \mathbf{x}_2, \ldots, \mathbf{x}_k$. A common question in applications of linear models is: Which explanatory variables (if any) help to explain $\mathbf{y}$? We addressed this question previously in Methodological Interlude #6, where some distribution-free model selection criteria were presented. When the distribution of $\mathbf{y}$ is multivariate normal, a collection of hypothesis testing strategies known as *stepwise variable selection procedures* are commonly used to address the same question. In the *forward selection* approach, one begins by fitting all k one-variable models (each with an intercept). For each such model, $H_0^{(j)}: \beta_{j+1} = 0$ versus $H_a^{(j)}: \beta_{j+1} \neq 0$ is tested using the decision rule (15.8) with a prespecified size ξ, which in this context can be written as

$$\frac{|\hat{\beta}_{j+1}|}{\hat{\sigma}_j\sqrt{c_{jj}}} > t_{\xi/2,n-2}$$

where $\hat{\sigma}_j^2$ is the residual mean square for the model with $\mathbf{x}_j$ (plus an intercept), and c_{jj} is the (2, 2) element of $[(\mathbf{1}_n, \mathbf{x}_j)^T(\mathbf{1}_n, \mathbf{x}_j)]^{-1}$. If none of the k null hypotheses are rejected, the procedure stops and one adopts the intercept-only model. If any of the null hypotheses are rejected, however, then a provisional model is formed from the intercept and the variable, denoted by $\mathbf{x}_{(1)}$, with the largest t test statistic. This completes the first step of the procedure. Next, one fits all $k - 1$ possible two-variable models formed by adding one of the remaining variables to the provisional first-stage model, and then tests the same hypotheses as in the first step (except for the one involving $\beta_{(1)}$) again using decision rule (15.8) with the same prespecified size, which in this context can be written as

$$\frac{|\hat{\beta}_{j+1}|}{\hat{\sigma}_{(1)j}\sqrt{c_{(1)jj}}} > t_{\xi/2,n-3}$$

where $\hat{\sigma}_{(1)j}^2$ is the residual mean square for the model $\mathbf{x}_{(1)}$ and $\mathbf{x}_j$ (plus an intercept), and $c_{(1)jj}$ is the (3, 3) element of $[(\mathbf{1}_n, \mathbf{x}_{(1)}, \mathbf{x}_j)^T(\mathbf{1}_n, \mathbf{x}_{(1)}, \mathbf{x}_j)]^{-1}$. If none of these

$k-1$ hypotheses are rejected, the procedure stops and one adopts the model with intercept and $\mathbf{x}_{(1)}$ only. Otherwise, a new provisional model is formed from the previous one plus the variable, denoted by $\mathbf{x}_{(2)}$, with the largest second-stage test statistic. The procedure continues in this fashion, adding variables one at a time until none of the null hypotheses corresponding to the remaining variables are rejected.

In the *backward elimination* approach, the model with all k explanatory variables (plus an intercept) is fitted first, and within the context of this model $H_0^{(j)} : \beta_{j+1} = 0$ versus $H_a^{(j)} : \beta_{j+1} \neq 0$ is tested at a prespecified size using (15.8) as it specializes when $\mathbf{c} = \mathbf{u}_j^{(k+1)}$ and $p^* = k+1$. If all of the hypotheses are rejected, then the procedure stops and one adopts the model with all k explanatory variables (plus an intercept). But if at least one hypothesis is not rejected, then the variable with the smallest t test statistic is removed from the model. This completes the first step. Next, within the context of the provisional $(k-1)$-variable model obtained via the first step, the same hypotheses are tested at the same size using (15.8) as it specializes to $\mathbf{X}_{-(1)}$, where $\mathbf{X}_{-(1)}$ is the model matrix corresponding to the provisional $(k-1)$-variable model. The procedure continues, deleting one variable at a time until every null hypothesis about the remaining variables is rejected.

Stepwise regression model selection procedures are very popular because they are computationally efficient; many fewer than the total number $\binom{k}{2}$ of possible models must be fit to the data. However, they have several shortcomings. First, they suffer from the "multiple testing" problem, meaning that the probability of incorrectly rejecting at least one of the stepwise null hypotheses is larger than the prespecified size ξ of the test at each step. Second, they only produce one "best" model, rather than a rank-ordering or other indication of how good other candidate models are. Finally, there is no guarantee that forward selection and backward elimination will select the same model, nor that the "best" model among all possible models will be obtained using either approach. Consequently, unless the number of variables is so large that fitting models corresponding to all possible subsets of the variables is not possible, other procedures for selecting the best model(s) are preferred. Some of those procedures were presented in Methodological Interlude #6, and others will be described in Methodological Interlude #12. ■

Next we give a result on the monotonicity, with respect to the noncentrality parameter, of the power function of the F test; the result is due to Ghosh (1973). Two theorems are required to reach the desired result.

Theorem 15.2.5 *Suppose that* $u \sim \chi^2(\nu, \lambda)$. *For fixed* ν *and* c, *with* $c > 0$, $Pr(u > c)$ *is a strictly increasing function of* λ.

Proof Note that

$$\Pr(u > c) = \int_c^\infty \sum_{j=0}^\infty \left(\frac{e^{-\lambda}\lambda^j}{j!}\right)\left(\frac{x^{(\nu+2j-2)/2}e^{-x/2}}{\Gamma(\frac{\nu+2j}{2})2^{(\nu+2j)/2}}\right) dx$$

$$= \int_c^\infty \left[e^{-\lambda}\left(\frac{x^{(\nu-2)/2}e^{-x/2}}{\Gamma(\frac{\nu}{2})2^{\nu/2}}\right) + \sum_{j=1}^\infty \left(\frac{e^{-\lambda}\lambda^j}{j!}\right)\left(\frac{x^{(\nu+2j-2)/2}e^{-x/2}}{\Gamma(\frac{\nu+2j}{2})2^{(\nu+2j)/2}}\right)\right] dx.$$

Thus, letting $f_{\chi^2}(x;\nu)$ denote the pdf of a central chi-square distribution with ν degrees of freedom, we have

$$\frac{d\Pr(u > c)}{d\lambda}$$

$$= \int_c^\infty \left[-e^{-\lambda}\left(\frac{x^{(\nu-2)/2}e^{-x/2}}{\Gamma(\frac{\nu}{2})2^{\nu/2}}\right) + \sum_{j=1}^\infty \left(\frac{e^{-\lambda}j\lambda^{j-1} - e^{-\lambda}\lambda^j}{j!}\right)\right.$$

$$\left.\times \left(\frac{x^{(\nu+2j-2)/2}e^{-x/2}}{\Gamma(\frac{\nu+2j}{2})2^{(\nu+2j)/2}}\right)\right] dx$$

$$= \int_c^\infty \left[\sum_{j=1}^\infty \left(\frac{e^{-\lambda}\lambda^{j-1}}{(j-1)!}\right) f_{\chi^2}(x;\nu+2j) - \sum_{j=0}^\infty \left(\frac{e^{-\lambda}\lambda^j}{j!}\right) f_{\chi^2}(x;\nu+2j)\right] dx$$

$$= \int_c^\infty \left[\sum_{j=0}^\infty \left(\frac{e^{-\lambda}\lambda^j}{j!}\right)[f_{\chi^2}(x;\nu+2j+2) - f_{\chi^2}(x;\nu+2j)]\right] dx$$

$$= \sum_{j=0}^\infty \left(\frac{e^{-\lambda}\lambda^j}{j!}\right)\left[\Pr\left(\sum_{i=1}^{\nu+2j+2} z_i^2 > c\right) - \Pr\left(\sum_{i=1}^{\nu+2j} z_i^2 > c\right)\right]$$

where $z_1, z_2, \ldots$ are iid N(0, 1) random variables.[1] In this last expression, the probability of the first event is clearly larger than that of the second; thus $\frac{d\Pr(u>c)}{d\lambda} > 0$, i.e., $\Pr(u > c)$ is an increasing function of λ. □

Theorem 15.2.6 *Suppose that $w \sim F(\nu_1, \nu_2, \lambda)$. For fixed ν_1, ν_2, and c, with $c > 0$, $Pr(w > c)$ is a strictly increasing function of λ.*

[1]Moving the derivative inside the integral can be justified by the dominated convergence theorem.

Proof Let u_1 and u_2 represent independent random variables such that $u_1 \sim \chi^2(\nu_1, \lambda)$ and $u_2 \sim \chi^2(\nu_2)$. Then by Definition 14.5.1 and the law of total probability for expectations,

$$\begin{aligned}
\Pr(w > c) &= \Pr\left(\frac{u_1/\nu_1}{u_2/\nu_2} > c\right) \\
&= \Pr\left(u_1 > \frac{\nu_1 c u_2}{\nu_2}\right) \\
&= \mathrm{E}[I(\{u_1 > \frac{\nu_1 c u_2}{\nu_2}\})] \\
&= \mathrm{E}\{\mathrm{E}[I(\{u_1 > \frac{\nu_1 c u_2}{\nu_2}\})|u_2]\} \\
&= \int_0^\infty \Pr\left(u_1 > \frac{\nu_1 c x}{\nu_2}\right) f_{\chi^2}(x; \nu_2)\, dx.
\end{aligned}$$

[Here $I(\cdot)$ represents the indicator function of the set specified within the parentheses.] The integrand depends on λ only through $\Pr(u_1 > \nu_1 c x/\nu_2)$, which by Theorem 15.2.5 strictly increases with λ. Because the rest of the integrand is positive, the entire integrand strictly increases with λ and therefore so does the integral. □

Corollary 15.2.6.1 *Suppose that* $t \sim t(\nu, \mu)$. *For fixed* ν *and* c, *with* $c > 0$, $Pr(|t| > c)$ *is a strictly increasing function of* $|\mu|$.

Proof By Theorem 14.5.1, $\Pr(|t| > c) = \Pr(t^2 > c^2) = \Pr(w > c^2)$ where $w \sim \mathrm{F}(1, \nu, \mu^2/2)$, which (by Theorem 15.2.6) is a strictly increasing function of $\mu^2/2$, or equivalently, of $|\mu|$. □

As an important application of Theorem 15.2.6, the power of the F test for $H_0 : \mathbf{C}^T\boldsymbol{\beta} = \mathbf{c}_0$ versus $H_a : \mathbf{C}^T\boldsymbol{\beta} \neq \mathbf{c}_0$ is an increasing function of $(\mathbf{C}^T\boldsymbol{\beta} - \mathbf{c}_0)^T[\mathbf{C}^T(\mathbf{X}^T\mathbf{X})^-\mathbf{C}]^{-1}(\mathbf{C}^T\boldsymbol{\beta} - \mathbf{c}_0)/(2\sigma^2)$, and the power of the t test for $H_0 : \mathbf{c}^T\boldsymbol{\beta} = c_0$ versus $H_a : \mathbf{c}^T\boldsymbol{\beta} \neq c_0$ is an increasing function of $\frac{|\mathbf{c}^T\boldsymbol{\beta} - c_0|}{\sigma\sqrt{\mathbf{c}^T(\mathbf{X}^T\mathbf{X})^-\mathbf{c}}}$. To the extent that the investigator using the linear model has control over the combinations of the explanatory variables where responses are observed, i.e., over $\mathbf{X}$, the investigator may control, or even maximize, the power of these tests.

Example 15.2-3. Maximizing the Power of the Tests for Nonzero Slope and Intercept in Normal Simple Linear Regression and for Nonzero Factor Effects in a Normal Gauss–Markov One-Factor Model

Consider the size-ξ F test for $H_0 : \beta_2 = 0$ versus $H_a : \beta_2 \neq 0$ considered in Example 15.2-1, which rejects H_0 if and only if

$$\frac{\hat{\beta}_2^2 SXX}{\hat{\sigma}^2} > F_{\xi,1,n-2}.$$

The power of this test is equal to $\Pr(w > F_{\xi,1,n-2})$ where $w \sim \mathrm{F}[1, n-2, \beta_2^2 SXX/(2\sigma^2)]$. Thus, to maximize this power for fixed n, β_2, and σ^2, the investigator should seek to maximize SXX, i.e., $\sum_{i=1}^n (x_i - \bar{x})^2$. Without any constraints, there is no solution to the problem of maximizing $f(\mathbf{x}) = \sum_{i=1}^n (x_i - \bar{x})^2$. If, however, the domain of the values of the explanatory variable is constrained to the interval $[a, b]$ and n is even, then $f(\mathbf{x})$ is maximized by taking half of the elements of $\mathbf{x}$ at a and the other half at b (see the exercises).

Next consider the size-ξ F test for

$$H_0 : \alpha_1 = \alpha_2 = \cdots = \alpha_q \quad \text{versus} \quad H_a : \text{not } H_0$$

considered in Example 15.2-2, which rejects H_0 if and only if

$$\frac{\sum_{i=1}^q n_i(\bar{y}_{i\cdot} - \bar{y}_{\cdot\cdot})^2}{(q-1)\hat{\sigma}^2} > F_{\xi,q-1,n-q}.$$

According to Example 14.5-2, the distribution of the F test statistic under H_a is

$$\mathrm{F}\left(q-1, n-q, \sum_{i=1}^q n_i \left(\alpha_i - \frac{\sum_{k=1}^q n_k \alpha_k}{n}\right)^2 \Big/ (2\sigma^2)\right).$$

Thus, by Theorem 15.2.6 the power of this test is a strictly increasing function of

$$\sum_{i=1}^q n_i \left(\alpha_i - \frac{\sum_{k=1}^q n_k \alpha_k}{n}\right)^2 \Big/ (2\sigma^2).$$

This reveals the intuitively reasonable result that for fixed σ^2, the power of the test increases as either the estimable functions $\{\mu + \alpha_1, \mu + \alpha_2, \ldots, \mu + \alpha_q\}$ become more disparate or the number of observations at each level grows. ■

Methodological Interlude #9: Testing for an Outlier

Consider the following modification of the normal Gauss–Markov model $\{\mathbf{y}, \mathbf{X}\boldsymbol{\beta}, \sigma^2\mathbf{I}\}$, which is called the normal Gauss–Markov mean shift outlier model. For this model, the model equation may be written as

$$\begin{pmatrix} y_1 \\ \mathbf{y}_{-1} \end{pmatrix} = \begin{pmatrix} \mathbf{x}_1^T \\ \mathbf{X}_{-1} \end{pmatrix} \boldsymbol{\beta} + \begin{pmatrix} e_1 \\ \mathbf{e}_{-1} \end{pmatrix}.$$

That is, we partition $\mathbf{y}$, $\mathbf{X}$, and $\mathbf{e}$ row-wise into two parts: the first part corresponds to the first observation, and the second part corresponds to the remaining $n-1$ observations. Furthermore, under this model,

$$\begin{pmatrix} e_1 \\ \mathbf{e}_{-1} \end{pmatrix} \sim \mathrm{N}_n\left(\begin{pmatrix} \Delta_1 \\ \mathbf{0}_{n-1} \end{pmatrix}, \sigma^2\mathbf{I} \right).$$

Here Δ_1 is an unknown, possibly nonzero, scalar parameter that allows the mean of the first observation to be shifted away from $\mathbf{x}_1^T\boldsymbol{\beta}$. Assume that $\text{rank}(\mathbf{X}_{-1}) = \text{rank}(\mathbf{X})$ $(= p^*)$, i.e., the rank of $\mathbf{X}$ is not affected by deleting its first row. Define $\hat{e}_{1,-1} = y_1 - \mathbf{x}_1^T\hat{\boldsymbol{\beta}}_{-1}$ where $\hat{\boldsymbol{\beta}}_{-1}$ is any solution to the normal equations $\mathbf{X}_{-1}^T\mathbf{X}_{-1}\boldsymbol{\beta} = \mathbf{X}_{-1}^T\mathbf{y}_{-1}$. Also define

$$\hat{\sigma}_{-1}^2 = \frac{(\mathbf{y}_{-1} - \mathbf{X}_{-1}\hat{\boldsymbol{\beta}}_{-1})^T(\mathbf{y}_{-1} - \mathbf{X}_{-1}\hat{\boldsymbol{\beta}}_{-1})}{n - p^* - 1}$$

and

$$p_{11,-1} = \mathbf{x}_1^T(\mathbf{X}_{-1}^T\mathbf{X}_{-1})^-\mathbf{x}_1.$$

Then by Theorems 4.2.1–4.2.3,

$$\mathrm{E}(\hat{e}_{1,-1}) = \mathbf{x}_1^T\boldsymbol{\beta} + \Delta_1 - \mathbf{x}_1^T(\mathbf{X}_{-1}^T\mathbf{X}_{-1})^-\mathbf{X}_{-1}^T\mathbf{X}_{-1}\boldsymbol{\beta} = \Delta_1$$

and

$$\text{var}(\hat{e}_{1,-1}) = \sigma^2 + \mathbf{x}_1^T(\mathbf{X}_{-1}^T\mathbf{X}_{-1})^-\mathbf{X}_{-1}^T(\sigma^2\mathbf{I})\mathbf{X}_{-1}(\mathbf{X}_{-1}^T\mathbf{X}_{-1})^-\mathbf{x}_1 = \sigma^2(1 + p_{11,-1}).$$

Thus $\hat{e}_{1,-1} \sim \mathrm{N}(\Delta_1, \sigma^2(1 + p_{11,-1}))$ and upon partial standardization,

$$\frac{\hat{e}_{1,-1}}{\sigma\sqrt{1 + p_{11,-1}}} \sim \mathrm{N}\left(\frac{\Delta_1}{\sigma\sqrt{1 + p_{11,-1}}}, 1 \right).$$

This quantity is often called the *externally studentized residual* (for the first observation). Clearly, because $\hat{\sigma}^2_{-1}$ is merely the residual mean square based on all observations save the first,

$$\frac{(n - p^* - 1)\hat{\sigma}^2_{-1}}{\sigma^2} \sim \chi^2(n - p^* - 1).$$

Furthermore,

$$(n - p^* - 1)\hat{\sigma}^2_{-1} = \begin{pmatrix} y_1 \\ \mathbf{y}_{-1} \end{pmatrix}^T \begin{pmatrix} 0 & \mathbf{0}^T \\ \mathbf{0} & \mathbf{I} - \mathbf{P}_{\mathbf{X}_{-1}} \end{pmatrix} \begin{pmatrix} y_1 \\ \mathbf{y}_{-1} \end{pmatrix}$$

and

$$\left(1 \; -\mathbf{x}_1^T(\mathbf{X}_{-1}^T\mathbf{X}_{-1})^-\mathbf{X}_{-1}^T\right)(\sigma^2\mathbf{I}) \begin{pmatrix} 0 & \mathbf{0}^T \\ \mathbf{0} & \mathbf{I} - \mathbf{P}_{\mathbf{X}_{-1}} \end{pmatrix} = \mathbf{0}^T,$$

which, by Corollary 14.3.1.2b, establishes that $\hat{e}_{1,-1}$ and $\hat{\sigma}^2_{-1}$ are independent. Therefore,

$$\frac{\hat{e}_{1,-1}}{\sigma\sqrt{1 + p_{11,-1}}} \Bigg/ \sqrt{\frac{(n - p^* - 1)\hat{\sigma}^2_{-1}}{\sigma^2(n - p^* - 1)}}$$

$$= \frac{\hat{e}_{1,-1}}{\hat{\sigma}_{-1}\sqrt{1 + p_{11,-1}}} \sim \mathrm{t}\left(n - p^* - 1, \frac{\Delta_1}{\sigma\sqrt{1 + p_{11,-1}}}\right).$$

By this last result, we see that $H_0 : \Delta_1 = 0$ could be tested versus $H_a : \Delta_1 \neq 0$ at the ξ level of significance by rejecting H_0 if and only if

$$\frac{|\hat{e}_{1,-1}|}{\hat{\sigma}_{-1}\sqrt{1 + p_{11,-1}}} > t_{\xi/2,n-p^*-1}.$$

Moreover, by Corollary 15.2.6.1 the power of this test is a monotone increasing function of $|\Delta_1|/(\sigma\sqrt{1 + p_{11,-1}})$. Thus larger $|\Delta_1|$, smaller σ, and/or smaller $p_{11,-1}$ yield a more powerful test.

Similarly, for any $i = 1, \ldots, n$, if we denote $\mathrm{E}(e_i)$ by Δ_i and assume that $\mathrm{E}(e_j) = 0$ for all $j \neq i$ but retain all of the other model assumptions, it is clear that $H_0 : \Delta_i = 0$ could be tested versus $H_a : \Delta_i \neq 0$ at size ξ by rejecting H_0 if and only if

$$\frac{|\hat{e}_{i,-i}|}{\hat{\sigma}_{-i}\sqrt{1 + p_{ii,-i}}} > t_{\xi/2,n-p^*-1}.$$

■

Methodological Interlude #10: Testing for Lack of Fit

Recall the scenario described in Sect. 9.5, in which some rows of the model matrix **X** are identical to each other, so that we may write the model matrix as

$$\mathbf{X} = \begin{pmatrix} \mathbf{1}_{n_1}\mathbf{x}_1^T \\ \mathbf{1}_{n_2}\mathbf{x}_2^T \\ \vdots \\ \mathbf{1}_{n_m}\mathbf{x}_m^T \end{pmatrix},$$

where $\mathbf{x}_1^T, \ldots, \mathbf{x}_m^T$ are distinct and n_i is the number of replications of the ith distinct row, and it is assumed that $p^* < m < n$. Recall further that in this scenario the residual sum of squares admits the decomposition

$$\mathbf{y}^T(\mathbf{I} - \mathbf{P_X})\mathbf{y} = \mathbf{y}^T(\mathbf{I} - \bar{\mathbf{J}})\mathbf{y} + \mathbf{y}^T(\bar{\mathbf{J}} - \mathbf{P_X})\mathbf{y},$$

where $\bar{\mathbf{J}} = \oplus_{i=1}^m (1/n_i)\mathbf{J}_{n_i}$. It was shown in Examples 14.2-4 and 14.3-4 that the scaled "pure error" sum of squares $\mathbf{y}^T(\mathbf{I} - \bar{\mathbf{J}})\mathbf{y}/\sigma^2$ and scaled "lack-of-fit" sum of squares $\mathbf{y}^T(\bar{\mathbf{J}} - \mathbf{P_X})\mathbf{y}/\sigma^2$ are distributed independently as $\chi^2(n-m)$ and $\chi^2(m-p^*)$ under the assumed Gauss–Markov model. Therefore,

$$F \equiv \frac{[\mathbf{y}^T(\bar{\mathbf{J}} - \mathbf{P_X})\mathbf{y}/\sigma^2]/(m - p^*)}{[\mathbf{y}^T(\mathbf{I} - \bar{\mathbf{J}})\mathbf{y}/\sigma^2]/(n - m)}$$

has an $\mathrm{F}(m - p^*, n - m)$ distribution under that model. Consequently, the decision rule for a size-ξ test of $H_0 : \mathrm{E}(y_{ij}) = \mathbf{x}_i^T\boldsymbol{\beta}$ (for all i and j) for some $\boldsymbol{\beta}$, versus $H_a : \mathrm{E}(y_{ij}) \neq \mathbf{x}_i^T\boldsymbol{\beta}$ is to reject H_0 if and only if $F > F_{\xi,m-p^*,n-m}$. This test is frequently presented in applied regression courses as a method for assessing the adequacy of the mean structure assumed for the model. Under the alternative hypothesis, if we write (as in Sect. 9.5) $\mathrm{E}(y_{ij}) = f(\mathbf{x}_i)$ where $f(\cdot)$ is an arbitrary function of p variables, then

$$F \sim \mathrm{F}(m - p^*, n - m, K/(2\sigma^2)),$$

where $K \geq 0$ was defined in the proof of Theorem 9.5.2. Generally speaking, K increases as $|f(\mathbf{x}_i) - \mathbf{x}_i^T\boldsymbol{\beta}|$ increases for any i, so by Theorem 15.2.6, the power of the test increases as the disparity between the model's assumed mean structure and its true, possibly nonlinear, mean structure increases. ∎

15.3 Simultaneous Confidence and Prediction Intervals

Suppose that we wish to make a joint, or *simultaneous*, confidence statement about $\mathbf{c}_1^T\boldsymbol{\beta}, \ldots, \mathbf{c}_s^T\boldsymbol{\beta}$. That is, we want to specify an s-dimensional region that contains all of these functions with a specified probability. If the functions are linearly independent, the $100(1-\xi)\%$ confidence ellipsoid given by (15.2) accomplishes the task; however, if the functions are not linearly independent, then it is unclear how to construct a confidence region. Furthermore, if s is larger than 3, the confidence ellipsoid given by (15.2) is not easy to visualize, describe, or interpret. A $100(1-\xi)\%$ confidence region that takes the form of an s-dimensional rectangle may be preferable. That is, we might prefer a set of s intervals, one for each function $\mathbf{c}_1^T\boldsymbol{\beta}, \ldots, \mathbf{c}_s^T\boldsymbol{\beta}$, for which the probability of simultaneous coverage equals (or exceeds) $1-\xi$. In this section, we describe several approaches for obtaining simultaneous confidence intervals for s estimable functions. We also extend these approaches to obtain simultaneous prediction intervals for s predictable functions.

For these purposes, it is helpful first to document four "facts" pertaining to confidence intervals that are seldom noted in presentations of confidence intervals made in mathematical statistics courses. Proofs of these facts, save the first, are left as exercises. For these facts, it is assumed that A and the A_i's, B and the B_i's, and τ and the τ_i's are random variables defined on a common sample space.[2] Also let $0<\xi<1$ and define $x^+ = \max\{x, 0\}$ and $x^- = \min\{x, 0\}$ for any real number x.

Fact 1 The statement, $\Pr(A_i \le \tau_i \le B_i) = 1-\xi$ for all $i = 1, \ldots, k$, neither implies nor is implied by the statement, $\Pr(A_i \le \tau_i \le B_i \text{ for all } i = 1, \ldots, k) = 1-\xi$.

Fact 2 $\Pr(A \le \tau \le B] = \Pr[c^+A + c^-B \le c\tau \le c^+B + c^-A \text{ for all } c \in \mathbb{R})$.

Fact 3 $\Pr(A_i \le \tau_i \le B_i \text{ for } i = 1, \ldots, k) \le \Pr(\sum_{i=1}^k A_i \le \sum_{i=1}^k \tau_i \le \sum_{i=1}^k B_i)$.

Fact 4

$$\Pr(A_i \le \tau_i \le B_i \text{ for } i = 1, \ldots, k)$$

$$= \Pr\left[\sum_{i=1}^k (c_i^+ A_i + c_i^- B_i) \le \sum_{i=1}^k c_i\tau_i \le \sum_{i=1}^k (c_i^+ B_i + c_i^- A_i) \text{ for all } c_i \in \mathbb{R}, i = 1, \ldots, k\right].$$

Fact 1 gets at the heart of the distinction between a set of simultaneous confidence intervals and a set of what we call "one-at-a-time" confidence intervals. Consider the special case of Fact 1 in which $k = 2$ and $\Pr(A_1 \le \tau_1 \le B_1) = \Pr(A_2 \le \tau_2 \le B_2) = 1-\xi$. Then $[A_1, B_1]$ and $[A_2, B_2]$ are $100(1-\xi)\%$ *one-at-a-time confidence intervals* for τ_1 and τ_2, meaning that each interval individually has

[2] More precisely, random variables on a common probability space.

coverage probability $1-\xi$. But their simultaneous coverage probability is generally not equal to $1-\xi$. For instance, if the bivariate random vectors $(A_1, B_1)^T$ and $(A_2, B_2)^T$ are independent, then

$$\begin{aligned}\Pr(A_1 \leq \tau_1 \leq B_1 \text{ and } A_2 \leq \tau_2 \leq B_2) &= \Pr(A_1 \leq \tau_1 \leq B_1) \cdot \Pr(A_2 \leq \tau_2 \leq B_2) \\ &= (1-\xi)^2,\end{aligned}$$

which is not equal to $1-\xi$ for $0 < \xi < 1$. Fact 2 implies that a $100(1-\xi)\%$ confidence interval for a given estimand or predictand yields (indeed, is equivalent to) an infinite collection of $100(1-\xi)\%$ simultaneous confidence intervals, one for each multiple of the original estimand or predictand. Fact 3 implies that a set of $100(1-\xi)\%$ simultaneous confidence intervals for a finite set of estimands/predictands can be used to generate (by summing endpoints) a single confidence interval for the sum of those estimands/predictands that has coverage probability *at least* $1-\xi$. Finally, Fact 4 reveals that a set of $100(1-\xi)\%$ simultaneous confidence intervals for a finite set of estimands/predictands can also be used to generate (indeed, is equivalent to) a set of $100(1-\xi)\%$ simultaneous confidence intervals for the infinite set of estimands/predictands consisting of all linear combinations of those estimands/predictands.

Until noted otherwise, take the model to be the normal Gauss–Markov model $\{\mathbf{y}, \mathbf{X}\boldsymbol{\beta}, \sigma^2\mathbf{I}\}$ with $n > p^*$, and let

$$\mathbf{C}^T\boldsymbol{\beta} = \begin{pmatrix} \mathbf{c}_1^T\boldsymbol{\beta} \\ \vdots \\ \mathbf{c}_s^T\boldsymbol{\beta} \end{pmatrix}$$

represent a vector of s estimable, but not necessarily linearly independent, functions. Let $s^* = \text{rank}(\mathbf{C}^T)$. Clearly, $s^* \leq s$. Also, let $\hat{\boldsymbol{\beta}}$ represent any solution to the normal equations $\mathbf{X}^T\mathbf{X}\boldsymbol{\beta} = \mathbf{X}^T\mathbf{y}$, and let

$$\mathbf{M} = \begin{pmatrix} m_{11} & \cdots & m_{1s} \\ \vdots & & \vdots \\ m_{s1} & \cdots & m_{ss} \end{pmatrix} = \mathbf{C}^T(\mathbf{X}^T\mathbf{X})^-\mathbf{C}.$$

Note (by Theorem 7.2.2) that $\sigma^2 m_{ij}$ is the covariance between $\mathbf{c}_i^T\hat{\boldsymbol{\beta}}$ and $\mathbf{c}_j^T\hat{\boldsymbol{\beta}}$.

The confidence intervals obtained by applying (15.1) to each of $\mathbf{c}_1^T\boldsymbol{\beta}, \ldots, \mathbf{c}_s^T\boldsymbol{\beta}$, i.e.,

$$[\,\mathbf{c}_i^T\hat{\boldsymbol{\beta}} - t_{\xi/2,n-p^*}\hat{\sigma}\sqrt{m_{ii}},\ \mathbf{c}_i^T\hat{\boldsymbol{\beta}} + t_{\xi/2,n-p^*}\hat{\sigma}\sqrt{m_{ii}}\,] \quad (i = 1, \ldots, s), \qquad (15.11)$$

each have a probability of coverage equal to $1-\xi$; hence they are one-at-a-time $100(1-\xi)\%$ confidence intervals for $\mathbf{c}_1^T\boldsymbol{\beta}, \ldots, \mathbf{c}_s^T\boldsymbol{\beta}$. However, the probability of simultaneous coverage by all s of them is in general not equal to $1-\xi$ (recall Fact 1); in fact, this probability is generally less than or equal to $1-\xi$. But by modifying the cutoff point of the $t(n-p^*)$ distribution used in the one-at-a-time confidence intervals, we can obtain intervals whose simultaneous coverage probability equals or exceeds $1-\xi$. One way to accomplish this modification is based on Bonferroni inequality.

Theorem 15.3.1 (The Bonferroni Inequality) *Let E_i $(i = 1,\ldots,k)$ be a collection of subsets of a sample space,*[3] *and let $\xi_i = Pr(E_i^c)$. Then,*

$$Pr\left(\bigcap_{i=1}^{k} E_i\right) \geq 1 - \sum_{i=1}^{k} \xi_i.$$

Proof By elementary laws of probability (including one of DeMorgan's laws),

$$\Pr\left(\bigcap_{i=1}^{k} E_i\right) = 1-\Pr\left[\left(\bigcap_{i=1}^{k} E_i\right)^c\right] = 1-\Pr\left(\bigcup_{i=1}^{k} E_i^c\right) \geq 1-\sum_{i=1}^{k}\Pr(E_i^c) = 1-\sum_{i=1}^{k}\xi_i.$$

□

Now, for real numbers $\xi_1,\ldots,\xi_s$ lying in the interval (0, 1), let E_i represent the event

$$\mathbf{c}_i^T\hat{\boldsymbol{\beta}} - t_{\xi_i/2,n-p^*}\hat{\sigma}\sqrt{m_{ii}} \leq \mathbf{c}_i^T\boldsymbol{\beta} \leq \mathbf{c}_i^T\hat{\boldsymbol{\beta}} + t_{\xi_i/2,n-p^*}\hat{\sigma}\sqrt{m_{ii}} \quad (i = 1,\ldots,s),$$

i.e., E_i is the event that the $100(1-\xi_i)\%$ one-at-a-time confidence interval for $\mathbf{c}_i^T\boldsymbol{\beta}$ contains that parametric function. By substituting ξ/s for ξ_i, Bonferroni's inequality implies that

$$\begin{aligned}\Pr[\,&\mathbf{c}_i^T\hat{\boldsymbol{\beta}} - t_{\xi/(2s),n-p^*}\hat{\sigma}\sqrt{m_{ii}} \\ &\leq \mathbf{c}_i^T\boldsymbol{\beta} \leq \mathbf{c}_i^T\hat{\boldsymbol{\beta}} + t_{\xi/(2s),n-p^*}\hat{\sigma}\sqrt{m_{ii}} \quad \text{for } i = 1,\ldots,s\,] \geq 1-\xi.\end{aligned}$$

Thus, we have the following theorem.

Theorem 15.3.2 *The rectangular region in $\mathbb{R}^s$ defined by*

$$\mathbf{c}_i^T\hat{\boldsymbol{\beta}} - t_{\xi/(2s),n-p^*}\hat{\sigma}\sqrt{m_{ii}} \leq \mathbf{c}_i^T\boldsymbol{\beta} \leq \mathbf{c}_i^T\hat{\boldsymbol{\beta}} + t_{\xi/(2s),n-p^*}\hat{\sigma}\sqrt{m_{ii}} \quad (i = 1,\ldots,s)$$

[3] More precisely, measurable subsets of a probability space.

comprises a confidence region for $\mathbf{c}_1^T\boldsymbol{\beta}, \ldots, \mathbf{c}_s^T\boldsymbol{\beta}$ *with confidence coefficient equal to or exceeding* $100(1-\xi)\%$. *These simultaneous confidence intervals are called the* $100(1-\xi)\%$ Bonferroni intervals *for* $\mathbf{c}_1^T\boldsymbol{\beta}, \ldots, \mathbf{c}_s^T\boldsymbol{\beta}$.

Plainly, the Bonferroni intervals for $\mathbf{c}_1^T\boldsymbol{\beta}, \ldots, \mathbf{c}_s^T\boldsymbol{\beta}$ are of exactly the same form as the one-at-a-time intervals for those functions but with $t_{\xi/(2s),n-p^*}$ in place of $t_{\xi/2,n-p^*}$.

The Bonferroni intervals are widely applicable to any *finite* set of estimable functions, but they are conservative, i.e., they are wider than is usually necessary, and they become ever more so as the number of estimable functions of interest (s) increases. Therefore, other less conservative (but less widely applicable) methods of constructing simultaneous confidence intervals have been proposed. The first alternative proposal we consider is the so-called Scheffé method.

The Scheffé method (Scheffé, 1959) is based on the following theorem, which may be obtained immediately from Corollary 2.16.1.1.

Theorem 15.3.3 *Let* $\mathbf{C}^T\boldsymbol{\beta}$ *be a vector of* s^* *linearly independent estimable functions. Under the normal Gauss–Markov model with* $n > p^*$,

$$\frac{(\mathbf{C}^T\hat{\boldsymbol{\beta}} - \mathbf{C}^T\boldsymbol{\beta})^T\mathbf{M}^{-1}(\mathbf{C}^T\hat{\boldsymbol{\beta}} - \mathbf{C}^T\boldsymbol{\beta})}{s^*\hat{\sigma}^2} = \left(\frac{1}{s^*\hat{\sigma}^2}\right)\max_{\mathbf{a}\neq\mathbf{0}}\left(\frac{[\mathbf{a}^T(\mathbf{C}^T\hat{\boldsymbol{\beta}} - \mathbf{C}^T\boldsymbol{\beta})]^2}{\mathbf{a}^T\mathbf{M}\mathbf{a}}\right).$$

Observe that the quantity on the left-hand side of the equality in Theorem 15.3.3 is identical to the quantity in Theorem 14.5.2 that has a central $\mathrm{F}(s^*, n-p^*)$ distribution. Thus

$$\begin{aligned}
1-\xi &= \Pr\left(\frac{(\mathbf{C}^T\hat{\boldsymbol{\beta}} - \mathbf{C}^T\boldsymbol{\beta})^T\mathbf{M}^{-1}(\mathbf{C}^T\hat{\boldsymbol{\beta}} - \mathbf{C}^T\boldsymbol{\beta})}{s^*\hat{\sigma}^2} \le F_{\xi,s^*,n-p^*}\right) \\
&= \Pr\left(\left(\frac{1}{s^*\hat{\sigma}^2}\right)\frac{[\mathbf{a}^T(\mathbf{C}^T\hat{\boldsymbol{\beta}} - \mathbf{C}^T\boldsymbol{\beta})]^2}{\mathbf{a}^T\mathbf{M}\mathbf{a}} \le F_{\xi,s^*,n-p^*} \quad \text{for all } \mathbf{a}\neq\mathbf{0}\right) \\
&= \Pr\Big(\mathbf{a}^T\mathbf{C}^T\hat{\boldsymbol{\beta}} - \sqrt{s^*F_{\xi,s^*,n-p^*}\hat{\sigma}^2(\mathbf{a}^T\mathbf{M}\mathbf{a})} \\
&\qquad \le \mathbf{a}^T\mathbf{C}^T\boldsymbol{\beta} \le \mathbf{a}^T\mathbf{C}^T\hat{\boldsymbol{\beta}} + \sqrt{s^*F_{\xi,s^*,n-p^*}\hat{\sigma}^2(\mathbf{a}^T\mathbf{M}\mathbf{a})} \quad \text{for all } \mathbf{a}\Big) \\
&= \Pr\Big((\mathbf{C}\mathbf{a})^T\hat{\boldsymbol{\beta}} - \sqrt{s^*F_{\xi,s^*,n-p^*}\hat{\sigma}^2\{(\mathbf{C}\mathbf{a})^T(\mathbf{X}^T\mathbf{X})^-(\mathbf{C}\mathbf{a})\}} \le (\mathbf{C}\mathbf{a})^T\boldsymbol{\beta} \\
&\qquad \le (\mathbf{C}\mathbf{a})^T\hat{\boldsymbol{\beta}} + \sqrt{s^*F_{\xi,s^*,n-p^*}\hat{\sigma}^2\{(\mathbf{C}\mathbf{a})^T(\mathbf{X}^T\mathbf{X})^-(\mathbf{C}\mathbf{a})\}} \quad \text{for all } \mathbf{a}\Big). \qquad (15.12)
\end{aligned}$$

The vector $(\mathbf{C}\mathbf{a})^T = \mathbf{a}^T\mathbf{C}^T$ in (15.12) is plainly an element of $\mathcal{R}(\mathbf{C}^T)$; in fact, $\{\mathbf{a}^T\mathbf{C}^T : \mathbf{a} \in \mathbb{R}^{s^*}\} = \mathcal{R}(\mathbf{C}^T)$. Thus, we have established the following theorem.

Theorem 15.3.4 *If* $\mathbf{C}^T\boldsymbol{\beta}$ *is a vector of* s^* *linearly independent estimable functions, then the probability of simultaneous coverage for the infinite collection of intervals*

$$\mathbf{c}^T\hat{\boldsymbol{\beta}} - \sqrt{s^* F_{\xi,s^*,n-p^*}\hat{\sigma}^2\{\mathbf{c}^T(\mathbf{X}^T\mathbf{X})^-\mathbf{c}\}} \leq \mathbf{c}^T\boldsymbol{\beta}$$
$$\leq \mathbf{c}^T\hat{\boldsymbol{\beta}} + \sqrt{s^* F_{\xi,s^*,n-p^*}\hat{\sigma}^2\{\mathbf{c}^T(\mathbf{X}^T\mathbf{X})^-\mathbf{c}\}} \quad \text{for all } \mathbf{c}^T \in \mathcal{R}(\mathbf{C}^T)$$

is $1-\xi$. *These are the* $100(1-\xi)\%$ Scheffé intervals *for* $\{\mathbf{c}^T\boldsymbol{\beta} : \mathbf{c}^T \in \mathcal{R}(\mathbf{C}^T)\}$.

The following properties of the Scheffé intervals are noteworthy:

1. In general, the Scheffé intervals are of the same form as the one-at-a-time t intervals except that $\sqrt{s^* F_{\xi,s^*,n-p^*}}$ replaces $t_{\xi/2,n-p^*}$ $(= \sqrt{F_{\xi,1,n-p^*}})$.
2. The simultaneous coverage probability for the Scheffé intervals is *exactly* $1-\xi$ for an *infinite* collection of estimable functions, in contrast to the simultaneous coverage probability for the Bonferroni intervals, which is *at least* $1-\xi$ for a *finite* set of functions.
3. The subset of $100(1-\xi)\%$ Scheffé intervals for $\{\mathbf{c}^T\boldsymbol{\beta} : \mathbf{c}^T \in \mathcal{R}(\mathbf{C}^T)\}$ corresponding to any finite set of functions $\mathbf{c}_1^T\boldsymbol{\beta}, \ldots, \mathbf{c}_s^T\boldsymbol{\beta}$ for which $\mathbf{c}_i^T \in \mathcal{R}(\mathbf{C}^T)$ $(i = 1, \ldots, s)$ may be used as simultaneous confidence intervals for this finite set of functions, but their simultaneous coverage probability is generally larger than the nominal level of $1-\xi$.
4. The Scheffé intervals require that the estimable functions that generate the space of functions of interest be linearly independent. If intervals are desired for a set of linearly dependent estimable functions, one must first obtain a basis for those functions and use that basis as $\mathbf{C}^T\boldsymbol{\beta}$ in Theorem 15.3.4. The Scheffé intervals generated using that basis will include intervals not only for the basis functions but also for all the functions in the original set of interest.
5. In the special case $s^* = 1$, the Scheffé intervals simplify to (and provide no more information than) the single one-at-a-time interval

$$\left(\mathbf{c}^T\hat{\boldsymbol{\beta}} - t_{\xi/2,n-p^*}\hat{\sigma}\sqrt{\mathbf{c}^T(\mathbf{X}^T\mathbf{X})^-\mathbf{c}},\ \mathbf{c}^T\hat{\boldsymbol{\beta}} + t_{\xi/2,n-p^*}\hat{\sigma}\sqrt{\mathbf{c}^T(\mathbf{X}^T\mathbf{X})^-\mathbf{c}}\right).$$

6. In the special case $s^* = p^*$, the Scheffé intervals are intervals for *all* estimable functions $\mathbf{c}^T\boldsymbol{\beta}$ [because in that case $\mathcal{R}(\mathbf{C}^T) = \mathcal{R}(\mathbf{X})$ by Theorem 2.8.5].
7. It can be shown (see the exercises) that

$$s^* F_{\xi,s^*,n-p^*} < (s^* + l)F_{\xi,s^*+l,n-p^*} \tag{15.13}$$

for every $l > 0$. It follows that the Scheffé intervals get wider as the dimensionality of the basis for $\mathcal{R}(\mathbf{C}^T)$ increases, being narrowest when $s^* = 1$ and widest when $s^* = p^*$.

8. The F test for testing $H_0 : \mathbf{C}^T\boldsymbol{\beta} = \mathbf{c}_0$ versus $H_a : \mathbf{C}^T\boldsymbol{\beta} \neq \mathbf{c}_0$ (assuming $s^* = s$) can be stated in terms of the Scheffé intervals: the F test consists of not rejecting (or "accepting," as some might say) H_0 if, for every s-vector $\mathbf{a}$, the Scheffé interval for $(\mathbf{Ca})^T\boldsymbol{\beta}$ includes $\mathbf{a}^T\mathbf{c}_0$.

Next we turn to the multivariate t method for constructing simultaneous confidence intervals. Recall from Sect. 14.6 that the multivariate t distribution $t(r, \nu, \mathbf{R})$ is the distribution of $\mathbf{t} = (t_i) = \left(\frac{w}{\nu}\right)^{-\frac{1}{2}}\mathbf{x}$ where w and $\mathbf{x}$ are independently distributed as $\chi^2(\nu)$ and $\mathrm{N}_r(\mathbf{0}, \mathbf{R})$, respectively, and $\mathbf{R}$ is a correlation matrix. We may define a particular scalar cutoff point involving this distribution, useful for simultaneous confidence interval estimation, as follows. Let $t_{\xi/2,r,\nu,\mathbf{R}}$ be the real number for which

$$1 - \xi = \Pr(-t_{\xi/2,r,\nu,\mathbf{R}} \le t_i \le t_{\xi/2,r,\nu,\mathbf{R}} \quad \text{for } i = 1, \ldots, r),$$

or equivalently by

$$1 - \xi = \Pr(\max_{i=1,\ldots,r} |t_i| \le t_{\xi/2,r,\nu,\mathbf{R}}).$$

This cutoff point may be obtained using the `qmvt` function in the `mvtnorm` package in `R`, with the `tail="both"` option specified. It should be noted that when $r > 2$, values returned by repeated calls to this function may vary slightly, due to numerical methods involved in their calculation.

Under the normal Gauss–Markov model, the s-vector having ith element

$$\frac{(\mathbf{c}_i^T\hat{\boldsymbol{\beta}} - \mathbf{c}_i^T\boldsymbol{\beta})/(\sigma\sqrt{m_{ii}})}{\sqrt{\frac{(n-p^*)\hat{\sigma}^2/\sigma^2}{n-p^*}}} = \frac{\mathbf{c}_i^T\hat{\boldsymbol{\beta}} - \mathbf{c}_i^T\boldsymbol{\beta}}{\hat{\sigma}\sqrt{m_{ii}}}$$

is distributed as $\mathrm{t}(s, n - p^*, \mathbf{R})$, where

$$\mathbf{R} = \mathrm{var}\left(\frac{\mathbf{c}_1^T\hat{\boldsymbol{\beta}}}{\sigma\sqrt{m_{11}}}, \ldots, \frac{\mathbf{c}_s^T\hat{\boldsymbol{\beta}}}{\sigma\sqrt{m_{ss}}}\right)^T.$$

Consequently, we have the following theorem.

Theorem 15.3.5 *The probability of simultaneous coverage for the s intervals*

$$\mathbf{c}_i^T\hat{\boldsymbol{\beta}} - t_{\xi/2,s,n-p^*,\mathbf{R}}\hat{\sigma}\sqrt{m_{ii}} \le \mathbf{c}_i^T\boldsymbol{\beta} \le \mathbf{c}_i^T\hat{\boldsymbol{\beta}} + t_{\xi/2,s,n-p^*,\mathbf{R}}\hat{\sigma}\sqrt{m_{ii}} \quad (i = 1, \ldots, s)$$

is $1 - \xi$. *These are the* $100(1 - \xi)\%$ multivariate t intervals *for* $\mathbf{c}_1^T\boldsymbol{\beta}, \ldots, \mathbf{c}_s^T\boldsymbol{\beta}$.

The following properties of the multivariate t intervals are noteworthy:

1. The multivariate t intervals are of the same form as the one-at-a-time intervals except that $t_{\xi/2,s,n-p^*,\mathbf{R}}$ replaces $t_{\xi/2,n-p^*}$.
2. In contrast to the Bonferroni and Scheffé intervals, the multivariate t intervals have simultaneous coverage probability equal to *exactly* $1 - \xi$ for a *finite* set of intervals.
3. In contrast to the Scheffé approach, the multivariate t approach does not require the estimable functions of interest to be linearly independent.
4. In the past, one practical shortcoming of the multivariate t method was that tables of cutoff points $t_{\xi/2,s,n-p^*,\mathbf{R}}$ were available for very few $\mathbf{R}$ matrices. One way that this problem was "finessed" used the fact, shown by Sidák (1968), that for any correlation matrix $\mathbf{R}$, $t_{\xi/2,r,\nu,\mathbf{R}} \leq t_{\xi/2,r,\nu,\mathbf{I}}$. This implies that, in applying the multivariate t method, conservative intervals can be obtained by replacing $\mathbf{R}$ with $\mathbf{I}$ or, in effect, by proceeding as though the estimators $\mathbf{c}_1^T\hat{\boldsymbol{\beta}}, \ldots, \mathbf{c}_s^T\hat{\boldsymbol{\beta}}$ are uncorrelated. Tables for $t_{\xi/2,r,\nu,\mathbf{I}}$ are available in some books, but nowadays, because $t_{\xi/2,r,\nu,\mathbf{R}}$ be obtained for any correlation matrix $\mathbf{R}$ using the aforementioned package in R, the conservative approach is no longer needed (unless one does not want to bother to determine $\mathbf{R}$).
5. It can be shown (see the exercises) that

$$t_{\xi/2,s,\nu,\mathbf{I}} \leq \sqrt{sF_{\xi,s,\nu}}. \tag{15.14}$$

In tandem with the fourth property listed above, this implies that the multivariate t intervals for a specific finite set of linearly independent estimable functions are narrower than the Scheffé intervals for those functions.
6. By Fact 4,

$$\begin{aligned}
1 - \xi &= \Pr(\mathbf{c}_i^T\hat{\boldsymbol{\beta}} - t_{\xi/2,s,n-p^*,\mathbf{R}}\hat{\sigma}\sqrt{m_{ii}} \leq \mathbf{c}_i^T\boldsymbol{\beta} \leq \mathbf{c}_i^T\hat{\boldsymbol{\beta}} \\
&\quad + t_{\xi/2,s,n-p^*,\mathbf{R}}\hat{\sigma}\sqrt{m_{ii}} \quad \text{for } i = 1, \ldots, s) \\
&= \Pr\Big[\sum_{i=1}^{s} c_i\mathbf{c}_i^T\hat{\boldsymbol{\beta}} - t_{\xi/2,s,n-p^*,\mathbf{R}}\hat{\sigma}\sum_{i=1}^{s}|c_i|\sqrt{m_{ii}} \leq \sum_{i=1}^{s} c_i\mathbf{c}_i^T\boldsymbol{\beta} \\
&\quad \leq \sum_{i=1}^{s} c_i\mathbf{c}_i^T\hat{\boldsymbol{\beta}} + t_{\xi/2,s,n-p^*,\mathbf{R}}\hat{\sigma}\sum_{i=1}^{s}|c_i|\sqrt{m_{ii}} \quad \text{for all } c_1, \ldots, c_s \in \mathbb{R}\Big].
\end{aligned}$$

The upshot of this is that the (finite set of) multivariate t intervals for s estimable functions $\mathbf{c}_1^T\boldsymbol{\beta}, \ldots, \mathbf{c}_s^T\boldsymbol{\beta}$ can be used to generate (an infinite set of) simultaneous intervals for all $\mathbf{c}^T\boldsymbol{\beta}$ such that $\mathbf{c}^T \in \mathcal{R}\begin{pmatrix} \mathbf{c}_1^T \\ \vdots \\ \mathbf{c}_s^T \end{pmatrix}$. (Note, however, that a sufficient condition for the intervals in the infinite set to be wider than the widest of the intervals for the original functions $\mathbf{c}_1^T\boldsymbol{\beta},\ldots,\mathbf{c}_s^T\boldsymbol{\beta}$ is: $|c_i| \geq 1$ for all i.) This extension of the multivariate t method can be viewed as a competitor to the Scheffé method for use when intervals are desired for an infinite number of estimable functions, hence we call it the "infinite-collection extension of the multivariate t method." The same idea could also be applied to generate an infinite collection of intervals from Bonferroni intervals for a finite number of estimable functions.

The final simultaneous confidence interval procedure we consider is known as Tukey's method, named after its originator (Tukey, 1949a). This method applies specifically to the setting where $\mathbf{c}_1^T\boldsymbol{\beta},\ldots,\mathbf{c}_s^T\boldsymbol{\beta}$ represent all possible differences among k linearly independent estimable functions $\psi_1,\ldots,\psi_k$. (This implies that $s = \frac{k(k-1)}{2}$, and that $s > s^*$ if $k \geq 3$.) One setting where this might be appropriate is the one-factor model, with ψ_i taken to be $\mu + \alpha_i$ for each i, so that each $\mathbf{c}^T\boldsymbol{\beta}$ is of the form $\alpha_i - \alpha_{i'}$.

Tukey's method is based on the studentized range distribution, defined in Sect. 14.6 as the distribution of

$$\frac{\max_{i=1,\ldots,k} z_i - \min_{i=1,\ldots,k} z_i}{\sqrt{w/\nu}}$$

where w and $\mathbf{z} = (z_1,\ldots,z_k)^T$ are distributed independently as $\chi^2(\nu)$ and $\mathrm{N}(\mathbf{0},\mathbf{I})$, respectively. Define $q^*_{\xi,k,\nu}$ by

$$\Pr\left(\frac{\max_{i=1,\ldots,k} z_i - \min_{i=1,\ldots,k} z_i}{\sqrt{w/\nu}} > q^*_{\xi,k,\nu}\right) = \xi.$$

These cutoff points may be obtained using the `qtukey` function in R.

Because

$$\frac{1}{\sigma c}\begin{pmatrix} \hat{\psi}_1 - \psi_1 \\ \vdots \\ \hat{\psi}_k - \psi_k \end{pmatrix} \sim \mathrm{N}_k(\mathbf{0},\mathbf{I}) \quad \text{and} \quad \frac{(n-p^*)\hat{\sigma}^2}{\sigma^2} \sim \chi^2(n-p^*),$$

and these two quantities are independent, we have

$$1-\xi = \Pr\left[\frac{\max_{i=1,\ldots,k}\left(\frac{\hat{\psi}_i-\psi_i}{\sigma c}\right) - \min_{i=1,\ldots,k}\left(\frac{\hat{\psi}_i-\psi_i}{\sigma c}\right)}{\sqrt{\frac{(n-p^*)\hat{\sigma}^2/\sigma^2}{n-p^*}}} \leq q^*_{\xi,k,n-p^*}\right]$$

$$= \Pr\left[\frac{\max_{i=1,\ldots,k}(\hat{\psi}_i - \psi_i) - \min_{i=1,\ldots,k}(\hat{\psi}_i - \psi_i)}{\hat{\sigma} c} \leq q^*_{\xi,k,n-p^*}\right]$$

$$= \Pr\left[\max_{i=1,\ldots,k} (\hat{\psi}_i - \psi_i) - \min_{i=1,\ldots,k} (\hat{\psi}_i - \psi_i) \le c\hat{\sigma} q^*_{\xi,k,n-p^*} \right]$$

$$= \Pr[\, |(\hat{\psi}_i - \psi_i) - (\hat{\psi}_{i'} - \psi_{i'})| \le c\hat{\sigma} q^*_{\xi,k,n-p^*} \quad \text{for } i' \neq i = 1, \ldots, k\,]$$

$$= \Pr[\, (\hat{\psi}_i - \hat{\psi}_{i'}) - c\hat{\sigma} q^*_{\xi,k,n-p^*} \le (\psi_i - \psi_{i'}) \le (\hat{\psi}_i - \hat{\psi}_{i'})$$

$$+ c\hat{\sigma} q^*_{\xi,k,n-p^*} \quad \text{for } i' \neq i = 1, \ldots, k\,].$$

Consequently, we have the following theorem.

Theorem 15.3.6 *The probability of simultaneous coverage for the* $\frac{k(k-1)}{2}$ *intervals*

$$(\hat{\psi}_i - \hat{\psi}_{i'}) - c\hat{\sigma} q^*_{\xi,k,n-p^*} \le (\psi_i - \psi_{i'}) \le (\hat{\psi}_i - \hat{\psi}_{i'}) + c\hat{\sigma} q^*_{\xi,k,n-p^*} \quad (i' \neq i = 1, \ldots, k)$$

is $1-\xi$. *These are the* $100(1-\xi)\%$ Tukey intervals *for* $\{\psi_i - \psi_{i'} : i' \neq i = 1, \ldots, k\}$.

The following are noteworthy facts about the Tukey intervals:

1. Although the Bonferroni, Scheffé, and multivariate t methods could also be used to obtain simultaneous confidence intervals for the $\frac{k(k-1)}{2}$ differences $\psi_i - \psi_{i'}$ $(i' \neq i = 1, \ldots, k)$, the Tukey intervals are known to be narrowest for these specific functions.
2. Tukey's method can be extended to give simultaneous confidence intervals for all functions of the form $\sum_{i=1}^k d_i \psi_i$ with $\sum_{i=1}^k d_i = 0$, i.e., for all contrasts (rather than merely all differences) among the linearly independent estimable functions $\psi_1, \ldots, \psi_k$; see the exercises.

Simultaneous confidence intervals for the differences $\psi_i - \psi_{i'}$ $(i' \neq i = 1, \ldots, k)$, whether obtained using the Bonferroni, Scheffé, multivariate t, or Tukey method, can be used to divide $\{\psi_1, \ldots, \psi_k\}$ into groups, within which the intervals for the differences $\psi_i - \psi_{i'}$ include zero. These groups need not be mutually exclusive. Two of the functions in $\{\psi_1, \ldots, \psi_k\}$ are said to differ significantly if they do not appear together in any group. Comparisons based on these groupings are called *multiple comparisons*.

The simultaneous confidence intervals presented here were derived under the assumption that the model is a normal Gauss–Markov model, but their extension to the normal positive definite Aitken model $\{\mathbf{y}, \mathbf{X}\boldsymbol{\beta}, \sigma^2\mathbf{W}\}$ is straightforward; one merely replaces $\hat{\boldsymbol{\beta}}$ with $\tilde{\boldsymbol{\beta}}$, $\hat{\sigma}^2$ with $\tilde{\sigma}^2$, $\mathbf{X}^T\mathbf{X}$ with $\mathbf{X}^T\mathbf{W}^{-1}\mathbf{X}$, m_{ii} with the ith diagonal element of $\mathbf{C}^T(\mathbf{X}^T\mathbf{W}^{-1}\mathbf{X})^-\mathbf{C}$, and $\hat{\psi}_1, \ldots, \hat{\psi}_k$ with $\tilde{\psi}_1, \ldots, \tilde{\psi}_k$. Verifying this is left as an exercise.

So far in this section, we have presented only simultaneous *confidence* intervals (for estimable functions of $\boldsymbol{\beta}$). As a consequence of Theorem 14.5.5, however, very similar approaches can be applied to the problem of obtaining simultaneous *prediction* intervals for a set of s predictable functions $\mathbf{c}_1^T\boldsymbol{\beta} +$

$u_1, \ldots, \mathbf{c}_s^T\boldsymbol{\beta} + u_s$ under the normal prediction-extended positive definite Aitken model $\left\{\begin{pmatrix}\mathbf{y}\\ \mathbf{u}\end{pmatrix}, \begin{pmatrix}\mathbf{X}\boldsymbol{\beta}\\ \mathbf{0}\end{pmatrix}, \sigma^2\begin{pmatrix}\mathbf{W} & \mathbf{K}\\ \mathbf{K}^T & \mathbf{H}\end{pmatrix}\right\}$. For all but the Scheffé intervals, such intervals are obtained merely by replacing $\mathbf{c}_i^T\hat{\boldsymbol{\beta}}$ with $\mathbf{c}_i^T\tilde{\boldsymbol{\beta}} + \tilde{u}_i$, $\hat{\sigma}$ with $\tilde{\sigma}$, and $\mathbf{M} = (m_{ij})$ with $\mathbf{Q} = (q_{ij})$ in the expressions for the simultaneous confidence intervals. Thus, the $100(1-\xi)\%$ one-at-a-time prediction intervals for $\mathbf{c}_i^T\boldsymbol{\beta} + u_i$ $(i = 1, \ldots, s)$ are

$$[\,(\mathbf{c}_i^T\tilde{\boldsymbol{\beta}} + \tilde{u}_i) - t_{\xi/2,n-p^*}\tilde{\sigma}\sqrt{q_{ii}},\ (\mathbf{c}_i^T\tilde{\boldsymbol{\beta}} + \tilde{u}_i) + t_{\xi/2,n-p^*}\tilde{\sigma}\sqrt{q_{ii}}\,] \quad (i = 1, \ldots, s), \tag{15.15}$$

and the $100(1-\xi)\%$ Bonferroni simultaneous prediction intervals and $100(1-\xi)\%$ multivariate t simultaneous prediction intervals for the same functions are obtained merely by replacing $t_{\xi/2,n-p^*}$ in (15.15) with $t_{\xi/(2s),n-p^*}$ and $t_{\xi/2,s,n-p^*,\mathbf{R}}$, respectively, where

$$\mathbf{R} = \mathrm{var}\left(\frac{\mathbf{c}_1^T\tilde{\boldsymbol{\beta}} + \tilde{u}_1 - u_1}{\sigma\sqrt{q_{11}}}, \ldots, \frac{\mathbf{c}_s^T\tilde{\boldsymbol{\beta}} + \tilde{u}_s - u_s}{\sigma\sqrt{q_{ss}}}\right)^T.$$

Furthermore, if: $\mathbf{c}_1^T\boldsymbol{\beta} + u_1, \ldots, \mathbf{c}_s^T\boldsymbol{\beta} + u_s$ represent all possible differences among k predictable functions $\psi_1, \ldots, \psi_k$; $\tilde{\psi}_1, \ldots, \tilde{\psi}_k$ are the BLUPs of those functions; $\mathbf{Q}$ is positive definite (as is the case under either condition of Theorem 13.2.2b with s in that theorem taken to coincide with k here); and $\mathrm{var}[(\tilde{\psi}_1 - \psi_1, \ldots, \tilde{\psi}_k - \psi_k)^T] = c^2\sigma^2\mathbf{I}$ for some $c > 0$; then the $100(1-\xi)\%$ Tukey simultaneous prediction intervals for those differences are

$$(\tilde{\psi}_i - \tilde{\psi}_{i'}) \pm c\tilde{\sigma}q^*_{\xi,k,n-p^*}.$$

We may obtain Scheffé-based simultaneous prediction intervals using the following theorem, the proof of which is left as an exercise.

Theorem 15.3.7 *Under the normal prediction-extended positive definite Aitken model*

$$\left\{\begin{pmatrix}\mathbf{y}\\ \mathbf{u}\end{pmatrix}, \begin{pmatrix}\mathbf{X}\boldsymbol{\beta}\\ \mathbf{0}\end{pmatrix}, \sigma^2\begin{pmatrix}\mathbf{W} & \mathbf{K}\\ \mathbf{K}^T & \mathbf{H}\end{pmatrix}\right\}$$

with $n > p^*$, if $\mathbf{C}^T\boldsymbol{\beta} + \mathbf{u}$ is an s-vector of predictable functions and $\mathbf{Q}$ is positive definite (as is the case under either condition of Theorem 13.2.2b), then the probability of simultaneous coverage of the infinite collection of intervals for $\{\mathbf{a}^T(\mathbf{C}^T\boldsymbol{\beta} + \mathbf{u}) : \mathbf{a} \in \mathbb{R}^s\}$ given by

$$\mathbf{a}^T(\mathbf{C}^T\tilde{\boldsymbol{\beta}} + \tilde{\mathbf{u}}) \pm \sqrt{sF_{\xi,s,n-p^*}\tilde{\sigma}^2(\mathbf{a}^T\mathbf{Q}\mathbf{a})} \quad \textit{for all } \mathbf{a}$$

is $1-\xi$.

There is an important practical difference between Scheffé confidence intervals and Scheffé prediction intervals. Recall that if the matrix $\mathbf{C}^T$ in (15.12) has p^* rows and if those rows are linearly independent, then the Scheffé confidence intervals given by Theorem 15.3.4 include an interval for every estimable function of $\boldsymbol{\beta}$ because every such function can be expressed as a linear combination of the elements of $\mathbf{C}^T\boldsymbol{\beta}$. In contrast, although the Scheffé prediction intervals given by Theorem 15.3.7 include an interval for every linear combination of the predictable functions $\mathbf{c}_1^T\boldsymbol{\beta} + u_1, \ldots, \mathbf{c}_s^T\boldsymbol{\beta} + u_s$, they do not include an interval for every predictable function, even if $s = p^*$ and $\mathbf{c}_1^T, \ldots, \mathbf{c}_s^T$ are linearly independent. For example, there is no interval for a "new" predictable function $\mathbf{c}_{s+1}^T\boldsymbol{\beta} + u_{s+1}$ among those given by Theorem 15.3.7, even if $\mathbf{c}_{s+1}^T$ is a linear combination of $\mathbf{c}_1^T, \ldots, \mathbf{c}_s^T$.

Intervals for estimable functions *and* predictable functions that have a specified simultaneous coverage probability may also be considered; an example is considered in the exercises.

Example 15.3-1. Simultaneous Confidence Intervals and Simultaneous Prediction Intervals in Normal Gauss–Markov Simple Linear Regression

Consider the normal Gauss–Markov simple linear regression model. According to Example 15.1-1, $100(1-\xi)\%$ one-at-a-time intervals for the slope and intercept are

$$\hat{\beta}_2 \pm t_{\xi/2,n-2}\hat{\sigma}/\sqrt{SXX}$$

and

$$\hat{\beta}_1 \pm t_{\xi/2,n-2}\hat{\sigma}\sqrt{(1/n) + (\bar{x}^2/SXX)},$$

respectively. Suppose that we want $100(1-\xi)\%$ simultaneous confidence intervals for these two parameters. The Bonferroni approach yields the intervals

$$\hat{\beta}_2 \pm t_{\xi/4,n-2}\hat{\sigma}/\sqrt{SXX}$$

and

$$\hat{\beta}_1 \pm t_{\xi/4,n-2}\hat{\sigma}\sqrt{(1/n) + (\bar{x}^2/SXX)},$$

while the multivariate t intervals are

$$\hat{\beta}_2 \pm t_{\xi/2,2,n-2,\mathbf{R}}\hat{\sigma}/\sqrt{SXX}$$

and

$$\hat{\beta}_1 \pm t_{\xi/2,2,n-2,\mathbf{R}}\hat{\sigma}\sqrt{(1/n) + (\bar{x}^2/SXX)}$$

where, from Example 7.2-1

$$\mathbf{R} = \begin{pmatrix} 1 & \frac{-\bar{x}}{\sqrt{(SXX/n)+\bar{x}^2}} \\ \frac{-\bar{x}}{\sqrt{(SXX/n)+\bar{x}^2}} & 1 \end{pmatrix}.$$

Now suppose that we want $100(1-\xi)\%$ simultaneous confidence intervals for $E(y|x) = \beta_1 + \beta_2 x$ over a finite collection of specified values of x, i.e., for $x \in \{x_{n+1}, \ldots, x_{n+s}\}$ (where the elements in this set are distinct but some or all of them could coincide with values of x in the dataset). Again from Example 15.1-1, one-at-a-time $100(1-\xi)\%$ confidence intervals for these quantities are

$$(\hat{\beta}_1 + \hat{\beta}_2 x_{n+i}) \pm t_{\xi/2,n-2}\hat{\sigma}\sqrt{\frac{1}{n} + \frac{(x_{n+i}-\bar{x})^2}{SXX}} \quad (i = 1, \ldots, s).$$

Thus, $100(1-\xi)\%$ Bonferroni simultaneous confidence intervals for these quantities are

$$(\hat{\beta}_1 + \hat{\beta}_2 x_{n+i}) \pm t_{\xi/(2s),n-2}\hat{\sigma}\sqrt{\frac{1}{n} + \frac{(x_{n+i}-\bar{x})^2}{SXX}} \quad (i = 1, \ldots, s).$$

The analogous $100(1-\xi)\%$ multivariate t simultaneous confidence intervals are

$$(\hat{\beta}_1 + \hat{\beta}_2 x_{n+i}) \pm t_{\xi/2,s,n-2,\mathbf{R}}\hat{\sigma}\sqrt{\frac{1}{n} + \frac{(x_{n+i}-\bar{x})^2}{SXX}} \quad (i = 1, \ldots, s),$$

where

$$\mathbf{R} = \mathrm{var}\left(\frac{\hat{\beta}_1 + \hat{\beta}_2 x_{n+1}}{\sigma\sqrt{(1/n) + (x_{n+1}-\bar{x})^2/SXX}}, \ldots, \frac{\hat{\beta}_1 + \hat{\beta}_2 x_{n+s}}{\sigma\sqrt{(1/n) + (x_{n+s}-\bar{x})^2/SXX}}\right)^T.$$

Scheffé intervals that include intervals for the quantities described in this paragraph are saved for another example to be presented later in this section.

Suppose instead that we want $100(1-\xi)\%$ simultaneous prediction intervals for unobserved y-values y_{n+i} $(= \beta_1 + \beta_2 x_{n+i} + e_{n+i})$ $(i = 1, \ldots, s)$ under a prediction-extended normal Gauss–Markov model. Building on Example 15.1-3, we find that $100(1-\xi)\%$ Bonferroni simultaneous prediction intervals for $\{y_{n+1}, \ldots, y_{n+s}\}$ are

$$(\hat{\beta}_1 + \hat{\beta}_2 x_{n+i}) \pm t_{\xi/(2s),n-2}\hat{\sigma}\sqrt{1 + \frac{1}{n} + \frac{(x_{n+i}-\bar{x})^2}{SXX}} \quad (i = 1, \ldots, s).$$

The corresponding multivariate t prediction intervals are

$$(\hat{\beta}_1 + \hat{\beta}_2 x_{n+i}) \pm t_{\xi/2,s,n-2,\mathbf{R}}\hat{\sigma}\sqrt{1 + \frac{1}{n} + \frac{(x_{n+i} - \bar{x})^2}{SXX}} \quad (i = 1, \ldots, s),$$

where

$$\mathbf{R} = \text{var}\left(\frac{\hat{\beta}_1 + \hat{\beta}_2 x_{n+1} - y_{n+1}}{\sigma\sqrt{1 + (1/n) + (x_{n+1} - \bar{x})^2/SXX}}, \ldots, \frac{\hat{\beta}_1 + \hat{\beta}_2 x_{n+s} - y_{n+s}}{\sigma\sqrt{1 + (1/n) + (x_{n+s} - \bar{x})^2/SXX}}\right)^T.$$

Determination of the elements of **R** for all of the intervals presented in this example is left as an exercise. ■

Example 15.3-2. Simultaneous Confidence Intervals for All Level Mean Differences in the Normal Gauss–Markov One-Factor Model

Consider the normal Gauss–Markov one-factor model and suppose initially that the data are balanced, with r observations at each of the q levels of the factor. Thus $n = qr$. Suppose that we wish to obtain $100(1 - \xi)\%$ simultaneous confidence intervals for all level mean differences $\alpha_i - \alpha_{i'}$ $(i' > i = 1, \ldots, q)$. In Example 15.1-2, we gave expressions for the $100(1 - \xi)\%$ one-at-a-time intervals for these functions, which specialize here (because of the assumed balance) to

$$(\bar{y}_{i\cdot} - \bar{y}_{i'\cdot}) \pm t_{\xi/2,q(r-1)}\hat{\sigma}\sqrt{2/r} \quad (i' > i = 1, \ldots, q).$$

Because there are $q(q - 1)/2$ distinct level mean differences, the $100(1 - \xi)\%$ Bonferroni intervals are

$$(\bar{y}_{i\cdot} - \bar{y}_{i'\cdot}) \pm t_{\xi/[q(q-1)],q(r-1)}\hat{\sigma}\sqrt{2/r} \quad (i' > i = 1, \ldots, q)$$

and the $100(1 - \xi)\%$ multivariate t intervals are

$$(\bar{y}_{i\cdot} - \bar{y}_{i'\cdot}) \pm t_{\xi/2,q(q-1)/2,q(r-1),\mathbf{R}}\hat{\sigma}\sqrt{2/r} \quad (i' > i = 1, \ldots, q),$$

where

$$\mathbf{R} = \text{var}\left(\frac{\bar{y}_{1\cdot} - \bar{y}_{2\cdot}}{\sigma\sqrt{2/r}}, \frac{\bar{y}_{1\cdot} - \bar{y}_{3\cdot}}{\sigma\sqrt{2/r}}, \ldots, \frac{\bar{y}_{(q-1)\cdot} - \bar{y}_{q\cdot}}{\sigma\sqrt{2/r}}\right)^T.$$

A second way to implement the multivariate t method is to begin with the $100(1 - \xi)\%$ multivariate t simultaneous confidence intervals for the $q - 1$ differences $\alpha_1 - \alpha_i$ $(i = 2, \ldots, q)$ only and then use the infinite-collection extension to construct intervals for all linear combinations of those differences, which of course include intervals for the remaining differences. The intervals for $\alpha_1 - \alpha_i$ $(i = 2, \ldots, q)$ are

$$(\bar{y}_{1\cdot} - \bar{y}_{i\cdot}) \pm t_{\xi/2,q-1,q(r-1),\mathbf{R}^*}\hat{\sigma}\sqrt{2/r} \quad (i = 2, \ldots, q),$$

where

$$\mathbf{R}^* = \begin{pmatrix} 1 & 0.5 & 0.5 & \cdots & 0.5 \\ 0.5 & 1 & 0.5 & \cdots & 0.5 \\ 0.5 & 0.5 & 1 & \cdots & 0.5 \\ \vdots & \vdots & \vdots & \ddots & \vdots \\ 0.5 & 0.5 & 0.5 & \cdots & 1 \end{pmatrix}.$$

Then, for any $i' \neq i = 2, \ldots, q$, $\alpha_i - \alpha_{i'} = (\alpha_1 - \alpha_{i'}) - (\alpha_1 - \alpha_i)$. Thus, the intervals for the remaining differences are

$$(\bar{y}_{i\cdot} - \bar{y}_{i'\cdot}) \pm 2t_{\xi/2,q-1,q(r-1),\mathbf{R}^*}\hat{\sigma}\sqrt{2/r} \quad (i' > i = 2, \ldots, q).$$

The simultaneous coverage probability of the intervals for all the differences is exactly $1 - \xi$. This set of intervals consists of $q - 1$ intervals of constant width and $[q(q - 1)/2] - (q - 1)$ intervals also of constant width, but the latter intervals are twice as wide as the former. This implementation of the multivariate t method may be of interest when there is a single factor level (such as a "control" treatment in an experimental design in which the factor levels are treatments) whose differences with other levels are of greater interest than the differences among other pairs of levels.

As for the Scheffé intervals, although the $q(q - 1)/2$ level mean differences are linearly dependent, any basis for them consists of $q - 1$ linearly independent differences. Thus, the Scheffé intervals are

$$(\bar{y}_{i\cdot} - \bar{y}_{i'\cdot}) \pm \sqrt{(q - 1)F_{\xi,q-1,q(r-1)}}\hat{\sigma}\sqrt{2/r} \quad (i' > i = 1, \ldots, q).$$

For the Tukey intervals, we take $\psi_i = \mu + \alpha_i$ and $\hat{\psi}_i = \bar{y}_{i\cdot}$ $(i = 1, \ldots, q)$. Because

$$\text{var}\begin{pmatrix} \bar{y}_{1\cdot} \\ \vdots \\ \bar{y}_{q\cdot} \end{pmatrix} = (\sigma^2/r)\mathbf{I}_q,$$

the Tukey intervals are

$$(\bar{y}_{i\cdot} - \bar{y}_{i'\cdot}) \pm q^*_{\xi,q,q(r-1)}\hat{\sigma}\sqrt{1/r} \quad (i' > i = 1, \ldots, q).$$

Expected squared lengths of 95% simultaneous confidence intervals of all of these types are given in the table below, for general q and r and also for the specific case $q = 5$ and $r = 4$. For this specific case, expressions for the variances and covariances between differences in level means given in Example 7.2-2 yield

$$\mathbf{R} = \begin{pmatrix} 1 & 0.5 & 0.5 & 0.5 & -0.5 & -0.5 & -0.5 & 0 & 0 & 0 \\ & 1 & 0.5 & 0.5 & 0.5 & 0 & 0 & -0.5 & -0.5 & 0 \\ & & 1 & 0.5 & 0 & 0.5 & 0 & 0.5 & 0 & -0.5 \\ & & & 1 & 0 & 0 & 0.5 & 0 & 0.5 & 0.5 \\ & & & & 1 & 0.5 & 0.5 & -0.5 & -0.5 & 0 \\ & & & & & 1 & 0.5 & 0.5 & 0 & -0.5 \\ & & & & & & 1 & 0 & 0.5 & 0.5 \\ & & & & & & & 1 & 0.5 & -0.5 \\ & & & & & & & & 1 & 0.5 \\ & & & & & & & & & 1 \end{pmatrix}.$$

Intervals	Expected squared length/σ^2 General case		$q = 5, r = 4$
One-at-a-time	$(8/r)t^2_{0.025,q(r-1)}$		9.09
Bonferroni	$(8/r)t^2_{0.05/[q(q-1)],q(r-1)}$		21.60
Scheffé	$(8/r)(q-1)F_{0.05,q-1,q(r-1)}$		24.44
Multivariate t (exact)	$(8/r)t^2_{0.025,q(q-1)/2,q(r-1),\mathbf{R}}$		19.07
Multivariate t (conservative)	$(8/r)t^2_{0.025,q(q-1)/2,q(r-1),\mathbf{I}}$		20.83
Multivariate t (infinite-collection)	$(8/r)t^2_{0.025,q-1,q(r-1),\mathbf{R}^*}$	for $\alpha_1 - \alpha_i$ $(i = 2, \ldots, q)$	14.88
	$(32/r)t^2_{0.025,q-1,q(r-1),\mathbf{R}^*}$	for others	59.54
Tukey	$(4/r)q^{*2}_{0.05,q,q(r-1)}$		19.07

Extensions to the case of unbalanced data are left as an exercise. ■

Example 15.3-3. Simultaneous Confidence "Bands" in Normal Full-Rank Multiple and Simple Linear Regression

Consider the normal Gauss–Markov full-rank multiple regression model with an intercept, in its centered parameterization. The model equation is

$$y_i = \beta_1 + \beta_2(x_{i2} - \bar{x}_2) + \cdots + \beta_p(x_{ip} - \bar{x}_p) + e_i$$
$$= (1, \mathbf{x}_{i,-1}^T)\begin{pmatrix} \beta_1 \\ \boldsymbol{\beta}_{-1} \end{pmatrix} + e_i \quad (i = 1, \ldots, n),$$

or equivalently,

$$\mathbf{y} = \mathbf{1}_n\beta_1 + \mathbf{X}_{-1}\boldsymbol{\beta}_{-1} + \mathbf{e},$$

where $\mathbf{x}_{i,-1}^T = (x_{i2} - \bar{x}_2, x_{i3} - \bar{x}_3, \ldots, x_{ip} - \bar{x}_p)$ and

$$\mathbf{X}_{-1} = \begin{pmatrix} \mathbf{x}_{1,-1}^T \\ \vdots \\ \mathbf{x}_{n,-1}^T \end{pmatrix}.$$

Suppose that we want $100(1-\xi)\%$ simultaneous confidence intervals for the infinite number of quantities $\{\mathbf{x}^T\boldsymbol{\beta} \equiv (1, \mathbf{x}_{-1}^T)\boldsymbol{\beta} : \mathbf{x}_{-1} \in \mathbb{R}^{p-1}\}$. Of the methods presented herein, only Scheffé's method is directly applicable to this problem. By choosing $\mathbf{C}^T$ in Theorem 15.3.4 to be $\mathbf{I}_p$, we see that the $100(1-\xi)\%$ Scheffé intervals for $\{\mathbf{c}^T\boldsymbol{\beta} = (c_1, \mathbf{x}_{-1}^T)\boldsymbol{\beta} : \text{all } c_1 \in \mathbb{R},\ \mathbf{x}_{-1} \in \mathbb{R}^{p-1}\}$ comprise a simultaneous confidence region given by

$$\left\{\mathbf{c}^T\hat{\boldsymbol{\beta}} \pm \sqrt{pF_{\xi,p,n-p}\hat{\sigma}^2\mathbf{c}^T(\mathbf{X}^T\mathbf{X})^{-1}\mathbf{c}} \text{ for all } c_1 \in \mathbb{R},\ \mathbf{x}_{-1} \in \mathbb{R}^{p-1}\right\}. \quad (15.16)$$

This infinite collection of intervals includes the subcollection

$$\left\{(\hat{\beta}_1 + \mathbf{x}_{-1}^T\hat{\boldsymbol{\beta}}_{-1}) \pm \sqrt{pF_{\xi,p,n-p}\hat{\sigma}^2[(1/n) + \mathbf{x}_{-1}^T(\mathbf{X}_{-1}^T\mathbf{X}_{-1})^{-1}\mathbf{x}_{-1}]} \text{ for all } \mathbf{x}_{-1} \in \mathbb{R}^{p-1}\right\} \quad (15.17)$$

comprising intervals for the desired quantities, $\{\mathbf{x}^T\boldsymbol{\beta} : \mathbf{x}_{-1} \in \mathbb{R}^{p-1}\}$. Observe that the domain for the intervals in the subcollection is $(p-1)$-dimensional rather than p-dimensional because the first element of $\mathbf{x}^T$ is equal to 1 for all $\mathbf{x}$. For the special case of simple linear regression (i.e., $p = p^* = 2$), the subcollection specializes to

$$\left\{[\hat{\beta}_1 + \hat{\beta}_2(x - \bar{x})] \pm \sqrt{2F_{\xi,2,n-2}\hat{\sigma}^2\left(\frac{1}{n} + \frac{(x-\bar{x})^2}{SXX}\right)} \text{ for all } x \in \mathbb{R}\right\},$$

which takes the form of a two-dimensional "band." This band is displayed in Fig. 15.2; the straight line represents the estimated least squares regression line, and the curved lines mark the boundaries of the band. The band is narrowest at $x = \bar{x}$, and its width is a monotone increasing function of $|x - \bar{x}|$.

It may seem that the simultaneous coverage probability of the confidence region given by (15.17) is merely bounded (below) by $1-\xi$, rather than exactly equal to $1-\xi$, because the region comprises a proper subcollection of the intervals given by (15.16) whose confidence coefficient is exactly $1-\xi$. However, it can be shown (see

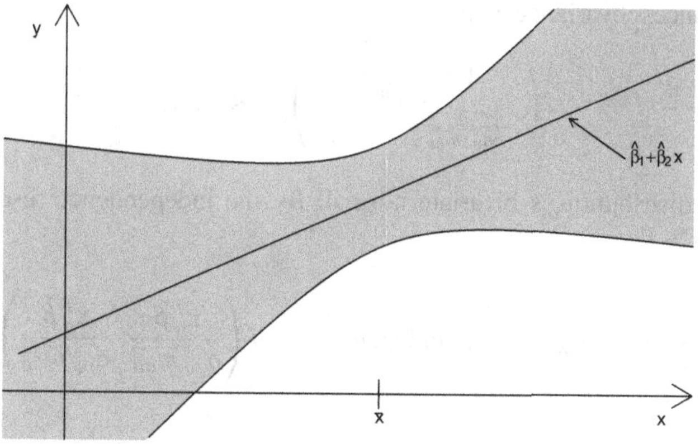

Fig. 15.2 Scheffé-based confidence band (shaded region) for $\{\beta_1 + \beta_2(x - \bar{x}) : x \in \mathbb{R}\}$ under a simple linear regression model

the exercises) that the simultaneous coverage probability of the subcollection is, in fact, exactly $1 - \xi$ despite this restriction.

In some situations we might be more interested in obtaining a simultaneous confidence band for $\mathbf{x}^T\boldsymbol{\beta}$ over a bounded region of $\mathbf{x}$-values. An extension of the multivariate t method can be used to obtain such a band, as is now described. We consider only the case $p = p^* = 2$, for which the form of the band is a trapezoid, or in some cases a parallelogram, rather than the shape depicted in Fig. 15.2. Note that a band that is shaped as a parallelogram has equal width for all x, which might be desirable in some situations. Let a and b represent the endpoints of a closed interval $[a, b]$, let $\mathbf{x}_a^T = (1, a - \bar{x})$, and let $\mathbf{x}_b^T = (1, b - \bar{x})$. Under a normal Gauss–Markov model, it is easily demonstrated that

$$\begin{pmatrix} \mathbf{x}_a^T\hat{\boldsymbol{\beta}} \\ \mathbf{x}_b^T\hat{\boldsymbol{\beta}} \end{pmatrix} \sim \mathrm{N}\left(\begin{pmatrix} \beta_1 + \beta_2(a - \bar{x}) \\ \beta_1 + \beta_2(b - \bar{x}) \end{pmatrix}, \sigma^2 \begin{pmatrix} m_{aa} & m_{ab} \\ m_{ab} & m_{bb} \end{pmatrix} \right),$$

where

$$m_{aa} = \frac{1}{n} + \frac{(a - \bar{x})^2}{SXX}, \quad m_{bb} = \frac{1}{n} + \frac{(b - \bar{x})^2}{SXX}, \quad m_{ab} = \frac{1}{n} + \frac{(a - \bar{x})(b - \bar{x})}{SXX}.$$

Consider the vector

$$\begin{pmatrix} \dfrac{\mathbf{x}_a^T\hat{\boldsymbol{\beta}}}{\sigma\sqrt{m_{aa}}} \\ \dfrac{\mathbf{x}_b^T\hat{\boldsymbol{\beta}}}{\sigma\sqrt{m_{bb}}} \end{pmatrix},$$

whose variance–covariance matrix is

$$\begin{pmatrix} 1 & \frac{m_{ab}}{\sqrt{m_{aa}m_{bb}}} \\ \frac{m_{ab}}{\sqrt{m_{aa}m_{bb}}} & 1 \end{pmatrix} \equiv \mathbf{R}, \text{ say,}$$

and whose distribution is bivariate normal. By the independence result noted in Example 14.3-1,

$$\frac{(n-2)\hat{\sigma}^2}{\sigma^2} \sim \chi^2(n-2) \text{ independently of } \left(\frac{\mathbf{x}_a^T\hat{\boldsymbol{\beta}}}{\sigma\sqrt{m_{aa}}}, \frac{\mathbf{x}_b^T\hat{\boldsymbol{\beta}}}{\sigma\sqrt{m_{bb}}}\right)^T.$$

Thus,

$$\begin{pmatrix} \frac{\mathbf{x}_a^T\hat{\boldsymbol{\beta}}-\mathbf{x}_a^T\boldsymbol{\beta}}{\hat{\sigma}\sqrt{m_{aa}}} \\ \frac{\mathbf{x}_b^T\hat{\boldsymbol{\beta}}-\mathbf{x}_b^T\boldsymbol{\beta}}{\hat{\sigma}\sqrt{m_{bb}}} \end{pmatrix} = \begin{pmatrix} \frac{[\mathbf{x}_a^T\hat{\boldsymbol{\beta}}-\mathbf{x}_a^T\boldsymbol{\beta}]/\sigma\sqrt{m_{aa}}}{\sqrt{(n-2)\hat{\sigma}^2/\sigma^2(n-2)}} \\ \frac{[\mathbf{x}_b^T\hat{\boldsymbol{\beta}}-\mathbf{x}_b^T\boldsymbol{\beta}]/\sigma\sqrt{m_{bb}}}{\sqrt{(n-2)\hat{\sigma}^2/\sigma^2(n-2)}} \end{pmatrix} \sim \mathrm{t}(2, n-2, \mathbf{R}).$$

Thus, $100(1-\xi)\%$ bivariate t confidence intervals for the expected responses at a and b are given via

$$1-\xi = \Pr(\mathbf{x}_a^T\hat{\boldsymbol{\beta}} - t_{\xi/2,2,n-2,\mathbf{R}}\hat{\sigma}\sqrt{m_{aa}} \le \mathbf{x}_a^T\boldsymbol{\beta} \le \mathbf{x}_a^T\hat{\boldsymbol{\beta}} + t_{\xi/2,2,n-2,\mathbf{R}}\hat{\sigma}\sqrt{m_{aa}}$$
$$\text{and } \mathbf{x}_b^T\hat{\boldsymbol{\beta}} - t_{\xi/2,2,n-2,\mathbf{R}}\hat{\sigma}\sqrt{m_{bb}} \le \mathbf{x}_b^T\boldsymbol{\beta} \le \mathbf{x}_b^T\hat{\boldsymbol{\beta}} + t_{\xi/2,2,n-2,\mathbf{R}}\hat{\sigma}\sqrt{m_{bb}}).$$

Now let ℓ_a and u_a represent the (lower and upper, respectively) endpoints of the above confidence interval for $\mathbf{x}_a^T\boldsymbol{\beta}$ and let ℓ_b and u_b represent the endpoints of the above confidence interval for $\mathbf{x}_b^T\boldsymbol{\beta}$. Let $\ell(x)$ represent the ordinate at x of the line segment connecting the point (a, ℓ_a) with the point (b, ℓ_b), and let $u(x)$ represent the ordinate of x at the line segment connecting (a, u_a) with (b, u_b); see Fig. 15.3. Define $\eta(x) = \beta_1 + \beta_2(x - \bar{x})$, $\hat{\eta}(x) = \hat{\beta}_1 + \hat{\beta}_2(x - \bar{x})$, $\eta_a = \mathbf{x}_a^T\boldsymbol{\beta}$, and $\eta_b = \mathbf{x}_b^T\boldsymbol{\beta}$. Using the same ideas that led to Fact 4 given earlier in this section, it can be established that the confidence band of values between $\ell(x)$ and $u(x)$, for all $x \in [a, b]$, has a simultaneous coverage probability of $1-\xi$, as follows:

$$\begin{aligned} 1-\xi &= \Pr(\ell_a \le \eta_a \le u_a \text{ and } \ell_b \le \eta_b \le u_b) \\ &= \Pr(k\ell_a \le k\eta_a \le ku_a \text{ and } (1-k)\ell_b \le (1-k)\eta_b \le (1-k)u_b \text{ for all } k \in [0,1]) \\ &= \Pr(k\ell_a + (1-k)\ell_b \le k\eta_a + (1-k)\eta_b \le ku_a + (1-k)u_b \text{ for all } k \in [0,1]). \end{aligned}$$

Now as k takes on values in $[0, 1]$, $k\ell_a + (1-k)\ell_b$ "traces out" $\{\ell(x); x \in [a, b]\}$ and $ku_a + (1-k)u_b$ traces out $\{u(x) : x \in [a, b]\}$. Thus

$$\Pr\Big(\ell(x) \le \eta(x) \le u(x) \text{ for all } x \in [a, b]\Big) = 1-\xi. \qquad (15.18)$$

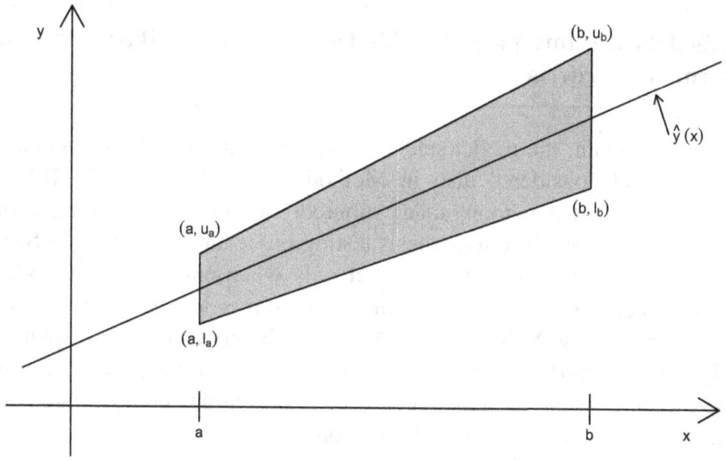

Fig. 15.3 Trapezoidal confidence band (shaded region) for $\{\beta_1 + \beta_2(x - \bar{x}) : a < x < b\}$ under a simple linear regression model

In general, the form of this confidence band is that of a trapezoid. It will have the form of a parallelogram if (and only if) the $100(1 - \xi)\%$ multivariate t intervals for η_a and η_b have equal width. This will happen if and only if $m_{aa} = m_{bb}$, which in turn holds if and only if a and b are chosen so that $\bar{x}$ is the midpoint of the interval $[a, b]$. ■

Methodological Interlude #11: Testing for Outliers

In Methodological Interlude #9 we considered the problem of testing whether a prespecified single case in the dataset was an outlier, under a given normal Gauss–Markov model. Now consider the related, but different, problem of testing whether there are *any* outliers in the dataset (under that same model). For this problem, account must be taken of the multiplicity of hypotheses being tested; equivalently, simultaneous confidence intervals, rather than one-at-a-time intervals, are needed for the expectations of model residuals. The Bonferroni method is applicable and reveals that we can perform a size-ξ test of

$$H_0 : \mathrm{E}(e_i) = 0 \text{ for all } i = 1, \ldots, n \qquad \text{versus} \qquad H_a : \text{not } H_0$$

by performing the t-test that a prespecified case is an outlier described in Methodological Interlude #9 for each case, one at a time, with significance level ξ/n. ■

Methodological Interlude #12: Model Selection via Likelihood-Based Information Criteria

Consider, once again, the model selection problem introduced in Methodological Interlude #6 and considered also in Methodological Interlude #8. Briefly, the problem is to select, from among many submodels of a linear model, the submodel that "best" fits the data. The hypothesis testing procedures presented in Sect. 15.2 may be used to address this problem, but only when one of the submodels is a further submodel of the other; they cannot be used to select between "non-nested" submodels, such as $\{\mathbf{y}, \mathbf{X}_1\boldsymbol{\beta}_1\}$ and $\{\mathbf{y}, \mathbf{X}_2\boldsymbol{\beta}_2\}$ in the context of a complete model $\{\mathbf{y}, \mathbf{X}_1\boldsymbol{\beta}_1 + \mathbf{X}_2\boldsymbol{\beta}_2\}$. Furthermore, hypothesis testing procedures, with their use of p-values as measures of the strength of evidence against null hypotheses, are gradually losing favor as a paradigm for model selection.

An alternative, more widely applicable, and preferred (by many) paradigm for model selection is the use of likelihood-based information criteria. A likelihood-based information criterion, as it applies to a Gauss–Markov model, has the general form

$$IC = -2\log f(\mathbf{y}; \hat{\boldsymbol{\beta}}, \dot{\sigma}^2) + (p^* + 1)c(n, p^*),$$

where $f(\cdot; \hat{\boldsymbol{\beta}}, \dot{\sigma}^2)$ is the pdf of $\mathbf{y}$ evaluated at maximum likelihood estimates $\hat{\boldsymbol{\beta}}$ and $\dot{\sigma}^2$, and $c(n, p^*)$ is a function of n and p^*. The first term, $-2\log f(\mathbf{y}; \hat{\boldsymbol{\beta}}, \dot{\sigma}^2)$, is called the "goodness-of-fit"; it decreases as the fit of the model to the data improves. The second term, $(p^* + 1)c(n, p^*)$, is called the "penalty term" and increases as the model becomes more complex, i.e., as p^* increases. Thus, models with smallest IC are best as they provide the best balance between model fit and model complexity. Under normality for $\mathbf{y}$,

$$\begin{aligned}
-2\log f(\mathbf{y}; \hat{\boldsymbol{\beta}}, \dot{\sigma}^2) &= -2\log \prod_{i=1}^{n} (2\pi\dot{\sigma}^2)^{-\frac{1}{2}} \exp[-(y_i - \mathbf{x}_i^T\hat{\boldsymbol{\beta}})^2/2\dot{\sigma}^2] \\
&= -2\log[(2\pi\dot{\sigma}^2)^{-n/2}\exp(-n/2)] \\
&= n\log(2\pi + 1) + n\log\dot{\sigma}^2.
\end{aligned}$$

Thus

$$IC = n\log(2\pi + 1) + n\log\dot{\sigma}^2 + (p^* + 1)c(n, p^*).$$

Several special cases of this general information criterion exist, which differ only with respect to $c(n, p^*)$. We consider three cases here; there is a large literature on these and other criteria, into which we do not delve. The interested reader may consult Burnham and Anderson (2002) for details. The case in which $c(n, p^*) = 2$ is the most popular and is known as Akaike's information criterion, AIC. A criterion that improves upon *AIC* when p^*/n is relatively large is the corrected Akaike's information criterion, *AICc*, and has $c(n, p^*) = 2n/(n - p^* - 2)$. And last, the Bayesian information criterion, *BIC*, sets $c(n, p^*) = \log n$. Because $2n/(n - p^* - 2) > 2$ and $\log n > 2$ for $n \geq 8$, *AICc* and *BIC* tend to favor more parsimonious models than those favored by *AIC*. However, *AICc* and *AIC* are asymptotically equivalent because $2n/(n - p^* - 2) \rightarrow 2$ as $n \rightarrow \infty$.

An information criterion may be used to select a best model and/or obtain a rank-ordering of all the models under consideration. The difference, Δ, between the value of *IC* for a candidate model and its minimum among all models under consideration can be used as a measure of the empirical support for that model. In conjunction with *AIC* or *AICc*, Burnham and Anderson (2002) suggest that models for which $0 < \Delta < 2$ have substantial empirical support; models for which $4 < \Delta < 7$ have considerably less support; and models for which $\Delta > 10$ have essentially no support. Similarly for *BIC*, Kass and Raftery (1995) describe the evidence against a model as not worth more than a bare mention if $0 < \Delta < 2$; positive if $2 < \Delta < 6$; strong if $6 < \Delta < 10$; and very strong if $\Delta > 10$.

There is a relationship between *AIC* and Mallow's C_p criterion (recall that the latter was introduced in Methodological Interlude #6). Specifically, *AIC* and C_p are asymptotically equivalent, i.e., as $n \rightarrow \infty$ the same model will minimize both of them. To see this, let $\hat{\sigma}^2$ represent the residual mean square for the model that includes all explanatory variables under consideration, and let $\dot{\sigma}_U^2$ represent the maximum likelihood estimator of σ^2 for an arbitrary rank-p^* submodel of that model. Now observe that the model that minimizes *AIC* will also minimize

$$n \log \dot{\sigma}_U^2 - n \log \hat{\sigma}^2 + 2p^* = n \log \left(\frac{\dot{\sigma}_U^2}{\hat{\sigma}^2} \right) + 2p^*.$$

A first-order Taylor series expansion of $\log \left(\frac{\dot{\sigma}_U^2}{\hat{\sigma}^2} \right)$ about 1 yields the approximation

$$\begin{aligned} n \log \left(\frac{\dot{\sigma}_U^2}{\hat{\sigma}^2} \right) &\doteq n \left[\log(1) + \left(\frac{\dot{\sigma}_U^2}{\hat{\sigma}^2} - 1 \right) \right] \\ &= \frac{n\dot{\sigma}_U^2}{\hat{\sigma}^2} - n. \end{aligned}$$

Thus, as $n \rightarrow \infty$, the model that minimizes *AIC* also minimizes

$$\frac{n\dot{\sigma}_U^2}{\hat{\sigma}^2} - n + 2p^* = p^* + \frac{(n - p^*)(\hat{\sigma}_U^2 - \hat{\sigma}^2)}{\hat{\sigma}^2} = C_p,$$

where $\hat{\sigma}_U^2$ is the residual mean square for the underspecified model. ■

15.4 Exercises

1. Consider the Gauss–Markov simple linear regression model for n observations, and suppose that in addition to the n responses that follow that model, there are m responses $y_1^*, y_2^*, \ldots, y_m^*$, all taken at an *unknown* value of x, say x^*, that also follow that model. Assume that the normal Gauss–Markov model holds for all the observations. Using the original observations $(x_1, y_1), \ldots, (x_n, y_n)$ and the additional y-observations, an estimate of the unknown x^* can be obtained. One estimator of x^* that has been proposed is

$$\hat{x}^* = \frac{\bar{y}^* - \hat{\beta}_1}{\hat{\beta}_2},$$

where $\bar{y}^* = \frac{1}{m}\sum_{i=1}^m y_i^*$. This is derived by equating $\bar{y}^*$ to $\hat{\beta}_1 + \hat{\beta}_2 x^*$ (because their expectations are both $\beta_1 + \beta_2 x^*$) and then solving for x^*. (Note that the estimator is not well defined when $\hat{\beta}_2 = 0$, but this is an event of probability zero).
   (a) Determine the distribution of $W \equiv \bar{y}^* - \hat{\beta}_1 - \hat{\beta}_2 x^*$.
   (b) The distribution of $W/(\hat{\sigma} c)$, where c is a nonrandom quantity, has a certain well-known distribution. Determine this distribution and the value of c.
   (c) Let $0 < \xi < 1$. Based on your solution to part (b), derive an exact $100(1-\xi)\%$ confidence interval for x^* whose endpoints depend on the unknown x^*, and then approximate the endpoints (by substituting $\hat{x}^*$ for x^*) to obtain an approximate $100(1-\xi)\%$ confidence interval for x^*. Is this confidence interval symmetric about $\hat{x}^*$?
2. Consider the normal Gauss–Markov quadratic regression model

$$y_i = \beta_1 + \beta_2 x_i + \beta_3 x_i^2 + e_i \quad (i = 1, \ldots, n),$$

where it is known that $\beta_3 \neq 0$. Assume that there are at least three distinct x_i's, so that $\mathbf{X}$ has full column rank. Let x_m represent the value of x where the quadratic function $\mu(x) = \beta_1 + \beta_2 x + \beta_3 x^2$ is minimized or maximized (such an x-value exists because $\beta_3 \neq 0$).
   (a) Verify that $x_m = -\beta_2/(2\beta_3)$.
   (b) Define $\tau = \beta_2 + 2\beta_3 x_m$ $(= 0)$ and $\hat{\tau} = \hat{\beta}_2 + 2\hat{\beta}_3 x_m$, where $(\hat{\beta}_1, \hat{\beta}_2, \hat{\beta}_3)$ are the least squares estimators of $(\beta_1, \beta_2, \beta_3)$. Determine the distribution of $\hat{\tau}$ in terms of x_m, σ^2, and $\{c_{ij}\}$, where c_{ij} is the ijth element of $(\mathbf{X}^T\mathbf{X})^{-1}$.
   (c) Based partly on your answer to part (b), determine a function of $\hat{\tau}$ and $\hat{\sigma}^2$ that has an F distribution, and give the parameters of this F distribution.
   (d) Let $\xi \in (0, 1)$. Use the result from part (c) to find a quadratic function of x_m, say $q(x_m)$, such that $\Pr[q(x_m) \leq 0] = 1 - \xi$. When is this confidence set an interval?

3. Consider the normal Gauss–Markov simple linear regression model

$$y_i = \beta_1 + \beta_2 x_i + e_i \quad (i = 1, \ldots, n).$$

In Example 12.1.1-1, it was shown that the mean squared error for the least squares estimator of the intercept from the fit of an (underspecified) intercept-only model is smaller than the mean squared error for the least squares estimator of the intercept from fitting the full simple linear regression model if and only if $\beta_2^2/(\sigma^2/SXX) \leq 1$. Consequently, it may be desirable to test the hypotheses

$$H_0 : \frac{\beta_2^2}{\sigma^2/SXX} \leq 1 \quad \text{versus} \quad H_a : \frac{\beta_2^2}{\sigma^2/SXX} > 1.$$

A natural test statistic for this hypothesis test is

$$\frac{\hat{\beta}_2^2}{\hat{\sigma}^2/SXX}.$$

Determine the distribution of this test statistic when

$$\frac{\beta_2^2}{\sigma^2/SXX} = 1.$$

4. By completing the three parts of this exercise, verify the claim made in Example 15.2-3, i.e., that the power of the F test for $H_0 : \beta_2 = 0$ versus $H_a : \beta_2 \neq 0$ in normal simple linear regression, when the explanatory variable is restricted to the interval $[a, b]$ and n is even, is maximized by taking half of the observations at $x = a$ and the other half at $x = b$.
   (a) Show that $\sum_{i=1}^n (x_i - \bar{x})^2 \leq \sum_{i=1}^n \left(x_i - \frac{a+b}{2}\right)^2$. [Hint: Add and subtract $\frac{a+b}{2}$ within $\sum_{i=1}^n (x_i - \bar{x})^2$.]
   (b) Show that $\sum_{i=1}^n \left(x_i - \frac{a+b}{2}\right)^2 \leq \frac{(b-a)^2}{4}$.
   (c) Show that $\sum_{i=1}^n \left(x_i - \frac{a+b}{2}\right)^2 = \frac{(b-a)^2}{4}$ for the design that takes half of the observations at $x = a$ and the other half at $x = b$.
5. Consider the following modification of the normal Gauss–Markov model, which is called the normal Gauss–Markov variance shift outlier model. For this model, the model equation $\mathbf{y} = \mathbf{X}\boldsymbol{\beta} + \mathbf{e}$ can be written as

$$\begin{pmatrix} y_1 \\ \mathbf{y}_{-1} \end{pmatrix} = \begin{pmatrix} \mathbf{x}_1^T \\ \mathbf{X}_{-1} \end{pmatrix} \boldsymbol{\beta} + \begin{pmatrix} e_1 \\ \mathbf{e}_{-1} \end{pmatrix},$$

where

$$\begin{pmatrix} e_1 \\ \mathbf{e}_{-1} \end{pmatrix} \sim \mathrm{N}_n\left(\mathbf{0}, \mathrm{diag}\{\sigma^2 + \sigma_\delta^2, \sigma^2, \sigma^2, \ldots, \sigma^2\}\right).$$

Here, σ_δ^2 is an unknown nonnegative scalar parameter. Assume that $\text{rank}(\mathbf{X}_{-1}) = \text{rank}(\mathbf{X})$ $(= p^*)$, i.e., the rank of $\mathbf{X}$ is not affected by deleting its first row, and define $\hat{e}_{1,-1}$, $\hat{\boldsymbol{\beta}}_{-1}$, $\hat{\sigma}^2_{-1}$, and $p_{11,-1}$ as in Methodological Interlude #9.

(a) Determine the distributions of $\hat{e}_{1,-1}$ and $(n - p^* - 1)\hat{\sigma}^2_{-1}/\sigma^2$.

(b) Determine the distribution of

$$\frac{\hat{e}_{1,-1}}{\hat{\sigma}_{-1}\sqrt{1 + p_{11,-1}}}.$$

(c) Explain how the hypotheses

$$H_0 : \sigma_\delta^2 = 0 \quad \text{vs} \quad H_a : \sigma_\delta^2 > 0$$

could be tested using the statistic in part (b), and explain how the value of $p_{11,-1}$ affects the power of the test.

6. Consider the normal positive definite Aitken model

$$\mathbf{y} = \mathbf{1}_n\mu + \mathbf{e}$$

where $n \geq 2$, $\mathbf{e} \sim \text{N}_n(\mathbf{0}, \sigma^2\mathbf{W})$, and $\mathbf{W} = (1-\rho)\mathbf{I}_n + \rho\mathbf{J}_n$. Here, σ^2 is unknown but $\rho \in \left(-\frac{1}{n-1}, 1\right)$ is a known constant. Let $\hat{\sigma}^2$ represent the sample variance of the responses, i.e., $\hat{\sigma}^2 = \sum_{i=1}^n (y_i - \bar{y})^2/(n-1)$.

(a) Determine the joint distribution of $\bar{y}$ and $\frac{(n-1)\hat{\sigma}^2}{\sigma^2(1-\rho)}$.

(b) Using results from part (a), derive a size-ξ test for testing $H_0 : \mu = 0$ versus $H_a : \mu \neq 0$.

(c) For the test you derived in part (b), describe how the magnitude of ρ affects the power.

(d) A statistician wants to test the null hypothesis $H_0 : \mu = 0$ versus the alternative hypothesis $H_a : \mu > 0$ at the ξ level of significance. However, rather than using the appropriate test obtained in part (b), (s)he wishes to ignore the correlations among the observations and use a standard size-ξ t test. That is, (s)he will reject H_0 if and only if

$$\frac{\bar{y}}{\hat{\sigma}/\sqrt{n}} > t_{\xi,n-1}.$$

The size of this test is not necessarily equal to its nominal value, ξ. What is the actual size of the test, and how does it compare to ξ?

7. Consider a situation where observations are taken on some outcome in two groups (e.g., a control group and an intervention group). An investigator wishes

to determine if exactly the same linear model applies to the two groups. Specifically, suppose that the model for n_1 observations in the first group is

$$\mathbf{y}_1 = \mathbf{X}_1\boldsymbol{\beta}_1 + \mathbf{e}_1$$

and the model for n_2 observations from the second group is

$$\mathbf{y}_2 = \mathbf{X}_2\boldsymbol{\beta}_2 + \mathbf{e}_2,$$

where $\mathbf{X}_1$ and $\mathbf{X}_2$ are model matrices whose p columns (each) correspond to the same p explanatory variables, and $\mathbf{e}_1$ and $\mathbf{e}_2$ are independent and normally distributed random vectors with mean $\mathbf{0}_{n_i}$ and variance–covariance matrix $\sigma^2\mathbf{I}_{n_i}$. Assume that $n_1 + n_2 > 2p$.

(a) Let $\mathbf{C}^T$ represent an $s \times p$ matrix such that $\mathbf{C}^T\boldsymbol{\beta}_1$ is estimable under the first model and $\mathbf{C}^T\boldsymbol{\beta}_2$ is estimable under the second model. Show that the two sets of data can be combined into a single linear model of the form

$$\mathbf{y} = \mathbf{X}\boldsymbol{\beta} + \mathbf{e}$$

such that the null hypothesis H_0: $\mathbf{C}^T\boldsymbol{\beta}_1 = \mathbf{C}^T\boldsymbol{\beta}_2$ is testable under this model and can be expressed as H_0: $\mathbf{B}^T\boldsymbol{\beta} = \mathbf{0}$ for a suitably chosen matrix $\mathbf{B}$.

(b) By specializing Theorem 15.2.1, give the size-ξ F test (in as simple a form as possible) for testing the null hypothesis of part (a) versus the alternative H_a: $\mathbf{C}^T\boldsymbol{\beta}_1 \neq \mathbf{C}^T\boldsymbol{\beta}_2$.

8. Consider a situation in which data are available on a pair of variables (x, y) from individuals that belong to two groups. Let $\{(x_{1i}, y_{1i}) : i = 1, \ldots, n_1\}$ and $\{(x_{2i}, y_{2i}) : i = 1, \ldots, n_2\}$, respectively, denote the data from these two groups. Assume that the following model holds:

$$y_{1i} = \beta_{11} + \beta_{21}(x_{1i} - \bar{x}_1) + e_{1i} \quad (i = 1, \ldots, n_1)$$
$$y_{2i} = \beta_{12} + \beta_{22}(x_{2i} - \bar{x}_2) + e_{2i} \quad (i = 1, \ldots, n_2)$$

where the e_{1i}'s and e_{2i}'s are independent $\mathrm{N}(0, \sigma^2)$ random variables, $n_1 \geq 2$, $n_2 \geq 2$, $x_{11} \neq x_{12}$, and $x_{21} \neq x_{22}$. Furthermore, define the following notation:

$$\bar{x}_1 = n_1^{-1}\sum_{i=1}^{n_1} x_{1i}, \quad \bar{x}_2 = n_2^{-1}\sum_{i=1}^{n_2} x_{2i}, \quad \bar{y}_1 = n_1^{-1}\sum_{i=1}^{n_1} y_{1i}, \quad \bar{y}_2 = n_2^{-1}\sum_{i=1}^{n_2} y_{2i},$$

$$SXX_1 = \sum_{i=1}^{n_1}(x_{1i} - \bar{x}_1)^2, \quad SXX_2 = \sum_{i=1}^{n_2}(x_{2i} - \bar{x}_2)^2,$$

$$SXY_1 = \sum_{i=1}^{n_1}(x_{1i} - \bar{x}_1)(y_{1i} - \bar{y}_1), \quad SXY_2 = \sum_{i=1}^{n_2}(x_{2i} - \bar{x}_2)(y_{2i} - \bar{y}_2).$$

Observe that this model can be written in matrix notation as the unordered 2-part model

$$\mathbf{y} = \begin{pmatrix} \mathbf{1}_{n_1} & \mathbf{0}_{n_1} \\ \mathbf{0}_{n_2} & \mathbf{1}_{n_2} \end{pmatrix} \begin{pmatrix} \beta_{11} \\ \beta_{12} \end{pmatrix} + \begin{pmatrix} \mathbf{x}_1 - \bar{x}_1 \mathbf{1}_{n_1} & \mathbf{0} \\ \mathbf{0} & \mathbf{x}_2 - \bar{x}_2 \mathbf{1}_{n_2} \end{pmatrix} \begin{pmatrix} \beta_{21} \\ \beta_{22} \end{pmatrix} + \mathbf{e}.$$

(a) Give nonmatrix expressions for the BLUEs of β_{11}, β_{12}, β_{21}, and β_{22} and for the residual mean square, in terms of the quantities defined above. Also give an expression for the variance–covariance matrix of the BLUEs.

(b) Give the two-part ANOVA table for the ordered two-part model that coincides with the unordered two-part model written above. Give nonmatrix expressions for the degrees of freedom and sums of squares. Determine the distributions of the (suitably scaled) sums of squares in this table, giving nonmatrix expressions for any noncentrality parameters.

(c) Give a size-ξ test for equal intercepts, i.e., for $H_0 : \beta_{11} = \beta_{12}$ versus $H_a : \beta_{11} \neq \beta_{12}$.

(d) Give a size-ξ test for equal slopes (or "parallelism"), i.e., for $H_0 : \beta_{21} = \beta_{22}$ versus $H_a : \beta_{21} \neq \beta_{22}$.

(e) Give a size-ξ test for identical lines, i.e., for $H_0 : \beta_{11} = \beta_{12}, \beta_{21} = \beta_{22}$ versus H_a: either $\beta_{11} \neq \beta_{12}$ or $\beta_{21} \neq \beta_{22}$.

(f) Let c represent a specified real number and let $0 < \xi < 1$. Give a size-ξ test of the null hypothesis that the two groups' regression lines intersect at $x = c$ and that the slope for the first group is the negative of the slope for the second group, versus the alternative hypothesis that at least one of these statements is false.

(g) Suppose that we desire to predict the value of a new observation $y_{1,n+1}$ from the first group corresponding to the value $x_{1,n+1}$ of x_1, and we also desire to predict the value of a new observation $y_{2,n+2}$ from the second group corresponding to the value $x_{2,n+2}$ of x_2. Give nonmatrix expressions for $100(1 - \xi)\%$ prediction intervals for these two new observations, and give an expression for a $100(1 - \xi)\%$ prediction ellipsoid for them. (In the latter expression, you may use vectors and matrices but if so, give nonmatrix expressions for each element of them.)

(h) Give confidence bands for the two lines for which the simultaneous coverage probability (for both lines and for all $x \in \mathbb{R}$) is at least $1 - \xi$.

9. Consider a situation similar to that described in the previous exercise, except that the two lines are known to be parallel. Thus the model is

$$y_{1i} = \beta_{11} + \beta_2(x_{1i} - \bar{x}_1) + e_{1i} \quad (i = 1, \ldots, n_1)$$

$$y_{2i} = \beta_{12} + \beta_2(x_{2i} - \bar{x}_2) + e_{2i} \quad (i = 1, \ldots, n_2),$$

where again the e_{1i}'s and e_{2i}'s are independent $\mathrm{N}(0, \sigma^2)$ random variables, $n_1 \geq 2$, $n_2 \geq 2$, $x_{11} \neq x_{12}$, and $x_{21} \neq x_{22}$. Adopt the same notation as in the previous exercise, but also define the following:

$$SXX_{12} = SXX_1 + SXX_2, \quad SXY_{12} = SXY_1 + SXY_2.$$

Observe that this model can be written in matrix notation as the unordered 2-part model

$$\mathbf{y} = \begin{pmatrix} \mathbf{1}_{n_1} & \mathbf{0}_{n_1} \\ \mathbf{0}_{n_2} & \mathbf{1}_{n_2} \end{pmatrix} \begin{pmatrix} \beta_{11} \\ \beta_{12} \end{pmatrix} + \begin{pmatrix} \mathbf{x}_1 - \bar{x}_1 \mathbf{1}_{n_1} \\ \mathbf{x}_2 - \bar{x}_2 \mathbf{1}_{n_2} \end{pmatrix} \beta_2 + \mathbf{e}.$$

(a) Give nonmatrix expressions for the BLUEs of β_{11}, β_{12}, and β_2, and for the residual mean square, in terms of the quantities defined above. Also give an expression for the variance–covariance matrix of the BLUEs.
(b) Give the two-part ANOVA table for the ordered two-part model that coincides with the unordered two-part model written above. Give nonmatrix expressions for the degrees of freedom and sums of squares. Determine the distributions of the (suitably scaled) sums of squares in this table, giving nonmatrix expressions for any noncentrality parameters.
(c) Suppose that we wish to estimate the distance, δ, between the lines of the two groups, defined as the amount by which the line for Group 1 exceeds the line for Group 2. Verify that

$$\delta = \beta_{11} - \beta_{12} - \beta_2(\bar{x}_1 - \bar{x}_2)$$

and obtain a $100(1 - \xi)\%$ confidence interval for δ.
(d) Obtain a size-ξ t test for $H_0 : \delta = 0$ versus $H_a : \delta \neq 0$. (This is a test for identical lines, given parallelism.)
(e) Suppose that enough resources (time, money, etc.) were available to take observations on a total of 20 observations (i.e. $n_1 + n_2 = 20$). From a design standpoint, how would you choose n_1, n_2, the x_{1i}'s and the x_{2i}'s in order to minimize the width of the confidence interval in part (c)?
(f) Give a size-ξ test for $H_0 : \beta_2 = 0$ versus $H_a : \beta_2 \neq 0$.
(g) Obtain the size-ξ F test of $H_0 : \begin{pmatrix} \delta \\ \beta_2 \end{pmatrix} = \mathbf{0}$ versus H_a : not H_0.
(h) Let $\xi \in (0, 1)$. Give an expression for a $100(1 - \xi)\%$ confidence ellipsoid for $\boldsymbol{\beta} \equiv (\beta_{11}, \beta_{12}, \beta_2)^T$.
(i) Let x_0 represent a specified x-value. Give confidence intervals for $\beta_{11} + \beta_2 x_0$ and $\beta_{12} + \beta_2 x_0$ whose simultaneous coverage probability is exactly $1 - \xi$.
(j) Give confidence bands for the two parallel lines whose simultaneous coverage probability (for both lines and for all $x \in \mathbb{R}$) is at least $1 - \xi$.

10. Consider the normal Gauss–Markov two-way main effects model with exactly one observation per cell, which has model equation

$$y_{ij} = \mu + \alpha_i + \gamma_j + e_{ij} \quad (i = 1, \ldots, q;\ j = 1, \ldots, m).$$

Define

$$f = \sum_{i=1}^{q} \sum_{j=1}^{m} b_{ij}(y_{ij} - \bar{y}_{i\cdot} - \bar{y}_{\cdot j} + \bar{y}_{\cdot\cdot}) = \mathbf{b}^T(\mathbf{I} - \mathbf{P_X})\mathbf{y},$$

where $\{b_{ij}\}$ is a set of real numbers, not all equal to 0, such that $\sum_{i=1}^{q} b_{ij} = 0$ for all j and $\sum_{j=1}^{m} b_{ij} = 0$ for all i.

(a) Show that

$$\frac{f^2}{\sigma^2 \mathbf{b}^T\mathbf{b}} \quad \text{and} \quad (1/\sigma^2)\sum_{i=1}^{q}\sum_{j=1}^{m}(y_{ij} - \bar{y}_{i\cdot} - \bar{y}_{\cdot j} + \bar{y}_{\cdot\cdot})^2 - \frac{f^2}{\sigma^2 \mathbf{b}^T\mathbf{b}}$$

are jointly distributed as independent $\chi^2(1)$ and $\chi^2(qm - q - m)$ random variables under this model.

(b) Let $\mathbf{y}_I$ represent the q-vector whose ith element is $\bar{y}_{i\cdot} - \bar{y}_{\cdot\cdot}$, let $\mathbf{y}_{II}$ represent the m-vector whose jth element is $\bar{y}_{\cdot j} - \bar{y}_{\cdot\cdot}$, and let $\mathbf{y}_{III}$ represent the qm-vector whose ijth element is $y_{ij} - \bar{y}_{i\cdot} - \bar{y}_{\cdot j} + \bar{y}_{\cdot\cdot}$. Show that $\mathbf{y}_I$, $\mathbf{y}_{II}$, and $\mathbf{y}_{III}$ are independent under this model.

(c) Now suppose that we condition on $\mathbf{y}_I$ and $\mathbf{y}_{II}$, and consider the expansion of the model to

$$y_{ij} = \mu + \alpha_i + \gamma_j + \lambda(\bar{y}_{i\cdot} - \bar{y}_{\cdot\cdot})(\bar{y}_{\cdot j} - \bar{y}_{\cdot\cdot}) + e_{ij} \quad (i = 1, \ldots, q;\ j = 1, \ldots, m).$$

Using part (a), show that conditional on $\mathbf{y}_I$ and $\mathbf{y}_{II}$,

$$\frac{\left(\sum_{i=1}^{q}\sum_{j=1}^{m}(\bar{y}_{i\cdot} - \bar{y}_{\cdot\cdot})(\bar{y}_{\cdot j} - \bar{y}_{\cdot\cdot})y_{ij}\right)^2}{\sigma^2 \sum_{i=1}^{q}(\bar{y}_{i\cdot} - \bar{y}_{\cdot\cdot})^2 \sum_{j=1}^{m}(\bar{y}_{\cdot j} - \bar{y}_{\cdot\cdot})^2}$$

and

$$(1/\sigma^2)\sum_{i=1}^{q}\sum_{j=1}^{m}(y_{ij} - \bar{y}_{i\cdot} - \bar{y}_{\cdot j} + \bar{y}_{\cdot\cdot})^2 - \frac{\left(\sum_{i=1}^{q}\sum_{j=1}^{m}(\bar{y}_{i\cdot} - \bar{y}_{\cdot\cdot})(\bar{y}_{\cdot j} - \bar{y}_{\cdot\cdot})y_{ij}\right)^2}{\sigma^2 \sum_{i=1}^{q}(\bar{y}_{i\cdot} - \bar{y}_{\cdot\cdot})^2 \sum_{j=1}^{m}(\bar{y}_{\cdot j} - \bar{y}_{\cdot\cdot})^2}$$

are jointly distributed as independent $\chi^2(1, \nu)$ and $\chi^2(qm - q - m)$ random variables, and determine ν. [Hint: First observe that $\sum_{i=1}^{q}\sum_{j=1}^{m}(\bar{y}_{i\cdot} - \bar{y}_{\cdot\cdot})(\bar{y}_{\cdot j} - \bar{y}_{\cdot\cdot})(y_{ij} - \bar{y}_{i\cdot} - \bar{y}_{\cdot j} + \bar{y}_{\cdot\cdot}) = \sum_{i=1}^{q}\sum_{j=1}^{m}(\bar{y}_{i\cdot} - \bar{y}_{\cdot\cdot})(\bar{y}_{\cdot j} - \bar{y}_{\cdot\cdot})y_{ij}$.]

(d) Consider testing $H_0 : \lambda = 0$ versus $H_a : \lambda \neq 0$ in the conditional model specified in part (c). Show that the noncentrality parameter of the chi-square distribution of the first quantity displayed above is equal to 0 under H_0, and then use parts (b) and (c) to determine the unconditional joint distribution of both quantities displayed above under H_0.

(e) Using part (d), obtain a size-ξ test of $H_0 : \lambda = 0$ versus $H_a : \lambda \neq 0$ in the unconditional statistical model (not a linear model according to our definition) given by

$$y_{ij} = \mu + \alpha_i + \gamma_j + \lambda(\bar{y}_{i\cdot} - \bar{y}_{\cdot\cdot})(\bar{y}_{\cdot j} - \bar{y}_{\cdot\cdot}) + e_{ij},$$

where the e_{ij}'s are independent $\mathrm{N}(0, \sigma^2)$ random variables.

Note: A nonzero λ in the model defined in part (e) implies that the effects of the two crossed factors are not additive. Consequently, the test ultimately obtained in part (e) tests for nonadditivity of a particular form. The test and its derivation (which is essentially equivalent to the steps in this exercise) are due to Tukey (1949b), and the test is commonly referred to as "Tukey's one-degree-of-freedom test for nonadditivity."

11. Prove Facts 2, 3, and 4 listed in Sect. 15.3.
12. Prove (15.13), i.e., prove that for any $\xi \in (0, 1)$ and any $s^* > 0$, $\nu > 0$, and $l > 0$,

$$s^* F_{\xi,s^*,\nu} \leq (s^* + l) F_{\xi,s^*+l,\nu}.$$

[Hint: Consider three independent random variables $U \sim \chi^2(s^*)$, $V \sim \chi^2(l)$, and $W \sim \chi^2(\nu)$.]

13. Prove (15.14).
14. Prove Theorem 15.3.7.
15. Consider the mixed linear model for two observations specified in Exercise 13.18, and suppose that the joint distribution of b, d_1, and d_2 is multivariate normal. Let $0 < \xi < 1$.
    (a) Obtain a $100(1 - \xi)\%$ confidence interval for β.
    (b) Obtain a $100(1 - \xi)\%$ prediction interval for b.
    (c) Obtain intervals for β and b whose simultaneous coverage probability is exactly $1 - \xi$.
16. Generalize all four types of simultaneous confidence intervals presented in Sect. 15.3 (Bonferroni, Scheffé, multivariate t, Tukey) to be suitable for use under a normal positive definite Aitken model.
17. Obtain specialized expressions for $\mathbf{R}$ for the multivariate t-based simultaneous confidence and simultaneous prediction intervals presented in Example 15.3-1.
18. Extend the expressions for simultaneous confidence intervals presented in Example 15.3-2 to the case of unbalanced data, if applicable.
19. Show that the simultaneous coverage probability of the classical confidence band given by (15.15) is $1 - \xi$ despite the restriction that the first element of $\mathbf{c}$ is equal to 1.

20. Consider the normal random-slope, fixed-intercept simple linear regression model

$$y_i = \beta + bz_i + d_i \quad (i = 1, \ldots, n),$$

described previously in more detail in Exercise 13.13, where var(**y**), among other quantities, was determined. The model equation may be written in matrix form as

$$\mathbf{y} = \beta\mathbf{1} + b\mathbf{z} + \mathbf{d}.$$

(a) Consider the sums of squares that arise in the following ordered two-part mixed-model ANOVA:

Source	df	Sum of squares
$\mathbf{z}$	1	$\mathbf{y}^T\mathbf{P}_\mathbf{z}\mathbf{y}$
$\mathbf{1}\|\mathbf{z}$	1	$\mathbf{y}^T(\mathbf{P}_{1,\mathbf{z}} - \mathbf{P}_\mathbf{z})\mathbf{y}$
Residual	$n-2$	$\mathbf{y}^T(\mathbf{I} - \mathbf{P}_{1,\mathbf{z}})\mathbf{y}$

Here $\mathbf{P}_\mathbf{z} = \mathbf{z}\mathbf{z}^T/(\sum_{i=1}^n z_i^2)$. Determine which two of these three sums of squares, when divided by σ^2, have noncentral chi-square distributions under the mixed simple linear regression model defined above. For those two, give nonmatrix expressions for the parameters of the distributions.

(b) It is possible to use the sums of squares from the ANOVA above to test $H_0 : \beta = 0$ versus $H_a : \beta \neq 0$ by an F test. Derive the appropriate F-statistic and give its distribution under each of H_a and H_0.

(c) Let $\boldsymbol{\tau} = (\beta, b)^T$ and let $\tilde{\boldsymbol{\tau}} = (\tilde{\beta}, \tilde{b})^T$ be the BLUP of $\boldsymbol{\tau}$. Let $\sigma^2\mathbf{Q} = \text{var}(\tilde{\boldsymbol{\tau}} - \boldsymbol{\tau}) = \sigma^2\begin{pmatrix} q_{11} & q_{12} \\ q_{12} & q_{22} \end{pmatrix}$ and assume that this matrix is positive definite. Starting with the probability statement

$$1 - \xi = \Pr\left(\frac{(\tilde{\boldsymbol{\tau}} - \boldsymbol{\tau})^T\mathbf{Q}^{-1}(\tilde{\boldsymbol{\tau}} - \boldsymbol{\tau})}{s\tilde{\sigma}^2} \leq F_{\xi,s,n-p^*}\right),$$

for a particular choice of s and p^*, Scheffé's method can be used to derive a $100(1-\xi)\%$ simultaneous prediction band for $\{\beta + bz : z \in \mathbb{R}\}$, where $\xi \in (0, 1)$. This band is of the form $(\tilde{\beta} + \tilde{b}z) \pm g(z)$, for some function $g(z)$. Determine $g(z)$; you may leave your answer in terms of q_{11}, q_{12}, and q_{22} (and other quantities) but determine numerical values for s and p^*.

21. Consider the normal Gauss–Markov simple linear regression model and suppose that $n = 22$ and $x_1 = x_2 = 5$, $x_3 = x_4 = 6, \ldots, x_{21} = x_{22} = 15$.

(a) Determine the expected squared lengths of the Bonferroni and multivariate t 95% simultaneous confidence intervals for β_1 and β_2. (Expressions for these intervals may be found in Example 15.3-1.)

(b) Let $x_{n+1} = 5$, $x_{n+2} = 10$, and $x_{n+3} = 15$, and suppose that the original responses and the responses corresponding to these three "new" values of x follow the normal Gauss–Markov prediction-extended simple linear regression model. Give an expression for the multivariate t 95% simultaneous prediction intervals for $\{\beta_1 + \beta_2 x_{n+i} i + e_{n+i} : i = 1, 2, 3\}$. Determine how much wider (proportionally) these intervals are than the multivariate t 95% simultaneous confidence intervals for $\{\beta_1 + \beta_2 x_{n+i} : i = 1, 2, 3\}$.

(c) Obtain the 95% trapezoidal (actually it is a parallelogram) confidence band described in Example 15.3-3 for the line over the bounded interval $[a = 5, b = 15]$, and compare the expected squared length of any interval in that band to the expected squared length of the intervals in the 95% classical confidence band at $x = 5$, $x = 10$, and $x = 15$.

22. Consider the two-way main effects model with equal cell frequencies, i.e.,

$$y_{ijk} = \mu + \alpha_i + \gamma_j + e_{ijk} \quad (i = 1, \ldots, q;\ j = 1, \ldots, m;\ k = 1, \ldots, r).$$

Here μ, the α_i's, and the γ_j's are unknown parameters, while the e_{ijk}'s are independent normally distributed random variables having zero means and common variances σ^2. Let $0 < \xi < 1$.

(a) Obtain an expression for a $100(1 - \xi)\%$ "one-at-a-time" (symmetric) confidence interval for $\alpha_i - \alpha_{i'} (i' > i = 1, \ldots, q)$. Obtain the analogous expression for a $100(1 - \xi)\%$ one-at-a-time confidence interval for $\gamma_j - \gamma_{j'}$ $(j' > j = 1, \ldots, m)$.

(b) Obtain an expression for a $100(1-\xi)\%$ confidence ellipsoid for the subset of Factor A differences $\{\alpha_1 - \alpha_i : i = 2, \ldots, q\}$; likewise, obtain an expression for a $100(1-\xi)\%$ confidence ellipsoid for the subset of Factor B differences $\{\gamma_1 - \gamma_j : j = 2, \ldots, m\}$. Your expressions can involve vectors and matrices provided that you give nonmatrix expressions for each of the elements of those vectors and matrices.

(c) Use each of the Bonferroni, Scheffé, and multivariate-t methods to obtain confidence intervals for all $[q(q - 1)/2] + m(m - 1)/2]$ differences $\alpha_i - \alpha_{i'}$ $(i' > i = 1, \ldots, q)$ and $\gamma_j - \gamma_{j'}(j' > j = 1, \ldots, m)$ such that the probability of simultaneous coverage is at least $1 - \xi$.

(d) Discuss why Tukey's method cannot be used directly to obtain simultaneous confidence intervals for $\alpha_i - \alpha_{i'}$ $(i' > i = 1, \ldots, q)$ and $\gamma_j - \gamma_{j'}(j' > j = 1, \ldots, m)$ such that the probability of simultaneous coverage is at least $1 - \xi$. Show, however, that Tukey's method could be used directly to obtain confidence intervals for $\alpha_i - \alpha_{i'}$ $(i' > i = 1, \ldots, q)$ such that the probability of simultaneous coverage is at least $1 - \xi_1$, and again to obtain confidence intervals for $\gamma_j - \gamma_{j'}$ $(j' > j = 1, \ldots, m)$ such that the probability of simultaneous coverage is at least $1 - \xi_2$; then combine these via the Bonferroni inequality to get intervals for the union of all of these differences whose simultaneous coverage probability is at least $1 - \xi_1 - \xi_2$.

(e) For each of the four sets of intervals for the α-differences and γ-differences obtained in parts (a) and (c), determine $\mathrm{E}(L^2)/\sigma^2$, where L represents the length of each interval in the set.

(f) Compute $[\mathrm{E}(L^2)/\sigma^2]^{\frac{1}{2}}$ for each of the four sets of confidence intervals when $q = 4$, $m = 5$, $r = 1$, and $\xi = 0.05$, and using $\mathbf{I}$ in place of $\mathbf{R}$ for the multivariate t intervals. Which of the three sets of simultaneous confidence intervals would you recommend in this case, assuming that your interest is in only the Factor A differences and Factor B differences?

23. Consider data in a 3×3 layout that follow a normal Gauss–Markov two-way model with interaction, and suppose that each cell contains exactly r observations. Thus, the model is

$$y_{ijk} = \mu + \alpha_i + \gamma_j + \xi_{ij} + e_{ijk} \quad (i = 1, \ldots, 3;\ j = 1, \ldots, 3;\ k = 1, \ldots, r)$$

where the e_{ijk}'s are independent $\mathrm{N}(0, \sigma^2)$ random variables.

(a) List the "essentially different" interaction contrasts $\xi_{ij} - \xi_{ij'} - \xi_{i'j} + \xi_{i'j'}$ within this layout, and their ordinary least squares estimators. (A list of interaction contrasts is "essentially different" if no contrast in the list is a scalar multiple of another.)

(b) Give expressions for one-at-a-time $100(1-\xi)\%$ confidence intervals for the essentially different interaction contrasts.

(c) Use each of the Bonferroni, Scheffé, and multivariate t methods to obtain confidence intervals for the essentially different interaction contrasts that have probability of simultaneous coverage at least $1-\xi$. For the multivariate t method, give the numerical entries of $\mathbf{R}$.

(d) Explain why Tukey's method cannot be used to obtain simultaneous confidence intervals for the essentially different interaction contrasts that have probability of simultaneous coverage at least $1 - \xi$.

(e) For each of the four sets of intervals for the essentially different error contrasts obtained in parts (b) and (c), determine $\mathrm{E}(L^2)/\sigma^2$, where L represents the length of each interval in the set.

(f) Compute $[\mathrm{E}(L^2)/\sigma^2]^{\frac{1}{2}}$ for each of the four sets of confidence intervals when $\xi = 0.05$. Which of the three sets of simultaneous confidence intervals would you recommend in this case?

24. Consider the two-factor nested model

$$y_{ijk} = \mu + \alpha_i + \gamma_{ij} + e_{ijk} \quad (i = 1, \ldots, q; \quad j = 1, \ldots, m_i; \quad k = 1, \ldots, n_{ij}),$$

where the e_{ijk}'s are independent $\mathrm{N}(0, \sigma^2)$ random variables. Suppose that $100(1 - \xi)\%$ simultaneous confidence intervals for all of the differences $\gamma_{ij} - \gamma_{ij'}$ $(i = 1, \ldots, q; j' > j = 1, \ldots, m_i)$ are desired.

(a) Give $100(1-\xi)\%$ one-at-a-time confidence intervals for $\gamma_{ij} - \gamma_{ij'}$ ($j' > j = 1, \ldots, m_i$; $i = 1, \ldots, q$).

(b) For each of the Bonferroni, Scheffé, and multivariate-t approaches to this problem, the desired confidence intervals can be expressed as

$$(\hat{\gamma}_{ij} - \hat{\gamma}_{ij'}) \pm a\hat{\sigma}\sqrt{v_{ijj'}},$$

where $\hat{\gamma}_{ij} - \hat{\gamma}_{ij'}$ is the least squares estimator of $\gamma_{ij} - \gamma_{ij'}$, $\hat{\sigma}^2$ is the residual mean square, $v_{ijj'} = \text{var}(\hat{\gamma}_{ij} - \hat{\gamma}_{ij'})/\sigma^2$, and a is a percentage point from an appropriate distribution. For each of the three approaches, give a (indexed by its tail probability, degree(s) of freedom, and any other parameters used to index the appropriate distribution). Note: You need not give expressions for the elements of the correlation matrix $\mathbf{R}$ for the multivariate t method.

(c) Consider the case $q = 2$ and suppose that n_{ij} is constant (equal to r) across j. Although Tukey's method is not directly applicable to this problem, it can be used to obtain $100(1-\xi_1)\%$ simultaneous confidence intervals for $\gamma_{1j} - \gamma_{1j'}$ ($j' > j = 1, \ldots, m_1$), and then used a second time to obtain $100(1-\xi_2)\%$ simultaneous confidence intervals for $\gamma_{2j} - \gamma_{2j'}$ ($j' > j = 1, \ldots, m_2$). If ξ_1 and ξ_2 are chosen appropriately, these intervals can then be combined, using Bonferroni's inequality, to obtain a set of intervals for all of the differences $\gamma_{ij} - \gamma_{ij'}(j' > j)$ whose simultaneous coverage probability is at least $1 - \xi_1 - \xi_2$. Give appropriate cutoff point(s) for this final set of intervals. (Note: The cutoff point for some of the intervals in this final set may not be the same as the cutoff point for other intervals in the set. Be sure to indicate the intervals to which each cutoff point corresponds).

(d) For each of the four sets of intervals for the γ-differences obtained in parts (a) and (b), determine an expression for $\text{E}(L^2/\sigma^2)$, where L represents the length of each interval in the set. Do likewise for the method described in part (c), with $\xi_1 = \xi_2 = \xi/2$. For the multivariate t approach, replace $\mathbf{R}$ with $\mathbf{I}$ to obtain conservative intervals.

(e) Evaluate $[\text{E}(L^2/\sigma^2)]^{\frac{1}{2}}$ for each of the four sets of confidence intervals when $q = 2$, $m_1 = m_2 = 5$, $n_{ij} = 2$ for all i and j, and $\xi = 0.05$. Based on these results, which set of simultaneous confidence intervals would you recommend in this case?

25. Under the normal Gauss–Markov two-way partially crossed model introduced in Example 5.1.4-1 with one observation per cell:
    (a) Give $100(1-\xi)\%$ one-at-a-time confidence intervals for $\alpha_i - \alpha_j$ and $\mu + \alpha_i - \alpha_j$.
    (b) Give $100(1-\xi)\%$ Bonferroni, Scheffé, and multivariate t simultaneous confidence intervals for $\{\alpha_i - \alpha_j : j > i = 1, \ldots, q\}$.

26. Under the normal Gauss–Markov Latin square model with q treatments introduced in Exercise 6.21:
    (a) Give $100(1-\xi)\%$ one-at-a-time confidence intervals for $\tau_k - \tau_{k'}$ ($k' > k = 1, \ldots, q$).

(b) Give $100(1-\xi)\%$ Bonferroni, Scheffé, and multivariate t simultaneous confidence intervals for $\{\tau_k - \tau_{k'} : k' > k = 1, \ldots, q\}$.

27. Consider the normal split-plot model introduced in Example 13.4.5-1, but with $\psi \equiv \sigma_b^2/\sigma^2 > 0$ known, and recall the expressions obtained for variances of various mean differences obtained in Exercise 13.25. Let $\tilde{\sigma}^2$ denote the generalized residual mean square.
   (a) Obtain $100(1-\xi)\%$ one-at-a-time confidence intervals for the functions in the sets $\{(\alpha_i - \alpha_{i'}) + (\bar{\xi}_{i\cdot} - \bar{\xi}_{i'\cdot}) : i' > i = 1, \ldots, q\}$ and $\{(\gamma_k - \gamma_{k'}) + (\bar{\xi}_{\cdot k} - \bar{\xi}_{\cdot k'}) : k' > k = 1, \ldots, m\}$. [Note that $\bar{y}_{i\cdot\cdot} - \bar{y}_{i'\cdot\cdot}$ is the least squares estimator of $(\alpha_i - \alpha_{i'}) + (\bar{\xi}_{i\cdot} - \bar{\xi}_{i'\cdot})$, and $\bar{y}_{\cdot\cdot k} - \bar{y}_{\cdot\cdot k'}$ is the least squares estimator of $(\gamma_k - \gamma_{k'}) + (\bar{\xi}_{\cdot k} - \bar{\xi}_{\cdot k'})$.]
   (b) Using the Bonferroni method, obtain a set of confidence intervals whose simultaneous coverage probability for all of the estimable functions listed in part (a) is at least $1-\xi$.
   (c) Using the Scheffé method, obtain a set of confidence intervals whose simultaneous coverage probability for the functions $\{(\alpha_i - \alpha_{i'}) + (\bar{\xi}_{i\cdot} - \bar{\xi}_{i'\cdot}) : i' > i = 1, \ldots, q\}$ and all linear combinations of those functions is $1-\xi$.

28. Consider the normal Gauss–Markov no-intercept simple linear regression model

$$y_i = \beta x_i + e_i \qquad (i = 1, \ldots, n)$$

where $n \geq 2$.
   (a) Find the Scheffé-based $100(1-\xi)\%$ confidence band for the regression line, i.e., for $\{\beta x : x \in \mathbb{R}\}$.
   (b) Describe the behavior of this band as a function of x (for example, where is it narrowest and how narrow is it at its narrowest point?). Compare this to the behavior of the Scheffé-based $100(1-\xi)\%$ confidence band for the regression line in normal simple linear regression.

29. Consider the normal prediction-extended Gauss–Markov full-rank multiple regression model

$$y_i = \mathbf{x}_i^T \boldsymbol{\beta} + e_i \quad (i = 1, \cdots, n+s),$$

where $\mathbf{x}_1, \ldots, \mathbf{x}_n$ are known p-vectors of explanatory variables (possibly including an intercept) but $\mathbf{x}_{n+1}, \ldots, \mathbf{x}_{n+s}$ $(s \geq 1)$ are p-vectors of the same explanatory variables that cannot be ascertained prior to actually observing $y_{n+1}, \cdots, y_{n+s}$ and must therefore be treated as unknown. This exercise considers the problem of obtaining prediction intervals for $y_{n+1}, \cdots, y_{n+s}$ that have simultaneous coverage probability at least $1-\xi$, where $0 < \xi < 1$, in this scenario.

(a) Apply the Scheffé method to show that one solution to the problem consists of intervals of form

$$\left\{\mathbf{x}_{n+i}^T\hat{\boldsymbol{\beta}} \pm \hat{\sigma}\sqrt{(p+s)F_{\xi,p+s,n-p}\mathbf{x}_{n+i}^T(\mathbf{X}^T\mathbf{X})^{-1}\mathbf{x}_{n+i}+1}\right.$$

$$\left.\text{for all } \mathbf{x}_{n+i} \in \mathbb{R}^p \text{ and all } i = 1,\ldots,s\right\}.$$

(b) Another possibility is to decompose $y_{n+i}(\mathbf{x})$ into its two components $\mathbf{x}_{n+i}^T\boldsymbol{\beta}$ and e_{n+i}, obtain confidence or prediction intervals separately for each of these components by the Scheffé method, and then combine them via the Bonferroni inequality to achieve the desired simultaneous coverage probability. Verify that the resulting intervals are of form

$$\left\{\mathbf{x}_{n+i}^T\hat{\boldsymbol{\beta}} \pm \hat{\sigma}[a(\mathbf{x}_{n+i})+b] \quad \text{for all } \mathbf{x}_{n+i} \in \mathbb{R}^p \text{ and all } i = 1,\ldots,s\right\},$$

where $a(\mathbf{x}_{n+i}) = \sqrt{pF_{\xi^*,p,n-p}\mathbf{x}_{n+i}^T(\mathbf{X}^T\mathbf{X})^{-1}\mathbf{x}_{n+i}}$, $b = \sqrt{sF_{\xi-\xi^*,s,n-p}}$, and $0 < \xi^* < 1$.

(c) Still another possibility is based on the fact that $(1/\hat{\sigma})(e_{n+1},\ldots,e_{n+s})^T$ has a certain multivariate-t distribution under the assumed model. Prove this fact and use it to obtain a set of intervals for $y_{n+1}(\mathbf{x}_{n+1}),\cdots,y_{n+s}(\mathbf{x}_{n+s})$ of the form

$$\left\{\mathbf{x}_{n+i}^T\hat{\boldsymbol{\beta}} \pm \hat{\sigma}[a(\mathbf{x}_{n+i})+c] \quad \text{for all } \mathbf{x}_{n+i} \in \mathbb{R}^p \text{ and all } i = 1,\ldots,s\right\}$$

whose simultaneous coverage probability is at least $1-\xi$, where c is a quantity you should determine.

(d) Are the intervals obtained in part (b) uniformly narrower than the intervals obtained in part (c), or vice versa? Explain.

(e) The assumption that $e_1,\ldots,e_{n+s}$ are mutually independent can be relaxed somewhat without affecting the validity of intervals obtained in part (c). In precisely what manner can this independence assumption be relaxed?

Note: The approaches described in parts (a) and (b) were proposed by Carlstein (1986), and the approach described in part (c) was proposed by Zimmerman (1987).

30. Consider the one-factor random effects model with balanced data

$$y_{ij} = \mu + b_i + d_{ij} \quad (i = 1,\ldots,q; \quad j = 1,\ldots,r),$$

described previously in more detail in Example 13.4.2-2. Recall from that example that the BLUP of $b_i - b_{i'}$ $(i \neq i')$ is

$$\widetilde{b_i - b_{i'}} = \frac{r\psi}{r\psi + 1}(\bar{y}_{i\cdot} - \bar{y}_{i'\cdot}).$$

Let $\tilde{\sigma}^2$ be the generalized residual mean square and suppose that the joint distribution of the b_i's and d_{ij}'s is multivariate normal.

(a) Give $100(1 - \xi)\%$ one-at-a-time prediction intervals for $b_i - b_{i'}$ $(i' > i = 1, \ldots, q)$.

(b) Obtain multivariate t-based prediction intervals for all $q(q-1)/2$ pairwise differences $b_i - b_{i'}$ $(i' > i = 1, \ldots, q)$ whose simultaneous coverage probability is $1 - \xi$. Determine the elements of the appropriate matrix $\mathbf{R}$.

(c) Explain why Tukey's method is not directly applicable to the problem of obtaining simultaneous prediction intervals for all $q(q-1)/2$ pairwise differences $b_i - b_{i'}$ $(i' > i = 1, \ldots, q)$.

31. Let $a_1, \ldots, a_k$ represent real numbers. It can be shown that $|a_i - a_{i'}| \leq 1$ for all i and i' if and only if $|\sum_{i=1}^k c_i a_i| \leq \frac{1}{2}\sum_{i=1}^k |c_i|$ for all c_i such that $\sum_{i=1}^k c_i = 0$.

(a) Using this result, extend Tukey's method for obtaining $100(1 - \xi)\%$ simultaneous confidence intervals for all possible differences among k linearly independent estimable functions $\psi_1, \ldots, \psi_k$, to all functions of the form $\sum_{i=1}^k d_i \psi_i$ with $\sum_{i=1}^k d_i = 0$ (i.e., to all contrasts).

(b) Specialize the intervals obtained in part (a) to the case of a balanced one-factor model.

32. Consider the normal Gauss–Markov model $\{\mathbf{y}, \mathbf{X}\boldsymbol{\beta}, \sigma^2\mathbf{I}\}$, and suppose that $\mathbf{X}$ is $n \times 3$ (where $n > 3$) and $\text{rank}(\mathbf{X}) = 3$. Let $(\mathbf{X}^T\mathbf{X})^{-1} = (c_{ij})$, $\boldsymbol{\beta} = (\beta_j)$, $\hat{\boldsymbol{\beta}} = (\hat{\beta}_j) = (\mathbf{X}^T\mathbf{X})^{-1}\mathbf{X}^T\mathbf{y}$, and $\hat{\sigma}^2 = (\mathbf{y} - \mathbf{X}\hat{\boldsymbol{\beta}})^T(\mathbf{y} - \mathbf{X}\hat{\boldsymbol{\beta}})/(n-3)$.

(a) Each of the Bonferroni, Scheffé, and multivariate t methods can be used to obtain a set of intervals for β_1, β_2, β_3, $\beta_1 + \beta_2$, $\beta_1 + \beta_3$, $\beta_2 + \beta_3$, and $\beta_1 + \beta_2 + \beta_3$ whose simultaneous coverage probability is equal to, or larger than, $1 - \xi$ (where $\xi \in (0, 1)$). Give the set of intervals corresponding to each method.

(b) Obtain another set of intervals for the same functions listed in part (a) whose simultaneous coverage probability is equal to $1 - \xi$, but which are constructed by adding the endpoints of $100(1 - \xi)\%$ multivariate t intervals for β_1, β_2, and β_3 only.

(c) Obtain a $100(1 - \xi)\%$ simultaneous confidence band for $\{\beta_1 x_1 + \beta_2 x_2$, i.e., an infinite collection of intervals $\{(L(x_1, x_2), U(x_1, x_2)): (x_1, x_2) \in \mathbb{R}^2\}$ such that

$$\Pr[\beta_1 x_1 + \beta_2 x_2 \in (L(x_1, x_2), U(x_1, x_2)) \text{ for all } (x_1, x_2) \in \mathbb{R}^2] = 1 - \xi.$$

33. Consider the normal Gauss–Markov model $\{\mathbf{y}, \mathbf{X}\boldsymbol{\beta}, \sigma^2\mathbf{I}\}$, let $\tau_i = \mathbf{c}_i^T\boldsymbol{\beta}$ $(i = 1, \ldots, q)$, and let $\tau_{q+1} = \sum_{i=1}^q \tau_i = (\sum_{i=1}^q \mathbf{c}_i)^T\boldsymbol{\beta}$. Give the interval for τ_{q+1} belonging to each of the following sets of simultaneous confidence intervals:
    (a) $100(1-\xi)\%$ Scheffé intervals that include intervals for $\tau_1, \ldots, \tau_q$.
    (b) The infinite collection of all intervals generated by taking linear combinations of the $100(1-\xi)\%$ multivariate t intervals for $\tau_1, \ldots, \tau_q$.
    (c) The finite set of multivariate t intervals for $\tau_1, \ldots, \tau_q, \tau_{q+1}$ that have exact simultaneous coverage probability equal to $1-\xi$.
34. Consider the normal Gauss–Markov no-intercept simple linear regression model

$$y_i = \beta x_i + e_i \qquad (i = 1, \ldots, n)$$

where $n \geq 2$. Let y_{n+1} and y_{n+2} represent responses not yet observed at x-values x_{n+1} and x_{n+2}, respectively, and suppose that

$$y_i = \beta x_i + e_i \qquad (i = n+1, n+2)$$

where

$$\begin{pmatrix} e_{n+1} \\ e_{n+2} \end{pmatrix} \sim \mathrm{N}_2\left(\mathbf{0}_2, \sigma^2\begin{pmatrix} x_{n+1}^2 & 0 \\ 0 & x_{n+2}^2 \end{pmatrix}\right),$$

and $(e_{n+1}, e_{n+2})^T$ and $(e_1, e_2, \ldots, e_n)^T$ are independent.
    (a) Obtain the vector of BLUPs of $(y_{n+1}, y_{n+2})^T$, giving each element of the vector by an expression free of matrices.
    (b) Obtain the variance–covariance matrix of prediction errors corresponding to the vector of BLUPs you obtained in part (a). Again, give expressions for the elements of this matrix that do not involve matrices.
    (c) Using the results of parts (a) and (b), obtain multivariate-t prediction intervals for $(y_{n+1}, y_{n+2})^T$ having simultaneous coverage probability $1-\xi$ (where $0 < \xi < 1$).
    (d) Obtain a $100(1-\xi)\%$ prediction interval for $(y_{n+1} + y_{n+2})/2$.
35. Suppose that observations (x_i, y_i), $i = 1, \ldots, n$ follow the normal Gauss–Markov simple linear regression model

$$y_i = \beta_1 + \beta_2 x_i + e_i$$

where $n \geq 3$. Let $\hat{\beta}_1$ and $\hat{\beta}_2$ be the ordinary least squares estimators of β_1 and β_2, respectively; let $\hat{\sigma}^2$ be the usual residual mean square; let $\bar{x} = (\sum_{i=1}^n x_i)/n$, and let $SXX = \sum_{i=1}^n (x_i - \bar{x})^2$. Recall that in this setting,

$$(\hat{\beta}_1 + \hat{\beta}_2 x_{n+1}) \pm t_{\xi/2,n-2}\hat{\sigma}\sqrt{1 + \frac{1}{n} + \frac{(x_{n+1} - \bar{x})^2}{SXX}}$$

is a $100(1 - \xi)\%$ prediction interval for an unobserved y-value to be taken at a specified x-value x_{n+1}. Suppose that it is desired to predict the values of three unobserved y-values, say y_{n+1}, y_{n+2}, and y_{n+3}, which are all to be taken at the same particular x-value, say x^*. Assume that the unobserved values of y follow the same model as the observed data; that is,

$$y_i = \beta_1 + \beta_2 x^* + e_i \quad (i = n+1,\ n+2,\ n+3)$$

where $e_{n+1}, e_{n+2}, e_{n+3}$ are independent $N(0, \sigma^2)$ random variables and are independent of $e_1, \ldots, e_n$.

(a) Give expressions for Bonferroni prediction intervals for $y_{n+1}, y_{n+2}, y_{n+3}$ whose simultaneous coverage probability is at least $1 - \xi$.
(b) Give expressions for Scheffé prediction intervals for $y_{n+1}, y_{n+2}, y_{n+3}$ whose simultaneous coverage probability is at least $1 - \xi$. (Note: The three requested intervals are part of an infinite collection of intervals, but give expressions for just those three.)
(c) Give expressions for multivariate t prediction intervals for $y_{n+1}, y_{n+2}, y_{n+3}$ whose simultaneous coverage probability is exactly $1 - \xi$. Note: The off-diagonal elements of the correlation matrix $\mathbf{R}$ referenced by the multivariate t quantiles in your prediction intervals are all equal to each other; give an expression for this common correlation coefficient.
(d) Give expressions for Bonferroni prediction intervals for all pairwise differences among $y_{n+1}, y_{n+2}, y_{n+3}$, whose simultaneous coverage probability is at least $1 - \xi$.
(e) Give expressions for multivariate t prediction intervals for all pairwise differences among $y_{n+1}, y_{n+2}, y_{n+3}$, whose simultaneous coverage probability is exactly $1 - \xi$. Give expressions for the off-diagonal elements of the correlation matrix $\mathbf{R}$ referenced by the multivariate t quantiles in your prediction intervals.
(f) Explain why Tukey's method is applicable to the problem of obtaining simultaneous prediction intervals for all pairwise differences among $y_{n+1}, y_{n+2}, y_{n+3}$, and obtain such intervals.

36. Consider the normal Gauss–Markov one-factor model with balanced data:

$$y_{ij} = \mu + \alpha_i + e_{ij} \qquad (i = 1, \ldots, q; \quad j = 1, \ldots, r)$$

Recall from Example 15.3-2 that a $100(1-\xi)\%$ confidence interval for a single level difference $\alpha_i - \alpha_{i'}$ ($i' > i = 1, \ldots, q$) is given by

$$(\bar{y}_{i\cdot} - \bar{y}_{i'\cdot}) \pm t_{\xi/2, q(r-1)} \hat{\sigma} \sqrt{2/r}$$

where $\hat{\sigma}^2 = \sum_{i=1}^{q} \sum_{j=1}^{r} (y_{ij} - \bar{y}_{i\cdot})^2 / [q(r-1)]$. Also recall that in the same example, several solutions were given to the problem of obtaining confidence intervals for the level differences $\{\alpha_i - \alpha_{i'} : i' > i = 1, \ldots, q\}$ whose simultaneous coverage probability is at least $1 - \xi$.

Now, however, consider a slightly different problem under the same model. Suppose that the factors represent treatments, one of which is a "control" treatment or placebo, and that the investigator has considerably less interest in estimating differences involving the control treatment than in estimating differences not involving the control treatment. That is, letting α_1 correspond to the control treatment, the investigator has less interest in $\alpha_1 - \alpha_2, \alpha_1 - \alpha_3, \ldots, \alpha_1 - \alpha_q$ than in $\{\alpha_i - \alpha_{i'} : i' > i = 2, \ldots, q\}$. Thus, rather than using any of the simultaneous confidence intervals obtained in Example 15.3-2, the investigator decides to use "new" confidence intervals for the treatment differences $\{\alpha_i - \alpha_{i'} : i' > i = 1, \ldots, q\}$ that satisfy the following requirements:

I. The new intervals for $\{\alpha_i - \alpha_{i'} : i' > i = 1, \ldots, q\}$, like the intervals in Example 15.3-2, have simultaneous coverage probability at least $1 - \xi$.

II. Each of the new intervals for $\{\alpha_i - \alpha_{i'} : i' > i = 2, \ldots, q\}$ obtained by a given method is narrower than the interval in Example 15.3-2 for the same treatment difference, obtained using the same method.

III. Each of the new intervals for $\alpha_1 - \alpha_2, \alpha_1 - \alpha_3, \ldots, \alpha_1 - \alpha_q$ is no more than twice as wide as each of the new intervals for $\{\alpha_i - \alpha_{i'} : i' > i = 2, \ldots, q\}$.

IV. All of the new intervals for $\{\alpha_i - \alpha_{i'} : i' > i = 2, \ldots, q\}$ are of equal width and as narrow as possible, subject to the first three rules.

(a) If it is possible to use the Scheffé method to obtain new intervals that satisfy the prescribed rules, give such a solution. Otherwise, explain why it is not possible.

(b) If it is possible to use the multivariate t method to obtain new intervals that satisfy the prescribed rules, give such a solution. Otherwise, explain why it is not possible.

Note: You may not use the Bonferroni method in formulating solutions, but you may use the notion of linear combinations of intervals.

37. Consider a normal prediction-extended linear model

$$\begin{pmatrix} \mathbf{y} \\ y_{n+1} \\ y_{n+2} \end{pmatrix} = \begin{pmatrix} \mathbf{X} \\ \mathbf{x}_{n+1}^T \\ \mathbf{x}_{n+2}^T \end{pmatrix} \boldsymbol{\beta} + \begin{pmatrix} \mathbf{e} \\ e_{n+1} \\ e_{n+2} \end{pmatrix}$$

where $\begin{pmatrix} \mathbf{e} \\ e_{n+1} \\ e_{n+2} \end{pmatrix}$ satisfies Gauss–Markov assumptions except that $\text{var}(e_{n+1}) = 2\sigma^2$ and $\text{var}(e_{n+2}) = 3\sigma^2$.

(a) Give a $100(1 - \xi)\%$ prediction interval for y_{n+1}.

(b) Give a $100(1 - \xi)\%$ prediction interval for y_{n+2}.

(c) Give the $100(1-\xi)\%$ Bonferroni simultaneous prediction intervals for y_{n+1} and y_{n+2}.

(d) The intervals you obtained in part (c) do not have the same width. Indicate how the Bonferroni method could be used to obtain prediction intervals for

y_{n+1} and y_{n+2} that have the same width yet have simultaneous coverage probability at least $1 - \xi$.
(e) Obtain the $100(1 - \xi)\%$ multivariate t simultaneous prediction intervals for y_{n+1} and y_{n+2}.

38. Consider the normal Gauss–Markov one-factor analysis-of-covariance model for balanced data,

$$y_{ij} = \mu + \alpha_i + \gamma x_{ij} + e_{ij} \quad (i = 1, \ldots, q, j = 1, \ldots, r),$$

where $r \geq 2$, $x_{ij} \neq x_{ij'}$ for $j \neq j'$ and for $i = 1, \ldots, q$. Let $\mathbf{y} = (y_{11}, y_{12}, \ldots, y_{qr})^T$, let $\mathbf{X}$ be the corresponding model matrix, and let $\boldsymbol{\beta} = (\mu, \alpha_1, \ldots, \alpha_q, \gamma)^T$. Let $\hat{\boldsymbol{\beta}}$ be a solution to the normal equations and let $\hat{\sigma}^2$ be the residual mean square. Furthermore, let $\mathbf{c}_0$, $\mathbf{c}_i$, and $\mathbf{c}_{ix}$ be such that

$$\mathbf{c}_0^T \boldsymbol{\beta} = \gamma, \quad \mathbf{c}_i^T \boldsymbol{\beta} = \mu + \alpha_i, \quad \text{and} \quad \mathbf{c}_{ix}^T \boldsymbol{\beta} = \mu + \alpha_i + \gamma x \quad \text{where } -\infty < x < \infty.$$

Also let $\mathbf{c}_{ii'} = \mathbf{c}_i - \mathbf{c}_{i'}$ for $i' > i = 1, \ldots, q$. Let $0 < \xi < 1$. For each of the following four parts, give a confidence interval or a set of confidence intervals that satisfy the stated criteria. Express these confidence intervals in terms of the following quantities: $\mathbf{X}$, $\hat{\boldsymbol{\beta}}$, $\hat{\sigma}$, σ, $\mathbf{c}_0$, $\mathbf{c}_i$, $\mathbf{c}_{ii'}$, $\mathbf{c}_{ix}$, ξ, q, r, and appropriate cut-off points from an appropriate distribution.
(a) A $100(1 - \xi)\%$ confidence interval for γ.
(b) A set of confidence intervals for $\{\mu + \alpha_i : i = 1, \ldots, q\}$ whose simultaneous coverage probability is exactly $1 - \xi$.
(c) A set of confidence intervals for $\{\mu + \alpha_i + \gamma x : i = 1, \ldots, q$ and all $x \in (-\infty, \infty)\}$ whose simultaneous coverage probability is at least $1 - \xi$.
(d) A set of confidence intervals for $\{\alpha_i - \alpha_{i'} : i' > i = 1, \ldots, q\}$ whose simultaneous coverage probability is exactly $1 - \xi$.

39. Consider a situation in which a response variable, y, and a single explanatory variable, x, are measured on n subjects. Suppose that the response for each subject having a nonnegative value of x is related to x through a simple linear regression model without an intercept, and the response for each subject having a negative value of x is also related to x through a simple linear regression without an intercept; however, the slopes of the two regression models are possibly different. That is, assume that the observations follow the model

$$y_i = \begin{cases} \beta_1 x_i + e_i, & \text{if } x_i < 0 \\ \beta_2 x_i + e_i, & \text{if } x_i \geq 0, \end{cases}$$

for $i = 1, \ldots, n$. Suppose further that the e_i's are independent $\mathrm{N}(0, \sigma^2)$ variables; that $x_i \neq 0$ for all i; and that there is at least one x_i less than 0 and at least one x_i greater than 0. Let n_1 denote the number of subjects whose x-value is less than 0, and let $n_2 = n - n_1$. Finally, let $0 < \xi < 1$.

(a) Give expressions for the elements of the model matrix **X**. (Note: To make things easier, assume that the elements of the response vector **y** are arranged in such a way that the first n_1 elements of **y** correspond to those subjects whose x-value is less than 0.) Also obtain nonmatrix expressions for the least squares estimators of β_1 and β_2, and obtain the variance–covariance matrix of those two estimators. Finally, obtain a nonmatrix expression for $\hat{\sigma}^2$.

(b) Give confidence intervals for β_1 and β_2 whose simultaneous coverage probability is exactly $1 - \xi$.

(c) Using the Scheffé method, obtain a $100(1 - \xi)\%$ simultaneous confidence band for $\mathrm{E}(y)$; that is, obtain expressions (in as simple a form as possible) for functions $a_1(x)$, $b_1(x)$, $a_2(x)$, and $b_2(x)$ such that

$$\Pr[a_1(x) \leq \beta_1 x \leq b_1(x) \text{ for all } x < 0 \quad \text{and}$$
$$a_2(x) \leq \beta_2 x \leq b_2(x) \text{ for all } x \geq 0] \geq 1 - \xi.$$

(d) Consider the confidence band obtained in part (c). Under what circumstances will the band's width at x equal the width at $-x$ (for all x)?

40. Consider a full-rank normal Gauss–Markov analysis-of-covariance model in a setting where there are one or more factors of classification and one or more regression variables. Let q represent the number of combinations of the factor levels. Suppose further that at each such combination the model is a regression model with the same number, p_c, of regression variables (including the intercept in this number); see part (c) for two examples. For $i = 1, \ldots, q$ let $\mathbf{x}_i$ represent an arbitrary p_c-vector of the regression variables (including the intercept) for the ith combination of classificatory explanatory variables, and let $\boldsymbol{\beta}_i$ and $\hat{\boldsymbol{\beta}}_i$ represent the corresponding vectors of regression coefficients and their least squares estimators from the fit of the complete model, respectively. Note that an arbitrary p-vector $\mathbf{x}$ of the explanatory variables consists of $\mathbf{x}_i$ for some i, padded with zeroes in appropriate places.

(a) For $i = 1, \ldots, q$ let $\mathbf{G}_{ii}$ represent the submatrix of $(\mathbf{X}^T\mathbf{X})^{-1}$ obtained by deleting the rows and columns that correspond to the elements of $\mathbf{x}$ excluded to form $\mathbf{x}_i$. Show that

$$\Pr\left[\mathbf{x}_i^T\boldsymbol{\beta}_i \in \mathbf{x}_i^T\hat{\boldsymbol{\beta}}_i \pm \hat{\sigma}\sqrt{p_c F_{\xi/q, p_c, n-p}\mathbf{x}_i^T\mathbf{G}_{ii}\mathbf{x}_i}\right.$$
$$\left.\text{for all } \mathbf{x}_i \in \mathbb{R}^{p_c} \text{ and all } i = 1, \ldots, q\right] \geq 1 - \xi. \qquad (15.19)$$

(Hint: First apply Scheffé's method to obtain a confidence band for $\mathbf{x}_i^T\boldsymbol{\beta}_i$ for fixed $i \in \{1, \ldots, q\}$.)

(b) Obtain an expression for the ratio of the width of the interval for $\mathbf{x}_i^T\boldsymbol{\beta}_i$ in the collection of $100(1 - \xi)\%$ simultaneous confidence intervals determined in part (a), to the width of the interval for the same $\mathbf{x}_i^T \boldsymbol{\beta}_i$ in the collection of standard $100(1-\xi)\%$ Scheffé confidence intervals

$$\mathbf{x}^T \hat{\boldsymbol{\beta}} \pm \hat{\sigma}\sqrt{pF_{\xi,p,n-p}\mathbf{x}^T(\mathbf{X}^T\mathbf{X})^{-1}\mathbf{x}}, \quad \text{for all } \mathbf{x} \in \mathbb{R}^p.$$

(c) For each of the following special cases of models, evaluate the ratio obtained in part (b) when $\xi = 0.05$:
   i. The three-group simple linear regression model

$$y_{ij} = \beta_{i1} + \beta_{i2}x_{ij} + e_{ij} \quad (i = 1, 2, 3;\ j = 1, \ldots, 10).$$

   ii. The three-group common-slope simple linear regression model

$$y_{ij} = \beta_{i1} + \beta_{2}x_{ij} + e_{ij} \quad (i = 1, 2, 3;\ j = 1, \ldots, 10).$$

Note: This exercise was inspired by results from Lane and Dumouchel (1994).

References

Burnham, K. P., & Anderson, D. R. (2002). *Model selection and multimodel inference: A practical information-theoretic approach* (2nd ed.). New York: Springer.

Carlstein, E. (1986). Simultaneous confidence regions for predictions. *The American Statistician, 40*, 277–279.

Ghosh, B. K. (1973). Some monotonicity theorems for χ^2, F and t distributions with applications. *Journal of the Royal Statistical Society, Series B, 35*, 480–492.

Kass, R. E., & Raftery, A. E. (1995). Bayes factors. *Journal of the American Statistical Association, 90*, 773–795.

Lane, T. P., & Dumouchel, W. H. (1994). Simultaneous confidence intervals in multiple regression. *The American Statistician, 48*, 315–321.

Lehmann, E., & Romano, J. P. (2005). *Testing statistical hypotheses* (3rd ed.). New York: Springer.

Scheffé, H. (1959). *The analysis of variance*. New York: Wiley.

Sidák, Z. (1968). On multivariate normal probabilities of rectangles: Their dependence on correlations. *Annals of Mathematical Statistics, 39*, 1425–1434.

Tukey, J. W. (1949a). Comparing individual means in the analysis of variance. *Biometrics, 5*, 99–114.

Tukey, J. W. (1949b). One degree of freedom for non-additivity. *Biometrics, 5*, 232–242.

Zimmerman, D. L. (1987). Simultaneous confidence regions for predictions based on the multivariate t distribution. *The American Statistician, 41*, 247.

Inference for Variance–Covariance Parameters

16

In Chaps. 11 and 13, we obtained BLUEs for estimable linear functions under the Aitken model and BLUPs for predictable linear functions under the prediction-extended Aitken model, and we noted that this methodology could be used to estimate estimable linear functions or predict predictable linear functions of $\boldsymbol{\beta}$, $\mathbf{b}$ and $\mathbf{d}$ in the mixed (and random) effects model, or more generally to estimate estimable and predict predictable linear functions in a general mixed linear model, provided that the variance–covariance parameters $\boldsymbol{\psi}$ (in the case of a mixed effects model) or $\boldsymbol{\theta}$ (in the case of a general mixed model) are known. Moreover, it was also noted at the ends of both chapters that the customary procedure for performing these inferences when the variance–covariance parameters are unknown is to first estimate those parameters from the data and then use BLUE/BLUP formulas with the estimates substituted for the unknown true values. It is natural, then, to ask how the variance–covariance parameters should be estimated. Answering this question is the topic of this chapter. We begin with an answer that applies when the model is a components-of-variance model, for which a method known as quadratic unbiased estimation can be used to estimate the variance–covariance parameters, which are variance components in that case. Then we give an answer, based on likelihood-based estimation, that applies to any normal general mixed linear model.

16.1 Quadratic Unbiased Estimation and Related Inference Methods

16.1.1 The General Components-of-Variance Model

Throughout this section we consider a special case of the mixed effects model introduced in Sect. 5.2.3. The model considered here is as follows:

$$\mathbf{y} = \mathbf{X}\boldsymbol{\beta} + \mathbf{Z}_1\mathbf{b}_1 + \mathbf{Z}_2\mathbf{b}_2 + \cdots + \mathbf{Z}_{m-1}\mathbf{b}_{m-1} + \mathbf{d},$$

D. L. Zimmerman, *Linear Model Theory*,
https://doi.org/10.1007/978-3-030-52063-2_16

where $\mathbf{X}$ is a specified $n \times p$ matrix of rank p^*, $\boldsymbol{\beta}$ is a p-vector of unknown parameters, $\mathbf{Z}_i$ is a specified $n \times q_i$ matrix, $\mathbf{b}_i$ is a random q_i-vector such that $\mathrm{E}(\mathbf{b}_i) = \mathbf{0}$ and $\mathrm{var}(\mathbf{b}_i) = \sigma_i^2\mathbf{I}$, and $\mathbf{d}$ is a random n-vector such that $\mathrm{E}(\mathbf{d}) = \mathbf{0}$ and $\mathrm{var}(\mathbf{d}) = \sigma^2\mathbf{I}$. Furthermore, it is assumed that $\mathrm{E}(\mathbf{d}\mathbf{b}_i^T) = \mathbf{0}$ and $\mathrm{E}(\mathbf{b}_i\mathbf{b}_j^T) = \mathbf{0}$ for $j \neq i = 1, \ldots, m-1$. The variances $\sigma_1^2, \sigma_2^2, \ldots, \sigma_{m-1}^2, \sigma^2$ are called *variance components*, and the model is called the *general components-of-variance model.* The parameter space for the vector $\boldsymbol{\theta} = (\sigma_1^2, \sigma_2^2, \ldots, \sigma_{m-1}^2, \sigma^2)^T$ of variance components is

$$\Theta = \{\boldsymbol{\theta} : \sigma_i^2 \geq 0 \text{ for } i = 1, \ldots, m-1,\ \sigma^2 > 0\},$$

and the parameter space for all model parameters is $\{\boldsymbol{\beta}, \boldsymbol{\theta} : \boldsymbol{\beta} \in \mathbb{R}^p, \boldsymbol{\theta} \in \Theta\}$.

Upon comparison, we see that the model equation for the general components-of-variance model is a multi-part decomposition of the same model equation for the mixed effects model, i.e., the components-of-variance model equation may be written as

$$\mathbf{y} = \mathbf{X}\boldsymbol{\beta} + \mathbf{Z}\mathbf{b} + \mathbf{d}$$

with $\mathbf{Z} = (\mathbf{Z}_1, \ldots, \mathbf{Z}_{m-1})$ and $\mathbf{b} = \begin{pmatrix} \mathbf{b}_1 \\ \vdots \\ \mathbf{b}_{m-1} \end{pmatrix}$. Although the model equations of the two models are essentially the same, the variance–covariance matrices of $\mathbf{d}$ and $\mathbf{b}$ are more specialized under the components-of-variance model; they are $\mathbf{R} = \mathbf{I}$ and $\mathbf{G} = \mathbf{G}(\boldsymbol{\psi}) = \oplus_{i=1}^{m-1}(\sigma_i^2/\sigma^2)\mathbf{I}_{q_i}$, respectively, where

$$\boldsymbol{\psi} = (\psi_i) = (\sigma_1^2/\sigma^2, \sigma_2^2/\sigma^2, \ldots, \sigma_{m-1}^2/\sigma^2)^T.$$

The parameter space for $\boldsymbol{\psi}$, derived from that for $\boldsymbol{\theta}$, is $\Psi = \{\boldsymbol{\psi} : \psi_i \geq 0 \text{ for } i = 1, \ldots, m-1\}$. Thus, for this model $\mathbf{R}$ is known and positive definite, but $\mathbf{G}$ is not fully known and merely nonnegative definite (not necessarily positive definite).

Although there are (inequality) constraints on the variance components, we call a components-of-variance model *unconstrained* if there are no constraints on the parameters in its mean structure. This is in keeping with terminology introduced in Chap. 5.

16.1.2 Estimability

Similar to the situation with $\boldsymbol{\beta}$ described in Chap. 6, it is not always the case that every element of $\boldsymbol{\theta}$ (i.e., every variance component) or function thereof may be sensibly estimated. For example, consider a Gauss–Markov model for which $n = p^*$. Under this model, there is no sensible estimator of the residual variance,

σ^2. The goal of this subsection is to obtain necessary and sufficient conditions under which a linear function $\mathbf{f}^T\boldsymbol{\theta}$ of the variance components may be sensibly (unbiasedly) estimated, where $\mathbf{f} = (f_i)$ is an m-vector.

We begin with three definitions.

Definition 16.1.1 An estimator $q(\mathbf{y})$ of a linear function of variance components $\mathbf{f}^T\boldsymbol{\theta}$ is said to be *quadratic* if $q(\mathbf{y}) = \mathbf{y}^T\mathbf{A}\mathbf{y}$ for some symmetric matrix $\mathbf{A}$ of constants.

Definition 16.1.2 Under an unconstrained components-of-variance model, an estimator $q(\mathbf{y})$ of a linear function of variance components $\mathbf{f}^T\boldsymbol{\theta}$ is said to be *unbiased* if $\mathrm{E}[q(\mathbf{y})] = \mathbf{f}^T\boldsymbol{\theta}$ for all $\boldsymbol{\beta}$ and all $\boldsymbol{\theta} \in \Theta$.

Definition 16.1.3 Under an unconstrained components-of-variance model, a linear function $\mathbf{f}^T\boldsymbol{\theta}$ is said to be (quadratically) *estimable* if a quadratic function $\mathbf{y}^T\mathbf{A}\mathbf{y}$ exists that estimates it unbiasedly. Otherwise, it is said to be *nonestimable*.

Under an unconstrained components-of-variance model, for any $n \times n$ symmetric matrix $\mathbf{A}$ we have, by Theorem 4.2.4,

$$\mathrm{E}(\mathbf{y}^T\mathbf{A}\mathbf{y}) = \boldsymbol{\beta}^T\mathbf{X}^T\mathbf{A}\mathbf{X}\boldsymbol{\beta} + \mathrm{tr}\left[\mathbf{A}\left(\sum_{i=1}^{m-1}\theta_i\mathbf{Z}_i\mathbf{Z}_i^T + \theta_m\mathbf{I}\right)\right]. \tag{16.1}$$

This leads us to the following theorem, which characterizes the collection of estimable linear functions of the variance components.

Theorem 16.1.1 *Under an unconstrained components-of-variance model with model matrix* $\mathbf{X}$ *and random effects coefficient matrix* $\mathbf{Z} = (\mathbf{Z}_1, \ldots, \mathbf{Z}_{m-1})$*, a linear function* $\mathbf{f}^T\boldsymbol{\theta}$ *is estimable if and only if a symmetric matrix* $\mathbf{A}$ *exists such that* $\mathbf{X}^T\mathbf{A}\mathbf{X} = \mathbf{0}$*,* $\mathrm{tr}(\mathbf{Z}_i^T\mathbf{A}\mathbf{Z}_i) = f_i$ $(i = 1, \ldots, m-1)$*, and* $\mathrm{tr}(\mathbf{A}) = f_m$*.*

Proof Observe that

$$\mathrm{tr}\left(\mathbf{A}\sum_{i=1}^{m-1}\theta_i\mathbf{Z}_i\mathbf{Z}_i^T\right) = \mathrm{tr}\left(\sum_{i=1}^{m-1}\theta_i\mathbf{A}\mathbf{Z}_i\mathbf{Z}_i^T\right) = \sum_{i=1}^{m-1}\theta_i\mathrm{tr}(\mathbf{A}\mathbf{Z}_i\mathbf{Z}_i^T) = \sum_{i=1}^{m-1}\theta_i\mathrm{tr}(\mathbf{Z}_i^T\mathbf{A}\mathbf{Z}_i).$$

The sufficiency follows directly upon applying the given conditions on $\mathbf{A}$ to (16.1). For the necessity, suppose that $\mathbf{f}^T\boldsymbol{\theta}$ is estimable, implying that an unbiased estimator $\mathbf{y}^T\mathbf{A}\mathbf{y}$ of $\mathbf{f}^T\boldsymbol{\theta}$ exists. Then it follows from (16.1) that $\boldsymbol{\beta}^T\mathbf{X}^T\mathbf{A}\mathbf{X}\boldsymbol{\beta} + \mathrm{tr}(\mathbf{A}\sum_{i=1}^{m-1}\theta_i\mathbf{Z}_i\mathbf{Z}_i^T) + \mathrm{tr}(\mathbf{A}\theta_m\mathbf{I}) = \mathbf{f}^T\boldsymbol{\theta}$ for all $\boldsymbol{\beta}$ and all $\boldsymbol{\theta} \in \Theta$. Thus $\boldsymbol{\beta}^T\mathbf{X}^T\mathbf{A}\mathbf{X}\boldsymbol{\beta} = 0 = \boldsymbol{\beta}^T\mathbf{0}_{p\times p}\boldsymbol{\beta}$ for all $\boldsymbol{\beta}$, implying by Theorem 2.14.1 that $\mathbf{X}^T\mathbf{A}\mathbf{X} = \mathbf{0}$. Choosing $\boldsymbol{\theta}$ to be each of the unit m-vectors yields the remaining conditions. □

16.1.3 Point Estimation

Suppose that the variance components of a components-of-variance model are estimable. How then may we estimate them? One general procedure consists of the following three steps:

1. Find m quadratic forms, say $\mathbf{y}^T\mathbf{A}_1\mathbf{y}, \ldots, \mathbf{y}^T\mathbf{A}_m\mathbf{y}$ (where $\mathbf{A}_1, \ldots, \mathbf{A}_m$ are known nonrandom matrices) whose expectations are linearly independent linear combinations of the variance components and are all free of $\boldsymbol{\beta}$;
2. Equate these quadratic forms to their expectations;
3. Solve the equations for the variance components.

This procedure produces estimators that are (a) quadratic functions of the data, and (b) unbiased. Consequently, the estimators produced by this procedure are called *quadratic unbiased estimators (QUEs)*. Because the expectation of a quadratic form in $\mathbf{y}$ depends, by Theorem 4.2.4, only on the mean vector and variance–covariance matrix but not the distribution of $\mathbf{y}$, normality need not be assumed to obtain quadratic unbiased estimators.

Example 16.1.3-1. Quadratic Unbiased Estimators of Variance Components for a Toy Example

The following toy example with $m = 3$ illustrates the procedure for quadratic unbiased estimation of variance components. Suppose that the model has three variance components σ_1^2, σ_2^2, and σ^2 to estimate, and suppose that we can find matrices $\mathbf{A}_1$, $\mathbf{A}_2$, and $\mathbf{A}_3$ such that

$$\begin{aligned}
\mathrm{E}(\mathbf{y}^T\mathbf{A}_1\mathbf{y}) &= \sigma^2 + a\sigma_1^2 + b\sigma_2^2, \\
\mathrm{E}(\mathbf{y}^T\mathbf{A}_2\mathbf{y}) &= \sigma^2 + c\sigma_1^2, \\
\mathrm{E}(\mathbf{y}^T\mathbf{A}_3\mathbf{y}) &= \sigma^2,
\end{aligned}$$

where a, b, and c are known nonzero scalars. Then

$$\begin{aligned}
\hat{\sigma}^2 &= \mathbf{y}^T\mathbf{A}_3\mathbf{y}, \\
\hat{\sigma}_1^2 &= \frac{1}{c}(\mathbf{y}^T\mathbf{A}_2\mathbf{y} - \mathbf{y}^T\mathbf{A}_3\mathbf{y}), \\
\hat{\sigma}_2^2 &= \frac{1}{b}\left[\mathbf{y}^T\mathbf{A}_1\mathbf{y} - \frac{a}{c}\mathbf{y}^T\mathbf{A}_2\mathbf{y} - \left(1 - \frac{a}{c}\right)\mathbf{y}^T\mathbf{A}_3\mathbf{y}\right]
\end{aligned}$$

are quadratic unbiased estimators of the variance components. ■

Generally, for any particular components-of-variance model, we can often find many sets of m quadratic forms whose expectations are linearly independent linear combinations of the variance components, and it is not clear which set (if any) is "best." In the case of a classificatory components-of-variance model with balanced data, the sequential (in any order) mixed-model ANOVA table (i.e., the sequential ANOVA table for the corresponding fixed effects model) contains m mean squares whose expectations are linearly independent linear combinations of the variance components. It can be shown (Graybill, 1954) that, in this case, those m quadratic forms constitute the best choice in the sense that, if the joint distribution of $\mathbf{b}_1, \ldots, \mathbf{b}_{m-1}, \mathbf{d}$ is multivariate normal, they produce variance component estimators that have minimum variance among all unbiased estimators. That is, the quadratic unbiased estimators obtained from those mean squares are uniformly minimum variance unbiased estimators (UMVUE's). In this context they are also referred to as MIVQUEs, for *minimum variance quadratic unbiased estimators.*

If the data are unbalanced, however, it is less clear how the quadratic forms should be chosen because in this case the sums of squares in the sequential ANOVA are not invariant to the order of partitioning. In this case, generally there is no uniformly best choice. One possibility is to use the mean squares from some sequential ANOVA table. Three different sequential ANOVA tables that might be used were proposed by Henderson (1953) and are referred to as Henderson's Methods 1, 2, and 3; the interested reader may examine (Searle, 1971, pp. 424–451) for details. Another possibility is to abandon quadratic unbiased estimation altogether and instead estimate the variance components by the method of maximum likelihood (under an assumption that the joint distribution of $\mathbf{b}_1, \ldots, \mathbf{b}_{m-1}, \mathbf{d}$ is multivariate normal). This alternative method of estimation will be considered in the next section.

Example 16.1.3-2. Quadratic Unbiased Estimators of Variance Components in a One-Factor Components-of-Variance Model with Balanced Data

Consider a situation in which data follow the model

$$y_{ij} = \mu + b_i + d_{ij} \quad (i = 1, \ldots, q;\ j = 1, \ldots, r),$$

where $\mathrm{E}(b_i) = \mathrm{E}(d_{ij}) = 0$ for all i and j, $\mathrm{var}(b_i) = \sigma_b^2$ for all i, b_i and b_j are uncorrelated for $i \neq j$, $\mathrm{var}(d_{ij}) = \sigma^2$ for all i and j, d_{ij} and d_{kl} are uncorrelated unless $i = k$ and $j = l$, and each b_i is uncorrelated with each d_{ij}. The parameter space for the model is $\{\mu, \sigma_b^2, \sigma^2 : \mu \in \mathbb{R}, \sigma_b^2 \geq 0, \sigma^2 > 0\}$. Suppose that we wish to obtain quadratic unbiased estimators of σ_b^2 and σ^2.

The corrected overall ANOVA table for the one-factor fixed effects model presented in Example 9.3.3-2 may be specialized here to the case of balanced data, as follows:

Source	df	Sum of squares	Mean square
Classes	$q-1$	$r\sum_{i=1}^{q}(\bar{y}_{i\cdot}-\bar{y}_{\cdot\cdot})^2$	S_1^2
Residual	$q(r-1)$	$\sum_{i=1}^{q}\sum_{j=1}^{r}(y_{ij}-\bar{y}_{i\cdot})^2$	S_2^2
Total	$qr-1$	$\sum_{i=1}^{q}\sum_{j=1}^{r}(y_{ij}-\bar{y}_{\cdot\cdot})^2$	

This can be regarded as a special case of the mixed-model ANOVA introduced in Sect. 13.4.4. Note that we have labeled the mean squares in this table as S_1^2 and S_2^2.

The next task is to obtain the expectations of the mean squares under the given components-of-variance model. To this end it is helpful to write the mean squares as quadratic forms in which the matrices of the quadratic forms are expressed as Kronecker products. Because

$$\bar{y}_{i\cdot}=\frac{1}{r}\mathbf{1}_r^T\begin{pmatrix}y_{i1}\\ \vdots\\ y_{ir}\end{pmatrix}\quad\text{and}\quad \bar{y}_{\cdot\cdot}=\frac{1}{qr}\mathbf{1}_{qr}^T\mathbf{y},$$

we see that $r\sum_{i=1}^{q}(\bar{y}_{i\cdot}-\bar{y}_{\cdot\cdot})^2$ is the sum of squares of the vector

$$(\mathbf{I}_q\otimes\frac{1}{r}\mathbf{J}_r)\mathbf{y}-(\frac{1}{q}\mathbf{J}_q\otimes\frac{1}{r}\mathbf{J}_r)\mathbf{y}=[(\mathbf{I}_q-\frac{1}{q}\mathbf{J}_q)\otimes\frac{1}{r}\mathbf{J}_r]\mathbf{y}$$

and $\sum_{i=1}^{q}\sum_{j=1}^{r}(y_{ij}-\bar{y}_{i\cdot})^2$ is the sum of squares of the vector

$$(\mathbf{I}_q\otimes\mathbf{I}_r)\mathbf{y}-(\mathbf{I}_q\otimes\frac{1}{r}\mathbf{J}_r)\mathbf{y}=[\mathbf{I}_q\otimes(\mathbf{I}_r-\frac{1}{r}\mathbf{J}_r)]\mathbf{y}.$$

Observe that $[(\mathbf{I}_q-\frac{1}{q}\mathbf{J}_q)\otimes\frac{1}{r}\mathbf{J}_r]$ and $[\mathbf{I}_q\otimes(\mathbf{I}_r-\frac{1}{r}\mathbf{J}_r)]$ are symmetric and idempotent because each term in each Kronecker product is so, implying that

$$(q-1)S_1^2=\mathbf{y}^T[(\mathbf{I}_q-\frac{1}{q}\mathbf{J}_q)\otimes\frac{1}{r}\mathbf{J}_r]\mathbf{y}\quad\text{and}\quad q(r-1)S_2^2=\mathbf{y}^T[\mathbf{I}_q\otimes(\mathbf{I}_r-\frac{1}{r}\mathbf{J}_r)]\mathbf{y}.$$

Also, observe that

$$\text{cov}(y_{ij},y_{i'j'})=\begin{cases}\sigma^2+\sigma_b^2 & \text{if } i=i' \text{ and } j=j'\\ \sigma_b^2 & \text{if } i=i' \text{ and } j\neq j'\\ 0 & \text{otherwise,}\end{cases}$$

or equivalently in matrix notation,

$$\mathrm{var}(\mathbf{y}) = \sigma^2\mathbf{I}_{qr} + \sigma_b^2(\mathbf{I}_q \otimes \mathbf{J}_r).$$

Then, using Theorems 4.2.4 and 2.17.6 we find that

$$\begin{aligned}
\mathrm{E}(S_1^2) &= \mathrm{E}\left\{\mathbf{y}^T\left[(\mathbf{I}_q - \frac{1}{q}\mathbf{J}_q)\otimes\frac{1}{r}\mathbf{J}_r\right]\mathbf{y}\Big/(q-1)\right\} \\
&= (\mu\mathbf{1}_q\otimes\mathbf{1}_r)^T\left[(\mathbf{I}_q - \frac{1}{q}\mathbf{J}_q)\otimes\frac{1}{r}\mathbf{J}_r\right](\mu\mathbf{1}_q\otimes\mathbf{1}_r)\Big/(q-1) \\
&\quad + \mathrm{tr}\left\{\left[(\mathbf{I}_q - \frac{1}{q}\mathbf{J}_q)\otimes\frac{1}{r}\mathbf{J}_r\right][\sigma^2\mathbf{I} + \sigma_b^2(\mathbf{I}_q\otimes\mathbf{J}_r)]\right\}\Big/(q-1) \\
&= \left\{(\mu\mathbf{1}_q\otimes\mathbf{1}_r)^T(\mu\mathbf{0}_q\otimes\mathbf{1}_r) + \sigma^2\mathrm{tr}\left[\left(\mathbf{I}_q - \frac{1}{q}\mathbf{J}_q\right)\otimes\frac{1}{r}\mathbf{J}_r\right]\right. \\
&\quad \left. + \sigma_b^2\mathrm{tr}\left[\left(\mathbf{I}_q - \frac{1}{q}\mathbf{J}_q\right)\otimes\mathbf{J}_r\right]\right\}\Big/(q-1) \\
&= \sigma^2 + r\sigma_b^2
\end{aligned}$$

and

$$\begin{aligned}
\mathrm{E}(S_2^2) &= \mathrm{E}\left\{\mathbf{y}^T\left[\mathbf{I}_q\otimes(\mathbf{I}_r - \frac{1}{r}\mathbf{J}_r)\right]\mathbf{y}\right\}\Big/[q(r-1)] \\
&= (\mu\mathbf{1}_q\otimes\mathbf{1}_r)^T\left[\mathbf{I}_q\otimes(\mathbf{I}_r - \frac{1}{r}\mathbf{J}_r)\right](\mu\mathbf{1}_q\otimes\mathbf{1}_r)\Big/[q(r-1)] \\
&\quad + \mathrm{tr}\left\{\left[\mathbf{I}_q\otimes(\mathbf{I}_r - \frac{1}{r}\mathbf{J}_r)\right][\sigma^2\mathbf{I}_{qr} + \sigma_b^2(\mathbf{I}_q\otimes\mathbf{J}_r)]\right\}\Big/[q(r-1)] \\
&= \left\{(\mu\mathbf{1}_q\otimes\mathbf{1}_r)^T(\mu\mathbf{1}_q\otimes\mathbf{0}_r) + \sigma^2\mathrm{tr}\left[\mathbf{I}_q\otimes(\mathbf{I}_r - \frac{1}{r}\mathbf{J}_r)\right]\right. \\
&\quad \left. + \sigma_b^2\mathrm{tr}(\mathbf{I}_q\otimes\mathbf{0}_{r\times r})\right\}\Big/[q(r-1)] \\
&= \sigma^2.
\end{aligned}$$

Thus, quadratic unbiased estimators of σ^2 and σ_b^2 are as follows:

$$\hat{\sigma}^2 = S_2^2, \qquad \hat{\sigma}_b^2 = \frac{S_1^2 - S_2^2}{r}.$$

Observe that if $\Pr(S_1^2 < S_2^2) > 0$ (as is true under normality, see the exercises), then $\hat{\sigma}_b^2$ as defined above could be negative. Because the parameter space constrains

σ_b^2 to be nonnegative, one might consider alternatively estimating σ_b^2 by the estimator $\max\{0, \hat{\sigma}_b^2\}$. This estimator is biased, however. ■

16.1.4 Interval Estimation

Suppose that we wish to have not merely point estimates but also interval estimates, i.e., confidence intervals, for the variance components or for functions of the variance components. How should we proceed? Obviously, to obtain parametric confidence intervals we must make some additional assumptions about the distributions of the random effects in the model. Here we assume that the joint distribution of $\mathbf{b}_1, \ldots, \mathbf{b}_{m-1}, \mathbf{d}$ is multivariate normal; more specifically, we assume that these vectors are independently distributed as $\mathbf{b}_i \sim \mathrm{N}(\mathbf{0}, \sigma_i^2 \mathbf{I}_{q_i})$ and $\mathbf{d} \sim \mathrm{N}(\mathbf{0}, \sigma^2 \mathbf{I})$.

Obtaining confidence intervals for the variance components or functions thereof can be simple, complicated but possible, or (if an interval is desired whose nominal coverage probability equals its actual coverage probability exactly) impossible, depending on the particular components-of-variance model and the function of the variance components for which the interval is desired. One situation that occurs frequently, however, leads very simply to an exact $100(1-\xi)\%$ confidence interval. If $g(\sigma_1^2, \ldots, \sigma_{m-1}^2, \sigma^2)$ is a positively valued function of the variance components of interest and a positive definite quadratic form $\mathbf{y}^T\mathbf{A}\mathbf{y}$ exists for which

$$\frac{\mathbf{y}^T\mathbf{A}\mathbf{y}}{g(\sigma_1^2, \ldots, \sigma_{m-1}^2, \sigma^2)} \sim \chi^2(\nu),$$

then for any $\xi \in (0, 1)$,

$$\begin{aligned} 1 - \xi &= \Pr\left(\chi^2_{1-(\xi/2),\nu} \leq \frac{\mathbf{y}^T\mathbf{A}\mathbf{y}}{g(\sigma_1^2, \ldots, \sigma_{m-1}^2, \sigma^2)} \leq \chi^2_{\xi/2,\nu}\right) \\ &= \Pr\left(\frac{\mathbf{y}^T\mathbf{A}\mathbf{y}}{\chi^2_{\xi/2,\nu}} \leq g(\sigma_1^2, \ldots, \sigma_{m-1}^2, \sigma^2) \leq \frac{\mathbf{y}^T\mathbf{A}\mathbf{y}}{\chi^2_{1-(\xi/2),\nu}}\right). \end{aligned}$$

Therefore, for such a function of the variance components, a $100(1-\xi)\%$ confidence interval is

$$\left[\frac{\mathbf{y}^T\mathbf{A}\mathbf{y}}{\chi^2_{\xi/2,\nu}}, \frac{\mathbf{y}^T\mathbf{A}\mathbf{y}}{\chi^2_{1-(\xi/2),\nu}}\right]. \tag{16.2}$$

Observe that this interval: (a) is not symmetric around the unbiased point estimator $\mathbf{y}^T\mathbf{A}\mathbf{y}/\nu$ of $g(\sigma_1^2, \ldots, \sigma_{m-1}^2, \sigma^2)$; and (b) has endpoints that are positive (with probability one).

Another situation that leads easily to an exact $100(1-\xi)\%$ confidence interval and occurs often enough to mention is as follows. Suppose that two independent positive definite quadratic forms, $\mathbf{y}^T\mathbf{A}_1\mathbf{y}$ and $\mathbf{y}^T\mathbf{A}_2\mathbf{y}$, exist for which

$$\frac{\mathbf{y}^T\mathbf{A}_1\mathbf{y}}{g_1(\sigma_1^2,\ldots,\sigma_{m-1}^2,\sigma^2)} \sim \chi^2(\nu_1),$$

$$\frac{\mathbf{y}^T\mathbf{A}_2\mathbf{y}}{g_2(\sigma_1^2,\ldots,\sigma_{m-1}^2,\sigma^2)} \sim \chi^2(\nu_2),$$

where $g_1(\sigma_1^2,\ldots,\sigma_{m-1}^2,\sigma^2)$ and $g_2(\sigma_1^2,\ldots,\sigma_{m-1}^2,\sigma^2)$ are two positively valued functions of the variance components. Then for any $\xi \in (0,1)$,

$$1-\xi = \Pr\left(F_{1-(\xi/2),\nu_1,\nu_2} \le \frac{\mathbf{y}^T\mathbf{A}_1\mathbf{y}/\nu_1}{g_1(\sigma_1^2,\ldots,\sigma_{m-1}^2,\sigma^2)} \frac{g_2(\sigma_1^2,\ldots,\sigma_{m-1}^2,\sigma^2)}{\mathbf{y}^T\mathbf{A}_2\mathbf{y}/\nu_2} \le F_{\xi/2,\nu_1,\nu_2}\right)$$

$$= \Pr\left(\frac{\mathbf{y}^T\mathbf{A}_1\mathbf{y}/\mathbf{y}^T\mathbf{A}_2\mathbf{y}}{(\nu_1/\nu_2)F_{\xi/2,\nu_1,\nu_2}} \le \frac{g_1(\sigma_1^2,\ldots,\sigma_{m-1}^2,\sigma^2)}{g_2(\sigma_1^2,\ldots,\sigma_{m-1}^2,\sigma^2)} \le \frac{\mathbf{y}^T\mathbf{A}_1\mathbf{y}/\mathbf{y}^T\mathbf{A}_2\mathbf{y}}{(\nu_1/\nu_2)F_{1-(\xi/2),\nu_1,\nu_2}}\right).$$

Therefore, a $100(1-\xi)\%$ confidence interval for the ratio $\frac{g_1(\sigma_1^2,\ldots,\sigma_{m-1}^2,\sigma^2)}{g_2(\sigma_1^2,\ldots,\sigma_{m-1}^2,\sigma^2)}$ is

$$\left[\frac{\mathbf{y}^T\mathbf{A}_1\mathbf{y}/\mathbf{y}^T\mathbf{A}_2\mathbf{y}}{(\nu_1/\nu_2)F_{\xi/2,\nu_1,\nu_2}}, \frac{\mathbf{y}^T\mathbf{A}_1\mathbf{y}/\mathbf{y}^T\mathbf{A}_2\mathbf{y}}{(\nu_1/\nu_2)F_{1-(\xi/2),\nu_1,\nu_2}}\right]. \tag{16.3}$$

In some situations that do not fall into the categories just mentioned, it is still possible to obtain an exact confidence interval by clever manipulation of intervals that do fall into those categories. A few such cases will be encountered in a forthcoming example and in the exercises. In other situations, it may not be possible to obtain an exact confidence interval in this way, but an approximate one may be derived with the aid of Bonferroni's inequality; again, an example of this is forthcoming.

In cases in which the function of the variance components for which an interval is desired is a linear combination of quantities for which an interval of the form (16.2) exists, we can obtain an approximate confidence interval by a method known as Satterthwaite's approximation (Satterthwaite, 1946). Suppose that $U_i = \frac{n_i S_i^2}{\alpha_i^2} \sim$ independent $\chi^2(n_i)$ for $i = 1,\ldots,m$, where $\alpha_i^2 = \mathrm{E}(S_i^2)$, and that a confidence interval is desired for $\alpha^2 \equiv \sum_{i=1}^m c_i\alpha_i^2$ where the c_i's are specified constants. If we were to act as though $\frac{t(\sum_{i=1}^m c_i S_i^2)}{\alpha^2} \sim \chi^2(t)$, where t is a constant to be determined, then a $100(1-\xi)\%$ confidence interval for α^2 would be given by (16.2), which in this context may be written as

$$\left[\frac{t(\sum_{i=1}^m c_i S_i^2)}{\chi^2_{\xi/2,t}}, \frac{t(\sum_{i=1}^m c_i S_i^2)}{\chi^2_{1-(\xi/2),t}}\right].$$

Note that

$$\mathrm{E}\left(\frac{t(\sum_{i=1}^m c_i S_i^2)}{\alpha^2}\right) = \frac{t}{\alpha^2}\left(\sum_{i=1}^m c_i\alpha_i^2\right) = \frac{t}{\alpha^2}\alpha^2 = t,$$

$$\mathrm{var}\left(\frac{t(\sum_{i=1}^m c_i S_i^2)}{\alpha^2}\right) = \frac{t^2}{\alpha^4}\sum_{i=1}^m c_i^2(2n_i)\left(\frac{\alpha_i^4}{n_i^2}\right) = \frac{2t^2}{\alpha^4}\sum_{i=1}^m \frac{(c_i\alpha_i^2)^2}{n_i}.$$

Equating the variance of $t(\sum_{i=1}^m c_i S_i^2)/\alpha^2$ to the variance of a $\chi^2(t)$ random variable and solving for t, we obtain

$$2t = \frac{2t^2}{\alpha^4}\sum_{i=1}^m \frac{(c_i\alpha_i^2)^2}{n_i},$$

or equivalently,

$$t = \frac{\alpha^4}{\sum_{i=1}^m \frac{(c_i\alpha_i^2)^2}{n_i}} = \frac{(\sum_{i=1}^m c_i\alpha_i^2)^2}{\sum_{i=1}^m \frac{(c_i\alpha_i^2)^2}{n_i}}.$$

In practice we do not know α_i^2, so we replace it by its unbiased estimator S_i^2 in the expression for t to obtain an estimate

$$\hat{t} = \frac{(\sum_{i=1}^m c_i S_i^2)^2}{\sum_{i=1}^m \frac{(c_i S_i^2)^2}{n_i}}. \tag{16.4}$$

Thus, an approximate $100(1-\xi)\%$ confidence interval for α^2 is

$$\left[\frac{\hat{t}\sum_{i=1}^m c_i S_i^2}{\chi^2_{\xi/2,\hat{t}}}, \frac{\hat{t}\sum_{i=1}^m c_i S_i^2}{\chi^2_{1-(\xi/2),\hat{t}}}\right]. \tag{16.5}$$

There are two "levels" of approximation here: first in acting as though $\frac{t(\sum_{i=1}^m c_i S_i^2)}{\alpha^2} \sim \chi^2(t)$ for t as given above, and second in replacing t by its estimate $\hat{t}$.

Example 16.1.4-1. Confidence Intervals for Variance Components of a One-Factor Components-of-Variance Model with Balanced Data

Consider the same model featured in the previous example, but suppose now that we desire confidence intervals for the variance components. To that end, assume that the joint distribution of the b_i's and d_{ij}'s is multivariate normal, in which case they are independent. We proceed by obtaining the distributions (not merely the expectations as in Example 16.1.3-2) of the mean squares (suitably scaled). First recall from the aforementioned example that

$$(q-1)S_1^2 = r\sum_{i=1}^{q}(\bar{y}_{i\cdot} - \bar{y}_{\cdot\cdot})^2 = \mathbf{y}^T[(\mathbf{I}_q - \frac{1}{q}\mathbf{J}_q) \otimes \frac{1}{r}\mathbf{J}_r]\mathbf{y},$$

$$q(r-1)S_2^2 = \sum_{i=1}^{q}\sum_{j=1}^{r}(y_{ij} - \bar{y}_{i\cdot})^2 = \mathbf{y}^T[\mathbf{I}_q \otimes (\mathbf{I}_r - \frac{1}{r}\mathbf{J}_r)]\mathbf{y},$$

$$\boldsymbol{\Sigma} \equiv \text{var}(\mathbf{y}) = \sigma_b^2(\mathbf{I}_q \otimes \mathbf{J}_r) + \sigma^2(\mathbf{I}_q \otimes \mathbf{I}_r).$$

Define

$$\mathbf{A}_1 = \left(\frac{1}{\sigma^2 + r\sigma_b^2}\right)[(\mathbf{I}_q - \frac{1}{q}\mathbf{J}_q) \otimes \frac{1}{r}\mathbf{J}_r] \quad \text{and} \quad \mathbf{A}_2 = \left(\frac{1}{\sigma^2}\right)[\mathbf{I}_q \otimes (\mathbf{I}_r - \frac{1}{r}\mathbf{J}_r)].$$

Then

$$\begin{aligned}\mathbf{A}_1\boldsymbol{\Sigma} &= \left(\frac{1}{\sigma^2 + r\sigma_b^2}\right)\{\sigma_b^2[(\mathbf{I}_q - \frac{1}{q}\mathbf{J}_q) \otimes \mathbf{J}_r] + \sigma^2[(\mathbf{I}_q - \frac{1}{q}\mathbf{J}_q) \otimes \frac{1}{r}\mathbf{J}_r]\} \\ &= (\mathbf{I}_q - \frac{1}{q}\mathbf{J}_q) \otimes \frac{1}{r}\mathbf{J}_r\end{aligned}$$

and

$$\begin{aligned}\mathbf{A}_2\boldsymbol{\Sigma} &= \left(\frac{1}{\sigma^2}\right)\{\sigma_b^2[\mathbf{I}_q \otimes \mathbf{0}_{r\times r}] + \sigma^2[\mathbf{I}_q \otimes (\mathbf{I}_r - \frac{1}{r}\mathbf{J}_r)]\} \\ &= \mathbf{I}_q \otimes (\mathbf{I}_r - \frac{1}{r}\mathbf{J}_r).\end{aligned}$$

Now $\mathbf{A}_1\mathbf{\Sigma}$ and $\mathbf{A}_2\mathbf{\Sigma}$ are idempotent (because each term in each Kronecker product is idempotent); $\text{rank}(\mathbf{A}_1) = q - 1$ and $\text{rank}(\mathbf{A}_2) = q(r - 1)$ by Theorem 2.17.8; and

$$\mathbf{A}_1[\mu(\mathbf{1}_q \otimes \mathbf{1}_r)] = \mu\left(\frac{1}{\sigma^2 + r\sigma_b^2}\right)(\mathbf{0}_q \otimes \mathbf{1}_r) = \mathbf{0}$$

and

$$\mathbf{A}_2[\mu(\mathbf{1}_q \otimes \mathbf{1}_r)] = \mu\left(\frac{1}{\sigma^2}\right)(\mathbf{1}_q \otimes \mathbf{0}_r) = \mathbf{0}.$$

Therefore, by Corollary 14.2.6.1,

$$\frac{(q-1)S_1^2}{\sigma^2 + r\sigma_b^2} \sim \chi^2(q-1), \tag{16.6}$$

$$\frac{q(r-1)S_2^2}{\sigma^2} \sim \chi^2(q(r-1)). \tag{16.7}$$

Furthermore, by Corollary 14.3.1.2a, these two (scaled) mean squares are independent because

$$(\mathbf{A}_1\mathbf{\Sigma})\mathbf{A}_2 = [(\mathbf{I}_q - \frac{1}{q}\mathbf{J}_q) \otimes \frac{1}{r}\mathbf{J}_r]\left(\frac{1}{\sigma^2}\right)[\mathbf{I}_q \otimes (\mathbf{I}_r - \frac{1}{r}\mathbf{J}_r)] = \left(\frac{1}{\sigma^2}\right)(\mathbf{I}_q - \frac{1}{q}\mathbf{J}_q) \otimes \mathbf{0}_{r\times r} = \mathbf{0}.$$

Now we may illustrate the derivation of confidence intervals for various functions of σ_b^2 and σ^2.

1. A $100(1-\xi)\%$ confidence interval for σ^2:
   By (16.7) we have, as a special case of (16.2), that
   $$\left[\frac{q(r-1)S_2^2}{\chi^2_{\xi/2,q(r-1)}}, \frac{q(r-1)S_2^2}{\chi^2_{1-(\xi/2),q(r-1)}}\right]$$
   is a $100(1-\xi)\%$ confidence interval for σ^2.
2. A $100(1-\xi)\%$ confidence interval for σ_b^2:
   An exact interval for σ_b^2 is unknown, but we will give two approximate intervals shortly.
3. A $100(1-\xi)\%$ confidence interval for $\psi \equiv \sigma_b^2/\sigma^2$. Note that ψ is the value of the main diagonal elements of $\mathbf{G}(\psi)$ in the model:

Define $U = \frac{S_1^2}{S_2^2}$ and $U^* = \frac{S_1^2/(\sigma^2+r\sigma_b^2)}{S_2^2/\sigma^2} = \frac{\sigma^2}{\sigma^2+r\sigma_b^2}U$. Clearly, (16.6) and (16.7) imply that $U^* \sim \text{F}(q-1, q(r-1))$. Thus,

$$\begin{aligned}
1-\xi &= \Pr(\, F_{1-(\xi/2),q-1,q(r-1)} \leq U^* \leq F_{\xi/2,q-1,q(r-1)}\,) \\
&= \Pr(\, F_{1-(\xi/2),q-1,q(r-1)} \leq \frac{\sigma^2}{\sigma^2+r\sigma_b^2}U \leq F_{\xi/2,q-1,q(r-1)}\,) \\
&= \Pr(\, F_{1-(\xi/2),q-1,q(r-1)} \leq \frac{1}{1+r\psi}U \leq F_{\xi/2,q-1,q(r-1)}\,) \\
&= \Pr\left(\frac{U}{F_{\xi/2,q-1,q(r-1)}} \leq 1+r\psi \leq \frac{U}{F_{1-(\xi/2),q-1,q(r-1)}} \right) \\
&= \Pr\left(\underbrace{\frac{U/F_{\xi/2,q-1,q(r-1)}-1}{r}}_{\text{call this } L} \leq \psi \leq \underbrace{\frac{U/F_{1-(\xi/2),q-1,q(r-1)}-1}{r}}_{\text{call this } R} \right).
\end{aligned} \quad (16.8)$$

Note: If the realized value of L is negative, in practice we may replace it by 0; if the realized values of both R and L are negative, we may replace each of them by 0, in which case the interval reduces to the single point $\{0\}$. This amounts to replacing the interval $[L, R]$ with the interval $[L^+, R^+]$, where $L^+ = \max\{L, 0\}$ and $R^+ = \max\{R, 0\}$. That the coverage probability of this second interval is still $1-\xi$ may be seen by the following probabilistic argument. Let Ω represent the sample space of $\mathbf{y}$-values, and partition it into the three disjoint subsets $\Omega_1 = \{\mathbf{y} : L > 0\}$, $\Omega_2 = \{\mathbf{y} : L \leq 0 < R\}$, and $\Omega_3 = \{\mathbf{y} : R \leq 0\}$. Also denote the event $\{L \leq \psi \leq R\}$, defined in (16.8), by E. Then for any $\xi \in (0, 1)$,

$$\begin{aligned}
1-\xi &= \Pr(E) \\
&= \Pr(E \cap \Omega) \\
&= \Pr(E \cap \Omega_1) + \Pr(E \cap \Omega_2) + \Pr(E \cap \Omega_3) \\
&= \Pr(L \leq \psi \leq R,\ L > 0) \\
&\quad + \Pr(L \leq \psi \leq R,\ L \leq 0 < R) + \Pr(L \leq \psi \leq R,\ R \leq 0) \\
&= \Pr(L^+ \leq \psi \leq R^+,\ L > 0) + \Pr(0 \leq \psi \leq R,\ L \leq 0 < R) \\
&\quad + \Pr(0 \leq \psi \leq 0,\ R \leq 0) \\
&= \Pr(L^+ \leq \psi \leq R^+,\ L > 0) + \Pr(L^+ \leq \psi \leq R^+,\ L \leq 0 < R) \\
&\quad + \Pr(L^+ \leq \psi \leq R^+,\ R \leq 0) \\
&= \Pr(L^+ \leq \psi \leq R^+).
\end{aligned} \quad (16.9)$$

4. A $100(1-\xi)\%$ confidence interval for $\frac{\sigma_b^2}{\sigma_b^2+\sigma^2}$. This quantity is the proportion of total variability in an observation that is attributable to the random factor effect:

First, note that

$$\frac{\sigma_b^2}{\sigma_b^2+\sigma^2} = \frac{\sigma_b^2/\sigma^2}{(\sigma_b^2/\sigma^2)+1} = \frac{\psi}{\psi+1} \equiv f(\psi),$$

say, where the domain of $f(\psi)$ is $[0,\infty)$. Now, $f(0)=0$ and f is continuous and strictly increasing on $[0,\infty)$. Thus

$$\{\mathbf{y}: L^+ \leq \psi \leq R^+\} = \left\{\mathbf{y}: \frac{L^+}{L^++1} \leq \frac{\psi}{\psi+1} \leq \frac{R^+}{R^++1}\right\},$$

which by (16.9) implies that

$$\left[\frac{L^+}{L^++1}, \frac{R^+}{R^++1}\right]$$

is a $100(1-\xi)\%$ confidence interval for $\frac{\sigma_b^2}{\sigma_b^2+\sigma^2}$.

5. Two approximate $100(1-\xi)\%$ confidence intervals for σ_b^2:
   (a) The first approach we describe is due to Williams (1962). It combines the previously derived confidence interval for σ_b^2/σ^2 with a confidence interval for $\sigma^2+r\sigma_b^2$ via the Bonferroni inequality and then does some further manipulation to get a conservative interval for σ_b^2.

   Using (16.6) and (16.2), we have

$$\Pr\left(\underbrace{\frac{(q-1)S_1^2}{\chi^2_{(\xi/2),q-1}}}_{\text{call this } L^*} \leq \sigma^2+r\sigma_b^2 \leq \underbrace{\frac{(q-1)S_1^2}{\chi^2_{1-(\xi/2),q-1}}}_{\text{call this } R^*}\right) = 1-\xi.$$

   Thus, by the Bonferroni inequality,

$$\Pr(L^+ \leq \frac{\sigma_b^2}{\sigma^2} \leq R^+ \text{ and } L^* \leq \sigma^2+r\sigma_b^2 \leq R^*) \geq 1-2\xi, \qquad (16.10)$$

   where L^+ and R^+ were defined in (16.9).

   The event in (16.10) is either a trapezoid (if $R^+>0$) or a line segment (if $R^+=0$) in the (σ^2,σ_b^2)-plane, as depicted in Fig. 16.1. In either case, this event is a subset of the set

$$\{(\sigma^2,\sigma_b^2): \sigma^2>0,\ L^{**} \leq \sigma_b^2 \leq R^{**}\},$$

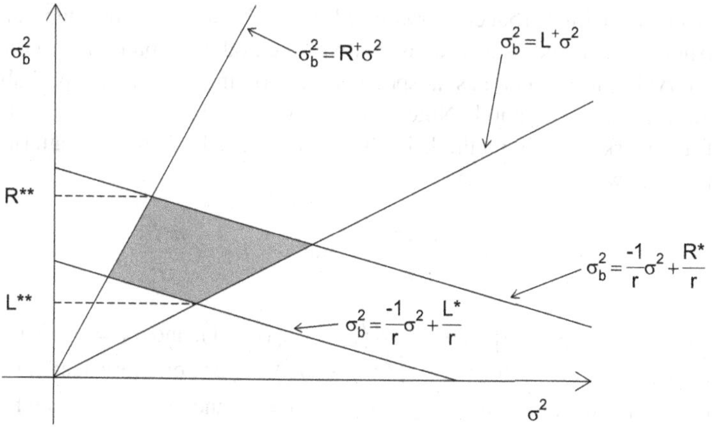

Fig. 16.1 The first of the two approximately $100(1-\xi)\%$ confidence intervals for σ_b^2 described in Example 16.1.3-1

where L^{**} is the σ_b^2-coordinate of the point where the lines $\sigma_b^2 = L^+\sigma^2$ and $\sigma_b^2 = (-\frac{1}{r})\sigma^2 + \frac{L^*}{r}$ intersect, and R^{**} is the σ_b^2-coordinate of the point where the lines $\sigma_b^2 = R^+\sigma^2$ and $\sigma_b^2 = (-\frac{1}{r})\sigma^2 + \frac{R^*}{r}$ intersect. Thus,

$$\Pr(L^{**} \le \sigma_b^2 \le R^{**}) \ge \Pr(L^+ \le \frac{\sigma_b^2}{\sigma^2} \le R^+ \text{ and } L^* \le \sigma^2 + r\sigma_b^2 \le R^*) \ge 1-2\xi.$$

Expressions for L^{**} and R^{**} (in terms of L^+, L^*, R^+, and R^*) may be determined by equating the lines at the two relevant intersection points and solving for σ_b^2:

$$\sigma_b^2 = L^+\sigma^2 = \left(-\frac{1}{r}\right)\sigma^2 + \frac{L^*}{r}$$
$$\Rightarrow \left(L^+ + \frac{1}{r}\right)\sigma^2 = \frac{L^*}{r} \Rightarrow \sigma^2 = \frac{L^*}{rL^+ + 1} \Rightarrow \sigma_b^2 = \frac{L^+L^*}{rL^+ + 1};$$
$$\sigma_b^2 = R^+\sigma^2 = \left(-\frac{1}{r}\right)\sigma^2 + \frac{R^*}{r}$$
$$\Rightarrow \left(R^+ + \frac{1}{r}\right)\sigma^2 = \frac{R^*}{r} \Rightarrow \sigma^2 = \frac{R^*}{rR^+ + 1} \Rightarrow \sigma_b^2 = \frac{R^+R^*}{rR^+ + 1}.$$

Thus

$$\left[\frac{L^+L^*}{rL^+ + 1}, \frac{R^+R^*}{rR^+ + 1}\right]$$

is an approximate (but conservative) $100(1-2\xi)\%$ confidence interval for σ_b^2.

(b) Another, not necessarily conservative, method for obtaining a confidence interval for σ_b^2 that has a specified approximate coverage probability is Satterthwaite's method. Note that (16.6) and (16.7) provide the required framework for this method; in the notation used to present Satterthwaite's method, we have

$$U_1 = \frac{n_1 S_1^2}{\alpha_1^2} \quad \text{and} \quad U_2 = \frac{n_1 S_2^2}{\alpha_1^2}$$

where $n_1 = q-1$, $\alpha_1^2 = \sigma^2 + r\sigma_b^2$, $n_2 = q(r-1)$, and $\alpha_2^2 = \sigma^2$. We desire a confidence interval for $\sigma_b^2 = \frac{1}{r}(r\sigma_b^2 + \sigma^2) - \frac{1}{r}\sigma^2$, or equivalently, using the notation above, for $c_1\alpha_1^2 + c_2\alpha_2^2$ where $c_1 = \frac{1}{r}$ and $c_2 = -\frac{1}{r}$. Then by (16.5) and (16.4),

$$\left[\frac{\hat{t}(\frac{1}{r}S_1^2 - \frac{1}{r}S_2^2)}{\chi^2_{\xi/2,\hat{t}}}, \frac{\hat{t}(\frac{1}{r}S_1^2 - \frac{1}{r}S_2^2)}{\chi^2_{1-(\xi/2),\hat{t}}}\right]$$

is an approximate $100(1-\xi)\%$ confidence interval for σ_b^2, where

$$\hat{t} = \frac{(\frac{1}{r}S_1^2 - \frac{1}{r}S_2^2)^2}{\frac{(\frac{1}{r}S_1^2)^2}{q-1} + \frac{(-\frac{1}{r}S_2^2)^2}{q(r-1)}} = \frac{(S_1^2 - S_2^2)^2}{\frac{S_1^4}{q-1} + \frac{S_2^4}{q(r-1)}}.$$ ■

Confidence intervals (exact and approximate) for variance components in a large variety of other random and mixed classificatory linear models are given by Burdick and Graybill (1992).

16.1.5 Hypothesis Testing

Results on distributions of mean squares and their ratios can be used not only to obtain confidence intervals for variance components, but also to derive formal procedures for testing hypotheses on them. For example, in Example 16.1.4-1, we found that

$$\frac{(q-1)S_1^2}{r\sigma_b^2 + \sigma^2} \sim \chi^2(q-1) \quad \text{and} \quad \frac{q(r-1)S_2^2}{\sigma^2} \sim \chi^2[q(r-1)],$$

and we noted that these two quantities are independent. Thus,

$$\frac{S_1^2}{S_2^2} \cdot \frac{\sigma^2}{r\sigma_b^2 + \sigma^2} \sim \mathrm{F}(q-1, q(r-1)),$$

or equivalently,

$$\frac{S_1^2}{S_2^2} \sim \frac{r\sigma_b^2 + \sigma^2}{\sigma^2} \mathrm{F}(q-1, q(r-1)).$$

Now consider testing H_0: $\sigma_b^2 = 0$ versus H_a: $\sigma_b^2 > 0$. (Note that the union of H_0 and H_a is the entire parameter space for σ_b^2 under the components-of-variance model.) Under H_0,

$$\frac{S_1^2}{S_2^2} \sim \mathrm{F}(q-1, q(r-1))$$

so a size-ξ test of H_0 versus H_a is obtained by rejecting H_0 if and only if

$$\frac{S_1^2}{S_2^2} > F_{\xi, q-1, q(r-1)}.$$

The power of this test is

$$\Pr\left(\frac{r\sigma_b^2 + \sigma^2}{\sigma^2} F > F_{\xi, q-1, q(r-1)}\right),$$

where $F \sim \mathrm{F}(q-1, q(r-1))$, which clearly increases with σ_b^2. Interestingly, the test statistic and the cutoff point for this test (under the one-factor random effects model) are identical to the test statistic and cutoff point for testing the equality of Factor A effects under the one-factor fixed effects model; see, for example, Examples 14.5-2 and 15.2-2. However, the distributions of the test statistic under the alternative hypotheses are different; in this case it is a multiple of a central F distribution, while in the fixed effects case it is a noncentral F distribution.

Similar manipulations may be used to derive hypothesis tests on the variance components of many balanced random and mixed effects models. As a somewhat more general example, suppose that for a certain model with three variance components σ_1^2, σ_2^2, and σ_3^2 and three independent mean squares S_1^2, S_2^2, and S_3^2, we have the following distributional results:

$$\frac{d_1 S_1^2}{\sigma_1^2 + a\sigma_2^2 + b\sigma_3^2} \sim \chi^2(d_1),$$

$$\frac{d_2 S_2^2}{\sigma_1^2 + a\sigma_2^2} \sim \chi^2(d_2),$$

$$\frac{d_3 S_3^2}{\sigma_1^2} \sim \chi^2(d_3),$$

where a and b are nonnegative constants. Then

$$\frac{S_2^2}{S_3^2} \sim \frac{\sigma_1^2 + a\sigma_2^2}{\sigma_1^2}\mathrm{F}(d_2, d_3) \quad \text{and} \quad \frac{S_1^2}{S_2^2} \sim \frac{\sigma_1^2 + a\sigma_2^2 + b\sigma_3^2}{\sigma_1^2 + a\sigma_2^2}\mathrm{F}(d_1, d_2),$$

implying that a size-ξ test of H_0: $\sigma_2^2 = 0$ versus H_a: $\sigma_2^2 > 0$ is obtained by rejecting H_0 if and only if

$$\frac{S_2^2}{S_3^2} > F_{\xi, d_2, d_3},$$

and a size-ξ test of H_0: $\sigma_3^2 = 0$ versus H_a: $\sigma_3^2 > 0$ is obtained by rejecting H_0 if and only if

$$\frac{S_1^2}{S_2^2} > F_{\xi, d_1, d_2}.$$

For further consideration of tests on variance components, we refer the reader to Khuri et al. (1998).

16.2 Likelihood-Based Estimation

The previous section described quadratic unbiased estimation of the variance components of a components-of-variance model. However, there are at least two problems with quadratic unbiased estimation:

1. It can produce negative estimates, and if a negative estimate is replaced with 0, which in effect modifies the estimator to $\max(0, QUE)$ where QUE represents the quadratic unbiased estimator, then the modified estimator will be biased;
2. It is not clear which quadratic forms should be used when the data are unbalanced (which is often the case in practice).

In this section we consider alternative approaches to the estimation of variance components, that are likelihood-based: the maximum likelihood (ML) approach and its close relative, the residual maximum likelihood (REML) approach. ML and REML estimators of variance components are inherently nonnegative and well defined, so the two problems with quadratic unbiased estimation noted above are not encountered with ML and REML estimation. And although likelihood-based estimation has its own limitations (for example, the form of the distribution of $\mathbf{y}$ must be specified), we will see that these limitations are not actually very restrictive.

Although variance components are the only parameters in the variance–covariance structure of a components-of-variance model, we present likelihood-based approaches to their estimation in the somewhat more general context of a general mixed linear model. This more general context allows us to also deal with estimation of variance–covariance structure parameters other than variances with little extra conceptual difficulty.

Recall from Chap. 5 that a *general mixed linear model* specifies that

$$\mathbf{y} = \mathbf{X}\boldsymbol{\beta} + \mathbf{e},$$

where var($\mathbf{e}$) $= \mathbf{V}(\boldsymbol{\theta})$ and the elements of $\mathbf{V}(\boldsymbol{\theta})$ are known functions of an m-dimensional parameter vector $\boldsymbol{\theta} = (\theta_i)$. The parameter space for the model is $\{\boldsymbol{\beta}, \boldsymbol{\theta} : \boldsymbol{\beta} \in \mathbb{R}^p, \boldsymbol{\theta} \in \Theta\}$, where Θ is the set of vectors $\boldsymbol{\theta}$ for which $\mathbf{V}(\boldsymbol{\theta})$ is nonnegative definite or a given subset of that set. Throughout this section we assume that Θ is the subset of vectors for which $\mathbf{V}(\boldsymbol{\theta})$ is positive definite. We also assume that $\mathbf{e}$ (hence also $\mathbf{y}$) has a multivariate normal distribution.

The ML approach to estimating the parameters $\boldsymbol{\beta}$ and $\boldsymbol{\theta}$ consists of maximizing the likelihood function, or equivalently maximizing the log-likelihood function, with respect to these parameters. Because of the assumed normality and Theorem 14.1.5, the log-likelihood function, apart from an additive constant not depending on $\boldsymbol{\beta}$ or $\boldsymbol{\theta}$, is

$$L(\boldsymbol{\beta}, \boldsymbol{\theta}; \mathbf{y}) = -\frac{1}{2}\log|\mathbf{V}| - \frac{1}{2}(\mathbf{y} - \mathbf{X}\boldsymbol{\beta})^T\mathbf{V}^{-1}(\mathbf{y} - \mathbf{X}\boldsymbol{\beta}). \tag{16.11}$$

Observe that the functional dependence of $\mathbf{V}$ on $\boldsymbol{\theta}$ has been suppressed in this expression. With a few exceptions, throughout this chapter we write $\mathbf{V}(\boldsymbol{\theta})$ as $\mathbf{V}$ because it makes the mathematical expressions much less cumbersome.

Definition 16.2.1 $(\hat{\boldsymbol{\beta}}_{ML}, \hat{\boldsymbol{\theta}}_{ML})$ is a *maximum likelihood (ML) estimate* of $(\boldsymbol{\beta}, \boldsymbol{\theta})$ under the normal general mixed linear model if $L(\boldsymbol{\beta}, \boldsymbol{\theta}; \mathbf{y})$ attains its maximum value, over $\{\boldsymbol{\beta}, \boldsymbol{\theta} : \boldsymbol{\beta} \in \mathbb{R}^p, \boldsymbol{\theta} \in \Theta\}$, at $(\boldsymbol{\beta}, \boldsymbol{\theta}) = (\hat{\boldsymbol{\beta}}_{ML}, \hat{\boldsymbol{\theta}}_{ML})$.

The likelihood equations associated with the full log-likelihood $L(\boldsymbol{\beta}, \boldsymbol{\theta}; \mathbf{y})$, whose solution may yield a maximum likelihood estimate, are

$$\frac{\partial L}{\partial \beta_i} = 0 \quad (i = 1, \ldots, p), \qquad \frac{\partial L}{\partial \theta_i} = 0 \quad (i = 1, \ldots, m),$$

assuming that the partial derivatives of L with respect to the elements of $\boldsymbol{\theta}$ exist; the partial derivatives of L with respect to the elements of $\boldsymbol{\beta}$ clearly exist.

From the form of $L(\boldsymbol{\beta}, \boldsymbol{\theta}; \mathbf{y})$ and results obtained in Example 14.1-1, it is clear that for any fixed value $\boldsymbol{\theta}^*$ of $\boldsymbol{\theta}$, $L(\boldsymbol{\beta}, \boldsymbol{\theta}; \mathbf{y})$ is maximized with respect to $\boldsymbol{\beta}$ at $\hat{\boldsymbol{\beta}}(\boldsymbol{\theta}^*)$, where $\hat{\boldsymbol{\beta}}(\boldsymbol{\theta}^*)$ is any solution to the Aitken equations

$$\mathbf{X}^T[\mathbf{V}(\boldsymbol{\theta}^*)]^{-1}\mathbf{X}\boldsymbol{\beta} = \mathbf{X}^T[\mathbf{V}(\boldsymbol{\theta}^*)]^{-1}\mathbf{y}$$

corresponding to $\mathbf{W} = \mathbf{V}(\boldsymbol{\theta}^*)$. Thus,

$$\begin{aligned}\max_{\boldsymbol{\beta}\in\mathbb{R}^p,\boldsymbol{\theta}\in\Theta} L(\boldsymbol{\beta},\boldsymbol{\theta};\mathbf{y}) &= \max_{\boldsymbol{\theta}\in\Theta}\max_{\boldsymbol{\beta}\in\mathbb{R}^p} L(\boldsymbol{\beta},\boldsymbol{\theta};\mathbf{y}) \\ &= \max_{\boldsymbol{\theta}\in\Theta} L(\hat{\boldsymbol{\beta}}(\boldsymbol{\theta}),\boldsymbol{\theta};\mathbf{y}).\end{aligned}$$

Therefore, ML estimates of $\boldsymbol{\beta}$ and $\boldsymbol{\theta}$ may be obtained by first maximizing the function

$$\begin{aligned}L_0(\boldsymbol{\theta};\mathbf{y}) &\equiv L(\hat{\boldsymbol{\beta}}(\boldsymbol{\theta}),\boldsymbol{\theta};\mathbf{y}) \\ &= -\frac{1}{2}\log|\mathbf{V}| - \frac{1}{2}[\mathbf{y}-\mathbf{X}\hat{\boldsymbol{\beta}}(\boldsymbol{\theta})]^T\mathbf{V}^{-1}[\mathbf{y}-\mathbf{X}\hat{\boldsymbol{\beta}}(\boldsymbol{\theta})] \\ &= -\frac{1}{2}\log|\mathbf{V}| - \frac{1}{2}[\mathbf{y}-\mathbf{X}(\mathbf{X}^T\mathbf{V}^{-1}\mathbf{X})^-\mathbf{X}^T\mathbf{V}^{-1}\mathbf{y}]^T \\ &\quad \times\mathbf{V}^{-1}[\mathbf{y}-\mathbf{X}(\mathbf{X}^T\mathbf{V}^{-1}\mathbf{X})^-\mathbf{X}^T\mathbf{V}^{-1}\mathbf{y}]\end{aligned}$$

(to obtain $\hat{\boldsymbol{\theta}}_{ML}$), after which $\hat{\boldsymbol{\beta}}_{ML}$ may be obtained as a solution to the "empirical" Aitken equations

$$\mathbf{X}^T[\mathbf{V}(\hat{\boldsymbol{\theta}}_{ML})]^{-1}\mathbf{X}\boldsymbol{\beta} = \mathbf{X}^T[\mathbf{V}(\hat{\boldsymbol{\theta}}_{ML})]^{-1}\mathbf{y}.$$

The function $L_0(\boldsymbol{\theta};\mathbf{y})$ is called the *profile (or concentrated) log-likelihood function for* $\boldsymbol{\theta}$. Because

$$\begin{aligned}&[\mathbf{y}-\mathbf{X}(\mathbf{X}^T\mathbf{V}^{-1}\mathbf{X})^-\mathbf{X}^T\mathbf{V}^{-1}\mathbf{y}]^T\mathbf{V}^{-1}[\mathbf{y}-\mathbf{X}(\mathbf{X}^T\mathbf{V}^{-1}\mathbf{X})^-\mathbf{X}^T\mathbf{V}^{-1}\mathbf{y}] \\ &= \mathbf{y}^T[\mathbf{I}-\mathbf{X}(\mathbf{X}^T\mathbf{V}^{-1}\mathbf{X})^-\mathbf{X}^T\mathbf{V}^{-1}]^T[\mathbf{V}^{-1}-\mathbf{V}^{-1}\mathbf{X}(\mathbf{X}^T\mathbf{V}^{-1}\mathbf{X})^-\mathbf{X}^T\mathbf{V}^{-1}]\mathbf{y} \\ &= \mathbf{y}^T[\mathbf{I}-\mathbf{V}^{-1}\mathbf{X}\{(\mathbf{X}^T\mathbf{V}^{-1}\mathbf{X})^-\}^T\mathbf{X}^T][\mathbf{V}^{-1}-\mathbf{V}^{-1}\mathbf{X}(\mathbf{X}^T\mathbf{V}^{-1}\mathbf{X})^-\mathbf{X}^T\mathbf{V}^{-1}]\mathbf{y} \\ &= \mathbf{y}^T[\mathbf{V}^{-1}-\mathbf{V}^{-1}\mathbf{X}(\mathbf{X}^T\mathbf{V}^{-1}\mathbf{X})^-\mathbf{X}^T\mathbf{V}^{-1}-\mathbf{V}^{-1}\mathbf{X}\{(\mathbf{X}^T\mathbf{V}^{-1}\mathbf{X})^-\}^T\mathbf{X}^T\mathbf{V}^{-1} \\ &\quad +\mathbf{V}^{-1}\mathbf{X}\{(\mathbf{X}^T\mathbf{V}^{-1}\mathbf{X})^-\}^T\mathbf{X}^T\mathbf{V}^{-1}\mathbf{X}(\mathbf{X}^T\mathbf{V}^{-1}\mathbf{X})^-\mathbf{X}^T\mathbf{V}^{-1}]\mathbf{y} \\ &= \mathbf{y}^T[\mathbf{V}^{-1}-\mathbf{V}^{-1}\mathbf{X}(\mathbf{X}^T\mathbf{V}^{-1}\mathbf{X})^-\mathbf{X}^T\mathbf{V}^{-1}]\mathbf{y},\end{aligned}$$

the profile log-likelihood function may be rewritten more simply as

$$L_0(\boldsymbol{\theta};\mathbf{y}) = -\frac{1}{2}\log|\mathbf{V}| - \frac{1}{2}\mathbf{y}^T[\mathbf{V}^{-1}-\mathbf{V}^{-1}\mathbf{X}(\mathbf{X}^T\mathbf{V}^{-1}\mathbf{X})^-\mathbf{X}^T\mathbf{V}^{-1}]\mathbf{y}. \qquad (16.12)$$

The likelihood equations associated with the profile log-likelihood $L_0(\boldsymbol{\theta};\mathbf{y})$ are

$$\frac{\partial L_0}{\partial\theta_i} = 0 \quad (i = 1,\ldots,m)$$

(again assuming that these derivatives exist).

We investigate how to actually maximize the log-likelihood and profile log-likelihood functions later in this section, but first we address one known shortcoming of the ML method for estimating $\boldsymbol{\theta}$, namely, that it does not account for the loss in degrees of freedom from estimating $\mathbf{X}\boldsymbol{\beta}$. To understand what is meant by the failure of ML to take into account the loss in degrees of freedom from estimating $\mathbf{X}\boldsymbol{\beta}$, consider the estimation of σ^2 under a normal Gauss–Markov model in which $\mathbf{X}$ is equal to $\mathbf{0}$, so that $p^* = 0$ and there is no $\mathbf{X}\boldsymbol{\beta}$ to estimate. In this case, the ML estimator of σ^2 is $\mathbf{y}^T\mathbf{y}/n$ as a special case of Example 14.1-1, and it is easily verified that this estimator is unbiased. If $\mathbf{X}$ is not equal to $\mathbf{0}$, however, then the ML estimator, as another special case of Example 14.1-1, is

$$\bar{\sigma}^2 = \frac{(\mathbf{y} - \mathbf{X}\hat{\boldsymbol{\beta}})^T(\mathbf{y} - \mathbf{X}\hat{\boldsymbol{\beta}})}{n}, \tag{16.13}$$

where $\hat{\boldsymbol{\beta}}$ is any solution to $\mathbf{X}^T\mathbf{X}\boldsymbol{\beta} = \mathbf{X}^T\mathbf{y}$. However, provided that $n > p^*$, the uniformly minimum variance unbiased estimator of σ^2 is

$$\hat{\sigma}^2 = \frac{(\mathbf{y} - \mathbf{X}\hat{\boldsymbol{\beta}})^T(\mathbf{y} - \mathbf{X}\hat{\boldsymbol{\beta}})}{n - p^*}. \tag{16.14}$$

Now, $\mathrm{E}(\bar{\sigma}^2) = \sigma^2(n - p^*)/n = \sigma^2[1 - (p^*/n)]$, so that the bias of $\bar{\sigma}^2$ is

$$\mathrm{E}(\bar{\sigma}^2) - \sigma^2 = -\sigma^2\left(\frac{p^*}{n}\right).$$

Thus, the ML estimator is unbiased when $p^* = 0$ (i.e., when $\mathbf{X} = \mathbf{0}$) but biased when $0 < p^* < n$ (i.e., when $\mathbf{X} \neq \mathbf{0}$), and the bias becomes worse as p^* increases towards an upper limit of $n - 1$. This is the sense in which the ML estimator of σ^2 fails to account for the loss in degrees of freedom from estimating $\mathbf{X}\boldsymbol{\beta}$.

As an aside, it is true that $\mathrm{var}(\bar{\sigma}^2) \leq \mathrm{var}(\hat{\sigma}^2)$; however, it can be shown (see the exercises) that the mean square error of $\hat{\sigma}^2$ is less than or equal to that of $\bar{\sigma}^2$ if $p^* > 5$ and $n > p^*(p^* - 2)/(p^* - 4)$. (If $p^* \geq 13$, then $n > p^* + 2$ suffices.)

The REML approach to the estimation of variance components seeks to eliminate or reduce the bias incurred by the ML approach when $\mathbf{X} \neq \mathbf{0}$. In the REML approach, the ML technique is applied to the likelihood function associated with a set of "error contrasts" derived from the response vector $\mathbf{y}$, rather than to the likelihood function associated with $\mathbf{y}$ itself. We now define precisely what is meant by the term "error contrast," and we assume henceforth that $\mathbf{X} \neq \mathbf{0}$.

Definition 16.2.2 Under the general mixed linear model, a linear combination $\mathbf{a}^T\mathbf{y}$ of the elements of $\mathbf{y}$ is said to be an *error contrast* if $\mathrm{E}(\mathbf{a}^T\mathbf{y}) = 0$ for all $\boldsymbol{\beta}$ (and all $\boldsymbol{\theta} \in \Theta$), i.e., if $\mathbf{X}^T\mathbf{a} = \mathbf{0}$.

How do we find error contrasts? Recall that $\mathbf{P_X}$ represents the orthogonal projection matrix onto the column space of $\mathbf{X}$. It turns out that each element of the vector $(\mathbf{I} - \mathbf{P_X})\mathbf{y}$ is an error contrast because

$$\mathrm{E}(\mathbf{I} - \mathbf{P_X})\mathbf{y} = (\mathbf{I} - \mathbf{P_X})\mathrm{E}(\mathbf{y}) = (\mathbf{I} - \mathbf{P_X})\mathbf{X}\boldsymbol{\beta} = \mathbf{0}.$$

Note, however, that $\mathbf{I} - \mathbf{P_X}$ has rank $n - p^*$ while its dimensions are $n \times n$, so there are some redundancies (linear dependencies) among its elements. So the question arises, "How many essentially different error contrasts can be included in a single set?" The following definition and theorem answer this question.

Definition 16.2.3 Error contrasts $\mathbf{a}_1^T\mathbf{y}, \ldots, \mathbf{a}_k^T\mathbf{y}$ are said to be *linearly independent* if $\mathbf{a}_1, \ldots, \mathbf{a}_k$ are linearly independent vectors.

Theorem 16.2.1 *Any set of error contrasts contains at most $n - p^*$ linearly independent error contrasts.*

Proof Suppose that $\mathbf{A}^T\mathbf{y}$ is a vector of error contrasts. Then $\mathbf{X}^T\mathbf{A} = \mathbf{0}$, implying that

$$\mathbf{A} = [\mathbf{I} - \mathbf{X}(\mathbf{X}^T\mathbf{X})^-\mathbf{X}^T]\mathbf{Z} \quad \text{for some matrix } \mathbf{Z}$$

(by Theorem 3.2.3 because $\mathbf{X}(\mathbf{X}^T\mathbf{X})^-$ is a generalized inverse of $\mathbf{X}^T$). Thus, $\mathrm{rank}(\mathbf{A}) \leq \mathrm{rank}(\mathbf{I} - \mathbf{P_X}) = n - p^*$. □

We can obtain a vector of $n - p^*$ linearly independent error contrasts that are derived from the matrix $\mathbf{I} - \mathbf{P_X}$ in the following way. By Theorem 2.15.15, an $n \times (n - p^*)$ matrix $\mathbf{B}$ exists such that $\mathbf{B}^T\mathbf{B} = \mathbf{I}_{n-p^*}$ and $\mathbf{B}\mathbf{B}^T = \mathbf{I} - \mathbf{P_X}$. Set $\mathbf{A} = \mathbf{B}$. Then

$$\mathbf{0} = \mathbf{X}^T(\mathbf{I} - \mathbf{P_X}) = \mathbf{X}^T\mathbf{A}\mathbf{A}^T,$$

implying (by Theorem 3.3.2) that $\mathbf{X}^T\mathbf{A} = \mathbf{0}$. Furthermore, $\mathrm{rank}(\mathbf{A}) = \mathrm{rank}(\mathbf{A}^T\mathbf{A}) = \mathrm{rank}(\mathbf{I}_{n-p^*}) = n - p^*$. Consequently, the $(n - p^*)$-vector $\mathbf{w}$ defined by $\mathbf{w} = \mathbf{A}^T\mathbf{y}$ is a vector of $n - p^*$ linearly independent error contrasts.

Under the normality assumption on $\mathbf{y}$ and positive definiteness assumption on $\mathbf{V}$ made earlier in this section, the pdf of the vector $\mathbf{w}$ constructed in the manner just described is

$$f_{\mathbf{w}}(\mathbf{w};\, \boldsymbol{\theta}) = (2\pi)^{-\frac{n-p^*}{2}} |\mathbf{A}^T\mathbf{V}\mathbf{A}|^{-\frac{1}{2}} \exp[-\frac{1}{2}\mathbf{w}^T(\mathbf{A}^T\mathbf{V}\mathbf{A})^{-1}\mathbf{w}], \quad \mathbf{w} \in \mathbb{R}^{n-p^*}, \tag{16.15}$$

where $\boldsymbol{\theta} \in \Theta$. However, this $\mathbf{w}$ is not necessarily the only vector of $n - p^*$ linearly independent error contrasts for a given model; there could be many other such vectors. A natural question is, "Would the estimate of $\boldsymbol{\theta}$ obtained by maximizing $f_{\mathbf{w}}(\mathbf{w};\ \boldsymbol{\theta})$ be the same as the estimate obtained by maximizing the likelihood function associated with any other vector of $n - p^*$ linearly independent error contrasts?" The next theorem pertains to this question.

Theorem 16.2.2 *Let* $\mathbf{u}$ *represent any vector of* $n - p^*$ *linearly independent error contrasts. Then the likelihood function associated with* $\mathbf{u}$ *is a positive scalar multiple of* $f_{\mathbf{w}}(\mathbf{w};\ \boldsymbol{\theta})$*, and that scalar is free of* $\boldsymbol{\theta}$.

Proof Because $\mathbf{u}$ is a vector of $n-p^*$ linearly independent error contrasts, $\mathbf{u} = \mathbf{C}^T\mathbf{y}$ for some $(n - p^*) \times n$ matrix $\mathbf{C}^T$ of full row rank for which $\mathcal{R}(\mathbf{C}^T) = \mathcal{R}(\mathbf{I} - \mathbf{P_X})$. Furthermore, because $\mathbf{A}^T$ has the same dimensions as $\mathbf{C}^T$ and also has full row rank $n - p^*$ and $\mathcal{R}(\mathbf{A}^T) = \mathcal{R}(\mathbf{I} - \mathbf{P_X})$, we see (by Theorem 2.8.5) that $\mathbf{C}^T = \mathbf{L}\mathbf{A}^T$ for some nonsingular $(n-p^*)\times(n-p^*)$ matrix of constants $\mathbf{L}$. Thus $\mathbf{u} = \mathbf{L}\mathbf{A}^T\mathbf{y} = \mathbf{L}\mathbf{w}$, implying that the pdf of $\mathbf{u}$ is

$$f_{\mathbf{u}}(\mathbf{u};\ \boldsymbol{\theta}) = |\mathbf{L}|^{-1} f_{\mathbf{w}}(\mathbf{L}^{-1}\mathbf{u};\ \boldsymbol{\theta}).$$

□

It follows immediately from Theorem 16.2.2 that, in estimating $\boldsymbol{\theta}$ from the likelihood function associated with a set of $n - p^*$ linearly independent error contrasts, it makes no difference which set of error contrasts is used; the same estimate(s) would be obtained from any such set.

In general, the expression for the pdf of $\mathbf{w}$ given by (16.15) is not in the most convenient computational form because some "pre-processing" is required to obtain $\mathbf{A}$ from the model matrix $\mathbf{X}$. It would be preferable to express $f_{\mathbf{w}}(\mathbf{w};\ \boldsymbol{\theta})$ explicitly in terms of $\mathbf{X}$ (and $\mathbf{V}$). We now proceed to derive such an expression.

Let $\breve{\mathbf{X}}$ represent any $n \times p^*$ matrix whose columns are linearly independent columns of $\mathbf{X}$, and define $\mathbf{M} = \mathbf{V}^{-1}\breve{\mathbf{X}}(\breve{\mathbf{X}}^T\mathbf{V}^{-1}\breve{\mathbf{X}})^{-1}$. Then $(\mathbf{A}, \mathbf{M})$ is an $n\times n$ matrix such that

$$\begin{aligned}
|(\mathbf{A}, \mathbf{M})| &= \left|(\mathbf{A}, \mathbf{M})^T(\mathbf{A}, \mathbf{M})\right|^{\frac{1}{2}} \\
&= \left|\begin{pmatrix} \mathbf{I} & \mathbf{A}^T\mathbf{M} \\ \mathbf{M}^T\mathbf{A} & \mathbf{M}^T\mathbf{M} \end{pmatrix}\right|^{\frac{1}{2}} \qquad (16.16)\\
&= \left(|\mathbf{I}| \cdot |\mathbf{M}^T\mathbf{M} - \mathbf{M}^T\mathbf{A}\mathbf{A}^T\mathbf{M}|\right)^{\frac{1}{2}} \\
&= |\mathbf{M}^T[\mathbf{I} - (\mathbf{I} - \mathbf{P_X})]\mathbf{M}|^{\frac{1}{2}} \\
&= |\mathbf{M}^T\mathbf{X}(\mathbf{X}^T\mathbf{X})^{-}\mathbf{X}^T\mathbf{M}|^{\frac{1}{2}} \\
&= |\mathbf{M}^T\breve{\mathbf{X}}(\breve{\mathbf{X}}^T\breve{\mathbf{X}})^{-1}\breve{\mathbf{X}}^T\mathbf{M}|^{\frac{1}{2}} \\
&= |(\breve{\mathbf{X}}^T\breve{\mathbf{X}})^{-1}|^{\frac{1}{2}} \\
&= |\breve{\mathbf{X}}^T\breve{\mathbf{X}}|^{-\frac{1}{2}}, \qquad (16.17)
\end{aligned}$$

where we used Theorems 2.11.3, 2.11.5, and 2.11.7, Corollary 8.1.1.1, and Theorem 2.11.6 (in that order). Besides giving an explicit expression for the determinant of $(\mathbf{A}, \mathbf{M})$, (16.17) together with Theorem 2.11.2 establishes that $(\mathbf{A}, \mathbf{M})$ is nonsingular because the determinant of $\breve{\mathbf{X}}^T\breve{\mathbf{X}}$ is positive. Also, by Theorems 2.11.3, 2.11.5, and 2.11.7,

$$\begin{aligned}|(\mathbf{A}, \mathbf{M})|^2|\mathbf{V}| &= |(\mathbf{A}, \mathbf{M})^T\mathbf{V}(\mathbf{A}, \mathbf{M})| \\ &= \left|\begin{pmatrix} \mathbf{A}^T\mathbf{VA} & \mathbf{A}^T\mathbf{VM} \\ \mathbf{M}^T\mathbf{VA} & \mathbf{M}^T\mathbf{VM}\end{pmatrix}\right| \\ &= \left|\begin{pmatrix} \mathbf{A}^T\mathbf{VA} & \mathbf{0} \\ \mathbf{0} & (\breve{\mathbf{X}}^T\mathbf{V}^{-1}\breve{\mathbf{X}})^{-1}\end{pmatrix}\right| \\ &= |\mathbf{A}^T\mathbf{VA}| \cdot |\breve{\mathbf{X}}^T\mathbf{V}^{-1}\breve{\mathbf{X}}|^{-1}.\end{aligned}$$

Combining these results gives

$$\begin{aligned}|\mathbf{A}^T\mathbf{VA}| &= |\breve{\mathbf{X}}^T\mathbf{V}^{-1}\breve{\mathbf{X}}| \cdot |\mathbf{V}| \cdot |(\mathbf{A}, \mathbf{M})|^2 \\ &= |\breve{\mathbf{X}}^T\mathbf{V}^{-1}\breve{\mathbf{X}}| \cdot |\mathbf{V}| \cdot |\breve{\mathbf{X}}^T\breve{\mathbf{X}}|^{-1}.\end{aligned} \tag{16.18}$$

Furthermore, by Theorem 2.9.4 and Corollary 2.9.5.1,

$$\begin{aligned}(\mathbf{A}, \mathbf{M})^{-1}\mathbf{V}^{-1}[(\mathbf{A}, \mathbf{M})^T]^{-1} &= [(\mathbf{A}, \mathbf{M})^T\mathbf{V}(\mathbf{A}, \mathbf{M})]^{-1} \\ &= \begin{pmatrix} \mathbf{A}^T\mathbf{VA} & \mathbf{0} \\ \mathbf{0} & (\breve{\mathbf{X}}^T\mathbf{V}^{-1}\breve{\mathbf{X}})^{-1}\end{pmatrix}^{-1} \\ &= \begin{pmatrix} (\mathbf{A}^T\mathbf{VA})^{-1} & \mathbf{0} \\ \mathbf{0} & \breve{\mathbf{X}}^T\mathbf{V}^{-1}\breve{\mathbf{X}}\end{pmatrix}.\end{aligned}$$

This implies further that

$$\begin{aligned}\mathbf{V}^{-1} &= (\mathbf{A}, \mathbf{M})\begin{pmatrix} (\mathbf{A}^T\mathbf{VA})^{-1} & \mathbf{0} \\ \mathbf{0} & \breve{\mathbf{X}}^T\mathbf{V}^{-1}\breve{\mathbf{X}}\end{pmatrix}(\mathbf{A}, \mathbf{M})^T \\ &= \mathbf{A}(\mathbf{A}^T\mathbf{VA})^{-1}\mathbf{A}^T + \mathbf{M}\breve{\mathbf{X}}^T\mathbf{V}^{-1}\breve{\mathbf{X}}\mathbf{M}^T \\ &= \mathbf{A}(\mathbf{A}^T\mathbf{VA})^{-1}\mathbf{A}^T + \mathbf{V}^{-1}\breve{\mathbf{X}}(\breve{\mathbf{X}}^T\mathbf{V}^{-1}\breve{\mathbf{X}})^{-1}\breve{\mathbf{X}}^T\mathbf{V}^{-1},\end{aligned}$$

yielding

$$\begin{aligned}\mathbf{A}(\mathbf{A}^T\mathbf{VA})^{-1}\mathbf{A}^T &= \mathbf{V}^{-1} - \mathbf{V}^{-1}\breve{\mathbf{X}}(\breve{\mathbf{X}}^T\mathbf{V}^{-1}\breve{\mathbf{X}})^{-1}\breve{\mathbf{X}}^T\mathbf{V}^{-1} \\ &= \mathbf{V}^{-1} - \mathbf{V}^{-1}\mathbf{X}(\mathbf{X}^T\mathbf{V}^{-1}\mathbf{X})^{-}\mathbf{X}^T\mathbf{V}^{-1},\end{aligned} \tag{16.19}$$

where we used Theorem 11.1.6c for the last equality. By substituting (16.18) and (16.19) into expression (16.15) for $f_{\mathbf{w}}(\mathbf{w}; \boldsymbol{\theta})$, we obtain the following theorem.

Theorem 16.2.3 *The log-likelihood function associated with any vector of $n - p^*$ linearly independent error contrasts is, apart from an additive constant that does not depend on $\boldsymbol{\theta}$,*

$$L_1(\boldsymbol{\theta};\, \mathbf{y}) = -\frac{1}{2}\log|\mathbf{V}| - \frac{1}{2}\log|\breve{\mathbf{X}}^T\mathbf{V}^{-1}\breve{\mathbf{X}}| - \frac{1}{2}\mathbf{y}^T[\mathbf{V}^{-1} - \mathbf{V}^{-1}\mathbf{X}(\mathbf{X}^T\mathbf{V}^{-1}\mathbf{X})^{-}\mathbf{X}^T\mathbf{V}^{-1}]\mathbf{y}.$$

Note that the REML log-likelihood function, L_1, differs from the profile log-likelihood function L_0 given by (16.12) only by having the additional term, $-\frac{1}{2}\log|\breve{\mathbf{X}}^T\mathbf{V}^{-1}\breve{\mathbf{X}}|$.

Definition 16.2.4 $\hat{\boldsymbol{\theta}}_{REML}$ is a *REML estimate* of $\boldsymbol{\theta}$ under the normal general mixed linear model if $L_1(\boldsymbol{\theta};\, \mathbf{y})$ attains its maximum value, over $\boldsymbol{\theta} \in \Theta$, at $\boldsymbol{\theta} = \hat{\boldsymbol{\theta}}_{REML}$.

In contrast to the situation with ML estimation, a REML estimate of $\boldsymbol{\beta}$ is not defined as a maximizer (or a component of a maximizer) of some log-likelihood function. Rather, the corresponding estimate of $\boldsymbol{\beta}$ typically is taken to be any solution to the empirical Aitken equations with $\hat{\boldsymbol{\theta}}_{REML}$ substituted for $\boldsymbol{\theta}$, i.e., any solution to

$$\mathbf{X}^T[\mathbf{V}(\hat{\boldsymbol{\theta}}_{REML})]^{-1}\mathbf{X}\boldsymbol{\beta} = \mathbf{X}^T[\mathbf{V}(\hat{\boldsymbol{\theta}}_{REML})]^{-1}\mathbf{y}.$$

We refer to the likelihood equations associated with $L_1(\boldsymbol{\theta};\, \mathbf{y})$ as the REML equations. Thus, the REML equations are

$$\frac{\partial L_1}{\partial \theta_i} = 0 \quad (i = 1, \ldots, m),$$

assuming that these derivatives exist.

It is rather easy (and is left as an exercise) to show that if $n > p^*$, the unique REML estimator of the residual variance σ^2 in the normal Gauss–Markov model is the unbiased estimator

$$\hat{\sigma}^2 = \frac{(\mathbf{y} - \mathbf{X}\hat{\boldsymbol{\beta}})^T(\mathbf{y} - \mathbf{X}\hat{\boldsymbol{\beta}})}{n - p^*}.$$

Thus, in this case REML completely accounts for the loss of degrees of freedom from estimating $\mathbf{X}\boldsymbol{\beta}$. Somewhat more generally, consider the case of a normal mixed effects model with $\theta_1, \ldots, \theta_m$ taken to be the variance components. It can be shown [see, e.g., Anderson (1979)] that, for cases of this model with crossed or nested data structures and balanced data, the REML equations have an explicit solution and this solution coincides with the quadratic unbiased estimator from the usual sequential ANOVA. However, this solution may lie outside the parameter space (e.g., the solution for a variance component may be negative), so the solution is not necessarily a REML estimator. Consequently, a REML estimator may not be unbiased under these more general models. Nor is the REML estimator necessarily unbiased under the still more general mixed linear model. However, simulation studies and the author's practical experience with various mixed linear models suggest that in cases where the REML and ML estimators exist and are unique, the bias of the REML estimator is usually less than that of the ML estimator.

That the REML equations have an explicit solution that coincides, at least in the case of balanced crossed and nested data structures, with the minimum variance quadratic unbiased estimator derived without assuming normality or any other form of distribution, suggests that REML estimation is reasonable even when the distribution of $\mathbf{y}$ is unspecified.

Although the REML estimators of the elements of $\boldsymbol{\theta}$ are often less biased than the ML estimators, they may or may not have smaller variance. In fact, one might suspect that the variances of the REML estimators would be larger than those of the ML estimators because it seems that some information may be lost by applying ML to a vector of $n - p^*$ transformed observations (error contrasts) rather than to the entire n-dimensional response vector $\mathbf{y}$. However, the following argument shows that when ML is applied to $\mathbf{y}$ itself, the ML estimator of $\boldsymbol{\theta}$ depends on $\mathbf{y}$ only through the values of $n - p^*$ linearly independent error contrasts; thus ML uses exactly the same information from the responses—no more and no less—that REML uses.

To make this argument, let us define

$$\mathbf{E} = \mathbf{V}^{-1} - \mathbf{V}^{-1}\mathbf{X}(\mathbf{X}^T\mathbf{V}^{-1}\mathbf{X})^-\mathbf{X}^T\mathbf{V}^{-1}.$$

(Note that $\mathbf{E}$, like $\mathbf{V}$, is functionally dependent on $\boldsymbol{\theta}$, but that again we suppress this dependence to make the mathematical expressions less cumbersome.) By Theorem 11.1.6d, $\mathbf{EX} = \mathbf{0}$, implying that $\mathbf{Ey}$ is an n-vector of error contrasts. It is easily verified that $\text{rank}(\mathbf{E}) = n - p^*$; furthermore, by Theorem 11.1.6d once more, $\mathbf{E}$ is symmetric and satisfies $\mathbf{EVE} = \mathbf{E}$. Therefore, we may re-express the profile log-likelihood function given by (16.12) as

$$\begin{aligned} L_0(\boldsymbol{\theta}; \mathbf{y}) &= -\frac{1}{2}\log|\mathbf{V}| - \frac{1}{2}\mathbf{y}^T\mathbf{EVEy} \\ &= -\frac{1}{2}\log|\mathbf{V}| - \frac{1}{2}(\mathbf{Ey})^T\mathbf{V}(\mathbf{Ey}), \end{aligned}$$

showing that the ML estimator of $\boldsymbol{\theta}$ depends on $\mathbf{y}$ only through the values of $n - p^*$ linearly independent error contrasts.

In general, neither the ML nor REML equations have an explicit solution. Various methods have been proposed for numerically maximizing the nonlinear function L with respect to $\boldsymbol{\beta}$ and $\boldsymbol{\theta}$, and the nonlinear functions L_0 and L_1 with respect to $\boldsymbol{\theta}$, subject to $\boldsymbol{\theta} \in \Theta$. Among these are grid search and iterative methods such as steepest ascent, Newton–Raphson, and the method of scoring. Most of the iterative methods require expressions for the first-order partial derivatives of the function with respect to the parameters, and some also require expressions for the second-order partial derivatives or their expectations. Consequently, we assume henceforth that L, L_0, and L_1 are twice-differentiable with respect to $\boldsymbol{\theta}$ for all $\boldsymbol{\theta} \in \Theta$. Using Theorems 2.19.1 and 2.21.1, we may then derive the required expressions. We do so here for the full log-likelihood function and leave similar derivations for the other log-likelihood functions as exercises. Differentiating (16.11), we obtain

$$\frac{\partial L}{\partial \boldsymbol{\beta}} = \frac{1}{2}\mathbf{X}^T\mathbf{V}^{-1}\mathbf{y} + \frac{1}{2}\mathbf{X}^T\mathbf{V}^{-1}\mathbf{y} - \frac{1}{2} \cdot 2\mathbf{X}^T\mathbf{V}^{-1}\mathbf{X} = \mathbf{X}^T\mathbf{V}^{-1}\mathbf{y} - \mathbf{X}^T\mathbf{V}^{-1}\mathbf{X}\boldsymbol{\beta}$$

and

$$\frac{\partial L}{\partial \theta_i} = -\frac{1}{2}\mathrm{tr}\left(\mathbf{V}^{-1}\frac{\partial \mathbf{V}}{\partial \theta_i}\right) + \frac{1}{2}(\mathbf{y} - \mathbf{X}\boldsymbol{\beta})^T\mathbf{V}^{-1}\frac{\partial \mathbf{V}}{\partial \theta_i}\mathbf{V}^{-1}(\mathbf{y} - \mathbf{X}\boldsymbol{\beta}).$$

Differentiating these expressions again yields

$$\begin{aligned}
\frac{\partial^2 L}{\partial \boldsymbol{\beta}\partial \boldsymbol{\beta}^T} &= -\mathbf{X}^T\mathbf{V}^{-1}\mathbf{X}, \\
\frac{\partial^2 L}{\partial \boldsymbol{\beta}\partial \theta_i} &= -\mathbf{X}^T\mathbf{V}^{-1}\frac{\partial \mathbf{V}}{\partial \theta_i}\mathbf{V}^{-1}\mathbf{y} + \mathbf{X}^T\mathbf{V}^{-1}\frac{\partial \mathbf{V}}{\partial \theta_i}\mathbf{V}^{-1}\mathbf{X}\boldsymbol{\beta}, \\
\frac{\partial^2 L}{\partial \theta_i\partial \theta_j} &= -\frac{1}{2}\mathrm{tr}\left(\mathbf{V}^{-1}\frac{\partial^2 \mathbf{V}}{\partial \boldsymbol{\theta}_i\partial \boldsymbol{\theta}_j}\right) + \frac{1}{2}\mathrm{tr}\left(\mathbf{V}^{-1}\frac{\partial \mathbf{V}}{\partial \theta_i}\mathbf{V}^{-1}\frac{\partial \mathbf{V}}{\partial \theta_j}\right) \\
&\quad + \frac{1}{2}(\mathbf{y} - \mathbf{X}\boldsymbol{\beta})^T\left(\mathbf{V}^{-1}\frac{\partial^2 \mathbf{V}}{\partial \theta_i\partial \theta_j}\mathbf{V}^{-1} - 2\mathbf{V}^{-1}\frac{\partial \mathbf{V}}{\partial \theta_i}\mathbf{V}^{-1}\frac{\partial \mathbf{V}}{\partial \theta_j}\mathbf{V}^{-1}\right)(\mathbf{y} - \mathbf{X}\boldsymbol{\beta}).
\end{aligned}$$

Taking expectations and using Theorem 4.2.4, we find that

$$\begin{aligned}
\mathrm{E}\left(\frac{\partial^2 L}{\partial \boldsymbol{\beta}\partial \boldsymbol{\beta}^T}\right) &= -\mathbf{X}^T\mathbf{V}^{-1}\mathbf{X}, \\
\mathrm{E}\left(\frac{\partial^2 L}{\partial \boldsymbol{\beta}\partial \theta_i}\right) &= -\mathbf{X}^T\mathbf{V}^{-1}\frac{\partial \mathbf{V}}{\partial \theta_i}\mathbf{V}^{-1}\mathbf{X}\boldsymbol{\beta} + \mathbf{X}^T\mathbf{V}^{-1}\frac{\partial \mathbf{V}}{\partial \theta_i}\mathbf{V}^{-1}\mathbf{X}\boldsymbol{\beta} = \mathbf{0}, \\
\mathrm{E}\left(\frac{\partial^2 L}{\partial \theta_i\partial \theta_j}\right) &= -\frac{1}{2}\mathrm{tr}\left(\mathbf{V}^{-1}\frac{\partial^2 \mathbf{V}}{\partial \boldsymbol{\theta}_i\partial \boldsymbol{\theta}_j}\right) + \frac{1}{2}\mathrm{tr}\left(\mathbf{V}^{-1}\frac{\partial \mathbf{V}}{\partial \theta_i}\mathbf{V}^{-1}\frac{\partial \mathbf{V}}{\partial \theta_j}\right) \\
&\quad + \frac{1}{2}\mathrm{tr}\left(\mathbf{V}^{-1}\frac{\partial^2 \mathbf{V}}{\partial \theta_i\partial \theta_j}\mathbf{V}^{-1}\mathbf{V} - 2\mathbf{V}^{-1}\frac{\partial \mathbf{V}}{\partial \theta_i}\mathbf{V}^{-1}\frac{\partial \mathbf{V}}{\partial \theta_j}\mathbf{V}^{-1}\mathbf{V}\right) \\
&= -\frac{1}{2}\mathrm{tr}\left(\mathbf{V}^{-1}\frac{\partial \mathbf{V}}{\partial \theta_i}\mathbf{V}^{-1}\frac{\partial \mathbf{V}}{\partial \theta_j}\right).
\end{aligned} \tag{16.20}$$

Observe that the expressions for the expectations of the second-order partial derivatives are simpler than those for the second-order partials themselves.

In a similar manner, it may be shown that

$$\frac{\partial L_0}{\partial \theta_i} = -\frac{1}{2}\text{tr}\left(\mathbf{V}^{-1}\frac{\partial \mathbf{V}}{\partial \theta_i}\right) + \frac{1}{2}\mathbf{y}^T\mathbf{E}\left(\frac{\partial \mathbf{V}}{\partial \theta_i}\right)\mathbf{E}\mathbf{y}, \tag{16.21}$$

$$\frac{\partial^2 L_0}{\partial \theta_i \partial \theta_j} = \frac{1}{2}\text{tr}\left[\mathbf{V}^{-1}\left(\frac{\partial \mathbf{V}}{\partial \theta_i}\right)\mathbf{V}^{-1}\left(\frac{\partial \mathbf{V}}{\partial \theta_j}\right)\right] - \frac{1}{2}\text{tr}\left[\mathbf{V}^{-1}\left(\frac{\partial^2 \mathbf{V}}{\partial \theta_i \partial \theta_j}\right)\right] \tag{16.22}$$

$$+\frac{1}{2}\mathbf{y}^T\mathbf{E}\left(\frac{\partial^2 \mathbf{V}}{\partial \theta_i \partial \theta_j}\right)\mathbf{E}\mathbf{y} - \mathbf{y}^T\mathbf{E}\left(\frac{\partial \mathbf{V}}{\partial \theta_i}\right)\mathbf{E}\left(\frac{\partial \mathbf{V}}{\partial \theta_j}\right)\mathbf{E}\mathbf{y}, \tag{16.23}$$

$$\text{E}\left(\frac{\partial^2 L_0}{\partial \theta_i \partial \theta_j}\right) = \frac{1}{2}\text{tr}\left[\mathbf{V}^{-1}\left(\frac{\partial \mathbf{V}}{\partial \theta_i}\right)\mathbf{V}^{-1}\left(\frac{\partial \mathbf{V}}{\partial \theta_j}\right)\right] - \frac{1}{2}\text{tr}\left[\mathbf{V}^{-1}\left(\frac{\partial^2 \mathbf{V}}{\partial \theta_i \partial \theta_j}\right)\right] \tag{16.24}$$

$$+\frac{1}{2}\text{tr}\left[\mathbf{E}\left(\frac{\partial^2 \mathbf{V}}{\partial \theta_i \partial \theta_j}\right)\right] - \text{tr}\left[\mathbf{E}\left(\frac{\partial \mathbf{V}}{\partial \theta_i}\right)\mathbf{E}\left(\frac{\partial \mathbf{V}}{\partial \theta_j}\right)\right], \tag{16.25}$$

and that

$$\frac{\partial L_1}{\partial \theta_i} = -\frac{1}{2}\text{tr}\left(\mathbf{E}\frac{\partial \mathbf{V}}{\partial \theta_i}\right) + \frac{1}{2}\mathbf{y}^T\mathbf{E}\left(\frac{\partial \mathbf{V}}{\partial \theta_i}\right)\mathbf{E}\mathbf{y}, \tag{16.26}$$

$$\frac{\partial^2 L_1}{\partial \theta_i \partial \theta_j} = \frac{1}{2}\text{tr}\left[\mathbf{E}\left(\frac{\partial \mathbf{V}}{\partial \theta_j}\right)\mathbf{E}\left(\frac{\partial \mathbf{V}}{\partial \theta_i}\right)\right] - \frac{1}{2}\text{tr}\left[\mathbf{E}\left(\frac{\partial^2 \mathbf{V}}{\partial \theta_i \partial \theta_j}\right)\right] \tag{16.27}$$

$$+\frac{1}{2}\mathbf{y}^T\mathbf{E}\left(\frac{\partial^2 \mathbf{V}}{\partial \theta_i \partial \theta_j}\right)\mathbf{E}\mathbf{y} - \mathbf{y}^T\mathbf{E}\left(\frac{\partial \mathbf{V}}{\partial \theta_i}\right)\mathbf{E}\left(\frac{\partial \mathbf{V}}{\partial \theta_j}\right)\mathbf{E}\mathbf{y}, \tag{16.28}$$

$$\text{E}\left(\frac{\partial^2 L_1}{\partial \theta_i \partial \theta_j}\right) = -\frac{1}{2}\text{tr}\left(\mathbf{E}\frac{\partial \mathbf{V}}{\partial \theta_i}\mathbf{E}\frac{\partial \mathbf{V}}{\partial \theta_j}\right). \tag{16.29}$$

How does one actually maximize L, L_0, or L_1 using these expressions? We briefly describe the three aforementioned iterative methods, each of which is a gradient algorithm. In a gradient algorithm for maximizing a twice-differentiable function f, one computes the $(k+1)$th iterate $\boldsymbol{\theta}^{(k+1)}$ by updating the kth iterate $\boldsymbol{\theta}^{(k)}$ according to the equation

$$\boldsymbol{\theta}^{(k+1)} = \boldsymbol{\theta}^{(k)} + \mathbf{M}^{(k)}\boldsymbol{\delta}^{(k)},$$

where $\mathbf{M}^{(k)}$ is a matrix that can be determined from $\boldsymbol{\theta}^{(k)}$ and $\boldsymbol{\delta}^{(k)}$ is the vector gradient of f evaluated at $\boldsymbol{\theta} = \boldsymbol{\theta}^{(k)}$, i.e.,

$$\boldsymbol{\delta}^{(k)} = \left.\frac{\partial f}{\partial \boldsymbol{\theta}}\right|_{\boldsymbol{\theta}=\boldsymbol{\theta}^{(k)}}.$$

The choice of $\mathbf{M}^{(k)}$ for the three methods is as follows:

- Method of steepest ascent: $\mathbf{M}^{(k)} = \mathbf{I}$ for all k.
- Newton–Raphson method: $\mathbf{M}^{(k)} = -\left(\frac{\partial^2 f}{\partial\theta_i \partial\theta_j}\Big|_{\boldsymbol{\theta}=\boldsymbol{\theta}^{(k)}}\right)^{-1}$.
- Method of scoring: $\mathbf{M}^{(k)} = \left\{-\mathrm{E}\left(\frac{\partial^2 f}{\partial\theta_i \partial\theta_j}\Big|_{\boldsymbol{\theta}=\boldsymbol{\theta}^{(k)}}\right)\right\}^{-1}$.

Note that the method of scoring is identical to the Newton–Raphson method except that the second-order partial derivatives are replaced by their expectations.

Which gradient algorithm is fastest depends on the situation. Generally, the method of steepest ascent is not as demanding computationally as the other methods on a per-iteration basis, but it may require many more iterations to converge than the others. Much of the computational burden is associated with inversion of the $n \times n$ matrix $\mathbf{V}$ in the expressions for the log-likelihood function and its partial derivatives. This large-scale inversion and other required computations can often be performed more efficiently or avoided altogether when $\mathbf{V}$ has more structure, such as under a mixed effects model.

In order to implement any of these methods, an initial estimate $\boldsymbol{\theta}^{(0)}$ must be supplied. This estimate can be chosen based on prior knowledge or on some rough data-based approach. It should also be emphasized that the above methods must often incorporate modifications to ensure that the final iterate (and often, in practice, every iterate) belongs to the specified parameter space Θ. Finally, there is no guarantee that the final iterate will be a global, rather than merely local, maximizer. Therefore, in practice it is wise to repeat the gradient algorithm starting from many widely separated initial estimates; if the algorithm converges to the same final iterate from all these initial estimates, one can be somewhat more (but still not 100%) assured that the final iterate corresponds to a global maximum. For a more thorough discussion of algorithms for obtaining ML and REML estimators of $\boldsymbol{\theta}$ in general mixed linear models, and in components-of-variance models in particular, the reader is referred to Harville (1977) and Callanan and Harville (1991).

Under suitable regularity conditions, asymptotically valid confidence intervals/regions and hypothesis testing procedures for the elements of $\boldsymbol{\theta}$ may be obtained using standard asymptotic theory of maximum likelihood estimation (e.g., Casella and Berger 2002, Sec. 10.1). According to this theory, $\hat{\boldsymbol{\theta}}_{ML}$ and $\hat{\boldsymbol{\theta}}_{REML}$ are consistent and asymptotically normal, with asymptotic variance–covariance matrices

$$\mathbf{B}_{ML}(\boldsymbol{\theta}) = \left[-\mathrm{E}\left(\frac{\partial^2 L_0}{\partial\theta_i \partial\theta_j}\right)\right]^{-1}$$

and

$$\mathbf{B}_{REML}(\boldsymbol{\theta}) = \left[-\mathrm{E}\left(\frac{\partial^2 L_1}{\partial\theta_i \partial\theta_j}\right)\right]^{-1}.$$

Therefore, asymptotically valid confidence intervals/regions and hypothesis testing procedures for a vector $\mathbf{C}^T\boldsymbol{\beta}$ of estimable functions and asymptotically valid prediction intervals/regions for a vector $\mathbf{C}^T\boldsymbol{\beta} + \mathbf{u}$ of predictable functions may be obtained by substituting $\hat{\boldsymbol{\beta}}_{ML}$ and $\begin{pmatrix} \mathbf{V}_{yy}(\hat{\boldsymbol{\theta}}_{ML}) & \mathbf{V}_{yu}(\hat{\boldsymbol{\theta}}_{ML}) \\ \mathbf{V}_{yu}(\hat{\boldsymbol{\theta}}_{ML})^T & \mathbf{V}_{uu}(\hat{\boldsymbol{\theta}}_{ML}) \end{pmatrix}$, or $\hat{\boldsymbol{\beta}}_{REML}$ and $\begin{pmatrix} \mathbf{V}_{yy}(\hat{\boldsymbol{\theta}}_{REML}) & \mathbf{V}_{yu}(\hat{\boldsymbol{\theta}}_{REML}) \\ \mathbf{V}_{yu}(\hat{\boldsymbol{\theta}}_{REML})^T & \mathbf{V}_{uu}(\hat{\boldsymbol{\theta}}_{REML}) \end{pmatrix}$, for $\tilde{\boldsymbol{\beta}}$ and $\tilde{\sigma}^2\begin{pmatrix} \mathbf{W} & \mathbf{K} \\ \mathbf{K}^T & \mathbf{H} \end{pmatrix}$ in the appropriate expressions in Chap. 15 [e.g., (15.4), (15.10), and (15.15)].

In practice, for a given dataset the parameters of several general mixed linear models, which differ with respect to either their mean structure or variance–covariance structure (or both), may be estimated by ML or REML, and the analyst often wants to select the best such model. If the models are nested, then the results of appropriate hypothesis testing procedures of the type described in the previous paragraph may be used to guide this choice. Alternatively, the likelihood-based information criteria introduced in Methodological Interlude #12 may be employed; for more specifics, see Gurka (2006).

16.3 Exercises

1. Consider a two-way layout with two rows and two columns. Suppose there are two observations, labeled as y_{111} and y_{112}, in the upper left cell; one observation, labeled as y_{121}, in the upper right cell; and one observation, labeled as y_{211}, in the lower left cell, as in the fourth layout displayed in Example 7.1-3 (the lower right cell is empty). Suppose that the observations follow the normal two-way mixed main effects model

$$y_{ijk} = \mu+\alpha_i+b_j+d_{ijk} \qquad (i, j, k) \in \{(1, 1, 1), (1, 1, 2), (1, 2, 1), (2, 1, 1)\},$$

where

$$\mathrm{E}\begin{pmatrix} b_1 \\ b_2 \\ d_{111} \\ d_{112} \\ d_{121} \\ d_{211} \end{pmatrix} = \mathbf{0} \quad \text{and} \quad \mathrm{var}\begin{pmatrix} b_1 \\ b_2 \\ d_{111} \\ d_{112} \\ d_{121} \\ d_{211} \end{pmatrix} = \begin{pmatrix} \sigma_b^2\mathbf{I}_2 & \mathbf{0} \\ \mathbf{0} & \sigma^2\mathbf{I}_4 \end{pmatrix}.$$

Here, the variance components $\sigma_b^2 \geq 0$ and $\sigma^2 > 0$ are unknown.

(a) Obtain quadratic unbiased estimators of the variance components.

(b) Let $\xi \in (0, 1)$. Obtain a $100(1 - \xi)\%$ confidence interval for $\psi \equiv \sigma_b^2/\sigma^2$, which depends on the data only through $y_{111} - y_{112}$ and $y_{121} - (y_{111} + y_{112})/2$. (Hint: First obtain the joint distribution of these two linear functions of the data, and then obtain the joint distribution of their squares, suitably scaled.)

2. Consider the normal two-way main effects components-of-variance model with only one observation per cell,

$$y_{ij} = \mu + a_i + b_j + d_{ij} \quad (i = 1, \ldots, q;\ j = 1, \ldots, m),$$

where $a_i \sim \mathrm{N}(0, \sigma_a^2)$, $b_j \sim \mathrm{N}(0, \sigma_b^2)$, and $d_{ij} \sim \mathrm{N}(0, \sigma^2)$ for all i and j, and where $\{a_i\}$, $\{b_j\}$, and $\{d_{ij}\}$ are mutually independent. The corrected two-part mixed-model ANOVA table (with Factor A fitted first) is given below, with an additional column giving the distributions of suitably scaled mean squares under the model.

Source	df	Sum of squares	Mean square
Factor A	$q-1$	$m\sum_{i=1}^{q}(\bar{y}_{i\cdot} - \bar{y}_{\cdot\cdot})^2$	S_1^2
Factor B	$m-1$	$q\sum_{j=1}^{m}(\bar{y}_{\cdot j} - \bar{y}_{\cdot\cdot})^2$	S_2^2
Residual	$(q-1)(m-1)$	$\sum_{i=1}^{q}\sum_{j=1}^{m} (y_{ij} - \bar{y}_{i\cdot} - \bar{y}_{\cdot j} + \bar{y}_{\cdot\cdot})^2$	S_3^2
Total	$qm-1$	$\sum_{i=1}^{q}\sum_{j=1}^{m}(y_{ij} - \bar{y}_{\cdot\cdot})^2$	

Here, $f_1 S_1^2 \sim \chi^2(q-1)$, $f_2 S_2^2 \sim \chi^2(m-1)$, and $f_3 S_3^2 \sim \chi^2[(q-1)(m-1)]$, where

$$f_1 = \frac{q-1}{\sigma^2 + m\sigma_a^2}, \quad f_2 = \frac{m-1}{\sigma^2 + q\sigma_b^2}, \quad f_3 = \frac{(q-1)(m-1)}{\sigma^2}.$$

Furthermore, S_1^2, S_2^2, and S_3^2 are mutually independent. Let $\xi \in (0, 1)$.

(a) Verify the distributions given in the last column of the table, and verify that S_1^2, S_2^2, and S_3^2 are mutually independent.

(b) Obtain minimum variance quadratic unbiased estimators of the variance components.

(c) Obtain a $100(1-\xi)\%$ confidence interval for $\sigma^2 + q\sigma_b^2$.

(d) Obtain a $100(1-\xi)\%$ confidence interval for σ_b^2/σ^2.

(e) Obtain a confidence interval for $(\sigma_a^2 - \sigma_b^2)/\sigma^2$ which has coverage probability at least $1-\xi$. [Hint: First combine the interval from part (d) with a similarly constructed interval for σ_a^2/σ^2 via the Bonferroni inequality.]

(f) Use Satterthwaite's method to obtain an approximate $100(1-\xi)\%$ confidence interval for σ_a^2.

(g) Obtain size-ξ tests of $H_0\colon \sigma_a^2 = 0$ versus $H_a\colon \sigma_a^2 > 0$ and $H_0\colon \sigma_b^2 = 0$ versus $H_a\colon \sigma_b^2 > 0$.

3. Consider the normal two-way components-of-variance model with interaction and balanced data:

$$y_{ijk} = \mu + a_i + b_j + c_{ij} + d_{ijk} \quad (i = 1, \ldots, q;\ j = 1, \ldots, m;\ k = 1, \cdots, r)$$

where $a_i \sim \mathrm{N}(0, \sigma_a^2)$, $b_j \sim \mathrm{N}(0, \sigma_b^2)$, $c_{ij} \sim \mathrm{N}(0, \sigma_c^2)$, and $d_{ijk} \sim \mathrm{N}(0, \sigma^2)$ for all i, j, and k, and where $\{a_i\}$, $\{b_j\}$, $\{c_{ij}\}$ and $\{d_{ijk}\}$ are mutually independent. The corrected three-part mixed-model ANOVA table (corresponding to fitting the interaction terms after the main effects) is given below, with an additional column giving the distributions of suitably scaled mean squares under the model.

Source	df	Sum of squares	Mean square
Factor A	$q-1$	$mr\sum_{i=1}^{q}(\bar{y}_{i\cdot\cdot} - \bar{y}_{\cdot\cdot\cdot})^2$	S_1^2
Factor B	$m-1$	$qr\sum_{j=1}^{m}(\bar{y}_{\cdot j\cdot} - \bar{y}_{\cdot\cdot\cdot})^2$	S_2^2
Interaction	$(q-1)(m-1)$	$r\sum_{i=1}^{q}\sum_{j=1}^{m}$ $(\bar{y}_{ij\cdot} - \bar{y}_{i\cdot\cdot} - \bar{y}_{\cdot j\cdot} + \bar{y}_{\cdot\cdot\cdot})^2$	S_3^2
Residual	$qm(r-1)$	$\sum_{i=1}^{q}\sum_{j=1}^{m}\sum_{k=1}^{r}(y_{ijk} - \bar{y}_{ij\cdot})^2$	S_4^2
Total	$qmr-1$	$\sum_{i=1}^{q}\sum_{j=1}^{m}\sum_{k=1}^{r}(y_{ijk} - \bar{y}_{\cdot\cdot\cdot})^2$	

Here $f_1 S_1^2 \sim \chi^2(q-1)$, $f_2 S_2^2 \sim \chi^2(m-1)$, $f_3 S_3^2 \sim \chi^2[(q-1)(m-1)]$, and $f_4 S_4^2 \sim \chi^2[qm(r-1)]$, where

$$f_1 = \frac{q-1}{\sigma^2 + r\sigma_c^2 + mr\sigma_a^2}, \quad f_2 = \frac{m-1}{\sigma^2 + r\sigma_c^2 + qr\sigma_b^2},$$

$$f_3 = \frac{(q-1)(m-1)}{\sigma^2 + r\sigma_c^2}, \quad f_4 = \frac{qm(r-1)}{\sigma^2}.$$

Furthermore, S_1^2, S_2^2, S_3^2, and S_4^2 are mutually independent. Let $\xi \in (0, 1)$.

(a) Verify the distributions given in the last column of the table, and verify that S_1^2, S_2^2, S_3^2, and S_4^2 are mutually independent.

(b) Obtain minimum variance quadratic unbiased estimators of the variance components.

(c) Obtain a $100(1-\xi)\%$ confidence interval for $\frac{\sigma_b^2}{\sigma^2 + r\sigma_c^2}$. (Hint: consider the distribution of S_2^2/S_3^2, suitably scaled.)

(d) Obtain a $100(1-\xi)\%$ confidence interval for $\frac{\sigma_a^2}{\sigma^2 + r\sigma_c^2}$. [Hint: exploit the similarity with part (c).]

(e) Using the results of part (c) and (d), obtain a confidence interval for $\frac{\sigma_a^2 + \sigma_b^2}{\sigma^2 + r\sigma_c^2}$ which has coverage probability at least $1-\xi$.

(f) Obtain size-ξ tests of H_0: $\sigma_a^2 = 0$ versus H_a: $\sigma_a^2 > 0$, H_0: $\sigma_b^2 = 0$ versus H_a: $\sigma_b^2 > 0$, and H_0: $\sigma_c^2 = 0$ versus H_a: $\sigma_c^2 > 0$.

4. Consider the normal two-factor nested components-of-variance model with balanced data:

$$y_{ijk} = \mu + a_i + b_{ij} + d_{ijk} \quad (i = 1, \ldots, q;\ j = 1, \ldots, m;\ k = 1, \ldots, r),$$

where $a_i \sim \mathrm{N}(0, \sigma_a^2)$, $b_{ij} \sim \mathrm{N}(0, \sigma_b^2)$, and $d_{ijk} \sim \mathrm{N}(0, \sigma^2)$ for all i, j, and k, and where $\{a_i\}$, $\{b_j\}$, and $\{d_{ijk}\}$ are mutually independent. The corrected two-part mixed-model ANOVA table (with Factor A fitted first) is given below, with an additional column giving the distributions of suitably scaled mean squares under the model.

Source	df	Sum of squares	Mean square
Factor A	$q-1$	$mr\sum_{i=1}^{q}(\bar{y}_{i\cdot\cdot} - \bar{y}_{\cdot\cdot\cdot})^2$	S_1^2
Factor B within A	$q(m-1)$	$r\sum_{i=1}^{q}\sum_{j=1}^{m}(\bar{y}_{ij\cdot} - \bar{y}_{i\cdot\cdot})^2$	S_2^2
Residual	$qm(r-1)$	$\sum_{i=1}^{q}\sum_{j=1}^{m}\sum_{k=1}^{r}(y_{ijk} - \bar{y}_{ij\cdot})^2$	S_3^2
Total	$qmr-1$	$\sum_{i=1}^{q}\sum_{j=1}^{m}\sum_{k=1}^{r}(y_{ijk} - \bar{y}_{\cdot\cdot\cdot})^2$	

Here $f_1 S_1^2 \sim \chi^2(q-1)$, $f_2 S_2^2 \sim \chi^2[q(m-1)]$, and $f_3 S_3^2 \sim \chi^2[qm(r-1)]$, where

$$f_1 = \frac{q-1}{\sigma^2 + r\sigma_b^2 + mr\sigma_a^2}, \quad f_2 = \frac{q(m-1)}{\sigma^2 + r\sigma_b^2}, \quad f_3 = \frac{qm(r-1)}{\sigma^2}.$$

Furthermore, S_1^2, S_2^2, and S_3^2 are mutually independent. Let $\xi \in (0, 1)$.

(a) Verify the distributions given in the last column of the table, and verify that S_1^2, S_2^2, and S_3^2 are mutually independent.

(b) Obtain minimum variance quadratic unbiased estimators of the variance components.

(c) Obtain a $100(1-\xi)\%$ confidence interval for σ^2.

(d) Obtain a $100(1-\xi)\%$ confidence interval for $\frac{\sigma_b^2 + m\sigma_a^2}{\sigma^2}$.

(e) Use Satterthwaite's method to obtain an approximate $100(1-\xi)\%$ confidence interval for σ_a^2.

(f) Obtain size-ξ tests of H_0: $\sigma_a^2 = 0$ versus H_a: $\sigma_a^2 > 0$ and H_0: $\sigma_b^2 = 0$ versus H_a: $\sigma_b^2 > 0$.

5. In a certain components-of-variance model there are four unknown variance components: σ^2, σ_1^2, σ_2^2, and σ_3^2. Suppose that four quadratic forms (S_i^2: $i = 1, 2, 3, 4$) whose expectations are linearly independent linear combinations of the variance components are obtained from an ANOVA table. Furthermore,

suppose that S_1^2, S_2^2, S_3^2, and S_4^2 are mutually independent and that the following distributional results hold:

$$\frac{f_1 S_1^2}{\sigma^2 + \sigma_1^2 + \sigma_2^2} \sim \chi^2(f_1)$$

$$\frac{f_2 S_2^2}{\sigma^2 + \sigma_1^2 + \sigma_3^2} \sim \chi^2(f_2)$$

$$\frac{f_3 S_3^2}{\sigma^2 + \sigma_2^2 + \sigma_3^2} \sim \chi^2(f_3)$$

$$\frac{f_4 S_4^2}{\sigma^2} \sim \chi^2(f_4)$$

Let $\xi \in (0, 1)$.

(a) Obtain quadratic unbiased estimators of the four variance components.
(b) Obtain a $100(1 - \xi)\%$ confidence interval for $\sigma^2 + \sigma_1^2 + \sigma_2^2$.
(c) Obtain a confidence interval for $\sigma_2^2 - \sigma_3^2$ which has coverage probability at least $1 - \xi$. [Hint: using the Bonferroni inequality, combine the interval from part (b) with a similarly constructed interval for another quantity.]
(d) Use Satterthwaite's method to obtain an approximate $100(1 - \xi)\%$ confidence interval for σ_1^2.

6. Consider once again the normal one-factor components-of-variance model with balanced data that was considered in Examples 16.1.3-2 and 16.1.4-1. Recall that $\hat{\sigma}_b^2 = (S_2^2 - S_3^2)/r$ is the minimum variance quadratic unbiased estimator of σ_b^2 under this model. Suppose that $\sigma_b^2/\sigma^2 = 1$.
   (a) Express $\Pr(\hat{\sigma}_b^2 < 0)$ as $\Pr(F < c)$ where F is a random variable having an $\mathrm{F}(\nu_1, \nu_2)$ distribution and c is a constant. Determine ν_1, ν_2, and c in as simple a form as possible.
   (b) Using the `pf` function in `R`, obtain a numerical value for $\Pr(\hat{\sigma}_b^2 < 0)$ when $q = 5$ and $r = 4$.
7. Consider once again the normal one-factor components-of-variance model with balanced data that was considered in Examples 16.1.3-2 and 16.1.4-1. Simulate 10,000 response vectors $\mathbf{y}$ that follow the special case of such a model in which $q = 4$, $r = 5$, and $\sigma_b^2 = \sigma^2 = 1$, and obtain the following:
   (a) the proportion of simulated vectors for which the quadratic unbiased estimator of σ_b^2 given in Example 16.1.3-2 is negative;
   (b) the average width and empirical coverage probability of the approximate 95% confidence interval for σ_b^2 derived in Example 16.1.4-1 using Satterthwaite's approach;
   (c) the average width and empirical coverage probability of the approximate 95% confidence intervals for σ_b^2 derived in Example 16.1.4-1 using Williams' approach.

Based on your results for parts (b) and (c), which of the two approaches for obtaining an approximate 95% confidence interval for σ_b^2 performs best? (Note: In parts (b) and (c), if any endpoints of the intervals are negative, you should modify your code to set them equal to 0.)

8. Consider the normal general mixed linear model $\{\mathbf{y}, \mathbf{X}\boldsymbol{\beta}, \mathbf{V}(\boldsymbol{\theta})\}$ with model matrix

$$\mathbf{X} = \begin{pmatrix} 1 & 0 \\ 1 & 0 \\ 1 & 0 \\ 0 & 1 \\ 0 & 1 \end{pmatrix}.$$

Specify, in terms of the elements $y_1, \ldots, y_5$ of $\mathbf{y}$, the elements of a vector $\mathbf{w}$ of error contrasts (of appropriate dimension) whose likelihood, when maximized over the parameter space for $\boldsymbol{\theta}$, yields a REML estimate of $\boldsymbol{\theta}$.

9. Consider the normal Gauss–Markov model with $n > p^*$, and define $\hat{\sigma}^2$ and $\bar{\sigma}^2$ as in (16.14) and (16.13), respectively.
   (a) Show that $\hat{\sigma}^2$ is the REML estimator of σ^2.
   (b) Show that the mean square error of $\hat{\sigma}^2$ is less than or equal to that of $\bar{\sigma}^2$ if $p^* > 5$ and $n > p^*(p^*-2)/(p^*-4)$, and that if $p^* \geq 13$, then $n > p^*+2$ suffices.

10. Show that, under the general mixed linear model, solving the REML equations

$$\frac{\partial L_1}{\partial \theta_i} = 0 \quad (i = 1, \cdots, m)$$

is equivalent to equating m quadratic forms in $\mathbf{y}$ to their expectations. Also show that the profile likelihood equations, defined as

$$\frac{\partial L_0}{\partial \theta_i} = 0 \quad (i = 1, \cdots, m),$$

do not have this property.

11. Verify the expressions for $\frac{\partial L_0}{\partial \theta_i}$, $\frac{\partial^2 L_0}{\partial \theta_i \partial \theta_j}$, and $\mathrm{E}\left(\frac{\partial^2 L_0}{\partial \theta_i \partial \theta_j}\right)$ given by (16.21)–(16.25).
12. Verify the expressions for $\frac{\partial L_1}{\partial \theta_i}$, $\frac{\partial^2 L_1}{\partial \theta_i \partial \theta_j}$, and $\mathrm{E}\left(\frac{\partial^2 L_1}{\partial \theta_i \partial \theta_j}\right)$ given by (16.26)–(16.29).

References

Anderson, R. D. (1979). Estimating variance components from balanced data: Optimum properties of REML solutions and MIVQUE estimators. In L. D. Van Vleck & S. R. Searle (Eds.), *Variance components and animal breeding* (pp. 205–216). Ithaca, NY: Animal Science Department, Cornell University.

Burdick, R. K., & Graybill, F. A. (1992). *Confidence intervals on variance components*. New York: Dekker.

Callanan, T. P., & Harville, D. A. (1991). Some new algorithms for computing restricted maximum likelihood estimates of variance components. *Journal of Statistical Computation and Simulation, 38*, 239–259.

Casella, G., & Berger, R. L. (2002). *Statistical inference* (2nd ed.). Pacific Grove, CA: Duxbury.

Graybill, F. A. (1954). On quadratic estimates of variance components. *Annals of Mathematical Statistics, 25*, 367–372.

Gurka, M. J. (2006). Selecting the best linear mixed model under REML. *The American Statistician, 60*, 19–26.

Harville, D. A. (1977). Maximum likelihood approaches to variance component estimation and to related problems. *Journal of the American Statistical Association, 72*, 320–338.

Henderson, C. R. (1953). Estimation of variance and covariance components. *Biometrics, 9*, 226–252.

Khuri, A. I., Mathew, T., & Sinha, B. K. (1998). *Statistical tests for mixed linear models*. New York: Wiley.

Satterthwaite, F. E. (1946). An approximate distribution of estimates of variance components. *Biometrics Bulletin, 2*, 110–114.

Searle, S. R. (1971). *Linear models*. New York: Wiley.

Williams, J. S. (1962). A confidence interval for variance components. *Biometrika, 49*, 278–281.

Empirical BLUE and BLUP

17

Recall from Sect. 13.5 that E-BLUP, the conventional procedure for predicting a predictable linear function $\tau = \mathbf{c}^T\boldsymbol{\beta} + u$ under the prediction-extended general mixed linear model

$$\left\{\begin{pmatrix}\mathbf{y}\\ u\end{pmatrix}, \begin{pmatrix}\mathbf{X}\boldsymbol{\beta}\\ 0\end{pmatrix}, \begin{pmatrix}\mathbf{V}_{yy}(\boldsymbol{\theta}) & \mathbf{v}_{yu}(\boldsymbol{\theta})\\ \mathbf{v}_{yu}(\boldsymbol{\theta})^T & v_{uu}(\boldsymbol{\theta})\end{pmatrix}\right\},$$

is to first obtain an estimate $\hat{\boldsymbol{\theta}}$ of $\boldsymbol{\theta}$ and then proceed as though this estimate was the true $\boldsymbol{\theta}$. Specifically, for the positive definite case of the model, the E-BLUP of τ is

$$\hat{\tilde{\tau}} \equiv \mathbf{c}^T\hat{\tilde{\boldsymbol{\beta}}} + \hat{\mathbf{v}}_{yu}^T\hat{\mathbf{E}}\mathbf{y} \tag{17.1}$$

where $\hat{\tilde{\boldsymbol{\beta}}}$ is any solution to the "empirical" Aitken equations

$$\mathbf{X}^T\hat{\mathbf{V}}_{yy}^{-1}\mathbf{X}\boldsymbol{\beta} = \mathbf{X}^T\hat{\mathbf{V}}_{yy}^{-1}\mathbf{y},$$

$\hat{\mathbf{V}}_{yy} = \mathbf{V}_{yy}(\hat{\boldsymbol{\theta}})$, $\hat{\mathbf{v}}_{yu} = \mathbf{v}_{yu}(\hat{\boldsymbol{\theta}})$, and $\hat{\mathbf{E}} = \hat{\mathbf{V}}_{yy}^{-1} - \hat{\mathbf{V}}_{yy}^{-1}\mathbf{X}(\mathbf{X}^T\hat{\mathbf{V}}_{yy}^{-1}\mathbf{X})^-\mathbf{X}^T\hat{\mathbf{V}}_{yy}^{-1}$. Furthermore, the conventional procedure for estimating the E-BLUP's prediction error variance is to substitute $\hat{\boldsymbol{\theta}}$ for $\boldsymbol{\theta}$ in the expression for $M(\boldsymbol{\theta})$, the BLUP's prediction error variance; that is, to estimate $\text{var}(\hat{\tilde{\tau}} - \tau)$ by $M(\hat{\boldsymbol{\theta}})$, where

$$M(\hat{\boldsymbol{\theta}}) = \hat{v}_{uu} - \hat{\mathbf{v}}_{yu}^T\hat{\mathbf{V}}_{yy}^{-1}\hat{\mathbf{v}}_{yu} + (\mathbf{c}^T - \hat{\mathbf{v}}_{yu}^T\hat{\mathbf{V}}_{yy}^{-1}\mathbf{X})(\mathbf{X}^T\hat{\mathbf{V}}_{yy}^{-1}\mathbf{X})^-(\mathbf{c}^T - \hat{\mathbf{v}}_{yu}^T\hat{\mathbf{V}}_{yy}^{-1}\mathbf{X})^T$$

where $\hat{v}_{uu} = v_{uu}(\hat{\boldsymbol{\theta}})$. If the sample size is sufficiently large (relative to the number of functionally independent parameters comprising $\boldsymbol{\beta}$ and $\boldsymbol{\theta}$) and $\boldsymbol{\theta}$ is estimated consistently, and the regularity conditions mentioned at the end of the previous chapter are satisfied, then the asymptotically valid inference procedures based on the E-BLUP (which were also described at the end of the previous chapter) may

D. L. Zimmerman, *Linear Model Theory*,
https://doi.org/10.1007/978-3-030-52063-2_17

perform well. If the sample size is not sufficiently large, however, then inference procedures based on the distribution of the E-BLUP for small samples may perform better. The objectives of this chapter are to derive some properties of the exact sampling distribution of the E-BLUP and $M(\hat{\boldsymbol{\theta}})$ and to use them to obtain inference procedures that perform better for small samples.

All results in this chapter are presented in terms of E-BLUP. However, because E-BLUE (empirical BLUE) is the special case of E-BLUP for which $u \equiv 0$, all results presented here for E-BLUP specialize to E-BLUE.

17.1 Properties of the E-BLUP

Except in special cases (such as an Aitken model), the E-BLUP is neither linear [as is clear from inspection of (17.1)] nor "best" in any sense. Remarkably, however, it is unbiased under fairly unrestrictive conditions. The first two theorems of this chapter pertain to the unbiasedness of the E-BLUP. The first theorem's relevance to this issue is not immediately obvious but will become apparent when it is used in the proof of the second theorem.

Theorem 17.1.1 *If* $\mathbf{w}$ *is a random n-vector whose distribution is symmetric about* $\mathbf{0}$ *(in the sense that the distributions of* $\mathbf{w}$ *and* $-\mathbf{w}$ *are identical) and* $f(\cdot)$ *is an odd function of n variables [meaning that* $f(-\mathbf{v}) = -f(\mathbf{v})$ *for all* $\mathbf{v} \in \mathbb{R}^n$*], then the distribution of* $f(\mathbf{w})$ *is symmetric about 0.*

Proof By the given two conditions, for any real number x,

$$\Pr[f(\mathbf{w}) \leq x] = \Pr[f(-\mathbf{w}) \leq x] = \Pr[-f(\mathbf{w}) \leq x].$$

□

For the remainder of this chapter, it is assumed that $\mathbf{y}$ follows the prediction-extended positive definite general mixed linear model specified in the preamble, and that $\mathbf{c}^T\boldsymbol{\beta} + u$ is a predictable function. The theorems to be presented impose a condition on the estimator $\hat{\boldsymbol{\theta}}$ of $\boldsymbol{\theta}$. The condition is that $\hat{\boldsymbol{\theta}}$ is an even, translation-invariant function of the response vector $\mathbf{y}$; that is, that $\hat{\boldsymbol{\theta}}(-\mathbf{y}) = \hat{\boldsymbol{\theta}}(\mathbf{y})$ and that $\hat{\boldsymbol{\theta}}(\mathbf{y} + \mathbf{X}\mathbf{m}) = \hat{\boldsymbol{\theta}}(\mathbf{y})$ for all $\mathbf{y}$ and all $\mathbf{m}$. Nearly all estimators of $\boldsymbol{\theta}$ that are used in practice are even, translation-invariant estimators. In particular, it may be shown (see the exercises) that each of the maximum likelihood and REML estimators of $\boldsymbol{\theta}$, when it exists and is unique, is even and translation invariant.

Theorem 17.1.2 *Suppose that the distribution of* $\mathbf{e}$ *is symmetric around* $\mathbf{0}$*, and that* $\hat{\boldsymbol{\theta}}$ *is an even, translation-invariant estimator of* $\boldsymbol{\theta}$*. Then, the E-BLUP is distributed symmetrically around* $\mathbf{c}^T\boldsymbol{\beta}$*, so if its expectation exists it is unbiased.*

Proof Observe that $\text{E}[\tilde{\hat{\tau}} - (\mathbf{c}^T\boldsymbol{\beta} + u)] = \text{E}(\tilde{\hat{\tau}} - \mathbf{c}^T\boldsymbol{\beta})$, and that

$$\begin{aligned}\tilde{\hat{\tau}} - \mathbf{c}^T\boldsymbol{\beta} &= \mathbf{c}^T(\mathbf{X}^T\hat{\mathbf{V}}_{yy}^{-1}\mathbf{X})^-\mathbf{X}^T\hat{\mathbf{V}}_{yy}^{-1}(\mathbf{X}\boldsymbol{\beta} + \mathbf{e}) + \hat{\mathbf{v}}_{yu}^T\hat{\mathbf{E}}(\mathbf{X}\boldsymbol{\beta} + \mathbf{e}) - \mathbf{c}^T\boldsymbol{\beta} \\ &= \mathbf{c}^T(\mathbf{X}^T\hat{\mathbf{V}}_{yy}^{-1}\mathbf{X})^-\mathbf{X}^T\hat{\mathbf{V}}_{yy}^{-1}\mathbf{e} + \hat{\mathbf{v}}_{yu}^T\hat{\mathbf{E}}\mathbf{e}\end{aligned}$$

where we used Theorem 11.1.6a, d. Because $\hat{\boldsymbol{\theta}}$ is translation invariant, $\hat{\boldsymbol{\theta}}(\mathbf{y}) = \hat{\boldsymbol{\theta}}(\mathbf{y} - \mathbf{X}\boldsymbol{\beta}) = \hat{\boldsymbol{\theta}}(\mathbf{e})$, and because $\hat{\boldsymbol{\theta}}$ is even, $\hat{\boldsymbol{\theta}}(-\mathbf{e}) = \hat{\boldsymbol{\theta}}(\mathbf{e})$. Consequently, $\tilde{\hat{\tau}} - \mathbf{c}^T\boldsymbol{\beta}$ is an odd function of $\mathbf{e}$. By Theorem 17.1.1, the distribution of $\tilde{\hat{\tau}} - \mathbf{c}^T\boldsymbol{\beta}$ is symmetric around 0. Thus, if the expectation of $\tilde{\hat{\tau}}$ exists, then the expectation of $\tilde{\hat{\tau}} - \mathbf{c}^T\boldsymbol{\beta}$ exists and is equal to 0, implying that the E-BLUP is unbiased in this case. □

Does the E-BLUP's expectation exist, as Theorem 17.1.2 requires for the E-BLUP to be unbiased? Jiang (1999, 2000) answered that question in the affirmative for general components-of-variance models, provided that $\text{E}(\mathbf{e})$ exists and the elements of $\hat{\boldsymbol{\theta}}$ are nonnegative (as is the case for ML and REML estimators, but not, in general, for quadratic unbiased estimators such as those obtained from an ANOVA). Multivariate normality of $\mathbf{e}$ is not required. It appears that conditions under which the E-BLUP's expectation exists under a general mixed linear model have not yet been determined.

Example 17.1-1. E-BLUPs in a One-Factor Components-of-Variance Model with Balanced Data

Recall from Example 13.4.2-2 that, under a one-factor mixed effects model with balanced data and known variance component ratio $\psi = \sigma_b^2/\sigma^2$, the BLUPs of $\mu + b_i$ and $b_i - b_{i'}$ $(i \neq i')$ are

$$\widetilde{\mu + b_i} = \left(\frac{r\psi}{r\psi + 1}\right)\bar{y}_{i\cdot} + \left(\frac{1}{r\psi + 1}\right)\bar{y}$$

and

$$\widetilde{b_i - b_{i'}} = \left(\frac{r\psi}{r\psi + 1}\right)(\bar{y}_{i\cdot} - \bar{y}_{i'\cdot}),$$

respectively. Also recall from Example 16.1.3-2 that

$$\hat{\sigma}^2 = S_2^2, \qquad \hat{\sigma}_b^2 = \frac{S_1^2 - S_2^2}{r}$$

are quadratic unbiased estimators of σ^2 and σ_b^2. Therefore, an E-BLUP of $\mu + b_i$ is

$$\widetilde{\widetilde{\mu + b_i}} = \left(\frac{r\hat{\sigma}_b^2}{r\hat{\sigma}_b^2 + \hat{\sigma}^2}\right)\bar{y}_{i\cdot} + \left(\frac{\hat{\sigma}^2}{r\hat{\sigma}_b^2 + \hat{\sigma}^2}\right)\bar{y}_{\cdot\cdot} \tag{17.2}$$

$$= \left(\frac{r[(S_1^2 - S_2^2)/r]}{r[(S_1^2 - S_2^2)/r] + S_2^2}\right)\bar{y}_{i\cdot} + \left(\frac{S_2^2}{r[(S_1^2 - S_2^2)/r] + S_2^2}\right)\bar{y}_{\cdot\cdot}$$

$$= \left(1 - \frac{S_2^2}{S_1^2}\right)\bar{y}_{i\cdot} + \left(\frac{S_2^2}{S_1^2}\right)\bar{y}_{\cdot\cdot}$$

and an E-BLUP of $b_i - b_{i'}$ is

$$\widetilde{\widetilde{b_i - b_{i'}}} = \left(\frac{r\hat{\sigma}_b^2}{r\hat{\sigma}_b^2 + \hat{\sigma}^2}\right)(\bar{y}_{i\cdot} - \bar{y}_{i'\cdot}) \tag{17.3}$$

$$= \left(1 - \frac{S_2^2}{S_1^2}\right)(\bar{y}_{i\cdot} - \bar{y}_{i'\cdot}).$$

Other E-BLUPs of the same predictands could be obtained by replacing the ANOVA-based estimators of σ^2 and σ_b^2 with ML or REML estimators in (17.2) and (17.3).

It may be verified easily that S_1^2 and S_2^2 are even, translation-invariant estimators of σ^2 and σ_b^2, respectively. However, because $\hat{\sigma}_b^2$ is not nonnegative, there is no guarantee (by the discussion which immediately follows Theorem 17.1.2) that the expectations of the E-BLUPs of $\mu + b_i$ and $b_i - b_{i'}$ exist, hence no guarantee that these E-BLUPs are unbiased—even if the distribution of $\mathbf{e}$ is multivariate normal. If σ_b^2 was estimated instead by $\max\{\hat{\sigma}_b^2, 0\}$, or if σ^2 and σ_b^2 were estimated by ML or REML, then the E-BLUPs obtained by substituting those estimators for $\hat{\sigma}^2$ and $\hat{\sigma}_b^2$ in (17.2) and (17.3), respectively, would indeed be unbiased. ■

Next, we consider the E-BLUP's prediction error variance, $\text{var}(\tilde{\tilde{\tau}} - \tau)$. For this purpose, it is important to note that the maximal invariant (Ferguson 1967, Section 5.6; Lehmann and Romano, 2005, Section 6.2) with respect to the transformation group mapping $\mathbf{y}$ to $\mathbf{y} + \mathbf{Xm}$ is $\mathbf{A}^T\mathbf{y}$, where $\mathbf{A}$ is the matrix defined in Sect. 16.2; that is, $\mathbf{A}^T(\mathbf{y} + \mathbf{Xm}) = \mathbf{A}^T\mathbf{y}$ for all p-vectors $\mathbf{m}$, and if $\mathbf{y}_1$ and $\mathbf{y}_2$ are two n-vectors such that $\mathbf{A}^T\mathbf{y}_1 = \mathbf{A}^T\mathbf{y}_2$, then $\mathbf{y}_2 = \mathbf{y}_1 + \mathbf{Xm}$ for some p-vector $\mathbf{m}$. Thus $\hat{\boldsymbol{\theta}}$ is translation-invariant if and only if $\hat{\boldsymbol{\theta}}$ is a function of $\mathbf{y}$ only through the value of $\mathbf{A}^T\mathbf{y}$, i.e., if and only if $\hat{\boldsymbol{\theta}}$ depends on $\mathbf{y}$ only through a vector of $n - p^*$ linearly independent error contrasts.

Intuitively, it seems reasonable to expect that the extra uncertainty associated with prediction under a model in which the variance–covariance parameters are unknown would cause the E-BLUP to have a larger prediction error variance than the BLUP. The following theorem gives sufficient conditions for this to be so.

Theorem 17.1.3 *If $\hat{\boldsymbol{\theta}}$ is an even, translation-invariant estimator of $\boldsymbol{\theta}$, and the joint distribution of $\mathbf{e}$ and u is multivariate normal, and the prediction error variance of the E-BLUP exists, then*

$$M(\boldsymbol{\theta}) \leq var(\tilde{\tilde{\tau}} - \tau). \tag{17.4}$$

In words, the prediction error variance of the E-BLUP, when it exists, is at least as large as the prediction error variance of the BLUP.

Proof First observe, by (16.19), that

$$\begin{aligned}
\tilde{\tilde{\tau}} - \tilde{\tau} &= \mathbf{c}^T(\mathbf{X}^T\hat{\mathbf{V}}_{yy}^{-1}\mathbf{X})^-\mathbf{X}^T\hat{\mathbf{V}}_{yy}^{-1}\mathbf{y} + \hat{\mathbf{v}}_{yu}^T\hat{\mathbf{E}}\mathbf{y} - \mathbf{c}^T(\mathbf{X}^T\mathbf{V}_{yy}^{-1}\mathbf{X})^-\mathbf{X}^T\mathbf{V}_{yy}^{-1}\mathbf{y} - \mathbf{v}_{yu}^T\mathbf{E}\mathbf{y} \\
&= \mathbf{c}^T(\mathbf{X}^T\hat{\mathbf{V}}_{yy}^{-1}\mathbf{X})^-\mathbf{X}^T\hat{\mathbf{V}}_{yy}^{-1}[\mathbf{I} - \mathbf{X}(\mathbf{X}^T\mathbf{V}_{yy}^{-1}\mathbf{X})^-\mathbf{X}^T\mathbf{V}_{yy}^{-1}]\mathbf{y} + \hat{\mathbf{v}}_{yu}^T\mathbf{A}(\mathbf{A}^T\hat{\mathbf{V}}_{yy}\mathbf{A})^{-1}\mathbf{A}^T\mathbf{y} \\
&\quad - \mathbf{v}_{yu}^T\mathbf{A}(\mathbf{A}^T\mathbf{V}_{yy}\mathbf{A})^{-1}\mathbf{A}^T\mathbf{y} \\
&= \mathbf{c}^T(\mathbf{X}^T\hat{\mathbf{V}}_{yy}^{-1}\mathbf{X})^-\mathbf{X}^T\hat{\mathbf{V}}_{yy}^{-1}\mathbf{V}_{yy}\mathbf{A}(\mathbf{A}^T\mathbf{V}_{yy}\mathbf{A})^{-1}\mathbf{A}^T\mathbf{y} + \hat{\mathbf{v}}_{yu}^T\mathbf{A}(\mathbf{A}^T\hat{\mathbf{V}}_{yy}\mathbf{A})^{-1}\mathbf{A}^T\mathbf{y} \\
&\quad - \mathbf{v}_{yu}^T\mathbf{A}(\mathbf{A}^T\mathbf{V}_{yy}\mathbf{A})^{-1}\mathbf{A}^T\mathbf{y}.
\end{aligned}$$

Thus, $\tilde{\tilde{\tau}} - \tilde{\tau}$ depends on $\mathbf{y}$ only through the value of $\mathbf{A}^T\mathbf{y}$. Now by an argument very similar to that used to obtain the independence of the BLUP's prediction error and the generalized least squares residual vector in Example 14.1-3, it can be shown that $\tilde{\tau} - \tau$ and $\mathbf{A}^T\mathbf{y}$ are independent. Thus $\tilde{\tau} - \tau$ and any function of $\mathbf{A}^T\mathbf{y}$ are independent; in particular, $\tilde{\tau} - \tau$ and $\tilde{\tilde{\tau}} - \tilde{\tau}$ are independent hence uncorrelated. Therefore, if $\text{var}(\tilde{\tilde{\tau}} - \tau)$ exists, it satisfies

$$\begin{aligned}
\text{var}(\tilde{\tilde{\tau}} - \tau) &= \text{var}(\tilde{\tilde{\tau}} - \tilde{\tau} + \tilde{\tau} - \tau) \\
&= \text{var}(\tilde{\tilde{\tau}} - \tilde{\tau}) + M(\boldsymbol{\theta}) \\
&\geq M(\boldsymbol{\theta}).
\end{aligned}$$

□

Under what conditions does the prediction error variance of the E-BLUP exist, as required by Theorem 17.1.3? Once again, Jiang (2000) gives such conditions in the case of a general components-of-variance model. The conditions are that (a) the variance–covariance matrix of the joint distribution of $\mathbf{e}$ and u exists, and (b) the estimators of the variance components are nonnegative. Condition (a) is satisfied under the normality assumed in Theorem 17.1.3, so the only additional condition needed for the E-BLUP's prediction error variance to exist, beyond those given in Theorem 17.1.3, is nonnegativity of the variance component estimators. Thus, for example, even if the joint distribution of the random effects in the one-factor random model considered in Example 17.1-1 is multivariate normal, the prediction error variances of the E-BLUPs featured in that example may not exist, in which case Theorem 17.1.3 does not hold for those E-BLUPs. However, prediction error variances of E-BLUPs of the same form except that ML or REML estimators are substituted for the ANOVA-based estimators do exist, and Theorem 17.1.3 tells us that (under normality of the random effects) they are as large or larger than $M(\boldsymbol{\theta})$.

Although Theorem 17.1.3 establishes that the prediction error variance (when it exists) of the E-BLUP is generally larger than that of the BLUP under the given conditions, it neither quantifies the magnitude of the difference nor provides any insight into how to approximate the E-BLUP's prediction error variance. Harville and Jeske (1992) suggested the approximation

$$\text{var}(\tilde{\hat{\tau}} - \tau) \doteq M(\boldsymbol{\theta}) + \text{tr}[\mathbf{A}(\boldsymbol{\theta})\mathbf{B}(\boldsymbol{\theta})], \tag{17.5}$$

where $\mathbf{A}(\boldsymbol{\theta}) = \text{var}[\partial\tilde{\tau}(\boldsymbol{\theta})/\partial\boldsymbol{\theta}]$ and $\mathbf{B}(\boldsymbol{\theta})$ is either equal to the matrix $MSE(\hat{\boldsymbol{\theta}}) \equiv \text{E}[(\hat{\boldsymbol{\theta}} - \boldsymbol{\theta})(\hat{\boldsymbol{\theta}} - \boldsymbol{\theta})^T]$ or [in the likely event that $MSE(\hat{\boldsymbol{\theta}})$ is intractable] an approximation to $MSE(\hat{\boldsymbol{\theta}})$; a justification for this approximation is left as an exercise. If $\hat{\boldsymbol{\theta}}$ is the ML or REML estimator of $\boldsymbol{\theta}$, then we may take $\mathbf{B}(\boldsymbol{\theta})$ to be the asymptotic variance–covariance matrix of $\hat{\boldsymbol{\theta}}_{ML}$ or $\hat{\boldsymbol{\theta}}_{REML}$ described in Sect. 16.2.

17.2 Using the E-BLUP to Perform Inference

In order to adapt methodology for hypothesis testing and interval prediction derived under prediction-extended Aitken models to perform small-sample inferences for prediction-extended general mixed linear models using the E-BLUP, it is necessary to estimate the E-BLUP's prediction error variance. As noted in the opening paragraph of this chapter, it is a common practice to estimate the E-BLUP's prediction error variance by $M(\hat{\boldsymbol{\theta}})$, i.e., to substitute $\hat{\boldsymbol{\theta}}$ for $\boldsymbol{\theta}$ in the expression for the BLUP's prediction error variance. Harville and Jeske (1992) showed that if the joint distribution of $\mathbf{y}$ and u is multivariate normal and $\text{E}[\mathbf{V}_{yy}(\hat{\boldsymbol{\theta}})] = \mathbf{V}_{yy}(\boldsymbol{\theta})$, $\text{E}[\mathbf{v}_{yu}(\hat{\boldsymbol{\theta}})] = \mathbf{v}_{yu}(\boldsymbol{\theta})$, and $\text{E}[v_{uu}(\hat{\boldsymbol{\theta}})] = v_{uu}(\boldsymbol{\theta})$ [as would be the case if $\hat{\boldsymbol{\theta}}$ was an unbiased estimator of $\boldsymbol{\theta}$ and $\mathbf{V}_{yy}(\boldsymbol{\theta})$, $\mathbf{v}_{yu}(\boldsymbol{\theta})$, and $v_{uu}(\boldsymbol{\theta})$ were linear functions of $\boldsymbol{\theta}$], then

$$\text{E}[M(\hat{\boldsymbol{\theta}})] \leq M(\boldsymbol{\theta}). \tag{17.6}$$

Zimmerman and Cressie (1992) showed that multivariate normality was not needed for this result. When combined with inequality (17.4), inequality (17.6) indicates that the bias of $M(\hat{\boldsymbol{\theta}})$ as an estimator of the E-BLUP's prediction error variance may be even worse than anticipated.

Example 17.2-1. Bias of $M(\hat{\boldsymbol{\theta}})$ Under a One-Factor Components-of-Variance Model with Balanced Data

Consider the prediction error variance of the E-BLUP of $b_i - b_{i'}$ under the one-factor components-of-variance model with balanced data described in Example 17.1-1, and suppose that the joint distribution of the random effects is multivariate normal. It can be shown (indeed, it was part of Exercise 13.16) that the prediction error variance of the BLUP of $b_i - b_{i'}$ is given by

$$M(\boldsymbol{\theta}) = \frac{2\psi\sigma^2}{r\psi + 1}.$$

Thus, taking $\hat{\boldsymbol{\theta}}$ to be the vector of ANOVA-based quadratic unbiased estimators of the variance components also described in Example 17.1-1, we obtain

$$M(\hat{\boldsymbol{\theta}}) = \frac{2\left(\frac{S_1^2 - S_2^2}{r}\right) S_2^2}{r\left(\frac{S_1^2 - S_2^2}{r}\right) + S_2^2} = \left(\frac{2}{r}\right)\left(S_2^2 - \frac{S_2^4}{S_1^2}\right).$$

Recall from Example 16.1.4-1 that

$$\frac{(q-1)S_1^2}{\sigma^2 + r\sigma_b^2} \sim \chi^2(q-1) \quad \text{and} \quad \frac{q(r-1)S_2^2}{\sigma^2} \sim \chi^2(q(r-1)).$$

These results imply that $\mathrm{E}(S_1^2) = \sigma^2 + r\sigma_b^2$, $\mathrm{E}(S_2^2) = \sigma^2$,

$$\mathrm{var}(S_2^2) = 2q(r-1)\frac{(\sigma^2)^2}{[q(r-1)]^2} = \frac{2\sigma^4}{q(r-1)},$$

and (using Theorem 14.2.7)

$$\mathrm{E}\left(\frac{1}{S_1^2}\right) = \left(\frac{q-1}{q-3}\right)\left(\frac{1}{\sigma^2 + r\sigma_b^2}\right)$$

provided that $q > 3$. (If $q \leq 3$ then the expectation of $1/S_1^2$ does not exist.) Furthermore, S_1^2 and S_2^2 are independent. Thus, if $q > 3$, then

$$\begin{aligned} \mathrm{E}[M(\hat{\boldsymbol{\theta}})] &= \left(\frac{2}{r}\right)\left\{\mathrm{E}(S_2^2) - [\mathrm{var}(S_2^2) + (\mathrm{E}(S_2^2))^2]\mathrm{E}\left(\frac{1}{S_1^2}\right)\right\} \\ &= \left(\frac{2}{r}\right)\left[\sigma^2 - \left(\frac{2\sigma^4}{q(r-1)} + \sigma^4\right)\left(\frac{q-1}{q-3}\right)\left(\frac{1}{\sigma^2 + r\sigma_b^2}\right)\right], \end{aligned}$$

which, after some algebra, may be shown to equal $M(\boldsymbol{\theta})[1 - (a/\psi)]$ where $a = \frac{[2+q(r-1)](q-1)}{qr(r-1)(q-3)}$. The table below shows values of a for a few combinations of q and r:

q	r	a
4	2	$\frac{7}{4}$
4	4	$\frac{7}{8}$
4	8	$\frac{45}{112}$
8	2	$\frac{7}{8}$
8	4	$\frac{91}{240}$
8	8	$\frac{29}{160}$

Thus, the negative bias of $M(\hat{\boldsymbol{\theta}})$ as an estimator of $M(\boldsymbol{\theta})$ can be substantial unless r or ψ are sufficiently large, and (by Theorem 17.1.3) its bias as an estimator of the E-BLUP's prediction error variance is potentially even more substantial. ■

Although (17.4) and (17.6) suggest that $M(\hat{\boldsymbol{\theta}})$ tends to underestimate the E-BLUP's prediction error variance, they give no indication as to which is larger, the amount by which $M(\hat{\boldsymbol{\theta}})$ tends to underestimate $M(\boldsymbol{\theta})$ or the amount by which $M(\boldsymbol{\theta})$ is less than the E-BLUP's prediction error variance. Under the additional condition that $\hat{\boldsymbol{\theta}}$ is a complete sufficient statistic for the family of distributions of $\mathbf{A}^T\mathbf{y}$, Harville and Jeske (1992) showed that these two components of the bias of $M(\hat{\boldsymbol{\theta}})$ as an estimator of the E-BLUP's prediction error variance are equal! This motivated Harville and Jeske (1992) to propose the estimator

$$M^*(\hat{\boldsymbol{\theta}}) = M(\hat{\boldsymbol{\theta}}) + 2\mathrm{tr}[\mathbf{A}(\hat{\boldsymbol{\theta}})\mathbf{B}(\hat{\boldsymbol{\theta}})] \tag{17.7}$$

of the E-BLUP's prediction error variance. The approximate unbiasedness of $M^*(\hat{\boldsymbol{\theta}})$ for the E-BLUP's prediction error variance can be established when the joint distribution of $\mathbf{y}$ and u is multivariate normal, $\hat{\boldsymbol{\theta}}$ is unbiased, and $\mathbf{V}_{yy}(\boldsymbol{\theta})$, $\mathbf{v}_{yu}(\boldsymbol{\theta})$ and $v_{uu}(\boldsymbol{\theta})$ are linear functions of $\boldsymbol{\theta}$. But $M^*(\hat{\boldsymbol{\theta}})$ may perform reasonably well as an estimator of the E-BLUP's prediction error variance even in cases where one or more of these requirements is not satisfied. Results from simulation studies reported by Hulting and Harville (1991), Harville and Jeske (1992), and Singh et al. (1998) suggest that, in the case of normal components-of-variance models, the bias of

$M^*(\hat{\boldsymbol{\theta}})$ is reasonably small unless $\boldsymbol{\theta}$ is close to a boundary of the parameter space. When $\boldsymbol{\theta}$ is close to a boundary, $M^*(\hat{\boldsymbol{\theta}})$ tends to have substantial positive bias. Results from another simulation study conducted by Zimmerman and Cressie (1992) for a spatial mixed linear model were not as favorable to $M^*(\hat{\boldsymbol{\theta}})$; its bias was positive and substantial over a rather large portion of the parameter space.

A prediction interval for a predictable τ under a general mixed linear model may be constructed by acting as though

$$\hat{t} = \frac{\tilde{\hat{\tau}} - \tau}{\sqrt{\hat{M}}}$$

is a pivotal quantity, where $\hat{M}$ is either $M(\hat{\boldsymbol{\theta}})$ or $M^*(\hat{\boldsymbol{\theta}})$. Under the conditions of Theorem 17.1.1, the distribution of $\hat{t}$ is symmetric around 0, in which case an approximate $100(1-\xi)\%$ prediction interval for τ is

$$\tilde{\hat{\tau}} \pm \hat{t}_{\xi/2}\sqrt{\hat{M}} \tag{17.8}$$

where $\hat{t}_{\xi/2}$ is the upper $\xi/2$ cutoff point of a suitable approximate distribution of $\hat{t}$. Several possibilities exist for the choice of this distribution. One possibility, based on asymptotic results, is a standard normal distribution. Another is a central t distribution with $n-p^*$ degrees of freedom which, when combined with taking $\hat{M}$ to be $M(\hat{\boldsymbol{\theta}})$, yields the prediction interval described near the end of Sect. 16.2. Yet another choice, proposed by Jeske and Harville (1988), is a central t distribution with degrees of freedom determined empirically via a Satterthwaite approximation. Simulation studies conducted by Hulting and Harville (1991) and Harville and Jeske (1992) suggest that, in the case of normal components-of-variance models at least, prediction intervals constructed using this last choice of distribution, when combined with taking $\hat{M} = M^*(\hat{\boldsymbol{\theta}})$, perform very well.

For making simultaneous inferences about a vector $\boldsymbol{\tau} = \mathbf{C}^T\boldsymbol{\beta} + \mathbf{u}$ of $s \geq 2$ predictable functions, one possibility is to "Bonferronize" the prediction intervals just described, by using $\xi/(2s)$ in place of ξ in the cutoff points of those intervals. Such an approach is somewhat conservative, especially if s is large. Another, possibly less conservative possibility, is to act as though

$$\hat{F} = [(\mathbf{C}^T\tilde{\hat{\boldsymbol{\beta}}} + \hat{\mathbf{V}}_{yu}^T\hat{\mathbf{E}}\mathbf{y}) - (\mathbf{C}^T\boldsymbol{\beta} + \mathbf{u})]^T\hat{\mathbf{M}}^{-1}[(\mathbf{C}^T\tilde{\hat{\boldsymbol{\beta}}} + \hat{\mathbf{V}}_{yu}^T\hat{\mathbf{E}}\mathbf{y}) - (\mathbf{C}^T\boldsymbol{\beta} + \mathbf{u})]/s$$

is a pivotal quantity (where $\hat{\mathbf{M}}$ is the multivariate version of $\hat{M}$ and it is assumed that $\hat{\mathbf{M}}$ is positive definite). Kenward and Roger (1997) considered (for the special case where $\mathbf{u} \equiv \mathbf{0}$) a version of this approach to test hypotheses on s linear functions of the fixed effects in mixed linear models. In their version, which has been implemented in widely used software for fitting such models, $\hat{\boldsymbol{\theta}}$ is taken to be the REML estimator of $\boldsymbol{\theta}$, $\hat{\mathbf{M}}$ is taken to be the multivariate generalization of (17.7), and the distribution of $\lambda\hat{F}$ is taken to be a central F distribution with s and ν degrees of freedom, where λ and ν are determined by equating the mean and variance of $\lambda\hat{F}$ to those of the F distribution. They reported good performance (empirical sizes of nominal size-0.05 tests close to 0.05) of this approach for hypothesis tests on fixed effects of several types of mixed linear models. Harville (2008) noted the possibility of extending this approach to make simultaneous inferences about predictands.

17.3 Exercises

1. Show that, when it exists and is unique, the maximum likelihood estimator of $\boldsymbol{\theta}$ under a normal positive definite general mixed linear model is even and translation invariant. Show that the same result also holds for the REML estimator (when it exists and is unique).
2. Consider the problem of E-BLUP under the normal positive definite general mixed linear model considered in this chapter. Let $\tau = \mathbf{c}^T\boldsymbol{\beta} + u$ be a predictable linear combination; let $\hat{\boldsymbol{\theta}}$ be an even, translation-invariant estimator of $\boldsymbol{\theta}$; and let $\tilde{\tau}$ and $\hat{\tilde{\tau}}$ be the BLUP and E-BLUP of τ. Assume that the joint distribution of $\mathbf{e}$ and u is multivariate normal and that the prediction error variance of $\hat{\tilde{\tau}}$ exists.
   (a) Expand $[\hat{\tilde{\tau}} - \tilde{\tau}(\boldsymbol{\theta})]^2$ in a Taylor series (in $\hat{\boldsymbol{\theta}}$) about $\boldsymbol{\theta}$ to obtain the second-order approximation
   $$[\hat{\tilde{\tau}} - \tilde{\tau}(\boldsymbol{\theta})]^2 \doteq [(\partial\tilde{\tau}(\boldsymbol{\theta})/\partial\boldsymbol{\theta})^T(\hat{\boldsymbol{\theta}} - \boldsymbol{\theta})]^2.$$
   (b) Use the result of part (a) to argue that an approximation to the prediction error variance of the E-BLUP is
   $$\text{var}(\hat{\tilde{\tau}} - \tau) \doteq \text{var}(\tilde{\tau} - \tau) + \text{tr}[\mathbf{A}(\boldsymbol{\theta})\mathbf{B}(\boldsymbol{\theta})],$$
   where $\mathbf{A}(\boldsymbol{\theta}) = \text{var}[\partial\tilde{\tau}(\boldsymbol{\theta})/\partial\boldsymbol{\theta}]$ and $\mathbf{B}(\boldsymbol{\theta}) = \text{E}[(\hat{\boldsymbol{\theta}} - \boldsymbol{\theta})(\hat{\boldsymbol{\theta}} - \boldsymbol{\theta})^T]$.

References

Ferguson, T. S. (1967). *Mathematical statistics: A decision-theoretic approach*. New York: Academic Press.

Harville, D. A. (2008). Accounting for the estimation of variances and covariances in prediction under a general linear model: An overview. *Tatra Mountains Mathematical Publications, 39*, 1–15.

Harville, D. A., & Jeske, D. R. (1992). Mean squared error of estimation or prediction under a general linear model. *Journal of the American Statistical Association, 87*, 724–731.

Hulting, F. L., & Harville, D. A. (1991). Some Bayesian and non-Bayesian procedures for the analysis of comparative experiments and for small-area estimation: Computational aspects, frequentist properties, and relationships. *Journal of the American Statistical Association, 86*, 557–568.

Jeske, D. R., & Harville, D. A. (1988). Prediction-interval procedures and (fixed-effects) confidence-interval procedures for mixed linear models. *Communications in Statistics - Theory and Methods, 17*, 1053–1087.

Jiang, J. (1999). On unbiasedness of the empirical BLUE and BLUP. *Statistics and Probability Letters, 41*, 19–24.

Jiang, J. (2000). A matrix inequality and its statistical application. *Linear Algebra and Its Applications, 307*, 131–144.

Kenward, M. G., & Roger, J. H. (1997). Small sample inference for fixed effects from restricted maximum likelihood. *Biometrics, 53*, 983–987.

Lehmann, E., & Romano, J. P. (2005). *Testing statistical hypotheses* (3rd ed.). New York: Springer.

Singh, A. C., Stukel, D. M., & Pfeffermann, D. (1998). Bayesian versus frequentist measures of error in small area estimation. *Journal of the Royal Statistical Society, Series B, 60*, 377–396.

Zimmerman, D. L., & Cressie, N. (1992). Mean squared prediction error in the spatial linear model with estimated covariance parameters. *Annals of the Institute for Statistical Mathematics, 44*, 27–43.

Index

D. L. Zimmerman, *Linear Model Theory*,
https://doi.org/10.1007/978-3-030-52063-2

Printed in the United States
by Baker & Taylor Publisher Services